410

mono
di
sex-linked
blood tests

Compare Mitosis/Meiosis - Quiz
Mitosis / Meiosis - Thursday

Scott, Foresman

Biology

Authors

Irwin L. Slesnick

LeVon Balzer

Alan J. McCormack

David E. Newton

Frederick A. Rasmussen

Editorial Advisor

Peter Volpe

Program Consultant

Abraham Flexer

Scott, Foresman and Company

Glenview, Illinois

Dallas, Texas   Palo Alto, California   Oakland, New Jersey   Tucker, Georgia

Reviewers and Contributors

Carl Bollwinkel
Science Department Chairman
Palatine Schools
Palatine, IL

Ann L. Collins
Former Biology Teacher
Concord-Carlisle High School
Concord, MA

R. Larsen Fargo
Biology Teacher
Evanston Township High School
Evanston, IL

Howard E. Feldmann
Biology-Chemistry-Physical Science Teacher
Willowbrook High School
Villa Park, IL

Phyllis L. Goodson
Biology Teacher & Vice Coordinator
Science Department
Percy L. Julian High School
Chicago, IL

Gretchen M. Hanson
Science Specialist
Avoca District #37
Wilmette, IL

Kara M. Hsu
Biology Teacher
James B. Conant High School
Hoffman Estates, IL

Linda M. Lambert
Biology Teacher
Lake Park High School
Roselle, IL

Lois Lauer
Biology Teacher
Hinsdale South High School
Darien, IL

Arnold I. Miller
Research Director
Teacher of Biology, Microbiology, and Medical Laboratory Techniques
Syosset High School
Syosset, New York

Robert Motzko
Biology Teacher
Deerfield High School
Deerfield, IL

Charles D. Oviatt,
Asst. Director, Pupil Personnel Services
Missouri Department of Elementary and Secondary Education
Jefferson City, MO

Ed Triner
Assistant Chairman, Science
Oak Park & River Forest High School
Oak Park, IL

Barry L. Wulff
Associate Professor of Biology
Eastern Connecticut State College
Willimantic, CT

ISBN: 0-673-13735-X

Printed in the United States of America

345678910 WAK 898887868584

Contents

Unit One
Life 1

Chapter 1
The Organisms 3

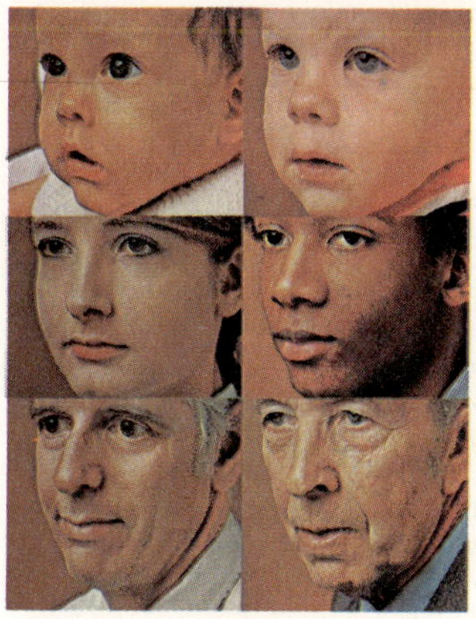

Chapter 2
The Humans 37

Unit Two
Cells 61

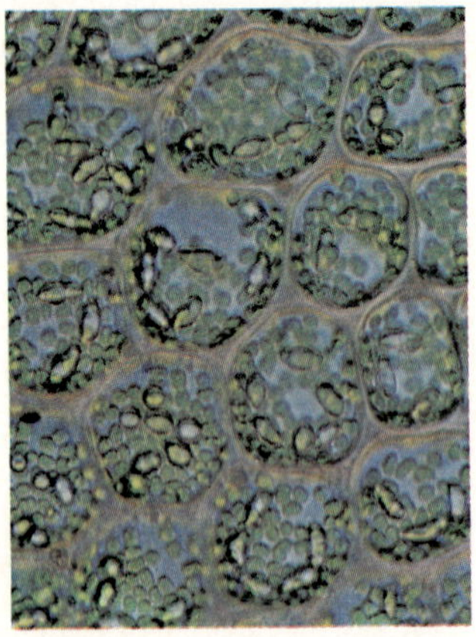

Chapter 3
Cell Structure 63

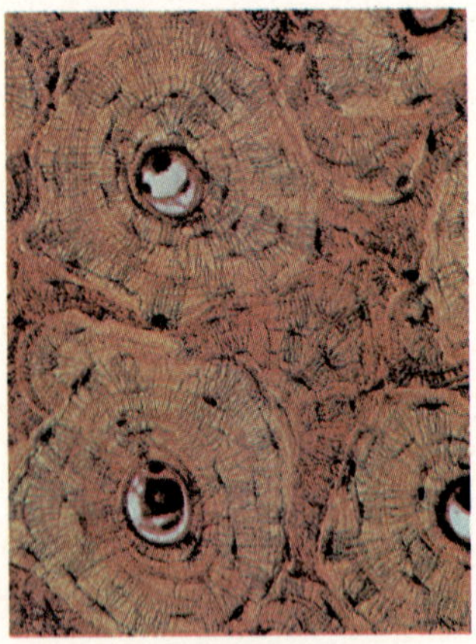

Chapter 4
Function of Cells 83

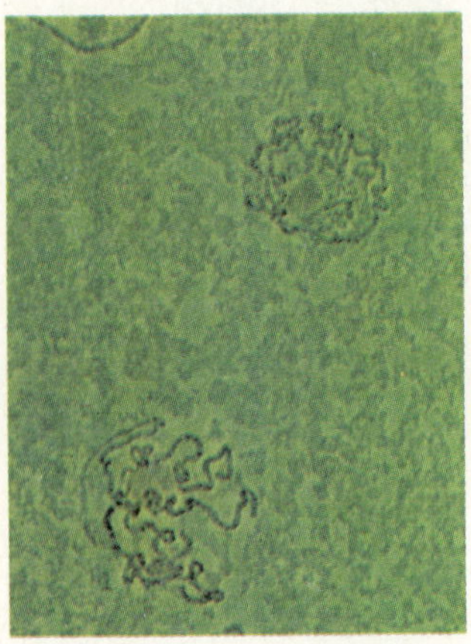

Chapter 5
Information Transfer by Cells 107

Unit Three
Genetics 137

Chapter 6
Inheritance of Traits 139

Chapter 7
Gene and Chromosome Changes 165

Unit Four
Evolutionary Theory 185

Chapter 8
Evidence of Change 187

Chapter 9
Mechanisms of Change 217

Unit Six
The Plants 293

Chapter 13
Survey of the Plants 295

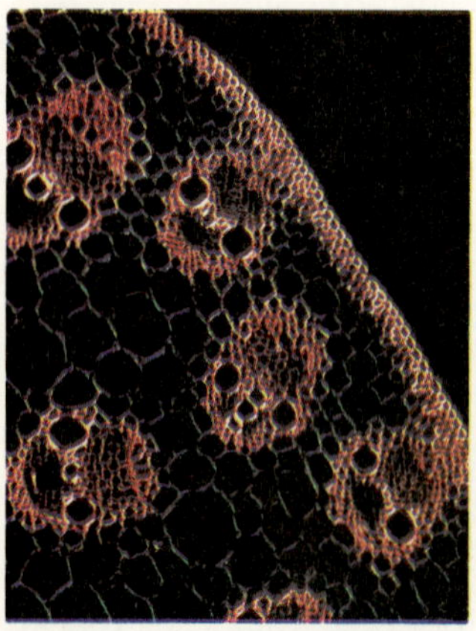

Chapter 14
Roots and Stems 313

Chapter 15
Leaves 333

Chapter 16
Flowers 355

Unit Seven
The Invertebrates 377

Chapter 17
Sponges to Starfish 379

Chapter 18 Arthropods 413

Unit Eight The Vertebrates 443

Chapter 19 Fish, Amphibians, and Reptiles 445

Chapter 20
Birds and Mammals 475

Unit Nine
Human Structure and Function 503

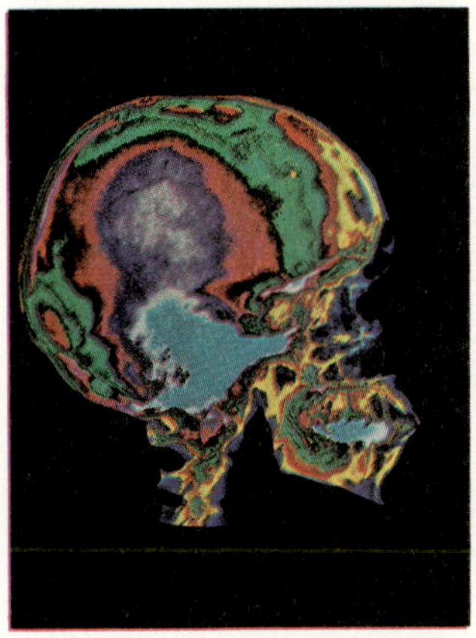

Chapter 21
Body Framework and Controls 505

Chapter 22
Nutrition and Digestion 535

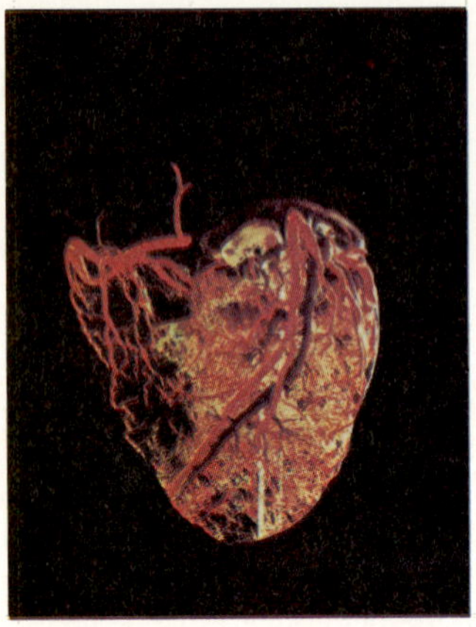

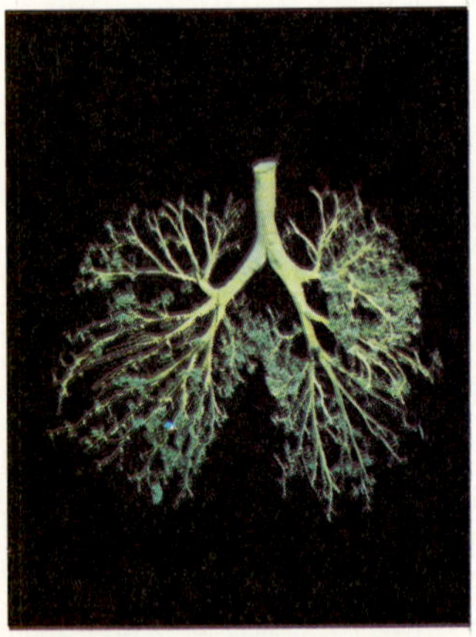

UNIT ONE

LIFE

Chapter 1

The Organisms

Our Earth supports more than two million different kinds of living things. Each kind of living thing, also called an *organism*, is suited to live in its own special place. The earth contains countless different places to live. So it should be no surprise that there is such a variety of life forms.

In this chapter you will think about what it means for something to be "alive." And you will examine some of the scientific explanations about the origin of life. Then you will look at the way major groups of organisms are classified. You will read about some of the special traits of each group. When you finish the chapter, you should have some ideas about why grouping or classifying organisms is such an important part of biology. And you should be aware of the difficulties in deciding how to put objects or organisms into groups.

Chapter Objectives

1. Describe some traits that often help distinguish living from nonliving things.
2. Compare the main ideas about the origin of life.
3. Explain the importance to biologists of having a worldwide system for the classification and naming of organisms.
4. Describe three traits used to distinguish among the five kingdoms of living things.
5. Describe traits of some of the major subgroups within each of the five kingdoms.
6. List at least two reasons why classification schemes in biology are continually changing.

The caption below refers, as throughout this book, to the chapter opener picture on the opposite page.

Pond water viewed under a microscope reveals an ameba in motion.

1–1
Traits of Living Things

Challenge!
A flame is often used as a symbol of life. What properties of living matter can you observe in a flame? What evidence do you have that a flame is not alive?

People often ask whether there is life in space. The question is difficult to answer for several reasons. First, much of the exploration of space must be done by machines instead of humans. Second, it is sometimes difficult for humans to distinguish living things from nonliving things right here on Earth. For example, you know that you are alive. And you know that a German shepherd and a geranium are alive. You also know that a rock, a cloud, and a candle flame are *not* alive. But a problem arises when you try to prove that you are alive and a candle flame is not alive.

Living things can usually be recognized by certain traits that they all share in common. Here are some of the major traits used to distinguish living from nonliving things:

1. Living things are organized as complex structures called **cells.** Some life forms are composed of a single cell. Others are made up of many different kinds of cells.
2. The cells of most living things contain certain kinds of complex and highly organized chemical compounds. Most nonliving matter contains compounds that are much less complex than those found in living matter.
3. All living things reproduce members of their own kind. Since living things eventually die, reproduction ensures survival of the group.
4. All living systems grow and develop by changing food into living cells and tissues.
5. Living things respond to stimuli in their environment. Organisms respond to certain specific stimuli. For example, a green plant bends toward light, an ameba moves toward food, and a rabbit runs away from sudden loud noises.
6. Living things have the ability to acquire and use energy by a process called **metabolism** (mə tab′ə liz′əm).
   a. Living systems obtain energy for metabolism either by turning the sun's energy into stored food energy or by taking in food energy from other life forms. For example, green plants manufacture their own food. Animals obtain energy by eating other forms of life.
   b. Living systems release energy during metabolism by "burning" food. They also get rid of excess heat and waste gases produced when the food is used in this way.
   c. Living systems regulate the use of energy released during metabolism. The energy is used for life functions such as growth, reproduction, and movement.

It is not always easy to decide whether or not something is alive. Evidence of most of the traits listed above is needed to be fairly sure that something is alive. Such evidence is sometimes difficult to obtain.

Traits of Living Things

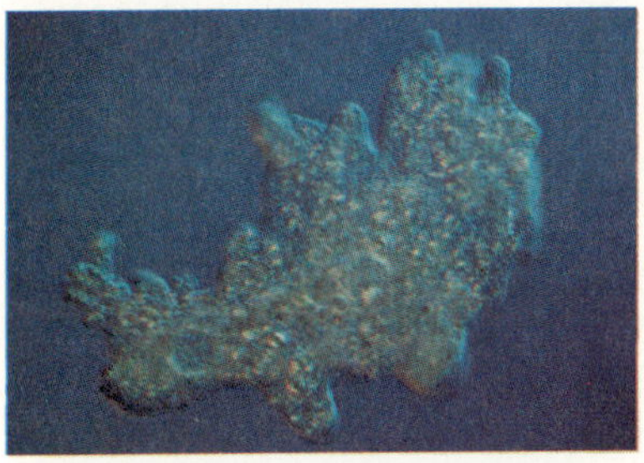

1. Living things are made up of cells

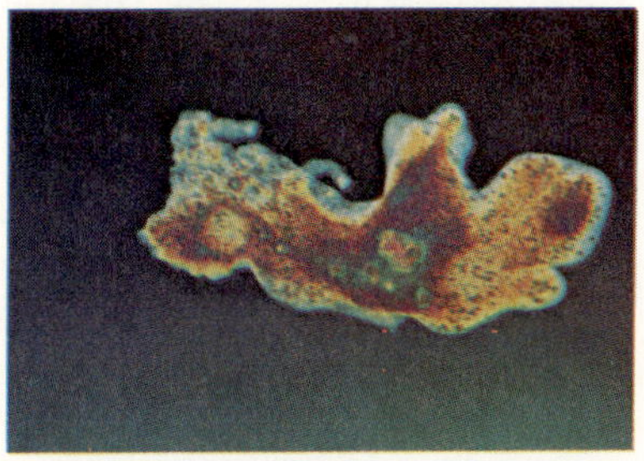

2. Living cells contain complex chemical compounds

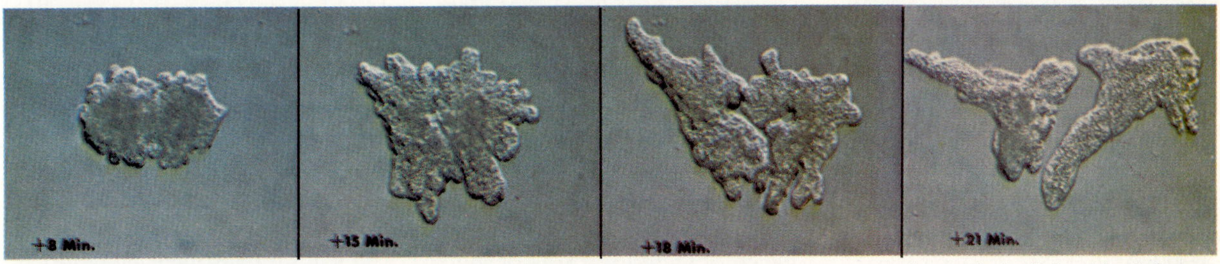

Courtesy Carolina Biological Supply Company

3. Living things reproduce

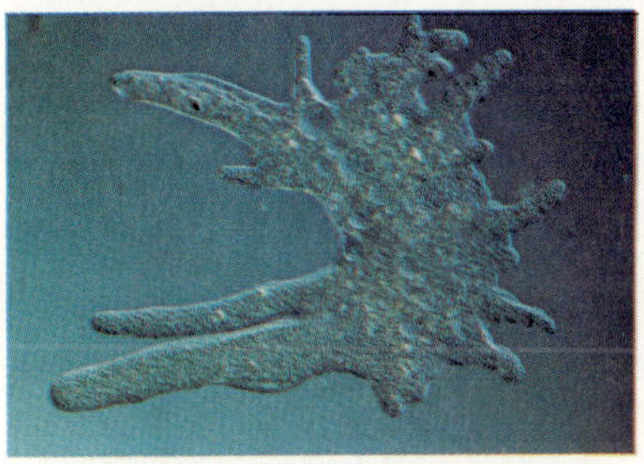

4. Living things grow and develop

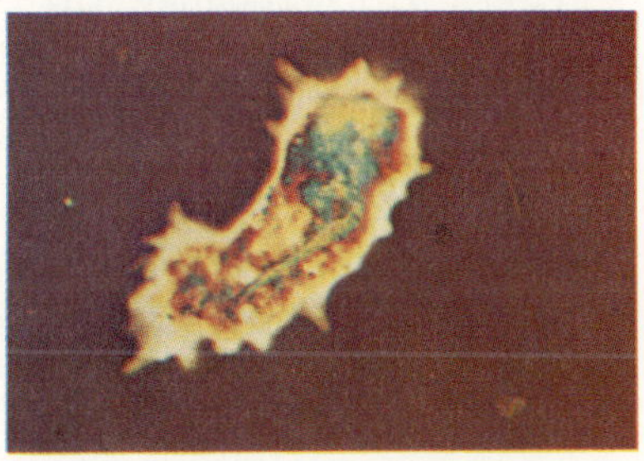

5. Living things respond to stimuli

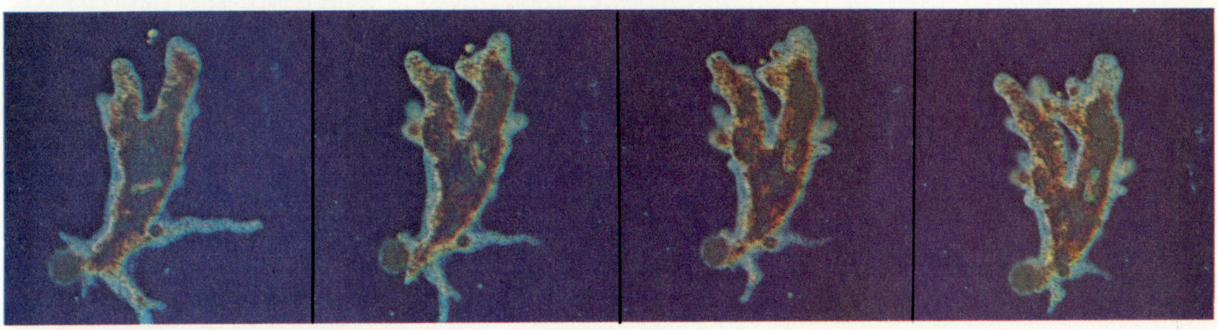

6. Living things acquire and use energy

Testing for Evidence of Life

Viking lander arm, about to scoop up a soil sample from the surface of Mars

Decisions about whether something is alive or that life exists are not always easy to make. The problem can be illustrated by using the tests performed aboard the Viking landers on the planet Mars. The Viking landers contained special laboratories and instruments that searched for life. Cameras scanned the landscape for objects that appeared to be alive. No living things were seen. However, the environment of Mars is harsh. The planet surface is cold, the air is thin, and the planet is exposed to deadly ultraviolet rays from the sun. But tiny living things might be able to survive in the Martian soil.

There is evidence that Mars might not always have been the freeze-dried planet it is today. The eroded landscape suggests that heavy rains may have once poured on the planet. At those times, Martian life may have been abundant. But today, the planet surface is dry, cold, and subjected to intense ultraviolet radiation. Large organisms may not have been able to survive. However, small organisms, such as bacteria and algae, may have been able to exist in the soil.

To find out, the Viking landers were equipped with digging arms which obtained samples of the soil. The soil samples were analyzed in three miniature laboratories that occupied 0.03 cubic meters of space and weighed 15 kilograms. The tiny laboratories were equipped to look for chemical signs of life. You might think that the labs would have used microscopes. However, even on Earth, it is often difficult to use a microscope to tell a microbe from a particle of soil.

Instead, each laboratory picked up a soil sample. Then it automatically heated the sample and added water to it. In one laboratory, artificial sunlight was shone on the soil. In the other two, food in the form of a nutritious "soup" was poured on the soil. The instruments in each lab measured how the soil reacted to the new conditions. Presumably, if organisms such as plants were in the soil, they might produce food under the right conditions. And if organisms such as bacteria were in the soil, they would consume the food, give off waste gases, and grow. Also, if living things lived in the soil, the sample should contain the chemical compounds of which living things are made. Such *organic compounds* are abundant in many soils on Earth.

A positive response to any of the tests performed aboard the Viking lander *might* have indicated the presence of life on Mars. However, it might also have meant that the equipment malfunctioned. Or it might have meant that some unknown chemicals existed in the Martian soil. A negative response, on the other hand, might have meant the landers settled in a sterile area. In a warmer or wetter area, life might be abundant. Thus, regardless of the test results, scientists would never be able to say that life definitely did or did not exist on Mars. On page 8 you can read about the three tests that were done.

Viking lander on the surface of Mars

Test #1: Is something in the soil making food?
Green plants use carbon dioxide gas to make food. This test detected whether anything in the Martian soil also used carbon dioxide to make food. *Radioactive* carbon dioxide was added to the soil. Radioactive carbon is unstable. The carbon atoms emit small particles that can be detected with special instruments. A light similar to Martian sunlight was also shone on the soil. Then the soil was heated to break down any food present and to release any radioactive carbon in the food. Special instruments tried to detect the presence of this gas.

Test #2: Is something in the soil using food?
On Earth, organisms release carbon dioxide gas as they "burn" food. The food is used as a fuel to stay alive. As the food is burned, carbon dioxide gas is given off. Therefore, in this test, a rich mixture of foods was added to a sample of Martian soil that had been warmed and humidified. The food contained radioactive carbon. If Martian organisms were present, they might eat this food and give off a radioactive carbon dioxide gas. Special instruments on the lander tried to detect the presence of this gas.

Test #3: Is something in the soil changing the air?
Active organisms on Earth change the make-up of the air. They take in certain gases and give off others as they "burn" food. In this final test, different amounts of food and water were added to the Martian soil. Such food and water might have given Martian organisms a chance to become active. Special instruments tried to detect changes in the make-up of the air. An increase or decrease of certain gases might have indicated that life processes were taking place.

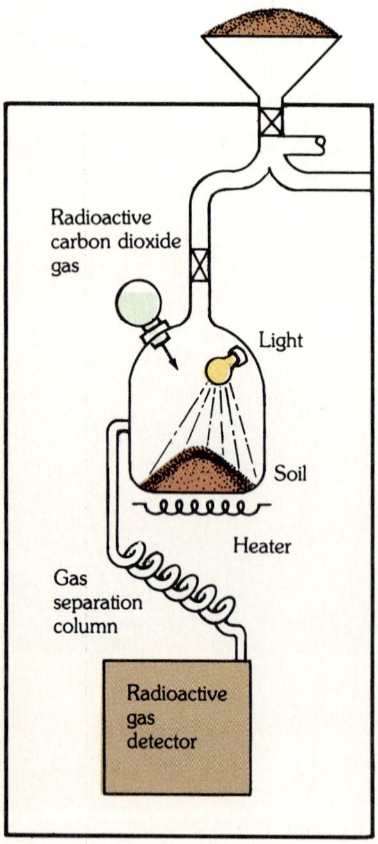

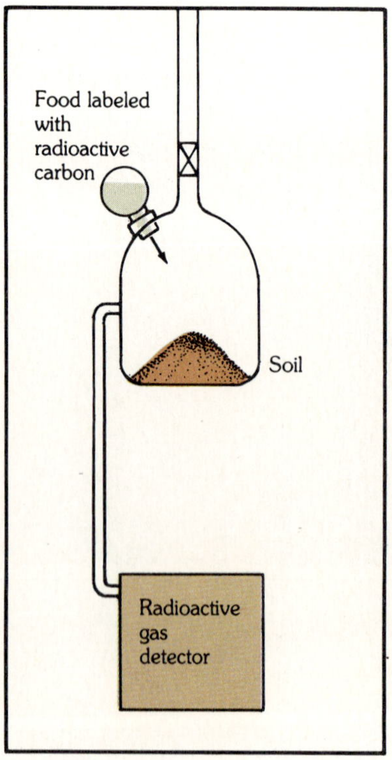

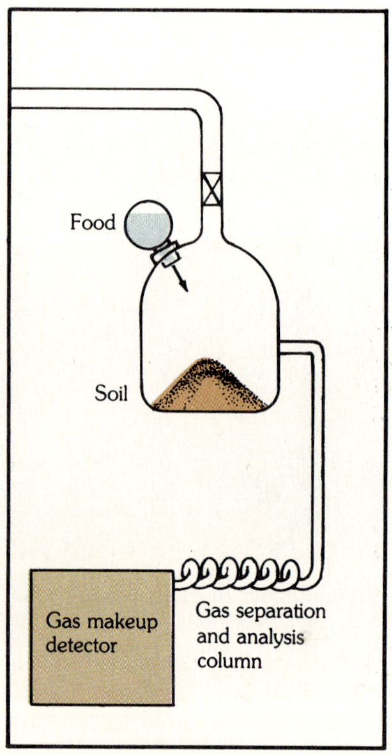

Conclusions

Results of the Viking lander test have been inconclusive. Several of the tests show that there are materials in the Martian soil that behave just like life forms on Earth. That is, something in the Martian soil appears to use Martian sunlight to make food out of carbon dioxide. And something else in the Martian soil appears to be able to burn food, releasing carbon dioxide. Both of these substances can be destroyed by heat. However, there is also the possibility that these substances in the Martian soil are unusual, nonliving chemical compounds which do not occur naturally on Earth.

A search on Mars for the chemical compounds characteristic of living things on Earth has failed so far. On Earth, the soil is full of organic compounds. And the air above soil on Earth contains organic compounds in the gaseous state. The Viking landers have not found any of these compounds. There may be too few organic molecules in soil samples for the lander's instruments to detect. Or organic molecules may not exist there at all.

So, while it appears that life on Mars was detected in several tests, the failure of instruments to find organic chemicals in the soil of Mars keeps the question of life on Mars unanswered.

Area on the rocky surface of Mars where scientists looked for evidence of life

The Origin of Life

The Scientific Method

- Make observations
- Describe observations
- Suggest explanation for observations
- Test explanation
- Change explanation, if necessary
- Test revised explanation
- Draw a conclusion

How did life originate? When and where did it originate? Did it happen only once? These questions have been asked by humans as far back as the beginning of recorded history. Now they are being asked with renewed interest as humans look for life on other planets.

Religious answers to questions about the origin of life vary. One common belief is that living things were made by a Divine Being. Since such acts of special creation would be supernatural events, they cannot be tested scientifically. Supernatural events are beyond the ability of humans to repeat, to control, and to detect with their senses.

Biology deals with the *scientific* study of life. This means that biologists gain knowledge by using controlled studies to observe and to test natural events. Other ways of knowing about life can be equally valid. But we will deal only with the scientific methods used to learn about life.

Spontaneous Generation

Early in the history of science, observers believed that some forms of life arose from nonliving materials. Some thought that rats formed from old rags, tadpoles grew out of the mud in ponds, or bacteria just appeared in old soup. This belief that living things could appear from nonliving matter, as if by magic, was known as **spontaneous generation** (spon tā′nē əs jen′ə rā′shən). One by one, the examples of spontaneous generation were shown to be incorrect by scientific tests. The scientific experiments on the next page are classic stories in the history of biology. These experiments helped to show that the theory of spontaneous generation was not correct.

Woodblocks showing early beliefs about spontaneous generation

Experiment #1: Does life arise from rotting meat?
In the 17th century an Italian scientist, Francesco Redi, provided evidence that flies did not arise from rotting meat. In his experiments Redi put meat inside two sets of jars. He left one set of jars open and sealed the other set. Flies landed on meat in the open jars and laid eggs. Maggots hatched from the eggs and developed into flies. But flies could not get into the sealed jars to lay eggs. The meat in the sealed jars remained free of flies.

Experiment #2: Does life arise from materials in soup broth?
By the 18th century the microscope was in wide use. People could see that many liquids swarmed with tiny living organisms called **microbes** (mī'krōbz). Some thought the microbes arose by spontaneous generation. But another Italian scientist, Lazzaro Spallanzani, demonstrated that microbes did not develop from nonliving materials in liquids. He put liquid broth in flasks and boiled the broth to kill any microbes present. Then he tightly sealed the flasks. No microbes appeared in these flasks. But microbes did appear in the flasks that had been left unsealed.

Experiment #3: Does life arise from materials in the air?
In the 19th century a French scientist, Louis Pasteur, demonstrated that life did not arise from nonliving materials in the air. Pasteur believed that microbes were carried about on dust particles in the air. So he designed a trap for the dust. He bent the necks of flasks into S shapes. Then he boiled broth in the flasks to force air out and to kill any microbes present. As the flasks cooled, air moved back inside. But the dust was trapped in the S-shaped necks. No microbes appeared in this broth for years. But as soon as broth touched the dust trapped in the necks, microbes began to grow in the broth.

Redi's Experiment

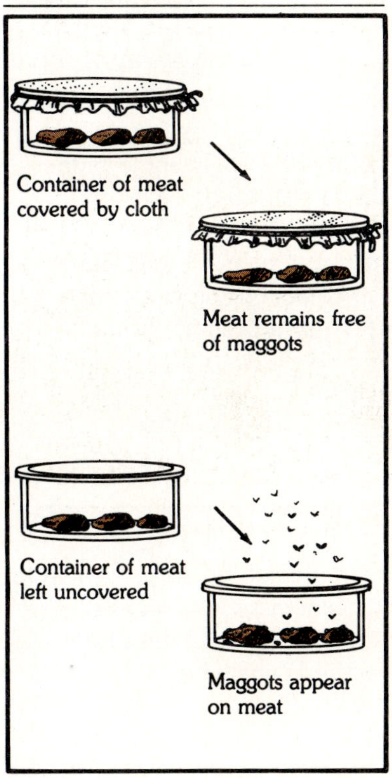

Spallanzani's Experiment

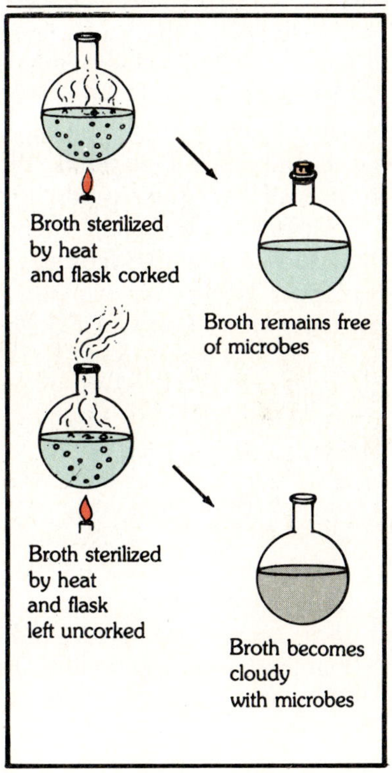

Pasteur's Experiment

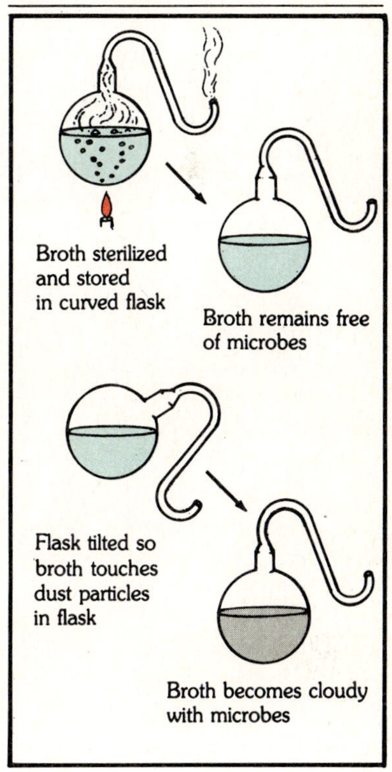

Chemosynthesis

A recent hypothesis about the origin of life suggests that life began on Earth billions of years ago when simple gases started to combine with each other. Scientists think the air of this early Earth was probably different from our present air. The early air contained mostly gases like methane, hydrogen, ammonia, and water vapor. Probably little oxygen, carbon dioxide, nitrogen, and ozone were present.

A Russian biochemist, Alexander Oparin, suggested in 1938 that energy from lightning and ultraviolet rays from the sun helped to combine these early gases. The gases eventually formed the chemical compounds of which living things are made. One group of such chemical compounds is the *amino acids*. Oparin guessed that as compounds like amino acids formed, they fell into hot seas. In time, the seas became a kind of rich "soup" of chemical compounds. These chemicals continued to form new and more complex compounds. Oparin suggested that eventually the complex compounds were able to reproduce themselves and to consume other chemicals for energy and food. Oparin's hypothesis on the origin of life is a good example of how ideas in science originate and grow.

Ever since Oparin presented his **Chemosynthesis** (kem′ō sin′thə-sis) **Hypothesis,** scientists have been trying to test it. In 1953, Stanley Miller tested the first part of Oparin's hypothesis. He tried to produce amino acids in the laboratory. Miller constructed a laboratory model of what the air might have been like billions of years ago. He mixed the gases methane, hydrogen, ammonia, and water vapor in a special chamber. Then he subjected the gas mixture to electrical sparks, to cooling, and to boiling. After one week of this treatment, Miller examined the "soup" in his chamber. He found amino acids, as well as sugars and other compounds. These compounds were some of the same ones that Oparin predicted would be produced.

Other parts of Oparin's hypothesis have also been tested. So far, the evidence produced indicates that part of Oparin's hypothesis might be correct. No one has yet, however, been able to produce living organisms in the laboratory. So the question of how life originated remains open to investigation.

Review It

1. Make a list of questions you would use to test if an object is living or nonliving.
2. How did the three early experiments described on page 11 help to show that the theory of spontaneous generation was incorrect?
3. What is the Chemosynthesis Hypothesis? Can scientists ever prove that the events described in the hypothesis really happened? Explain your answer.

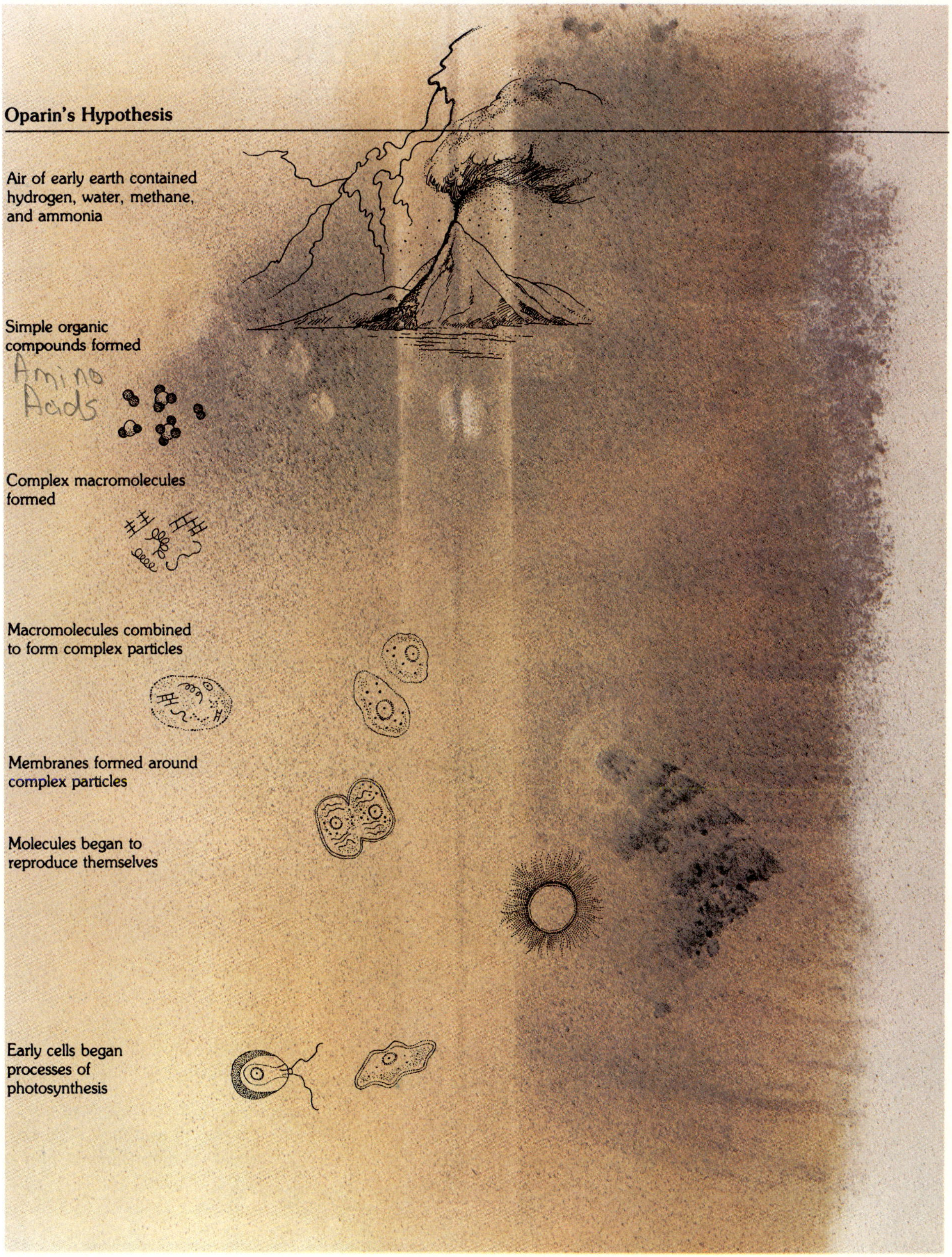
Oparin's Hypothesis
Air of early earth contained hydrogen, water, methane, and ammonia
Simple organic compounds formed
Complex macromolecules formed
Macromolecules combined to form complex particles
Membranes formed around complex particles
Molecules began to reproduce themselves
Early cells began processes of photosynthesis

Breakthroughs in Biology

Producing Artificial Proteins in the Laboratory
About the time Stanley Miller was completing his work, Sidney Fox had begun testing the next step in the Chemosynthesis Hypothesis. The next step was to show that amino acids could be combined to produce *proteins*. Proteins are complex chemical compounds that form a basic part of all living things.

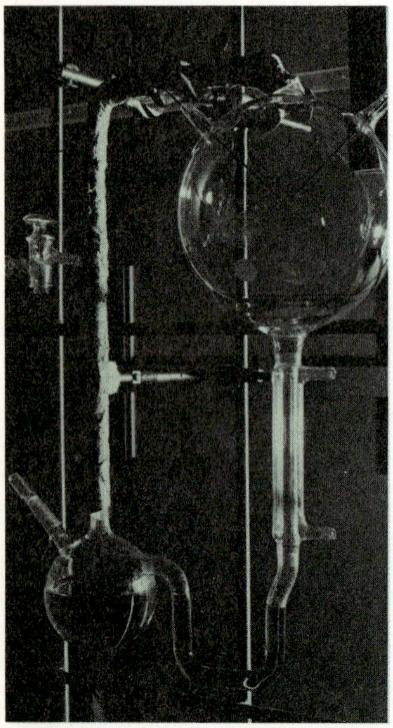

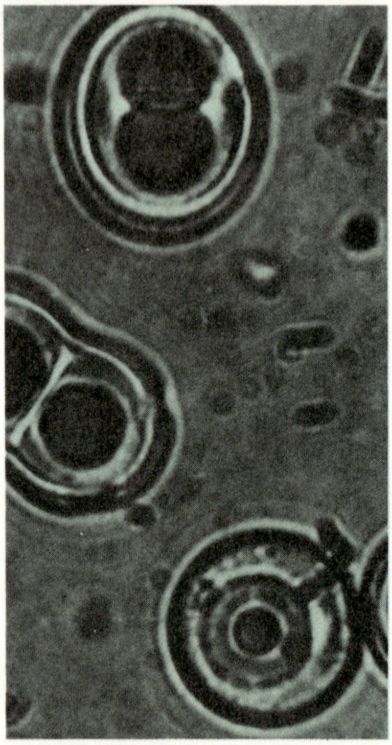

Fox reasoned that heat on the early earth might have caused the amino acids to join and form proteins. So Fox heated a mixture of 18 different amino acids. Later experiments showed that 60°C was sufficient. The mild heating produced large molecules that resembled the proteins of modern organisms. These proteins were highly ordered and had enzyme activity. When Fox brought artificial proteins into contact with water, he observed under the microscope that they formed small spheres. These *microspheres* looked like round bacteria.

In 1967 Fox and his co-workers reported another development.

When the microspheres were allowed to stand in the solution from which they had formed, small buds appeared. The buds broke off from the parent bud, grew, and the new parents had buds of their own.

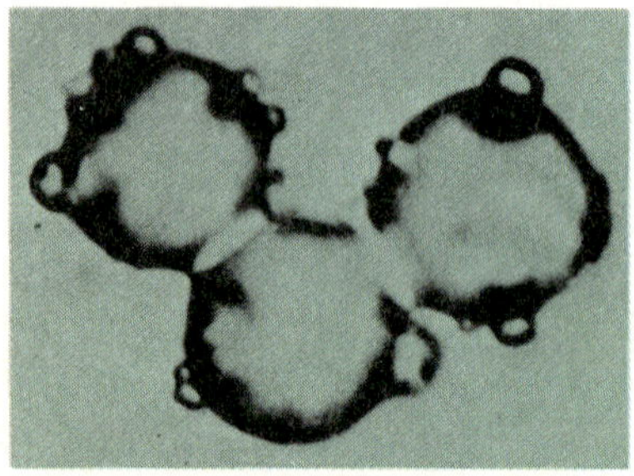

Fox's microspheres have several traits of modern living organisms. But they lack the nucleic acids found in all living things today. And they use food or energy in ways that are partly different from living things today. However, these microspheres are believed to represent an intermediate stage in the development of life as we know it.

Analysis

1. Compare the experiments of Sidney Fox to those of Stanley Miller.
2. What traits of living things do Fox's microspheres have? How do they differ from living systems of today?
3. Look up the terms *hypothesis*, *theory*, and *fact*. Why do you think chemosynthesis is a hypothesis and not a fact or theory?

1–2
Classifying Organisms

Until about 250 years ago people had no satisfactory way to organize and to name living things. Anyone who wanted to learn more about a certain organism had no easy way to find out about it. The way people named plants and animals was confusing. People in different places had different names for the same organism. For example, the mountain lion might be known by different groups of people as a cougar, puma, catamount, or panther.

You might compare the situation to shopping in a grocery store with all of the food scattered and mixed up on the shelves. Think of how hard it would be to find a certain kind of cereal in such a store. And imagine trying to find out if the store even carried a brand that you wanted!

In 1735 Carolus Linnaeus invented a system for classifying organisms. He based his classification system on physical similarities among organisms. He began by grouping organisms together that shared a few of the same physical traits. He called the largest of these groups **kingdoms.** Then he broke the large groups into smaller and smaller ones. The smaller the group, the more physical traits the members shared in common. Thus, kingdoms were divided into **phyla** (fī′lə). Phyla were grouped into **classes.** Classes were subdivided into **orders.** Orders were split into **families.** Families were broken down into **genera** (jen′ər ə). Each **genus** (jē′nəs) was divided into **species.** Thus, the members of a species are more alike than members of any other group of organisms.

A species is what we usually think of as a kind of organism, such as a dog, a horse, or an oak tree. Members of one species are usually physically similar in structure and can mate and produce fertile offspring. Sometimes members of a species differ somewhat. But they do not differ enough to be separate species. Such groups of organisms that differ slightly from other groups of the same species are called *varieties* or *subspecies.*

System of Classifying Organisms

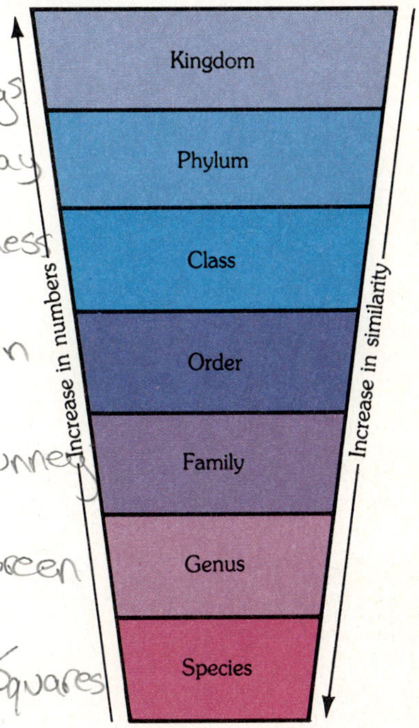

Naming Organisms
Linnaeus also invented a system for naming organisms. He named each organism by combining its genus and species names. These names were given in Latin because scientists everywhere understood the language. And the Latin names would not change because the language was no longer in use. Thus, all scientists today know the mountain lion by its genus and species names, *Felis concolor.* All varieties of house cats are *Felis catus.* All readers of this book are *Homo sapiens.* And the dandelion is called *Taraxacum officinale.*

A Five-Kingdom Classification System

At first, biologists grouped all organisms into two kingdoms. Organisms either belonged to the plant kingdom or to the animal kingdom. Basically, plants make their own food. Animals take in food from an outside source.

But as biologists learned more, they agreed that some groups did not fit well into either kingdom. For example, organisms such as bread mold do not make their own food as plants do. But they are not animals either. And large groups of one-celled organisms exist that have some of the traits of both plants and animals.

Biologists also discovered that two kinds of one-celled organisms exist. One group has a nucleus without a covering, or **membrane,** around it. The nuclear material floats freely about the cell. Such a cell is called a **prokaryote** (prō'kar'ē ōt). However, organisms in the second group do have an organized nucleus surrounded by a membrane. And the structures inside these cells are much more complex than those in prokaryotic cells. The second kind of cell is called a **eukaryote** (yü'kar'ē ōt).

Prokaryotic Cell

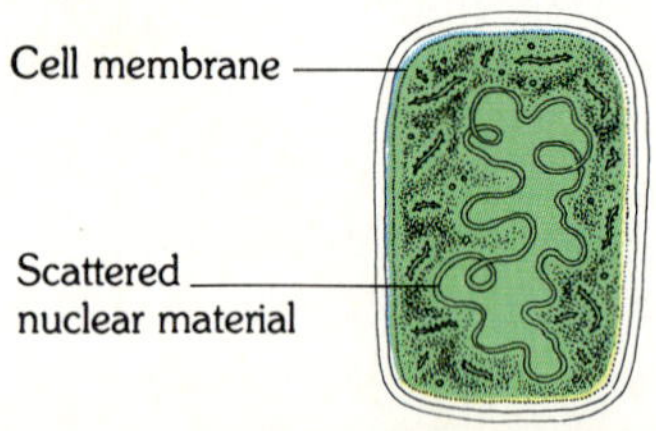

Eukaryotic Cell

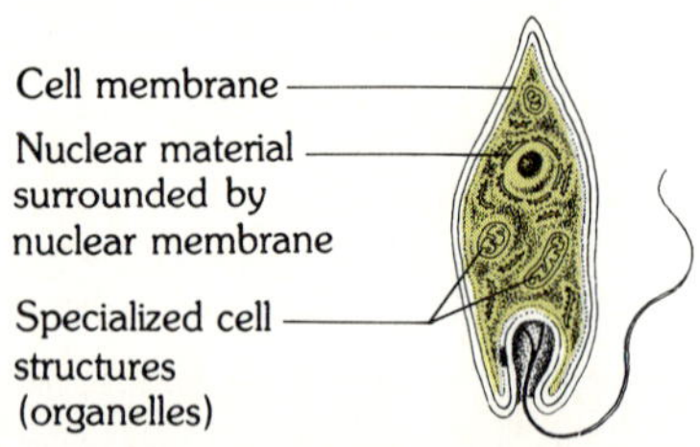

As a result of these findings, biologists decided that other kingdoms should be added to help distinguish among the different kinds of organisms. Today, many biologists believe that a five-kingdom classification system is best. Quite likely, the five-kingdom system will change as more information about organisms is discovered. But for now, the following three traits are used to help place organisms into one of five kingdoms:

1. The kind of cells that make up the organism.
2. The number of cells that make up the organism.
3. The method by which the organism obtains its food.

Review It

1. Describe the Linnaean system for classifying and grouping organisms. What are the advantages of such a system?
2. What is the difference between prokaryotic and eukaryotic cells?
3. Describe the special traits that distinguish members of each of the five kingdoms.

Five Kingdom Classification System

Kingdom Plantae

Members of the plant kingdom are also multicellular, but the cells are *specialized* for different tasks—such as support or transport

Plants use chlorophyll to *make their own food,* and so lack the ability to move about

Kingdom Fungi

These organisms are made up of many cells, but lack the ability to move about

They obtain food by *absorbing* it from dead or living organisms

Kingdom Animalia

Members of this kingdom are multicellular, but they obtain food from *outside sources*

Animals move from place to place to obtain food, swallow the food, and digest it inside the body

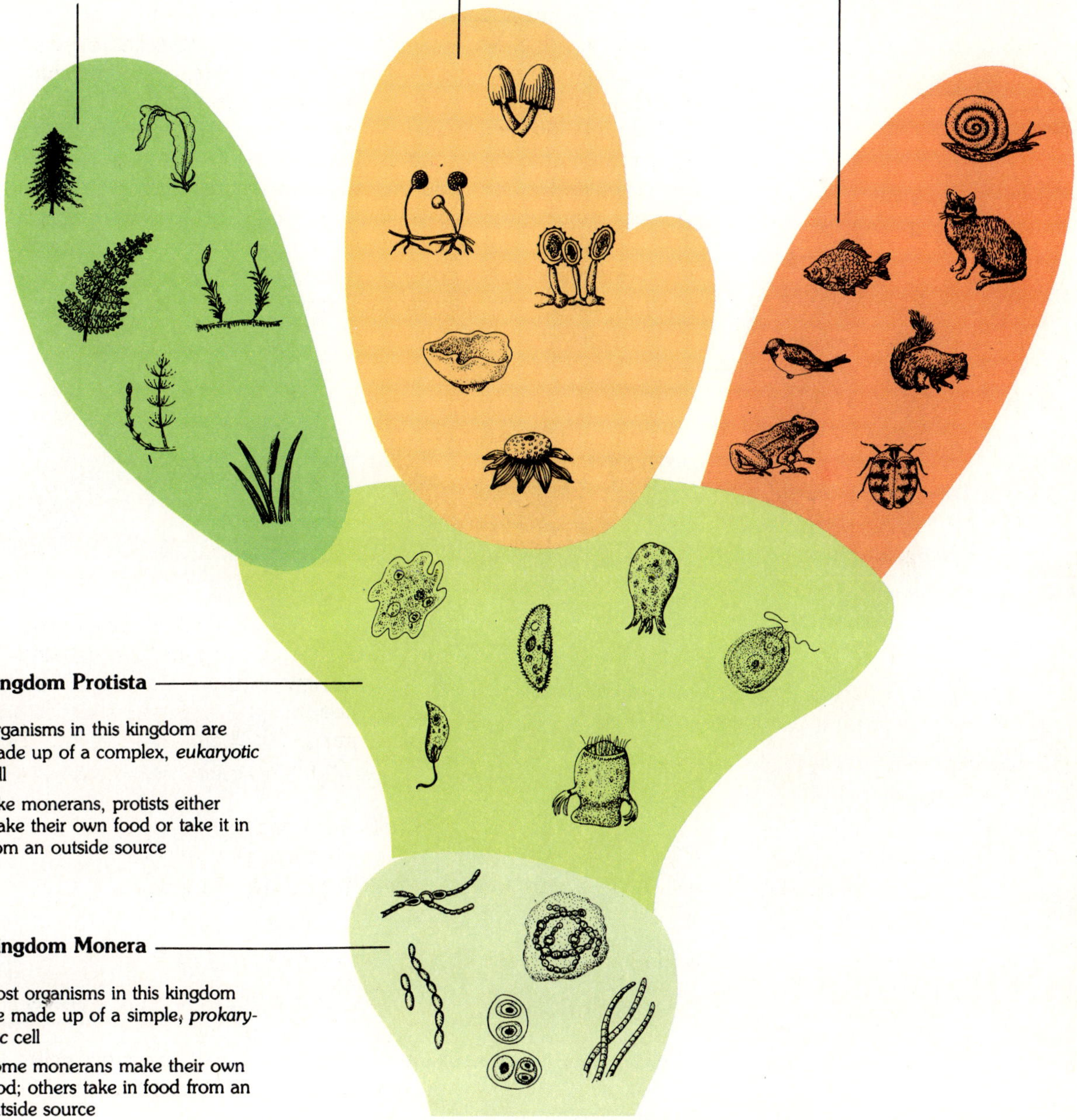

Kingdom Protista

Organisms in this kingdom are made up of a complex, *eukaryotic* cell

Like monerans, protists either make their own food or take it in from an outside source

Kingdom Monera

Most organisms in this kingdom are made up of a simple, *prokaryotic* cell

Some monerans make their own food; others take in food from an outside source

1–3
Kingdom Monera

The *monerans* include the smallest, the simplest, and the most widespread groups of organisms on Earth. Except for one species of bacteria that lives in our intestines, less is known about this kingdom than all of the others. There are two major phyla in this kingdom. You can study them below.

Bacteria

The smallest organisms in the five kingdoms are **bacteria** (bak tir′ē ə). Most bacteria are either round-, rod-, or spiral-shaped. Some kinds form colonies of chains, others form clusters. Some bacteria are capable of dividing in half every 20 minutes. At this rate, one bacterium can be the ancestor of over 1 million bacteria in just six hours.

Bacteria live almost everywhere on Earth. This includes the surface of the human body and the inside of the human digestive tract. Most bacteria feed on dead organisms and cause them to rot. Without bacteria, important materials would be locked up in dead organisms. Humans use bacteria to make industrial chemicals, to treat sewage, and to make foods like yogurt and cheese. But bacteria are also the cause of many human infections like strep throat, and tooth decay.

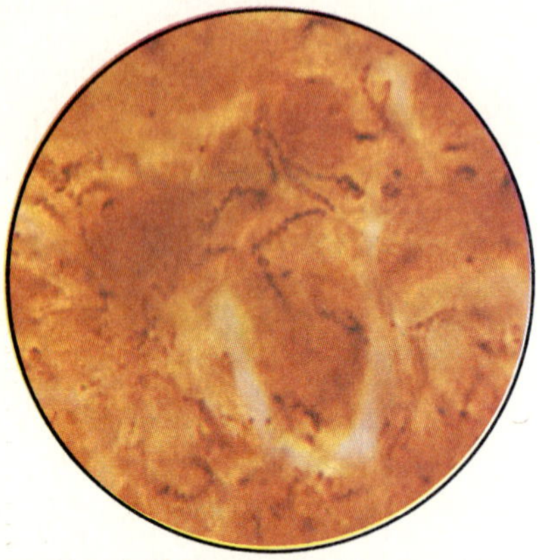

Spiral-shaped bacteria

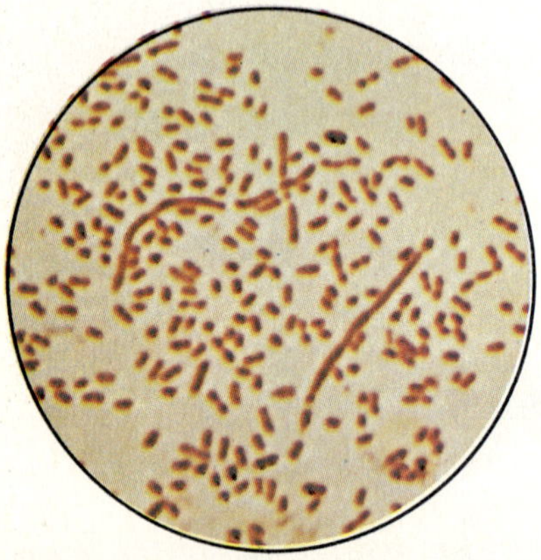

Rod-shaped bacteria

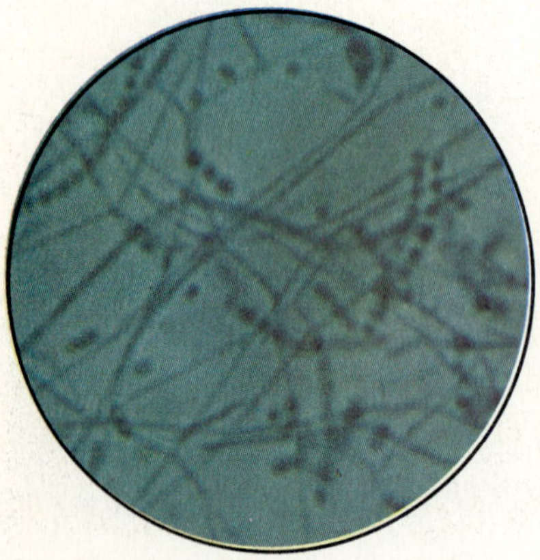

Round-shaped bacteria in chains

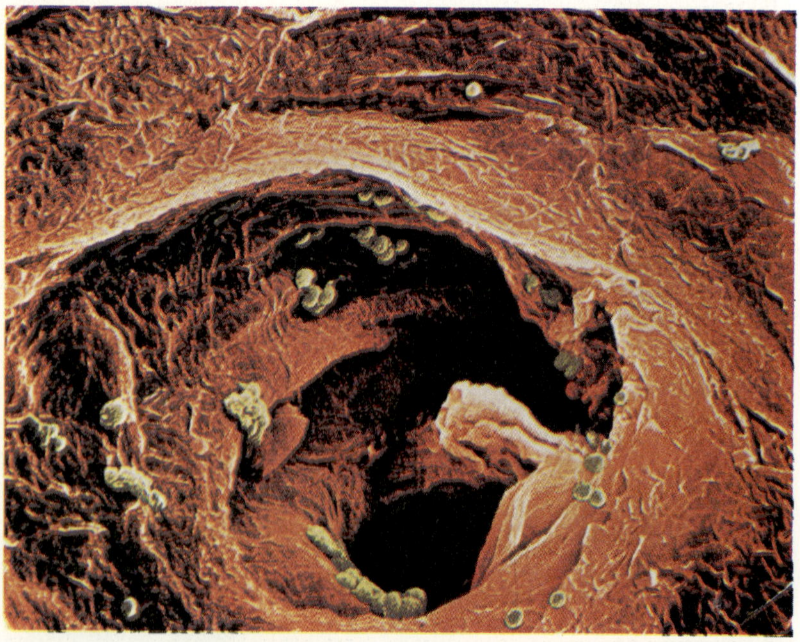

Round-shaped bacteria in pore

Blue-Green Bacteria

The cells of **blue-green bacteria** are bigger than most bacterial cells. They also contain colored chemicals which absorb sunlight and enable the organisms to make their own food. Blue-green bacteria are commonly found growing in water and on surfaces that stay wet.

Occasionally, some blue-green bacteria reproduce in large numbers. The Red Sea was named after a blue-green bacterium that actually contains large quantities of red pigments. When these bacteria increase in great numbers, the water looks as if it is red in color.

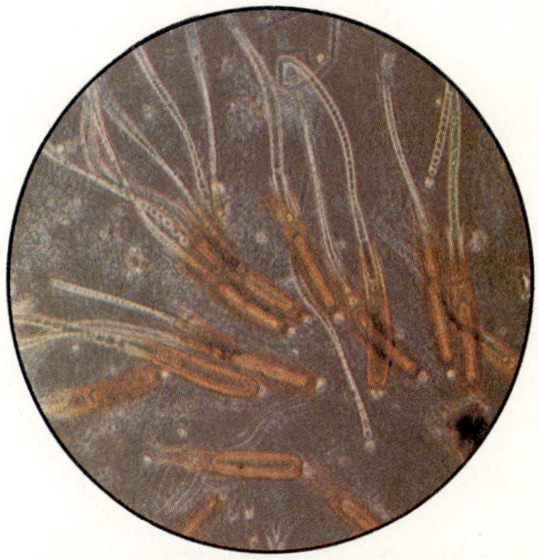

Blue-green bacteria

Water coloration due to the presence of blue-green bacteria

Kingdom Protista

Protists are much more complicated than monerans. A protist cell is somewhat like the whole body of a plant or an animal "stuffed" into a single cell. A protist cell has many specialized parts designed to carry out various life functions. For example, you have legs for walking that are made of many kinds of cells. But a protist uses special parts of its *single* cell as "legs." Protist cells are so complex that they are often thousands of times larger than one of your own body cells. You can study four phyla in the protist kingdom below and at the right.

Diatoms

Diatoms

Some protists have bodies that are covered by two "shells" which fit together like a pill box. Members of this phylum are called **diatoms** (dī′ə tomz). They float on the surfaces of fresh and salt water. Food is stored within their cells as small drops of oil. Much of our petroleum may be the stored food of diatoms that lived millions of years ago.

Dinoflagellates

Dinoflagellates

Other protists have two whiplike tails for movement. Members of this phylum are called **dinoflagellates** (din′ə flaj′ə lāts). Like diatoms, they have the ability to make their own food. And they live at the surface of water. Dinoflagellates are important to other forms of life because they capture and store energy from the sun. But an overpopulation can cause water to smell and taste bad.

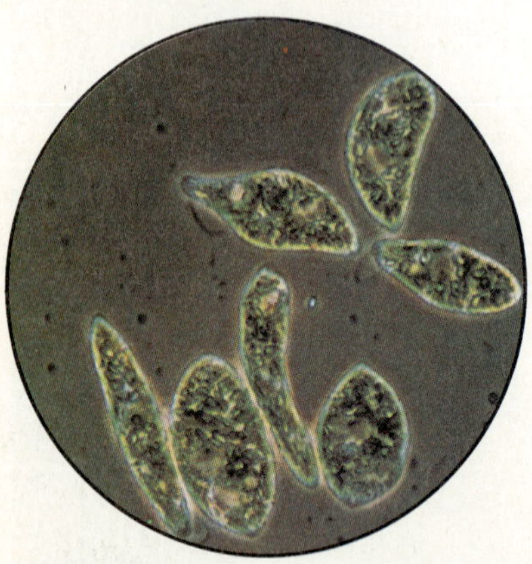

Euglena

Euglena

For years, biologists argued about a group of protists called **euglena** (yü glē′nə). They couldn't decide whether euglena were plants or animals. Euglena contain green pigments and make their own food. So some biologists argued that euglena were plants. However, euglena also have a kind of "eye," swim, and appear to have a mouthlike structure. So others argued that euglena were animals. But for now the arguments are over. Euglena are protists.

Protozoans

Another group of protists pursues, catches, and swallows its food. Members of this phylum are called **protozoans** (prō tə zō'anz). They are grouped into classes according to the way they move about. Four of these classes are shown and described below.

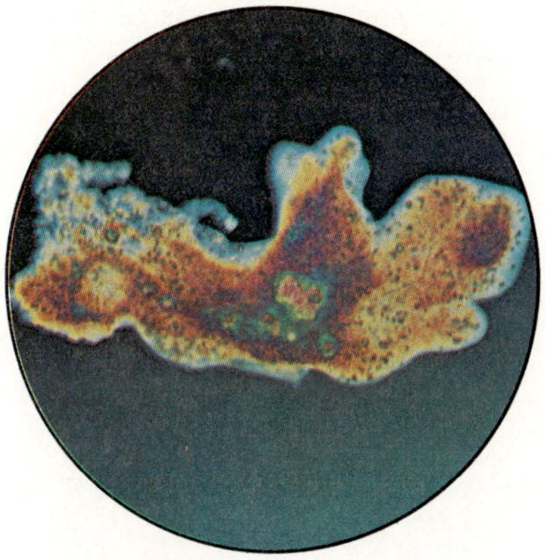

Sarcodines (sär'kō dinz) resemble small masses of cloudy gelatin. They move by pushing out parts of their cell membranes. Fluids in these parts of the cell flow in the direction that the organism is moving.

Ciliates (sil'ē āts) are covered with short, hairlike structures called **cilia.** The cilia contract in waves to push the organism along. Cilia may also help sweep food into mouthlike structures.

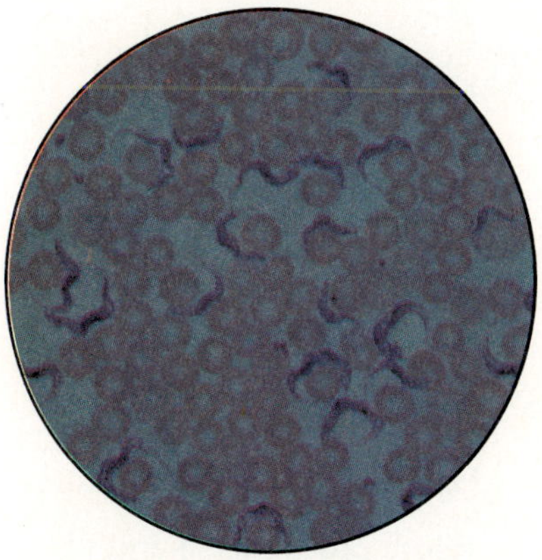

Flagellates (flaj'ə lāts) usually move by using one or two whiplike tails, or flagella. Most live independently, but some live inside other organisms. A few cause diseases such as African sleeping sickness.

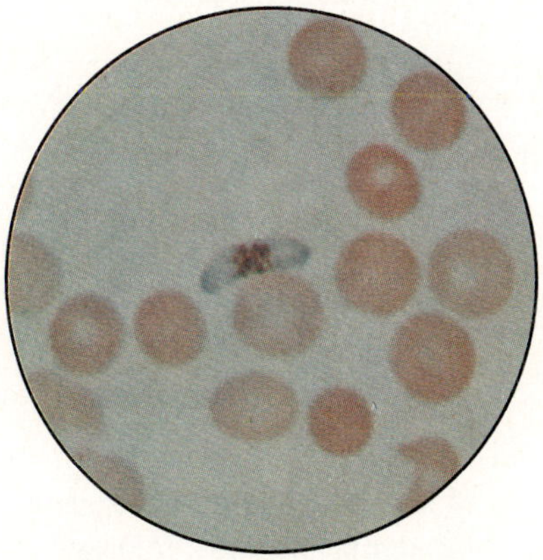

Sporozoans (spôr ə zō'anz) live inside other organisms. Most have *no* structures for moving about. They must depend on their hosts for transportation. For example, malaria sporozoans are carried about in the human body by the bloodstream.

Review It

1. Describe the kingdom Monera. What are some major differences between the bacteria phylum and the blue-green bacteria phylum in this kingdom?
2. What are the most important differences between members of kingdom Protista and members of kingdom Monera?
3. Describe members of one of the phyla in kingdom Protista.
4. Describe the method of locomotion used by each class in the phylum Protozoa.

Activity

Exploring Pond Water

Purpose
To find organisms in pond water, describe them, and compare their methods of movement and food gathering.

Materials
Water sample from pond or aquarium
Medicine dropper
Microscope slide
Cover slip
Microscope
Light source
Drawing materials

Protozoans and blue-green bacteria

Procedures
1. Add one or two drops of pond water to the center of a microscope slide. Try to include in the drops tiny amounts of mud or leaves. Organisms often collect around these things.
2. Gently place a cover slip over the drop. Look at the drop using the low power of your microscope.
3. Try to find some one-celled organisms in the drop. Then try to find places where these organisms occur in the largest numbers.
4. Use the high power of your microscope to examine the smallest organisms. Center some of the smallest organisms in the field of view and then switch to high power to examine them. Also examine details of the largest organisms.
5. Make drawings of the organisms you find most interesting. Use the pictures to help you identify some of these organisms. Use other reference books to help identify organisms that you cannot find here.

Analysis
1. How many different kinds of organisms were you able to identify?
2. Which kinds of organisms were the most common?
3. Which of the organisms probably make their own food? How can you tell?
4. Which organisms, if any, chased and then "swallowed" their food? In what other ways did you observe organisms obtaining food?
5. How many ways did you observe organisms moving about in the pond water? Describe each of these ways.

1–4 Kingdom Fungi

More than 100,000 kinds of *fungi* have been described and named. Biologists estimate that at least another 10,000 remain to be discovered.

A special trait of members of this kingdom is their method of obtaining food. Fungi feed by *absorbing* the dead or living material of other organisms. They produce threadlike structures that grow deep into their food materials. Chemicals from the threads dissolve the food. The fungi then absorb the dissolved food through these threads. When the food is part of a living organism, the fungus may cause disease to its host. Two examples of such diseases are athlete's foot in humans and Dutch elm disease in American elm trees.

You can see examples of the five major classes of fungi below.

Penicillium digitatum **on citrus**

Dragonfly larva parasitized by water mold

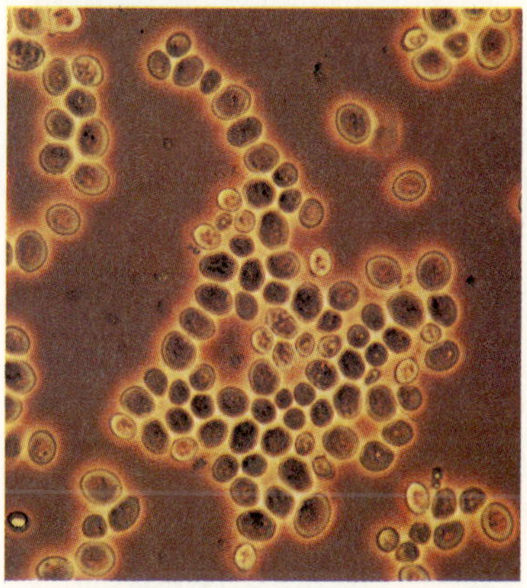

Ascomycetes sac fungi

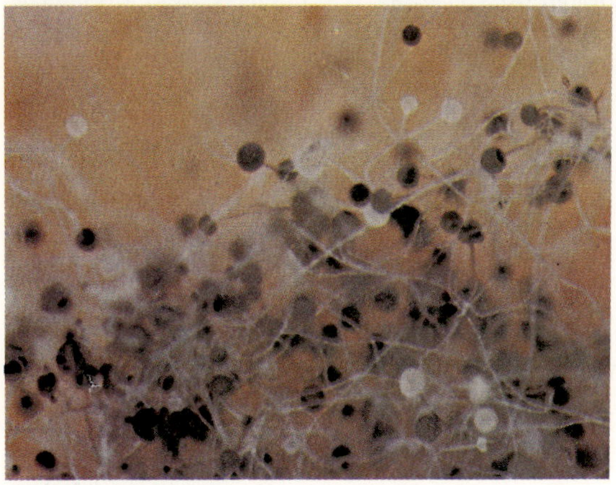

Rhizopus on bread

Lepiota rachodes **mushroom**

Kingdom Plantae

This kingdom contains mostly green, multicellular organisms. Their cells are of various types and form special organs. Most of these organs assist in the production of food and in the reproduction of the plant.

The members of the five phyla in this kingdom can be told apart by (1) where they live, (2) their color, and (3) whether or not they have an internal transport system for carrying food materials. You can study some members of the five phyla below and on the next page.

Brown algae—Fucus

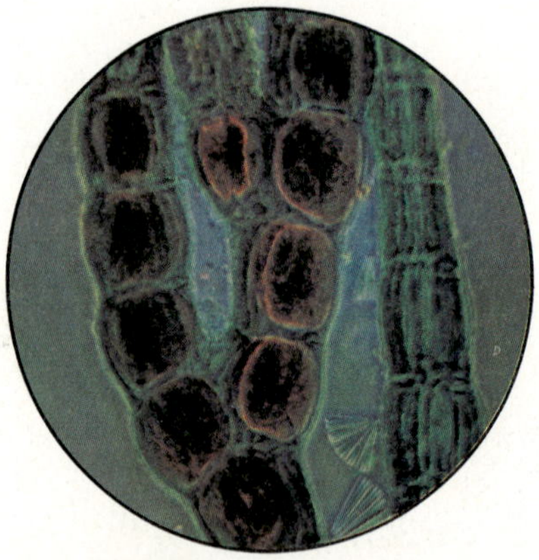

Red algae—Polysyphonia

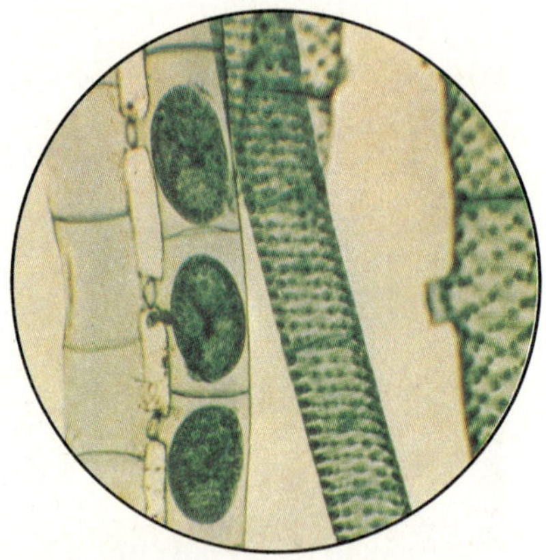

Green algae—Spirogyra

Brown Algae

Members of the **brown algae** phylum live mostly in cool seawater. They contain gold and green pigments that give them their brown colors. Those that live in shallow water have rootlike organs called **holdfasts.** They also have stemlike organs called **stipes** and leaflike organs called **blades.** The brown algae provide food and shelter for many ocean animals.

Red Algae

Another phylum in the plant kingdom is the **red algae.** They usually live in warm tropical seas. Sometimes they live at depths as great as 200 meters. Their color is due to a red pigment. The red pigment absorbs the blue light that penetrates deep water. They use the energy in this blue light to make food. Gelatin substitutes and other foods are obtained from red algae.

Green Algae

The group of algae that shows the most variety is the **green algae.** They live on the surfaces and near the shores of oceans. And they are found on the surfaces of fresh water. On land, they cover most of the surfaces that receive sunlight and stay damp. In some ways the green algae are like land plants. They are green, store food, and have cell walls.

Bryophytes

There are two major phyla of land plants. The first of these phyla is the **bryophytes** (brī′ə-fīts). They have no transport system inside their bodies to carry materials. They are small green plants that live only in moist places and rarely grow larger than 15 centimeters. Of the 25,000 different kinds of bryophytes, mosses and liverworts are the most common.

Tracheophytes

The other major phylum of land plants is the **tracheophytes** (trā′-kē ə fīts). They have vessels that transport food, water, and minerals to all parts of the plant body. The phylum is divided into three classes according to the kinds of seeds each plant forms. *Ferns* have no seeds. *Cone-bearing* plants have seeds that are uncovered. And *flowering plants* produce covered seeds that are enclosed by a fruit.

Liverwort

Fern

Moss

Cone-bearing plant

Flower-bearing plant

Kingdom Animalia

The animal kingdom is the most complex of all five kingdoms. Biologists have never been able to agree on how many phyla there should be in this kingdom. They also disagree over how animals in each phylum are related to one another.

At every moment in the history of modern biology there has been more than one classification system for the same organisms. And as more information about the relationships among organisms becomes available, the classification systems change. For example, all living fish were at first grouped together into one large class of fish. Now the fish are divided into two smaller classes. The fish in one class have skeletons made of bone. The fish in the other class have skeletons made of **cartilage** (kär′tl ij), or gristle.

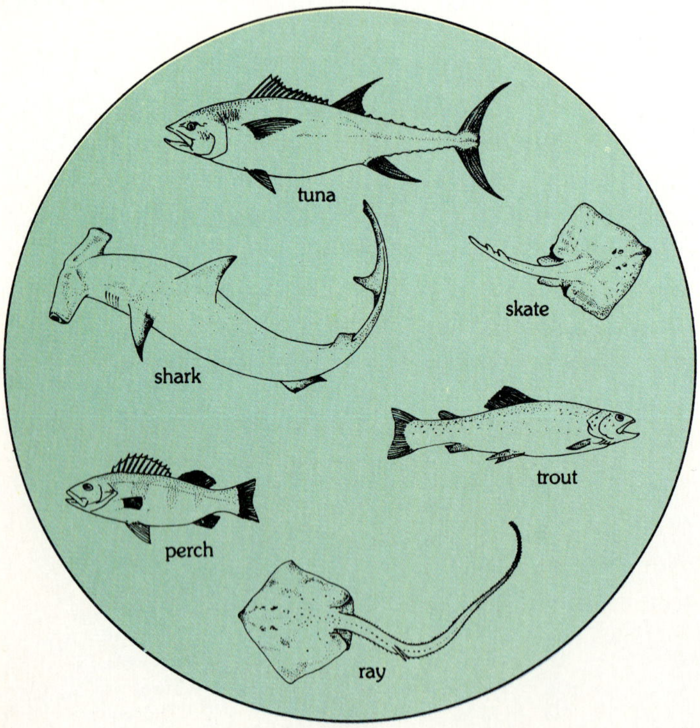

Fish

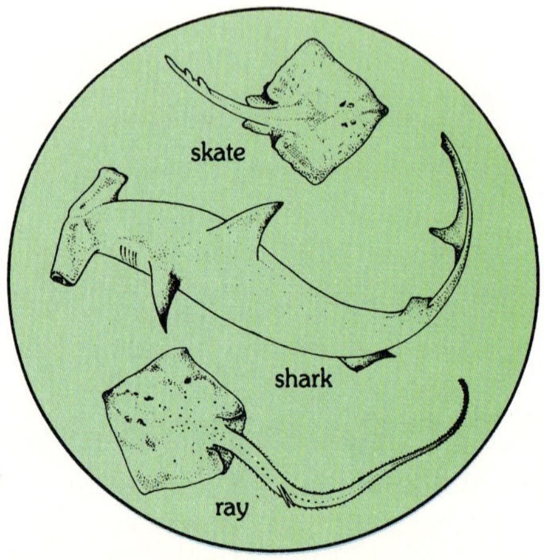

Cartilaginous Fish

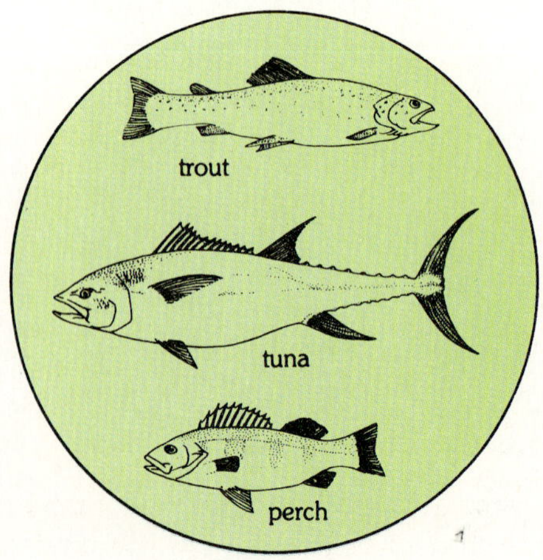

Bony Fish

Early Classification

Later Classification

Species in the Animal Kingdom

The pie chart below shows the number of species that are currently recognized in the different groups of the animal kingdom. You can see that there are more species in the arthropod phylum than in any other phylum. And there are more species of insects than there are species of all other groups combined!

Today there are at least nineteen recognized phyla in the animal kingdom. You can study some of the major groups on the following pages.

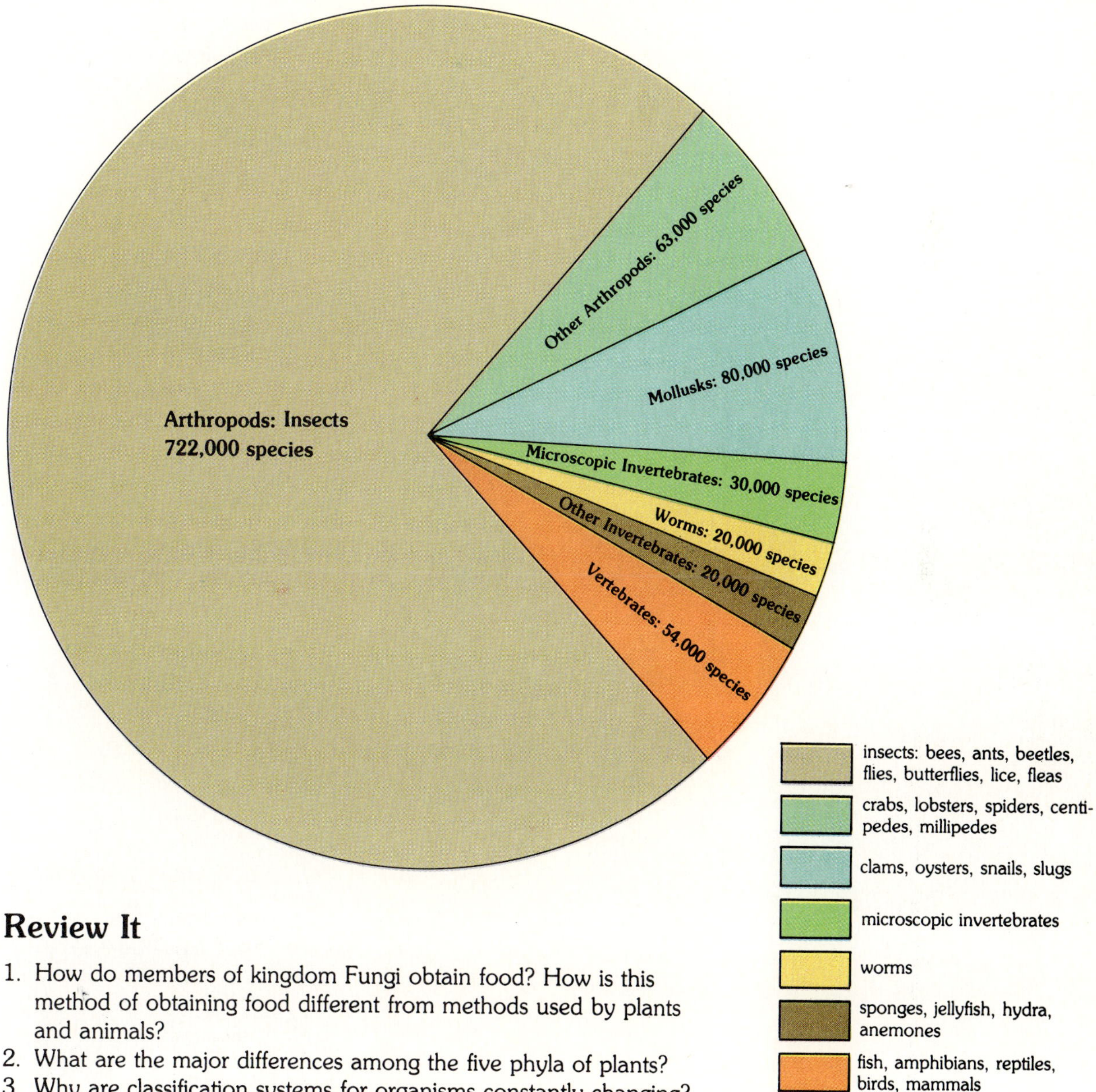

Review It

1. How do members of kingdom Fungi obtain food? How is this method of obtaining food different from methods used by plants and animals?
2. What are the major differences among the five phyla of plants?
3. Why are classification systems for organisms constantly changing?

Classification of Kingdom Animalia

Phylum Porifera

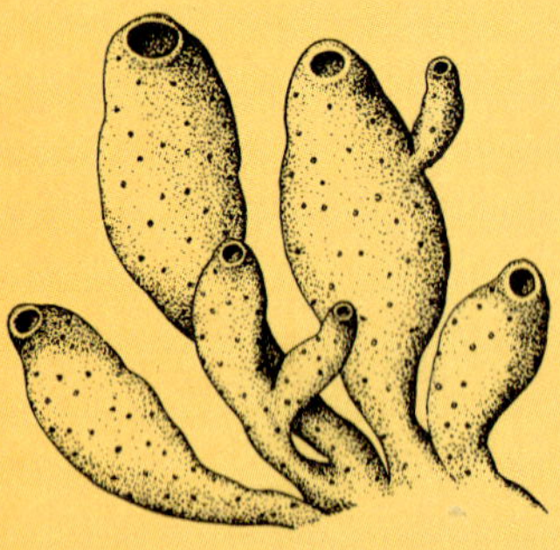

Sponge

Phylum Coelenterata

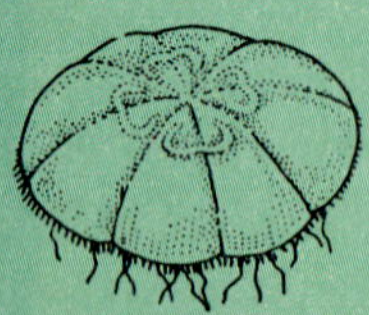

Jellyfish

Phylum Platyhelminthes

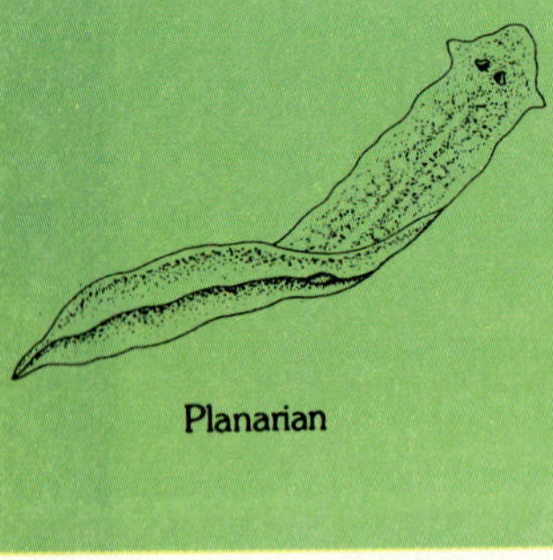

Planarian

Phylum Nematoda

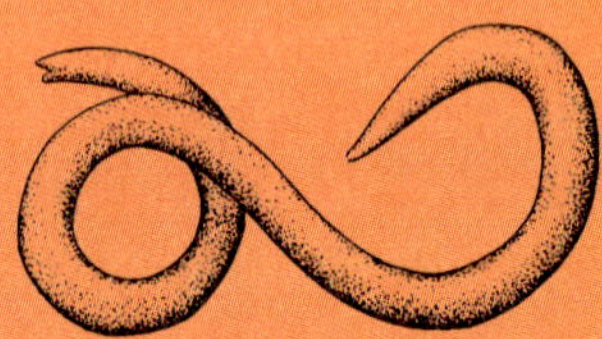

Roundworm

Phylum Rotifera

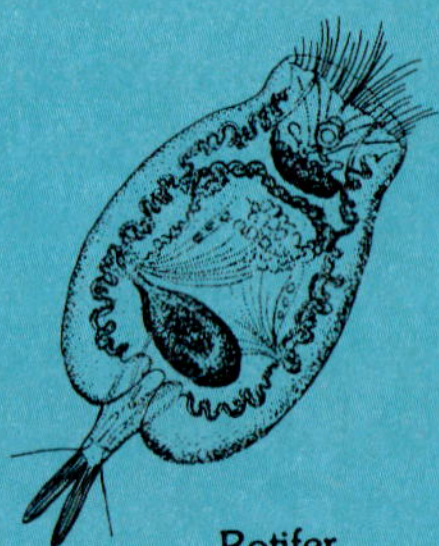

Rotifer

Phylum Annelida

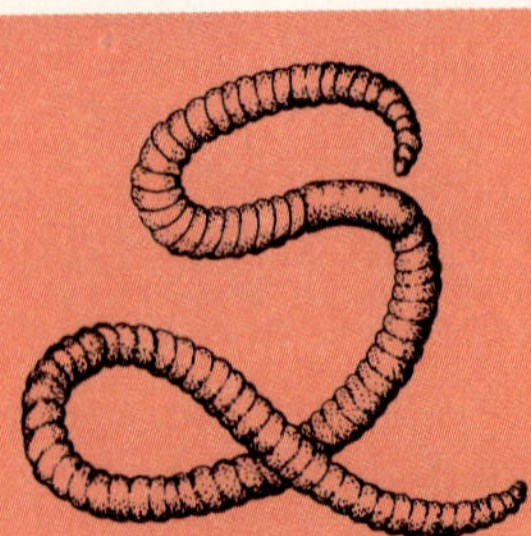

Earthworm

Phylum Echinodermata

Starfish

Sea urchin

Phylum Mollusca

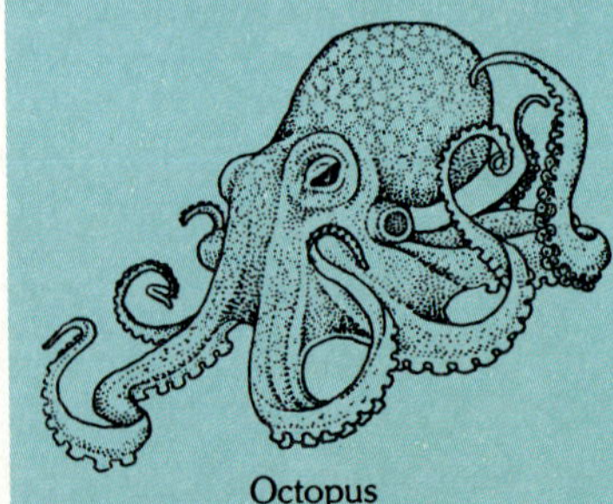

Octopus

Clam

Phylum Arthropoda

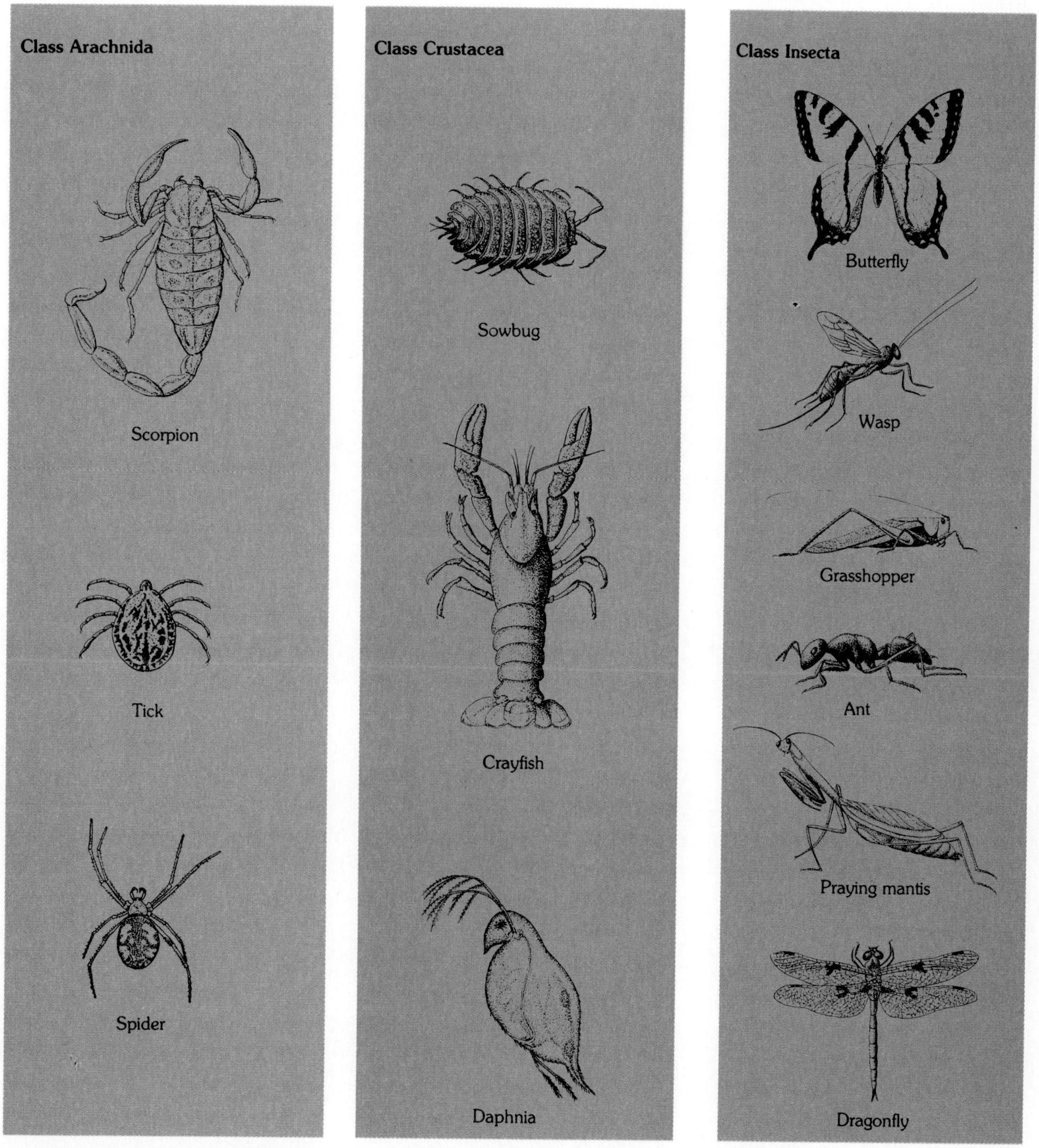

Phylum Chordata

Class Mammalia
Subclass Prototheria
(egg-laying mammals)
Platypus
Subclass Metatheria
(marsupials; mammals with pouches)
Kangaroo
Subclass Eutheria (mammals with placentas)
Rabbit
Wolf
Bat
Squirrel
Pangolin
Mole
Gorilla
Porpoise

Activity

Using a Taxonomy Key

Purpose
To be able to use a taxonomy key to identify organisms correctly.

Materials
Paper and pencil

Procedure
You are going to use a **taxonomy key** to identify some species of poisonous snakes from Africa. A *key* is a guide that helps people identify different kinds of organisms. To use a key, you must match up the physical traits of the organism you are trying to identify with the physical traits listed in the key.

Use the key at the right to find out the species name and the common name of each snake on the next page. Then answer the questions below.

Analysis

1. What physical traits were used in the key to help you identify the snakes?
2. How many *genera* of snakes did you identify? How many *species* did you identify?
3. Why might it be important for you to know how to use a snake key? When might you need to know how to use a *plant* key?

Key to Species of African Snakes

The key below describes some of the physical traits of the snakes at the right. The key always gives you *two* choices. If choice A does not describe the snake, then choice B will. After each choice, the snake is either named or a number tells you where to look further in the key. It is important to start at the *beginning* of the key. You will be dividing the snakes into smaller and smaller groups as you go along.

1. a. Crown of head covered with small, irregular scales; head arrow-shaped	Go to #2
b. Crown of head covered with large plates; head pointed or rounded	Go to #3
2. a. Body green with black markings on back; belly plates lightly colored	Sedge Viper, *Atheris nitschei*
b. Body green to yellow with no black markings on back; belly plates lightly colored	Green Bush Viper, *Atheris squamigera*
3. a. Head narrow and pointed; body speckled brown	Bird Snake, *Thelotornis kirtlandii*
b. Head and nose rounded	Go to #4
4. a. Black markings on head	Go to #5
b. Head unmarked	Go to #6
5. a. Black, "Y"-shaped markings on head; body light gray to dark brown	Rhombic Night Adder, *Causus rhombeatus*
b. Black bands on head; body tan or brown	Banded Water Cobra, *Boulengerina annulata*
6. a. Scales have ridges; body green with light belly plates	Boomslang, *Dispholidus typus*
b. Scales *smooth;* body dark with dark belly plates.	Go to #7
7. a. Body glossy and darkly colored	Forest Cobra, *Naja melanoleuca*
b. Body shiny, but not glossy	Spitting Cobra, *Naja nigricollis*

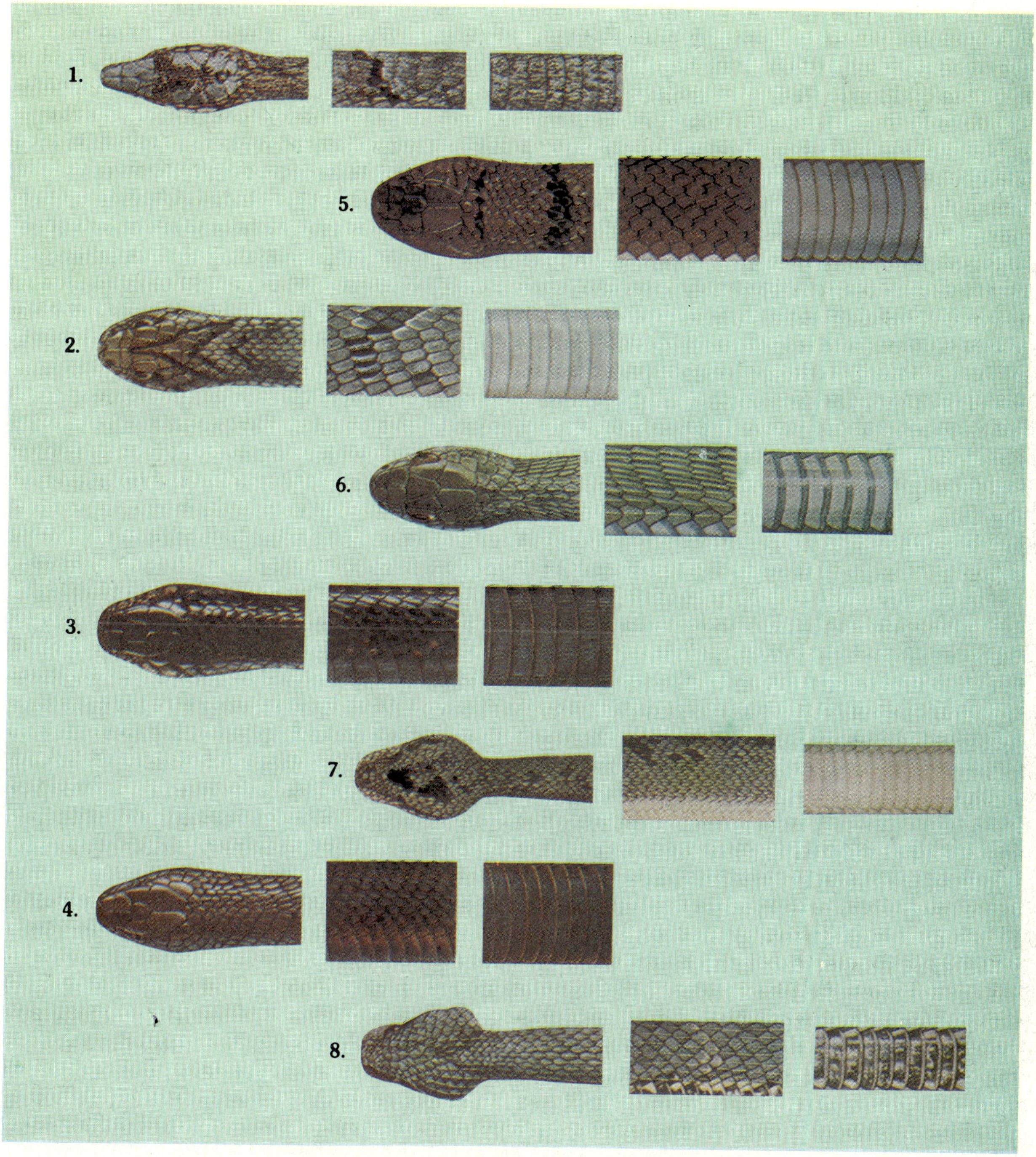
1.
5.
2.
6.
3.
7.
4.
8.

Chapter Summary

There are many ways to define life. But no single definition seems completely satisfactory. However, most living things do share the following traits: (1) they are organized as cells; (2) they are made up of complex chemical compounds; (3) they reproduce; (4) they grow and develop; (5) they respond to stimuli; and (6) they acquire and use energy by the process of metabolism.

People have been asking questions about how life originated for thousands of years. However, in recent years scientists have been testing Oparin's hypothesis of the origin of life. A few scientists have produced some of the basic chemicals of which organisms are made. Thus far, however, no one has created life in the laboratory.

There is a tremendous variety of living species. Organisms seem to exist in every size, shape, and color. Biologists keep track of their knowledge of organisms by classifying them. They group them into kingdoms, phyla, classes, orders, families, genera, and species. Groupings are made mostly on the basis of similarities in physical structure. Once organisms are classified, they can be given scientific names and identified by using keys.

The millions of kinds of organisms are grouped by most biologists into five kingdoms: monerans, protists, fungi, plants, and animals.

Questions to Think About

1. Suppose that Linnaeus had not created a special system for naming and classifying organisms. What kinds of problems would biologists be having today in working with organisms?
2. The chemosynthesis hypothesis states that life developed from nonliving materials. Does this support the idea of spontaneous generation? Why or why not?
3. Biologists estimate that there may be as many as ten million more undiscovered species. What might be some reasons why so many species remain to be discovered?
4. Class Insecta includes over 700,000 species, but class Mammalia has only 4,500 species. Why do you think there might be so many more insects than mammals?

Extra Research

1. Find out what additional research projects are being planned and carried out by NASA in its search for life on other planets.
2. Obtain a key to the trees, flowers, or insects in your part of the country. Practice using the key with organisms in your area.
3. Dutch elm disease destroyed American elm trees in many parts of North America. Now a disease called yellow palm is destroying palm trees in Florida. What organism is causing this disease, and what is being done to control it?
4. Find a specimen of a *lichen*. Examine it under a microscope. See if you can figure out what a lichen is and how it is put together.
5. A one-celled organism called a *methanogen* survives in places free of oxygen. Some researchers think this organism does not belong to *any* of the five kingdoms. Find out more. How might this discovery change the way organisms are classified?

Chapter Test

Part I

Copy the number of each test item. If the item is *true*, write *T* after the number; if the item is *false*, write *F*.

1. Some monerans make their own food. (19)
2. The cells of prokaryotes have a membrane around the nucleus. (16)
3. The classification of organisms into five kingdoms may change. (16)
4. Pasteur provided evidence that life does not arise from rotting meat. (11)
5. Protozoans can be grouped on the basis of how they move. (21)
6. One test for life on another planet is to scan the landscape with cameras. (6)
7. Stanley Miller provided evidence that cells were produced from the gases in the air of the early earth. (12-13)
8. Members of a species can mate but do not produce fertile offspring. (15)

Part II

Copy the number of each sentence below. After each number, write the word or words that complete the sentences correctly.

9. _____ is a protist that can make its own food, has an eye spot, and can swim. (20)
10. Mosses and liverworts belong to the phylum _____. (25)
11. Ferns differ from other tracheophytes because they produce no _____. (25)
12. The short hairs used for motion in a class of protozoans are called _____. (21)
13. Gases on the early earth might have consisted of water, methane, hydrogen, and _____.(12)
14. The _____ have transport vessels that carry food, water, and minerals to all parts of the plant. (25)
15. The scientific name of an organism is made up of its genus and its _____ names. (15)

Part III

Copy each item from List A. After each item, write the letter and words from List B that best describe the item.

List A

16. prokaryote (16)
17. eukaryote (16)
18. Stanley Miller (12)
19. spontaneous generation (10)
20. metabolism (5)
21. chemosynthesis (12)
22. Linnaeus (15)
23. phylum (15)
24. key (32)
25. tracheophyte (25)

List B

a. tool used to identify different kinds of organisms
b. researcher who created amino acids in the laboratory
c. hypothesis that life could arise from nonliving materials, such as meat
d. kind of cell that contains a membrane around its nucleus
e. one of the groups in Linnaeus's classification system
f. phylum in the kingdom Plantae
g. person who developed a system for naming and classifying organisms
h. hypothesis that life originated on Earth billions of years ago when gases in the air started to combine with each other
i. a trait of living things
j. kind of cell that contains no membrane around the nucleus.

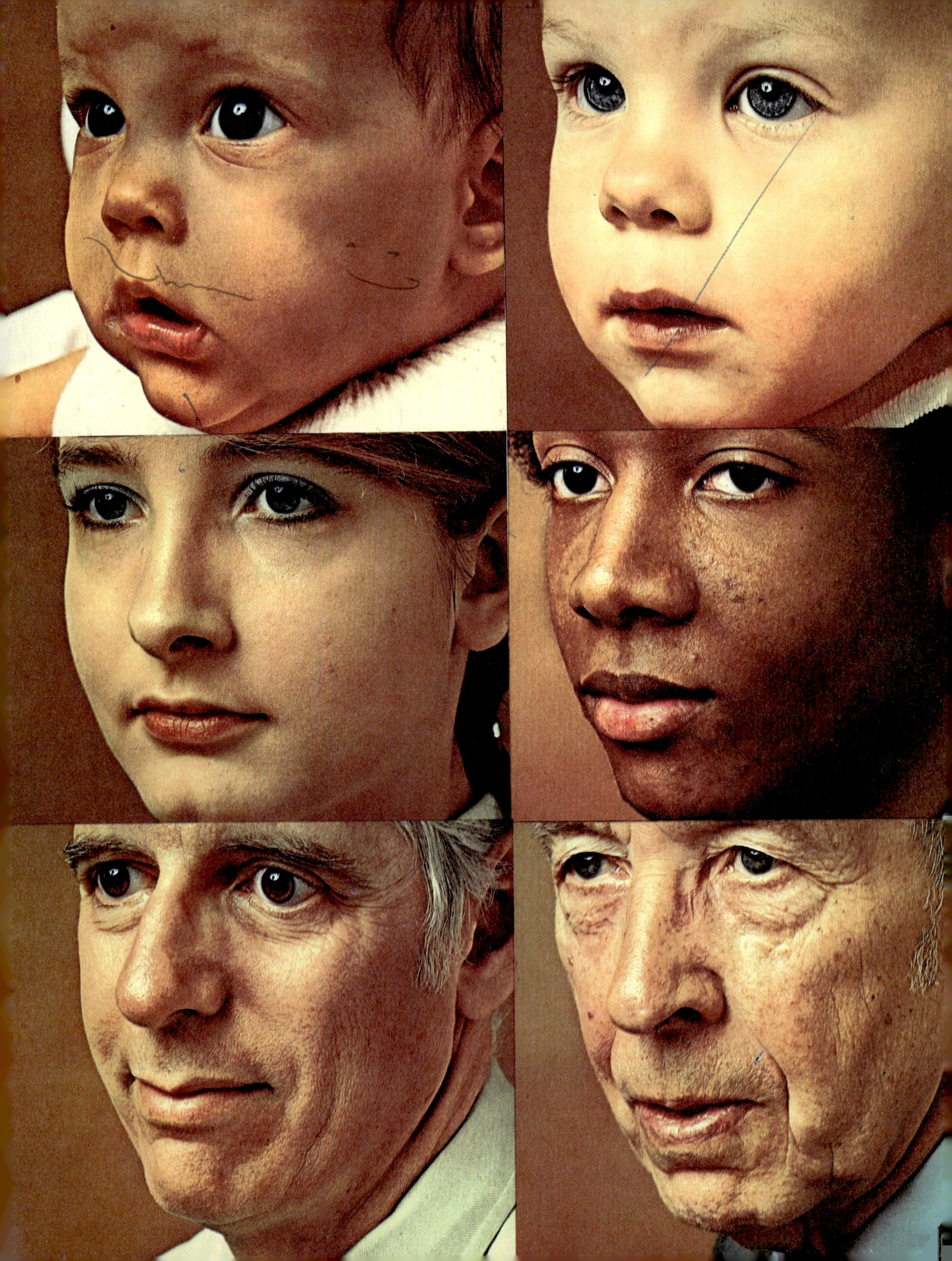

Chapter 2

The Humans

Humans are more like members of a group called *primates* than any other group of animals. However, we humans also possess traits that make us different from all other groups of living things. In this chapter you will examine some of the physical similarities and differences among humans and other primates. You will learn how the primate body is organized into tissues, organs, and systems. And you will survey some of the ways human behavior is different from that of other primates. You will also have a chance to recognize some of the ways individual humans vary in body structure and behavior.

Some of the human traits you will learn about have given us the power to control much of the environment. As a result of this ability to control the environment, humans must constantly make decisions that affect the future. In the last part of the chapter, you will have a chance to practice making decisions about a future event that might affect human survival.

Chapter Objectives

1. Describe how humans are physically similar to and different from other primates.
2. Identify some of the major physical differences between humans and apes.
3. Describe how cells in the human body are organized into tissues, organs, and systems.
4. Give examples of how human behavior is different from the behavior of other animals.
5. Recognize that within all groups of humans there is a wide range of variation in structure and behavior.
6. Explain why decision making is a trait important to the future survival of humans.

Although humans share many traits as members of the class of mammals, each individual is unique. Some of that uniqueness is expressed by facial characteristics.

2–1
Physical Traits of Primates

Humans are members of the order of mammals known as the **Primates** (prī′māts). Monkeys, apes, lemurs, and tarsiers are also included in this group. All primates share several major traits which make them different from other mammals.

1. The eyes of primates face forward. This allows the animals to have overlapping views of their surroundings. Such **stereovision** (ster′ē ō-vizh′ən) helps primates to judge distances and to perceive depth. Primates are the only mammals known to have stereovision. They are also the only mammals known to have full color vision.
2. The hands and arms of primates are different from all other groups of animals. Each hand has five digits. The first digit, or thumb, is able to touch the tips of the other four digits. This arrangement allows primates to grasp objects easily. And primates have nails instead of claws for grasping.
3. The forearm of primates is made up of two long bones. These are arranged so that the arm can turn at the elbow. Thus, primates can use their arms and hands for such tasks as peeling fruit and playing violins.
4. The brains of primates are more highly developed than the brains of any other group of animals.

Traits of primates

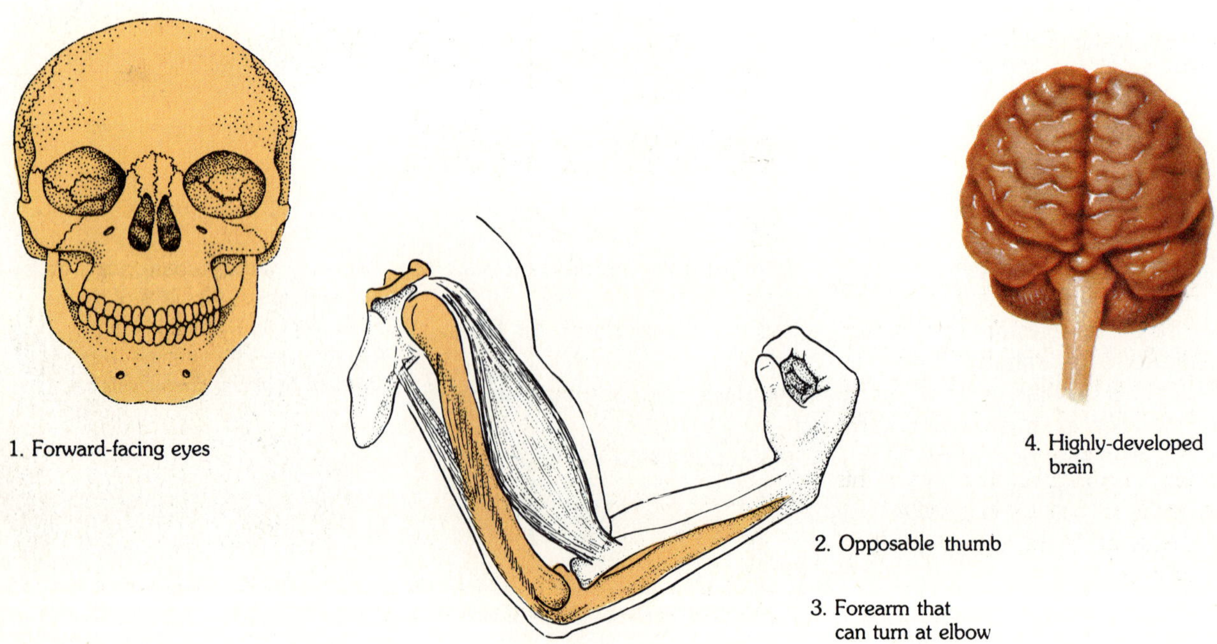

Groups of Primates

The order Primates is divided into five families. You can study each of these groups below and on the next page.

Lemurs, Tarsiers, and Their Relatives
Members of this group usually have pointed faces. Their face muscles do not allow them to make "funny faces" as monkeys, apes, and humans do. These primates feed mainly on insects And they drink by lapping liquids with their tongues. The more advanced primates drink by sucking liquids.

Pointed face

New World Monkeys (North and South America)
New World monkeys live in trees. In many species, the long tail can be used as another arm or leg to grasp objects. These monkeys have nostrils that are widely separated and open to the sides.

Nostrils widely separated

Old World Monkeys (Asia and Africa)
Old World monkeys are also tree dwellers. But some species, such as baboons, live on the ground. In general, the nostrils are close together and face forward. The tails are not able to grasp objects. An outstanding trait of this family is their **sitting pads.** These pads are often brightly colored.

Nostrils closely spaced

Brachiation in apes

Life Nature Library/*The Primates,* Published by Time-Life Books, Inc.

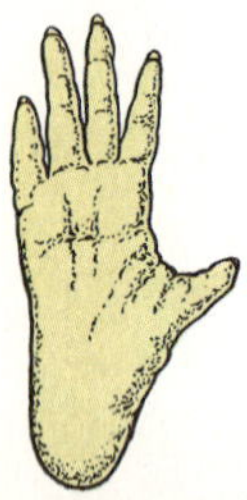

Chimpanzee hand

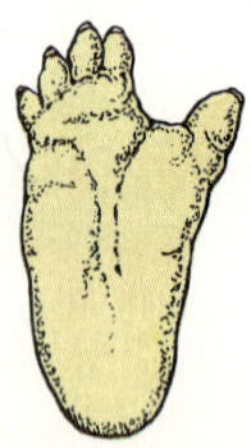

Chimpanzee foot

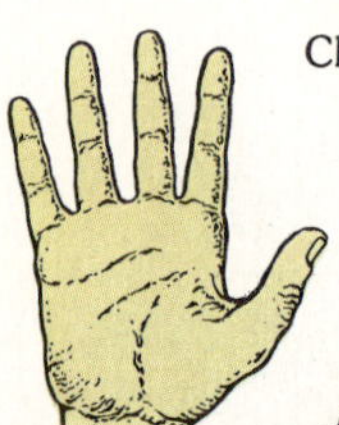

Human hand

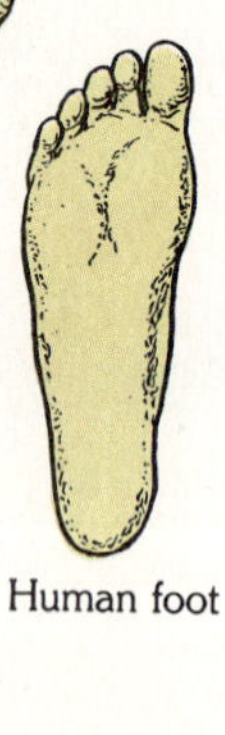

Human foot

The Apes

This group includes the orangutans, gibbons, chimpanzees, and gorillas. Members of this group have no tails. Although apes are often photographed while standing up, their most natural position for walking is on all four limbs. Apes live in forests. And they travel by using their long arms to swing from tree branch to tree branch. This arm-swinging technique is called **brachiation** (brak′ē ā shən). Apes are mainly vegetarians, but chimpanzees do eat some meat. Chimpanzees are also able to use simple tools.

Humans

Homo sapiens is the only living species in this group. Humans have a natural, upright posture for walking. Since humans can stand for long periods of time, they are able to use their hands for operating tools.

The human hand is different from that of other primates because it has a strong, precise grip. This grip is due to the large, muscular thumb. The human foot is also unlike that of other primates. It is completely adapted to heel-to-toe walking.

Humans can live in a variety of environments and eat a variety of foods. And since humans can make and use tools, they can live most places on Earth. They are not limited to the forests. Humans are also the only primates that have a voicebox suited for speaking a complex language.

Special Human Physical Traits

Humans share more traits in common with apes than any of the other primates. For example, both apes and humans have arms that can be raised above the head and used for swinging. And both groups have faces that are flat below the eyes. This enables apes and humans to examine an object close up with two eyes at the same time. However, we humans have many special physical structures that make us different from apes. One of our most important physical traits is the ability to stand and to walk in an upright position. Many of the structures described below and on the next page help us to maintain this upright position.

Pelvis
The human pelvis is broad and bowl-shaped. This shape helps to support the weight of the internal organs. And the shape also helps to keep the backbone upright over the legs.

Apes have a long, narrow pelvis. This type of pelvis tends to throw the backbone forward, causing the animal to appear stooped.

Legs
Human leg bones are long, straight, and held in place by knees that lock. Large buttock muscles pull the body upright.

Apes have knees that do not lock, and the leg bones are somewhat bowed. The buttock muscles are small and cannot pull the body completely upright.

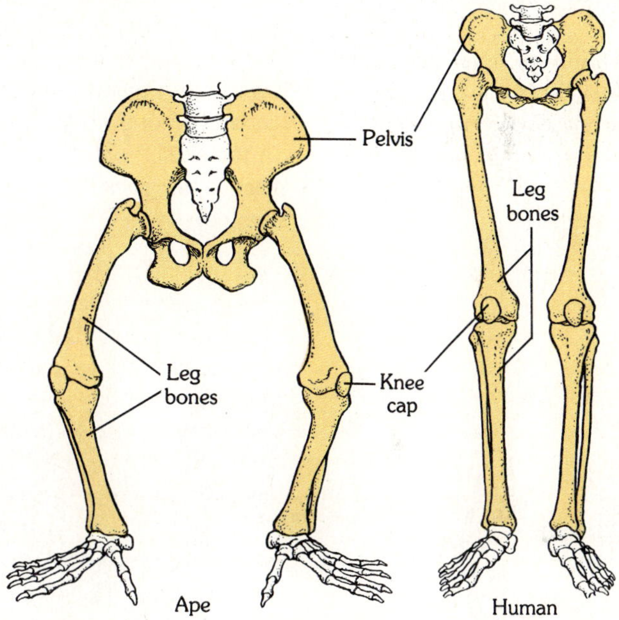

Backbone
The human backbone is S-shaped. This gives the upper body spring and strength while standing. The S-shape also helps to balance the upper body over the pelvis.

Apes have a backbone that is curved like a bow. Standing upright for long periods would probably give an ape a terrible backache!

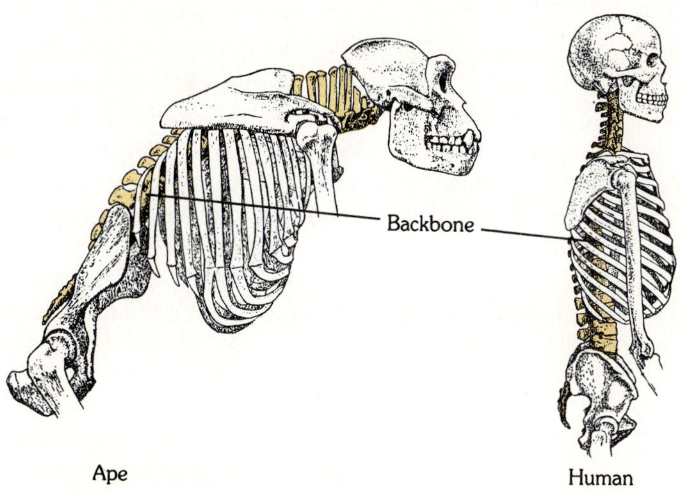

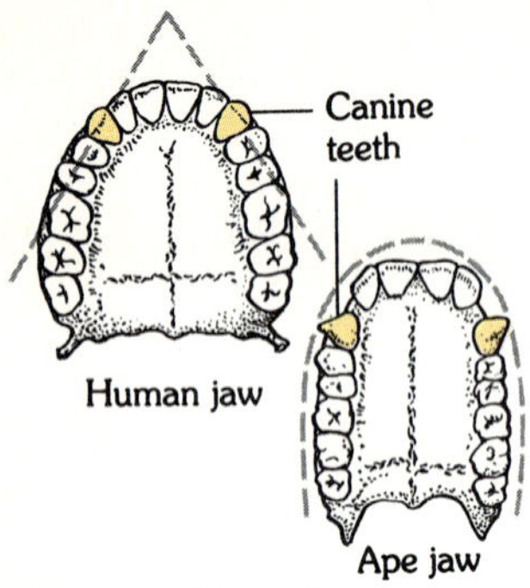

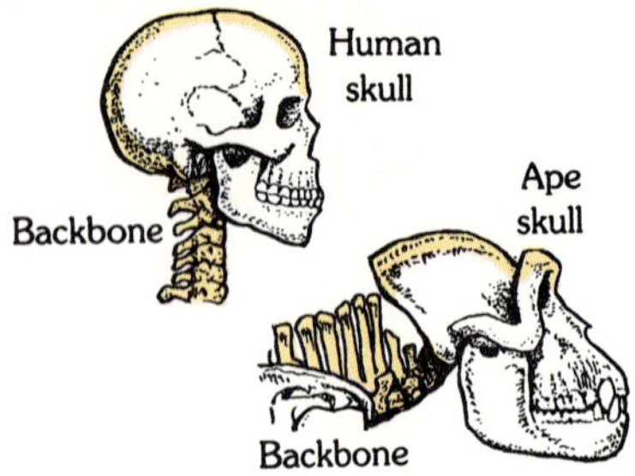

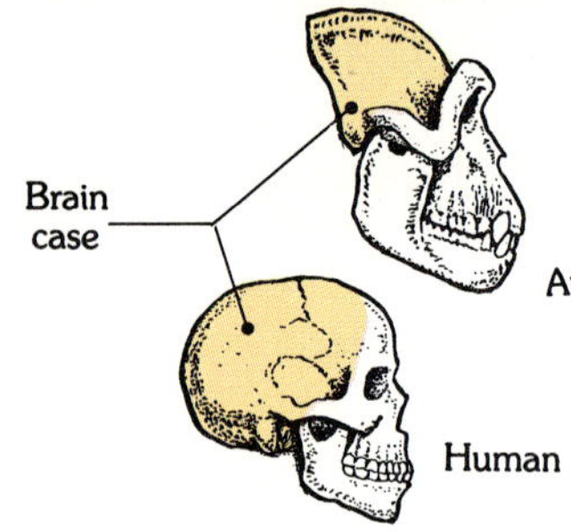

Jaws and Teeth
Humans have a jaw shaped like a wide letter *V*. Human teeth are all about the same size and length.

Apes, however, have a U-shaped jaw. And the **canine teeth,** or eye teeth, are much larger than the other teeth.

Skull
The human skull is balanced on top of the backbone. Face and neck muscles are small. So the face looks smooth and rounded. The ape skull lies forward on the backbone. Powerful muscles, attached to bony ridges and knobs on the skull, hold the skull in place. This gives the face a massive look.

Brain Case
The human brain is larger and better developed than the brain of any other primate. The **brain case** of the human is about twice the size of that of the ape.

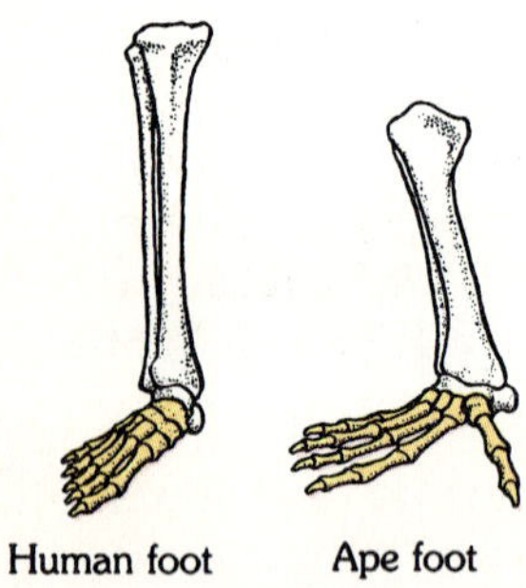

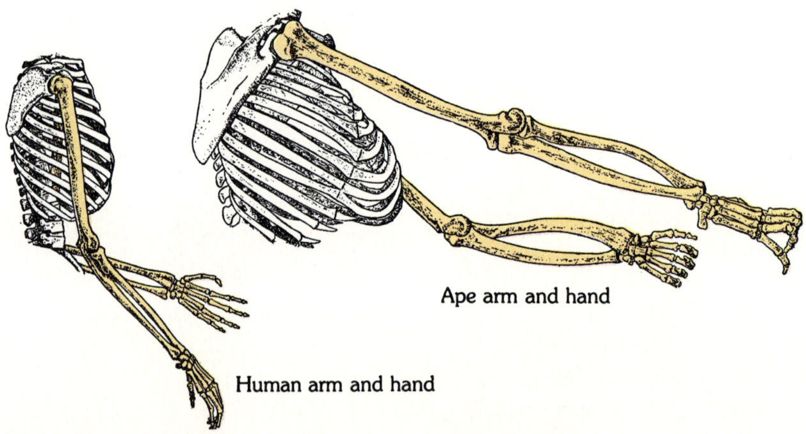

Foot
The human foot is much different from the feet of other primates. It is adapted for walking upright. The toe bones are short, and the first toe is not opposable to the other toes. However, the ape has a handlike foot. The first toe can be used for grasping.

Arm and Hand
The ape uses its hands and arms for walking and swinging through trees. The human arms and hands are specialized for using tools. Therefore, human arm bones are shorter and the arms are more muscular than those of apes. The bones of the human thumb are also longer in comparison to those of the ape. And the human thumb can be used with all the other fingers.

Review It

1. What are three traits of primates that make them different from other kinds of mammals?
2. Make a list of the physical differences and similarities between apes and humans.

Activity

Distinguishing Between the Skeletons of Gorillas and Humans

Purpose
To measure and compare the differences and similarities between the skeletons of two primates—a human and a gorilla.

Materials
Ruler
Pencil and paper

Procedure

1. The skeletons of modern gorillas and humans are alike in many ways. They are also different in numerous ways. Carefully examine the drawings of human and gorilla skeletons on this page.
2. Now make two columns on a sheet of paper. In Column I list the similarities between the two skeletons. In Column II list the differences.

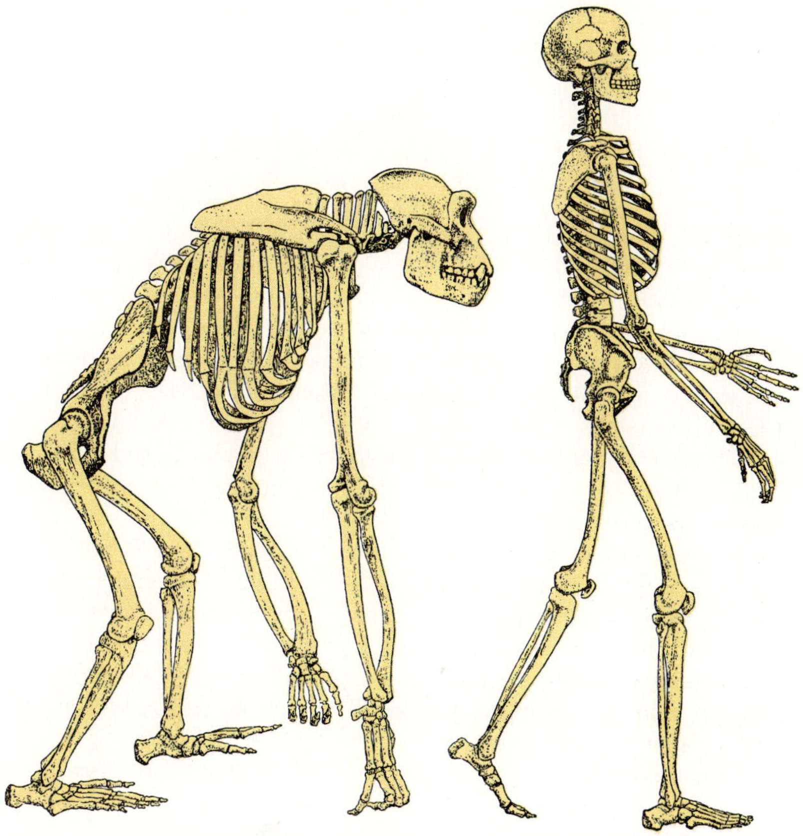

Ape skeleton　　Human skeleton

Analysis

1. What differences do you see between the shape and structure of the gorilla pelvis and the human pelvis? How do these differences help humans to walk upright while gorillas cannot?
2. How does the arm spread of the gorilla compare with the arm spread of the human? In what ways might differences in arm spread be related to differences in behavior of gorillas and humans?
3. How is the gorilla skull different from the human skull? How do these differences affect the appearance of the gorilla face and the human face?
4. Examine the differences in posture, and arm and hand structure of humans and gorillas. How might these differences make it easier for humans to use tools than for gorillas to use them?

2–2
Other Human Traits

Human Body Systems

- Circulatory System
- Digestive System
- Respiratory System
- Excretory System
- Nervous System
- Endocrine System
- Skeletal System
- Muscular System
- Reproductive System

The outside appearance and the skeleton of the human body tell only a little about what we are really like. Inside the body are many parts that perform activities to keep us alive. The parts of the body are organized from the simplest to the most complex as cells, tissues, organs, and systems. In many ways, the inside parts of the human body are like the inside parts of other organisms. Nevertheless, a human heart or a human muscle is physically and chemically *different* from the heart or muscles of other organisms. For example, an expert can easily tell the difference between a human liver and a calf liver.

Body Systems

The bodies of humans and many other complex organisms are organized into **systems** that perform certain jobs. The *circulatory system* carries digested food, gases, and other materials to all parts of the body. The *digestive system* receives food, digests some of this food, and gets rid of undigested wastes. The *respiratory system* collects certain gases in the air and expels waste gases. The *excretory system* eliminates wastes that are produced inside the body. The *nervous system* collects information from inside and outside of the body. This information is used to help regulate body functions. The *endocrine system* is made up of glands that help to regulate the chemical substances in the body. The *skeletal system* provides a framework for the body and protects many vital organs. The *muscular system* acts on the skeleton to allow body movement. The *reproductive system* makes the sex cells necessary for producing offspring.

Organs

Each system within the body is made up of **organs.** Each organ in a system performs certain tasks that help the entire system to function. For example, the heart, arteries and veins, and blood are organs of the circulatory system. The heart pumps blood to all parts of the body. The arteries and veins carry the blood to different parts of the body and return the blood to the heart. The blood carries food and oxygen to the cells and transports waste substances to the lungs and kidneys.

Tissues

Each organ is made up of groups of different kinds of *cells*. Such groups of cells are called **tissues.** Cells in each kind of tissue look alike and do the same kind of job in the body. For example, arteries are organs of the circulatory system. But the arteries are made up of three kinds of tissues shown on the facing page: a thin, inside lining of cells; an inner and outer layer of tough connecting cells; and a thick layer of muscle cells.

The organs and tissues of the human circulatory system are shown at the right.

Organs of Human Circulatory System

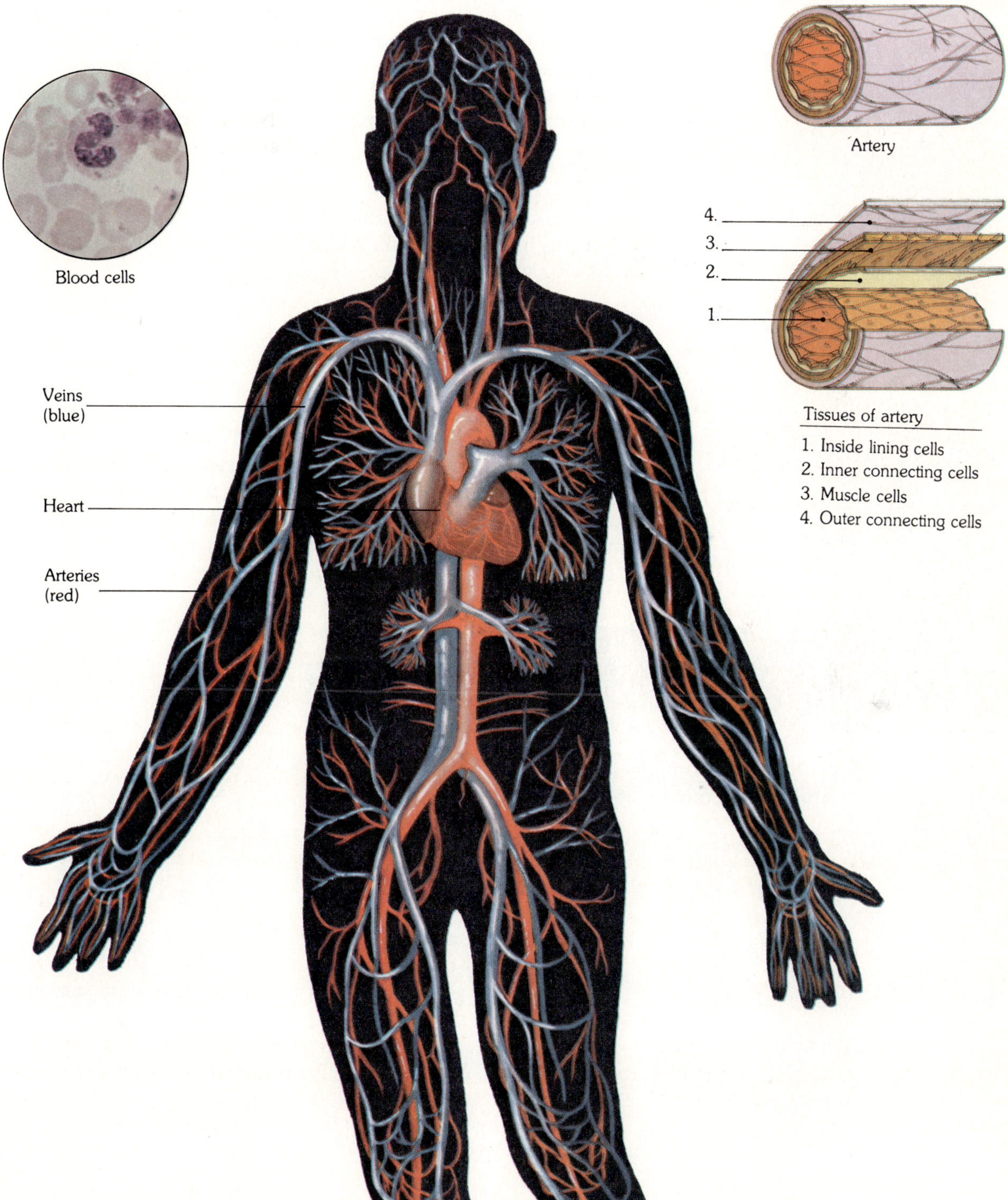

Tissues of artery

1. Inside lining cells
2. Inner connecting cells
3. Muscle cells
4. Outer connecting cells

Human Behavior

Humans have special physical traits that make them appear different from all other organisms, even the most advanced primates. But humans are not different just because of their physical traits. Humans are different because of the way their physical traits work together to produce a species that *behaves* much differently from all other species. Here are some of the ways humans behave differently.

Humans Can Be Educated

Dogs can be taught to sit up, and worms can be taught to turn left. Some animals can learn from others by copying their behavior. But only humans have created educational systems that teach people "to learn how to learn."

Humans Are Aware of Themselves

Individual humans know that they were born and that they will eventually die. *Homo sapiens* is the only species that has an awareness of itself.

Humans Communicate Through Space and Time

Humans have developed many kinds of languages that are made up of spoken and written words. These words are *symbols* that stand for objects and ideas. Language allows humans to communicate complex information and ideas in the present and into the future.

Humans Control the Environment

Humans control the temperature, food supply, and water supply where they live. They control temperature by making clothes and by building houses, fireplaces, furnaces, and air conditioners. Humans control the water supply by building dams. They use dams to make lakes, produce electricity, and prevent floods. Humans control the food supply by using large land areas to grow and to harvest their own special breeds of plants and animals. Humans are constantly using tools and machines to change the environment. Hardly any place on Earth is left where humans have not changed the natural surroundings. Now humans are exploring the possibility of living in areas other than Earth.

Humans Can Affect the Future

For thousands of years humans have decided which dogs, horses, cabbages, or apples to breed and grow. As a result, the plants and animals we use for agriculture and recreation are different from their wild ancestors. Other human decisions have ended or nearly ended the future of other species. Early in this century, the passenger pigeon became extinct. Humans had decided to kill and eat these birds in large numbers. They did not allow the birds enough time to reproduce and to replace themselves. Every day we must make decisions that will have an impact on our own future and the future of other organisms. To make the right decisions, we should understand the world.

Human Variation

Humans share many traits. But each one of us is *different* from everyone else. In fact, no two persons are exactly alike, not even "identical" twins. Some differences among people may be easily seen. For example, height, weight, and body build may vary. The shape of the nose, eyes, or mouth may also differ. There are many variations in skin tone. And there are variations in the skin of people of different ages. You can see some of these differences below.

Internal Differences

Differences among people may also be found *inside* the human body. Body organs, such as the heart, liver, kidneys, and brain, differ in size. People also differ in the way their bodies work. Blood pressure, pulse rate, and muscle strength vary among people. People even differ in the kinds of chemicals their bodies produce and release. Dogs can tell identical twins apart just by the way they smell. Some of these differences are shown below.

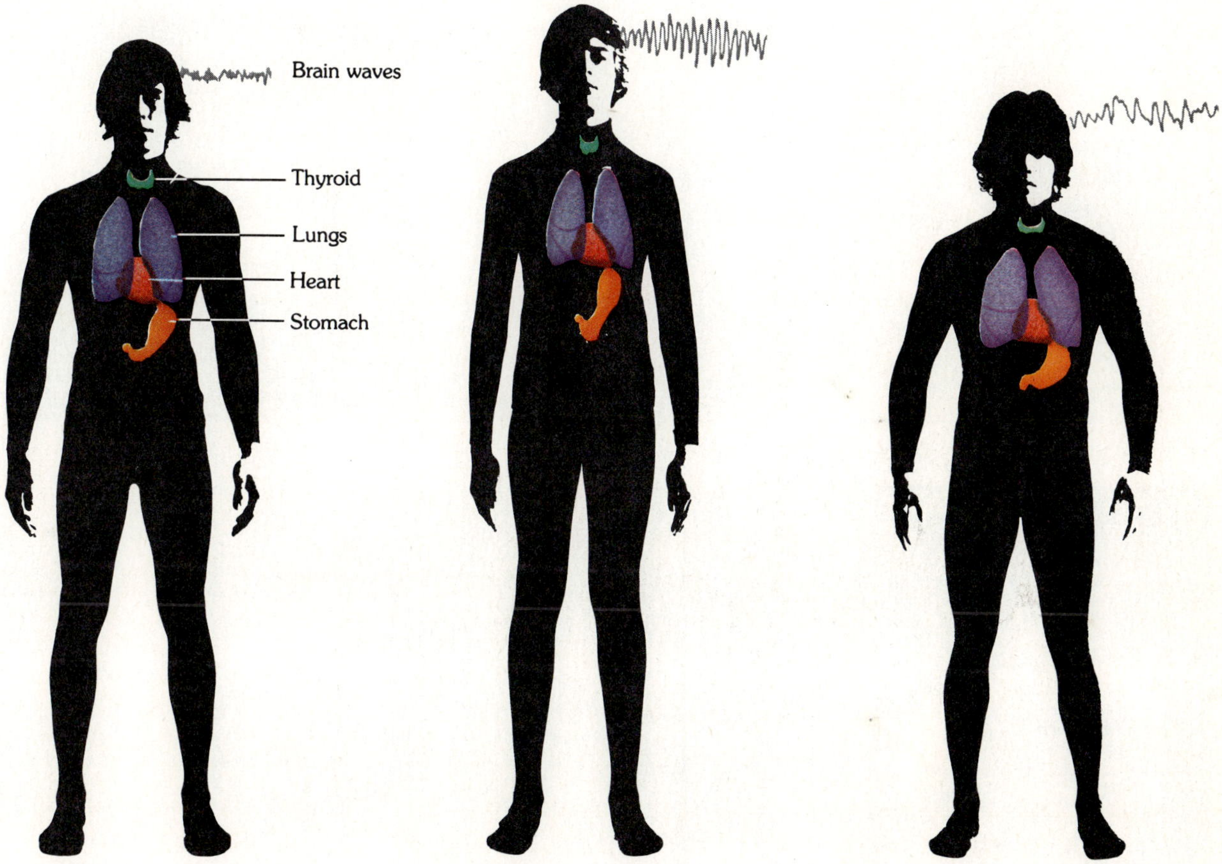

Behavioral Differences

Some differences among people relate to how their brains work. Such variations produce differences in personality, interests, and behavior. Sometimes such differences can be seen in the kinds of hobbies people choose. Or differences show up in the kinds of jobs people select. Some differences are shown in the pictures below.

Review It

1. List examples of organs and tissues in one human body system.
2. How is human behavior different from that of other organisms?
3. Describe some ways humans differ from each other.

Activity

Measuring Human Physical Traits

Purpose
To be able to measure, record, and graph difference in a human trait.

Materials
25 or more human subjects
Measuring tape
Paper and pencil
Graph paper

Procedure

1. In this activity you will measure, record, and graph differences in one human trait. Begin by selecting one human trait that you can measure. You might choose one of the traits listed below:
   diameter of head
   diameter of neck
   length of ear
   length of index finger
   diameter of ankle
   length of foot
2. Measure at least twenty-five of your classmates for the trait you have selected. Write down all of the measurements on a piece of paper.
3. Now make a table. In one column, order all of the measurements from smallest to largest. In the second column, write the total number of students who had each of the measurements in column one.

Length of Foot	Number of Students
17 cm	1
18 cm	3
19 cm	4
20 cm	6
21 cm	4
22 cm	2

4. Prepare a bar graph of your measurements. Draw a frame for the graph. The numbers along the bottom should be the smallest to the largest measurements you have for the trait. The numbers along the side should be the smallest to the largest groups of students you have for each measurement.
5. Draw bars on the graph. The length of each bar should show the total number of people who had a certain measurement for the trait you chose.

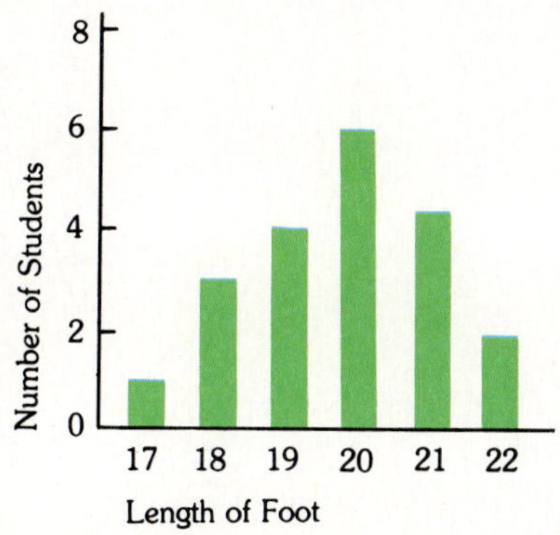

Analysis

1. How many *different* measurements do you have for the same trait?
2. What are your smallest and your largest measurements? These two numbers are the **range** for your trait.
3. Which measurement does the largest group of students have? This is called the **mode.** Find the mode by locating the highest bar on your bar graph.
4. What is the *average* measurement for your trait? Find the average by adding up all of the measurements and dividing this number by the total number of students you measured.
5. How does the average compare with the mode?
6. Sometimes people use the term *normal* when they are really talking about the mode or the average. How might such use of the term *normal* be misleading?

Challenge!
People also differ in the way their bodies work. Find the pulse rates of 25 classmates. What is their average pulse? What do doctors consider to be the normal range *for the human pulse? How many people fall above and below this range? What do you think this means?*

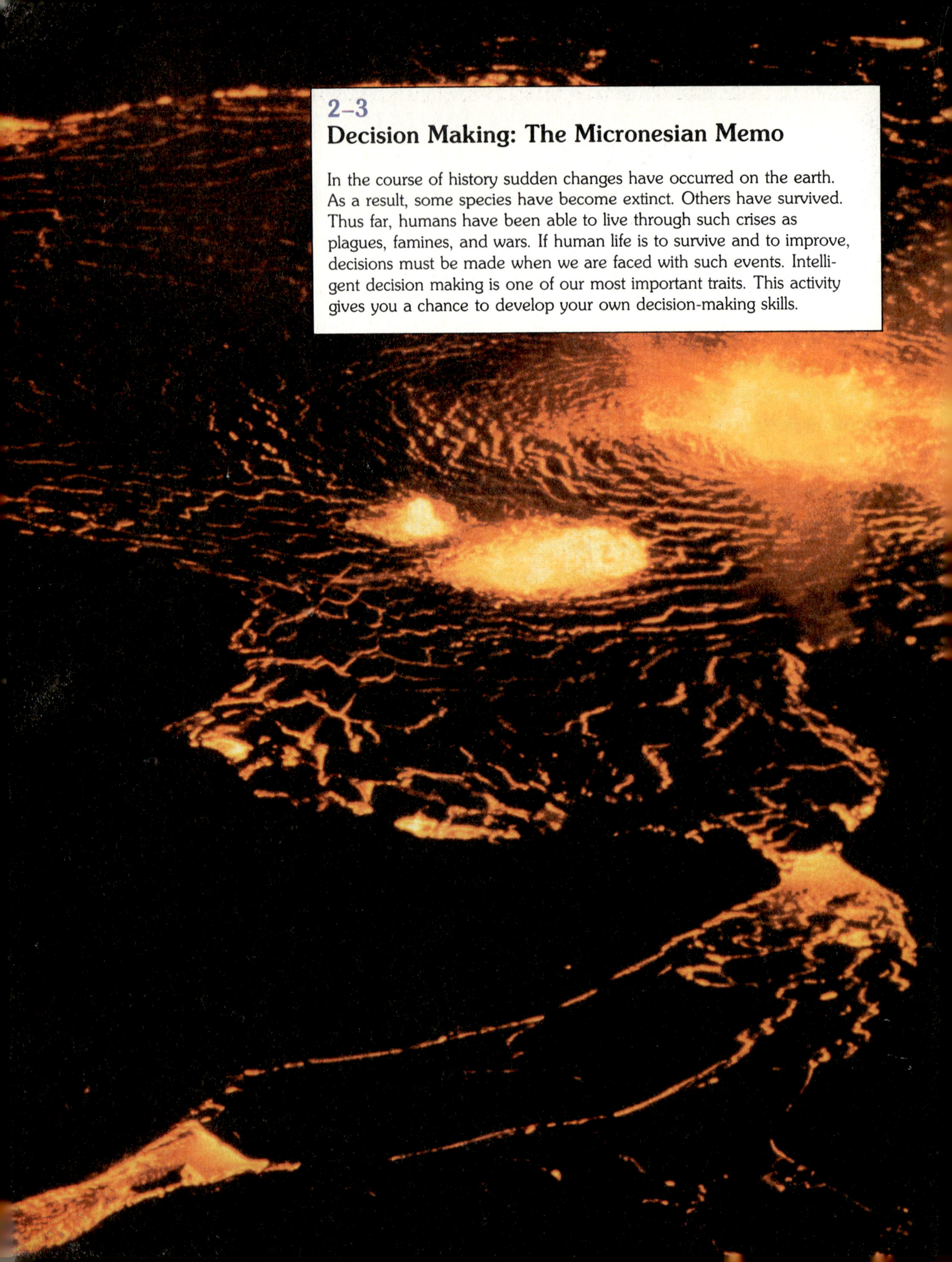

2–3
Decision Making: The Micronesian Memo

In the course of history sudden changes have occurred on the earth. As a result, some species have become extinct. Others have survived. Thus far, humans have been able to live through such crises as plagues, famines, and wars. If human life is to survive and to improve, decisions must be made when we are faced with such events. Intelligent decision making is one of our most important traits. This activity gives you a chance to develop your own decision-making skills.

Memo

TO:	**President's Advisory Panel**
FROM:	**Joint Advisory Staff**
DATE:	**October 8, 1989**
SUBJECT:	**Explosion of volcano Aruba in Micronesia**

We may be facing a disaster that will affect the entire world. We feel that all decisions must consider the worst possible consequences. We recommend that plans be drawn up immediately to deal with the problems listed under "Action To Be Taken." Failure to do so may result in the destruction of not only our society but, perhaps, most other societies.

Background

As we all know, the giant volcano Aruba exploded on the evening of October 2. Clouds of smoke and ash could be seen for over 50 miles. Aruba continues to give off large amounts of smoke and ash. A cloud of particles that blots out the sun is becoming larger each day. Our experts estimate that more ash has been put into the atmosphere than from the explosion of the volcano Tambora in 1815. The dust and ash from Tambora, as you may recall, circulated all over the world and had severe effects on climate. Weather records show there were frosts every month of the year in Europe and North America. The skies were always gray, and low rainfalls caused drought. The average annual temperature in Europe dropped almost 2° C. Thus, the growing season was severely shortened. The price of food doubled as a result of widespread food shortages. Weather conditions did not return to normal until 1818. In short, the explosion of Tambora was a disaster felt around the world.

The explosion of Aruba is already larger than that of Tambora. So its effects on world climate will be greater. And there are two other factors which you must consider: (1) The world population was less than one billion in 1815. The present world population is about five billion. (2) The atmosphere already has a pollution load that was not present in 1815.

Action To Be Taken

As you all know, decisions that affect all parts of our society must be made now. Whatever you decide, there will be those who consider your decisions unfair and even undemocratic. And since no changes have occurred yet, many people will question your judgment. But you alone must make important decisions *now*.

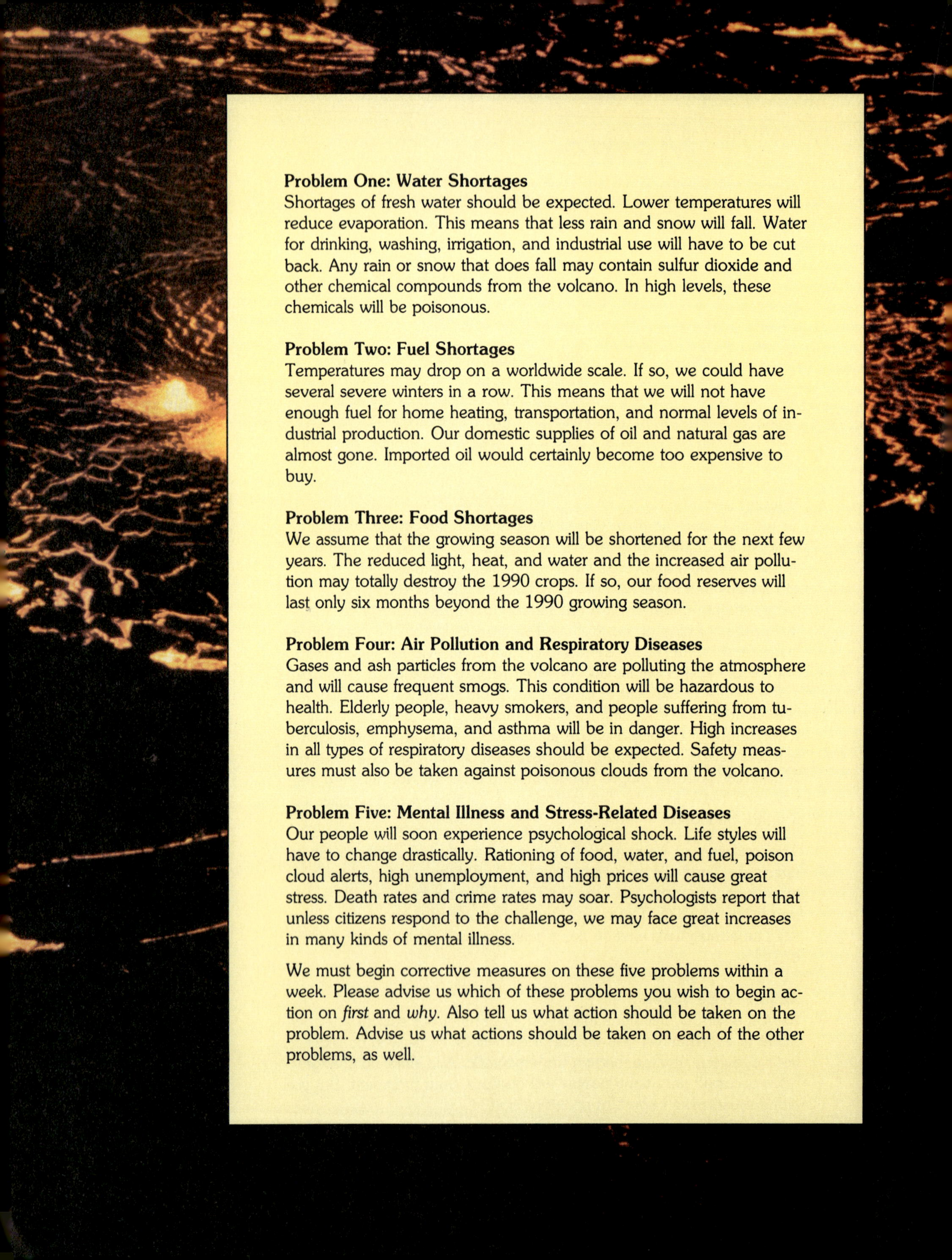

Problem One: Water Shortages
Shortages of fresh water should be expected. Lower temperatures will reduce evaporation. This means that less rain and snow will fall. Water for drinking, washing, irrigation, and industrial use will have to be cut back. Any rain or snow that does fall may contain sulfur dioxide and other chemical compounds from the volcano. In high levels, these chemicals will be poisonous.

Problem Two: Fuel Shortages
Temperatures may drop on a worldwide scale. If so, we could have several severe winters in a row. This means that we will not have enough fuel for home heating, transportation, and normal levels of industrial production. Our domestic supplies of oil and natural gas are almost gone. Imported oil would certainly become too expensive to buy.

Problem Three: Food Shortages
We assume that the growing season will be shortened for the next few years. The reduced light, heat, and water and the increased air pollution may totally destroy the 1990 crops. If so, our food reserves will last only six months beyond the 1990 growing season.

Problem Four: Air Pollution and Respiratory Diseases
Gases and ash particles from the volcano are polluting the atmosphere and will cause frequent smogs. This condition will be hazardous to health. Elderly people, heavy smokers, and people suffering from tuberculosis, emphysema, and asthma will be in danger. High increases in all types of respiratory diseases should be expected. Safety measures must also be taken against poisonous clouds from the volcano.

Problem Five: Mental Illness and Stress-Related Diseases
Our people will soon experience psychological shock. Life styles will have to change drastically. Rationing of food, water, and fuel, poison cloud alerts, high unemployment, and high prices will cause great stress. Death rates and crime rates may soar. Psychologists report that unless citizens respond to the challenge, we may face great increases in many kinds of mental illness.

We must begin corrective measures on these five problems within a week. Please advise us which of these problems you wish to begin action on *first* and *why*. Also tell us what action should be taken on the problem. Advise us what actions should be taken on each of the other problems, as well.

Class Activity

What would you do? How would you react? The situation described is not just science fiction. There is evidence that catastrophes of such proportions have occurred in the past. There is no reason to believe that they will not occur in the future.

Procedure
Your task as a class is to play the role of the President's Advisory Panel. You are to make decisions regarding the action to be taken in this emergency. Your goals are described in the last paragraph of the memo.

1. Divide your class into five groups. Each group selects a different problem. Discuss both the facts of the situation and the possible effects on human life.
2. Decide what should be done to help solve the problem chosen by your group. Make a list of the positive and negative effects of your actions. Write the decision of your group, along with the positive and negative effects, on the chalkboard.
3. Meet and decide as a class what action the President's Advisory Panel should take in this crisis. Prepare a list of recommendations to submit to the president for action.

Analysis
1. How do you think nonhuman organisms would behave under the threat of a worldwide cloud of volcanic ash and gas?
2. What does this activity have to do with what you have been learning in this chapter?
3. How does knowledge about science and technology affect the way people react to crises of this kind?
4. Which of the decisions that have to be made depend on knowledge of science? Which of the decisions involve human values?

Chapter Summary

The human is a member of the order Primates. This group includes the families of apes, monkeys, lemurs, and tarsiers. Humans share with all primates such physical traits as stereovision, fingernails, and a well-developed brain. Humans are most like apes. However, careful study shows many differences which place humans in a separate family from the apes.

The human body not only differs from other primates in its outward appearance, but it has equally special internal structures. The internal parts of the body are organized into systems, organs, tissues, and cells.

The most important trait of humans is our ability to create tools and language. These inventions have enabled humans to control their environment, "learn how to learn," communicate facts and ideas throughout the world, develop an awareness of self, and to control the future of other organisms.

But no two humans are alike. Considerable variation among all humans can be observed and measured. Possibly most important of all, each human has the ability and responsibility to make decisions that affect his or her future and the future of other organisms.

Questions to Think About

1. Trees in Asia and Africa are being removed for lumber and to turn the land into farms. How do you think these changes are affecting the other primates that live in the area?
2. Human organ transplants of the heart and the kidneys have occurred. What kinds of organ transplant problems resulted because of physical and chemical differences among humans?
3. Why is it important for humans to develop their decision-making skills?
4. How do humans affect the future of their own species as well as many other organisms?

Extra Research

1. Make an appointment with the person in charge of the primate house in a nearby zoo. Prepare a list of questions to ask about the care and behavior of primates in captivity.
2. Make plaster models of a human and an ape skull. These models should show basic similarities and differences between the two primate skulls.
3. Obtain microscope slides of human organs that have been stained. Make drawings of the different kinds of cells and tissues that you can see.
4. Both apes and humans have a range of facial expressions. Find out how humans and apes use these expressions to communicate ideas. Which expressions mean the same thing? Which convey different ideas?
5. Look for items in the newspaper which report decisions people are making and which you think will affect the future of life on Earth. Pick one item and describe the positive and negative effects of the decision.
6. Attend an open meeting of a group in your community that is studying an environmental problem. Compare the way this group makes decisions to the way your class made decisions in response to the "Micronesian Memo."

Chapter Test

Part I

Copy the number of each test item. If the item is *true*, write *T* after the number; if the item is *false*, write *F*.

1. One primate trait that is shared by lemurs, monkeys, and apes is fingernails. (38)
2. Tissues are made up of organs. (44)
3. The most natural walking position for apes is on two legs. (40)
4. The pelvis of an ape is shaped so that it can support the internal organs while the ape is in an upright position. (41)
5. All humans, including twins, produce chemicals that dogs can use to tell people apart. (49)
6. Only humans are able to "learn how to learn." (46)
7. The heart is a tissue found in the circulatory system. (44)
8. Each species of ape can live in a variety of natural environments and eat a variety of foods. (40)
9. The digestive system collects information from inside and outside the body. (44)
10. The ability to make intelligent decisions is one of the most important human traits. (52)

Part II

Copy the number of each sentence below. After each number, write the word or words that complete the sentences correctly.

11. ____ World monkeys are more likely to have grasping tails than ____ World monkeys. (39)
12. Primates are able to turn their palms up and down with the aid of two long bones in the ____. (38)
13. Human legs are held straight by ____ that lock in place. (41)
14. A primate that drinks by lapping is the ____. (39)
15. The parts of the human body are organized from simplest to most complex as cells, ____, ____, and ____. (44)
16. Only humans can communicate complex ideas about the past or future by using ____. (46)
17. Humans are the only species that uses tools to control their ____. (47)
18. Humans share many traits, but no two humans are exactly ____. (48)
19. Decisions humans make about how they will control the environment may affect other ____. (47)
20. If human life is to survive and to improve, we must make intelligent ____. (52)

Part III

Copy each item from List *A*. After each item, write the letter and words from List *B* that best describe it.

List A

21. brachiation (40)
22. excretory system (44)
23. respiratory system (44)
24. stereovision (38)
25. tissues (44)

List B

a. ability to judge distances and perceive depth
b. collects gases in the air and expels waste gases
c. specialized groups of cells
d. ability to swing by the arms
e. eliminates wastes produced in the body

Focus on Careers

Crime Laboratory Technician
"There really are very few perfect crimes. Today we have the technology to link almost every kind of criminal to the scene of the crime or to the victim. Fingerprints, blood, clothing, even teeth marks found at the scene of a crime can be matched to those of the suspect. We can match a chip of paint or a piece of glass found in the clothing of a hit-and-run victim to the suspect's car. By studying pieces of burned rubble, we can tell if a fire was deliberately started and how it was set. If explosives were used, we can tell about what time they were planted."

Gerilyn is describing a few of the things she does as a crime laboratory technician. In short, a crime lab technician searches for, collects, photographs, and preserves evidence. He or she determines if a crime has been committed, matches pieces of evidence to a suspect, tests alibis, and, at times, helps protect an innocent person from possible misinterpretation of evidence.

The crime lab technician works at the crime scene and in the laboratory. At the crime scene, the technician tries to reconstruct the crime, looks for evidence, and takes photographs. He or she may preserve tire treads, footprints, and other marks by making plaster casts. By collecting fingerprints, hair, and blood

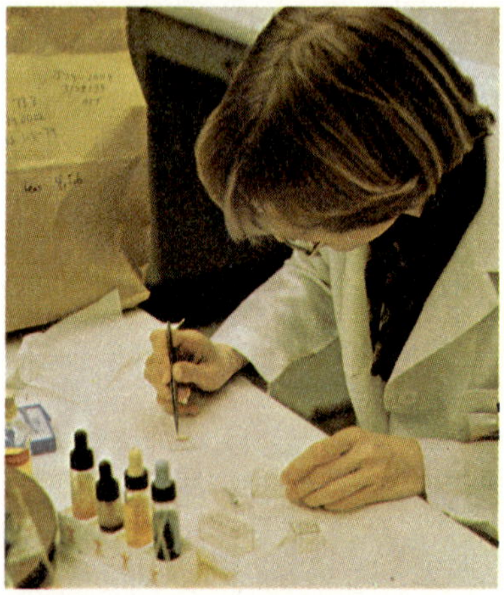

samples and locating weapons, the technician helps piece together the incident.

In the laboratory, the technician analyzes evidence using instruments such as microscopes, spectroscopes, and infrared and ultraviolet light. He or she performs and interprets laboratory results regarding substances such as poisons, alcohol, and narcotics that may have been eaten or injected into a victim's or a suspect's body.

The technician prepares careful reports of all findings. He or she may testify as an expert witness in court on evidence or laboratory techniques used.

Crime lab technicians must be curious, conscientious, responsible, and have good judgment. They must keep up with new developments in techniques and equipment.

Two-year programs in crime laboratory technology are available at some junior colleges. These programs include courses in biology, chemistry, physics, zoology, physiology, criminal law, investigative photography, fingerprint science, and crime-scene procedure. A person who successfully completes the course of study is better able to find a job than one who must rely only on on-the-job training. Advanced degrees and specialization in an area such as documents, ballistics, handwriting, medicine, or drugs qualify a person as a criminologist. Such a person supervises crime lab technicians.

Career Information
Biologist: American Institute of Biological Sciences, 1401 Wilson Blvd., Arlington, VA 22209.

Crime Laboratory Technician: Civil Service Commission, 1900 E Street NW, Washington, DC 20415.

Emergency Medical Technician: U.S. Department of Transportation, Emergency Medical Service Branch, N-42-13, Washington, DC 20540.

Medical Illustrator: Association of Medical Illustrators, Medical College of Georgia, Augusta, GA 30902

Biologist/Life Scientist

Duties: Studies living things and their life processes. One-third of all biologists conduct research in laboratories, botanical gardens, and outdoors in various parts of the world. Many others teach in high schools, colleges, and universities. Life scientists very often specialize: a botanist studies plants, a zoologist studies animals, an embryologist studies the development of an animal from a fertilized egg through hatching or gestation.

Training: A bachelor's degree qualifies a person for entry-level jobs in testing, inspecting, technical sales, or advanced technician positions. Advanced degrees are necessary for college teaching and independent or applied research positions.

Opportunities: Many opportunities are available in laboratory technology, health care, research, and education.

Medical Illustrator

Duties: Makes detailed sketches of the human body or of certain body organs and then makes finished paintings, sculptures, or models. An illustrator's work may be used in medical or science books, television programs, or films. Life-size models of the human body, created by medical illustrators, are often used in anatomy classes and in museums.

Training: Requires a bachelor's or master's degree in medical illustration. An ability to draw well and a good background in science are needed to be admitted to these degree programs.

Opportunities: Artists who specialize in medical illustration face less competition than commercial artists. Employment opportunities are best for those who have a variety of techniques.

Emergency Medical Technician (EMT)

Duties: Provides immediate on-site care to victims of such emergencies as automobile accidents, heart attacks, near drownings, poisonings, or gunshot wounds. EMTs drive emergency vehicles to accident scenes and give such care as is needed: restore breathing, stop bleeding, treat for shock, apply bandages, or assist in childbirth. EMTs then transfer the victims to hospital emergency rooms, and provide full reports of treatments they have given.

Training: Formal training, not previously required, is becoming necessary. A standard 81-hour training course is offered by police, fire, and health departments, in hospitals, and in some medical schools.

Opportunities: EMTs, sometimes known as paramedics, are hired by police and fire departments and private and hospital-based ambulance squads. Many EMTs are volunteers for fire departments.

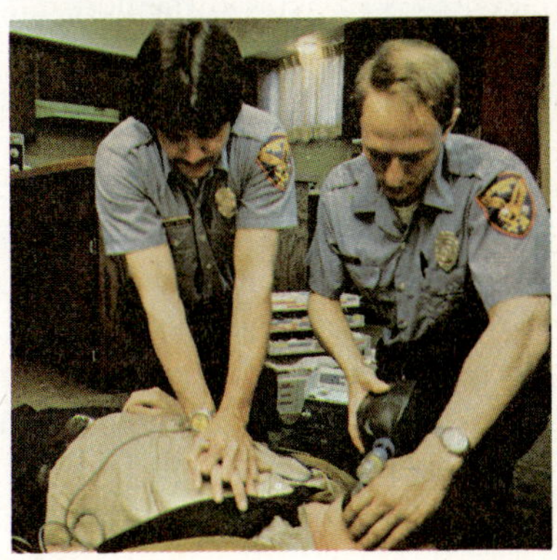

Bibliography: Unit One

Adler, Irving. *How Life Began.* The John Day Co., 1977. Discusses the nature of life and origin of life.

Ardrey, Robert. *The Hunting Hypothesis.* Atheneum Press, 1976. Describes human ability to overcome hardships of nature.

Asimov, Isaac. *A Short History of Biology.* Natural History Press, 1965. Discusses leading biologists and their ideas.

Aylesworth, Thomas. *The World of Microbes.* Franklin Watts, Inc., 1975. Interesting highlights of microbiologists, including Pasteur.

Bates, Marston. *The Forest and the Sea.* Random House, 1960. Overview of different forms of life and importance of some domestic animals.

Beck, William. *Modern Science and the Nature of Life.* Doubleday and Co., Inc., 1961. Discusses characteristics of living things.

Berger, Melvin. *Tools of Modern Biology.* Thomas Y. Crowell Co., 1970. Good overview of topics such as scientific method, classification, and instruments used in the study of biology.

Beveridge, W. I. *The Art of Scientific Investigation.* Random House, Inc., 1957. Applications of the scientific method.

Broms, Allan. *Thus Life Began.* Doubleday and Co., Inc., 1968. Story of the origin and growth of life on earth, and of the men who unraveled its mystery.

Carson, Rachel. *The Sea Around Us.* Oxford University Press, 1970. Describes life in the ocean.

Case, James, and Stiers, Vernon. *Biology: Observation and Concept.* Macmillan Co., 1971. Discusses scientific method, characteristics of life, origin of life, and more.

Dembeck, Hermann. *Animals and Men.* Doubleday and Co., 1965. Story of the relations between people and animals since biblical times.

De Kruif, Paul H. *Microbe Hunters.* Pocket Books, 1959. Describes lives of Leeuwenhoek and other notable scientists.

Dubos, Rene. *Pasteur and Modern Science.* Doubleday and Co., Inc., 1960. Discusses contributions of Pasteur.

Gallant, Roy. *How Life Began: Creation versus Evolution.* Four Winds, 1975. Explores various theories of creation and evolution. Contributions of Darwin, Lamarck, and others are included.

The Living Organism. Doubleday and Co., Inc., 1965. Illustrations and information on origin of life and cell structure and function.

Medawar, P. B., and Medawar, J. S. *The Life Science.* Harper and Row, 1977. Discusses current ideas about biogenesis, evolution, cell biochemistry, and other topics.

Oparin, A. I. *Life, Its Nature, Origin, and Development.* Academic Press, 1962. Presents a hypothesis explaining how life began.

Sears, Paul. *Living Landscape.* Basic Books, 1966. Discusses problems encountered by populations.

Shippen, Katherine. *Men, Microscopes, and Living Things.* Viking Press, 1963. Good account of events in the lives of early scientists, including Pasteur, Darwin, and others.

Stilley, Frank. *The Search.* G. P. Putnam's Sons, 1977. Interesting information for the average reader about the search for intelligent life in outer space.

Sullivan, Walter. *We Are Not Alone: The Search for Intelligent Life on Other Worlds.* McGraw-Hill, 1966. Exciting probe into life on other planets.

Articles

Asimov, Isaac. "Is Mars Alive? Viking Flight." *Saturday Review,* vol. 3 (May 1, 1976), pp. 14–16.

Frantel, H. "Amazing Search for Life on Mars; Viking Lander Flight." *Popular Mechanics,* vol. 145 (June, 1976), pp. 66–67.

"New Thoughts on Mars; Results of Viking Experiments." *Time,* vol. 109 (January 24, 1977), pp. 83–84.

UNIT TWO

CELLS

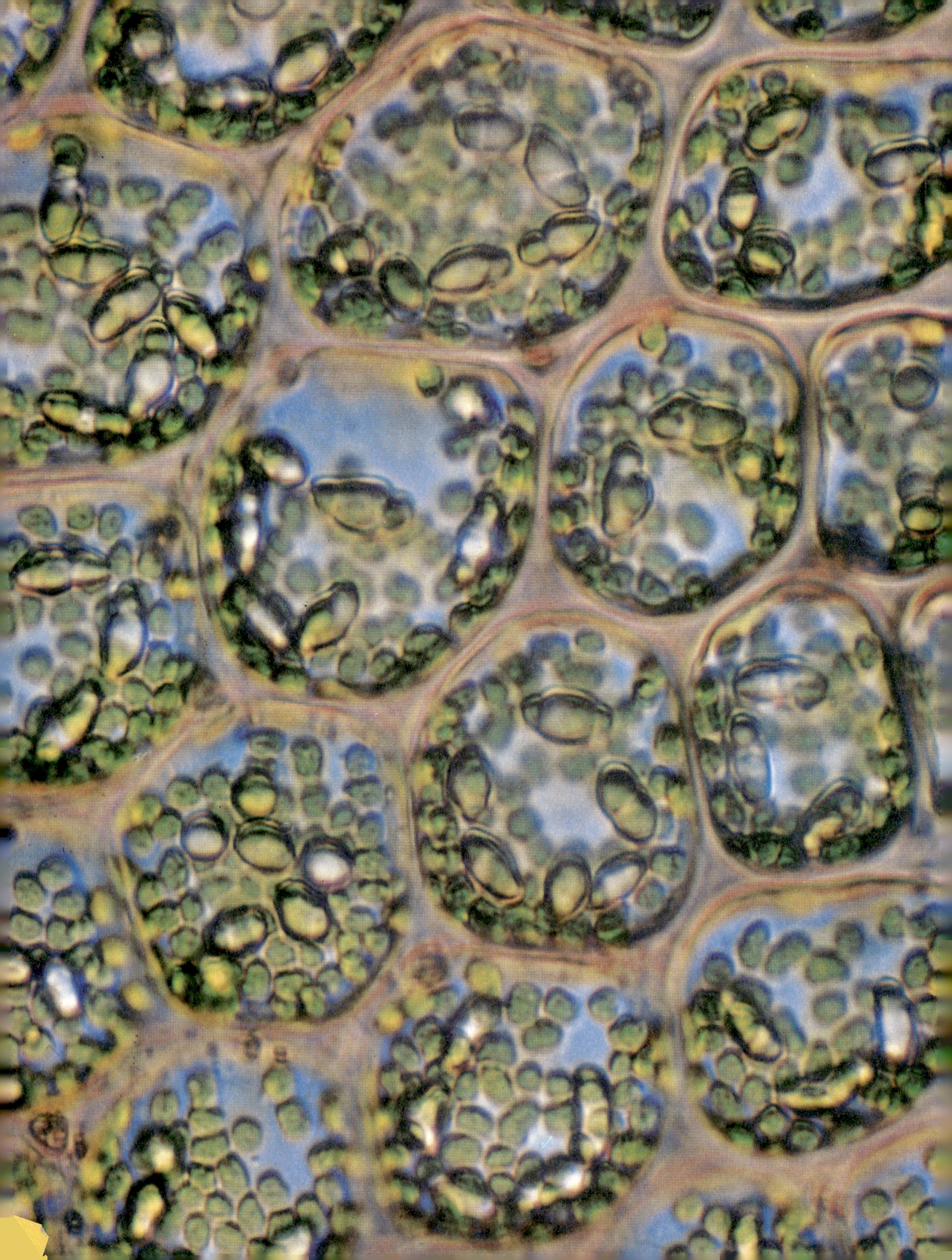

Chapter 3

Cell Structure

Cells are the building blocks from which living things are made. Learning how cells are put together and how they work is the key to understanding what life itself is like. We can find out a lot about living things by studying each organism alone. However, we can also find answers to the more basic question of what it means to be "alive" by looking beyond the whole organism. We can look at the one thing most organisms have in common: *cells.*

Chapter Objectives

1. Explain some of the ways in which cells differ from each other.
2. State the basic ideas of the cell theory.
3. Describe two techniques used to study cells.
4. Describe the major functions of at least four cell organelles.
5. List the major differences among protist cells, moneran cells, fungi cells, plant cells, and animal cells.

The chloroplasts of a moss as seen through a light microscope. Chloroplasts contain chlorophyll, which is essential for photosynthesis.

3–1
Cells of Many Kinds

Cells differ from each other in many different ways. They can be almost as large as a baseball or smaller than the thickness of a hair. They can be shaped like balls, boxes, rods, spirals, tubes, snowflakes, or starfish. A single cell may make up an entire organism. Or groups of cells may be loosely organized and live together. Other cells may be only one of millions of units that make up the tissues of larger organisms.

Cell Size

The largest known living cells are the egg cells of birds. Each cell is actually the yolk part of the egg. The largest of these cells is the yolk of an ostrich egg. It may measure eight centimeters in diameter.

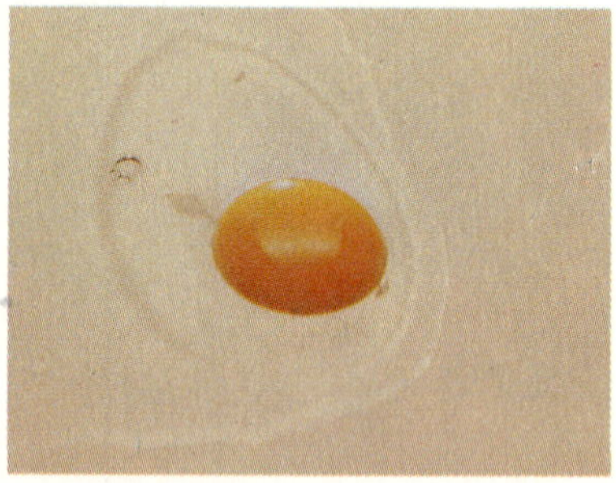

At the other extreme, some cells are so small that they can just barely be seen with an ordinary light microscope. A bacterium, like the one shown here, is only one ten-thousandth of a millimeter in diameter. This bacterium is found in the human intestines.

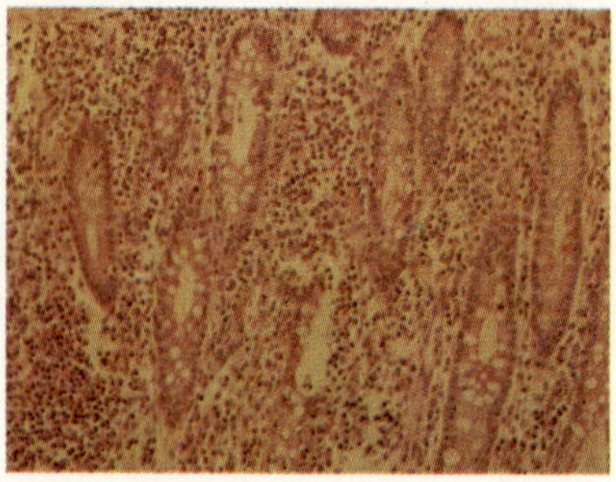

Cell Shape

The shape of a cell may be determined by the job it has to do. For example, a nerve cell like this one has to carry messages from one part of the human body to another. Such "long distance messages" may require a nerve cell one meter long.

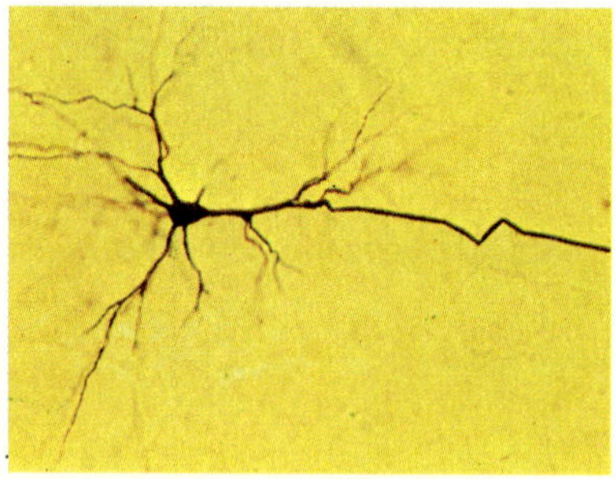

Courtesy Carolina Biological Supply Company

Cell shape may also be determined by the place the cell is located. The cells in this plant stem are shaped to fit tightly together. They are hollow, like straws. This shape enables them to transport water and food through the plant stem to the leaves and roots.

Cell Organization

Many cells live by themselves. Such cells often develop attractive and unusual shapes, such as these diatoms. Cells that live by themselves perform a variety of functions. Often they carry out all of their life functions without the aid of other cells.

Other cells live together. The structure and functions of such cells are often highly specialized. For example, the cells you see here are the units from which muscle tissue is made up.

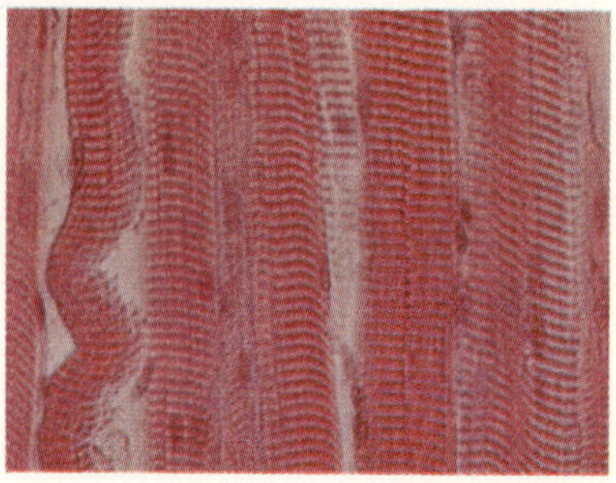

Cell Theory

Scientists have known about cells for a long time. The first record of cells being observed was made in the last half of the seventeenth century by the Dutch merchant Anton van Leeuwenhoek and the British scientist Robert Hooke. Hooke examined thin slices of ordinary cork under a microscope. He saw many tiny cavities surrounded by walls. The structures he saw in the cork looked like compartments in a honeycomb, so he named them cells.

It was almost 200 years before biologists fully realized that the cell is the basic unit of all living organisms. Another twenty years passed before the **cell theory** was stated in its present form by two German scientists, Matthias Schleiden and Theodor Schwann.

When we talk about the cell theory, we're referring to a set of ideas about what a cell is and how it functions. The major ideas listed below make up the cell theory.

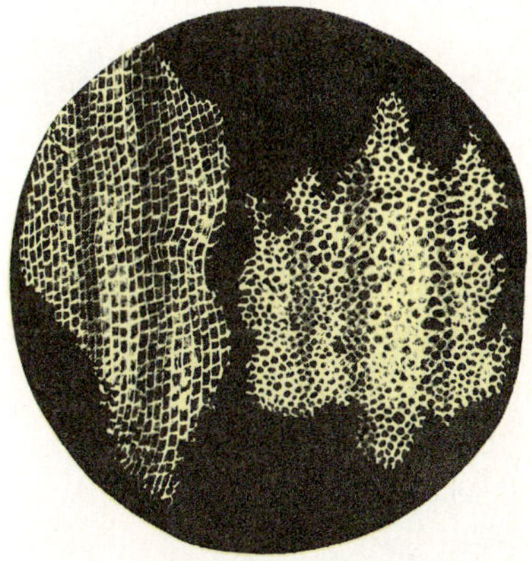

Hooke's drawing of cork cells as seen through an early light microscope

1. Living things are made up of cells and cell products.
2. Cells are very much alike in their structure and composition.
3. Cells have a set of functions that they carry out in order to stay alive.
4. New cells arise from old cells by cell reproduction. Healthy cells normally divide into two equal parts at regular intervals. This is the primary way in which new cells are formed.

The cell theory did not appear suddenly. The ideas developed slowly and with much controversy over a long period of time. Today we believe that the cell theory is a good description of the basic ideas about cells. Cell theory gives us a base from which we can continue to learn more about cells. Some of the special tools and techniques used to increase our knowledge of cells are described on the following pages.

Early cell drawings made by Schleiden and Schwann

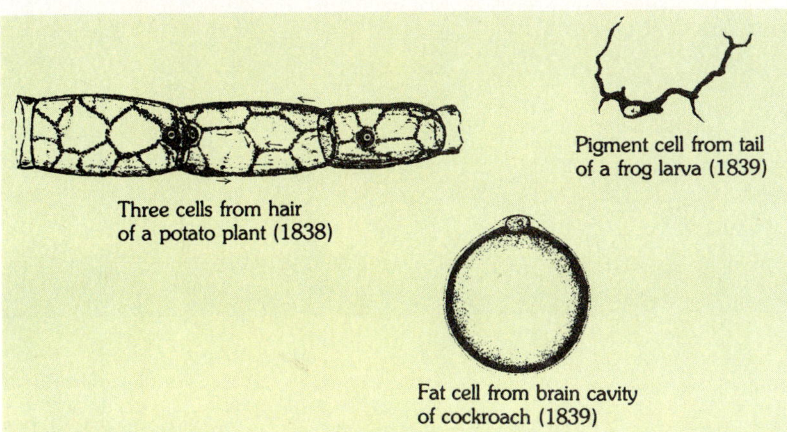

Challenge!
The development of ideas that now make up the cell theory was slow and complicated. Research the work of each of the following biologists on cell theory: Robert Hooke, Robert Brown, Theodor Schwann, Matthias Schleiden, Rudolph Virchow. What were some of the disagreements that arose regarding cell theory in the early stages of its development? How were they settled?

Studying Cells

Using a light microscope

How do scientists find out what cells are like? Cells are so small that almost nothing can be learned by simply looking at them with the unaided eye. For nearly 300 years, everything people knew about cells came from studies with simple **light microscopes.** Samples of biological material were cut into thin slices, stained with various dyes, and examined under a light microscope.

The stains that were used all made some parts of the cell show up more clearly than other parts. But stains also tended to kill the cell once it had been treated. Therefore, most of what is known about cells has come from studies of dead cells.

In the past 25 years, cell study has changed. Biologists and other scientists have invented many new tools and techniques for studying cells. Here is a review of some of those tools and techniques.

The Phase-Contrast Microscope

A different kind of light microscope is now used to study *living* cells. It is called a **phase-contrast microscope** and contains a special optical system. This system enables the viewer to see slight differences in the light that passes through different parts of the cell. Therefore, it is not necessary to use stains to emphasize cell parts. The phase-contrast microscope makes it possible to study cells while they are still alive. The photograph below shows a cell viewed through a phase-contrast microscope. Compare it with the one on the next page taken with a scanning electron microscope.

Pollen as seen through a phase-contrast microscope

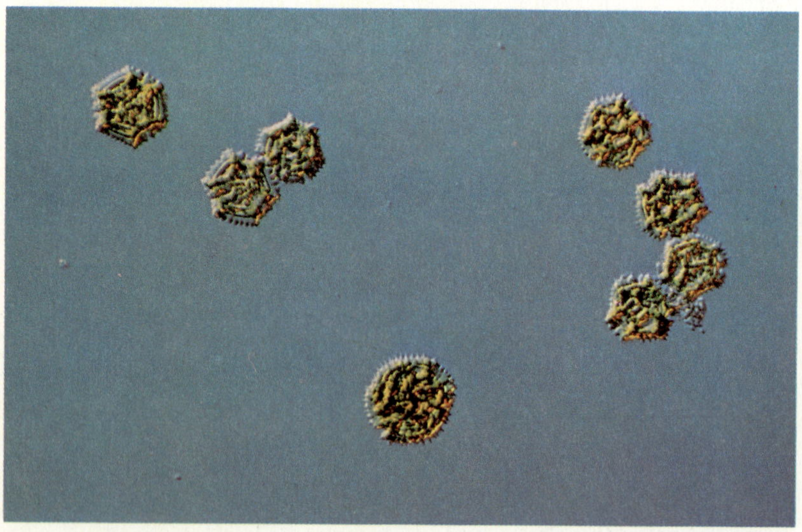

The Transmission Electron Microscope

The **transmission electron** (i lek'tron) **microscope** has enabled us to learn more about the detailed structure of the cell than any other tool. An electron microscope operates on the same principle as a light microscope. A light microscope enlarges an object by using *glass* lenses to bend beams of light. An electron microscope uses magnetic lenses to bend beams of electrons. Electrons are the tiny, negatively charged particles found in atoms.

Dead cells studied with a transmission electron microscope must be sliced very thinly. They are then stained and placed in a vacuum chamber. A beam of electrons is focused on the cell slice. Electrons that pass through the thin slice are transmitted to a televisionlike screen. Here they form a picture that can be seen by the operator. This picture shows part of the *inside* of the cell.

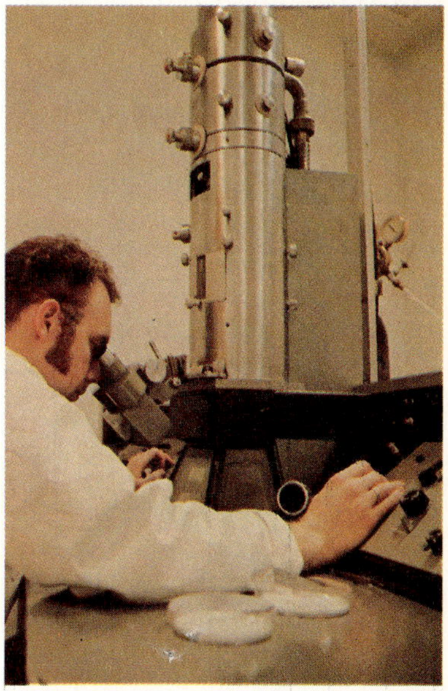

Using an electron microscope

The Scanning Electron Microscope

A **scanning electron microscope** is used to examine the *surface* of whole cells. Even living cells can be studied. The surface of the cell is scanned in a regular pattern by a beam of electrons. The viewing screen records electrons released from the cell surface. This provides a very detailed, three-dimensional picture of the *outside* of the cell.

Electron microscopes have two advantages over light microscopes. First, they can enlarge a cell 200,000 times. The best light microscope can enlarge a cell only 2,000 times. Second, electron microscopes have much better **resolving power** than light microscopes. Resolving power describes the ability of a microscope to distinguish between two objects that are very close together. Two objects seen as a single object by a light microscope would be seen as two separate objects by an electron microscope.

Pollen grain as seen through a scanning electron microscope

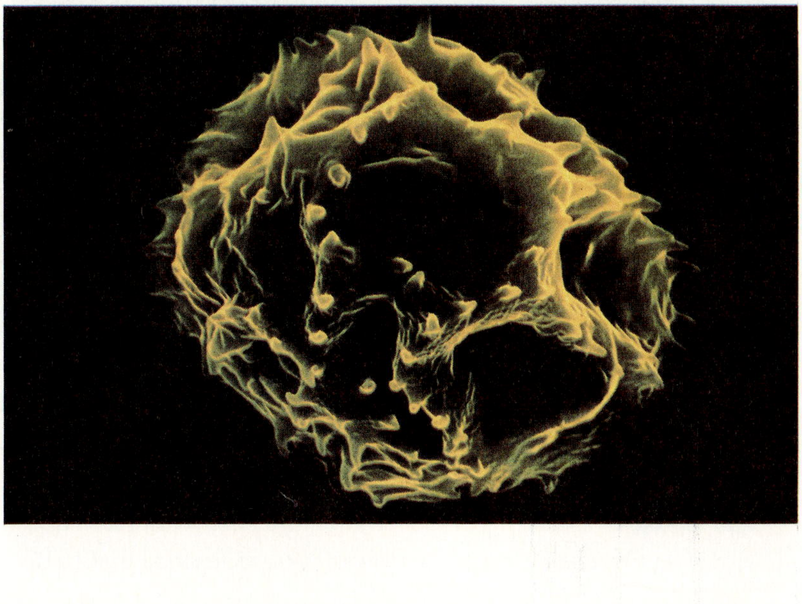

Ultracentrifugation: Taking Cells Apart

The tools mentioned so far are all devices for simply looking at cells the way they are. Biologists are also interested in studying cells in other ways. Much research depends on taking a cell apart and studying its separate parts. Researchers have developed several methods for taking cells apart. One method involves spinning cells very rapidly. This is called **ultracentrifugation** (ul′trə sen trif′yə-gā′shən).

The principle behind this technique is simple. First, a suspension of whole cells in a sugar solution is placed in an ordinary kitchen blender. The solution is mixed in the blender. This causes the cell membrane to break down, and the cell parts are set free in the sugar solution. Then, the mixture is placed in a test tube and the tube spun rapidly for a short time. The most dense parts of the cell, such as the nucleus, are thrown farthest away from the center. They settle in the bottom of the test tube. After these parts are removed, the solution can be spun again at higher speeds. Then the next most dense parts can be removed. Eventually, most of the major parts of the cell can be separated and studied. The diagram below shows how ultracentrifugation works.

How an Ultracentrifuge Works

1. Cells put in centrifuge

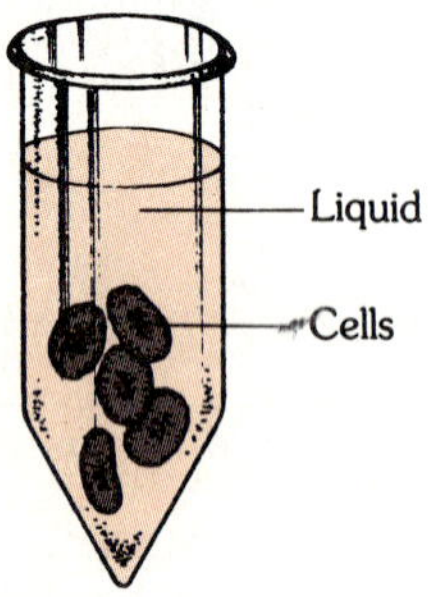

2. Cells spun at low speed and low force (less than 1,000 times gravity)

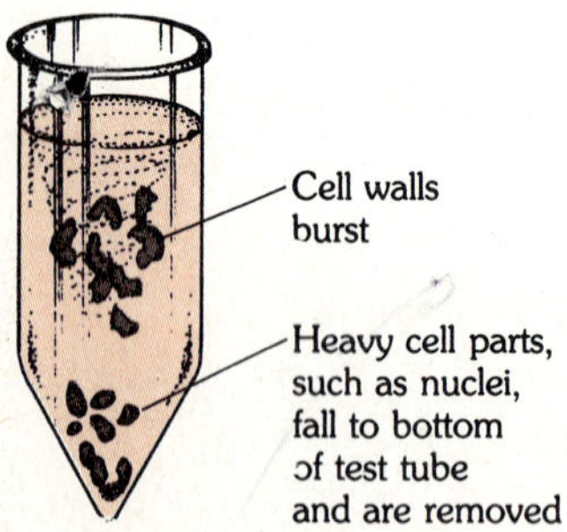

3. Cell contents transferred to new test tube and spun at higher speed and greater force (10,000 times gravity)

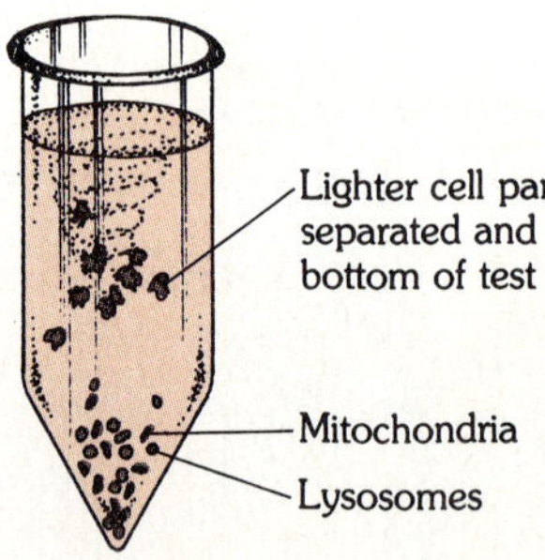

4. Cell contents spun at highest speed and greatest force (100,000 times gravity)

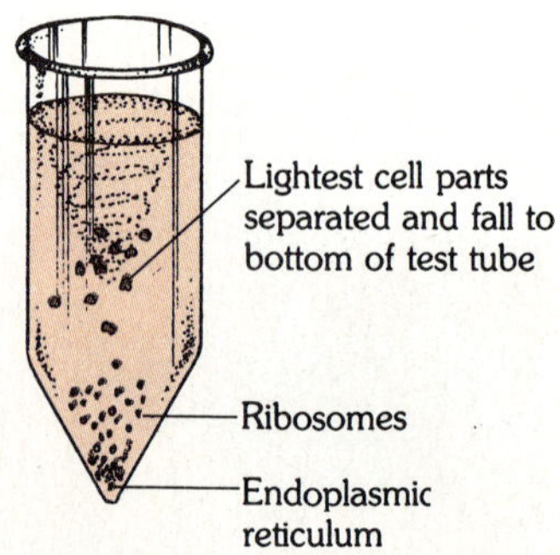

Review It

1. Define the word *cell.*
2. Describe the major ideas that make up the cell theory.
3. List some advantages of using each of the following microscopes to study cells: (1) phase-contrast, (2) transmission electron, (3) scanning electron.
4. How has ultracentrifugation improved our understanding of cells?

Breakthroughs in Biology

Performing Microsurgery on Cells

Do you know of any cells that need surgery? That may seem like a foolish question, because cells are such tiny things that it's hard to imagine how anyone could operate on them. Normal scalpels, clamps, and other operating tools are thousands of times larger than a cell.

Yet, for almost a century, biologists have been developing techniques that will allow them to dissect a cell the way you may have dissected a frog. Special tools are now available for such things as cutting into a cell, removing certain parts, injecting chemicals, transferring cell parts, and analyzing the contents of the cell.

Tools like these often have special attachments which reduce the operator's movements by 20 times or more. When the operator moves his or her arm 10 centimeters, the microtool may move as little as 0.5 millimeter.

Microsurgery has made possible some amazing research in biology. One new experiment is the **cloning** (klōn′ing) of cells. Using two microprobes, a biologist can carefully remove the nucleus from the egg cell of a frog. With the same two probes and a *micropipette* (mī′krō pī′ pet), the biologist can also remove the nucleus from a normal body cell of another frog. This nucleus is transferred to the egg cell. The egg cell then begins to divide, following directions stored in its new nucleus. The egg cell eventually develops into an exact copy of the frog from which the body cell was taken.

Analysis

1. What is microsurgery?
2. Why is it necessary to reduce the hand movements of the person operating the tools during microsurgery?
3. What other uses can you think of for microsurgery in addition to cloning?

Using a micropipette on a fertilized mouse egg

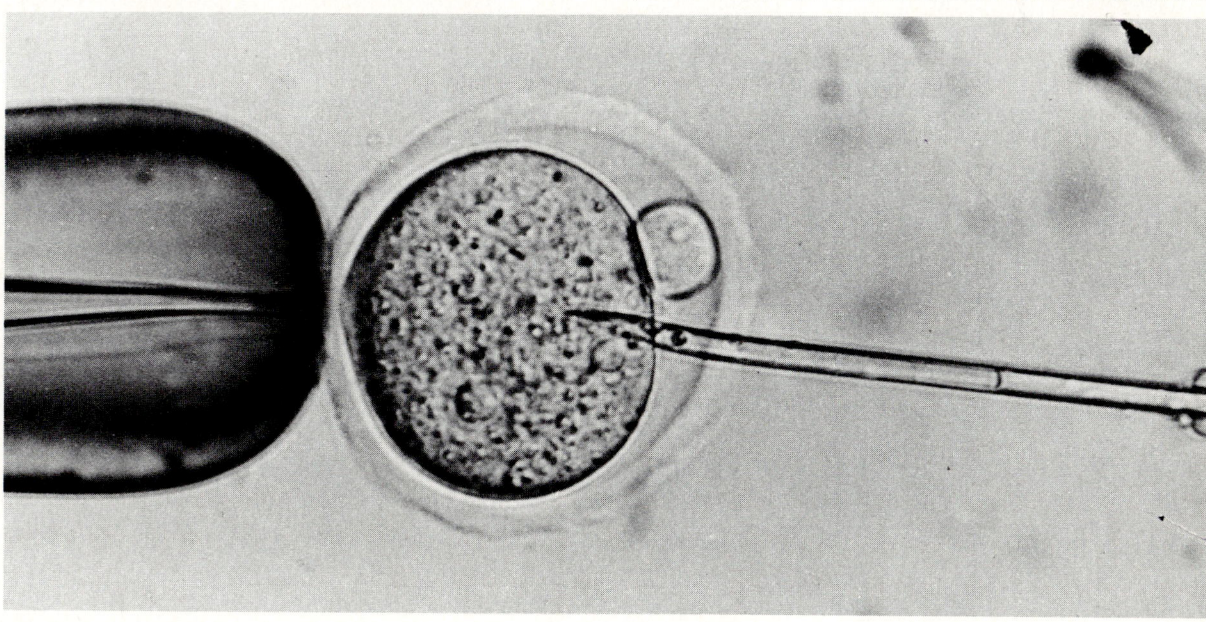

3–2
The Parts of a Cell

The electron microscope has revealed many specialized structures inside the cell. Normally these structures are not visible. They are surrounded by a cell fluid that is in constant motion. However, certain kinds of staining techniques can be used to help make most of the structures visible.

The term **organelle** (or′gə nel′), or "small organ," is often used to describe the specialized structures found inside a cell. The term is very appropriate because each structure does function somewhat like a miniature body organ. Each organelle performs one or more special tasks to help keep the cell alive.

The drawing on the opposite page is an attempt to show you what the organelles in a typical animal cell might look like. The function of each organelle is summarized in the table below. You can read about each structure in detail on the pages that follow.

Cell Organelle	Major Function(s)
Cell membrane	Controls passage of substances into and out of cell Encloses cell organelles and all other cell substances
Cytoplasm	Serves as a fluid container for cell organelles and other cell substances Assists in transport of substances within cell
Nucleus	Directs all cell activities Carries hereditary information
Nucleolus	Produces certain kinds of cell nucleic acid (RNA)
Endoplasmic Reticulum (ER)	Serves as a transport network and storage area for substances within cell
Ribosome	Manufactures different kinds of cell proteins
Golgi apparatus	Packages protein for storage or transport out of cell
Lysosome	Digests or breaks down food materials into simpler parts Removes waste materials from cell
Mitochondria	Serves as powerhouse of the cell—produces ATP, which is the source of energy for all cell activities
Vacuole	Supports cell wall of plant cells; may also digest food materials, remove wastes, and store substances
Microtubule	Serves as support system or skeleton of the cell
Microfilament	Assists in cell locomotion

Diagram of a "Typical" Animal Cell

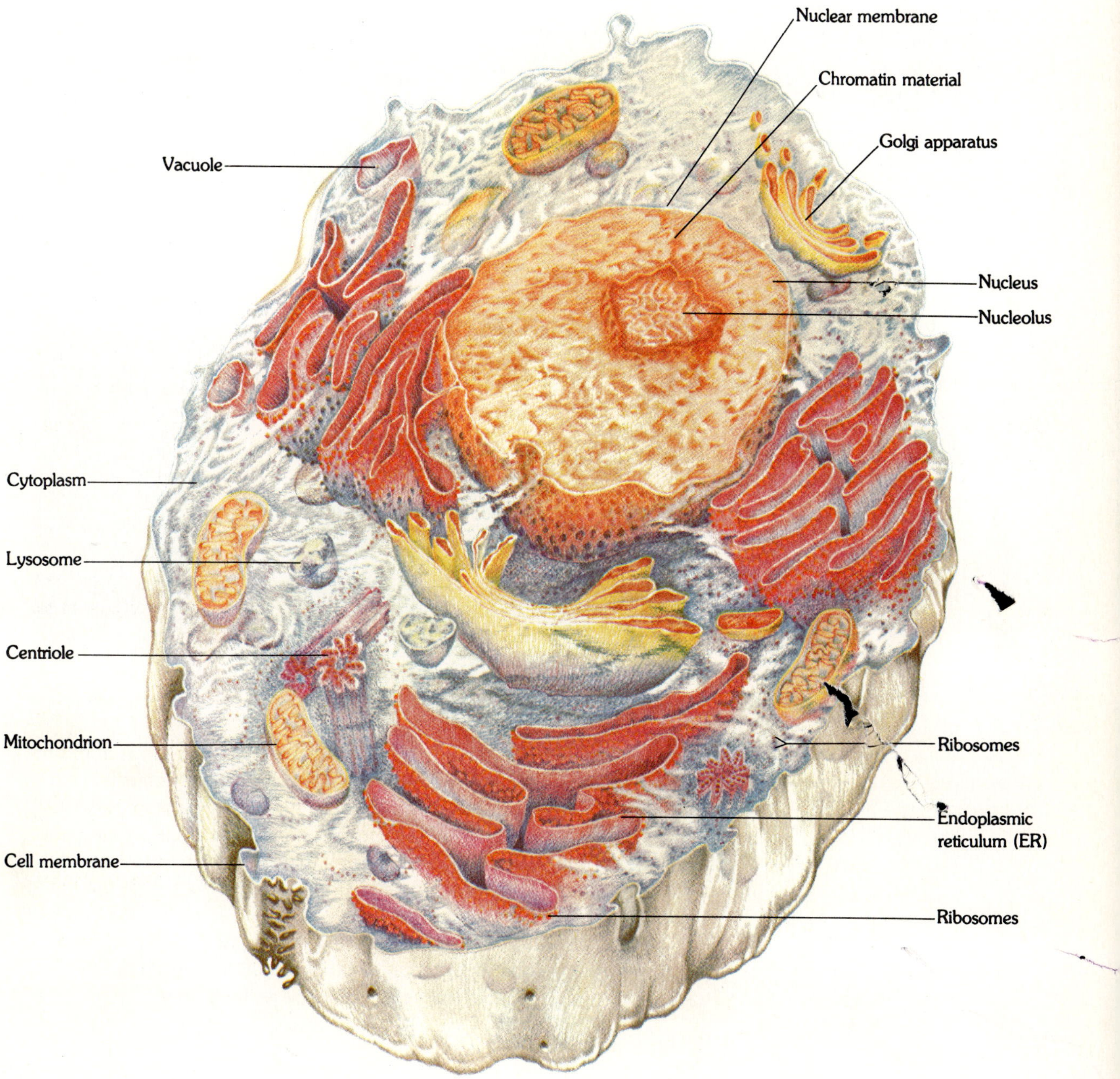

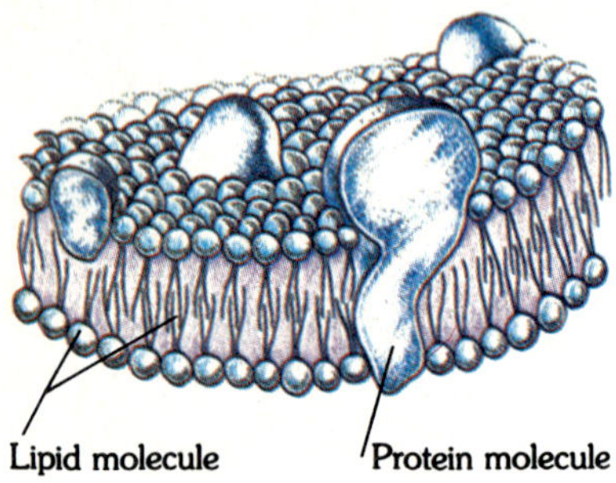

Cell Membrane

The outer boundary of a cell is a very thin, two-layered **membrane.** It is less than 1/100,000 of a millimeter thick. For years, scientists wondered if this membrane really existed. However, not only is it real, but the cell membrane is a vitally important organelle. It controls everything that passes into and out of the cell. It relays important messages between cells. And it encloses all of the other organelles within the cell.

The cell membrane is made up almost entirely of protein and lipid (fat) molecules. One current view is that the protein molecules are folded into compact globules. The globules float on a sea of lipid molecules. The globules act as pores or gates for substances entering or leaving the cell. However, the pores are not just holes in the membrane. They are adjustable portals that allow some substances to enter and prevent others from leaving the cell.

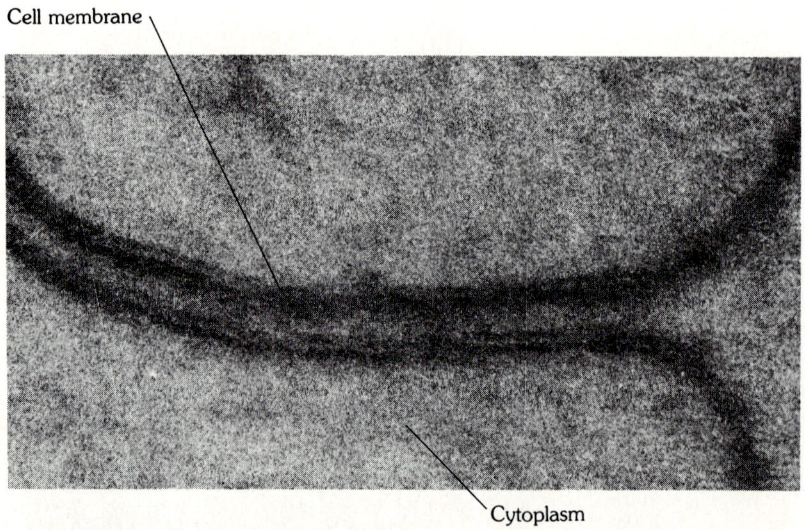

Cytoplasm

Within the cell membrane, different kinds of organelles move about in a watery environment. The clear, jellylike fluid that fills the cell is the **cytoplasm** (sī'tə plaz'əm). It is not really an organelle, but contains all of the other cell organelles. Depending on the type of cell, it may also contain other specialized materials such as pigments, oil droplets, and starch granules. In other words, the cytoplasm contains all of the functioning substances of the cell.

Materials are transported within the cell by the flowing action of the cytoplasm. Some organelles also move by flowing with the cytoplasm. These organelles may change shape as they move.

Nucleus

The easiest part of the cell to recognize is the **nucleus** (nü'klē əs). Most cells contain a single nucleus. However, some cells may have more than one. The nucleus is the command center of the cell. Near its center lies a mass of fibers called **chromatin** (krō'mə tən) **material.** These fibers seem to be the active forms of the **chromosomes** (krō'mə sōms). Chromosomes store instructions for all cell functions and direct all cell activities. They also control the reproduction of the cell and the inheritance of traits.

The nucleus is held together by a double-layered **nuclear membrane** which has many openings, or pores, in it. When studied through an electron microscope, it is possible to see a space between the two layers of the membrane. These pores allow materials to pass back and forth between nucleus and cytoplasm.

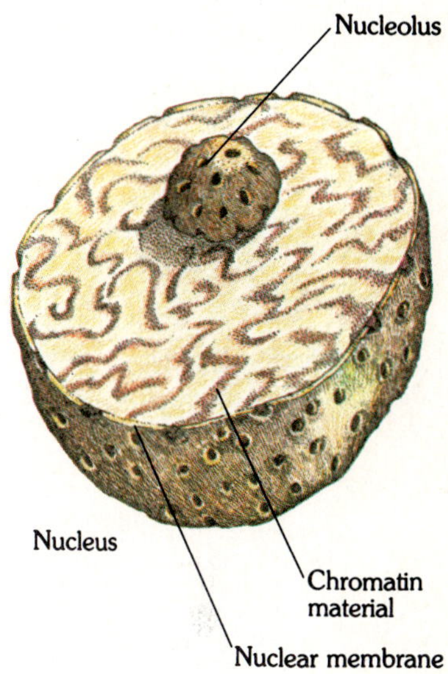

Nucleolus

The nucleus of most cells contains large numbers of granules called **nucleoli** (nü klē'ə lī). Such granules look like dark, round spots inside the nucleus when the cell is stained. The granules are rich in a particular type of nucleic acid known as **ribonucleic** (rī'bō nü klē'ik) **acid,** or **RNA.** The RNA in the granules is important in the production of proteins by the cell.

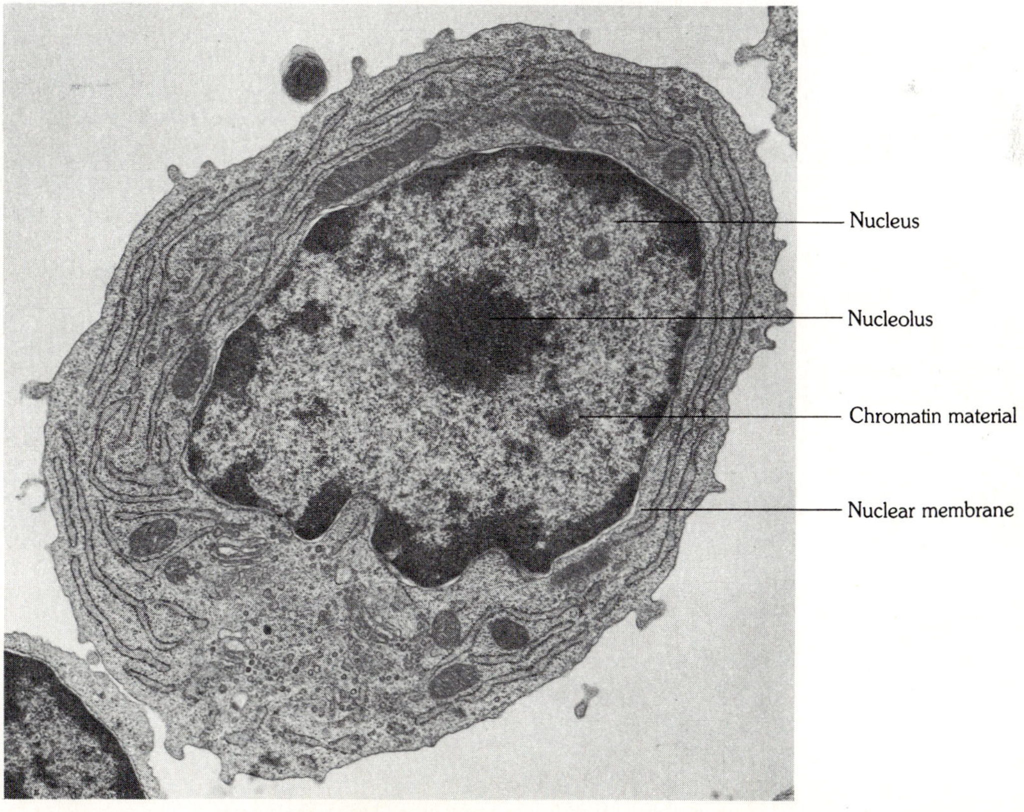

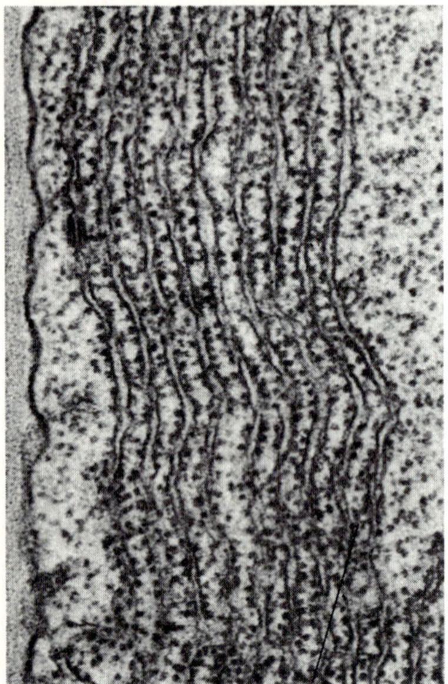

Endoplasmic reticulum (ER)

Endoplasmic Reticulum

A vast network of channels, bounded by membranes, runs through many parts of the cytoplasm. This is the **endoplasmic reticulum** (en′dō plaz′mik ri tik′yə ləm), or **ER.** The ER looks something like a vast highway system. All of the membranes in the ER interconnect. They form a system of tubes and flattened sacs which are connected to the nuclear membrane.

Some of the membranes in the ER are smooth. Others are rough. The rough ER is dotted with tiny organelles called *ribosomes.* These form granules on the outer surface of the membrane. It is the ribosomes that make the membrane appear rough.

All of the substances made in the cell must be transported to different sections of the cell for use. The main jobs of the ER are storage, separation, and transport of these substances (mostly proteins) within the cell.

Ribosomes

The cell contains a large number of tiny, grainy structures called **ribosomes** (rī′bə sōms). The ribosomes are so small that their internal structure cannot be seen even under an electron microscope. Most cells contain thousands, sometimes millions of them. Ribosomes are the protein factories of the cell. They may make up as much as one quarter of the mass of a cell. The large number of ribosomes allows many proteins to be made in the cell at the same time. It takes about one or two minutes for each protein to be completed.

There are two kinds of ribosomes. One floats free in the cytoplasm. The other is bound to membranes, such as the ER. Both kinds manufacture protein. But the free ribosomes release protein directly into the cytoplasm. The attached ribosomes transfer their protein to the membranes of the endoplasmic reticulum. From here the protein is transported to all parts of the cell.

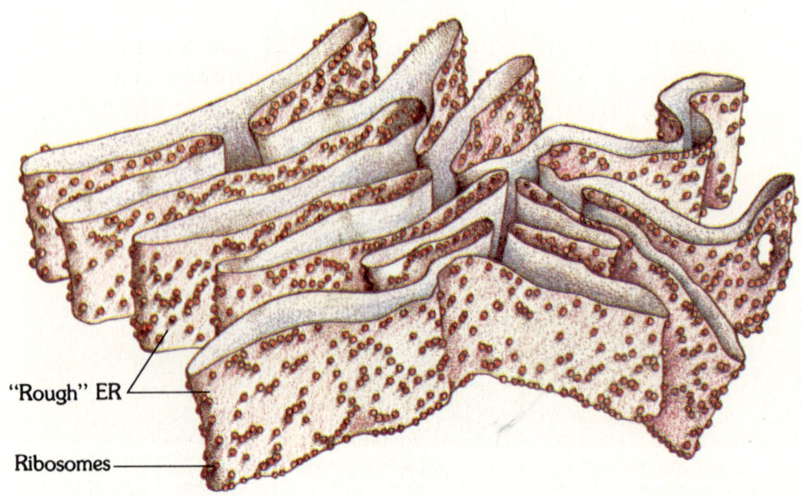

Golgi Apparatus

Near the nucleus is a collection of flat, saclike membranes that look like a pile of envelopes. This is called the **Golgi apparatus** (gol′jē). The Golgi apparatus packages proteins for export from the cell. Protein molecules in the ER are sent to the Golgi apparatus. Here large numbers of the protein molecules are wrapped together in a membrane package. This package is sent to the surface of the cell, where it moves out of the cell. The wrapping may also be used to store certain kinds of proteins within the cell.

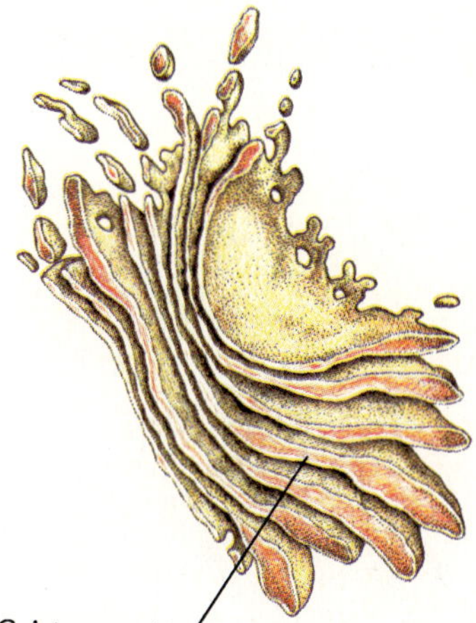

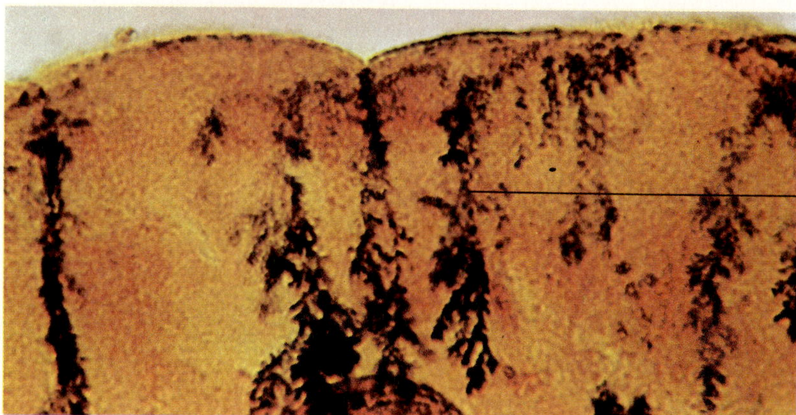

Courtesy Carolina Biological Supply Company

Lysosome

Another organelle found in the cell is a tiny package of fluid, held together by a membrane. It appears to change size and shape constantly, depending on its activity. This organelle is a **lysosome** (lī′sə sōm). It contains digestive enzymes, and its main job is to break down material in a cell into simpler parts. For example, protein, carbohydrate, and fat molecules are broken down into simpler substances within the lysosome. These can then be used as the building blocks for substances in other parts of the cell.

The lysosome also acts as a kind of disposal system for the cell. Undigested materials are left in the lysosome. The lysosome body moves to the cell surface, and the waste material is ejected into the external environment.

Did You Know?
In some cells, the membrane of the lysosome may weaken. Then digestive enzymes from the lysosome may leak into the cell. If enough of these enzymes are released, the cell may actually digest itself! The pain and swelling of arthritis may be due to the leakage of enzymes from the lysosomes of certain white blood cells.

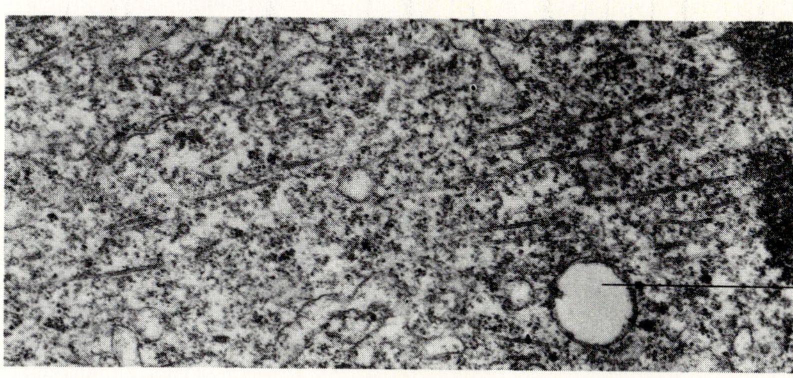

Mitochondrion

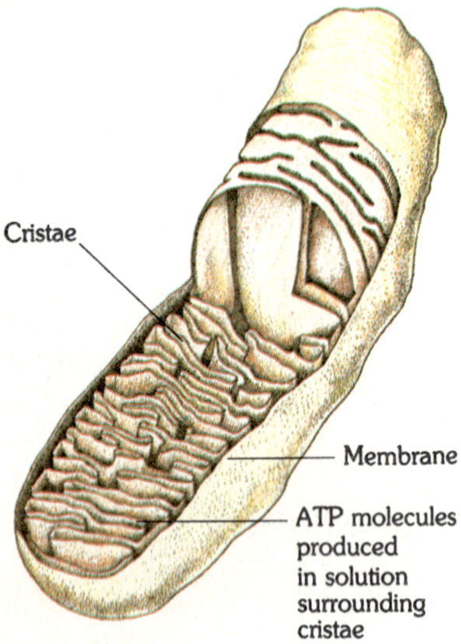

Mitochondria

The source of most body heat and energy can be traced to sausage-shaped organelles called the **mitochondria** (mī'tə kon'drē ə). This is the next largest organelle in the cell after the nucleus. The faint outline of the mitochondria can be seen with a good light microscope.

The mitochondrion is the powerhouse of the cell. Its outer surface is covered by a double-layered membrane. The inside membrane is folded to form a system of parallel ridges called **cristae** (kris'tē). Chemical reactions take place along these ridges to produce a chemical compound called **ATP.** ATP is the substance from which cells get the energy they need to carry out their life functions.

Mitochondria also help to control the amount of water, calcium, and other substances present in the cell. They break down and recycle fats, proteins, and carbohydrates. And they form *urea,* a waste product, which is secreted by the kidneys in multicellular animals.

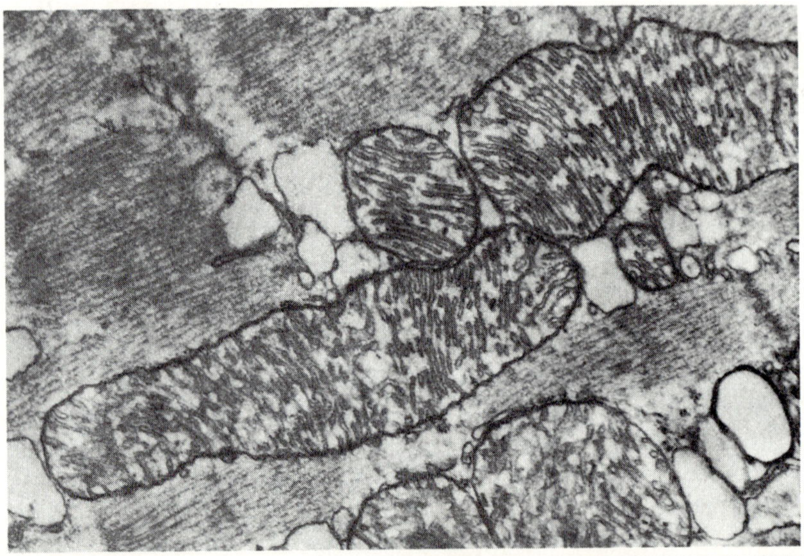

Mitochondria

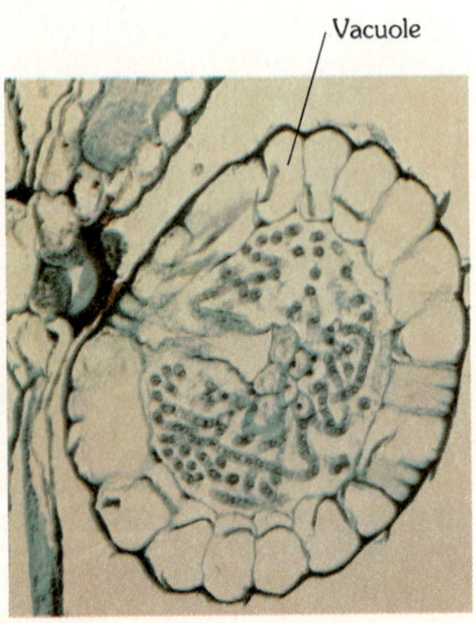

Courtesy Carolina Biological Supply Company

Vacuoles

Inside many cells are one or more saclike structures of various sizes. These structures consist only of a membrane filled with water and various other substances. Such structures are called **vacuoles** (vak'yü ōlz).

Vacuoles perform a variety of jobs. For example, some vacuoles digest food, some store food materials, others contain pigments, and still others dispose of wastes. One major function of vacuoles in plant cells is to form fluid-filled structures that help to support the tissues of the stem and leaves.

Microtubules

Another set of organelles provides a system of support, or a "skeleton," for the cell. This support system is made up of a series of very long, thin cylinders called **microtubules** (mī'krə tü'byülz). The microtubules help to stiffen and to shape such things as the round body of the red blood cell and the long fibers of nerve cells.

Microtubules also assist in the transport of substances into and out of the cell. And they play a role in the movement of cells. For example, microtubules form the cilia and flagella that wave back and forth on the surface of some cells. And microtubules also play an important role in sorting and separating chromosomes during cell division.

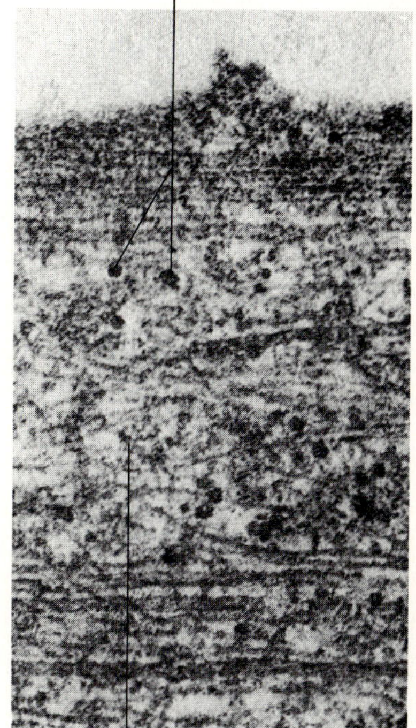

Microfilaments

Biologists wondered for a long time about the process by which cells move from place to place. Only recently have they been able to study living cells while they are moving. Inside the cell, biologists found tiny, threadlike fibers made of protein. These fibers are called **microfilaments** (mī'krə fil'ə mənts). They seem to have the ability to contract like muscle. In fact, microfilaments are found in large numbers in muscle cells.

The microfilaments seem to be involved in two kinds of cell movement. In one kind, special structures on the cell surface, such as cilia, wave back and forth. The cilia are attached to microfilaments inside the cell. The action of the filaments causes the waving action of the cilia. In another kind of movement, cells crawl along surfaces. They do this by forming *pseudopods* ("false feet"). The cytoplasm flows in the direction of the pseudopods. Microfilaments also seem to be involved somehow in the formation of these pseudopods.

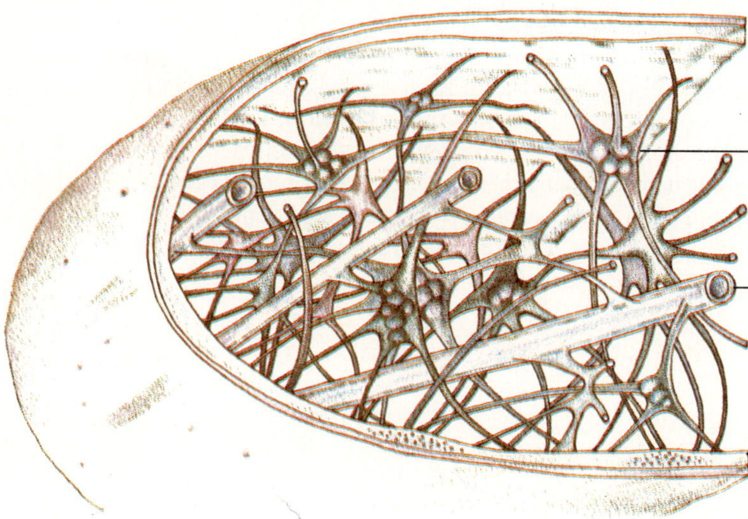

Differences Among Cells

You should be familiar now with the main organelles found in most animal cells. It is quite remarkable that all animal cells, as well as those of monerans, protists, fungi, and plants, are made up of the same basic building blocks. These building blocks are protein, lipids, carbohydrates, nucleic acid, water, and salts. All cell organelles contain these substances. Even the smallest and simplest cell contains enough nucleic acid to make at least 1,000 different kinds of proteins.

However, there are some basic differences among cells of the five kingdoms. Most of these differences are related to the presence or absence of certain cell organelles. You can study the differences below and on the next page.

A Moneran Cell (Bacteria)

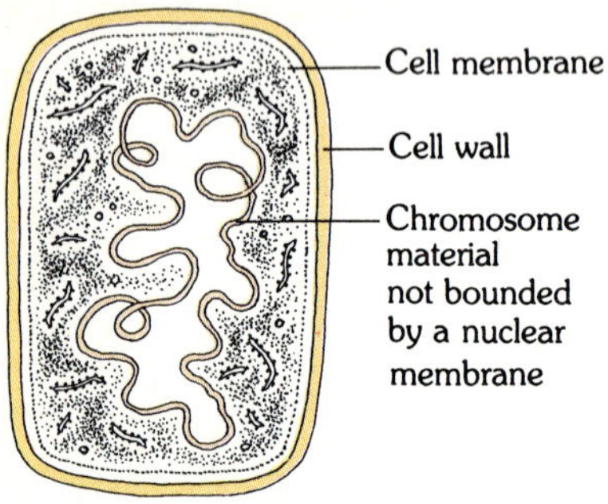

Moneran Cells

You can see the structure of a typical moneran cell on the left. Members of kingdom Monera are one-celled *prokaryotic* cells. Such cells have no nucleus. The chromosome material floats freely throughout the cytoplasm. And the cell contains no complex organelles bounded by membranes (such as mitochondria). Prokaryotic cells cannot combine with each other to form multicellular organisms.

A Protist Cell (Protozoan)

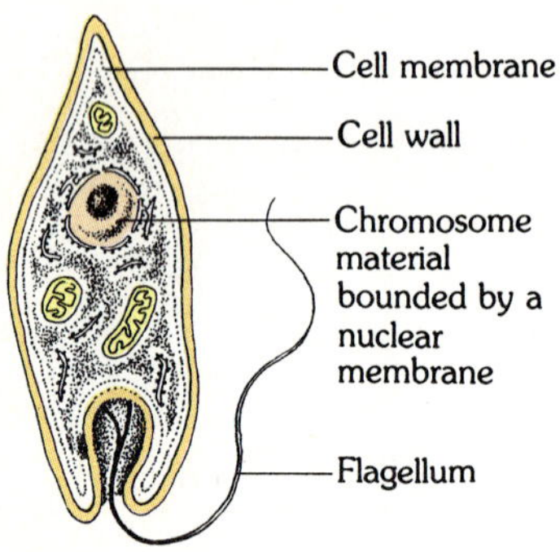

Protist Cells

One-celled protozoans, such as *euglena,* are members of kingdom Protista. Members of this kingdom have *eukaryotic* cells. Such cells have a nucleus bounded by a nuclear membrane. They also have many different kinds of complex organelles, also bounded by membranes that separate them from the cytoplasm. The organelles found in protist cells can be found in both plant cells and animal cells.

A Fungus Cell

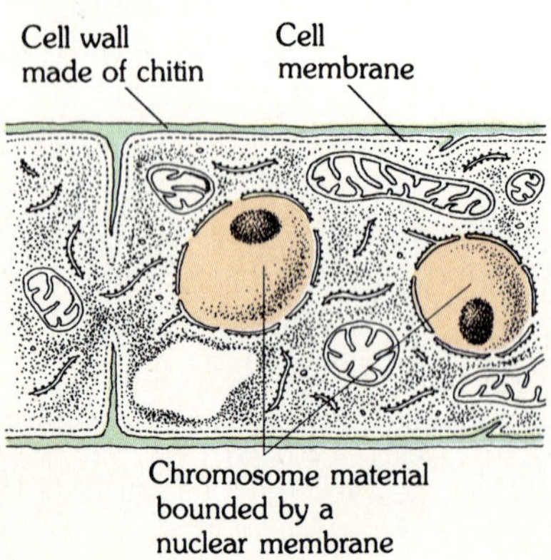

Fungus Cells

All members of kingdom Fungi have eukaryotic cells, as well. Such cells can combine to form simple, multicellular organisms. Fungus cells are bounded by a cell wall made of **chitin** (kīt′n). This material is the same as that which forms the outer skeleton of insects. Each fungus cell usually contains many nuclei. When fungal cells grow and develop, the cell walls do not completely close. Cytoplasm, along with all of the cell contents, can circulate freely from one cell to the next.

Plant Cells

Members of the plant kingdom are multicellular and are made up of eukaryotic cells. Plant cells have a rigid cell wall made of **cellulose** (sel'yə lōs). Each cell contains a single nucleus and many complex organelles, bounded by a membrane. One important group of organelles found only in plant cells and some protist cells are **plastids** (plas'tidz).

Each plastid is bounded by two outer membranes and has a system of internal membranes. There are three kinds of plastids: leucoplasts, chromoplasts, and chloroplasts. **Leucoplasts** are colorless. They are the sites at which glucose is converted to starch and where fats or proteins are stored. **Chromoplasts** manufacture and store pigments. These organelles give fruits, vegetables, and leaves their bright colors. **Chloroplasts** are the organelles in which photosynthesis takes place. Each plant cell also has a very large vacuole containing water and salt. This vacuole helps to keep the cell wall rigid and gives support to the leaves and stem of the plant.

Animal Cells

All members of the animal kingdom are also multicellular and made up of eukaryotic cells. However, animal cells have *no* cell wall and *no* plastids. They do have a special set of organelles called **centrioles** (sen'trē olz). These are active during the reproduction of the cell. Centrioles contain microtubules which help to separate and to sort the chromosome material during cell division.

A Plant Cell

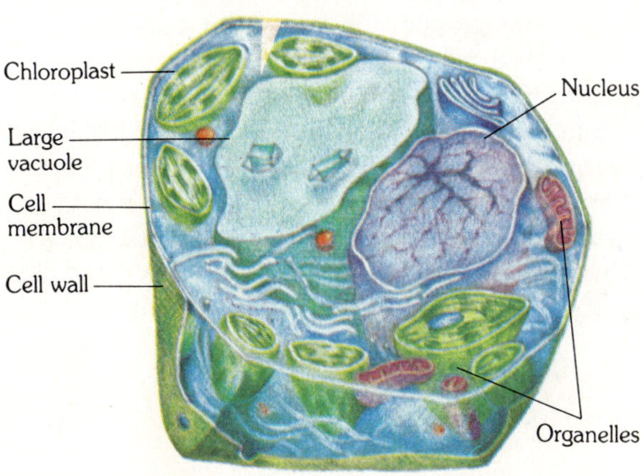

An Animal Cell

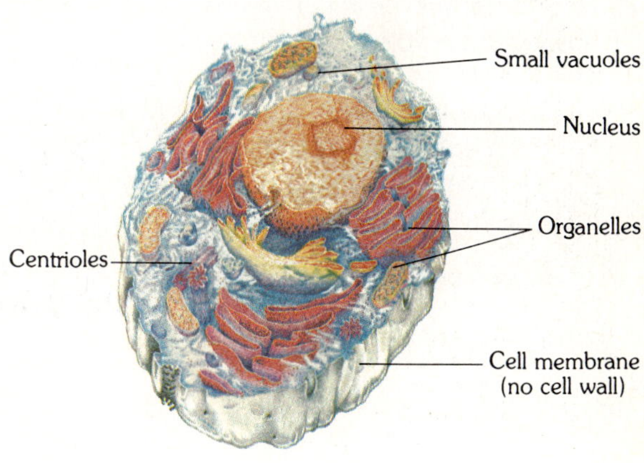

Review It

1. What is an organelle? Describe the functions of at least four organelles found in a cell.
2. List some of the major differences among moneran, protist, fungus, plant, and animal cells.

Chapter Summary

Cells are the basic units of living things. They come in many sizes and shapes. They may live alone or with other cells in more complex organisms. The appearance of a cell is determined by its environment and by its functions.

Our early understanding of the cell came from the study of cells with the light microscope. From this study, scientists slowly developed a cell theory. Today, this theory provides us with a good description of basic ideas about the cell.

Recently, the electron microscope has revealed additional details about cells. Other new techniques are also helping scientists learn more about cells. Some of these techniques involve taking cells apart and microsurgery.

The cell is not a simple structure. It contains many smaller structures called organelles. Every organelle has a specific job or set of jobs to do within the cell. All of the cell organelles work together to carry out life functions and to keep the cell alive.

Questions to Think About

1. What are some of the differences you might find in cells that live (1) by themselves, (2) as parts of larger, more complex organisms?
2. How might microsurgery be used to help diagnose and treat diseases such as cancer?
3. Think of the cell as an industrial complex. Now think of the organelles found in a cell as factories in this complex. Explain how the nucleus, ER, ribosomes, and other organelles might operate as factory divisions in the cell complex.

Extra Research

1. An instrument important in biological research is the **microtome** (mī'krə tōm). Find out what a microtome is, how it works, and why it is important.
2. A number of specialized microscopes have been developed in recent years. Find out what special job each of the following is designed to do: (1) darkfield microscope; (2) ultraviolet microscope; (3) infrared microscope; (4) polarizing microscope; (5) interference microscope; (6) field emission microscope.
3. The preparation of a sample for viewing with an electron microscope is somewhat different from the preparation of a sample for viewing with a light microscope. Find out what these differences in technique involve.
4. Sometimes cells have commercial importance or practical applications in our everyday lives. Egg yolks, for example, are used as a food by people in some parts of the world. Find out some other ways cells or their parts are used by humans. See what you can find out about diatoms, yeast, and bacteria.
5. Biologists are excited about the newest organelle found in cells: *membrane receptors*. Receptors are attached to the outer surface of cell membranes. Their job seems to be to watch for a certain substance out of all the materials that flow by a cell. Then they seize that substance and draw it into the cell. Biologists are fairly certain that some receptors are keyed for insulin molecules. People who don't have enough insulin receptors are unable to get insulin into their cells. Find out more about the research being done on receptors.

Chapter Test

Part I

Copy the number of each test item. If the item is *true*, write *T* after the number; if the item is *false*, write *F*.

1. The shape of a cell may be related to the job it has to do. (64)
2. The basic ideas of the cell theory were developed in the last 25 years. (65)
3. Living cells can be studied by using either the phase-contrast microscope or ultracentrifugation. (66–68)
4. The instructions that tell a cell what functions it is to perform are stored in the chromosomes. (73)
5. Protein synthesis occurs in the lysosomes. (75)
6. Another name for organelle is "small organ." (70–71)
7. The surface of cells can be studied with the transmission electron microscope. (67)
8. Plant cells have organelles called centrioles. (79)
9. ATP is produced in ribosomes. (76)
10. Plants, vegetables, and leaves get their bright color from leucoplasts. (79)

Part II

Copy the number of each sentence below. After each number, write the word or words that complete the sentence correctly.

11. The _____ help to shape the cell and give it structure. (77)
12. The beating action of cilia is caused by the _____ in the cell. (77)
13. Cells differ from each other primarily in three ways: _____, _____, and _____. (64)
14. One method for studying the parts of a cell is by _____. (68)
15. The basic unit of structure and function in all living things is the _____. (65)
16. Two structures found in plant cells but not animal cells are _____ and _____. (79)
17. New cells are produced by _____. (65)
18. Cell membranes appear to be made up of two kinds of molecules: _____ and _____. (72)
19. The large vacuole found in _____ cells contains water and salt which keep the cell wall rigid. (79)
20. The electron microscope can _____ objects thousands of times greater than the light microscope, and it has better _____ power than the light microscope. (67)

Part III

Copy each item from List A. After each item, write the letter and words from List B that best describe the item.

List A

21. ribosomes (74)
22. ER (74)
23. ATP (76)
24. mitochondria (76)
25. Golgi apparatus (75)

List B

a. a chemical that stores energy for use by the cell
b. major points at which protein is made in the cell
c. powerhouse of the cell; produce ATP
d. saclike membrane that packages protein and moves it out of the cell
e. "highway system" in the cell

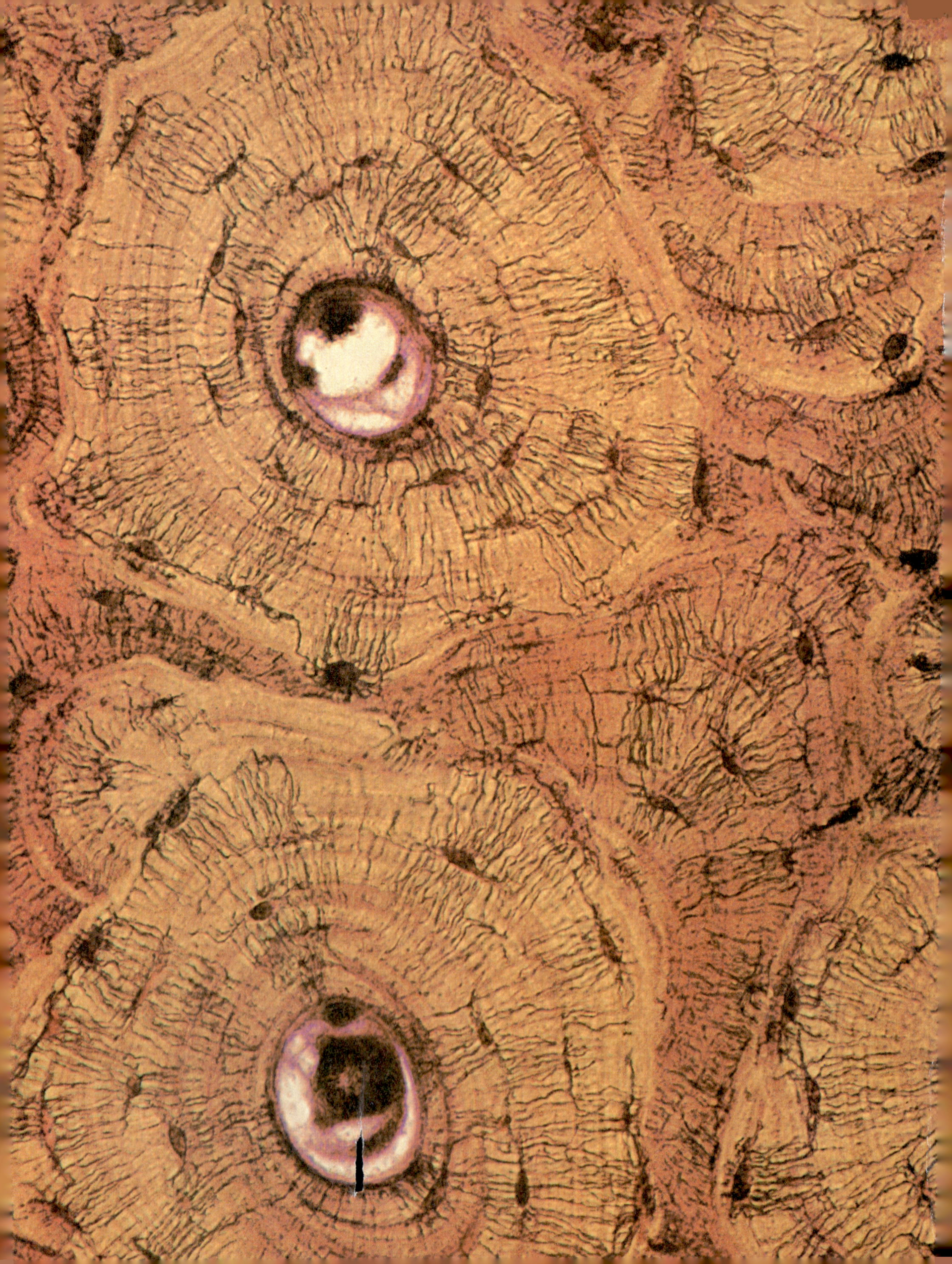

Chapter 4

Function of Cells

Cells are busy places. Whether they exist alone or as units in a larger organism, every cell performs many different functions. They take in substances from the outside environment and get rid of waste materials. They convert nutrients into energy and new cell parts. Some cells, such as those of green plants, convert chemical compounds and the energy from sunlight into food materials. The many jobs a cell has to do are related to each other. Changing one function also has an effect on other cell functions. In spite of the activity going on inside their membranes, cells are stable environments. Self-correcting mechanisms help them to maintain their essential functions even when drastic changes take place in the world around them.

Chapter Objectives

1. Explain how materials pass through the cell membrane.
2. Describe some of the special problems of cellular transport.
3. Give examples of the most common elements and compounds found in living systems.
4. Define a chemical reaction. Distinguish between covalent bonds and ionic bonds formed during chemical reactions.
5. List two important properties of chemical reactions in living things.
6. Describe the processes involved in respiration.
7. Distinguish between anaerobic and aerobic respiration.
8. Describe the processes involved in photosynthesis.
9. Distinguish between the light and dark reactions of photosynthesis.
10. Define homeostasis and explain how it occurs in the cell.

Bone cells in the thigh bone. The hollow circles contain blood vessels.

4–1
Cellular Transport

Cells do many things. They take in food, make changes in that food, build new cell parts, release materials to the outside environment, react to outside stimuli, and reproduce. The materials a cell needs to do its work come from outside the cell. Most of the substances a cell produces also return to the outside environment. In either case, they have to pass through a barrier to get into and out of the cell. This barrier is the *cell membrane,* and the transport is either *passive* or *active.*

The raw materials that pass through the cell membrane are in the form of **molecules.** Molecules are the smallest particles a substance can be divided into without changing it into another substance. Cells often take in partially digested molecules of fats, carbohydrates, and proteins as food. To understand how this is done, you need to know something about how molecules behave.

Random movement of molecules in a gas or liquid

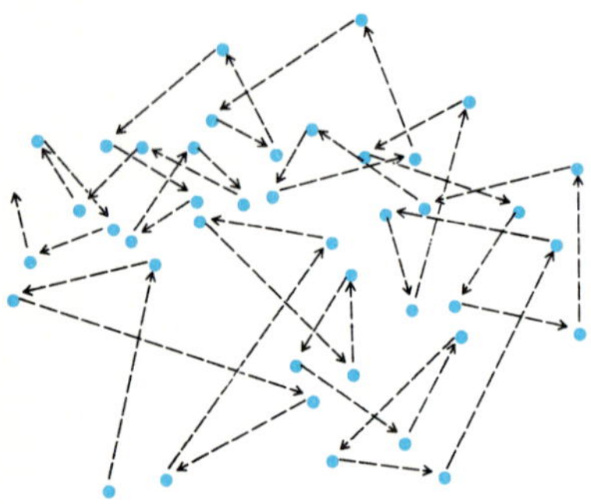

Diffusion

The molecules that make up any substance are constantly in motion. Molecules that make up a *solid* move slowly so the solid appears to stay in one place. However, molecules that make up a *liquid* or a *gas* are free to wander from place to place. As they move, they bump into each other. The result is that these molecules are bounced about in a random pattern. You can see the random paths of molecules in a liquid at the left.

Suppose you carefully add a drop of ink to a glass of water. At first, the ink molecules are bunched close together in a single drop. Then the molecules of ink begin to spread out until all of the water is colored by the ink. This spreading of color is evidence that the ink molecules are in motion. As ink molecules bump into water molecules, they slowly move out and away from each other. Eventually the ink molecules are evenly mixed throughout the water. The process by which molecules move from an area of higher concentration to an area of lower concentration is called **diffusion.**

Diffusion of ink molecules in a beaker of water

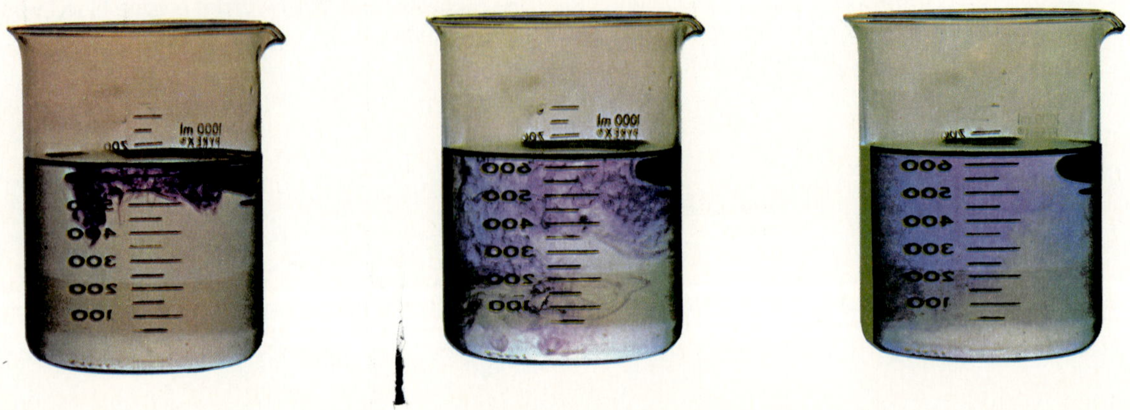

Semipermeable Membranes

Now suppose you could pour a layer of pure water on top of a layer of salt water. After a while, if you tasted the liquid, you would find that the layers of salt water and pure water have mixed completely. The reason for this is that the molecules of salt would have diffused from a region of higher concentration to a region of lower concentration. At the same time, molecules of water would have diffused from a region of higher concentration to one of lower concentration.

Suppose you repeat the experiment. But this time you put a *barrier* between the two liquids. The results you get will depend on the kind of barrier that you choose. For example, if you choose a barrier of plain cellophane, it will allow both water and salt molecules to pass back and forth. The cellophane is permeable to both water molecules and salt molecules. But if you choose another kind of membrane, such as *copper ferrocyanide,* it will produce a different result. This membrane will allow only water molecules to pass back and forth. The salt molecules will remain trapped on one side of the membrane. Membranes through which some substances can pass but others cannot are called **semipermeable** (sem′i pėr′mē ə bəl) **membranes.**

If you could examine a semipermeable membrane with an electron microscope, you would see tiny openings or pores in it. The size of the pores usually determines the size of the molecules that can pass through the membrane. You can see how a semipermeable membrane works by studying the diagram below.

Challenge!
Drop a raisin into a beaker of water. Observe the raisin over a period of about 10 minutes. What do you see happening? How can you explain your observations in terms of what you now know about diffusion?

Diffusion of molecules across a semipermeable membrane

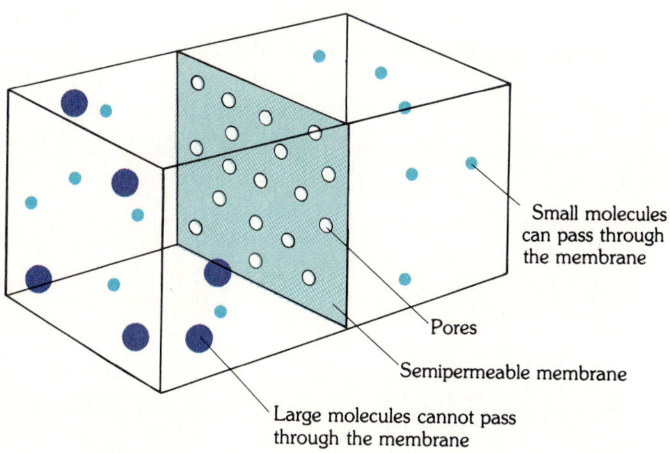

Transport Across Cell Membranes

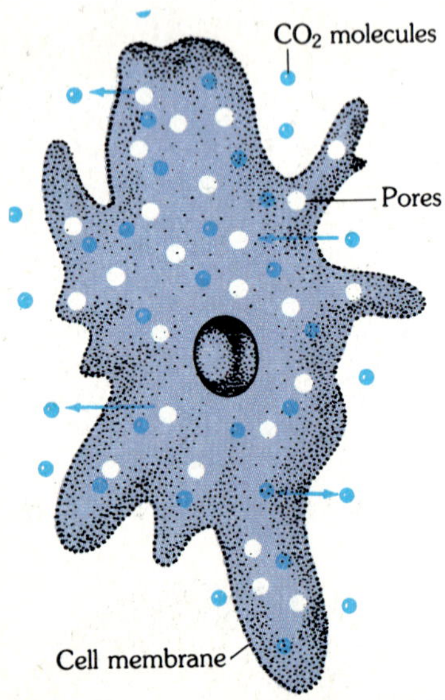

Transport of carbon dioxide molecules out of a cell by diffusion

Cell membranes are semipermeable membranes. Water, carbon dioxide, oxygen, and other small molecules can pass through the pores in the cell membrane easily. The larger molecules that make up simple sugars pass through the membrane slowly. And still larger molecules, such as those that make up proteins and some fats, can't get through the membranes at all. However, other factors besides size may affect the rate at which molecules pass through a cell membrane. For example, temperature of the outside environment seems to determine how easily some molecules are carried into the cell.

It sometimes helps to think of the cell contents as a water solution that contains many different substances. The cell is held together by a semipermeable cell membrane that separates it from the outside environment. And the outside environment is also a water solution of many different substances.

If you think of a cell in this way, it is possible to explain how things get into and out of the cell by diffusion. For example, carbon dioxide gas is produced in a cell during respiration. Over a period of time, the concentration of carbon dioxide molecules becomes greater inside the cell than outside. Then carbon dioxide molecules move through the cell membrane and out of the cell by diffusion. In this way, the concentration of carbon dioxide inside the cell stays nearly the same as the concentration outside the cell.

Osmosis

Water is the most abundant material in living things. Therefore its movement back and forth across a cell membrane is of special interest. In fact, the term **osmosis** (oz mō′sis) is used to describe the diffusion of water through a semipermeable membrane.

Let's take a simple example of osmosis first. Imagine a blood cell has been placed in an **isotonic** (ī′sə ton′ik) solution. This means that the concentration of water molecules in the solution is *equal* to the concentration of water molecules inside the cell. Water will flow into and out of the cell by osmosis. Since the concentration of water molecules is the same on both sides of the membrane, there will be no change in the cell.

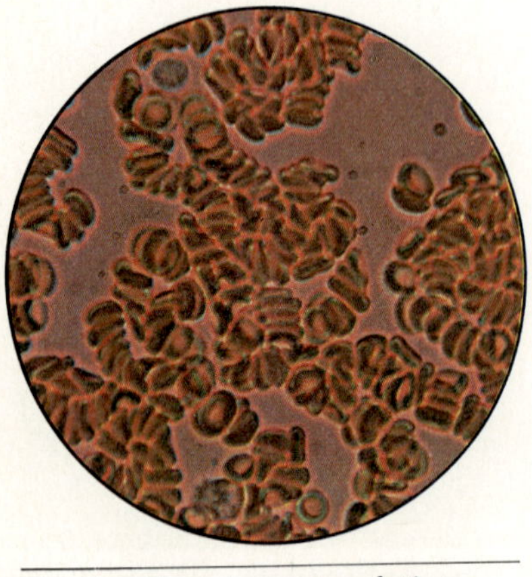

Blood cells in isotonic solution

Turgor

Sometimes the movement of water by osmosis can produce striking changes in a cell. Suppose a blood cell is placed in a **hypotonic** (hī′-pō ton′ik) solution. This means the concentration of dissolved particles outside the cell is *less* than the concentration of dissolved particles inside the cell. Water will flow into the cell faster than it will flow out. If the difference in concentrations is great enough, water will flow into the cell until the cell explodes.

One-celled organisms have developed a special method for handling this situation. Excess water is collected into a **contractile** (kən-trak′təl) **vacuole.** Then the water is pumped out of the vacuole and back into the surrounding environment.

Plant cells have a *cell wall* that keeps the excess water from exploding the cell. As water enters the cell, the plant cell becomes very stiff. Water pressure forces the cell contents against the cell wall. The stiffness of plant cells due to water pressure inside the cell is called **turgor** (tėr′gər). Turgor pressure plays an important part in keeping the leaves and stems of plants stiff and upright.

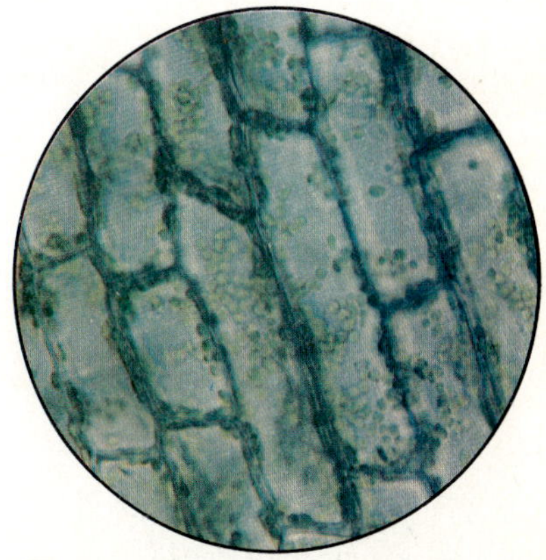

Plant cells in hypotonic solution

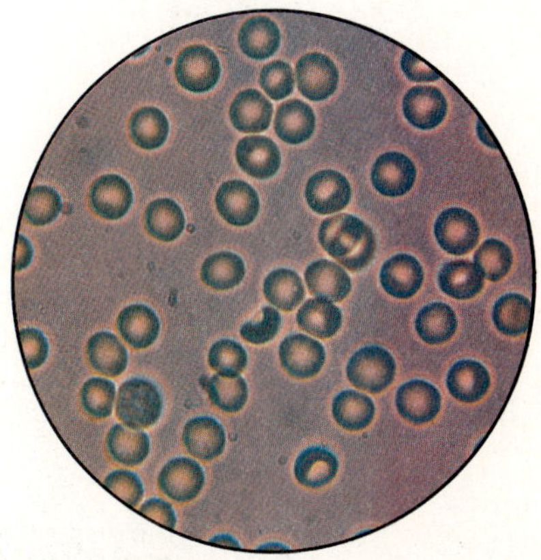

Blood cells in hypotonic solution

Plasmolysis

Now imagine just the reverse of the situation above. Suppose the blood cell is placed in a **hypertonic** (hī′pər ton′ik) solution. This means the concentration of dissolved particles outside of the cell is *greater* than the concentration of dissolved particles inside the cell. In this case, water will flow out of the cell faster than it will flow in. If the difference in concentrations of water molecules is great enough, the water loss can be very great. The cell may shrivel and die.

When cells shrink due to loss of water, the condition is called **plasmolysis** (plaz mol′ə sis). Plant cells can withstand such loss of water better than animal cells can. The rigid cell walls of plants help prevent the cells from collapsing when water is scarce. However, plasmolysis can occur if too much fertilizer is placed on the soil where plants are growing.

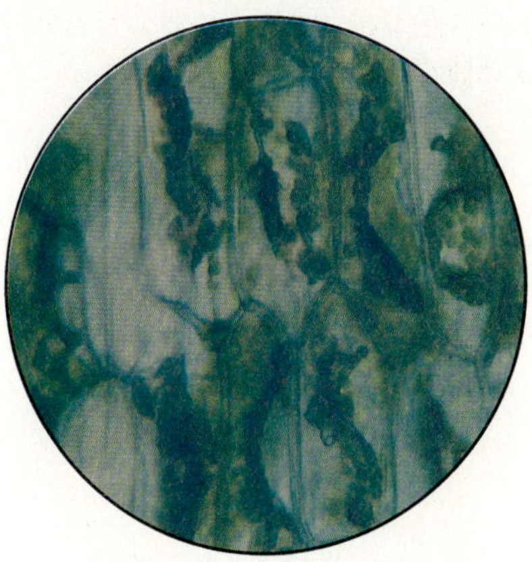

Plant cells in hypertonic solution

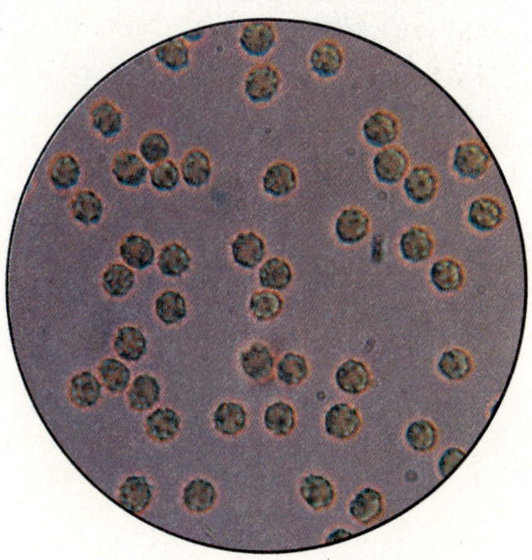

Blood cells in hypertonic solution

Active Transport

All the processes described so far have involved the movement of molecules from regions of higher concentration to regions of lower concentration. There are cases, however, when cells transport molecules in the *opposite* direction. The molecules pass in or out of cells *against* the direction of diffusion. This means that cells must use energy to pull molecules from a region of lower concentration to one of higher concentration. The process is called **active transport.**

Active transport requires the cell to use energy to force molecules in or out. This process is not clearly understood. But researchers believe that certain molecules along the cell membrane pick up the substance to be transported, and release the substance on the other side of the cell membrane. The process seems to work something like a revolving door.

Active transport allows cells to collect certain food substances in concentrations hundreds of times higher than those outside the cell. And it allows cells to get rid of substances that would be harmful in high concentrations. For instance, cells of the intestine use active transport to *force in* food molecules during digestion. But blood cells use active transport to *force out* sodium ions. The ions would be harmful to the human body in high concentrations.

Active Transport

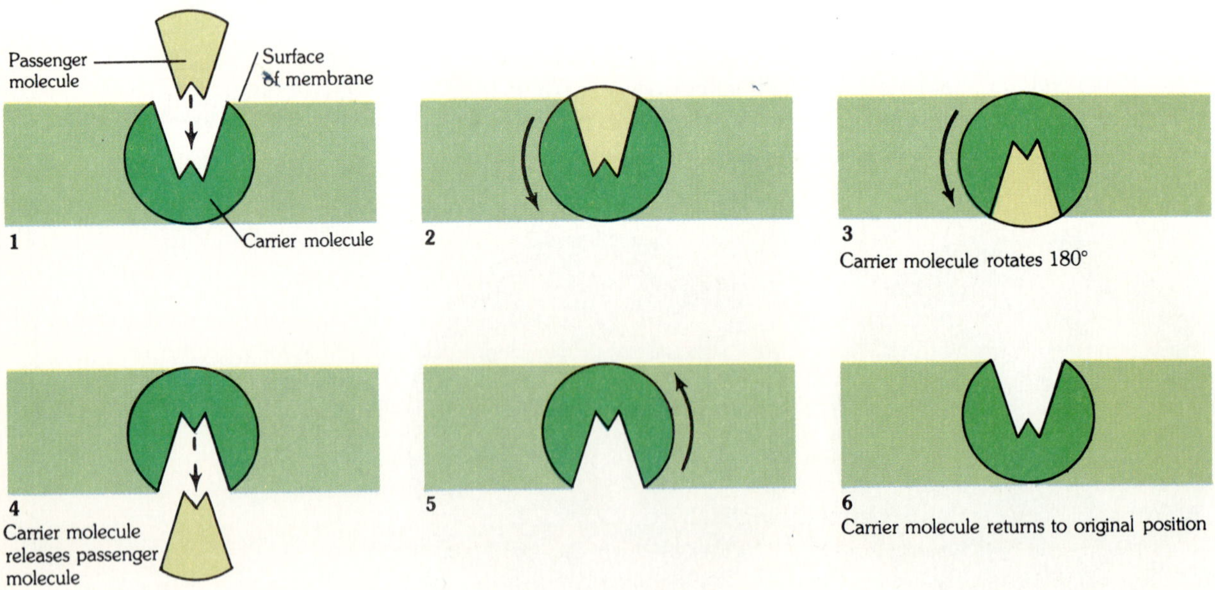

Special Problems of Cell Transport

Some food materials are too large to pass through the cell membrane. When an animal cell comes into contact with such materials, it forms a food pocket. The pocket surrounds the substance, seals itself, and becomes a vacuole within the cell. Eventually the food material is broken down and absorbed into the cell.

When large *solids* are taken in by a cell, the process is called **phagocytosis** (fag′ə sī tō′sis). White blood cells in our bodies surround and capture bacteria by the process of phagocytosis. This is one of the ways infection is controlled in the body. One-celled organisms, such as ameba, also capture their food by the process of phagocytosis.

When large droplets of *liquid* are engulfed and absorbed by cells, the process is called **pinocytosis** (pi′nō sī tō′sis). The process operates in the same way as phagocytosis. For example, large dissolved molecules of fats and certain proteins cannot pass through cell membranes. However, they can be absorbed by pinocytosis. Fats and proteins pass from the human bloodstream into body cells by this process.

Phagocytosis

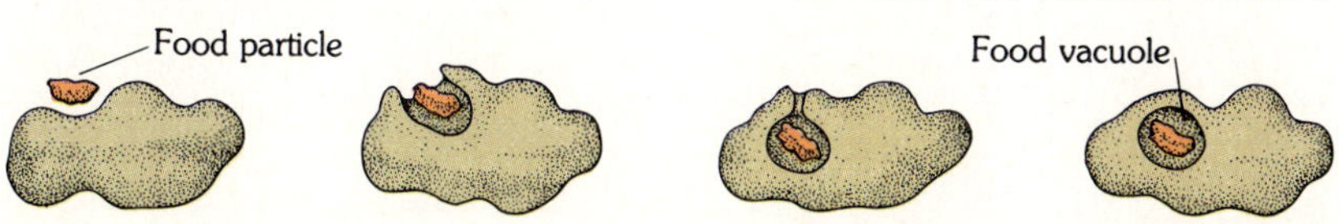

Pinocytosis

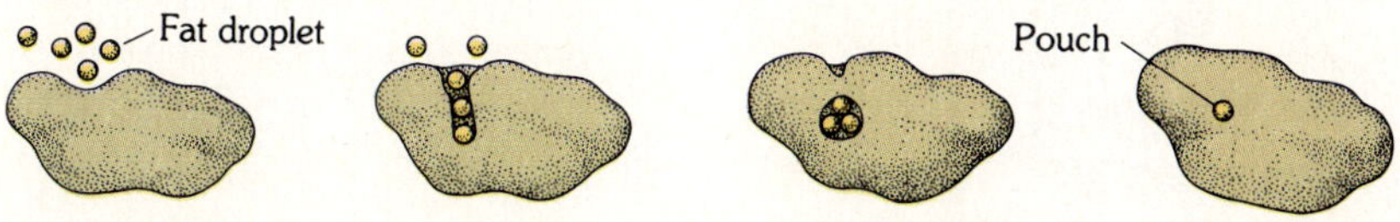

Review It

1. Distinguish between osmosis and diffusion.
2. What is a semipermeable membrane? Explain how a cell membrane functions as a semipermeable membrane.
3. How does active transport differ from osmosis? Give an example.
4. How do phagocytosis and pinocytosis differ from other methods of cellular transport?

Activity

Diffusion or Osmosis?

Materials
Dialysis membrane
10% sugar solution
20% sugar solution
10% salt solution
Silver nitrate solution
Glass container
Funnel tube or glass tubing
Marking pencil
Medicine dropper
Ring stand and clamp
Metric ruler

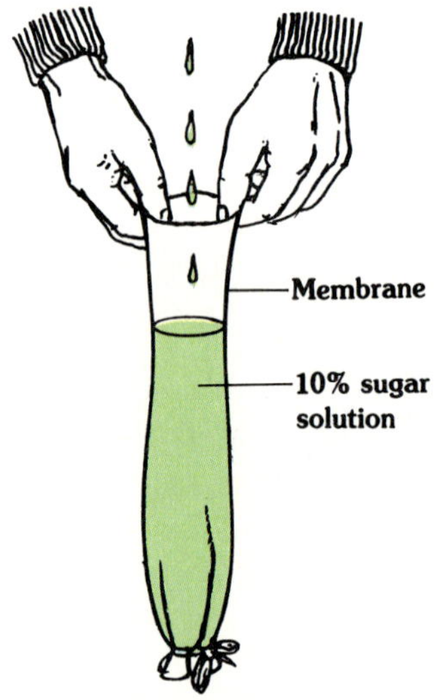

Procedure

1. Get a membrane. Fill it with 10% sugar solution and rinse off the outside of it with water. Tie the end of the membrane to a funnel tube.
2. Place the membrane in a container of water. Mark the beginning level of liquid in the tube with a marking pencil.
3. Record the level of the liquid every five minutes, for a total of 20 minutes.
4. Rinse out the membrane and repeat the procedure with a 20% sugar solution.
5. Rinse out the membrane again and fill it with a 10% salt solution. Repeat steps 1–3.

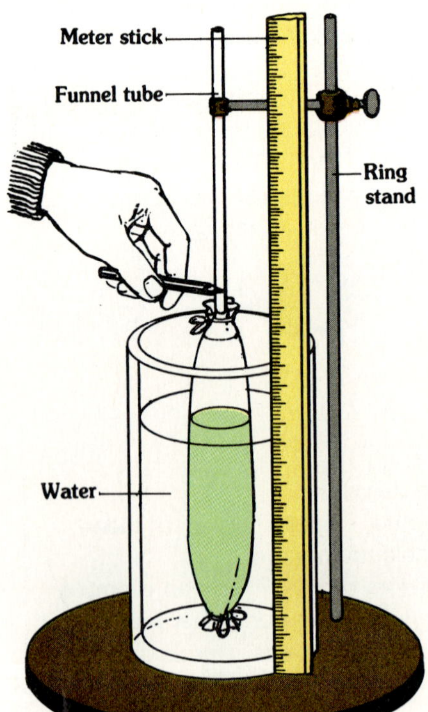

6. Test the water outside the membrane for the presence of salt. Do this by adding a few drops of silver nitrate to the water. If you see a cloudiness, then there is salt in the water. Record your observations.

Analysis

1. Compare the rates of movement of the liquid in the tubes with the different concentrations of sugar. What is the best explanation you can give for these results?
2. Compare the rates of liquid movement using 10% sugar solution and 10% salt. What is the best explanation you can give for these results?
3. Which of the above situations probably represents the results from diffusion? Why?
4. What can you infer about the nature of the membrane and the relative size of the salt and sugar molecules?

4–2 Basic Chemical Elements of Life

Each cell runs on energy. In order to do all the jobs it has to do, a cell needs a source of energy to power these activities. A cell is a very delicate thing. It cannot pump in heat or electricity, the way you would use energy in a home. Instead, a cell stores energy in the form of *chemical compounds.* Then these compounds are broken down and energy is released as it is needed.

All of the chemical processes by which a cell makes and uses energy-rich compounds is called **metabolism** (mə tab′ə liz′əm). This section is about compounds involved in cell metabolism.

Percentage of Chemical Elements

Earth's Crust		Living Matter	
Oxygen	47%	Oxygen	62%
Silicon	28%	Carbon	20%
Aluminum	8%	Hydrogen	10%
Iron	5%	Nitrogen	3%
Calcium	4%	Calcium	3%
Sodium	3%	Phosphorus	1%
Potassium	3%	Chlorine	0.2%
Magnesium	2%	Sulfur	0.1%

Elements

Both living and nonliving matter is formed from 92 natural chemical **elements.** Cells differ from nonliving things primarily in the kinds of elements of which they are made. The table at the right shows that oxygen, silicon, iron, and aluminum are the most abundant elements on Earth. However, cells are made up almost entirely of carbon, hydrogen, oxygen, nitrogen, phosphorus, and sulfur. As you can see from the table, the first four of these elements account for about 95 percent of all living matter.

Compounds

From the 92 elements that exist, more than three million different **compounds** have been formed. Compounds are formed when two or more different kinds of elements combine chemically in fixed amounts. Almost 95 percent of these compounds are **organic.** This means that the compounds contain the element carbon.

The majority of compounds found in cells are organic. Water and hydrochloric acid are two exceptions. Most organic compounds in living systems can be put into one of four "families": (1) carbohydrates: sugars, starches, and cellulose; (2) lipids: fats, oils, and waxes; (3) proteins; and (4) nucleic acids. You can see examples of some of these organic compounds to the right and below.

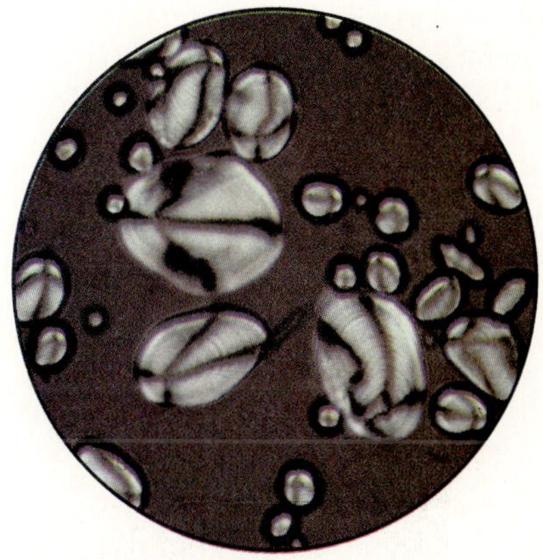

Carbohydrates

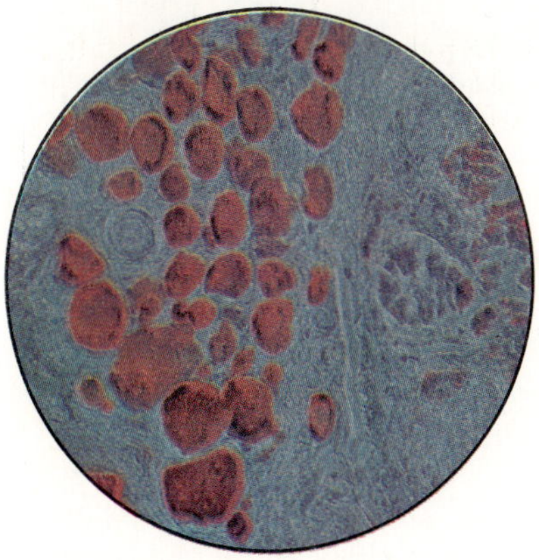

Fats

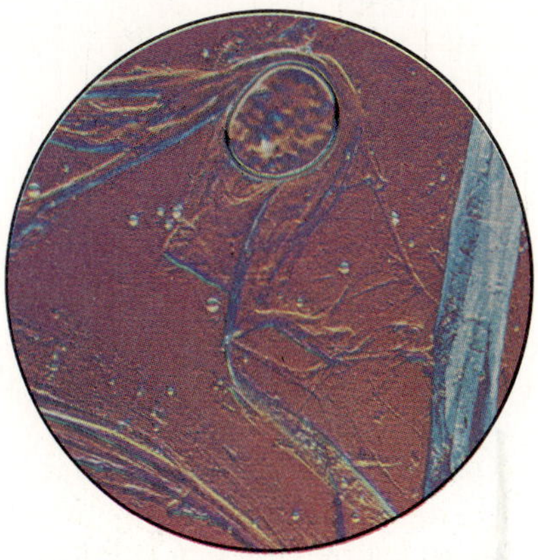

Proteins

Nucleic acids

Atoms

All of the elements that make up living matter are composed of units called **atoms.** No one has ever seen an atom. Yet scientists have been able to describe the structure and behavior of atoms through indirect means. The atom has an inner core, or *nucleus*, that contains small particles called **protons.** Each proton has a positive electrical charge. The nucleus may also contain **neutrons.** Each neutron has no electrical charge. Spinning around the nucleus is anywhere from one **electron** to more than a hundred electrons. The number of electrons depends on the kind of atom. Each electron is extremely small and has a negative electrical charge which is equal but opposite to the charge of the proton. Atoms are identified by the number of protons in the nucleus. An atom is neutral if it has equal numbers of protons and electrons. You can see the structure of hydrogen, oxygen, and carbon atoms below.

Only a certain number of electrons can occupy any one electron shell, or level. The first, or inner, shell will hold only two electrons. The second will hold eight and the third will hold eight or eighteen. One of the most important properties of an atom is that it tends to form a stable electron structure. Usually this means that an atom tends to complete its outer electron shell. Most **chemical reactions** between elements are nothing more than exchanges or sharing of electrons between atoms. The end result of such reactions is that each atom ends up with a completed outer electron shell. Now let's look at some examples.

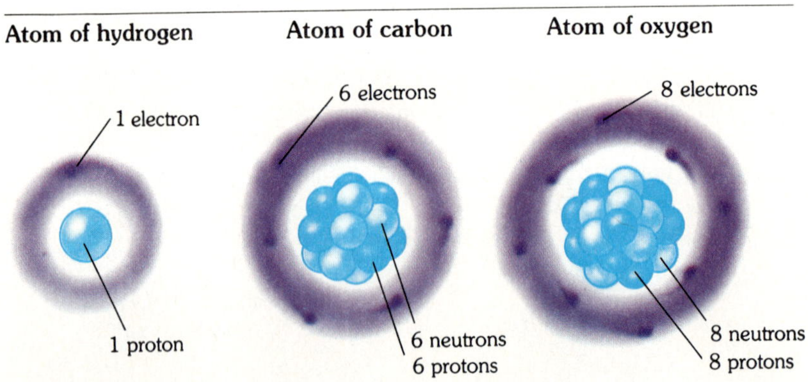

Element	Electrons in each shell 1	2	3	Protons in nucleus
Hydrogen	1	0	0	1
Carbon	2	4	0	6
Oxygen	2	6	0	8
Sodium	2	8	1	11
Chlorine	2	8	7	17

Covalent Bonds

Molecules of hydrogen and water are formed when two or more atoms *share* electrons. Sharing of electron pairs is called **covalent** (kō vā'lənt) **bonding.** For example, look at the table in the margin that shows the electron structure of hydrogen. Notice that hydrogen has only one electron in its electron shell. The hydrogen atom can complete its outer shell by sharing its electron with the electron from another hydrogen atom. When two hydrogen atoms share electron pairs, a molecule of hydrogen is formed. H_2 is the symbol.

Structure of Water

Now look at the table. Oxygen has six electrons in its outer shell. A molecule of water can be formed when two atoms of hydrogen share electrons with oxygen. The symbol for water is H_2O. You can see how these covalent bonds are formed by studying the diagram.

Ionic Bonds

Another kind of bonding involves the *transfer* of electrons, rather than the sharing of electrons. This is called **ionic** (ī on′ik) **bonding.** Sodium chloride, or table salt, is formed by ionic bonding. Notice in the table that sodium has only one electron in its third shell. Sodium tends to give up this electron (rather than sharing it) to form a stable electron structure. Now look at chlorine. It has seven electrons in its third shell. Chlorine tends to add an electron (rather than sharing one) to complete its shell.

When chlorine and sodium atoms meet, the chlorine atom captures the one electron in the outer shell of the sodium atom. This completes the outer shells of both atoms. However, since sodium has lost an electron, it now has a positive charge. But chlorine has added an electron, so it has a negative charge. Since opposites attract, the two charged atoms, or **ions** (ī′ənz), are held together by their opposite charges. The result is a molecule of sodium chloride or table salt. The diagram below shows how ionic bonds are formed.

Ionic Bonding

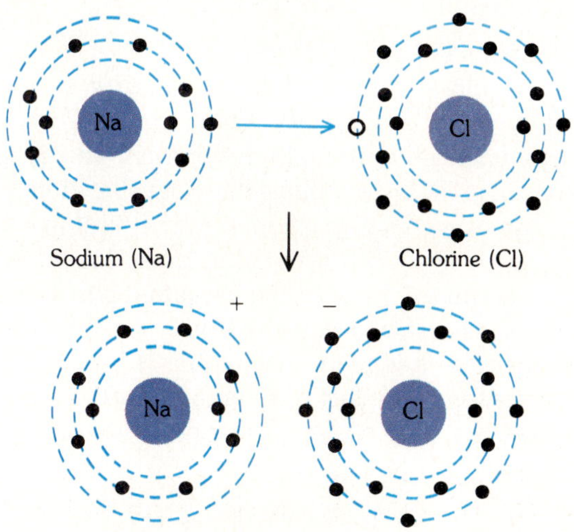

Sodium chloride molecule (Na^+Cl^-)

Covalent Bonding

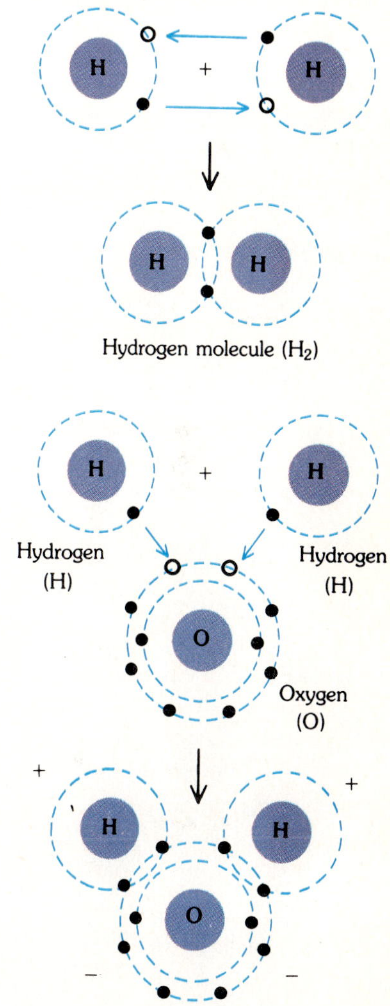

Water molecule (H_2O)

Carbon Atoms

Carbon is a somewhat unusual element. It has four electrons in its second shell. Therefore, it must share four electrons with other atoms to complete its outer shell. Carbon completes the shell by forming covalent bonds with other atoms or with ions. Therefore, carbon can combine with many different kinds of atoms and combinations of atoms. In fact, carbon has so many bonding possibilities that it is the backbone for the wide variety of organic compounds found in living things. Some of the possibilities are:

Methane is the simplest organic compound formed by carbon atoms

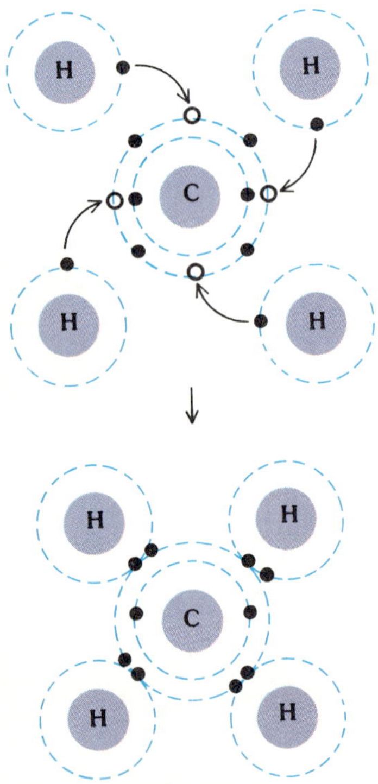

Methane (CH_4)

1. Carbon atoms can form covalent bonds with atoms or ions.

2. Carbon molecules can join to form long chains of molecules.

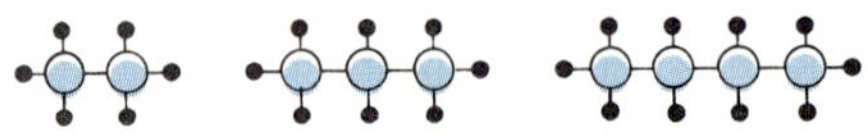

3. Carbon chains can join in various ways to form complex branches.

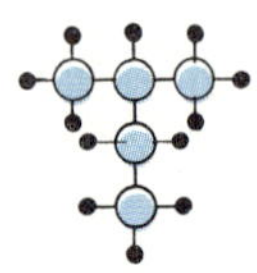

4. Carbon chains can join to form rings.

Chemical Reactions in Living Things

The forces which hold atoms together in a molecule are known as **chemical bonds.** When a chemical reaction occurs, existing bonds are broken and new ones are formed. In the process, energy may either be given off or taken in. Living things have developed very efficient ways to make use of the energy released in chemical reactions.

One property of chemical reactions in living things is that they usually occur in *sequence.* Most of the important chemical changes that occur in cells take place in a series of many steps (often 20 or more). This conserves energy and raw materials because the energy and products released in the first reaction are used to run a second reaction, and so on. A typical sequence reaction in a cell might look like this:

A + B ⟶ Product 1 + Energy ⟶ Product 2 + Energy ⟶
Product 3 + Energy ⟶ End Product

Such a sequence of reactions enables the cell to make use of most of the energy and compounds it produces. Very little is wasted as excess heat or is left over as waste products.

Enzymes

Another property of chemical reactions in living things is that they do not involve any large or sudden changes in heat. Chemical reactions that you may have observed in a laboratory would produce enough heat to destroy a cell. However, special compounds called **catalysts** (kat'l ists) can be used to control the amount of energy released in a chemical reaction. Catalysts speed up or slow down a reaction without being used up. They can make it possible for a reaction to take place by lowering the amount of energy needed to get things started.

The special catalysts found in living cells are called **enzymes** (en'-zīmz). All enzymes are complex proteins. Each enzyme has a surface that roughly fits the shape of the compound upon which it will act. However, it is believed that the enzyme doesn't fit the substance exactly. This slight difference in shape causes the molecular structure of the substance to be slightly out of shape. As a result, part of the molecule is weakened and exposed to other substances in the cell. The molecule reacts with these substances much more rapidly with the enzyme than without it. After the reaction, the new substance or substances that have been formed break away from the enzyme. But the enzyme remains unchanged. It is available to take part in the same kind of chemical reaction over and over again. You can see how this process works by studying the diagram below.

Enzyme Action in Cell

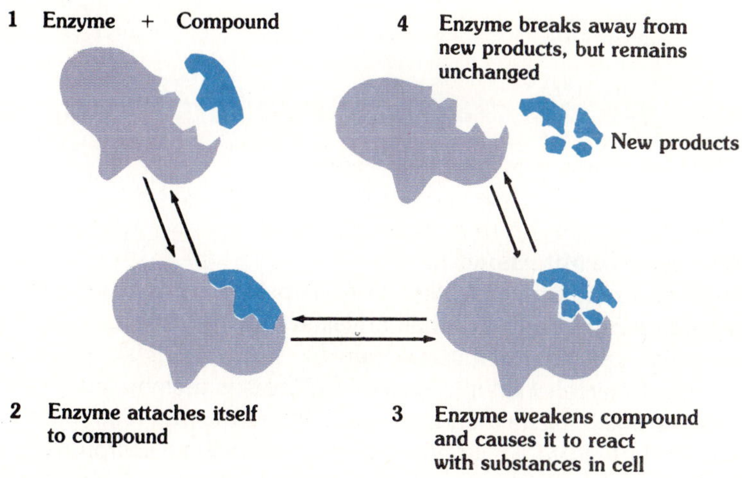

Review It

1. Distinguish between ionic and covalent bonds.
2. Why is carbon such a versatile element?
3. Explain how an enzyme works. Describe two ways chemical reactions in cells differ from those in nonliving things.
4. Define the following terms: metabolism, element, atom, compound, chemical reaction.

4–3
Cell Respiration

Molecule of ATP

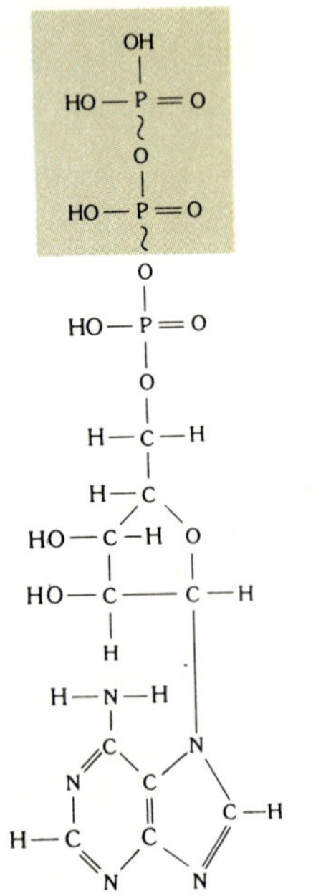

(~denotes an energy-rich bond)

The main source of energy for living systems is a compound called **glucose** (glü'kōs). The starch and sugar an organism eats is in the form of large, complex molecules. An organism digests or breaks down these large molecules into simpler glucose molecules. During **cell respiration,** the cell takes the glucose molecules apart. Energy is released and stored in a kind of chemical storage molecule called **ATP** (adenosine triphosphate). The energy in this molecule is always available for cell use.

The process can be shown by the following equation:

$$\text{Glucose} + \text{Oxygen} \longrightarrow \text{Carbon Dioxide} + \text{Water} + \text{ATP}$$
$$C_6H_{12}O_6 + 6\ O_2 \longrightarrow 6\ CO_2 + 6\ H_2O + \text{Energy}$$

ATP and Respiration

Look at the diagram of ATP. At one end, you will see a chain of three phosphorus groups. This is the storage part of the compound. The last two phosphates in this chain are high energy phosphates. Their chemical bonds can be easily broken, releasing chemical energy.

When a cell needs energy for any purpose, it obtains the energy by breaking the last phosphate bond of ATP. The reaction releases a certain amount of energy which the cell can use. When the bond is broken, the ATP molecule is left with only two phosphate groups. The new molecule is called **ADP** (adenosine diphosphate).

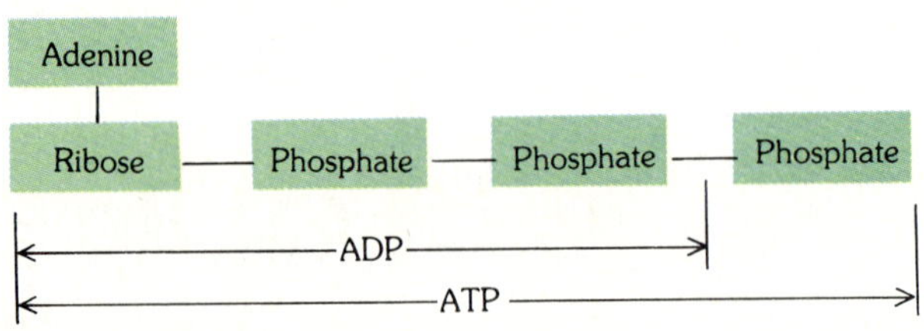

Mitochondria

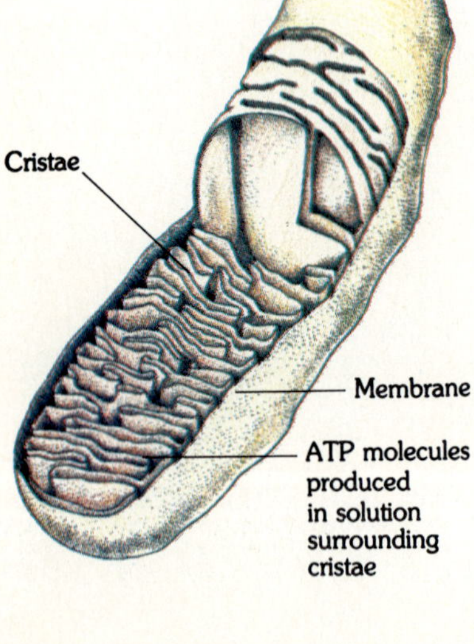

ATP and the Mitochondria

Inside each cell are organelles called *mitochondria.* Cells may contain anywhere from 20 to over half a million of these structures. Respiration takes place in the mitochondria.

Each mitochondrion is surrounded by two membranes. The outer one is smooth, while the inner one is folded over on itself many, many times. These folds are called *cristae,* as you may recall from Chapter 3. Surrounding the folds is a dense solution of compounds involved in respiration. The energy compound ATP is produced in this solution.

During respiration, energy released from glucose is used to restore the last phosphate group. The ADP is changed back to ATP. ATP holds this energy until it is needed. When work needs to be done, the ATP molecules leave the mitochondria. They travel to the part of the cell where energy is needed and release the stored energy. The whole process takes place in a fraction of a second and can be repeated over and over again, each time respiration occurs.

Cell Respiration

1. Glucose and oxygen enter the cell cytoplasm.

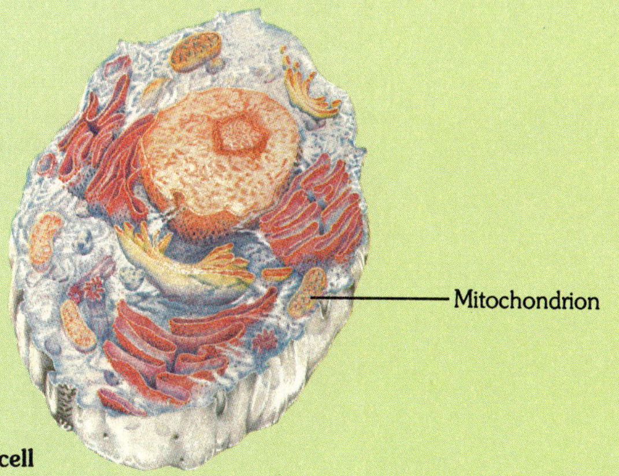

The cell

2. Glucose and oxygen from the cytoplasm enter the mitochondrion. Energy that is stored in the chemical bonds of glucose is released in the mitochondrion during respiration.

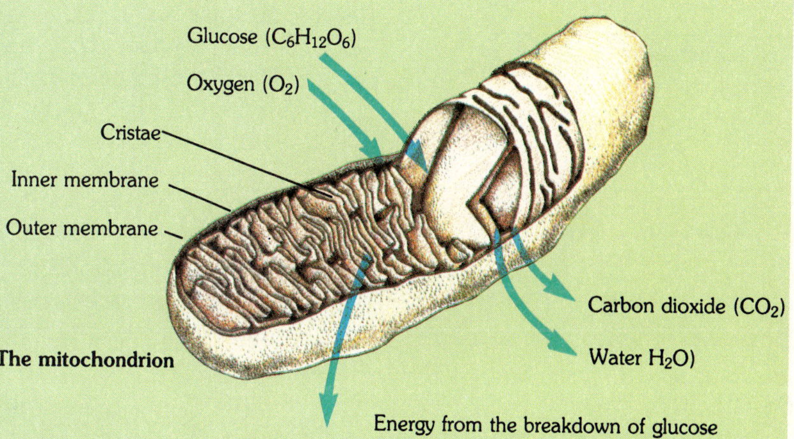

The mitochondrion

3a. The energy released by the breakdown of glucose is stored in the chemical bond that holds P to ADP, forming ATP.
3b. When energy is needed by the cell, that energy can be obtained by breaking the bond that holds P to ATP. When energy is released, ADP is formed.

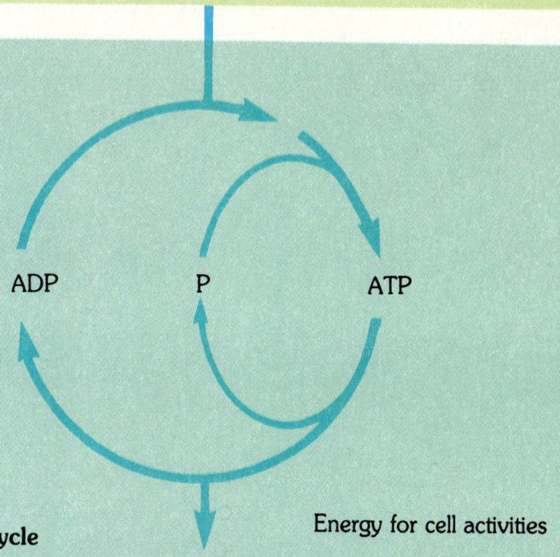

The ADP-ATP cycle

Anaerobic Respiration

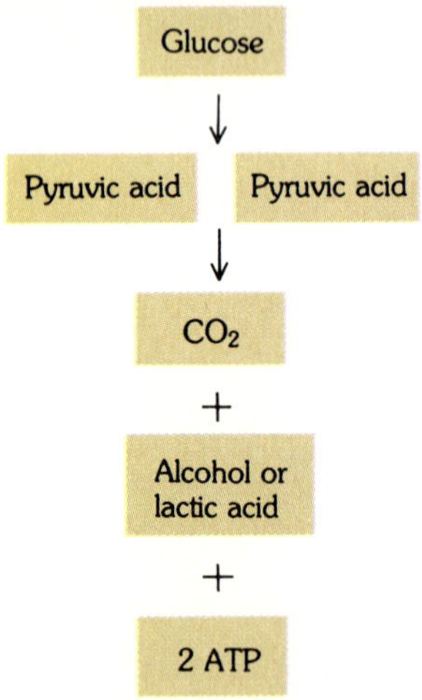

Aerobic Respiration

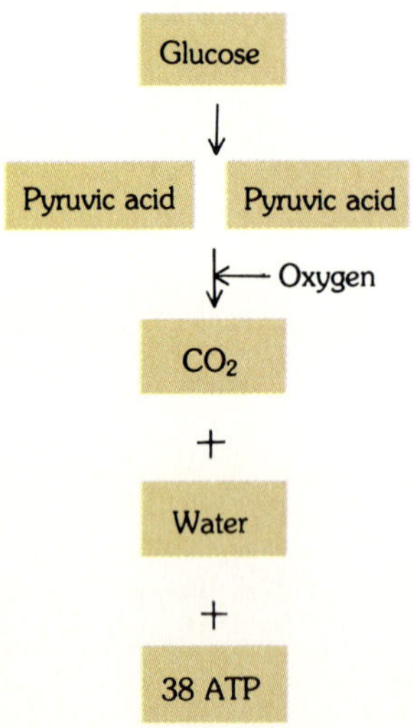

Anaerobic Respiration

The process of respiration involves many complex steps. The first series of steps always occurs in the absence of oxygen. Respiration that occurs without oxygen is called **anaerobic** (an′ə rō′bik) **respiration.** Another name for respiration without oxygen is **fermentation.** During fermentation, glucose is changed to carbon dioxide, alcohol, or various organic acids.

Some organisms, such as certain bacteria, obtain all of their energy by fermentation. Other cells, such as muscle cells, revert to this kind of respiration only when oxygen is in short supply. For example, muscle cells rely on anaerobic respiration during intense exercise. During this time, the oxygen supply from the bloodstream becomes depleted. You may have noticed an aching sensation in your muscles after heavy exercise. This is due partly to the presence of lactic acid produced during anaerobic respiration.

Fermentation is an inefficient means of extracting the energy present in glucose. For each molecule of glucose used, only two molecules of ATP are produced.

Aerobic Respiration

Most cells also conduct another series of reactions which do require oxygen. Respiration that takes place in the presence of oxygen is called **aerobic** (er′ō′bik) **respiration.** The last steps by which your own body makes energy available are aerobic stages. When oxygen is present, glucose is changed to carbon dioxide and water.

Cells use the oxygen you breathe in to transfer energy from the glucose molecule to the ATP molecule. Aerobic respiration is much more efficient than anaerobic respiration. For each molecule of glucose used up, a maximum of 38 molecules of ATP are formed. You can compare the two kinds of respiration by studying the diagrams in the margin.

Review It

1. Why is respiration important to the cell?
2. Distinguish between aerobic and anaerobic respiration.
3. What is ATP? What role does it play in respiration?

4–4
Photosynthesis

Plants are able to capture energy from the sun in the process of **photosynthesis** (fō'tō sin'the sis). During photosynthesis, plants change light energy into chemical-bond energy. And in this process, plants use carbon dioxide and water to build energy-containing compounds, such as glucose.

The process can be shown by the following equation:

$$\text{Carbon Dioxide} + \text{Water} + \text{Light Energy} \xrightarrow{\text{Chlorophyll}} \text{Glucose} + \text{Oxygen}$$
$$6\,CO_2 \qquad 6\,H_2O \qquad \longrightarrow C_6H_{12}O_6 \quad 6\,O_2$$

One of the important differences between plants and animals is in how they obtain their energy. Unlike plants, animals are unable to capture energy from the sun. Instead, animals obtain energy by consuming energy-containing compounds that come either directly or indirectly from plants.

But regardless of where the energy is obtained, all organisms need energy to carry on their activities. This energy is made available in the cells by the breakdown of energy-containing compounds in respiration. The process of respiration can be shown by the following equation:

$$\text{Glucose} + \text{Oxygen} \longrightarrow \text{Carbon Dioxide} + \text{Water} + \text{ATP}$$
$$C_6H_{12}O_6 + 6\,O_2 \longrightarrow 6\,CO_2 + 6\,H_2O + \text{Energy}$$

Notice that the equation of respiration is just the reverse of the equation of photosynthesis.

The Relation of Photosynthesis and Respiration

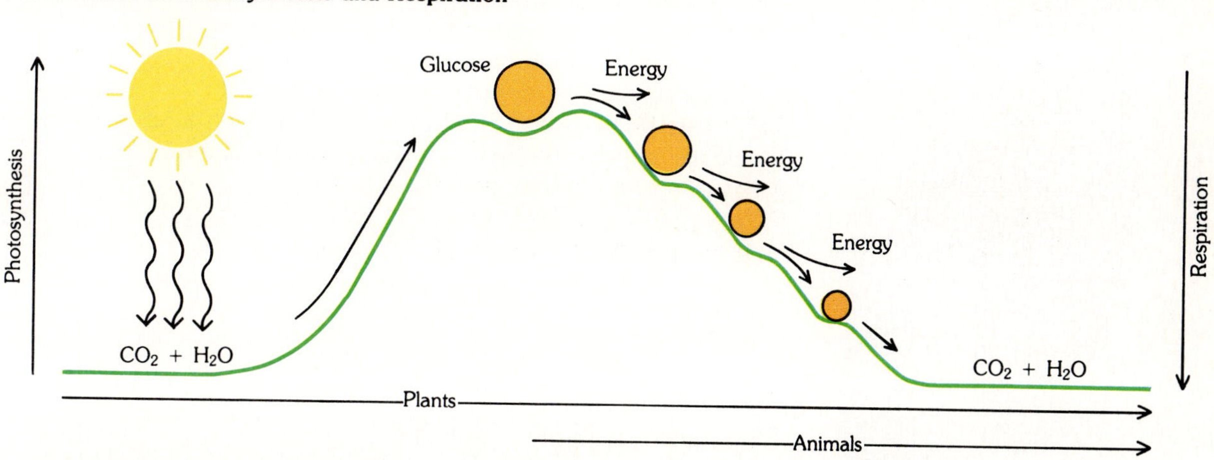

In photosynthesis, the energy of sunlight is used to push simple molecules up the energy "hill" to a position of high energy.

Energy is stored in the chemical bonds of complex molecules, such as glucose.

In respiration, energy from the "downhill" roll of glucose is made available in the cell.

Chloroplasts

Inside plant cells are organelles called *chloroplasts.* They contain the pigment, chlorophyll, which gives plants their green color. Each cell in a leaf may contain as many as 40 to 50 chloroplasts. Photosynthesis begins with the absorption of light by chlorophyll in the chloroplast.

The activities that occur in the chloroplast along with activities that occur in the mitochondrion make life possible. Inside the chloroplast, energy from the sun is captured. And that energy is stored in glucose molecules. Inside the mitochondrion, the energy stored in glucose is used to carry on other cell activities. In addition, the oxygen produced in the chloroplast can be used in the mitochondrion during the breakdown of glucose. Similarly, the carbon dioxide and water produced in the mitochondrion can be used in the chloroplast during photosynthesis.

The Cycle of Life

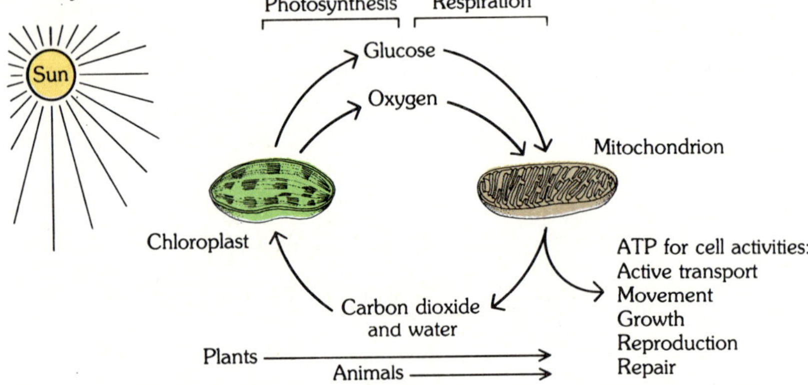

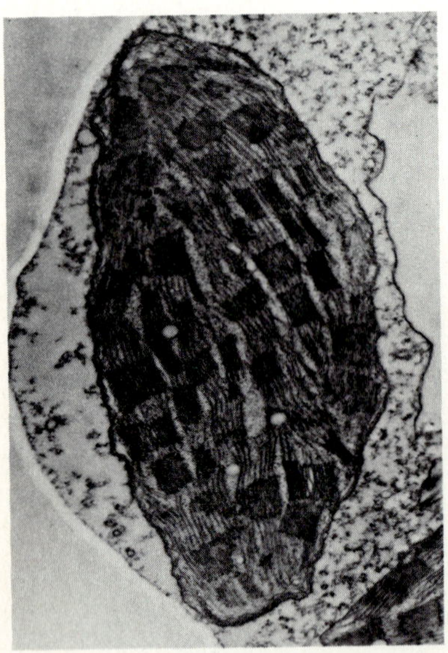

Chloroplast

A chloroplast is surrounded by two outer membranes. Inside there is another complex system of membranes. These membranes fold on top of each other to form a stack called a **granum** (grä′nəm). Chlorophyll pigments and other molecules that help to capture energy from sunlight are located here. Surrounding the granum is a dense solution of proteins and other compounds called **stroma** (strō′mə).

Inside a chloroplast

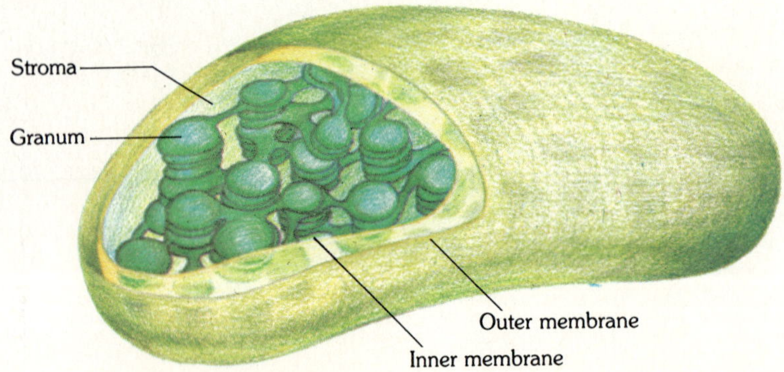

Light and Dark Reactions

Photosynthesis takes place in two major steps. The first step occurs in the presence of light such as sunlight and is called the **light reaction.** During this step, chlorophyll pigment in each granum absorbs light in red wavelengths of the spectrum. The energy trapped by chlorophyll is used to split water molecules. Chemical energy released by splitting the water molecules is stored in two special compounds called NADPH and ATP. Energy stored in the chemical bonds of these separate compounds is used to power the next step of photosynthesis.

The second step of photosynthesis does not require sunlight or chlorophyll, so it is called the **dark reaction.** This reaction occurs in the stroma of the chloroplasts. Energy stored in ATP and NADPH bonds is used to add carbon dioxide molecules to small organic molecules to form larger organic molecules. The main organic compound formed during photosynthesis is glucose. However, other carbohydrates, fats, and proteins may also be formed.

The Light and Dark Reactions of Photosynthesis

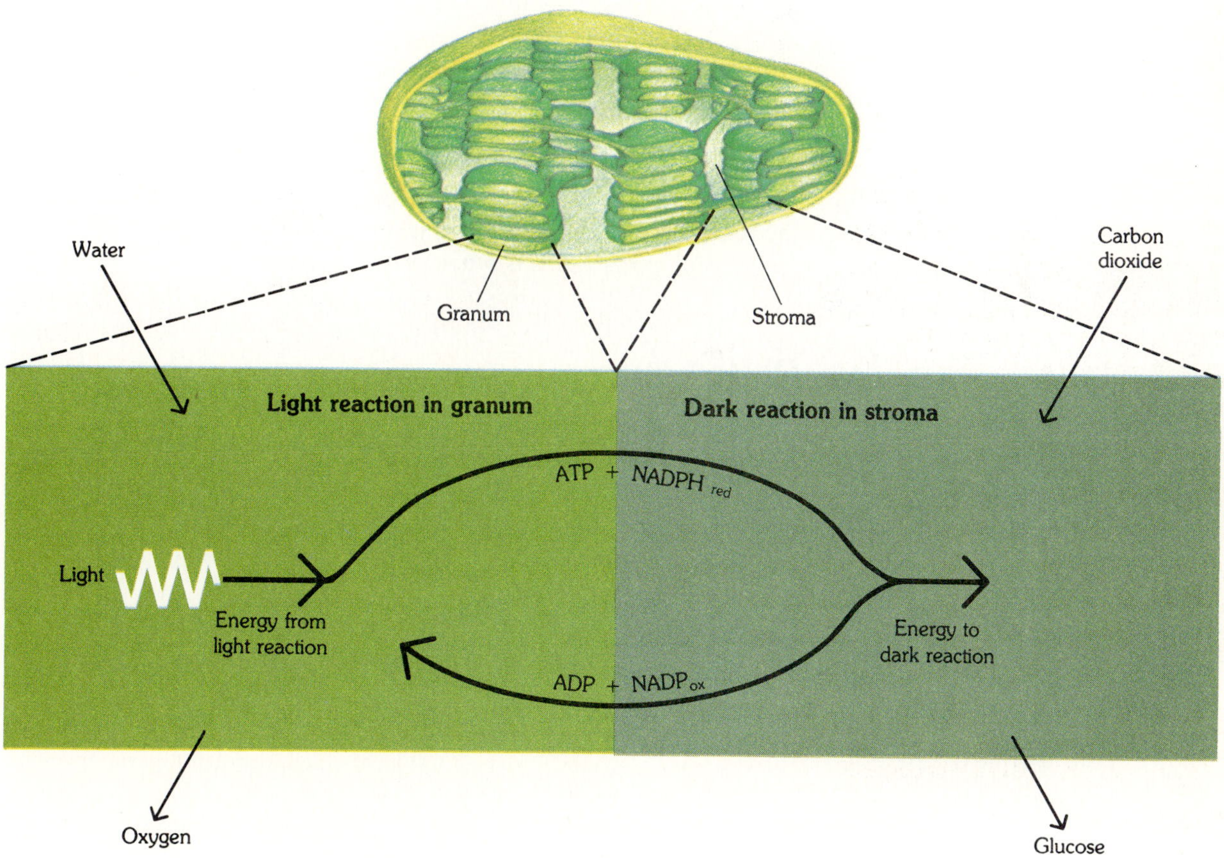

Cell Regulation: Homeostasis

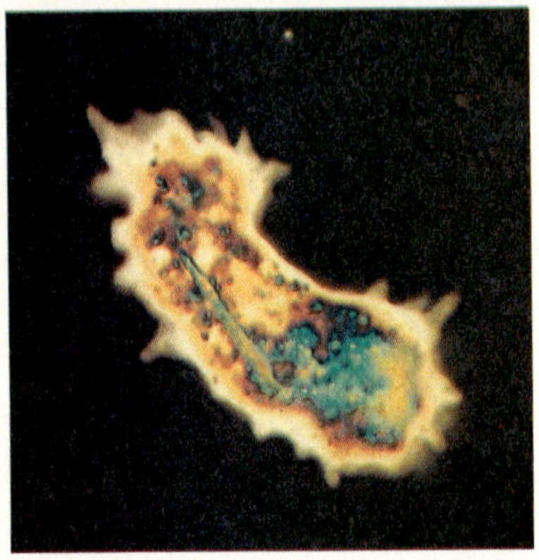

Feedback mechanisms regulate all cell functions and keep the cell environment stable

Our study of cell metabolism has left out one important point: the many jobs a cell has to do are closely related to each other. A change in one cell function affects what happens somewhere else in the cell. For example, suppose a cell needs to produce more energy. Then respiration in the cell has to speed up. But respiration depends on the presence of certain chemicals such as oxygen. Faster respiration can take place only if there is an increase in the number of these elements present. Therefore, the rate of *absorption*, or diffusion into the cell, has to increase.

And that's not all. With more chemicals in a cell and a faster rate of respiration, more waste products will form. A build-up of these wastes could damage or kill the cell. The rate of *excretion*, or diffusion out of the cell, must also increase.

The situation described here is called a *feedback mechanism*. Changing one part of a system brings about changes in other parts of the same system. If, for any reason, these changes cannot occur, the system may break down. Feedback mechanisms in the cell give it the ability to respond to changes from the outside. Greater demands for energy or new body parts, an excess or lack of food, outside stimuli—all cause the cell to change the rate at which it carries out various functions. But the cell makes these changes in such a way that the internal workings of the cell are not disrupted. The cell does not break down or fall apart just because some outside demands are made on it. *The ability of a living thing to maintain a stable internal environment when outside demands are placed on it is called* **homeostasis** (hō′mē ə stā′sis). Homeostasis is made possible by feedback mechanisms. And homeostasis is a trait of all living things.

Review It

1. Describe the process of photosynthesis. Distinguish between the light and dark reaction.
2. What is the importance of chlorophyll to photosynthesis?
3. Define homeostasis. Why is homeostasis important to the survival of the cell?

Breakthroughs in Biology

Electricity from Plants

Photosynthesis captures solar energy and uses it. The chlorophyll in green plants collects the energy of sunlight and changes it into a different form: the energy of chemical bonds.

It is interesting to speculate about the possibility of using photosynthesis for other energy conversion reactions. Imagine using chlorophyll to change solar energy into electricity. This might provide an exciting alternative for the world's diminishing fossil fuel supply.

Some steps in this direction have already been taken. Francis Fong and Nicholas Winograd at Purdue University have built a system by which photosynthetic reactions change sunlight directly into electricity.

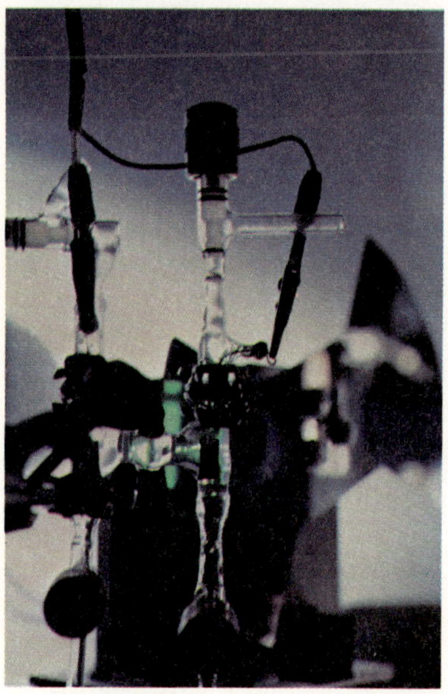

A sample of pure chlorophyll from spinach was attached to one electrode of a chemical cell. The other electrode was a rod of platinum. The whole system was set up something like the battery in a car. When the chlorophyll electrode was exposed to sunlight, an electrical current flowed from the chlorophyll electrode to the chlorophyll-free side of the cell. Sunlight was converted directly to electricity!

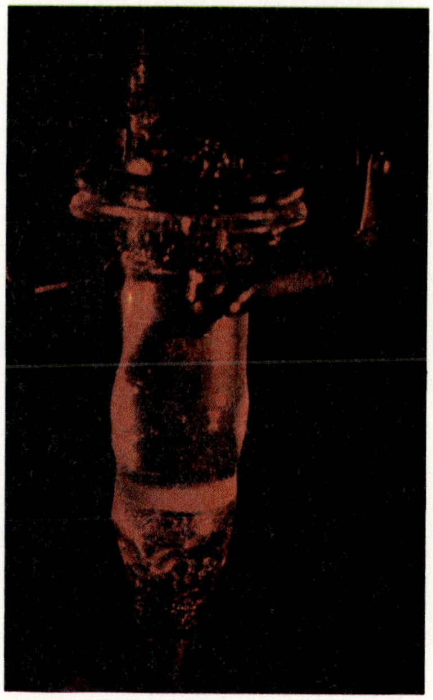

With a similar apparatus, Fong and his co-workers Lory Galloway and Daniel Fruge produced clean-burning hydrogen by splitting water. Water is split into hydrogen and oxygen using red light only.

Building a large-scale plant that uses chlorophyll and light to produce electricity or hydrogen and operates at reasonable cost will not be possible for some time. However, finding a method by which humans can do what green plants have done for millions of years is an important breakthrough in biology.

Analysis

1. We already use solar energy that has been collected and stored by plants for heating our homes and running our factories. What are some of the ways in which this is done?
2. What might be the next step in improving a photosynthetic electric generator?

Chapter Summary

Cells perform many functions. They take in and digest food. They use that food to make new body parts and to produce energy. But cells can't use the nutrients they need unless they can get them through the cell membrane and into the cell's interior. The natural flow of materials through the membrane is called passive transport. Diffusion and osmosis are the most common forms of passive transport. Cells can also force materials through the cell membrane by a variety of methods classified as active transport.

Cells produce the energy they need by respiration. Energy stored in the chemical bonds of glucose molecules is transferred to the energy storehouse compound, ATP. Green plants do not rely on food supplied by other organisms. They make their own food from carbon dioxide and water by the process of photosynthesis. Energy for this process comes from sunlight, and the whole reaction is triggered by chlorophyll.

Cells have the ability to maintain a balance among all of these cell activities. This balance is known as homeostasis.

Questions to Think About

1. Jellyfish normally live in salt water. What do you think would happen if a jellyfish were put into a freshwater pond? What would happen if a fish that lived in fresh water were placed in the salt water of the ocean? Base your answers on what you have learned about transport of materials across a cell membrane.
2. In what sense could you say that pure water is more concentrated than salt water?
3. Fat molecules can sometimes pass through cell membranes when protein molecules of the same size cannot. How would you explain this fact, based on what you know about the structure of the cell membrane.
4. The way in which cells release energy and the way an automobile engine releases energy are different. Explain the difference between the two. Why is it essential that cells work differently from the way car engines work?
5. When food is oxidized during cellular respiration, is any heat produced? How do you know? What would happen if a cell began to convert *all* of the energy in glucose into heat energy? What if a cell converted *none* of the energy in glucose into heat energy?
6. It is possible to destroy or to remove only a single molecule in a cell. When this happens, the whole cell may die. Explain how the survival of a whole cell can depend so critically on an individual molecule. Do you see any practical significance in this fact?

Extra Research

1. Workers in one field of science often borrow ideas from other fields of science. For example, the *kinetic theory of matter*, developed in physics, helps to explain diffusion and osmosis in living systems. What is the kinetic theory of matter? How does it apply to the transport of materials across cell membranes? What does the kinetic theory of matter say about factors which might affect the rate of diffusion?
2. *Phagocytes* are cells that help to rid the human body of materials which are harmful or foreign. White blood cells are examples of phagocytes. Where do you think the name phagocyte came from? Find out how phagocytes help to protect the body from disease.
3. Nothing has been said in this chapter about the death of cells. Do they live forever, or do they experience "natural death" the way larger organisms do? What does it mean to say that a cell "dies"?

Chapter Test

Part I

Copy the number of each test item. If the item is *true*, write *T* after the number; if the item is *false*, write *F*.

1. Cells maintain an internal balance by the process known as homeostasis. (102)
2. Pinocytosis is an example of passive transport. (89)
3. Cell membranes are semipermeable membranes. (85)
4. Aerobic respiration is less efficient than anaerobic respiration. (98)
5. Energy needed to power cell activities is stored in chemical compounds called mitochondria. (96)
6. The elements carbon, hydrogen, oxygen, and nitrogen account for 95 percent of all living matter. (91)
7. One important property of atoms is that they tend to leave their outer electron shells unfilled. (92)
8. Photosynthesis takes place in the chloroplasts. (100)

Part II

Copy the number of each sentence below. After each number, write the word or words that complete the sentence correctly.

9. The two kinds of chemical bonds that may be formed during a chemical reaction are _____ bonds and _____ bonds. (92)
10. All of the processes by which a cell makes and uses energy-rich compounds is called cell _____. (91)
11. Each atom is made up of three kinds of particles; _____, _____, and _____. (92)
12. The special catalysts found in living cells are called _____. (95)
13. The movement of materials through a cell membrane from regions of low concentration to regions of high concentration is known as _____. (88)
14. The special name given to the diffusion of water through a semipermeable membrane is _____. (86)
15. Some organisms take care of extra water that enters their cells by storing the water in _____. (87)
16. Cellular respiration which takes place without oxygen is known as _____ respiration. (98)
17. Photosynthesis takes place in two steps: the _____ reaction and the _____ reaction. (101)

Part III

Copy each numbered item from List A. After each item, write the letter and words from List B that best describe it.

List A

18. photosynthesis (99)
19. turgor (87)
20. plasmolysis (87)
21. respiration (96)
22. ATP (96)
23. catalyst (95)
24. stroma (100)
25. grana (100)

List B

a. process by which energy is released for use within the cell
b. substance which triggers or speeds up a chemical reaction, but remains unchanged itself
c. energy-storage compound
d. place where chlorophyll is located in the chloroplasts
e. process by which sunlight, water, and carbon dioxide are converted to glucose
f. place where chemical reactions of photosynthesis take place
g. stiffness of plant cell due to water pressure inside the cell
h. shrinkage of cell due to movement of water out of cell

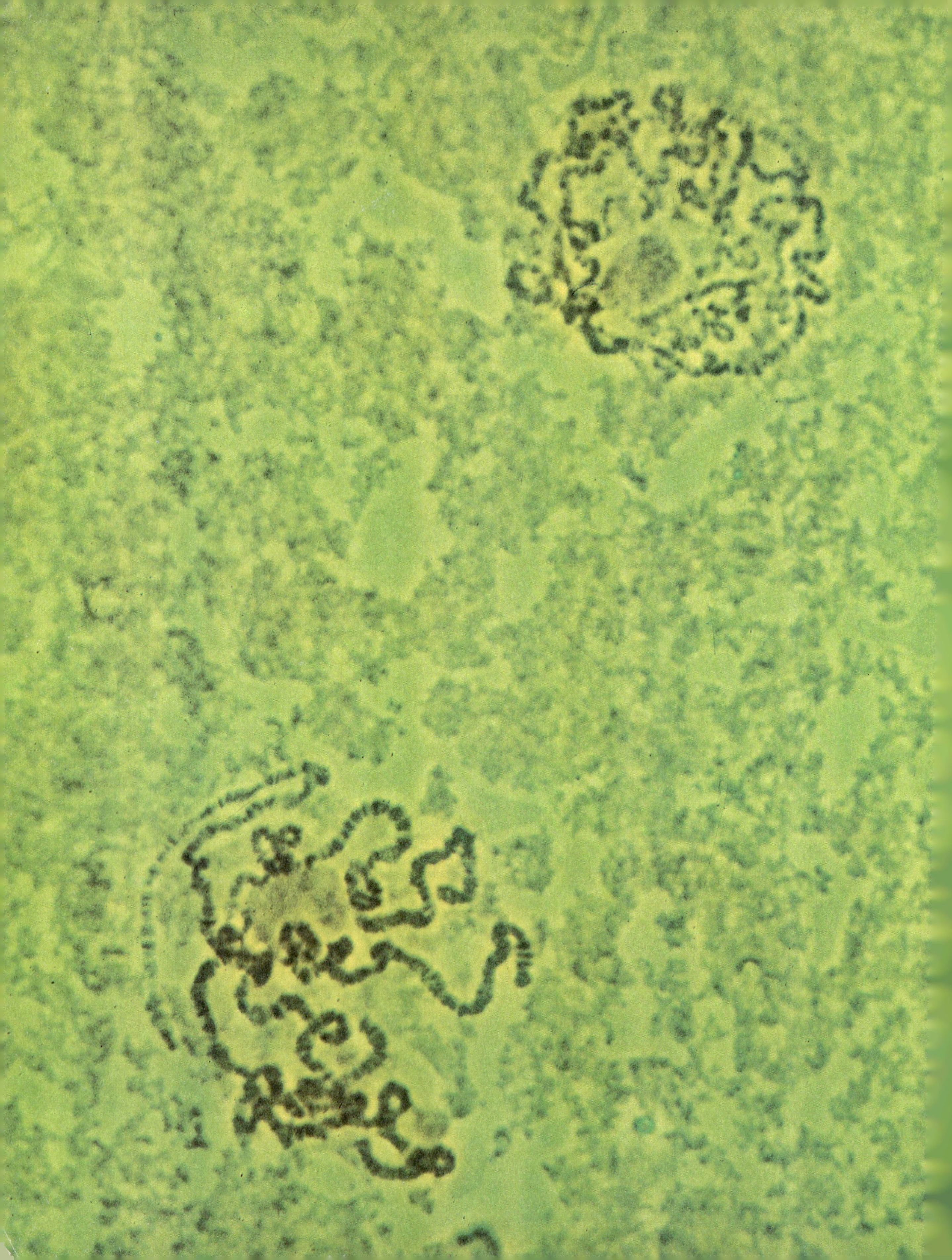

Chapter 5

Information Transfer by Cells

All the information directing every cell function must be stored somewhere in every cell. This information must be accessible for use as needed. And the information must also be accurately transmitted to new cells. Nearly every piece of information your cells need to function properly is stored in giant molecules found in the nucleus of your cells. *Deoxyribonucleic acid,* or *DNA* for short, is the information storage molecule of the cell.

A cell cannot function without DNA. The information it contains must in some way be available for use by the cell and the information must somehow be transferred to all new cells. By itself, DNA is like a recording tape. However, the information recorded must be decoded to be useful.

This chapter will introduce you to the structure and function of DNA and to the process of decoding and using the information stored within it. You will also learn how the information contained in DNA is passed from one generation to the next.

Chapter Objectives

1. Explain how DNA controls the functions in a cell.
2. Sketch the basic structures of DNA and RNA.
3. Describe how the DNA molecule replicates itself.
4. Describe how a segment of DNA controls the synthesis of a protein.
5. Explain how errors in the transfer of genetic information can occur.
6. Tell how genetic information can be transferred from parent cells to daughter cells.
7. Compare mitosis with meiosis.

Chromosomes of the common fruit fly, *Drosophila melanogaster.*

5–1
The Transfer of Information Within Cells

The key to the way your body functions lies within the molecules inside the cells of your body. Possibly the most important of these molecules is **deoxyribonucleic** (dē ok'sə rī'bō nü klē'ik) **acid,** or **DNA.** This molecule is thought to mastermind nearly all the processes involved in cellular functioning, including the most critical process of **protein synthesis.**

Most of the time, DNA is found loosely coiled within the nucleus of a cell, like spaghetti inside a tiny balloon. At certain times, the DNA constricts, or shortens, and is plainly visible with the aid of a microscope. Within the DNA molecule is coded the information that controls the functioning of your cells.

Chromosomal DNA

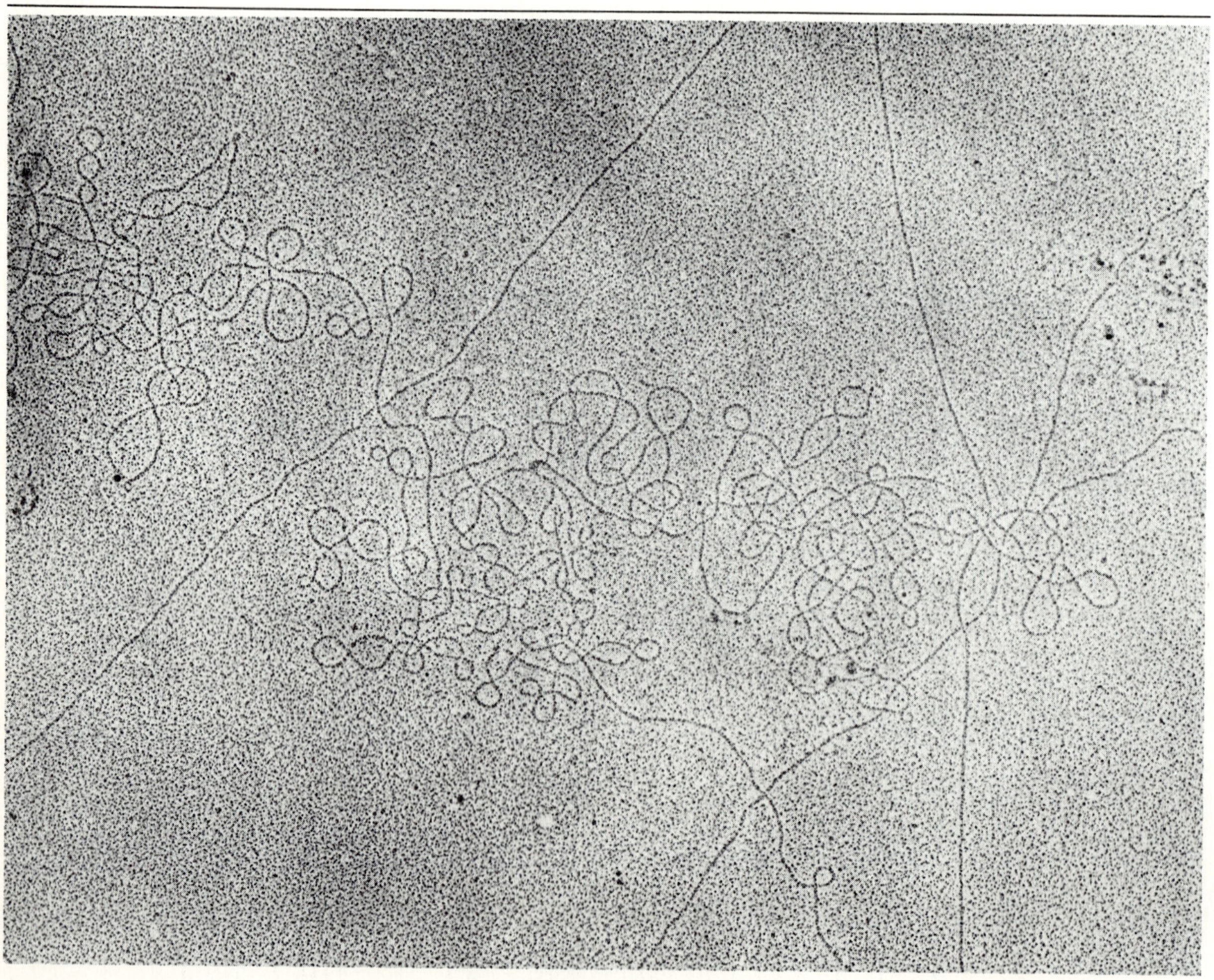

The Carrier of Information

Protein synthesis and most other cell functions occur outside the nucleus of the cell in the cytoplasm. The information encoded within the DNA molecule must somehow be carried outside the nucleus. The carrier of DNA's messages is a molecule known as **messenger ribonucleic** (rī'bō nü klē'ik) **acid,** or **mRNA.** Messenger RNA picks up the messages encoded within the DNA molecule and carries them out into the cytoplasm to the sites of protein synthesis, the **ribosomes** (rī'bə sōmz).

Ribosomes read messages carried by the mRNA. With the help of another RNA molecule called **transfer RNA,** or **tRNA,** the new protein molecules are built on ribosomes. The tRNAs help by bringing to the ribosomes the building blocks of proteins, which are called **amino** (ə mē'nō) **acids.** The entire transfer of information from the DNA inside the nucleus to the ribosomes in the cytoplasm can be summed up as follows:

1. The code, or message, of the DNA is transmitted to mRNA inside the cell nucleus.
2. The mRNA carries the message outside the nucleus to the sites of protein synthesis, the ribosomes.
3. Transfer RNA molecules connect the building blocks of proteins, amino acids, to the ribosomes.
4. Each ribosome moves along a messenger RNA building a new protein molecule. As the mRNA is decoded, the correct amino acid (brought by the tRNA) is attached to a growing protein molecule.

mRNA

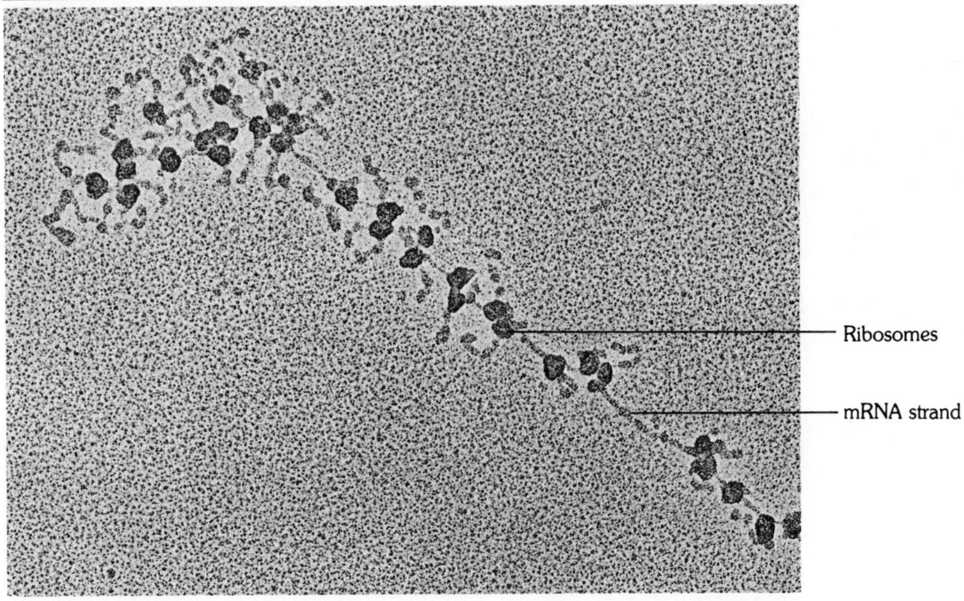

DNA Structure

To understand fully how DNA controls a cell function, such as protein synthesis, you must know something of the structure of DNA. It will help as you read about the structure of DNA if you visualize the molecule as a twisted ladder, or **double helix** (hē'liks). The double helix is made up of smaller units called **nucleotides** (nü'klē ə tīdz).

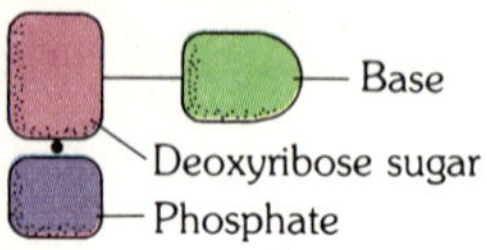

Nucleotide

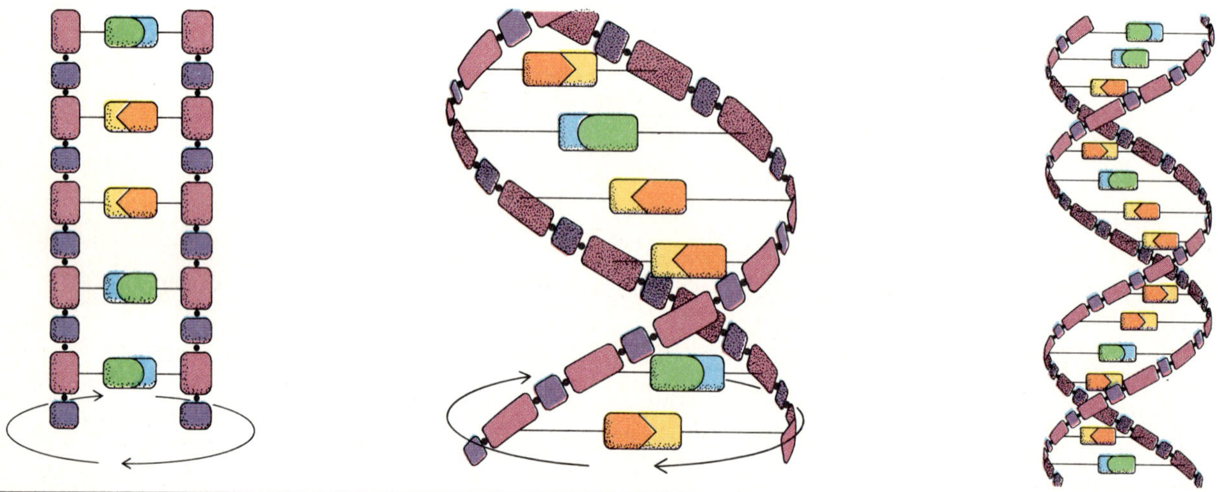

Double Helix of DNA

Nucleotides

The DNA in each one of your body's cells contains about eight billion nucleotides, the building blocks of DNA. Each nucleotide consists of a base, a sugar, and a phosphate group. The sugar is always **deoxyribose** (dē ok'sə rī'bōs). The base may be either **adenine** (ad'n ēn'), **guanine** (gwä'nēn'), **thymine** (thī'mēn'), or **cytosine** (sī'tə sēn').

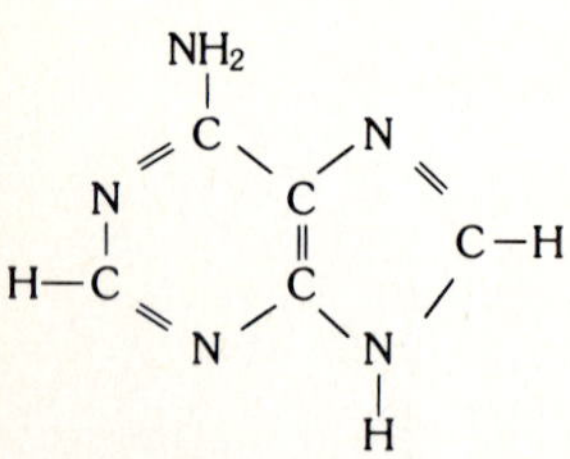

Adenine

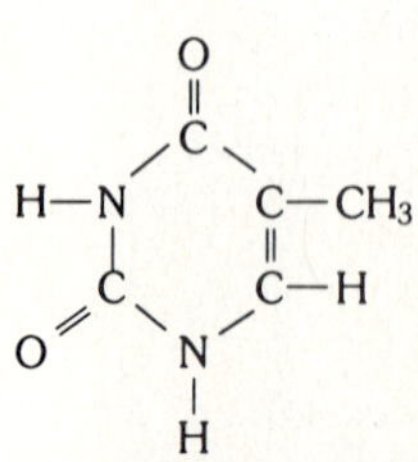

Thymine

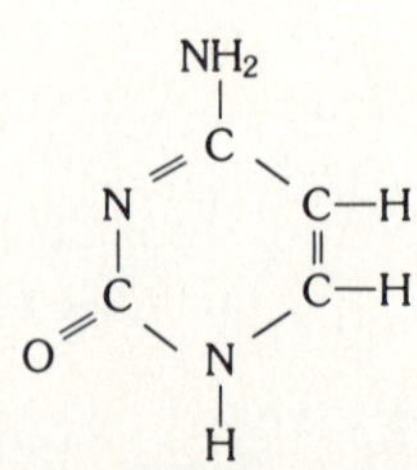

Cytosine

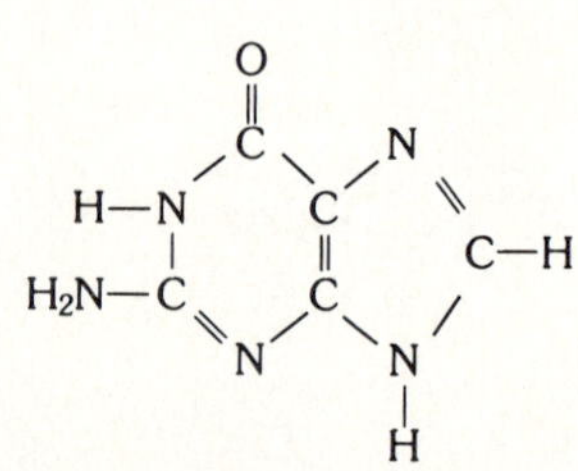

Guanine

Components of the DNA molecule

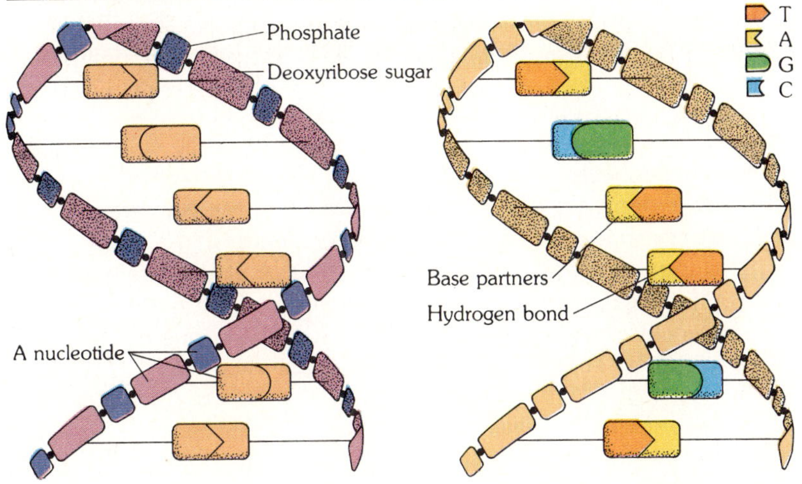

The Sides of the Double Helix

Thousands of nucleotides join together to form each DNA molecule. The sugar and the phosphate group of each nucleotide link together firmly by strong **phosphate bonds** to form the sides of the ladderlike DNA molecule. If you removed a piece of one side of the molecule, you would see something like this: deoxyribose–phosphate–deoxyribose–phosphate–deoxyribose–phosphate and so on.

The Rungs of Double Helix

The "rungs" of the ladderlike DNA molecule are formed by the bases of the nucleotides. The bases bond together in a set fashion. Nucleotides containing adenine (A) bond with nucleotides containing thymine (T). Similarly, nucleotides containing guanine (G) pair with nucleotides carrying cytosine (C). The structures of the bases ensure that A always bonds to T and the G always bonds to C. The order of the partner bases, or *complementary* bases, may be listed as G-C or C-G, or A-T or T-A.

The bases of each rung of the DNA molecule are joined by weak hydrogen bonds. These bonds allow the DNA molecule to split apart down the middle of each rung for a given length of the molecule. When the molecule splits, two strands with complementary bases are formed. It is the bases on the complementary strands that code for the formation of proteins.

DNA "unzipping"

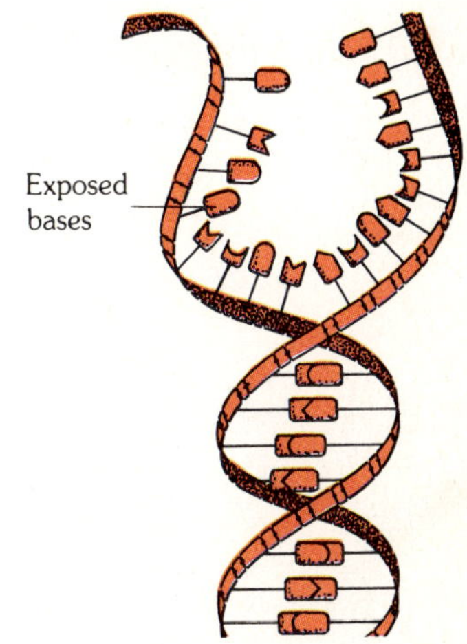

From Double Helix to Two Strands

The weak hydrogen bonds between the base pairs allow the DNA molecule to split apart down the middle of each "rung." When the DNA molecule splits, or unzips, two separate strands of DNA are formed. Each of these exposed strands of DNA can make a duplicate molecule of itself. This is a crucial feature of DNA. This process of duplication is known as **DNA replication** (rep′lə kā′shən).

DNA Replication

The coiled double helix molecule of DNA must make a molecule identical to itself. The specific base pairings (A-T and G-C) allow the replication process to occur smoothly and efficiently. The weak hydrogen bonds between the base pairs can easily be broken allowing the molecule to unzip. This exposes the bases of each strand of DNA. Free DNA nucleotides float into the exposed area and match up to complementary bases, with the help of enzymes.

New hydrogen bonds are formed between base partners. As the molecule continues to uncoil, a complementary, or mirror image, strand is formed on each original strand of DNA. It is in this manner that two identical molecules of DNA are produced. Note that each new molecule contains one original DNA strand and one newly formed complementary strand.

DNA Replication

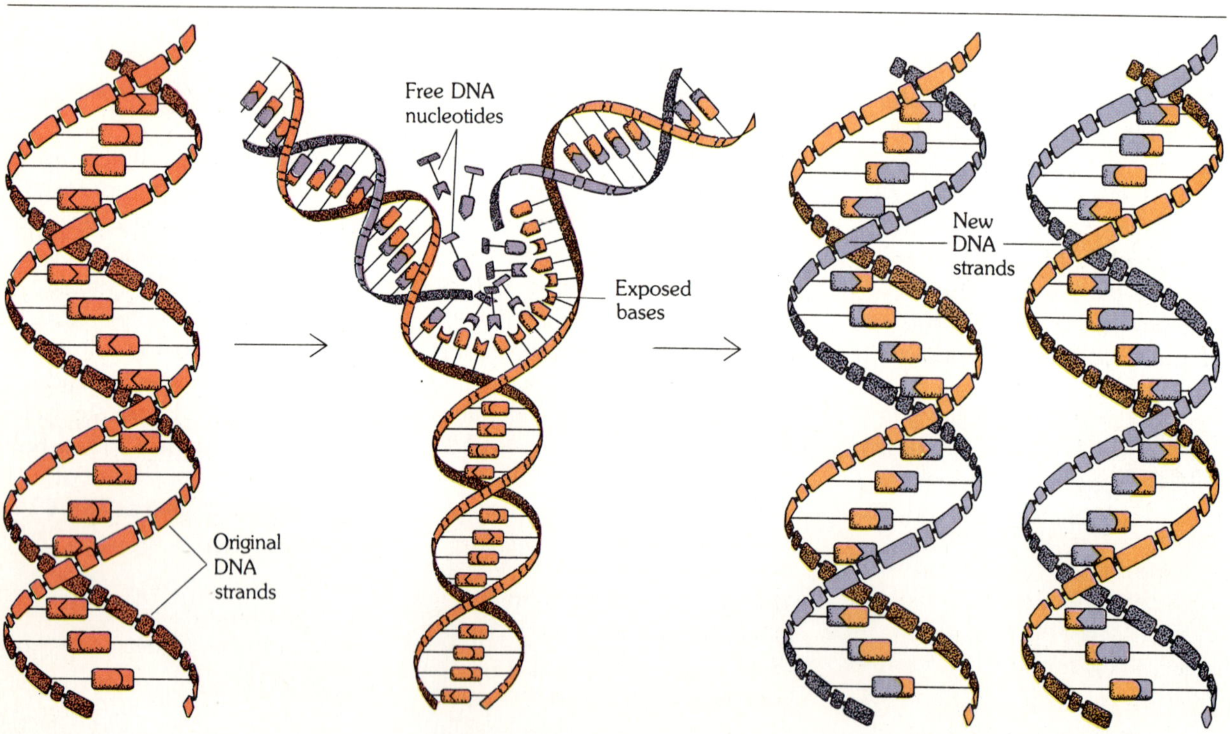

Review It

1. How are the bases paired in the DNA molecule?
2. One strand of DNA is known to be CCGATGCAAGGT. Give the complementary strand of DNA.
3. Describe how the DNA molecule can make an exact replica of itself. Explain why this is important.

Pioneers in Biology

Francis H. C. Crick, James D. Watson, Maurice H. F. Wilkins: Pioneers in the discovery of the structure of deoxyribonucleic acid, or DNA.

James Watson and Francis Crick with DNA model.

The discovery of the structure of DNA has been called the single most important development in biology in this century. Watson, Crick, and Wilkins shared the Nobel Prize in 1962 for this achievement.

Based on evidence compiled by Wilkins, Rosalind Franklin, and clues from other scientists, Watson and Crick perceived the double helix structure of the DNA molecule. In honor of their discovery, the double helix structure of DNA is often referred to as the *Watson-Crick model.*

Crick, a British biochemist, was born June 8, 1916, in Northampton, England. He attended University College in London, and studied at Strangeways Laboratory, Cambridge, from 1947 to 1949. As a physicist during World War II, he worked in radar research. Later he turned to biochemistry and worked with nucleic acids. In 1949, Crick and other physics-oriented scientists at Cambridge were developing the science of molecular biology, combining the disciplines of biology, physics, and chemistry. Crick received his Ph.D. from Cambridge in 1953 for his work in molecular biology.

Watson, an American biochemist, was born April 6, 1928, in Chicago. He was a "quiz kid" on the radio, and entered the University of Chicago at age 15. He graduated in 1947. At Indiana University, his graduate work was on bacterial viruses. In 1951 he joined Crick at the Cavendish Laboratory. There, in a shack known as The Hut, they worked out the three-dimensional model of the structure of DNA.

Wilkins, a biophysicist, was born December 15, 1916, of Irish parents in New Zealand. He attended St. John's College, Cambridge. His graduate research was on the improvement of radar screens. After working as one of the physicists on the Manhattan Project, which developed the atomic bomb, he changed his field to biophysics. At King's College, he studied the DNA molecules under the microscope by illuminating them with one color of transmitted light and another color of reflected light. Noticing the crystalline structure, he decided that X-ray diffraction techniques would be perfect for the study of DNA.

Rosalind Franklin: Pioneer in X-ray diffraction study of DNA.

Rosalind Franklin provided important evidence used by Watson and Crick to support their conclusion that DNA had the shape of a double helix. Franklin was a skilled scientist who used *X-ray diffraction,* or *crystallography,* techniques.

Franklin was born in London on July 25, 1920. The daughter of banking and artistic families, her interest in scientific research was not encouraged. She persisted, and graduated from Newnham College, Cambridge. In 1942, as a physical chemist, she worked on microstructures of coal. Later, she learned the art of crystallography in Paris.

Franklin went to King's College, London, to set up an X-ray diffraction unit. There, from 1951 to 1953, she accomplished her DNA work.

In 1958, at age 37, Dr. Rosalind Franklin died of cancer. She had gained international acclaim for her work with microcarbons, DNA, and later, viruses.

5–2
DNA and Protein Synthesis

Amino acid structure

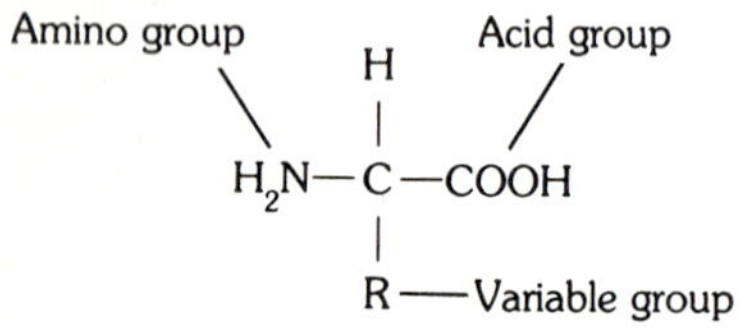

Proteins are an essential group of molecules. Protein is necessary in the formation and repair of an organism's tissues. Most of your tissues are composed of protein.

Enzymes, a special group of proteins, control and regulate all chemical reactions in the cell by acting as catalysts. This means that the enzyme changes (usually speeds up) the rate of the chemical reaction without itself being changed.

All proteins, including enzymes, are made up of hundreds of smaller units called amino acids. The general formula for an amino acid includes an acid group (COOH), an amino group (NH_2), a hydrogen atom (H), and a group that will vary in chemical formula, symbolically represented by the letter *R*. The table below lists the 20 commonly occurring amino acids.

Two common amino acids

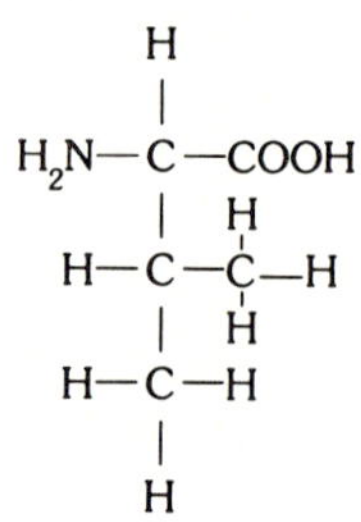

Valine

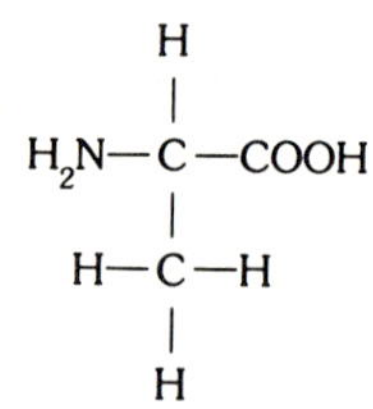

Alanine

Common Amino Acids

Amino Acid	Abbreviation	Amino Acid	Abbreviation
1. Alanine	Ala	11. Leucine	Leu
2. Arginine	Arg	12. Lysine	Lys
3. Asparagine	Asn	13. Methionine	Met
4. Aspartic Acid	Asp	14. Phenylalanine	Phe
5. Cysteine	Cys	15. Proline	Pro
6. Glutamic Acid	Glu	16. Serine	Ser
7. Glutamine	Gln	17. Threonine	Thr
8. Glycine	Gly	18. Tryptophane	Trp
9. Histidine	His	19. Tyrosine	Tyr
10. Isoleucine	Ileu	20. Valine	Val

Polypeptide Chains

Amino acids form proteins by linking together to form long chains, which in their functional form are usually folded or twisted. The chemical bonds linking the amino acids are called **peptide** (pep′tīd) **bonds.** The chains of amino acids that make up a given protein are called **polypeptides** (pol′ē pep′tīdz), after the many peptide bonds linking the amino acids.

The Base Triplet Code

For years, scientists were baffled by how DNA coded for the 20 kinds of amino acids that are repeated thousands of times to make up each protein molecule. Researchers realized that the code must lie in the arrangement of the four kinds of bases in the DNA molecule because these were the only parts that varied from one DNA molecule to the next. The puzzle was how could the four bases—cytosine, guanine, thymine, and adenine—code for 20 different amino acids?

If each base along a single strand of DNA served as a code for the synthesis of one amino acid that would account for only 4 of the 20 acids. Similarly, if pairs of bases acted as codes for the addition of different amino acids then only 16 pair combinations would be possible. Again this would not be enough to code for the 20 amino acids.

Scientists found that a code of 3 bases was theoretically possible. This hypothesis has now been tested sufficiently to indicate that it is probably correct. This would allow for 64 possible combinations, far more than enough to code for 20 amino acids. For example, the base triplet, CCA (cytosine–cytosine–adenine) along the length of a DNA strand means the amino acid "glycine." CGA (cytosine–guanine–adenine) means the amino acid "alanine." With 64 combinations for 20 amino acids, some amino acids will be coded for by more than one combination.

Challenge!
If two DNA bases in sequence were required to code for an amino acid, only sixteen combinations of two bases at a time would be possible. List the sixteen possible two-base combinations. Now list the combinations possible if three bases in sequence are used to code for an amino acid.

Polypeptide segment

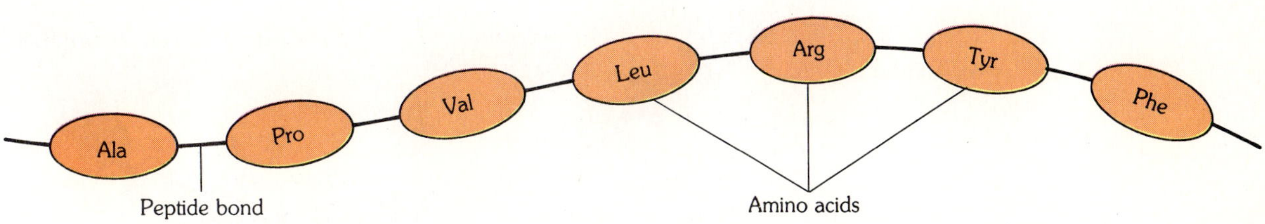

Comparing uracil with thymine

Uracil

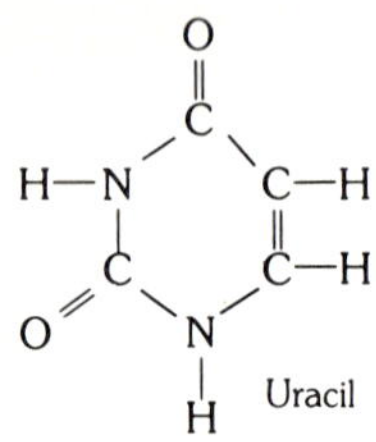

Thymine

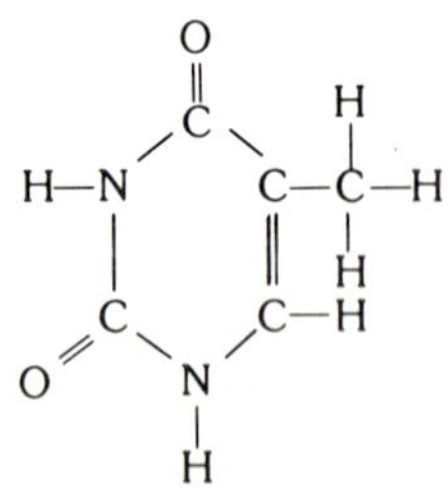

Transfer of Information from DNA to mRNA

Remember that the DNA is contained in the nucleus. For the code of the DNA to be useful, it must be carried out of the nucleus into the cytoplasm. This job of carrying the message outside of the nucleus to the ribosomes, where protein synthesis occurs, belongs to the messenger RNA, or mRNA. To understand how mRNA fulfills its function, you must learn something about its structure.

mRNA Structure

The structure of mRNA is very much like that of DNA with three major exceptions:

1. RNA normally appears as a single strand rather than a double-stranded molecule like DNA.
2. The sugar which forms the sugar-phosphate backbone of the mRNA molecule is a **ribose** (rī'bōs) instead of a deoxyribose as in the DNA molecule.
3. One of the four bases making up the mRNA molecule is different. Instead of thymine, RNA substitutes the base called **uracil** (yůr'ə-səl). Uracil will pair up with the base adenine in the same manner that the DNA base thymine would.

Single-stranded mRNA

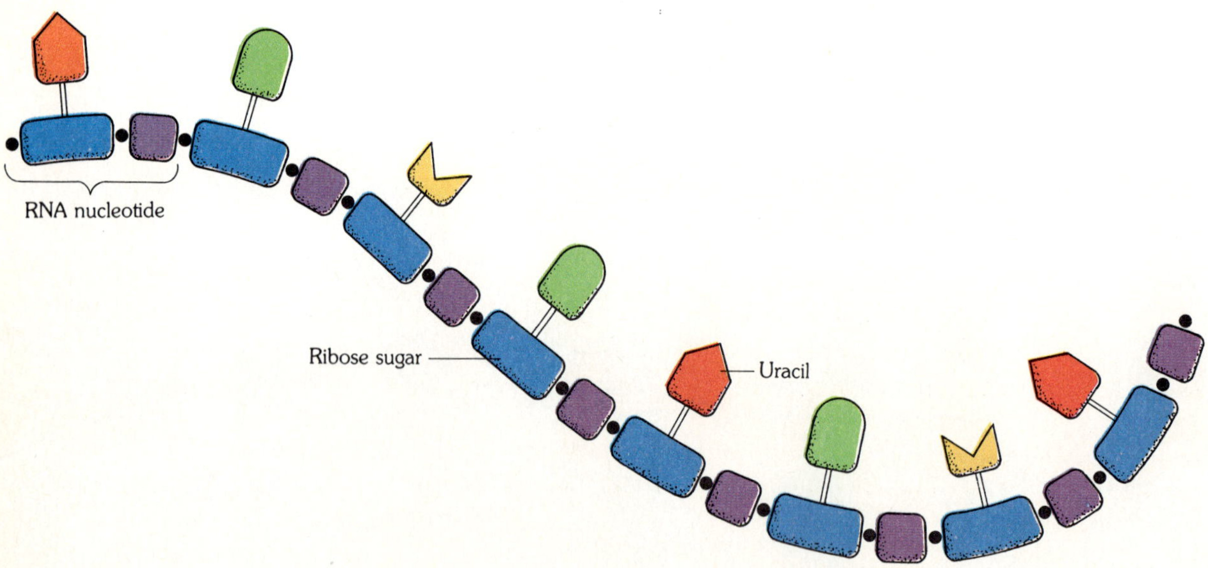

The RNA language

RNA molecules carry messages in much the same way as DNA molecules do: by means of base triplets. The "language" of the mRNA message is, however, different. The mRNA has a different base, uracil. Furthermore, the mRNA language is the complement of the DNA language. For example, CCA (cytosine–cytosine–adenine) on the DNA molecule codes for the amino acid glycine. During the decoding, or transcription, process, the DNA code CCA is translated into GGU (guanine–guanine–uracil) on the RNA molecule. If you study the table below, you can see the correspondence between the DNA code and the RNA code for each amino acid. RNA base triplets are often referred to as **codons** (kō′donz).

Amino Acid	DNA Code	mRNA Codons
Alanine	CGA, CGG, CGT, CGC	GCU, GCC, GCA, GCG
Arginine	TCT, TCC, GCA, GCG, GCT, GCC	AGA, AGG, CGU, CGC, CGA, CGG
Asparagine	TTA, TTG	AAU, AAC
Aspartic Acid	CTA, CTG	GAU, GAC
Cysteine	ACA, ACG	UGU, UGC
Glutamic Acid	CTT, CTC	GAA, GAG
Glutamine	GTT, GTC	CAA, CAG
Glycine	CCA, CCG, CCT, CCC	GGU, GGC, GGA, GGG
Histidine	GTA, GTG	CAU, CAC
Isoleucine	TAA, TAG, TAT	AUU, AUC, AUA
Leucine	AAT, AAC, GAA, GAG, GAT, GAC	UUA, UUG, CUU, CUC, CUA, CUG
Lysine	TTT, TTC	AAA, AAG
Methionine	TAC	AUG
Phenylalanine	AAA, AAG	UUU, UUC
Proline	GGA, GGG, GGT, GGC	CCU, CCC, CCA, CCG
Serine	AGA, AGG, AGT, AGC, TCA, TCG	UCU, UCC, UCA, UCG, AGU, AGC
Threonine	TGA, TGG, TGT, TGC	ACU, ACC, ACA, ACG
Tryptophane	ACC	UGG
Tyrosine	ATA, ATG	UAU, UAC
Valine	CAA, CAG, CAT, CAC	GUU, GUC, GUA, GUG

Transcription and Protein Synthesis

The code is passed from the DNA molecule to the mRNA molecule by a copying process known as **transcription.** Before the code can be transcribed on an RNA molecule, the two strands of DNA must separate, much in the same way it does for DNA replication. However, instead of DNA nucleotides floating in to pair with the exposed DNA bases, nucleotides of *RNA* float into the area. The RNA nucleotide molecules, with the help of enzymes, pair with corresponding partner bases of DNA. The exception is of course uracil, which is substituted everywhere thymine would normally pair up with adenine. The pairing of RNA nucleotides continues until a complete, complementary strand of RNA is formed. The DNA strand acts as a pattern, or **template** (tem'plit), for the mRNA. It is believed that the pairing of mRNA nucleotides with DNA nucleotides occurs on only one side of the DNA molecule. Thus only a single strand of mRNA is formed.

Once formed, the mRNA leaves the DNA molecule, allowing the two DNA strands to re-form into the double helix as before. The mRNA strand moves from the nucleus out into the cytoplasm to the ribosomes, the site of protein manufacture.

You will recall that to code for an amino acid, a sequence of three bases, or one codon, is necessary. The message for the amino acid sequence is decoded at the ribosome. Here, the mRNA attaches itself by one end to the ribosome, and the ribosome then begins to move along the length of the mRNA. As the ribosome moves and translates each codon of the mRNA, it, with the help of enzymes, adds amino acids to a forming polypeptide chain.

The amino acids are brought to the ribosome by another form of RNA called transfer RNA. For each of the 20 amino acids there is a tRNA molecule which will temporarily bond with that specific amino acid out in the cytoplasm and carry it to the ribosome. The tRNA is thought to have three exposed nucleotide bases, sometimes referred to as the **anti-codon** (an'tī kō'don), which will pair up with a complementary codon of mRNA, as it is read by the ribosome. This positions the amino acid for addition to the polypeptide chain.

It is in this fashion that the amino acids are brought to the ribosome in the correct order. Amino acids are joined together by peptide bonds in sequence as they are brought to the ribosome as it reads each codon of mRNA. The tRNA releases each amino acid as it is added to the polypeptide chain, and returns to the cytoplasm in search of another molecule of the same amino acid. The ribosome continues down the length of the mRNA, and amino acids are continually added to the growing chain until the polypeptide is complete.

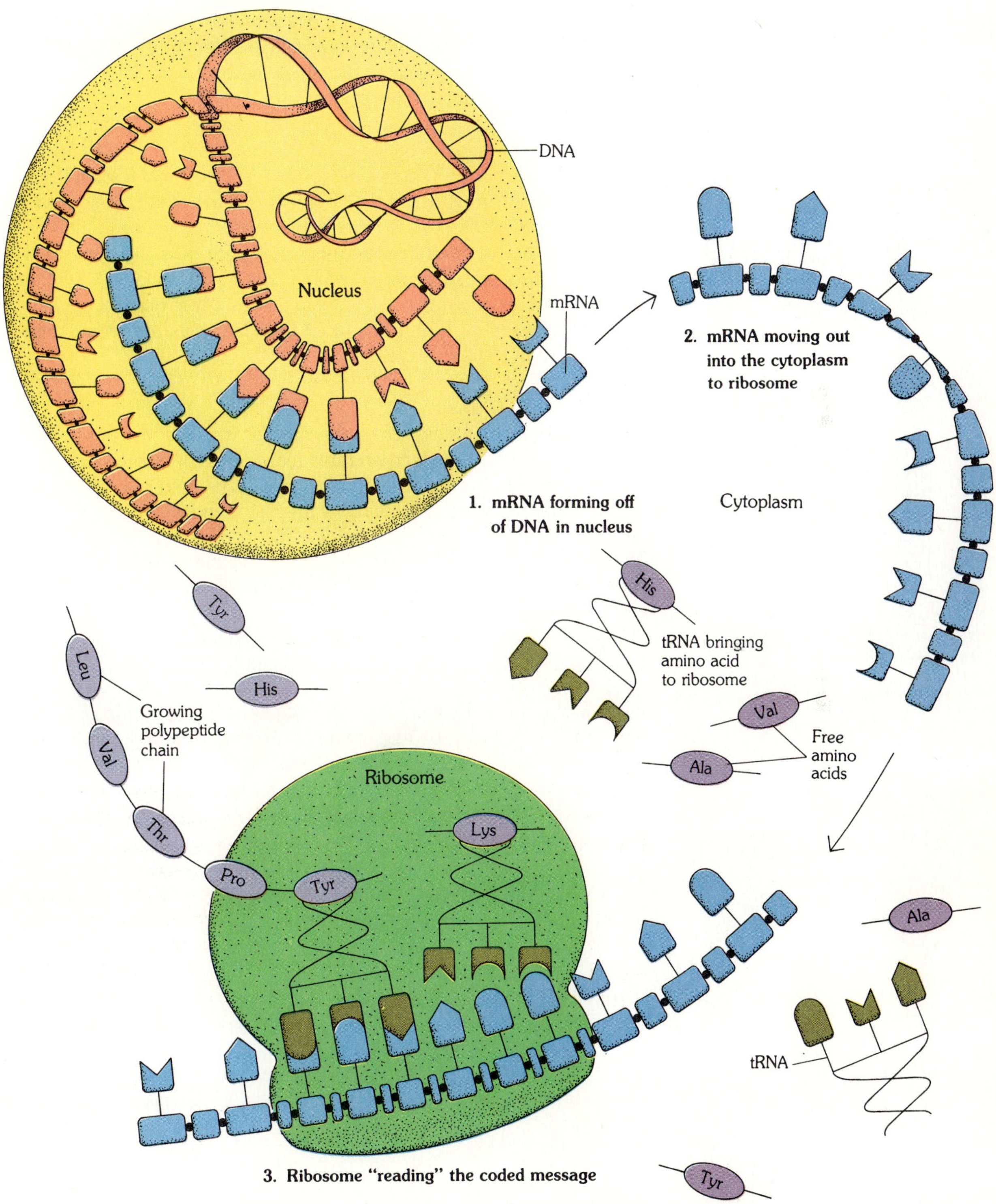
DNA
Nucleus
mRNA
2. mRNA moving out into the cytoplasm to ribosome
1. mRNA forming off of DNA in nucleus
Cytoplasm
His
Tyr
Leu
His
tRNA bringing amino acid to ribosome
Growing polypeptide chain
Val
Val
Free amino acids
Ala
Ribosome
Thr
Lys
Pro
Tyr
Ala
tRNA
3. Ribosome "reading" the coded message
Tyr

Errors in the Transfer of Information

Sickled Red Blood Cell

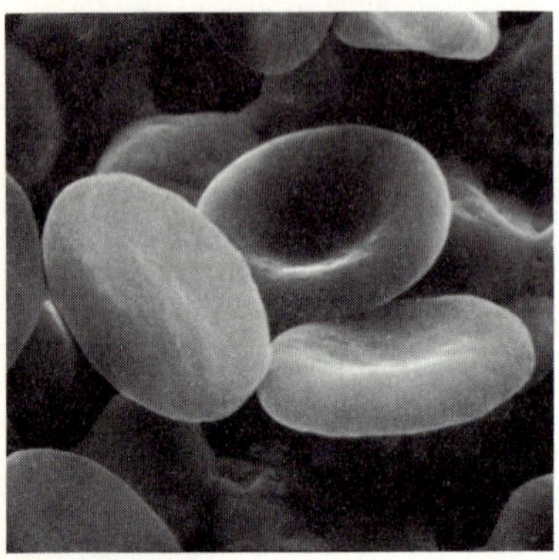

Normal Red Blood Cell

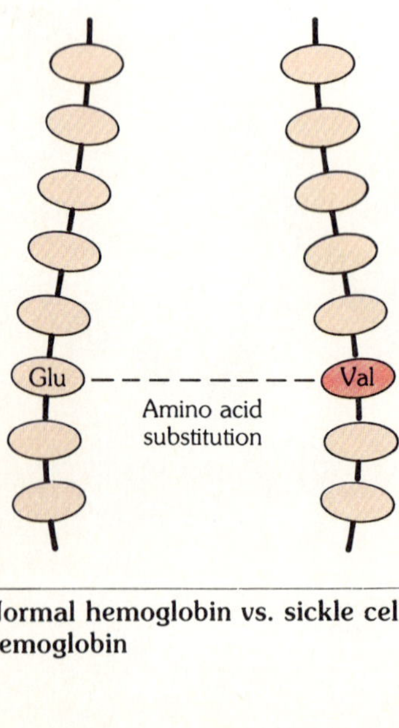

Normal hemoglobin vs. sickle cell hemoglobin

You have seen how DNA codes for the synthesis of proteins. The code for each protein is carried by a particular length of codons along a strand of DNA. The piece of DNA that carries the code for a particular polypeptide is called a **gene** (jēn).

There are hundreds of genes along the length of a single strand of your DNA. Each gene regulates indirectly the synthesis of a particular protein in your body. Proteins are constantly being synthesized inside the cells of your body. Right this moment you are synthesizing new proteins for new hair cells, for example.

It takes about three minutes for one of your cells to synthesize a large protein molecule, such as *hemoglobin*. Most of the time synthesis goes on without any mistakes. But sometimes errors occur. One kind of error is caused by **mutations** (myü tā'shənz), random, permanent changes in the DNA molecule.

Mutations

Mutations occur naturally. They also may be induced, or caused artificially. Exposure to high temperatures may cause mutations. Certain chemicals and forms of radiation, such as X rays, also may cause mutations. Agents that cause mutations are called **mutagens** (myü'tə-jənz). Mutations occur all the time. It appears, however, that some parts of the DNA molecule are more prone to mutations than others. Also, the *mutation rate,* or number of mutations per period of time, varies with age and other factors.

Most mutations in living organisms are not obvious. They occur in body cells and are not transmitted to offspring. Other mutations of the DNA of sex cells remain hidden, or unexpressed, in offspring. If a mutation is serious or harmful, it usually results in the death of an organism. A few mutations are beneficial and are transmitted from generation to generation.

The condition known as **sickle cell anemia** is believed to be the result of a mutation. People afflicted with the disease have blood cells that may be shaped like sickles, or crescents, under certain conditions. Sickled red blood cells cannot carry the normal supply of oxygen. The odd shape of sickled cells also causes them to clog tiny blood vessels in the body.

The disease is caused by an unusual form of hemoglobin, the molecule that carries oxygen in the blood. Researchers have found that this unusual hemoglobin differs by only one amino acid from normal hemoglobin. Instead of glutamic acid, the abnormal hemoglobin has valine in one of its polypeptide chains. Scientists assume this is because one base triplet in the DNA molecule was changed, or *mutated.* Changes in a single segment of a DNA or an RNA molecule are called *point mutations.*

Genetic Changes Produced by Viruses

Mutations are not the only causes of changes in the genetic code. A small particle called a *virus* (vī′rəs) can also bring about changes in the DNA of cells.

Viruses consist of bits of DNA or RNA wrapped in a protein coat. Viruses are *parasites* (par′ə sīts): they cannot reproduce outside of some organism. To live, they must invade a *host* cell. The figure shows what happens when a virus takes over a cell. In the example shown, the host is a simple one-celled organism called a **bacterium** (bak tir′ē əm).

The virus attaches itself to the outer wall of the bacterium by its "tail." With enzymes, the virus makes a hole in the wall of the bacterium. The virus then injects its RNA or DNA into the host cell, leaving the protein coat outside the bacterium. Occasionally the viral genetic material will be incorporated with the host cell's genetic material and replicate with it. In this manner, the bacterium obtains additional genetic material it did not have prior to invasion by the virus.

Challenge!

Once the transfer of genetic information is fully understood, there is no technical reason why humans could not affect the process. It might soon be possible, for example, to correct the genetic code for an individual who has a genetic disease, such as sickle cell or phenylketonuria. Tampering with the transfer of genetic information is called genetic engineering.

Find out if scientists think it might be possible eventually to "build babies to order." Do you think this would be a good idea? Why or why not? Try to find out the arguments people are offering for and against genetic engineering.

Bacterium obtaining new genetic material from virus

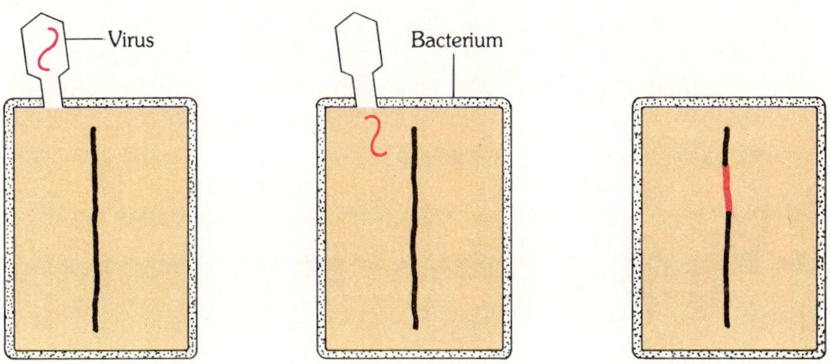

Review It

1. How is the structure of DNA different from RNA?
2. Use the table on p. 117 to design a strand of DNA that would code for a polypeptide whose amino acid sequence is: alanine-leucine-tyrosine-valine-lysine-alanine. Now name the complementary mRNA strand. Is there only one sequence of DNA bases possible? Explain.
3. Describe the process of transcription and protein synthesis.
4. What are two causes of errors in the transfer of genetic information?
5. What are mutations? What causes them?

5–3
The Transfer of Information Beyond the Cell

If your cells hadn't grown and **differentiated** (dif′ə ren′shē āt əd), or become specialized, you would still be one cell! If *all* cells stopped growing and differentiating, life would soon die out. To grow and differentiate, cells must somehow transmit genetic information beyond themselves to new generations of cells.

All cells come from other cells. In 1858, Rudolf Virchow, a German biologist, came to this conclusion after observing many kinds of cells. Since 1858, biologists have been trying to figure out exactly *how* cells produce new cells. The process is still not fully understood. But, biologists do know that there are two different ways by which cells produce more cells. With minor differences, the cells of plants and animals reproduce in the same way. Body cells are produced by a kind of cell division called **mitosis** (mī tō′sis). Sex cells are produced by **meiosis** (mī ō′sis). During both kinds of cell division, the DNA is replicated and transmitted to the new **daughter cells.**

Did You Know?
For a long time, scientists have been trying to discover why organisms age and die. Some scientists think the key to aging and death lies in cells.

Some researchers think that when the cells of an organism begin to die, the organism itself begins to age and die. Research shows that cells reproduce only a limited number of times. Human cells, for example, divide about 50 times before dying. Chicken cells and rat cells divide even fewer times before death. After a set number of divisions, chromosomes seem to lose the ability to duplicate themselves. Mitosis cannot occur. Certain biologists are trying to revive aging cells with drugs. If the biologists are successful, how do you think life on Earth will be affected?

Cells in various stages of division

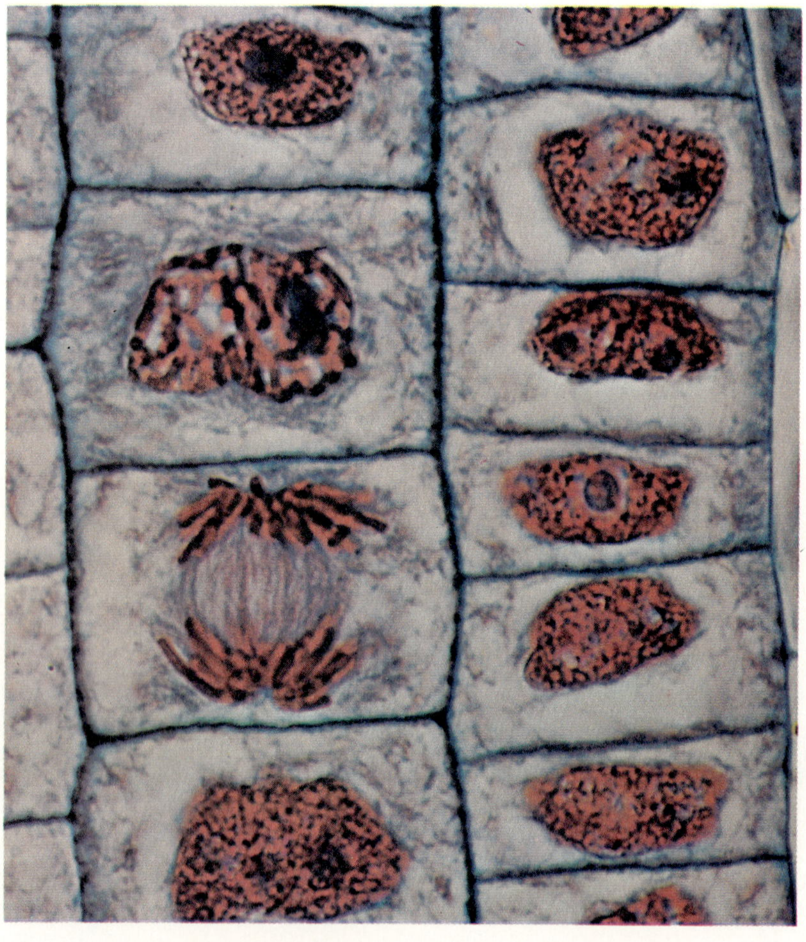

Interphase

Cells do not spend all their time dividing. The time in between cell divisions is called **interphase** (in′tər fāz). Interphase encompasses all the events occurring from the close of one cell division to the start of the next. The amount of time a cell spends in interphase depends on the cell's species, its age, its temperature, and so on. Human body cells grown in the laboratory typically spend about one day between divisions. Some cells, such as human brain cells, stop dividing altogether when humans reach the age of about sixteen. Most cells, however, divide on a fairly regular basis in order to repair or replace other cells.

Interphase begins immediately after cell division occurs. Each new daughter cell starts to grow to its mature size. This usually is a time of rapid synthesis of proteins and RNA. After the cell reaches a certain size, it begins to duplicate its DNA. The loosely coiled DNA, or **chromatin** (krō′mə tən), constricts, or coils up, to form rodlike **chromosomes** (krō′mə sōmz). In human body cells there are 46 chromosomes. The number varies in different species.

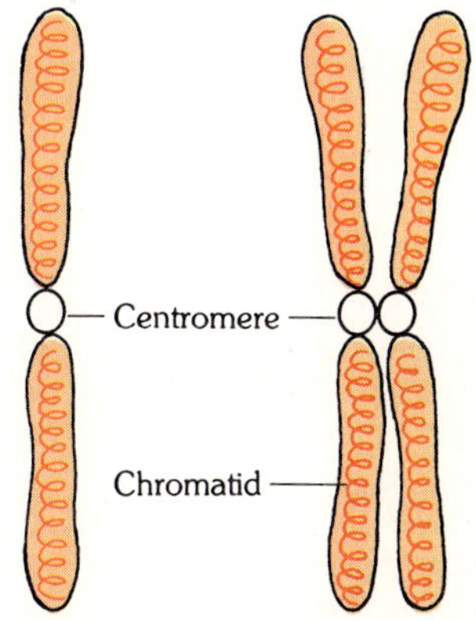

Chromosome

Replicated chromosome

Each chromosome duplicates itself to form two identical sister chromosomes. The sister chromosomes remain attached at a beadlike region called the **centromere** (sen′trə mir). As long as the sister chromosomes remain attached by their centromeres, they are referred to as **chromatids** (krō′mə tidz). Thus, where one chromosome was present early in interphase, later two chromatids appear. Generally, one cannot see the chromatin material under a light microscope when a cell is in interphase.

After the DNA is duplicated, the cell continues to increase in size. More proteins are synthesized. At this point, the cell is ready to divide and interphase ends. The exact factors that bring on cell division are not known.

Cell in interphase

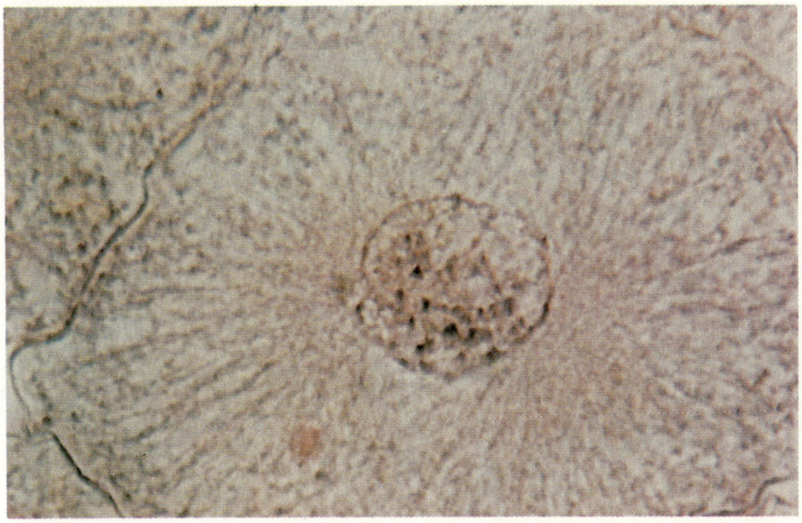

Courtesy Carolina Biological Supply Company

Mitosis

Cell division occurs in a smooth, continuous fashion. It is usually over quickly. For convenience, however, biologists generally divide the process of cell division into a series of steps, or *stages.* Certain events typically occur during each stage. As mentioned before, minor differences occur between plant and animal cells.

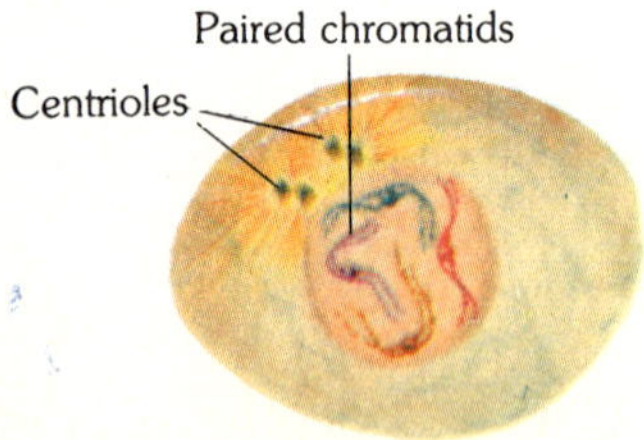

Early Prophase

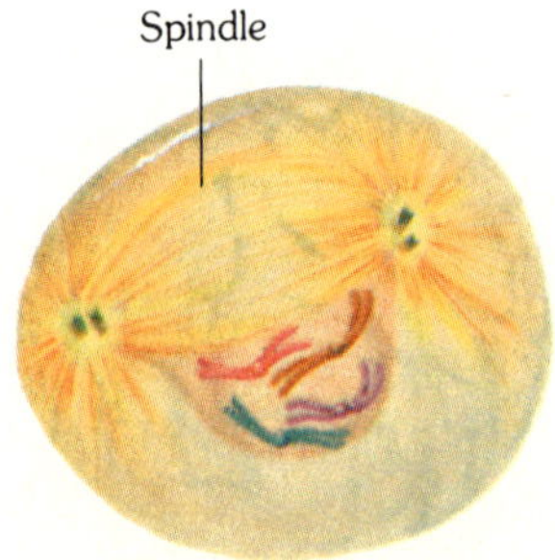

Prophase

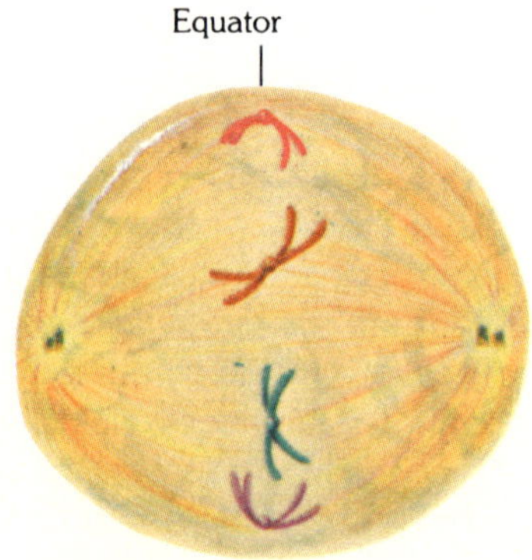

Metaphase

Prophase

The first stage of mitosis is called **prophase** (prō′fāz). A noticeable shortening and thickening of the chromatids marks the start of prophase. The membrane around the nucleus breaks down and disappears. A football-shaped structure called the **spindle** forms. The spindle is made up of tiny tubelike structures and it often extends the full length of the cell. The far ends of the spindle are called **poles.** In animal cells, a pair of structures located outside of the nucleus known as **centrioles** (sen′trē ōlz) each migrate to an opposite pole. The spindle fibers form between the centrioles. In plant cells, there are no centrioles. The spindle forms without them. Prophase ends with the movement of the chromosomes toward the middle of the spindle. This middle is sometimes referred to as the **equator** (i kwā′tər).

Metaphase

The stage during which the chromatids are lined up along the middle of the spindle is called **metaphase** (met′ə fāz). During the metaphase, sister chromatids face opposite poles. Some of the spindle fibers attach to the centromere of each chromatid.

Anaphase

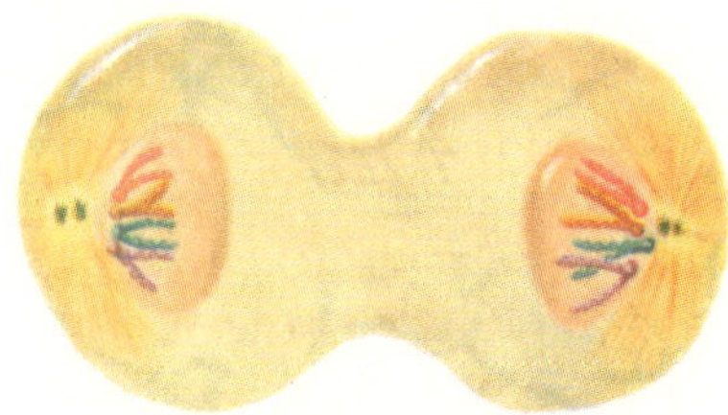

Telophase

Anaphase

Next, the centromeres divide separating the chromatids. The separation of the sister chromatids marks the start of **anaphase** (an′ə fāz). Once the chromatids separate, they are referred to as *daughter chromosomes.* One daughter chromosome migrates to one pole, while its twin travels to the opposite pole. This way, each new, still unformed cell will have the same genetic make-up.

Telophase

The daughter chromosomes arrive at the poles of the spindle. This event marks the onset of the final stage of mitosis, **telophase** (tel′ə fāz). The enlarged cell contains twice its normal number of chromosomes in the form of two identical sets clumped at the two poles of the spindle. Once the chromosomes have reached the poles of the spindle, the spindle starts to break down. The reverse of prophase begins. The individual chromosomes get longer, thinner, less visible. The nuclear membrane re-forms around each set of chromosomes. The centrioles, if present, replicate. It only remains for the cell to divide into two daughter cells.

The means of separation differs in plants and animals. In plants, a **cell plate** forms in the middle of the dividing cell. The cell plate expands outward until a wall is formed between the two daughter plant cells. In animal cells, the cell membrane constricts or furrows inward, until a complete membrane separates the two daughter cells.

Different methods in separating daughter cells

Plant cell

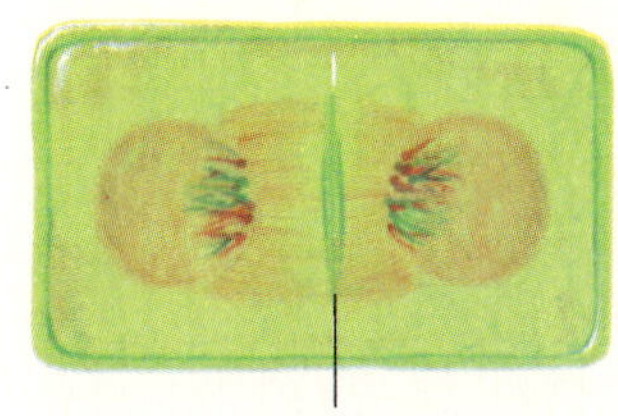

Animal cell

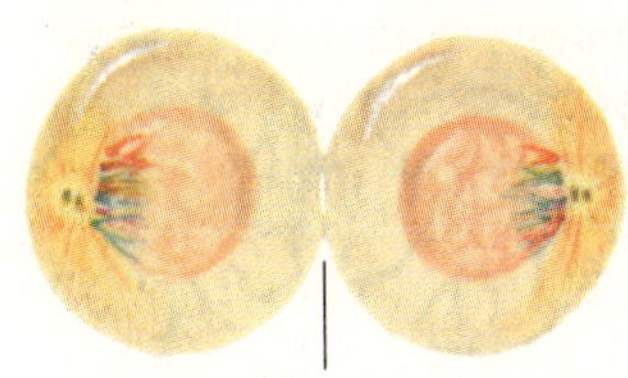

The Importance of Mitosis

Mitosis ensures that each new body cell has the same genetic make-up as its parent. Mutations can and do occur, but most mitotic divisions occur without error. Because of mitosis, all your body cells have identical molecules of DNA.

Meiosis

Organism	Diploid Chromosome Number	Haploid Chromosome Number
Human	46	23
Giant Sequoia	22	11
Goldfish	94	47
Rice	24	12
Fruit fly	8	4
Guinea pig	64	32
House fly	12	6

Sex cells in an organism are known as **gametes** (gam'ēts). These are usually known as the egg and sperm. The transfer of information to gametes is different from the transfer of information to body cells. The major difference lies in the number of chromosomes transmitted to the sex cell.

Chromosomes occur in most organisms in pairs. In the human, there are 46 chromosomes, or 23 pairs. The total number of chromosomes is called the **diploid** (dip'loid) number. Therefore, in the human, the diploid number is 46.

Gamete cells have half as many chromosomes as the diploid number because it usually takes two sex cells to form a new organism. This half number is called the **haploid** (hap'loid) number. In the human egg or sperm cell, there are 23 chromosomes. The haploid number in this example is 23. When two gametes unite—a process known as fertilization—the diploid number of chromosomes is restored.

Did You Know?

A lot of time, money, and effort are being spent trying to find a cure for cancer. Although there are many kinds of cancer, they all have one thing in common: rapid, uncontrolled growth of cells. If scientists could find a way to slow cell growth, they might be able to slow down cancer also.

One possible solution is a group of compounds called chalones *(kal'ōnz). Chalones prevent cell growth in two ways. First, some chalones prevent the duplication of chromosomes during interphase. Second, some chalones prevent chromosomes from separating during anaphase. If biologists can find out more about the way chalones work, they may be able to slow down the runaway growth of cancer cells.*

Human gamete cells prior to fertilization

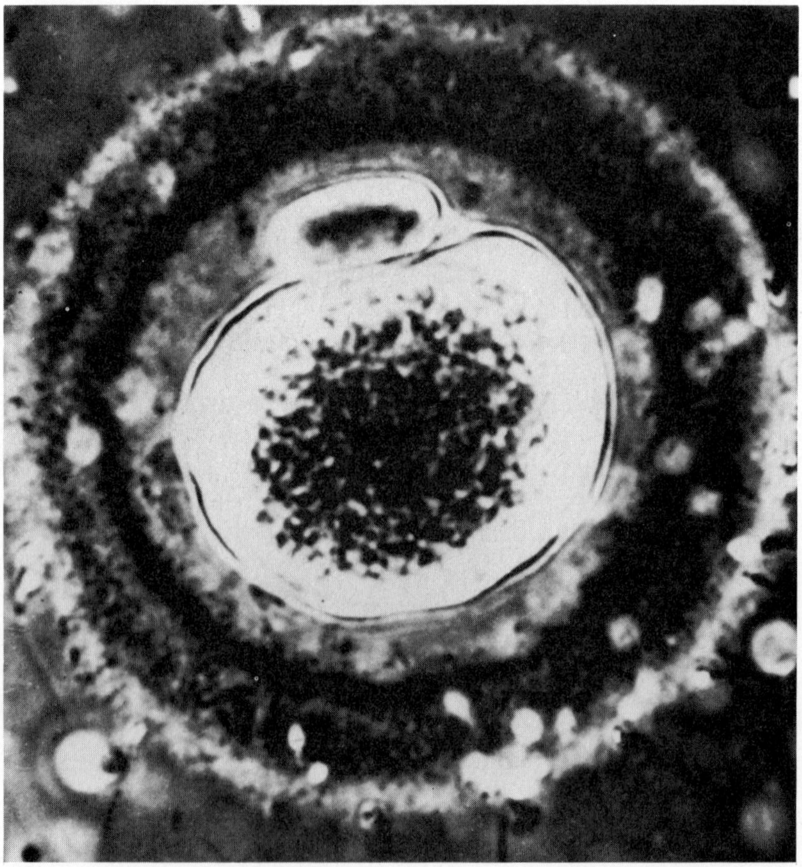

Comparison of Meiosis with Mitosis

Gamete cells are produced by a special kind of cell division called reduction division, or **meiosis** (mī ō′sis). In some ways, meiosis is like mitosis. But, there are important differences. First, two cell divisions occur in meiosis instead of one as in mitosis. As a result, four daughter cells are produced by meiosis instead of two. Second, during meiosis the chromatids may exchange similar parts. This exchange normally does not occur in mitosis. Third, the chromosomes duplicate only one time during meiosis even though there are two cell divisions. Fourth, the chromosomes line up on either side of the equator and randomly move to either pole. Because of the differences between meiosis and mitosis, each gamete is genetically unique. In other words, the gametes formed will have half the diploid number of chromosomes, and the DNA in one gamete will probably be different from the DNA in another.

To avoid confusion between the stages of each division, a summary of events for each division is given. The stages, as they occur, will be labeled with the accompanying diagrams.

Sperm formation in mouse testis

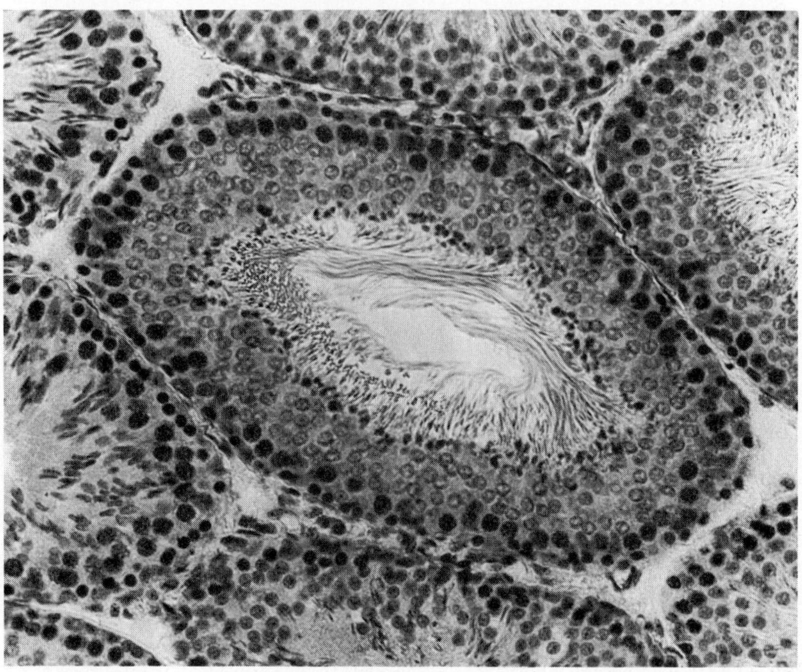

The Process of Meiosis

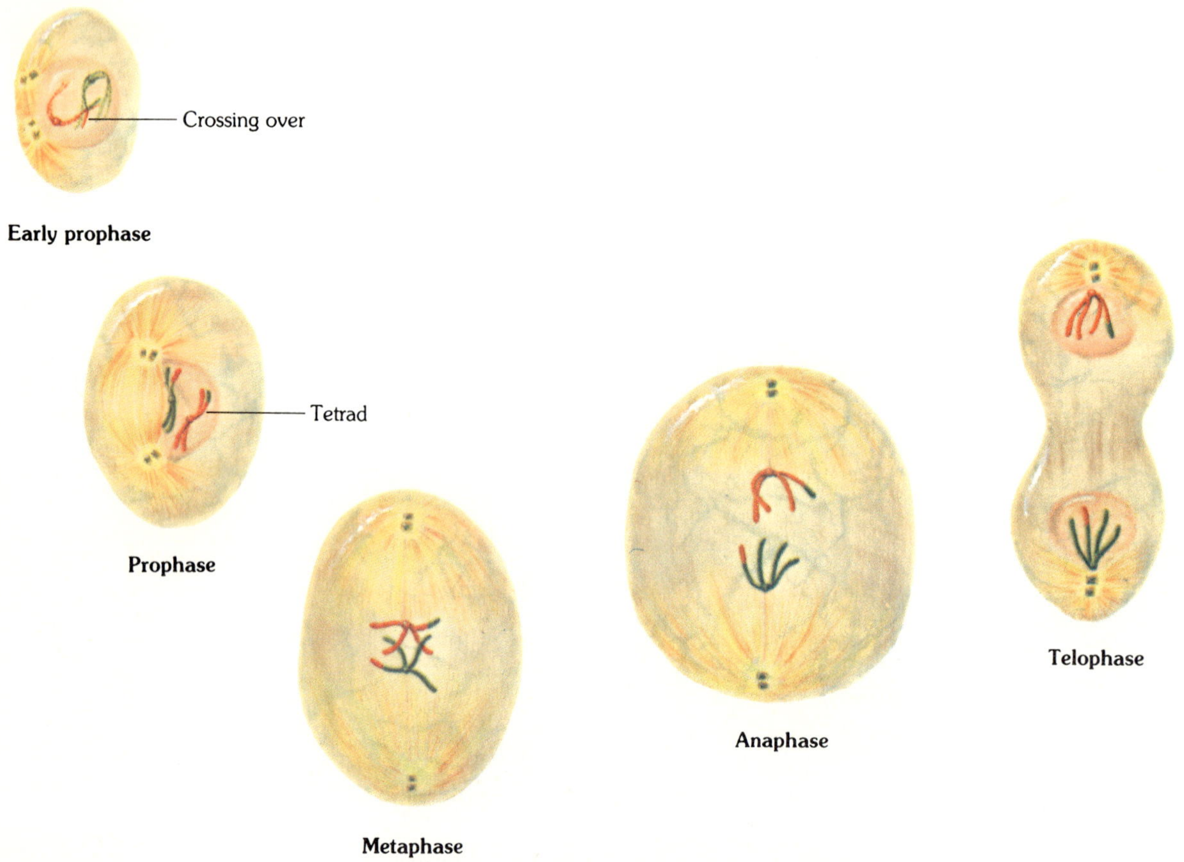

First Division

In the first meiotic division, each chromosome has already replicated itself and consists of two chromatids. Each replicated chromosome seeks out its sister chromosome (remember, chromosomes usually exist in pairs). Frequently, chromatids of partner chromosome pairs become entangled and will exchange corresponding sections of DNA. This process is called **crossing over.** This helps explain the variation of genetic material between parent and daughter cells. Crossing over does not usually occur in mitosis. The duplicated chromosomes pair up forming a four-stranded group called a **tetrad** (tet'rad).

While the tetrads are forming, the nuclear membrane is breaking down and the spindle fibers are forming. The tetrads migrate toward the equator, lining up in random fashion. It is equally likely that one pair of chromosomes will be oriented toward one pole as the other. Separation of the tetrad occurs. Note that the joined pair of chromatids move *together* toward the poles. This is different from mitosis where the chromatids separated from each other.

Complete chromosomes (consisting of two chromatids each) gather at each pole. The cytoplasm divides, either by furrowing or by forming a cell plate.

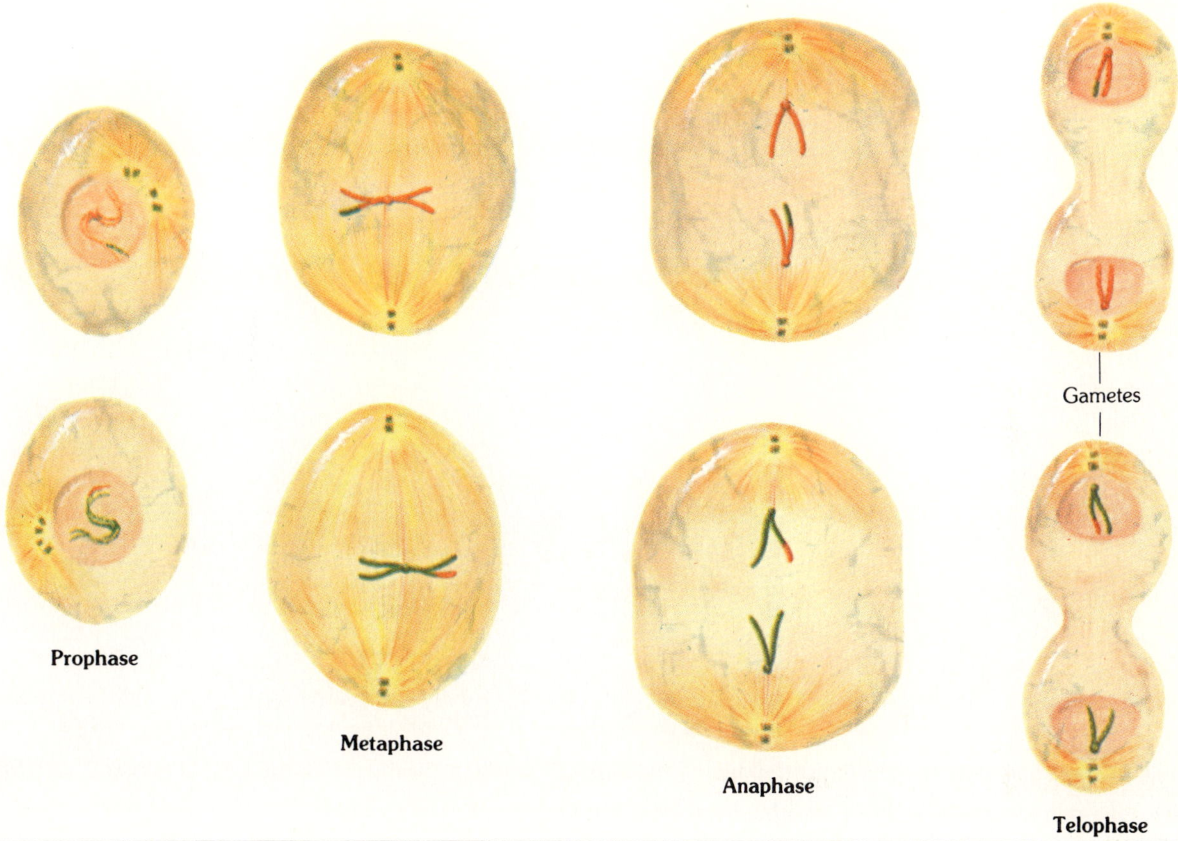

Second Division

The second meiotic division is very similar to mitosis. The spindle fibers form, and the paired chromatids line up at the equator. A spindle fiber attaches to each centromere and begins to draw the chromatids apart. The chromatids (now called chromosomes) migrate and group together at each pole. The cytoplasm again divides. Four daughter haploid cells, each with one-half the original number of chromosomes of the diploid parent cell, have now been formed. Not all daughter cells formed during meiosis become viable gametes. Usually, in male organisms, the cytoplasm divides evenly, and all four cells will function as sperm cells. But often in the female, the cytoplasm is divided unequally, with one cell receiving the majority of the cytoplasm. This one cell will then function as the female gamete cell. So, from one parent diploid cell, there may be one or four haploid gamete cells resulting from the meiotic process.

The Importance of Meiosis

Meiosis is the process by which the chromosome number is reduced by one-half. This results in the formation of as many as four gamete cells. The diploid number is restored upon the joining of two sex cells during fertilization. Organisms, such as amebas, that reproduce by mitosis, are genetically the same. Barring mutation, one generation of amebas is just like another. By contrast, meiosis produces organisms that are genetically different. Crossing over and random assortment of chromosomes into daughter cells ensure that each gamete is genetically unique. Further genetic mixing occurs during fertilization. This genetic recombination may enhance a species' chances for survival.

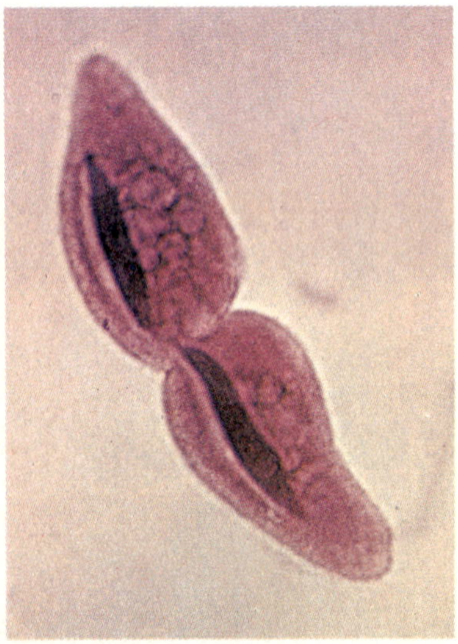

Courtesy Carolina Biological Supply Company

Protozoan reproducing by mitosis

Mammal offspring variation due to meiosis

Review It

1. How long does a cell normally spend in interphase?
2. What are the stages of mitosis?
3. Why is mitosis important?
4. What are the stages of meiosis?
5. Why is meiosis important?
6. How are mitosis and meiosis different?

Activity

Demonstrating the Transfer of Information Within and Beyond the Cell

Part A
Making and Breaking the Codes

Materials

One sheet of red paper
One sheet of blue paper
Scissors
Tape

Procedure

1. Cut the red paper into strips.
2. Cut each strip into two parts that fit together. Let each strip stand for a rung of the DNA molecule.
3. Label one part of one strip A (for adenine). Label the other part T (for thymine). Repeat step 2, using the labels T and A, G and C, and C and G until you run out of strips.

4. Using the DNA strips as patterns, cut the blue paper into complementary RNA bases. The RNA U (for uracil) should match up with the DNA piece labeled A. The RNA G (for guanine) should match up with the DNA C (for cytosine) and so on until you run out of strips.

5. Tape together a DNA base triplet of your choice.
6. Show how the DNA code translates into the RNA code by placing the proper complementary RNA base (e.g., G) next to the DNA base (e.g., C).

Analysis

1. Look at the chart on page 117. Which amino acid does your DNA and your RNA code stand for?
2. How many bases long would a strand of DNA have to be to code for all twenty amino acids? Describe the bases in the corresponding strand of RNA.

Part B
Making Models of Chromosomes During Mitosis and Meiosis

Materials

Red and white pipe cleaners
Tape
Yarn or thread
Cardboard

Procedure

1. Look back at pages 124–129. Study the diagrams of mitosis and meiosis carefully.
2. Let the two different colored pipe cleaners be models of chromosomes.
3. Use the tape and the yarn to show what happens to chromosomes during the stages of mitosis and meiosis.
4. Tape the pipe-cleaner "chromosomes" to the cardboard. Label each stage clearly.

Analysis

1. What happens to the pipe-cleaner "chromosomes" during mitosis? during meiosis?
2. How do your chromosome models illustrate the difference between the genetic material in non-sex cells and sex cells?

Chapter Summary

Directions for carrying out all cell functions are stored in the DNA molecules found in the nuclei of most cells. The arrangement of four different bases along the DNA molecule provides the genetic code that regulates cell functions.

The genetic code is transcribed from the DNA molecule onto a complementary strand of messenger RNA. The mRNA carries its message—in the form of codons—from inside the nucleus to the ribosomes, where protein synthesis takes place. Another kind of RNA, transfer RNA, assists in protein synthesis by bringing to the ribosomes amino acids, the building blocks of proteins.

Many of the proteins that are synthesized are enzymes, which stimulate chemical reactions throughout the cell. The segment of DNA that codes for the synthesis of a particular polypeptide is called a gene.

Errors sometimes occur in the genetic code or in the transfer of genetic information. Mutations that may be caused by radiation or chemicals are one source of errors. Viruses are another source of error.

Genetic information is transmitted to new generations of cells in two ways: mitosis and meiosis. Mitosis refers to the way in which new body cells, or non-sex cells, are formed. Meiosis is the process by which sex cells, or gametes, are formed. During mitosis, exact copies of the DNA are transmitted to new cells. By contrast, meiosis ensures that the DNA of each sex cell is different from its parent cell.

Questions to Think About

1. Some people have suggested that people should be tested to see if they are carriers of genes that cause diseases such as sickle cell anemia. What would your arguments pro and con be?
2. DNA has the same basic molecular structure and chemical make-up in nearly all living species. How then would you explain the diversity of living things?
3. Why are proteins, and enzymes in particular, important molecules in your body? What functions do they have?
4. All people, except for identical twins, are genetically unique. Identical twins are genetically identical because the fertilized egg from which they developed split and developed into two embryos. Explain why most people are genetically unique.

Extra Research

1. Make an appointment to talk with a genetic counselor in your town or community. Find out what advice a genetic counselor can offer.
2. Obtain prepared slides of mitosis in plant cells and in animal cells. Make drawings of the stages you see. Highlight the differences between dividing plant and animal cells.
3. One of the most thoroughly studied genetic diseases is *phenylketonuria*, or PKU. Find out what scientists know about this disease. Check books and encyclopedias in your library.
4. Bacteria are widely used in experiments on the biochemistry of genetics. What are the advantages of using bacteria for such studies? Find some descriptions of some bacterial studies that have added to our knowledge of molecular genetics.
5. Some state legislatures are considering laws re quiring that newborn babies be tested for genetic diseases. See what you can find out about existing and planned legislation in this area. What arguments would you offer for and against such legislation?
6. Obtain a kit for constructing a model of a DNA molecule or make up your own kit. Build a model of the DNA molecule for display in your classroom.

Chapter Test

Part I

Copy the number of each test item. If the item is *true,* write *T* after the number; if the item is *false,* write *F.*

1. The genetic code consists of the arrangement of the nucleotide bases along the length of a strand of DNA. (115)
2. Complementary strands of protein are transcribed from pieces of the DNA molecule. (118)
3. Codons govern the formation of polypeptide chains. (117)
4. Amino acids are the building blocks of RNA. (116)
5. Most proteins are synthesized inside the nucleus. (118)
6. The sickle cell disease is caused by an abnormal form of hemoglobin. (120)
7. When a virus infects a cell, the genetic material of the virus may be incorporated into the genetic material of the cell. (121)
8. The end results of mitosis and meiosis are the same. (127)
9. Cells spend most of their time in interphase. (123)
10. Crossing over usually occurs during mitosis. (128)

Part II

Copy the number of each sentence below. After each number, write the word or words that complete the sentences correctly.

11. The sides of the ladderlike DNA molecule are made up of alternating units of ____ and ____. (111)
12. The two strands of the DNA molecule are held together by ____ bonds. (111)
13. RNA molecules are like DNA molecules except for three things: (1) ____, (2) ____, and (3) ____. (116)
14. Each base triplet, or ____, codes for one ____, which the tRNA transfers to the sites of protein synthesis. (115)
15. Special molecules called ____ catalyze chemical reactions inside cells. (114)
16. Random changes that occur at a single location on a DNA or RNA molecule are called ____. (120)
17. An agent that causes a change in the genetic material is called a ____. (120)
18. Non-sex cells form by ____; sex cells form by ____. (122)
19. Gametes contain ____ the number of chromosomes found in body cells. (126)
20. Meiosis differs from mitosis in four major ways: (1) ____, (2) ____, (3) ____, and (4) ____. (127)

Part III

Copy each numbered item from List A. After each item, write the letter and words from List B that best describe it.

List A

21. metaphase (124)
22. anaphase (125)
23. interphase (123)
24. telophase (125)
25. prophase (124)

List B

a. spindle forms and chromosomes move toward middle
b. spindle breaks down and membrane re-forms
c. the period between cell divisions
d. chromatids arrange around middle of spindle
e. daughter chromosomes move to opposite ends of spindle

Focus on Careers

Cytotechnologist

In modern medical laboratories and cancer research centers, accurate diagnoses and complex procedures require the attention of well-trained personnel. Thus, there is an expanding need for trained cytotechnologists.

A cytotechnologist is a specialized medical technician who is trained to recognize abnormalities in cells. For example, minute abnormalities in the color, size, and shape of human cell organelles may signal the presence of cancer.

When a tissue sample is taken from a patient, the cells in the tissues are prepared on slides by a histologist, another medical technician. The histologist slices the tissue into thin sections and mounts the sections on slides. He or she uses stain techniques to make cell parts visible. The cells in the tissue section are examined under a microscope by the cytotechnologist.

If abnormal cells are detected, the cytotechnologist records the results on special medical forms which are then studied by the patient's physician. If the abnormality is a sign of cancer, cellular spread may be prevented with drugs or surgery. In some emergency situations, the cytotechnologist examines cells while a surgical team waits for the results.

To become a cytotechnologist, you need at least two years of college with some courses in biology. An individual must also spend twelve months training in a school of cytotechnology which is approved by the American Medical Association. The last six months is on-job training in a cytology laboratory. The cytotechnologist generally works under the supervision of a pathologist, a medical doctor in charge of the laboratory.

Most cytotechnology schools charge no tuition or may offer scholarships or stipends if tuition is charged. After training, a student takes an exam given by the Board of Registry of Medical Technologists. The successful student is designated CT(ASCP), which means Registered Cytotechnologist, as approved by the American Society of Clinical Pathologists.

Opportunities for cytotechnologists are excellent. There is a continuing need for qualified cytotechnologists in hospitals and clinics.

Career Information

Biomedical Engineer Technician: Biomedical Engineering Society, P.O. Box 2399, Culver City, CA 90230.

Cytotechnologist: American Society of Cytology, Jefferson University, 130 S. 9th Street, Suite 1006, Philadelphia, PA 19107.

Nuclear Medical Technologist: Society of Nuclear Medicine, 475 Park Avenue S., New York, NY 10016.

Pathologist: American Medical Association, Department of Pathology, 535 N. Dearborn Street, Chicago, IL 60612.

Biomedical Engineer Technician
Duties: Assists researchers in developing instruments which will speed up laboratory tests and aid people with certain handicaps. For example, biomedical technicians helped design a blood analyzer that can perform tests on hundreds of blood samples in just a few minutes. Biomedical engineers also design mechanical limbs for amputees, hearing aids for patients with hearing losses, and pacers for heart patients.

Training: Requires two years in a junior college or professional school. Courses in engineering and mathematics are essential.

Opportunities: Jobs are available in pharmaceutical companies, research laboratories, universities, and medical instrumentation companies. Advances in technology make job opportunities excellent.

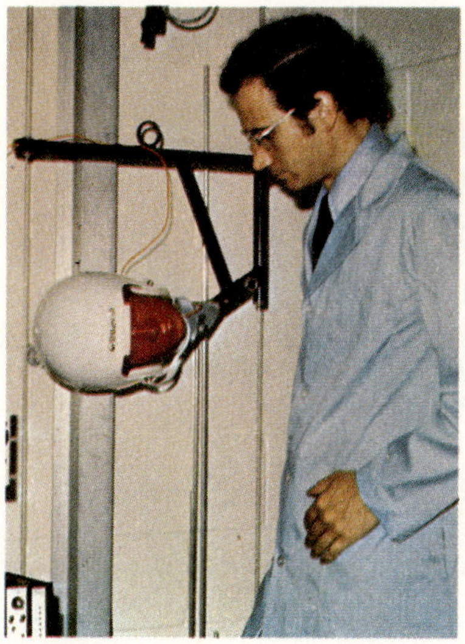

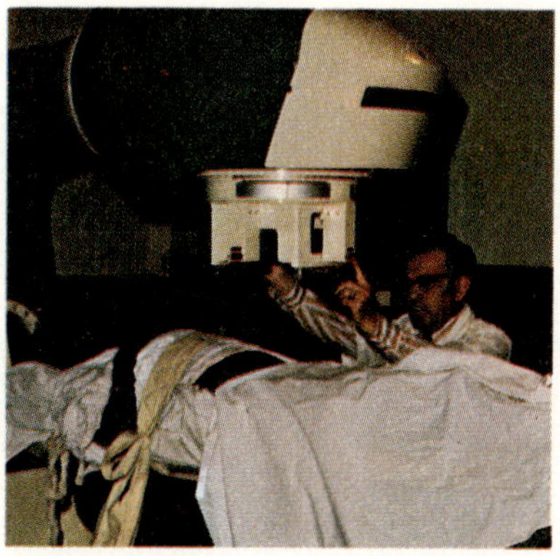

Nuclear Medical Technologist
Duties: Performs clinical tests with radioactive materials to diagnose or control various diseases. For example, small amounts of radioactive material may be introduced and traced through certain organs of a patient's body. This allows the patient's physician to get a "picture" of a specific area of the body and may help in the treatment of such diseases as thyroid cancer.

Training: Requires special training in a four-year college or in a specialized school for nuclear medical technologists. On-job training is also necessary.

Opportunities: Prospects are excellent. The Society of Nuclear Medicine claims there will not be enough qualified people to fill all the job openings in the near future.

Pathologist
Duties: Studies changes in organs and tissues caused by diseases. Working with the attending physician, a pathologist tries to identify a patient's disease and evaluate treatment. This is done by studying and testing cells, tissues, and fluids of the human body in the laboratory. A pathologist is also in charge of a medical laboratory in a hospital or research center. He or she supervises various kinds of medical technologists.

Training: Requires college, medical school, and internship, followed by three or four years as a pathology resident.

Opportunities: Profitable career as supervisor of medical laboratory, professor in a medical school, or as a consultant.

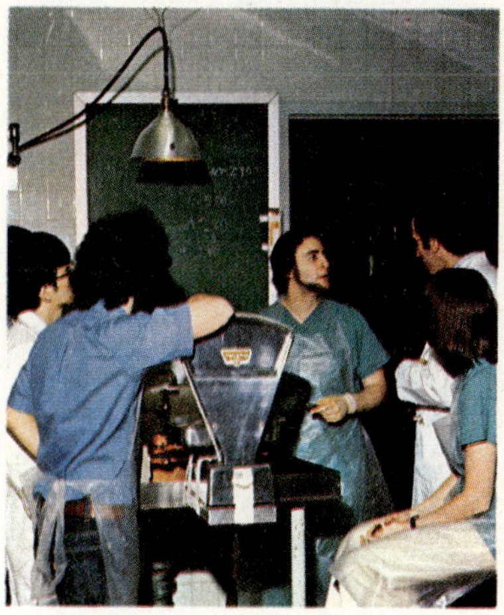

Bibliography: Unit Two

Asimov, Isaac. *The Genetic Code.* New American Library, Inc., 1963. Account of the role of DNA in cells. Above-average reading level.

Baker, J. J. W., and Allen, G. E. *Matter, Energy, and Life.* Addison-Wesley Publishing Co., Inc., 1965. Background information on the chemistry and physics needed to understand biology.

Beadle, George, and Beadle, Muriel. *The Language of Life.* Doubleday and Co., Inc., 1966. An introduction to genetics, including an account of the role of DNA.

Butler, J. A. V. *The Life of the Cell.* Basic Books, 1964. A nontechnical account of how secrets of life at the cellular level have been discovered.

Case, James, and Stiers, Vernon. *Biology: Observation and Concept.* The Macmillan Co., 1971. Well-written discussion of cell structure and function.

The Chemical Basis of Life: Readings from Scientific American. W. H. Freeman, 1973. Collection of articles on cell structure and function.

Cosslett, V. E. *Modern Microscopy or Seeing the Very Small.* Cornell, 1966. Contains recent information on the electron microscope.

Galston, Arthur. *The Life of the Green Plant,* 2nd ed. Prentice-Hall, Inc., 1964. Provides an account of photosynthesis as well as structure and function of plants.

Gillie, Oliver. *The Living Cell.* Funk and Wagnalls, 1971. Discusses cell structure and function, evolution of microorganisms, genetic abnormalities, and other topics.

Jenkins, Marie. *Embryos and How They Develop.* Holiday House, 1975. Contains photographs of a wide range of organisms and discusses cell division and development.

Kennedy, D., ed. *The Living Cell.* W. H. Freeman and Sons, 1965. Collection of articles on cell structure and function.

Klein, Aaron. *Threads of Life.* Natural History Press, 1970. History of research in genetics from the time of Aristotle to the present.

Lessing, Lawrence. *DNA, at the Core of Life Itself.* The Macmillan Co., 1967. Information about DNA and RNA written for the average reader.

The Living Organism. Doubleday and Co., Inc., 1965. Illustrations and information on cell structure and function.

McElroy, William D. *Cell Physiology and Biochemistry,* 3rd ed. Prentice-Hall, Inc., 1971. Describes biochemical and physiological processes.

Medawar, P. B., and Medawar, J. S. *The Life Science.* Harper and Row, 1977. Current ideas about cell biology, genetics, aging, and more.

Mercer, E. H. *Cells, Their Structure and Function.* Natural History Press, 1962. Provides general information about cell structure and function.

Morrison, John. *Functional Organelles.* Reinhold Publishing Corp., 1966. Describes functions of cell organelles.

Pfeiffer, John. *The Cell.* Time Inc., 1964. Beautifully illustrated book covering all aspects of structure and function of cells.

Stehli, G. *The Microscope and How to Use It.* Sterling, 1963. Describes the use of the microscope.

Swanson, C. P. *The Cell,* 3rd ed. Prentice-Hall, Inc., 1969. General information about cell structure and function.

Tufty, Barbara. *Cells–Units of Life.* G. P. Putnam's Sons, 1973. Good overview of cell structure and function.

Watson, J. D. *The Double Helix.* New American Library, 1969. Watson's own account of the discovery of DNA structure.

Wilson, G. B. *The Elements of Cytogenetics.* Reinhold Book Corp., 1968. Information on chromosomes, mitosis, and meiosis.

Articles

Arehart-Treichel, Joan. "Sickle Cells: The Molecular Attack." *Science News,* February 16, 1974.

Proujan, Carl. "Genes Made to Order." *Science World,* November 17, 1977.

Samz, Jane. "But . . . What Is Cloning?" *Science World,* November 2, 1978.

Schuman, Joan. "Chalones—Do They Say 'Stop' to Cell Division?" *Science World,* March 5, 1973.

UNIT THREE

GENETICS

Chapter 6

Inheritance of Traits

Genetics is the science that concerns heredity. It involves the study of how parents pass biological traits to their offspring. Before the 20th century, people believed that biological traits moved from one generation to the next through the blood. Many people still speak of "blood relatives." We now know that biological traits are passed as chemical messages in the chromosome. The chemical messages are coded on DNA molecules in the nucleus.

In this chapter, you will examine the ways in which various traits are passed from parents to offspring. You will begin the chapter by studying the basic principles of genetics. These principles were first discovered in the 19th century by Gregor Johann Mendel. Then you will learn about the more recent advances in the field of genetics. Finally, you will learn how these recent discoveries have added to our knowledge of *human* genetics.

Chapter Objectives

1. Describe the rules of dominance, segregation, and independent assortment.
2. Construct a Punnett square and describe how to use it.
3. Describe how crossing over provides new gene combinations in the sex cells.
4. Explain how the sex of an organism is determined.
5. Describe ways in which genes express themselves differently from the way Mendel described.
6. List some patterns of human inheritance and give examples of each.

The reason for the strong resemblance of the child to his parents is heredity. Genetics is the science that deals with the principles of heredity.

6–1 Mendelian Genetics

Gregor Johann Mendel

Gregor Johann Mendel was born to peasant parents in 1822. He became a priest at the age of 25 and studied math and science at the University of Vienna. He then returned to a monastery in Brünn, which is now a part of Czechoslovakia. At the monastery, he worked on plant breeding experiments. He also helped to start a science magazine for the Brünn Natural Science Society.

Mendel was different from other scholars who lived in Europe during the 19th century. He approached problem-solving in new and different ways. The brilliance of Gregor Mendel's work can be seen in the way he conducted his experiments. For example, he used garden peas for his studies. These plants contain both male and female parts. Mendel could remove the male, pollen-bearing structures from each plant. This prevented female egg cells on those plants from being fertilized. Therefore, Mendel could *control the way the plants were fertilized.* He could select the plants he wanted as parents and the traits he wanted to study.

Second, Mendel worked with *only one trait* at a time. He studied only those traits that were easy to see or to measure. For example, some of the traits Mendel selected included flower color, seed shape, and stem length.

Finally, Mendel kept *detailed records* of his crosses. He counted offspring and looked for mathematical answers, based on how often specific traits appeared.

In one experiment, Mendel crossed garden peas that had round seeds with those that had wrinkled seeds. Then he figured out the percentage of offspring that had round seeds and the percentage that had wrinkled seeds. He learned that about 75 percent of the offspring of such a cross had round seeds, while only about 25 percent had wrinkled seeds. From the results of experiments like these, Mendel was able to explain some basic principles of heredity. Today, these basic principles are referred to as **Mendelian** (men dē′lē ən) **genetics.**

Unfortunately, all of Mendel's findings were reported in the little-known magazine of the Brünn science group. So his work did not become well known until 35 years later. Unfortunately Mendel's genius was not recognized until after his death.

Garden pea flower

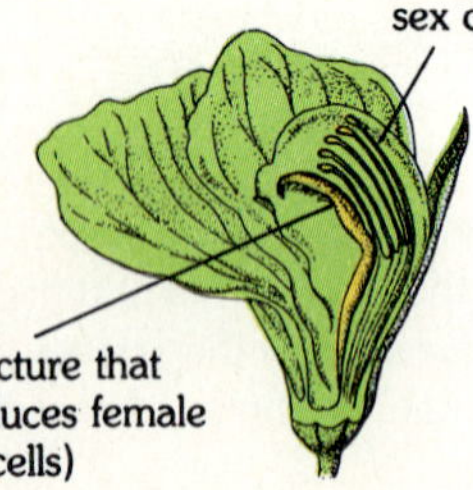

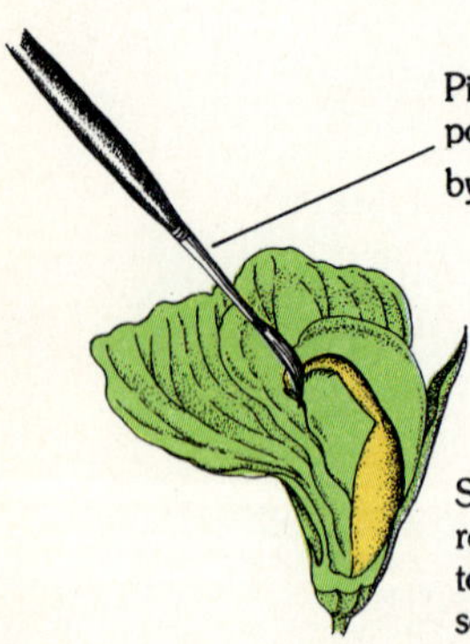

Unit Factors

One of Mendel's first experiments was a cross between two garden peas that had different colored flowers. Mendel believed that *traits are determined by individual units called factors.* Today these units are called genes. Mendel believed that each offspring received a factor from each parent. He guessed that each trait, such as flower color, is determined by a *pair* of genes.

Therefore, plants that inherit a gene from each parent for red flower color will be a pure strain, or **purebred,** for red flower color. Offspring will be purebred for white flower color if they inherit one factor from each parent for white flower color. However, garden peas that inherit a factor for red color from one parent and a factor for white color from the other parent will be mixed, or **hybrid,** for flower color.

You can see how this process works by studying the diagram below.

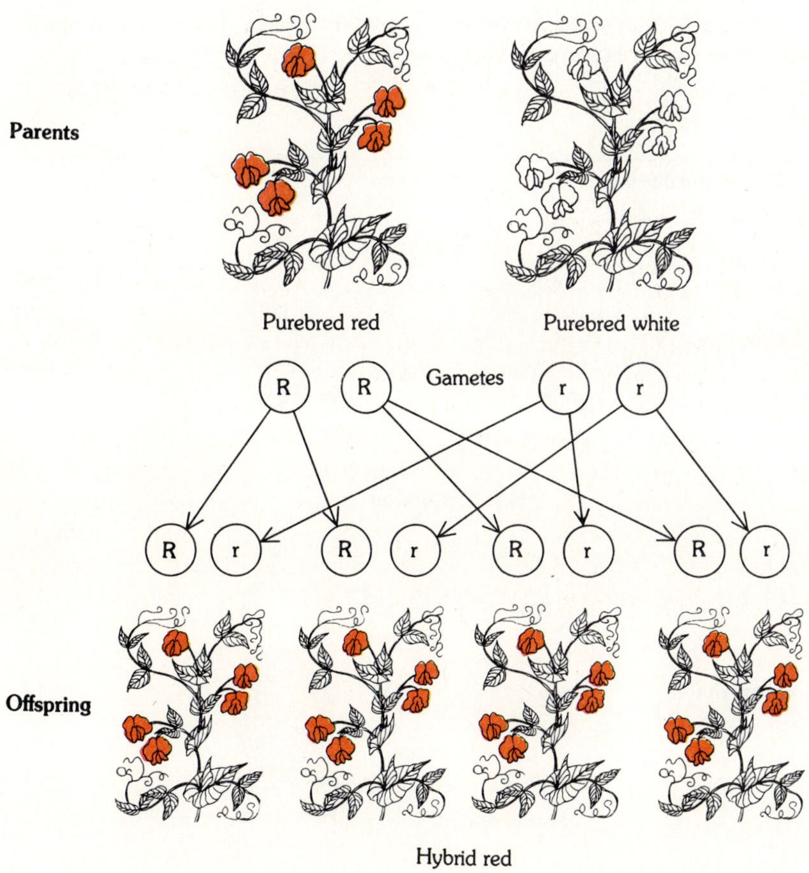

Mendel's Rule of Dominance

Mendel wanted to test his ideas. He began by crossing garden peas that were purebred for red flower color (RR) with plants that were purebred for white flower color (rr). Mendel found that all the offspring in the first, or **F_1 generation,** had red flowers. The flowers were red, even though each offspring was hybrid for flower color (Rr). In other words, each offspring received a gene for red color (R) from one parent and a gene for white color (r) from the other parent. Mendel concluded that the gene for red color is **dominant** over the gene for white color, since only red flowers appeared in the first generation. Mendel also concluded that the gene for white color is **recessive** because white flowers *did not* appear in the first generation.

Mendel's **Rule of Dominance** can be stated as follows: *When an individual is hybrid for a certain trait, only one of the two genes expresses itself. The gene that expresses itself is the dominant gene.*

Note that the symbol for the dominant gene is written as a capital letter and the symbol for the recessive gene is a small letter. Thus, the dominant gene for red flower color is written as *R.* The recessive gene for white flower color is written as *r.*

Rule of Dominance

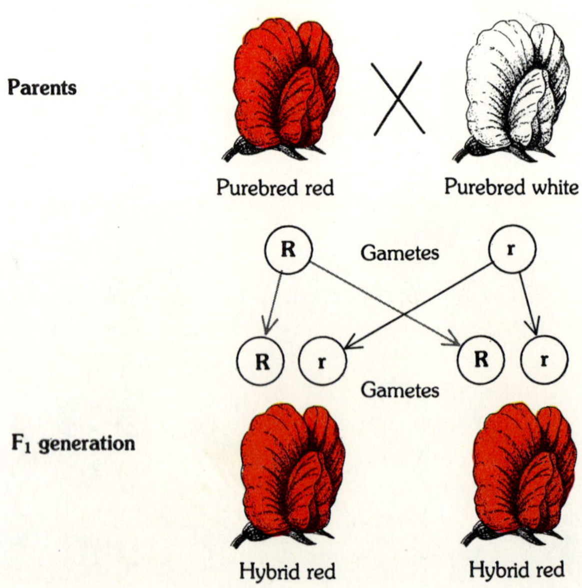

Mendel's Rule of Segregation

Mendel did not observe any white flowers in his first generation garden peas. So he wondered what had happened to the gene for white flower color. Mendel guessed that the *two* genes for flower color separate, or **segregate,** when sex cells form.

In the case of hybrid pea flowers (Rr), each F_1 plant would have one gene for red color (R) and one gene for white color (r). These two genes would separate from each other during the formation of sex cells. About one-half of the sex cells would receive a gene for red color. The other half would receive a gene for white color. When male and female sex cells unite, the offspring would inherit any one of four combinations for flower color: RR (purebred red), Rr (hybrid red), rR (hybrid red), or rr (purebred white). Because the dominant R gene appears in three groups of offspring, 75 percent of the flowers should be red. The recessive rr gene appears in only one group. So 25 percent of the flowers of the offspring should be white.

Monohybrid Cross

Mendel tested his idea by crossing two plants that were hybrid only for red flower color (Rr × Rr). Plants with one hybrid trait are called **monohybrids** (mon′ə hī′bridz). After examining the offspring of the second, or **F_2 generation,** Mendel found that approximately 75 percent of the plants did have red flowers. And approximately 25 percent had white flowers. A 75 percent to 25 percent ratio can also be written as a 3 to 1 ratio. This 3 to 1 ratio indicated that the genes separated the way Mendel thought they did.

Mendel's **Rule of Segregation** can be stated as follows: *The two genes that determine a particular trait will separate when sex cells form. Half the sex cells will receive one gene from each gene pair, the other half will receive the other gene of the pair.*

Rule of Segregation

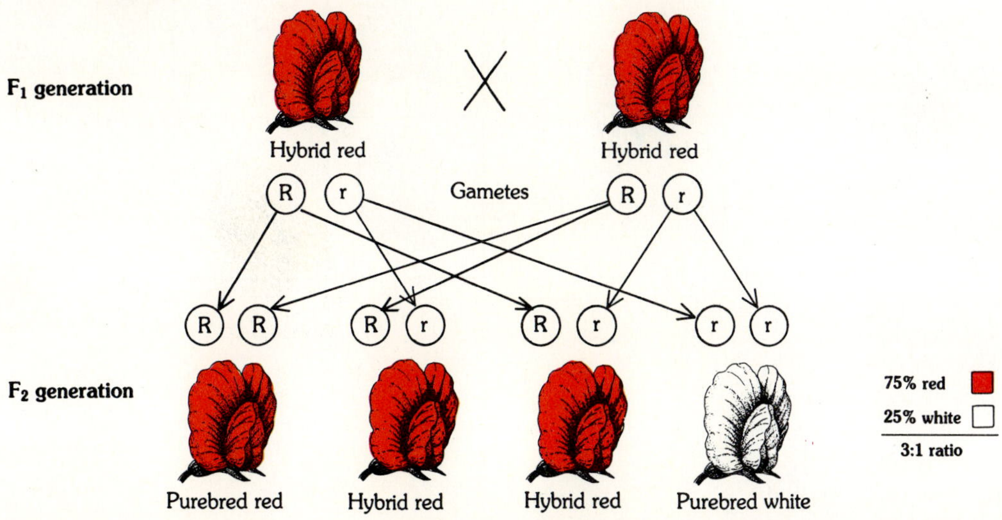

Punnett Squares

One way to picture how genes from each parent separate and recombine is to make a **Punnett** (pun′it) **square.** To make a Punnett square, do the following:

1. Select the trait to be studied.
2. List all gene possibilities for the female sex cells along the top of the square.
3. List all gene possibilities for the male sex cells along the side of the square.
4. Fill in the square by placing one male and one female gene possibility in each box of the square. When you are finished, each box will contain a male-female gene combination.

The finished Punnett square below shows the expected ratio of offspring from a monohybrid cross. The square shows the results of a cross between two garden peas that are hybrid for red flower color (Rr × Rr). As you can see, there is a 75 percent chance to get red flowers (Rr, rR, RR). And there is a 25 percent chance to get white flowers (rr). These percentages can also be written as a 3 to 1 ratio of red to white flowers.

Monohybrid Cross

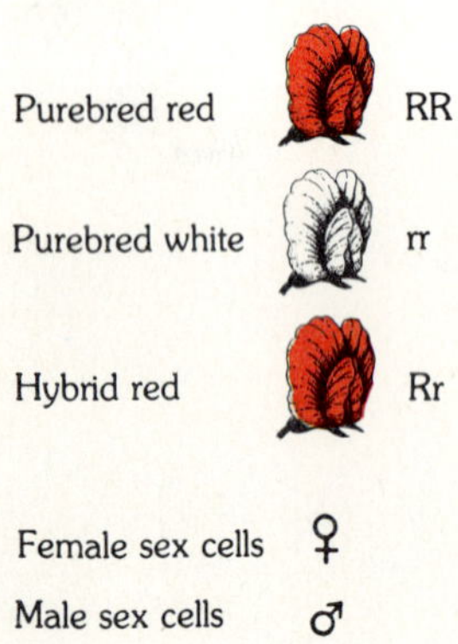

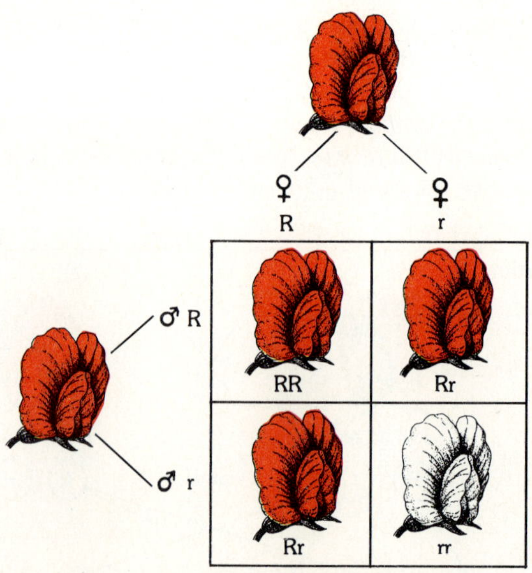

Mendel's Rule of Independent Assortment

Mendel studied seven traits in garden peas. In each case, he found there was a dominant and a recessive gene (factor). For example, he used varieties in which yellow seeds (Y) were dominant over green seeds (y). And round seeds (R) were dominant over wrinkled seeds (r).

Mendel wondered whether the inheritance of genes (factors) for one trait influenced the inheritance of genes (factors) for another trait. For example, is the inheritance of yellow seeds affected by the inheritance of round seeds? To find out, Mendel crossed a pure strain of pea plants that had round, yellow seeds (RRYY) with a pure strain of plants that had wrinkled, green seeds (rryy).

The offspring of this cross had round, yellow seeds. Based on his previous results, Mendel suspected that the offspring were hybrids. Their genetic make-up was probably RrYy. Plants with *two* hybrid traits are called **dihybrids** (dī hī'bridz).

The Formation of Dihybrid Offspring

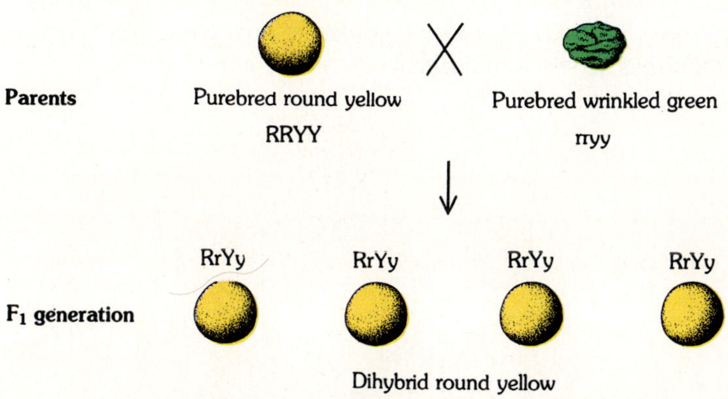

Dihybrid Cross

Mendel crossed two plants that were hybrid for seed color and seed shape (RrYy). He wanted to see if the inheritance of one trait was influenced by the inheritance of a second trait. Each of his parent plants had the genes RrYy.

If the genes for seed color and shape stayed together, each parent could produce only two kinds of sex cells: RY or ry. However, if the genes for seed color and shape sorted independently, each parent could produce four kinds of sex cells: RY, Ry, rY, or ry. Thus, sixteen different combinations of traits would be possible. You can see the possible gene combinations by studying the Punnett square below.

After making the dihybrid cross, Mendel sorted and counted the seeds that were produced. He found 315 round, yellow seeds; 101 wrinkled, yellow seeds; 108 round, green seeds; and 32 wrinkled, green seeds. The offspring fell into about a 9:3:3:1 ratio for seed color and type. Mendel expected such a ratio if the inheritance of seed shape was not affected by the inheritance of seed color. Since the actual experimental results were close to the results he expected, Mendel concluded that genes for different traits separate independently of one another.

Mendel's **Rule of Independent Assortment** can be stated as follows: *The inheritance of genes that control one trait is not affected by the inheritance of genes that control a different trait. Genes that control different traits separate independently of one another.*

Dihybrid Cross

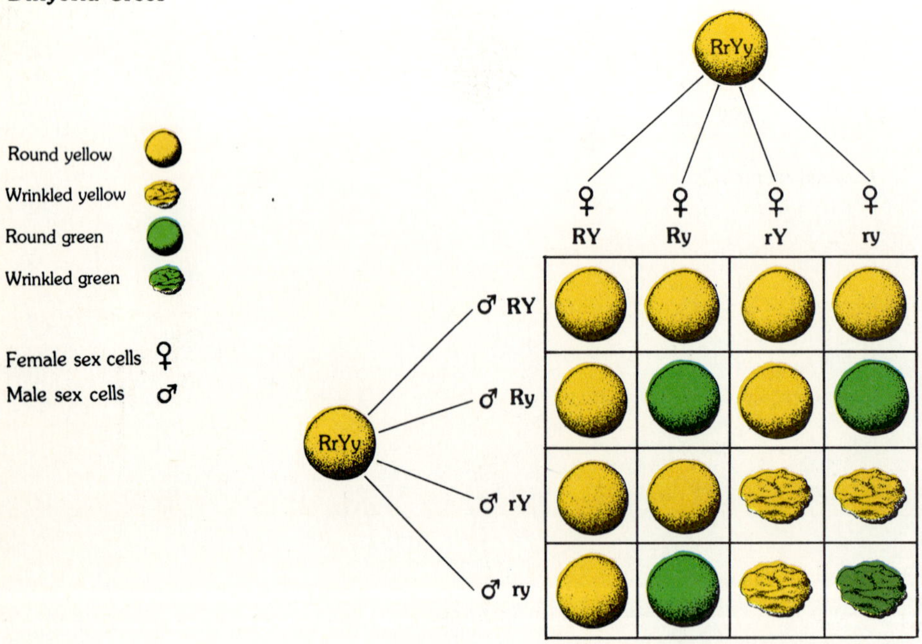

Incomplete Dominance

Today, we know that there are many exceptions to Mendel's Rule of Dominance. For example, red four o'clock flowers (RR) can be crossed with white four o'clocks (rr). However, the flowers of the first generation are a *blend* of red and white (Rr). They appear *pink*. When each of two genes in a gene pair expresses itself, the condition is called **incomplete dominance.**

If the hybrid pink flowers are crossed (Rr × Rr), the offspring of the F_2 generation would appear in percentages like those shown in the diagram below. One-half would be hybrid pink, one-fourth pure red, and one-fourth pure white. Note that the trait for pure white and for pure red flowers shows up in the F_2 generation. This proves that the genes did not mix in some way to produce pink flowers.

Challenge!
Make a Punnett square to show how flower color would be distributed in the F_2 generation if hybrid pink four o'clocks and purebred red four o'clocks are crossed (Rr × RR). What percentage of the offspring would be pink? What percentage would be red or white? How could these percentages be expressed as ratios?

Incomplete Dominance in Four-O'-Clocks

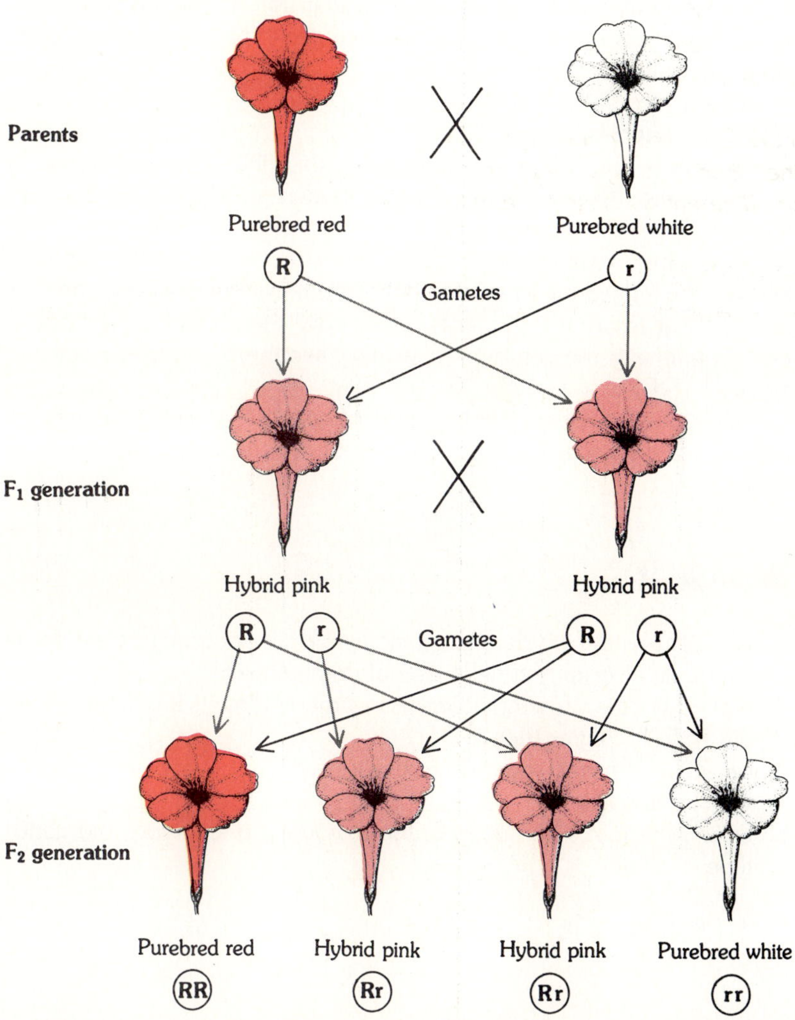

The Language of Genetics

Since the time of Mendel, much progress has been made in the science of genetics. As a result, a special language has developed to describe the many ideas in this field. For example, you have already learned that Mendel's unit factors of heredity are now called genes. However, there can be two forms of the same gene. The different forms of the same gene are called **alleles** (ə lēlz'). In garden peas, the two alleles for flower color are red (R) and white (r).

Alleles (genes) for flower color

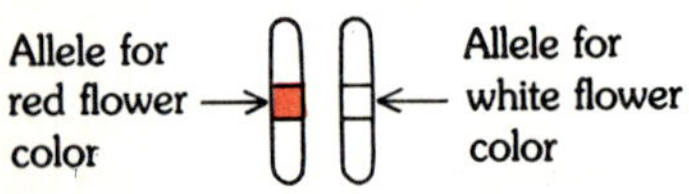

Pair of alleles homozygous for flower color

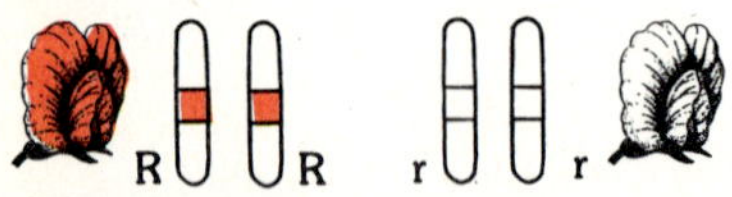

Pair of alleles heterozygous for flower color

Genotypes for flower color

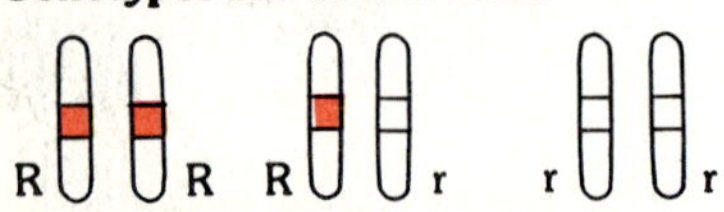

Phenotypes for flower color

Homozygote and Heterozygote

Today a purebred is called a **homozygote** (hō'mə zī'gōt). A homozygote has a pair of *identical* alleles for some trait. For example, in garden peas, a homozygote for red flower color has two identical alleles for red color (RR). A homozygote for white flower color has two identical alleles for white color (rr).

Hybrids are called **heterozygotes** (het'ər ə zī'gōts). A heterozygote has two *different* alleles for a particular trait. In garden peas, a heterozygote for red flower color has one allele for red color and one allele for white color (Rr).

Genotype and Phenotype

The **genotype** (jen'ə tīp) of an individual describes the *combination of genes* inherited for a particular trait. In garden peas, an individual can have the following genotype for flower color: RR, Rr, or rr.

The **phenotype** (fē'nə tīp) of an individual describes the appearance of the individual. In garden peas, an individual can have the genotype RR or Rr. But both plants would have the phenotype *red flowers*. A plant with the genotype rr would have the phenotype *white flowers*.

Do you understand the language of genetics? Do you understand how to use a Punnett square? To find out, try the activity on the next page.

Review It

1. What is Mendel's Rule of Dominance? How does incomplete dominance differ from Mendel's idea of dominance?
2. State the Rule of Independent Assortment. How is it different from the Rule of Segregation?
3. What is a Punnett square? How is it used?
4. What is an allele?
5. Define the terms homozygote, heterozygote, phenotype, and genotype.

Activity

Studying Mendelian Inheritance in Humans

The basic rules of Mendelian genetics apply to many human traits. The rules of dominance, segregation, and independent assortment can be used to explain human inheritance. You are going to study a few examples of dominant and recessive human traits in this activity. You will also have a chance to work with Punnett squares and with the vocabulary of genetics.

Materials
Paper and pencil

A. Procedure
Humans have *ear lobes* of two general shapes: attached or free. The gene for free ear lobes (E) is dominant over the gene for attached lobes (e).

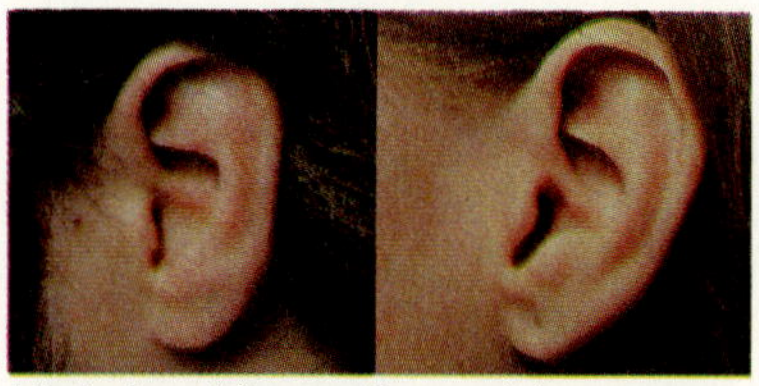

Attached ear lobe　　Free ear lobe

1. Assume that two human parents have attached ear lobes. Determine the genotype of each parent.
2. Make a Punnett square. Fill in the boxes to show all possible genotypes of the offspring of these parents.
3. List the possible genotypes and phenotypes of the offspring.

Analysis
1. What are the two alleles for ear lobes? Which is dominant?
2. What is the genotype of each parent? Are the parents homozygous or heterozygous for attached ear lobes? How do you know?
3. What are the genotypes of the offspring? Phenotypes?
4. What percentage of the offspring are homozygous for attached ear lobes? Explain.

B. Procedures
Humans also differ in *ability to roll the tongue.* Ability to roll the tongue is dominant (T) over inability to roll the tongue (t).

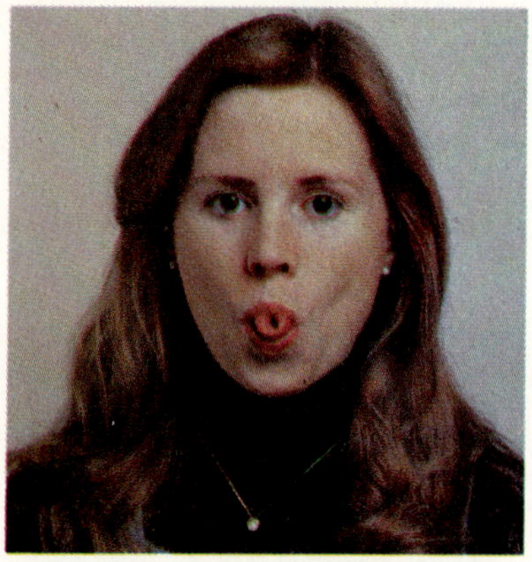

Tongue roller

1. Assume both parents are heterozygous for tongue rolling. Determine the genotype of each parent.
2. Make a Punnett square and show all possible genotypes of the offspring.
3. List the genotypes and phenotypes of the offspring.

Analysis
1. What is the genotype of the parents? How is their genotype different from the genotype of the parents with attached ear lobes?
2. How many different genotypes do the offspring have? How many phenotypes?
3. What is the ratio of tongue rollers to the others?
4. What proportion of children would be homozygous for tongue rolling? What proportion would be heterozygous?

Challenge!
Make a Punnett square to show a dihybrid cross for the two traits you have studied. Assume the parents are heterozygous for free ear lobes and for tongue rolling (EeTt × EeTt).

What are the four possible sex cell combinations for each parent? What are the possible genotypes and phenotypes of the children? What is the ratio of possible phenotypes? Ratio of genotypes?

6–2
Chromosome Theory of Heredity

Chromosomes dividing and separating during cell division

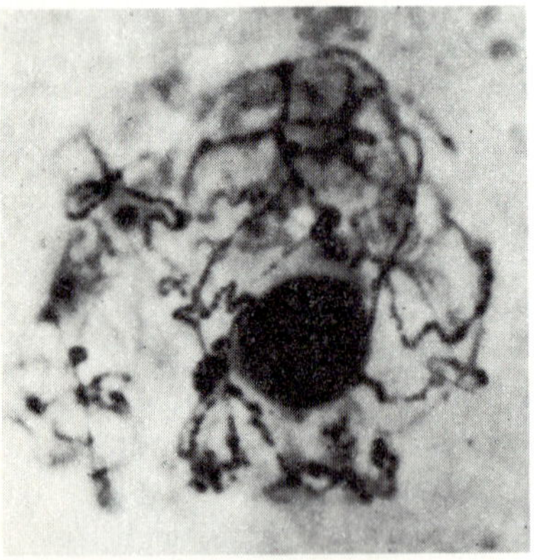

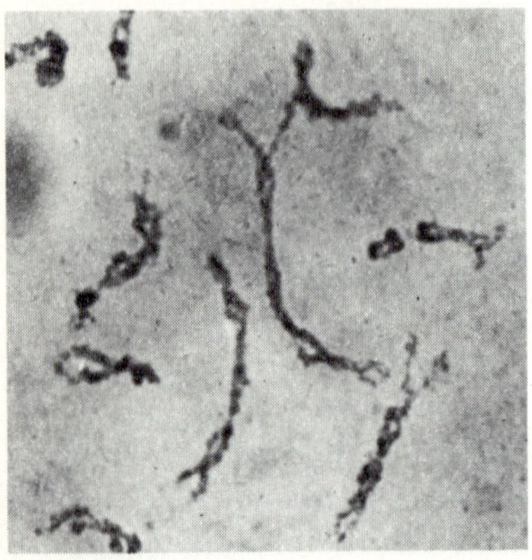

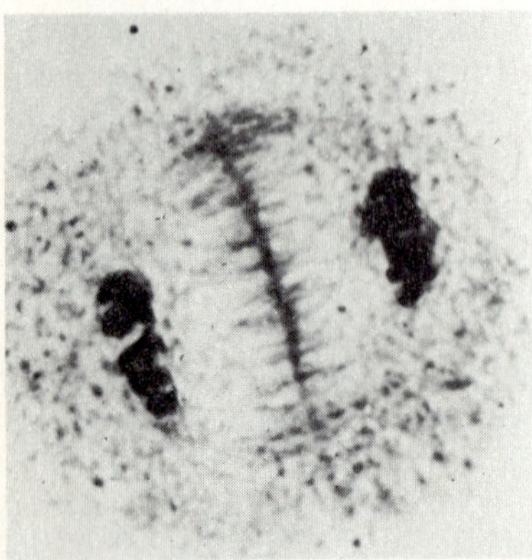

Mendel described the basic rules of heredity without really knowing what was happening inside the cells of his garden peas. He guessed there were factors, or genes, that controlled the traits of the garden peas. However, Mendel did not know exactly where these factors were located. And he did not know how the factors were passed from parent to offspring.

In 1902, Mendel's paper was read by Walter S. Sutton. Sutton provided the first clue as to where Mendel's factors were located. At the time, Sutton was a student at Columbia University in New York. He was observing chromosomes in the cells of grasshoppers. Sutton and others were trying to learn more about what chromosomes do during regular cell division and during the formation of sex cells.

Sutton noticed that the behavior of chromosomes during meiosis (sex cell formation) was much like the behavior of Mendel's factors. He observed the following similarities:

1. Chromosomes occur in pairs.	Mendel's factors occur in pairs.
2. Paired chromosomes separate during meiosis. Each sex cell receives only one chromosome of the pair.	Mendel's paired factors separate. Each sex cell receives only one factor of the pair.
3. Chromosome pairs separate independently of other chromosome pairs during meiosis.	Mendel's factor pairs separate independently of other factor pairs.
4. Individual chromosomes return to their original paired numbers in the fertilized egg.	Mendel's individual factors become paired factors in the offspring.

Putting all this together, Sutton guessed that Mendel's factors must be carried on the chromosomes. The factors, or genes, are then passed to the offspring on chromosomes during fertilization. Sutton's findings formed the basis for the **chromosome theory of heredity.**

Gene Linkage

Sutton also realized that there must be *more than one gene* on each chromosome. When Mendel ran his experiments, he studied only seven traits of garden peas. Each trait was controlled by a different gene. And each gene must have been located on a *different* chromosome. But garden peas obviously have more than seven traits. So Sutton suggested that each chromosome is made up of many genes that are linked together somewhat like a beaded necklace. The idea of a string of genes on the same chromosome is known as **gene linkage.**

Sutton recognized that linked genes on the same chromosome cannot separate independently during meiosis. They must stay together. So Mendel's Rule of Independent Assortment had to be changed slightly. Mendel's rule is only true when genes are located on *different* chromosomes. This change in Mendel's rule shows how scientific models can change as more facts are gathered.

Giant fruit fly larva chromosomes

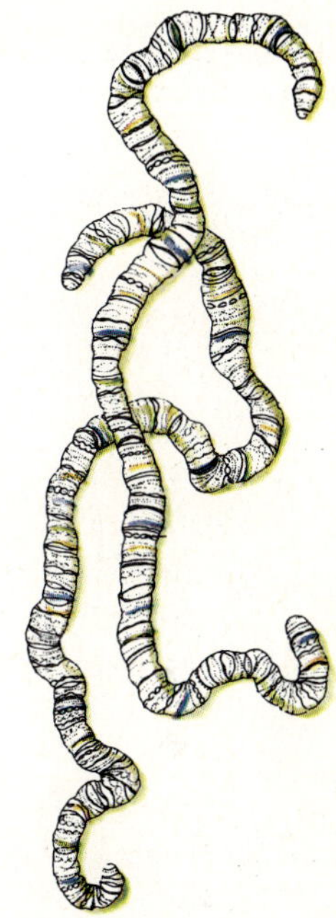

Crossing Over

Studies have shown that linked genes on the same chromosome can sometimes separate. You learned in Chapter 5 that this process is called *crossing over.* Crossing over occurs at prophase during the first part of meiosis. Sometimes, the chromosomes twist around each other. Parts of each chromosome may break off and become attached to the other chromosome. In this way, each chromosome receives some genes that are different from the ones it originally carried.

The advantage of crossing over is that it provides new combinations of genes in male and female sex cells. For example, suppose the genes for blue eyes and blond hair were linked on the same human chromosome. Also suppose the genes for dark eyes and dark hair were linked on the matching chromosome. When sex cells form, they would either have genes for blue eyes and blond hair—or for dark eyes and dark hair. However, if crossing over occurs, new combinations would be possible. Some sex cells would receive genes for blue eyes and dark hair. Others would receive genes for dark eyes and blond hair.

Crossing over

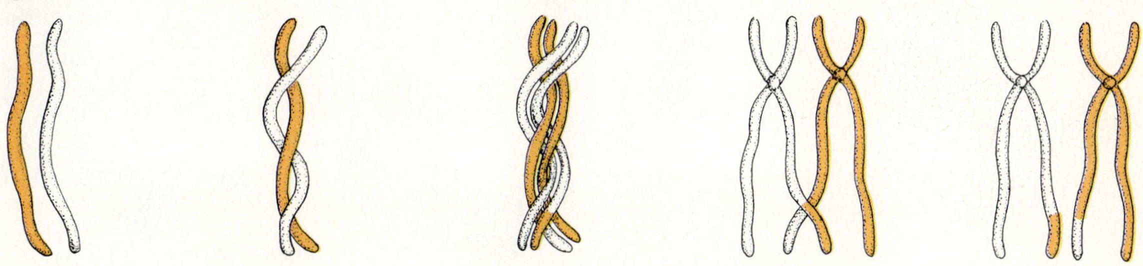

1. Chromosomes pair. Each is double. (Doubleness does not show.)
2. Chromosome pairs wind around each other
3. Chromosomes break and exchange equal amounts of chromosome material (Doubleness is shown.)
4. Chromosomes move away from each other.
5. Chromosomes separate, each with new genetic information

Morgan's Chromosome Studies

After Sutton's discovery of Mendel's paper, Columbia University became the center for research in genetics. An important researcher in this laboratory was Thomas Hunt Morgan. At the very start of his experiments, Morgan made a major decision. Rather than use garden peas as the experimental organism, Morgan chose the fruit fly. The fruit fly has several advantages over garden peas for genetic research.

1. The fruit fly has only four pairs of chromosomes.
2. Certain cells in the salivary glands of fruit flies have *giant chromosomes*. They are especially easy to see because they are 200 times larger than chromosomes in other cells. These giant chromosomes have been useful in finding the location of genes on the chromosome.
3. Female and male fruit flies are easy to tell apart. And females hatch earlier than males. So it is easy to remove females and keep them separate from the males. This enables the researcher to select the parents and the traits to be studied.
4. Fruit flies reproduce quickly. Adults appear just two weeks after the eggs are laid. Many generations can be raised and their traits studied in just a few months.
5. Thousands of flies can be kept in small bottles that take up very little space.

Giant fruit fly chromosomes

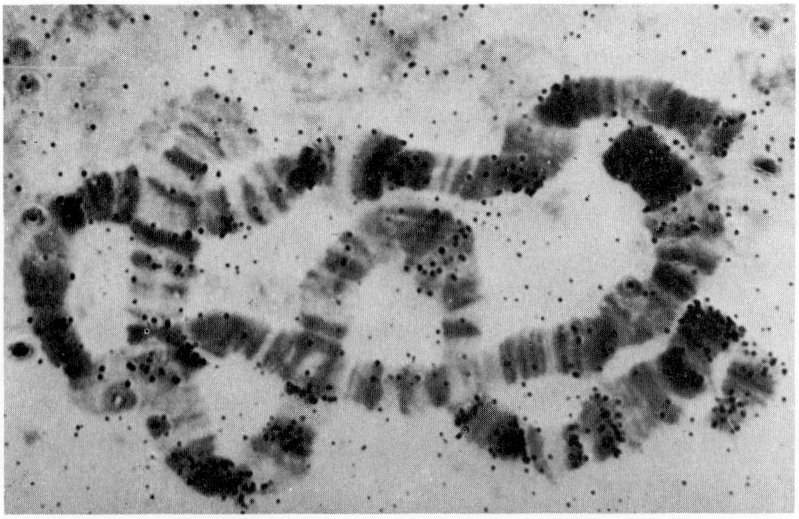

Sex Chromosomes

Morgan could clearly see the chromosomes of the fruit fly with a simple microscope. He saw eight chromosomes. Six chromosomes could be arranged into pairs so that each member of the pair was the same size and shape as the other member. These chromosome pairs are now known as **autosomes** (ô'tə sōmz). However, the remaining two chromosomes *did not* match. These are known as **sex chromosomes.** Some of the genes on the sex chromosomes determine whether the fly will be male or female.

In the female fruit fly, the two sex chromosomes appear to be the same. Each is called an X chromosome. The female is described as being XX. However, only one of the sex chromosomes in the male looks like the X chromosome in the female. The other chromosome has a hook at one end. This is called a Y chromosome. Males are described as being XY.

When sex cells are produced in females, each egg cell receives one X chromosome. In males, each sperm cell receives either an X chromosome or a Y chromosome. If a sperm with an X chromosome fertilizes an egg, which has an X chromosome, the offspring will be a female (XX). If a sperm with a Y chromosome fertilizes an egg, the offspring will be a male (XY). Therefore, the male (XY) parent determines the sex of the offspring.

The sex of mammals is also determined by X and Y chromosomes. In humans, the X chromosome is larger than the Y chromosome. Human females have the sex chromosomes XX. Males have the sex chromosomes XY.

Fruit fly chromosomes

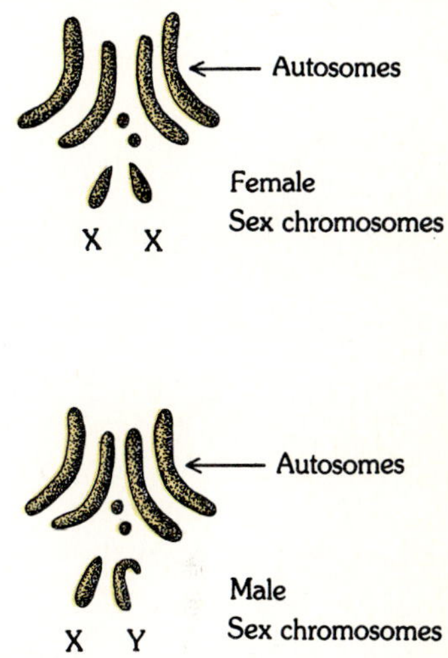

Inheritance of Sex in Fruit Flies

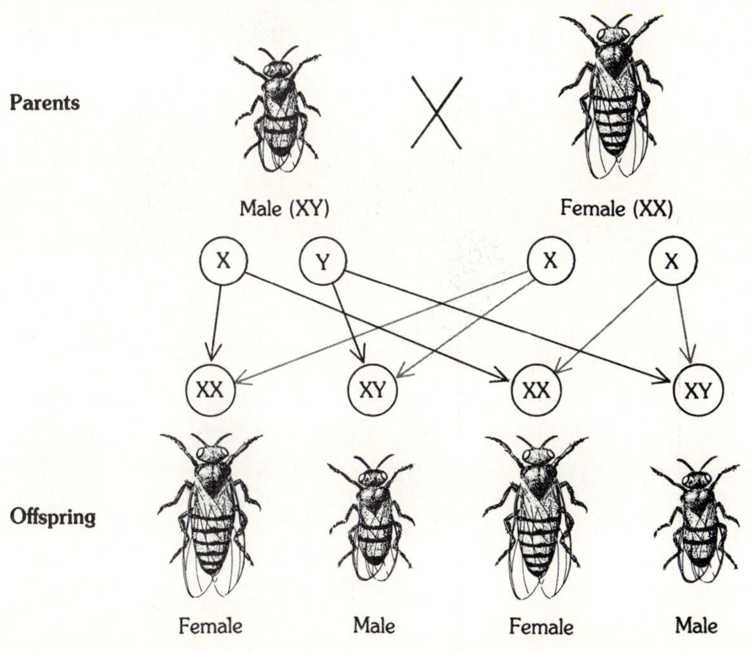

Sex Linkage

Walter Sutton had suggested that chromosomes contain genes, or the hereditary information that is passed from parent to offspring. But Sutton did not experiment to show that his idea was correct. However, Thomas Hunt Morgan did many experiments with fruit flies to show that chromosomes carry the genes that determine traits.

In one experiment, Morgan crossed a white-eyed male fly with a red-eyed female fly. The F_1 generation offspring of this cross had red eyes. When the hybrid males and females of the F_1 generation were crossed, about 75 percent of the F_2 generation had red eyes. The other 25 percent had white eyes. At first glance, this appeared to be an example of Mendelian segregation. Red eye color seemed to be a dominant trait. White eye color seemed to be recessive.

Now take a closer look at the F_2 generation. Notice that *only males have white eyes.* None of the females have white eyes! Inheritance of the gene for eye color in fruit flies does not appear to follow Mendel's Rule of Segregation.

Inheritance of a Sex-linked Trait

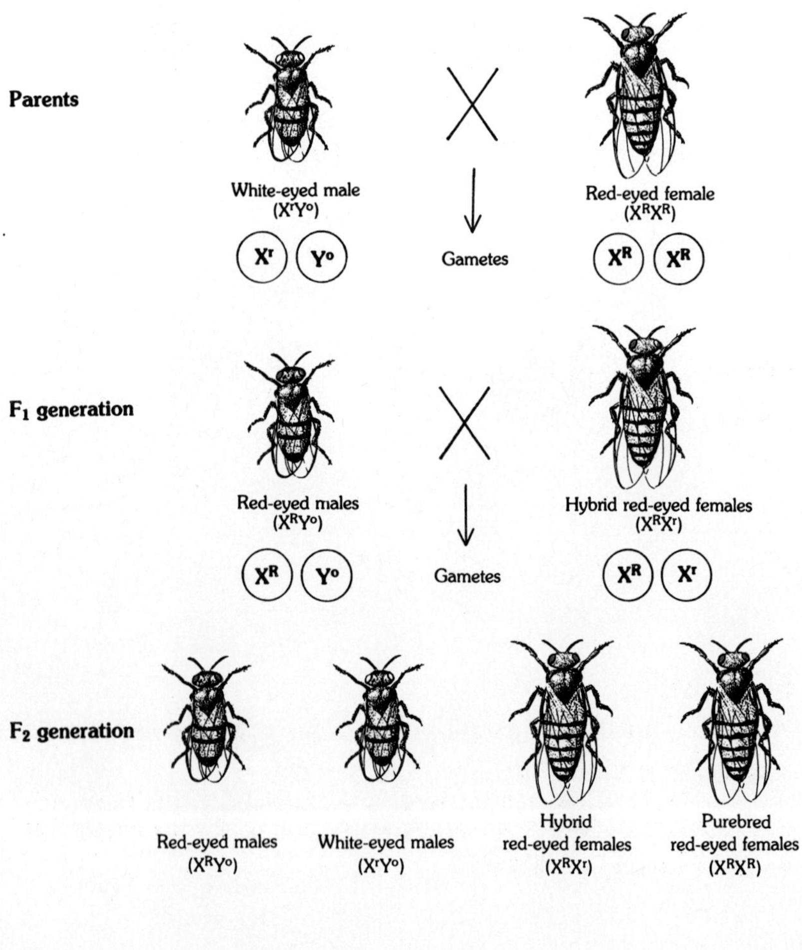

Test Cross

Morgan guessed that the trait for eye color was sex linked. That is, the genes for eye color were carried only on the X chromosomes. Genes for eye color were not carried on the Y chromosome. Therefore, the color of the female's eyes would be determined by two genes, one on each X chromosome. The male's eye color would be determined by only one gene located on the single X chromosome.

To test his idea, Morgan crossed white-eyed males (X^rY^o) with heterozygous, red-eyed females (X^RX^r). If the eye color gene were on the Y chromosome, all the males would be white eyed. All of the females would be red eyed. However, if the eye color gene were only on the X chromosome, half the female offspring would be white eyed. And half of the males would also be white eyed. In the results, about half the females produced by Morgan's test cross were white eyed. And about half the males were white eyed, as well. You can see the results by studying the diagram below.

Test Cross for Eye Color

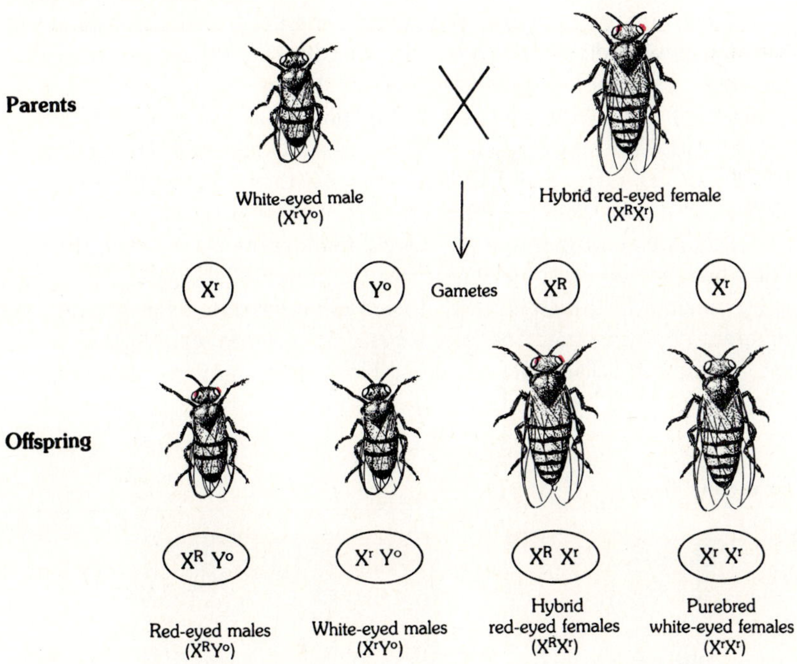

Review It

1. Describe two ways that Sutton's discoveries added information to the science of genetics.
2. How does crossing over provide new combinations in the sex cells?
3. How do the X and Y chromosomes determine the sex of fruit flies?
4. What is a sex-linked trait?

6–3 Human Genetics

Challenge!
Choose one human trait from the table. Find out as much as you can about how this trait is inherited. You may need to check reference books on human genetics.

Inherited Human Traits

Dominant	Recessive
Dark hair	Blond hair
Brown eyes	Blue, gray eyes
Hazel, green eyes	Blue, gray eyes
Nearsightedness	Normal vision
Farsightedness	Normal vision
Free ear lobes	Attached ear lobes
Long eyelashes	Short eyelashes
Six fingers	Five fingers

Since the beginning of the 20th century, many new things have been learned about genes. These findings have added to Mendel's ideas. They have also enriched the science of genetics. Many recent discoveries have helped us to better understand how *humans* inherit traits.

Mendelian genetics assumes that one gene controls one trait. And each trait occurs in at least two different forms. For example, human ear lobes are either attached or free. And humans either have the ability to roll their tongues or they do not.

Today we know of over a thousand human traits that are inherited strictly according to Mendelian rules. Many of the traits we know about are responsible for disease. Others are not important to human health or behavior as far as we know. The table in the margin lists a few human traits that are inherited. Some are controlled by a single gene. Others are controlled by many genes.

Many Genes, One Trait

Human traits, such as height, weight, intelligence, and skin color are determined by many genes. When more than one gene controls a single trait, the condition is called **polygenic** (pol′ē jēn′ik) **inheritance.** In the case of skin color, geneticists believe that as many as six sets of genes may control the trait. Thus, the lightest skin color would have the genotype aa bb cc dd ee ff. The darkest would have the genotype AA BB CC DD EE FF. An individual that is totally heterozygous for skin color would have the genotype Aa Bb Cc Dd Ee Ff.

You can see the range of skin colors possible by studying the skin color tiles below. Another way to see the range of skin colors possible is by making a Punnett square. If each gene for skin color was on a different chromosome, how many gene possibilities would there be for the female sex cells? Male sex cells? How many gene combinations for skin color would be possible in the offspring?

Skin color tiles

One Gene—Many Traits

Sometimes a single gene may influence more than one trait. This condition is known as **pleiotropy** (plī ot'rə pē). For example, in Marfan's syndrome, one gene causes abnormally long legs, fingers, and toes. Often, people with Marfan's syndrome have heart trouble. And the lenses of the eyes are in abnormal positions. Another example is the gene responsible for sickle cell anemia. The sickle cell gene affects such organs as the blood, bones, kidney and brain.

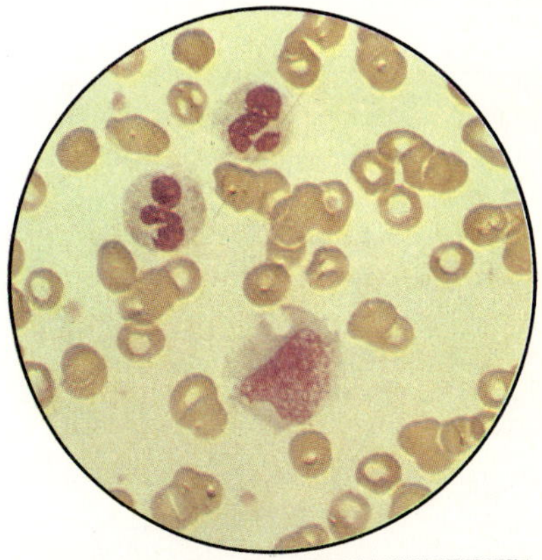

Normal red blood cells

Sickle Cell Anemia

You will recall from Chapter 5 that sickle cell anemia is a disease caused by a recessive gene. Its main characteristic is fragile red blood cells that collapse into a sickle shape when the oxygen concentration is low. These cells cannot carry enough oxygen to the body's cells. The sickle-shaped cells also tend to clump together in the small blood vessels. This cuts off the blood supply to a particular part of the body. Severe pain in the joints, kidney failure, heart failure, and blindness can result.

If a person is *homozygous* for sickle cell anemia, all of the red blood cells are affected. Such people usually do not survive very long. However, if a person is *heterozygous* for the condition, the red blood cells are able to carry enough oxygen to meet the body's requirements. Such people are generally healthy. But they are carriers of the disease, since they can transmit the sickling gene.

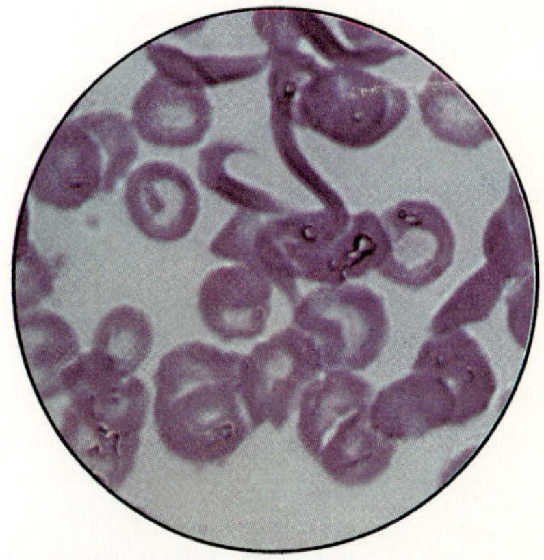

Sickled red blood cells

Early detection of the disease and proper care can help to keep the condition stable. Scientists are working with a chemical that has been found to prevent the sickling of human blood cells in test tubes. But more work must be done before the chemical can be widely used on human patients.

The heterozygous condition does have one advantage in certain parts of the world. People who are heterozygous for sickle cell anemia are resistant to malaria. So the trait tends to be maintained among people in parts of Africa where malaria still occurs.

Polygenic effect

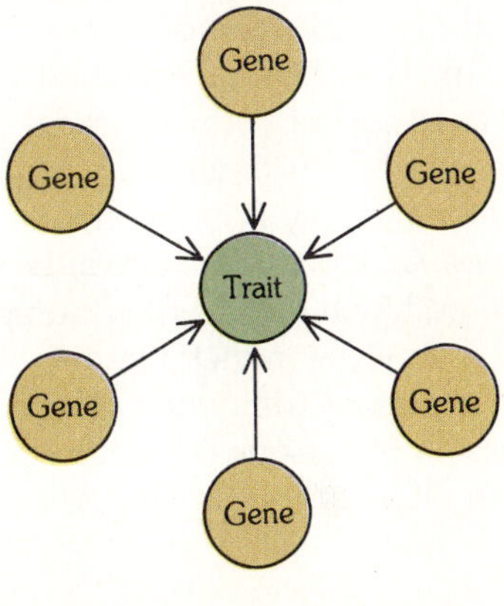

Pleiotropic effect

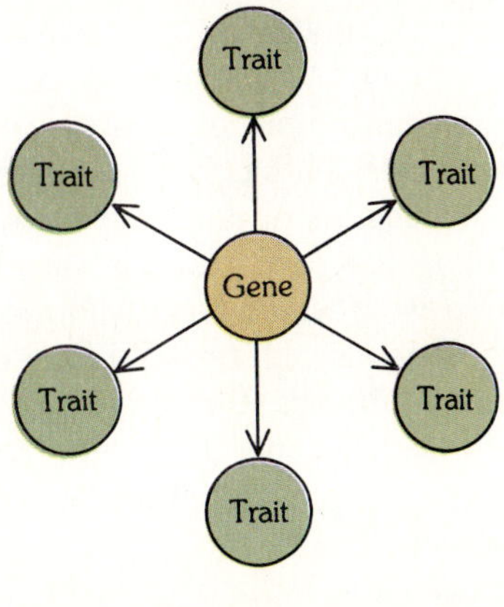

Challenge!
Under emergency conditions, a transfusion might have to be given without knowing the blood type of the person receiving the blood. Which of the blood types in the chart would not cause blood cells to clump if mixed together?

Chart showing which major blood types can be mixed safely

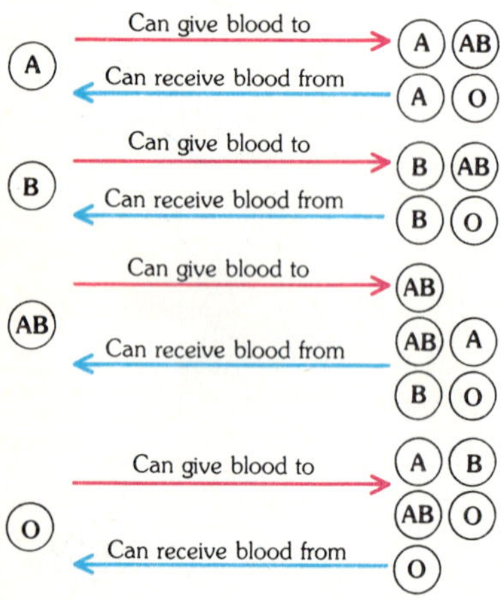

Multiple Alleles

So far we have assumed that at any one place on a chromosome there are only two possible alleles for each trait. For example, garden peas can have red (R) or white (r) flowers. And their seeds can be yellow (Y) or green (y). Geneticists now know that at many locations there can be *three or more forms of the same gene.* In sweet clover, there are 200 alleles that prevent flowers from fertilizing themselves. And in cattle, there are 250 forms of the gene that determines a blood group. The term **multiple allele** is used to describe the condition in which each of three or more alleles can produce a distinct trait.

Human Blood Types

Human blood type is determined by the inheritance of two of the different alleles: A, B, and O. An allele is inherited from each parent. And the two inherited alleles determine whether the blood type will be A, B, AB, or O. The diagram below shows you how alleles combine to determine blood type.

How alleles combine to determine blood types

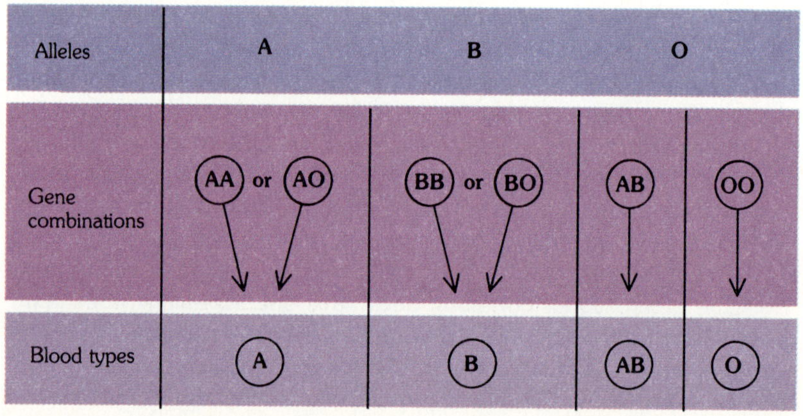

The alleles for blood type control the kind of chemical, called an **antigen** (an′tə jən), that appears on the surface of the red blood cell. Allele A causes the production of antigen A. Allele B causes the production of antigen B. And allele O causes the production of *no* antigen. In the blood plasma of one's body exists a chemical called **antibody** (an′ ti bod′ ē). It reacts with any antigens not normally found on one's own red blood cells. For example, a person with type A blood has antigen A on their red blood cells, and antibody B in the blood plasma. If type B blood is mixed with type A blood, the antibody B in the plasma would react with the type B red blood cells causing clumping. Persons with type B blood contain antibody A in their plasma. Both antibodies A and B are produced in persons with type O blood. And neither antibody is found in the plasma of persons with type AB blood. The chart on the left shows which blood types can be safely combined.

Sex Linkage in Humans

The genes that determine whether an individual will be a male or a female are found on the sex chromosomes. However, some of the genes on the sex chromosomes determine traits that are not related to the sex of the individual. A trait controlled by genes on the sex chromosomes is called a **sex-linked trait.** Almost every sex-linked trait is found on the X chromosome rather than on the Y chromosome. One example of a sex-linked trait you have already studied is red and white eye color in fruit flies.

Red-Green Color-Blindness

Red-green color-blindness is an example of a sex-linked trait found in humans. People with this trait cannot see the colors red or green. About eight percent of the males in the United States have the trait. However, color-blindness is rare in females. The way color-blindness is inherited explains why more males than females are red-green color-blind.

Challenge!
Two parents produce a female child who has red-green color-blindness. What is the genotype of each parent?

The allele for normal color vision is dominant (C). It is paired with the recessive allele for red-green color-blindness (c). Both alleles are located only on the X chromosome. The alleles are not found on the Y chromosome.

Therefore, if a male child inherits a dominant allele for normal vision, he will have normal vision. However, a male child who inherits a recessive allele for color-blindness will be color-blind. He will have the trait because no allele for normal vision can be carried on the Y chromosome.

A female child will be color-blind only if she inherits two recessive alleles for color-blindness. But she will have normal vision if the dominant allele for normal vision occurs on either one of her X chromosomes. Therefore, females are less likely to show a sex-linked recessive trait than are males.

The diagram shows you how color-blindness is inherited in males and females.

Inheritance of red-green color blindness

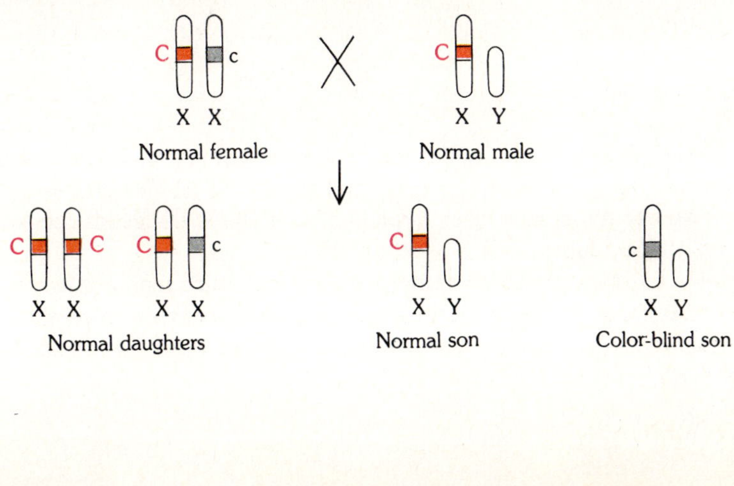

Color-Blindness Charts

The charts below are used to find out if a person is suffering from different kinds of color-blindness. If you have normal color vision, you will see a 15 in the large circle, a 97 in the top small circle, nothing in the middle small circle, and a 26 in the bottom small circle.

If you have red-green color-blindness, you will see a 17 in the large circle, a 5 in the middle small circle, and nothing in the other circles. If you are red color-blind, you will see only a 6 in the bottom small circle. If you are green color-blind, you will see only a 2 in this circle. And if you are totally color-blind, you will see *no* numbers in any of the circles.

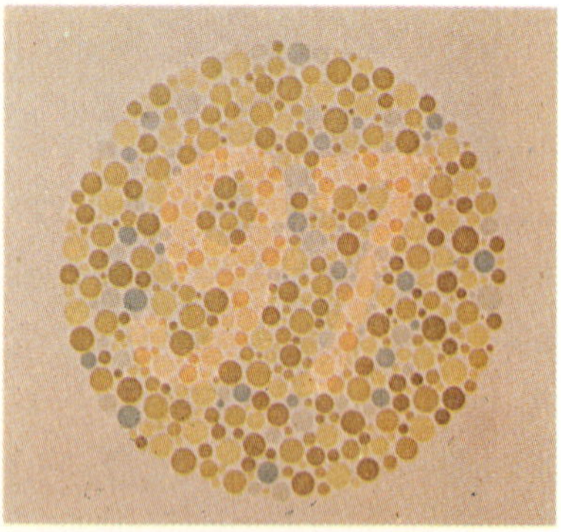

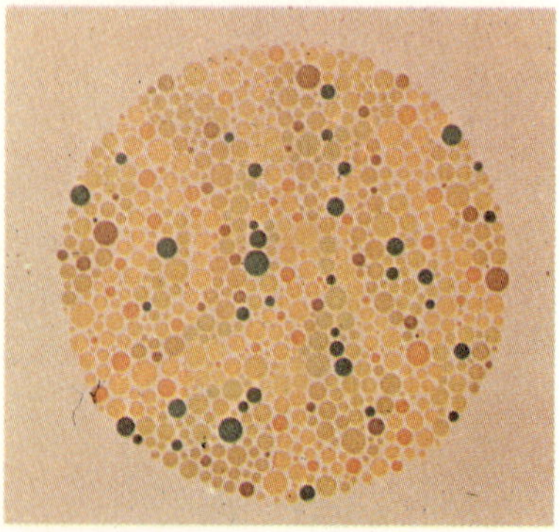

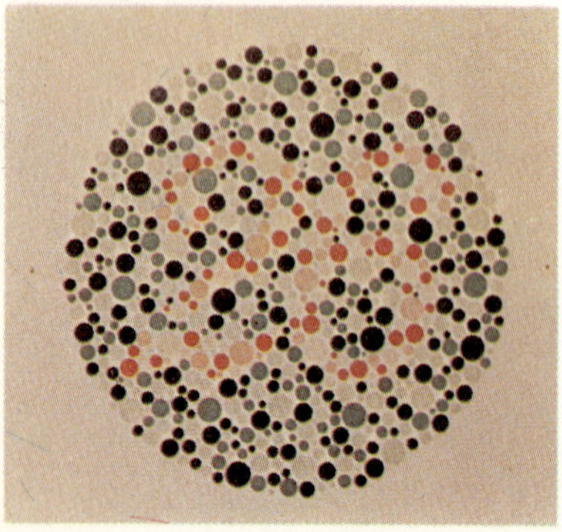

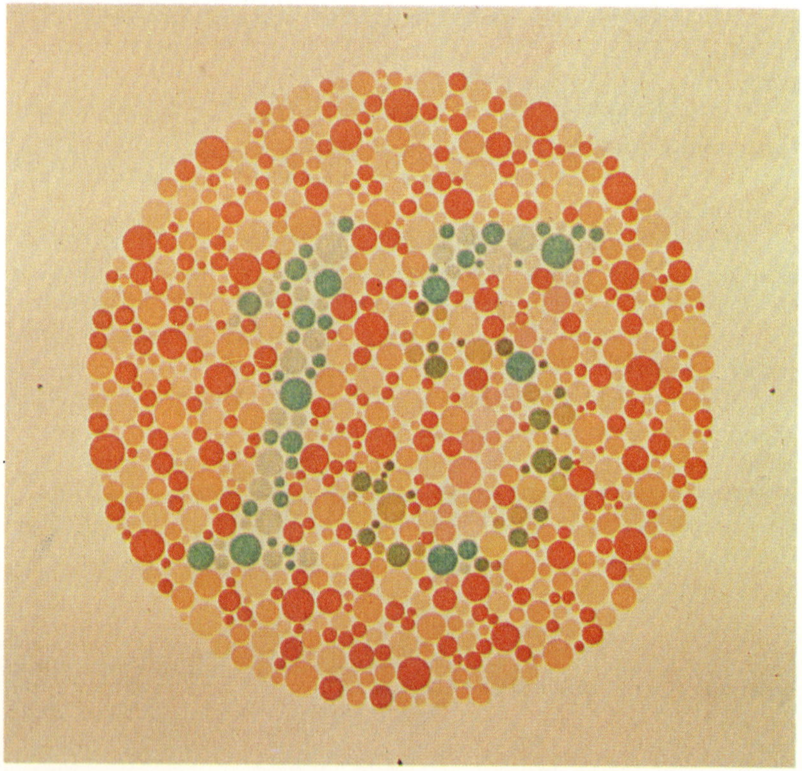

Review It

1. Explain how multiple genes can influence a single trait. Give an example.
2. What are multiple alleles? Explain how multiple alleles influence human blood type.
3. Give an example of how one gene can influence many traits.
4. Are females or males more likely to show a sex-linked recessive trait? Why?

Breakthroughs in Biology

Chromosome Mapping

Today, geneticists can construct **genetic maps** to show the position of genes on a chromosome. To do this, they begin by observing the traits of an organism such as the fruit fly. Then they carefully record the effects of crossing over and chromosome breakage on the traits of the fly. From these observations, they can locate and plot distances between genes on a chromosome.

The most complete genetic maps are those of fruit fly chromosomes. Many traits of this organism have been mapped to specific positions. Opposite, you can see a genetic map of the X chromosome of the fruit fly. Notice the great number of traits that have already been mapped.

Next to the map is a *photograph* of fruit fly chromosomes. The light and dark bands on each chromosome correspond almost exactly with the position of sets of genes on the genetic maps.

Analysis

1. What is a genetic map?
2. Why do you think it might be important to make maps of fruit fly chromosomes before mapping human chromosomes?
3. What value might genetic maps of human chromosomes have in medicine?

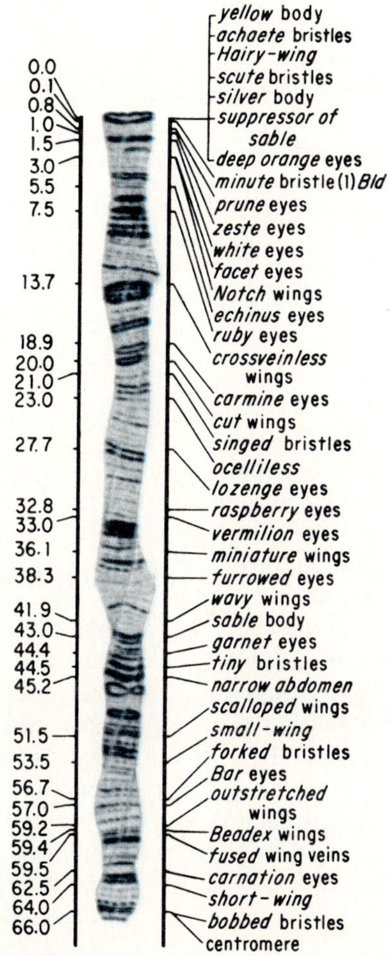

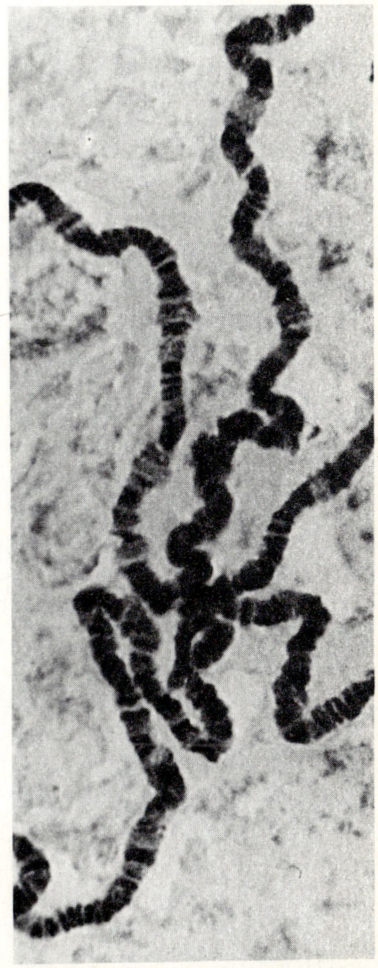

Chapter Summary

Genetics is the science of biological communication between generations. The traits of an organism are to a large extent determined by the DNA molecules that are present in every cell of the body and that come from the parents.

During meiosis, pairs of alleles separate. The alleles then recombine during fertilization. In this way, two alleles for a trait may combine to give the genotypes AA, Aa, or aa. Frequently, one allele is dominant, so individuals with the genotypes AA and Aa appear to be identical for the trait. When pairs of alleles occur on separate chromosomes, each pair will be sorted independently of the other by chance. Factors (genes), dominance, segregation, and independent assortment were discovered by Gregor Mendel.

Since the beginning of this century, much additional information has been learned about chromosomes and genes. For example, genes that are linked on one chromosome of a chromosome pair can be exchanged with genes on the other chromosome of the pair by crossing over. And sex in many species is determined by X and Y chromosomes. Certain traits are located on the sex chromosomes.

Recent discoveries have also helped us to better understand how humans inherit traits. Several patterns of inheritance can occur. In Mendelian inheritance, one gene pair influences one trait. In polygenic inheritance, many genes influence a single trait. And in pleiotropic inheritance, one gene influences many traits. Traits can also be determined by multiple alleles or they can be sex linked.

Questions to Think About

1. In horses, the color black is due to a dominant gene B, and chestnut color to its recessive allele b. The trotting gait is due to a dominant gene T, and pacing to its recessive allele t. A homozygous black trotter (BBTT) is crossed with a chestnut pacer (bbtt).

   What will be the genotypes and phenotypes of offspring from a cross of these horses? If two of the offspring from this cross mate, what would be the genotypes and phenotypes of horses in the next generation? In what ratios would you expect the phenotypes to appear?
2. Sex-linked traits usually occur on the X chromosome. However, some sex-linked traits may occur on the Y chromosome. One example is the gene for "hairy" ears. The trait is common among mature men of southern India.

   Assume "hairy" ears is a dominant sex-linked trait on the Y chromosome. If a man had the trait for hairy ears, how many of his sons would have this trait? Daughters? Explain your answer.

Extra Research

1. Much progress is being made in understanding inherited disorders. As a result, it is possible to diagnose and provide earlier treatment for genetic diseases such as hemophilia, sickle cell anemia, and diabetes. Investigate and report on some of the ways in which genetics is being applied to medical problems today.
2. Interview someone in your community who breeds animals such as dogs, guppies, or horses. You might also interview someone who breeds flowers or agricultural crops. Write a report that explains how the science of genetics is used in breeding plants and animals.
3. Outline the steps that occur during meiosis. Show how a pair of alleles on the chromosomes segregate during meiosis.

Chapter Test

Part I

Copy the number of each test item. If the item is *true*, write *T* after the number; if the item is *false*, write *F*.

1. Genes on the same chromosome are sorted independently when sex cells are formed. (151)
2. Mendel's pea plants had traits that showed incomplete dominance. (142)
3. RR and rr are purebred genotypes. (148)
4. A pea plant with one recessive gene for white color and one dominant gene for red color will have red flowers. (142)
5. A four o'clock plant with one gene for white color and one gene for red color shows incomplete dominance. It will have red flowers. (147)
6. In humans, the sex chromosome inherited from the father determines the sex of the child. (153)
7. Sickle cell anemia is an example of one gene influencing many traits. (157)
8. A sex-linked recessive trait shows up in females more often than in males. (154)

Part II

Copy the number of each sentence below. After each number, write the word or words that complete the sentence correctly.

9. One advantage of using fruit flies for genetic experiments is that fruit flies have only four pairs of ____. (152)
10. Offspring of blood genotype AA × BB parents will have blood type ____. (158)
11. A person with type B blood could have a genotype of BB or ____. (158)
12. A dihybrid cross of parent genotypes RrYy will usually produce offspring with RY, Ry, rY, ry cells in a ____ ratio. (146)
13. Different forms of the same kind of gene are called ____. (148)
14. The phenotype of a pea plant that is heterozygous for flower color (Rr) is ____. (148)
15. Human females have the sex chromosomes ____. (153)
16. Genes that are ____ do not separate independently unless crossing over occurs during meiosis. (151)
17. Unit factors of inheritance are called ____. (141)

Part III

Copy each item from List A. After each item, write the letter and words from List B that best describe the item.

List A

18. crossing over (151)
19. Morgan (154)
20. Mendel (140)
21. phenotype (148)
22. pleiotropy (157)
23. heterozygote (148)
24. genotype (148)
25. homozygote (148)

List B

a. researcher who discovered the basic rules of genetics
b. description of the effects of a gene pair on the appearance of an organism
c. condition in which one gene influences many traits
d. hybrid
e. researcher who used fruit flies to study chromosomes
f. description of the type of gene pair inherited by an organism
g. organism with a pair of identical alleles
h. process by which matching chromosomes exchange genetic material during meiosis

Chapter 7

Gene and Chromosome Changes

Genes on chromosomes carry the messages that the body uses to produce traits. Sometimes, however, traits appear that the parents did not have. This chapter will provide information that explains how changes in genes and chromosomes can cause the appearance of new traits. It will also examine how the appearance of new traits may cause genetic diseases.

The chapter explains some methods that physicians use to detect genetic diseases and the importance of detection in early treatment. The influence of the environment on the way a trait expresses itself is also discussed.

Chapter Objectives

1. Describe how gene mutations differ from chromosome mutations.
2. List three kinds of structural changes in chromosomes.
3. Describe two kinds of mutations that involve changes in the number of chromosomes.
4. Explain two methods scientists use to detect genetic diseases.
5. Describe one treatment for a genetic disease.
6. Explain how gene expression can be influenced by the environment and give an example.

This tiger has a gene mutation that has made him an albino.

7–1 Kinds of Genetic Changes

Where do new genetic traits come from? For example, how can a white-eyed fruit fly suddenly appear in a pure culture of red-eyed flies? Or how can a short-legged sheep appear in a population of sheep with legs of normal length? In the early 1900s, Hugo de Vries attempted to answer this question. At the time, he was working on the genetics of the primrose plant. He noticed that several new traits appeared in the offspring of his plants. He guessed the cause of the new traits to be permanent changes in the genetic material. Such changes are now called **mutations.** To be a true mutation, the trait should be capable of being passed on to all the descendants of the mutated cell. Today, we know that two kinds of mutations exist. One kind is a *gene* mutation, and the other is a *chromosome* mutation.

Gene Mutations

A chemical change that occurs in the DNA of a cell is called a **gene mutation.** Such a mutation may alter the sequence of nucleotides within a part of the DNA molecule. This alteration changes the information on the DNA chain and results in differences in the proteins being produced. These differences, in turn, may affect the way a trait is expressed.

For example, in sickle cell anemia, the mutation of a single gene causes a slight change in the structure of the protein molecule of hemoglobin. Because of that slight change, the blood cell that carries the hemoglobin takes a sickle shape.

Many scientists believe that mutations may be caused naturally by the radiation that constantly enters the earth's atmosphere. For example, gene mutations are probably caused when the sex cells of an organism are exposed to X rays, gamma rays, cosmic rays, and ultraviolet rays. In addition to this radiation, certain chemicals called **mutagens** (myü'tə jəns) can change nucleotides within DNA molecules. Such mutagens are often present in food additives and in substances that contribute to air and water pollution.

The chance for a particular human gene to mutate in one generation is between 1 in 10,000 to 1 in 1,000,000. Since humans have at least 30,000 genes, it is likely that each person carries at least one mutation. Thus, you may well have genetic information that is not present in the genes of either one of your parents.

Chromosome Mutations

Another way for the genetic traits of an organism to be altered is through changes involving *whole* chromosomes or *parts* of chromosomes. Structural changes in chromosomes are also caused by radiation, chemicals, and even by some virus infections.

Changes occur in the structure of chromosomes during cell division. When homologous chromosomes pair up, linked genes on the chromosomes may break apart. The genes may join another chromosome. Or they may be lost. The result is one of three kinds of chromosome mutations you can read about below.

A **deletion** involves the loss of a piece of chromosome. If a chromosome breaks and the parts do not reattach, the pieces may be lost. Deletion is the most serious kind of chromosome mutation. Bits of genetic information are no longer available to the offspring.

A **duplication** occurs when an extra, but identical piece of chromosome is added to the normal chromosome. Duplications are not always harmful to the offspring, and in some cases, may be an advantage.

Finally, an **inversion** occurs when pieces of chromosome break apart and the pieces rejoin the same chromosome in a different order. Usually inversions have no harmful effects on the offspring.

Did You Know?
***Transduction** is a process by which a virus carries a piece of DNA from one bacterium to another. In this way, new DNA is added to the chromosomes of the bacterium. Bacteria that have received new genetic instructions are considered to be recombinants.*

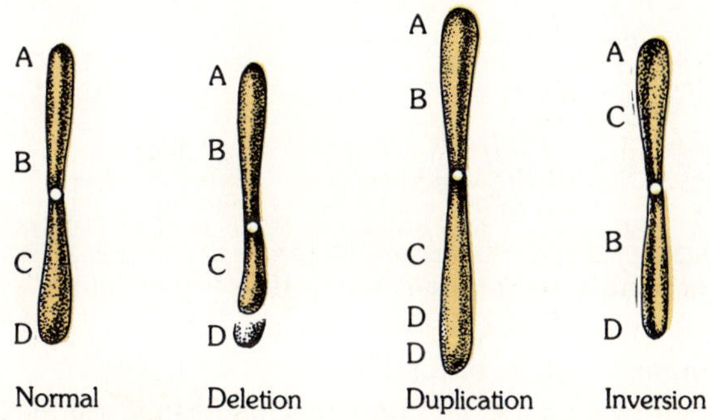

Types of chromosome mutations

Not all mutations are bad. In fact, mutations are the source of the great variety we see among members of a population. Such variations often help some individuals to fit into the environment better than other individuals. However, in general, random mutations are of no help. In fact, most cause the death of the organism. Mutations that survive are usually recessive and remain hidden for many generations. Mutations are like making random changes in a piece of music. The chances of improving the music in this way are very small.

Changes in Number of Chromosomes

Sometimes chromosome mutations involve the addition or loss of an entire chromosome. The loss of a whole chromosome is called **monosomy.** The addition of a chromosome is termed **trisomy.** These two kinds of mutations often involve the X and Y chromosomes. Instead of having the normal XX or XY make-up, an organism may have a sex chromosome make-up of XXX, XYY, XXY, or XO (one X only).

Whole chromosome mutations usually occur during meiosis. When the pairs of chromosomes are supposed to separate, certain pairs occasionally do *not* separate. As a result, an egg may contain two X chromosomes or no sex chromosomes at all. A sperm cell may also contain both an X and a Y chromosome or no sex chromosomes at all. The failure of chromosome pairs to separate during either meiosis or mitosis is called **nondisjunction.** The causes of nondisjunctional mutations are not well understood.

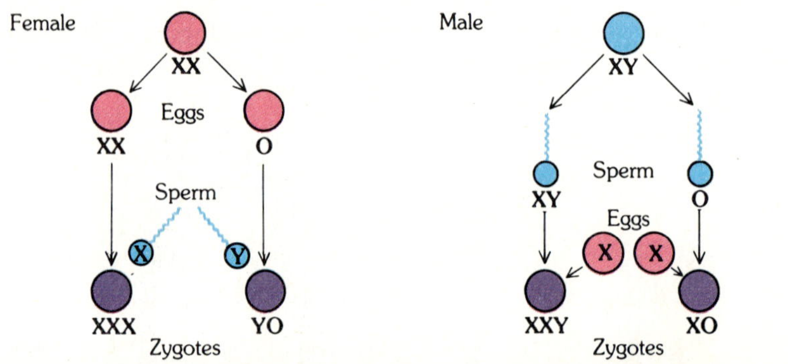

Whole chromosome mutations in sex cells (nondisjunction)

Studying Changes in Chromosomes

In recent years, methods have been developed to study the full set of human chromosomes. Such methods have enabled geneticists to identify various chromosome defects in cells, such as those discussed above.

In one technique, a specimen of human blood, bone marrow, or skin is used. A certain chemical is added to the cells to stop all cell division during metaphase. At this stage, the chromosomes are easiest to see. Then the chromosomes are photographed through a microscope, and the photo is enlarged. Each chromosome is cut out of the photo and arranged in pairs, according to size and shape. The chromosome arrangement that results is called a **karyotype** (kar′ē ō tīp).

Careful study of cell karyotypes shows that each chromosome pair has a particular size, shape, and form. In humans, there are 22 chromosome pairs and two sex chromosomes. Geneticists can use karyotypes to see if any chromosomes are missing or if extra chromosomes are present. You can see how karyotypes are prepared by studying the diagram on the right.

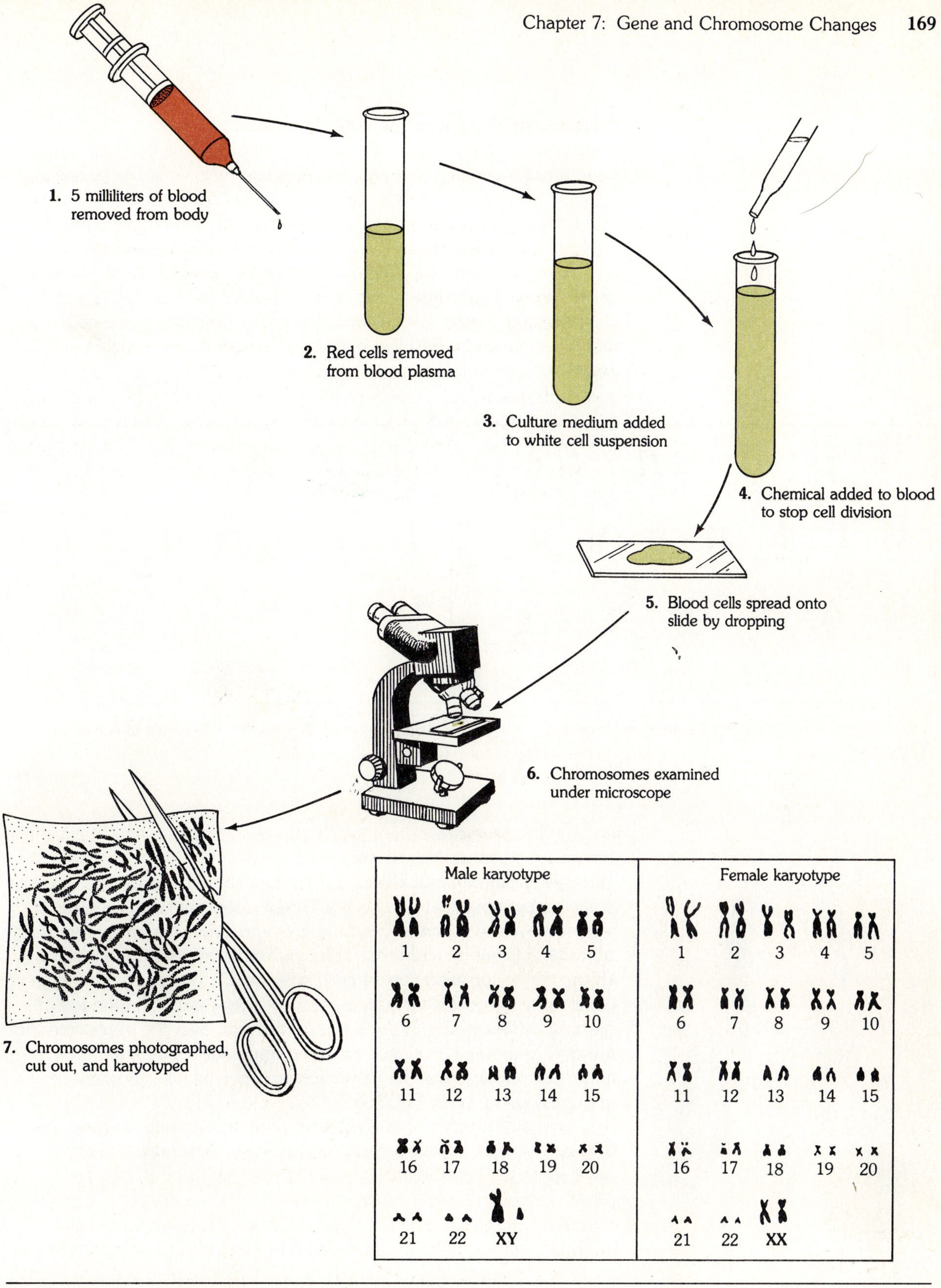

Preparation of a Cell Karyotype

Human Sex Chromosome Mutations

Variations of the normal XX or XY sex chromosome pair can be discovered by examining karyotypes. A person with only one X has the genetic make-up XO and is diagnosed as having *Turner's syndrome.* XO females occur in one of about every 2,500 live female births. Such women are sterile, short of stature, and have other distinct traits.

A person with one Y chromosome and more than one X chromosome generally has the genetic make-up XXY or even XXXY and is diagnosed as having *Klinefelter's syndrome.* This syndrome occurs in one of about every 800 live male births. Such men are sterile and may have secondary female traits.

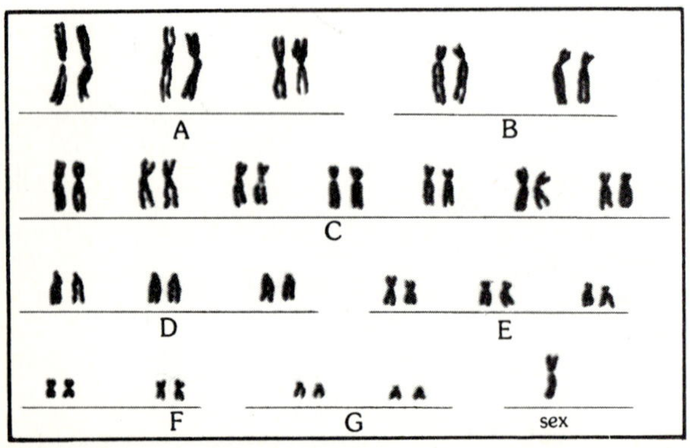

Karyotype of a person with Turner's syndrome (45, XO)

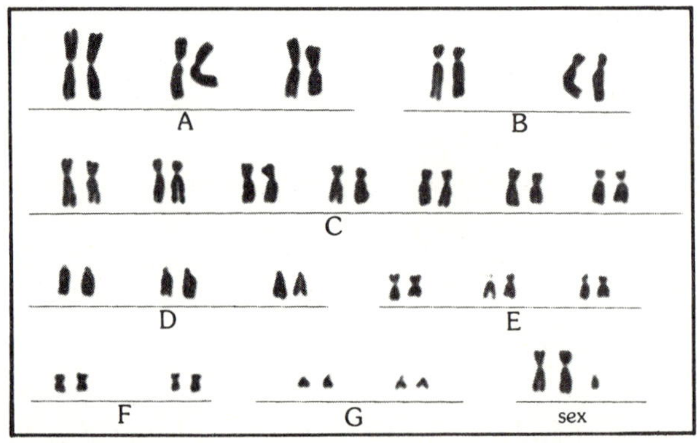

Karyotype of a person with Klinefelter's syndrome (47, XXY)

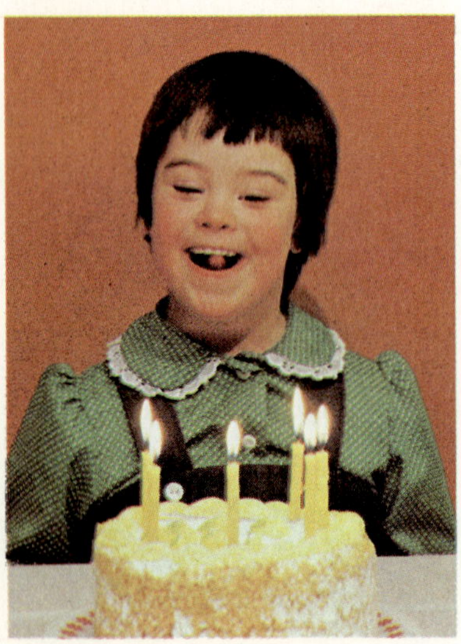

Child with Down's syndrome

Human Autosome Mutations

Changes in human autosomes, or non-sex chromosomes, can also be revealed by looking at karyotypes. Some disorders have been described in which a person has an extra autosome. The most common of these is *Down's syndrome.* This syndrome results when an extra chromosome, or extra part of chromosome 21, is present. The extra genes on chromosome 21 seem to affect many traits. The bridge of the nose is flattened and there is a fold of skin over the inner corner of the eye. Typically, the hands are short and wide. In general, there is abnormal physical and mental development. Down's syndrome occurs in one of about every 900 live births.

Live births occur where one arm of an autosome is missing, but they are rare. Such deletions, or partial monosomy, result in severe developmental retardation.

Changes in Sets of Chromosomes

In a third kind of chromosome mutation, one or more whole sets of chromosomes may be added. Mutants with an extra set of chromosomes are known as **polyploids.** Instead of the normal two sets of chromosomes, the organism may have three, four, five, or more sets. Polyploidy is common in plants, but rare in animals. Polyploid plants are often larger and hardier than plants that are not polyploids. Many of these plants are valued for their food or beauty.

Polyploidy occurs during meiosis if the chromosomes fail to separate. For example, the ordinary pea plant has 14 chromosomes in each cell. However, if the chromosomes fail to separate when sex cells form, one sex cell will receive 14 chromosomes and the other will receive none. The sex cell with 14 chromosomes may be fertilized by a sex cell that has 7 chromosomes. The offspring that is formed would have 21 chromosomes, instead of 14. Such a plant would survive and grow well, even though it has an extra set of chromosomes.

Plant breeders have even developed techniques to produce polyploid mutant plants. A common procedure is to treat sex cells with *colchicine.* This chemical prevents the chromosomes from separating during meiosis. By treating pea plants with colchicine, offspring with 21, 42, or even 56 chromosomes can be produced. You can see some examples of other kinds of polyploid plants below.

Some plants that exhibit polyploidy

Review It

1. What is a gene mutation? How does a gene mutation differ from a chromosome mutation?
2. Explain how a substitution in the DNA molecule might change a genetic trait.
3. Describe the three kinds of chromosome mutations that can occur. Give an example of each.

Activity

Finding Your Blood Type

Human blood types are determined by the alleles A, B, and O. The three alleles combine in various ways to produce the blood types A, AB, B, and O. One or more of these alleles is probably the result of a gene mutation that occurred at some time in the distant past.

Blood types have often been used by geneticists to learn more about the relationship between gene mutations and disease. In this activity, you and your classmates will find out how blood types are determined by finding each of your own blood types.

Materials

ABO blood typing sera (anti-a and anti-b)
Blood lancet
Alcohol
Cotton
Microscope slide
Grease pencil
Toothpicks

Procedure

1. Prepare a clean glass slide by drawing a line down the middle with a grease pencil. On one side write *anti-a,* on the other write *anti-b.* Add one drop of each serum to its space.

2. Clean the tip of one of your fingers with rubbing alcohol. Puncture the spot with a sharp jab of the blood lancet. You may need someone to do this for you.

3. Squeeze out one drop of blood into each of the drops of serum. Stir each drop of blood into the serum with a toothpick. Use different ends of the toothpick to stir the different mixtures.

4. Look for clumping of blood in the mixtures. Use the photos to help you determine which blood type you have. Compare your blood type with those of your classmates.

Anti-A

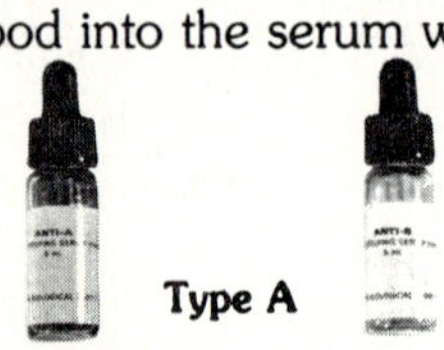

Anti-B

Type A

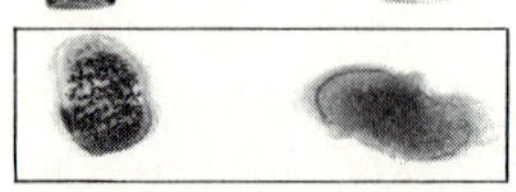

Clumping No clumping

Type B

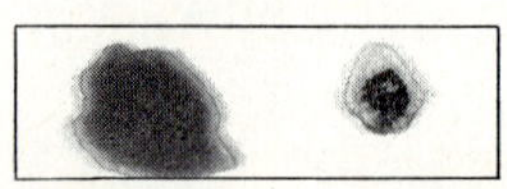

No clumping Clumping

Type AB

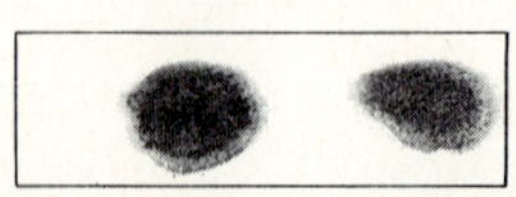

Clumping Clumping

Type O

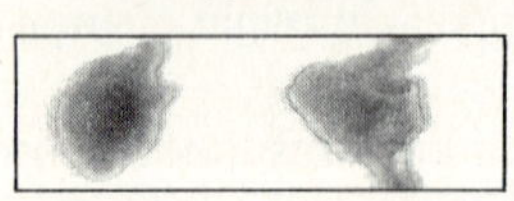

No clumping No clumping

Analysis

1. What blood type do you have?

2. How many members of your class have each of the blood types A, B, AB, and O? What *percentage* have each of these blood types?

3. In the United States, the percentage of blood types is:

	A	B	AB	O
White	39.5	11.2	4.2	45.2
Black	28.3	20.0	4.1	47.0
Indian	22.5	0.0	0.0	77.7

How do these data compare with the class average? Why do you think such differences exist among members of different populations?

4. Almost a dozen different blood types exist in addition to A, AB, B, and O. What are some of these blood types and how important are they to our health?

7–2
Detecting and Treating Human Genetic Mutations

The different kinds of mutations that you have been reading about are known to cause about 2,000 human genetic disorders. Many of these disorders occur only when both chromosomes in a pair carry the mutation. This means that both parents must pass on the mutation before the trait will be expressed in the offspring. However, some genetic disorders will show in the offspring if only one parent carries the trait and passes it on to the child.

Mutations are usually present at the time the sperm fertilizes the egg cell. Therefore, every cell of the developing zygote will contain the mutation. Any genetic disorder will be a part of the whole organism. It is not possible to remove the trait caused by the mutation, just as there is no way to remove the trait for tallness or for brown eyes. No genetic mutation can actually be corrected until a way is discovered to replace or to repair the faulty genes.

The trait of albinism is caused by gene mutation.

However, not all mutations are harmful. For example, color-blindness is due to a mutation. But color-blindness by itself does not affect the health of the individual. In other cases, a minor genetic disorder can be easily treated. For example, some kinds of nearsightedness and farsightedness are due to mutations. However, eye glasses will treat both conditions. Webbed fingers and toes are also caused by a gene mutation. But surgery can be used to separate fingers and toes that are attached.

Chemical Treatment of Genetic Disorders

Today, even some serious genetic disorders can be treated if detected early enough. For example, the disease *PKU (phenylketonuria)* occurs in one of every 14,000 live births. It is an example of a disorder caused by a single recessive gene. People who have inherited PKU are not able to break down the chemical *phenylalanine* which is present in milk and other foods containing certain proteins. The chemical products that remain are poisonous. They collect in the blood and slow down the growth of the brain. This causes moderate to severe mental retardation, as well as other side effects. If PKU is detected within three weeks after birth, the child can be fed a special diet. The diet simply restricts the intake of proteins that contain phenylalanine. The special diet prevents the collection of poisons in the blood. And mental retardation usually does not occur.

PKU can be detected by running a chemical test on the blood or urine of the newborn child. Starting in the early 1960s, many states began to require that this test be run on all newborns. Before that time, there was little hope for PKU children. Today, this test is just one of many being developed to detect and allow treatment of children who have serious genetic disorders.

Amniocentesis

Another way to detect genetic disorders is to study the cells of a baby before it is born. This is done by a technique called **amniocentesis** (am′nē ō sen tē′sis). A doctor trained for this procedure inserts a hollow needle through the abdominal wall of the mother. A sample of the fluid that surrounds the unborn child is withdrawn. The fluid contains skin cells that have been lost by the baby. Chromosomes in the skin cells are genetically the same as other cells in the baby's body.

The skin cells are separated from the fluid and grown in special laboratory dishes. Then the cells may be studied in several ways. First, karyotypes of the cells can be made. These will help determine if the baby has a chromosome disease, such as Down's syndrome. The karyotypes can also be used to tell if the baby is male or female.

Second, chemical tests can be carried out on the cells. These tests determine if the cells are making substances necessary for normal growth and development. For example, if a child has *Tay Sachs disease,* he or she cannot make an enzyme necessary for normal development of the brain. A chemical study of the child's cells will show if the enzyme is being made.

You can see how amniocentesis is done by studying the diagram below.

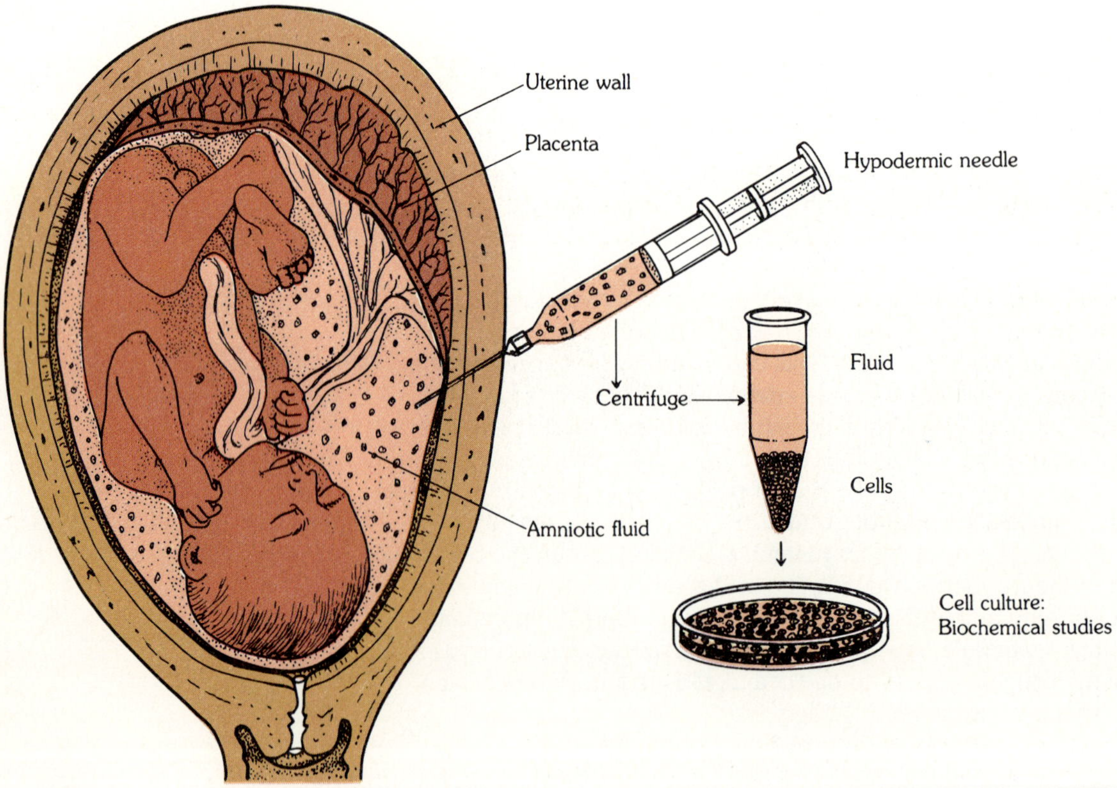

Amniocentesis

Ultrasound

Before delivery of a baby, doctors may use X rays to learn the exact position of the baby in the mother's womb. However, X rays can cause mutations in the rapidly dividing cells of the baby. Now a safe method of examination during pregnancy is available. Doctors can use **ultrasound** to make a picture of the baby's organs. Ultrasound consists of sound waves that are too high for the human ear to hear. The ultrasound waves used in medical testing do not cause mutations of the genes and chromosomes.

Ultrasound waves are made by a small cylindrical machine placed over the area of the body to be examined. The instrument sends ultrasound waves through the body. These waves are reflected from the surface of the organs. The reflected sound waves are recorded as a picture called an **echogram.** The whole process is similar to what happens when your voice bounces off a wall or a cliff and returns as an echo.

As ultrasound equipment is improved, it is hoped that early detection of genetic and other diseases will be possible. It is already possible to record the heartbeat of the unborn baby. The echogram of a beating heart is recorded as a series of waves. The number of times a heart beats in a minute can show if the heart is under too much stress. A heart with a genetic defect may not be able to withstand the stress during delivery. The doctor can take emergency action during delivery if the echogram shows a heart problem.

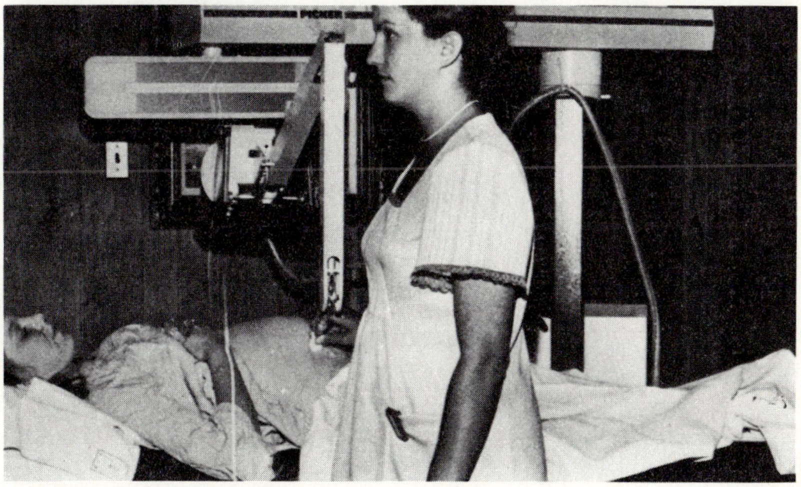

Ultrasound equipment in use

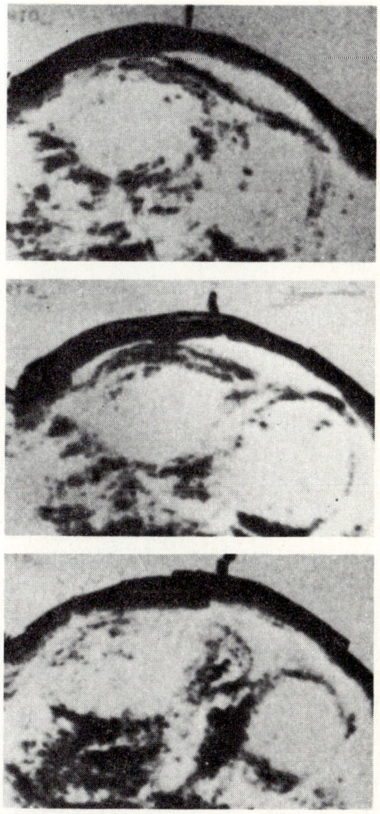

Echogram

Review It

1. Why can genetic diseases be treated, but not cured?
2. Describe two methods used to detect genetic diseases.

7–3
Influence of Environment on Gene Expression

How much of our physical, mental, and personality traits are determined by our genes? How much of these same traits are determined by where we were raised and how we were raised? There is much controversy over the answers to these questions. However, it is known that from the moment fertilization occurs, an organism is influenced by the environment. The way a gene expresses, or shows, itself is the result of the way the gene and the environment interact. This interaction can be with the environment *inside* the body of the organism. Or the interaction can be with the environment *outside* the organism.

Internal Environment

An example of a gene influenced by the internal environment is *early pattern baldness* in humans. The gene for baldness is a dominant gene (B), usually expressed only in males. Pattern baldness is affected by the amount of male sex hormone produced. Since females normally have lower levels of this hormone than males, they do not show the trait in the same way as males. Females must inherit two dominant genes (BB) to show pattern baldness. However, males will show the trait with either one dominant gene (Bb) or with two dominant genes (BB).

Male with early pattern baldness (BB or Bb)

Male without pattern baldness (bb)

External Environment

There are many examples of ways in which the external environment can affect the traits of an organism. In humans, height is determined by a number of genes. You may recall that height is an example of polygenic inheritance. Tall parents tend to have tall children. And short parents tend to have short children. However, environmental factors, such as nutrition and disease, also influence the trait. For example, the population of the United States has grown taller over the past several generations. Most scientists believe the increase in height has been due to a general improvement in the diet and health conditions of the population.

Twin Studies

Studies of identical and fraternal twins often provide valuable information about the effect of genes and environment on the development of traits. **Identical twins** develop from a single fertilized egg that splits shortly after fertilization. So the two new individuals have the same genetic information. **Fraternal twins,** however, develop from two eggs that are fertilized by *different* sperms. Each twin develops into an individual with different genetic information. You can see how this works by studying the diagram below.

Twin studies are used to examine a trait, such as height or weight, for twins raised together and for those raised apart. Identical twins separated early in life and raised apart share the same genes, but grow up in different environments. Studies of such twins are used to help determine which traits are influenced more strongly by genes and which are influenced more strongly by the environment. You can see a summary of such a study in the margin.

Trait Differences	Identical Twins—Reared Together	Identical Twins—Reared Apart
Height		
Less than 1″	82%	48%
1″ or more	18%	42%
Intelligence Test Scores		
0–9 points	65%	62%
10	35%	38%

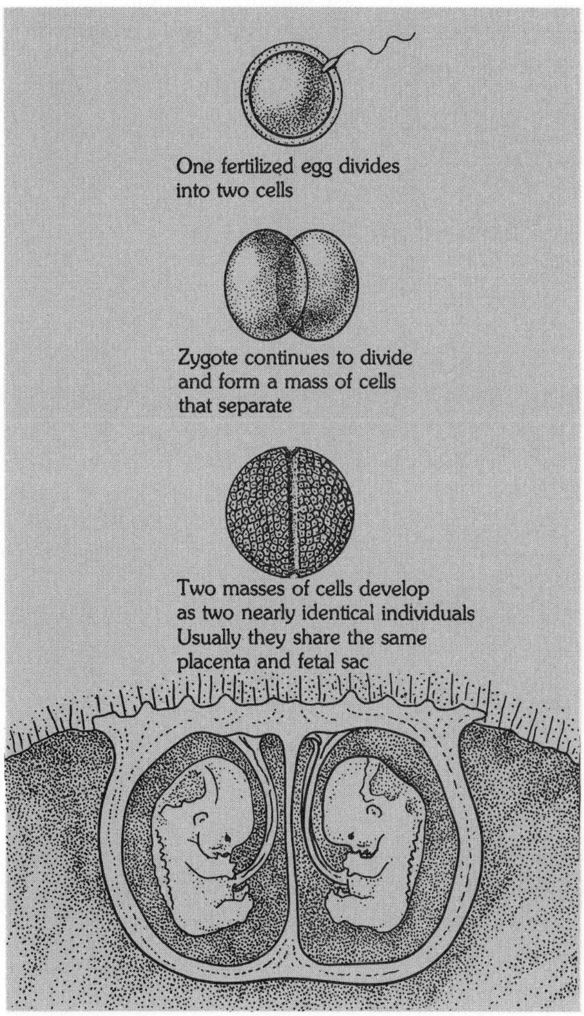

Identical twins

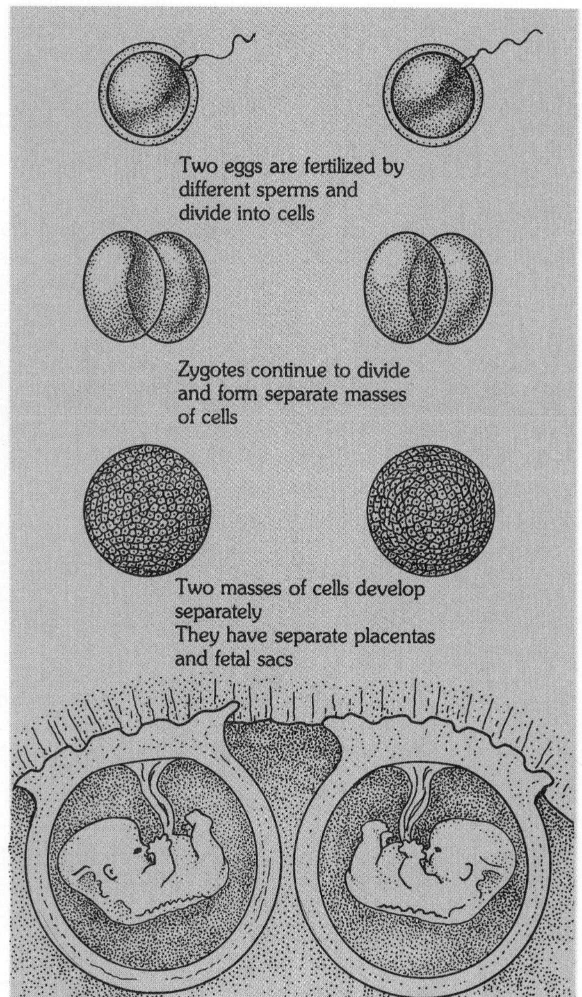

Fraternal twins

Interaction of Genes and the Environment

In general, no trait is inherited in a simple manner. The picture is almost always complicated by environmental factors. For example, the tendency to become infected with pulmonary tuberculosis is inherited. However, only a small portion of such people actually develop the disease because nutrition, general health, and exposure to infection are more important than genetic tendency to develop the disease. It is uncommon for heredity or environment to be entirely responsible for any particular trait or disease. In most cases, both factors are responsible. However, sometimes one may appear more important than the other.

For example, at one extreme there are diseases, such as Down's syndrome or muscular dystrophy, that are almost completely genetic in origin. The environment seems to play no direct part in the development of these disorders. At the other extreme, we have infectious diseases that are almost completely the result of environmental factors. Scurvy and tuberculosis are examples of diseases due almost entirely to environmental factors. In between these extremes are such conditions as diabetes mellitus, hypertension, peptic ulcer, and certain cancers. Both environmental and genetic factors are involved in the development of such conditions.

You can see below how diseases range from those largely environmental in cause to those which are almost entirely genetic.

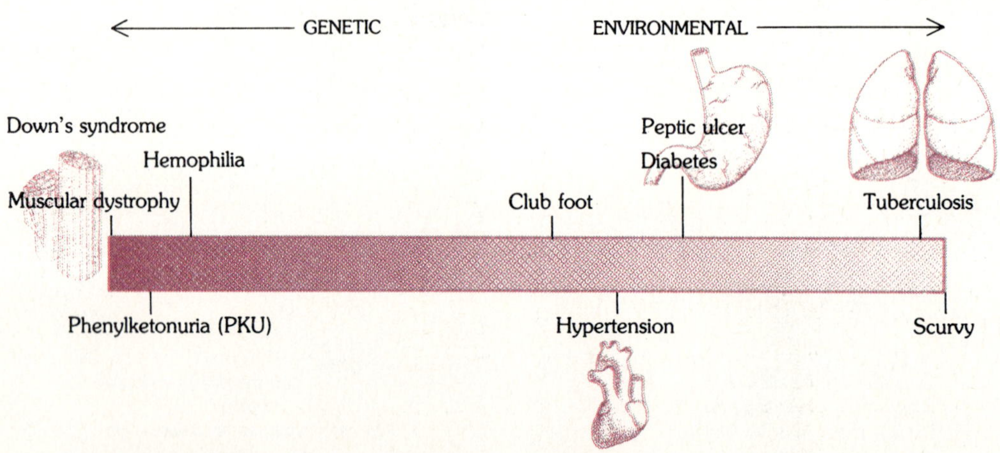

Spectrum of human diseases influenced by genes and the environment

Review It

1. How does the internal and external environment of an organism influence gene expression?
2. Explain how twin studies might be used to determine the effect of environment on gene expression.
3. How do genes and the environment interact to influence development of a trait?

Breakthroughs in Biology

Making Insulin

DIAGRAM OF A HUMAN INSULIN MOLECULE

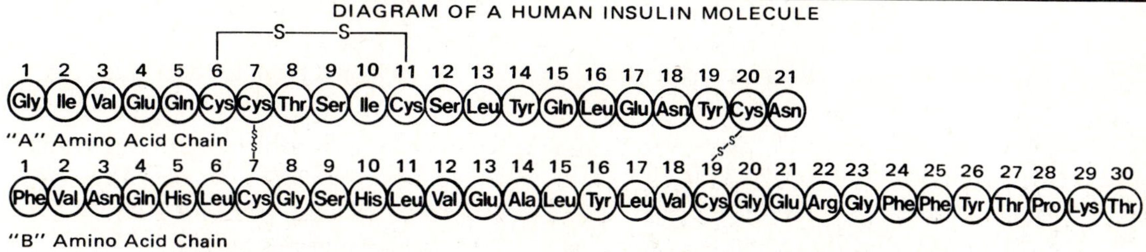

COMPARISON OF ANIMAL INSULIN AMINO ACID CHAINS WITH THAT OF HUMANS

	A_8	A_9	A_{10}	B_{30}
Human	THR	SER	ILE	THR
Bovine	ALA	SER	VAL	ALA
Swine	THR	SER	ILE	ALA

People with one form of diabetes have excess sugar in their blood. This can be due to a lack of insulin, a hormone made in the pancreas. Over 1.5 million diabetics in the United States depend on getting insulin from animals.

Notice the human insulin molecule's two amino acid chains, and the differences between animal and human insulin. About 5% of the diabetics who need insulin are allergic or resistant to insulin from other mammals.

Making insulin involves splicing synthesized amino acid chains into a bacterial ring of DNA known as a *plasmid.* The bacteria's protein-making mechanism is given new instructions: to make human insulin.

DIAGRAM OUTLINING HOW THE INSULIN WAS MADE

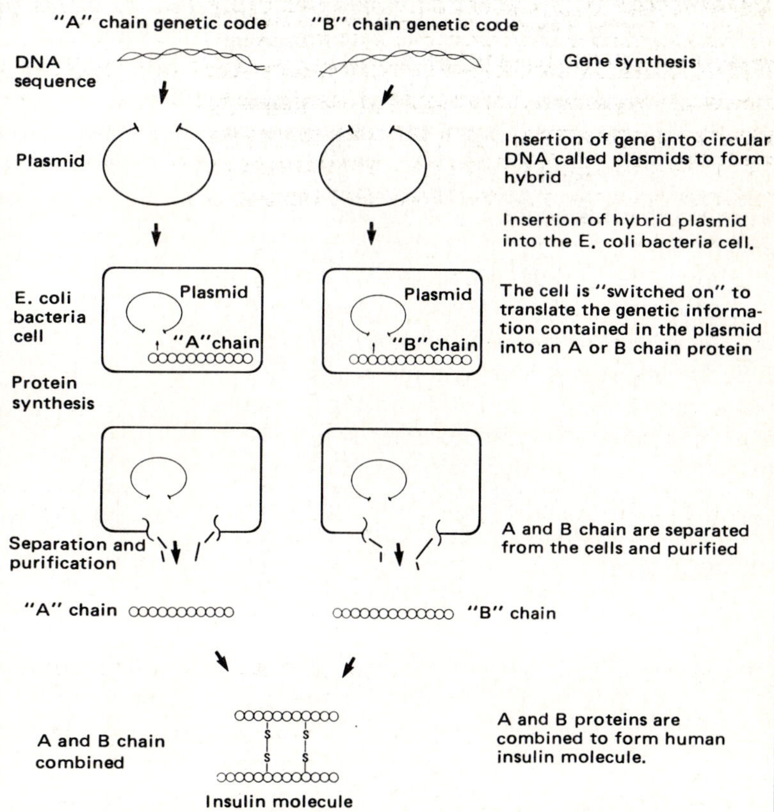

Analysis

1. What is the difference between swine and human insulin?
2. What is a plasmid?

Chapter Summary

Early in the twentieth century, Hugo de Vries guessed that the new traits that appear in an organism are due to permanent changes in its genetic material. Today, it is believed that such changes, or mutations, are due to changes in either the genes or chromosomes of the sex cells.

Gene mutations are due to a chemical change in the DNA of a cell. Sometimes the change is slight and cannot be noticed. Other times, a major change in a trait is produced by a single gene mutation. Sickle cell anemia is the result of such a mutation.

Changes in whole chromosomes or parts of chromosomes can also alter the traits of an organism. Deletions, duplications, and inversions are examples of mutations that occur in parts of chromosomes. Mutations may also involve the addition or loss of single chromosomes, or the addition of entire *sets* of chromosomes.

Mutations are the source of the great variety we see among members of a population. However, many times mutations are harmful. They either result in the death of an organism or are recessive and remain hidden for many generations. When they do appear, the mutations can cause various genetic disorders. Some of these, such as colorblindness or webbed toes, are minor and can be treated. Others, such as PKU or Down's syndrome, are major and can cause serious problems.

Many techniques have been developed to help detect such disorders. Karyotyping is used to study the size, shape, form, and number of chromosomes in an organism. This technique is used to detect abnormalities in chromosomes or to detect the addition or loss of chromosomes. Amniocentesis and echograms are other techniques used to determine if any abnormalities are present in an unborn child.

Even though many disorders are produced by chromosome and gene mutations, the environment plays an important role in modifying such mutations. The way each trait is expressed is due to the way the gene and the environment interact. Much remains to be learned about how such interactions result in the production of a unique individual.

Questions to Think About

1. Explain why genetic diseases are incurable and can only be treated. What must be done to truly *cure* a genetic disease?
2. In what ways can the study of blood types provide important information in the field of human genetics? How can the study of identical twins also supply valuable information?
3. Discuss how environmental factors influence the study of human genetics.
4. *Genetic counseling* is the guidance a geneticist provides couples or individuals about the risks of having defective offspring. Why might individuals want to seek the advice of a genetic counselor before having children?

Extra Research

1. What are some of the current ideas about the causes of cancer? What evidence, if any, is there of a connection between cancer and genetics?
2. Find out how triplets, quadruplets, and quintuplets are conceived and develop. Do they originate from one egg, many eggs, or a combination of both? How does their conception and development determine their genetic make-up?
3. Make a list of five human traits that you think are (a) entirely determined by genes, (b) entirely determined by the environment, (c) determined by both genes and the environment. Look up information on the traits you have listed. Write a report about your findings.
4. Look up the term *eugenics*. What do you believe are the differences between positive and negative eugenics? On what bases do some people oppose eugenics?

Chapter Test

Part I

Copy the number of each test item. If the item is *true*, write *T* after the number; if the item is *false*, write *F*.

1. Gene mutations are due to additions or deletions of parts of chromosomes. (166)
2. Most mutations are harmful to the organism. (167)
3. Chromosome changes usually occur after a cell has divided. (167)
4. Mutations are responsible for the great variety you see among individuals in a population. (167)
5. Whole chromosome mutations are due to the failure of a chromosome pair to separate during cell division. (168)
6. The deletion of one arm of an autosome results in severe developmental retardation. (170)
7. Some serious genetic disorders, such as PKU, can be treated if detected early enough. (173)
8. A gene can interact with the environment inside and outside the body to produce a certain trait. (176)
9. X rays cause a decrease in the number of mutations. (166)
10. Disorders range from those largely environmental in cause to those which are almost entirely genetic. (178)

Part II

Copy the number of each sentence below. After each number, write the word or words that complete the sentence correctly.

11. A_____ _____ is a chemical change within one gene. (166)
12. Sex _____ can influence how genes are expressed in traits, as illustrated by pattern baldness in humans. (176)
13. Chemicals that cause changes in genes are called _____. (166)
14. _____ syndrome is a genetic disorder due to the presence of an extra autosome. (170)
15. A technique used to show the size, shape, and form of chromosomes is called a cell _____. (169)
16. Mutants with one or more extra sets of chromosomes are known as _____. (171)
17. _____ and _____ are techniques used to detect genetic disorders in the unborn child. (174 and 175)
18. Doctors use _____ to make a picture of the organs of an unborn baby. (175)
19. The loss of a whole chromosome is called _____. (168)
20. _____ twins develop from a single fertilized egg that splits shortly after fertilization. (177)

Part III

Copy each item from List A. After each item, write the letter and words from List B that best describe the item.

List A

21. inversion (167)
22. trisomy (168)
23. polyploidy (171)
24. duplication (167)
25. deletion (167)

List B

a. addition of a whole chromosome
b. loss of a piece of chromosome
c. addition of an extra, identical piece of chromosome
d. breaking and rejoining of chromosome parts in a different order
e. addition of one or more whole chromosome sets

Focus on Careers

Geneticist

The study of genetics is not new. Scientists have long used Mendel's discoveries to improve crops and breed better farm animals. The genetic control of insects by sterilization is helping to control several crop pests. A person who studies and experiments with inherited characteristics of living things is known as a geneticist.

Geneticists work in many fields: agriculture, plant and animal science, and medicine. In each field, a geneticist studies the inheritance and variation of characteristics in living things. They try to develop strains, varieties, breeds, and hybrids of plants and animals that are better suited for various human needs than those available. For example, in agriculture, a geneticist crossbreeds animals to obtain desirable characteristics. Cattle, chickens, or swine may be bred for quick maturity rate, resistance to disease, or quality of meat.

A plant geneticist experiments with plants, especially crop plants, to develop new methods of growing crops or to develop and improve varieties of crops. Fruits and vegetables can be bred for flavor, tenderness, yield, vigor, ease in harvesting, and resistance to disease.

One of the goals of medical geneticists is the prevention of inherited diseases. A medical geneticist is often part of a counseling team that includes a cytologist, a pathologist, and a social worker. Together, this team may help guide people who have a high risk of transmitting genetic diseases to their children. A medical geneticist might also be involved in research to discover how certain environmental substances affect human genes. For example, it is known that some compounds in smog cause mutations in cells. However, it is not known if such materials affect the reproductive cells.

Geneticists are well-educated scientists with advanced degrees in their field of specialization.

Much of the work of a geneticist is carried on in a laboratory. Geneticists work in private and government research centers and hospitals. They must be conscientious, and willing to do painstaking, detailed work.

Career Information

Animal Breeder Technician: American Veterinary Medical Association, 930 N. Meacham Road, Schaumburg, IL 60196.

Geneticist: American Genetic Association, 1028 Connecticut Avenue NW, Washington, DC 20036.

Plant Propagator: Institute of Agricultural Technology, Michigan State University, East Lansing, MI 48823.

Statistician: American Statistical Association, 806 15th Street NW, Washington, DC 20005.

Animal Breeder Technician
Duties: Breeds and raises mice, rats, guinea pigs, monkeys, and other animals for use in experimental and biological research, such as in a research center for studying animal tumors. The technician keeps records of "family trees," weight, diet, and other data. He or she may breed animals selectively to produce offspring with specific genes or traits. The technician builds animal housing and provides everything the animals need to be healthy.

Training: A two-year course of study in laboratory animal care qualifies a person for many jobs in this field. Many science courses are required and laboratory work is stresssed.

Opportunities: Jobs are available at research centers, drug companies, medical schools, and commercial breeders. The outlook for job opportunities is good.

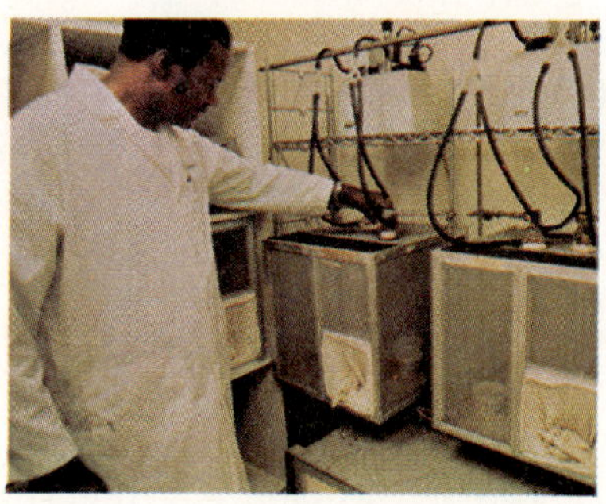

Plant Propagator
Duties: Propagates plants by various methods, for example, root and stem cuttings and air layering, to produce new rootstocks. In an orchard or vineyard, a propagator selects the best seeds or cuttings to start new plants. A plant propagator has knowledge of plant germination, soil content, and growing habits of plants. Work may be inside or outside depending on the climate and the kind of plants he or she works with.

Training: On-the-job training is necessary. Better paying jobs may require technical training, for example, in orchard and vineyard or farm crop technology.

Opportunities: Trained people have the best chance of being hired.

Statistician
Duties: Devises, carries out, and interprets the numerical results of surveys and experiments. A statistician decides where to get data and what size samples should be used. He or she may develop questionnaires or reporting forms. The statistician summarizes results in tables, charts, mathematical models, and written reports.

Training: Requires at least a bachelor's degree in statistics or mathematics.

Opportunities: Outlook is favorable. Statisticians work in many fields: science, engineering, astronomy, economics, and psychology.

Bibliography: Unit Three

Auerbach, Charlotte. *The Science of Genetics.* Harper and Row, 1961. Clear account of topics including chromosomes, genes, mutations, heredity and environment, and more.

Beadle, George, and Beadle, Muriel. *The Language of Life.* Doubleday and Co., Inc., 1966. Clear account of topics covering DNA, mechanisms of inheritance, and genetic changes.

Bonner, D. M., and Mills, S. E. *Heredity.* Prentice-Hall, Inc., 1964. Information on gene structure and inheritance.

Brewbaker, J. L. *Agricultural Genetics.* Prentice-Hall, Inc., 1964. Contains information on basic Mendelian inheritance and data relating to inheritance in domestic plants and animals.

Dunn, L. C., and Dobzhansky, Theodore. *Heredity, Race, and Society.* New American Library, 1952. Good discussions of heredity and environment, inheritance of traits, and eugenics.

Fast, Julius. *Blueprint for Life.* St. Martin's Press, 1965. Interesting discussions of a wide range of topics, such as mutations, genetic diseases, and laws of inheritance.

Fried, John J. *The Mystery of Heredity.* The John Day Co., 1971. Explains some of the problems faced by scientists as they try to correct genetic defects, regenerate lost limbs, and create human replicas.

Goldstein, Philip. *Genetics Is Easy,* 2nd ed. Viking, 1961. Brief introduction to the subject of genetics.

Karp, Laurence. *Genetic Engineering: Threat or Promise?* Nelson-Hall, 1976. An up-to-date account of breakthroughs in genetic screening, genetic engineering, cloning, and other fascinating topics. Above-average reading level.

McKusick, V. A. *Human Genetics.* Prentice-Hall, Inc., 1964. Methods and results of human genetic studies.

Montagu, Ashley. *Human Heredity,* 2nd ed. World Publishing Co., 1963. Interesting account of laws of inheritance, effects of environment on heredity, and a brief description of numerous genetic disorders.

Olby, Robert. *Origins of Mendelism.* Schocken Books, Inc., 1966. Describes the growth of ideas about inheritance and variation which eventually led to the Mendelian solution. Famous scientists in the field of genetics are mentioned.

Papazian, Haig. *Modern Genetics.* W. W. Norton and Co., Inc., 1967. Describes how chromosome and gene combination and mutations are the basis for evolution. Discusses natural selection and acquired characteristics.

Pedder, I. J., and Wynne, E. G. *Genetics: A Basic Guide.* W. W. Norton and Co., Inc., 1972. Includes topics such as cell division, Mendelian and non-Mendelian inheritance, and genetic experiments that can be performed.

Rogers, Michael. *Biohazard.* Alfred A. Knopf, 1977. Good discussion of the controversy over recombinant DNA research and the possible hazards.

Scheinfeld, Amram. *Human Heredity Handbook.* Lippincott, 1956. Good basic information on genetic disorders.

Scheinfeld, Amram. *Your Heredity and Environment.* Lippincott, 1965. Good information about human traits.

Sootin, Harry. *Gregor Mendel: Father of the Science of Genetics.* Vanguard Press, Inc., 1959. Biography of Mendel and his discovery of the laws of inheritance.

Sturtevant, A. H. *A History of Genetics.* Harper and Row, 1965. Traces the history of genetics from the time of Aristotle. Describes contributions of famous geneticists.

Sullivan, Navin. *The Message of the Genes.* Basic Books, Inc., 1967. Clearly written account of DNA structure and function, genes and chromosomes, and the mechanism of heredity.

Tanner, James M., and others. *Growth.* Time Inc., 1971. Beautifully illustrated book containing several chapters on genes, environment, and genetic flaws.

Winchester, A. M. *Heredity and Your Life,* 2nd ed. Dover Publications, Inc., 1960. Interesting variety of topics covering genetic abnormalities, heredity influenced by sex and environment, eugenics, and more.

UNIT FOUR

EVOLUTIONARY THEORY

Chapter 8

Evidence of Change

Most biologists think that the theory of evolution is the most important theory in biology. It is important because it explains so many observations that have been made about organisms alive today and those that have lived in the past. In fact, this chapter is mainly about the kinds of observations that the theory of evolution explains.

Evolution itself is defined as a change in gene frequencies of a population over a period of time. The change does not refer to changes within an *individual,* but to changes within *populations* across successive generations. There is much evidence of such change.

What kind of evidence is there about evolution? Is any of it visible around us? Can we observe evolution? The chapter will attempt to provide answers to some of these questions. Then you will have an opportunity to draw some conclusions of your own.

Chapter Objectives

1. Describe how fossils form.
2. Describe how the age of a fossil is measured.
3. Discuss the pattern of simple and complex fossils in the fossil record, using examples.
4. Explain how homologous structures provide evidence about evolution.
5. Describe how different kinds of vertebrate embryos are similar to each other. Give examples of how they differ.
6. List examples of chemical similarities and differences among organisms. Tell how they are used to explain evolutionary relationships.
7. Describe an animal "race circle" and tell how it implies evolution.
8. List examples of observed evolution. Describe how one of these examples provides evidence of evolution.

The fossil remains of the *Archaeopteryx,* which lived 130 million years ago. This bird may be an important link in the evolution of reptiles and birds.

8–1
Evidence: Fossils

Any sign of earlier life, usually laid down in soil or rock, is a **fossil.** A fossil may be an animal track, or it may be an imprint of an organism (or a part of an organism) that has been replaced by soil or rock. A fossil may also be an organism that has actually changed into rock.

Fossils give us *indirect* evidence of evolution. For example, you may have seen or handled fossils, so you know they are real objects. But seeing a fossil is not the same as seeing evolution. So the fossil record provides scientists with indirect evidence in support of the evolution theory.

The preservation of organisms as fossils is called **fossilization.** In a few cases, complete organisms that are now extinct have been preserved. Woolly mammoths have been found frozen solid in Siberia. Specimens of the furry rhinoceros, now extinct, have also been found. These were found preserved in oily and waxy deposits in Poland.

Usually, fossilization is not so complete. Most organisms die and decay without being fossilized. When fossil formation does occur, it is usually the hard parts that are preserved. Skeletons, shells, and tree trunks resist decay long enough for this process to take place.

Fossil reconstruction

Reconstructing Fossils

Often, only a few parts of a skeleton have been fossilized. Sometimes these parts do not fit any living species. Then scientists must compare the fossil parts with skeletons of living forms. They do this to find the most similar living species. The skeleton of the similar species gives clues to the structure of the missing parts of the fossil. In other cases, parts missing in one fossil may be present in a similar fossil elsewhere. The parts are then combined to rebuild the organism. An artist may also be asked to sketch in muscle and skin to show what the fossil might have looked like. The process of reconstructing fossils through careful observation and comparison is called **fossil reconstruction.**

Examples of fossils

Measuring the Age of Fossils

Fossils are found mainly in **sedimentary rock.** Such rock is formed as soil particles are laid down, forming new layers over old ones. After a long time, the layers harden. Most often, this happens on the bottoms of lakes, seas, and oceans. Over long periods of time, such areas may be lifted up and become visible on the surface of the earth.

If you compare fossils from different layers of sedimentary rock, you often find small differences among them. Since the different layers were laid down at different times, the differences observed among the fossils are evidence of change across time.

Sedimentary rock layers

Sedimentation

It is important to be able to estimate the age of a fossil and of a layer of sedimentary rock. If scientists can do this, they can construct a likely history of the changes in the populations represented by the fossils.

Several methods are used for telling the age of fossils. One method is based on **rock sedimentation.** Younger sedimentary rocks usually rest on top of older sedimentary rocks. So fossils in lower layers would be older than fossils in upper layers. Therefore, layers with similar fossils in them, but found in different places, can be compared to each other.

Fossil formation in stream bed

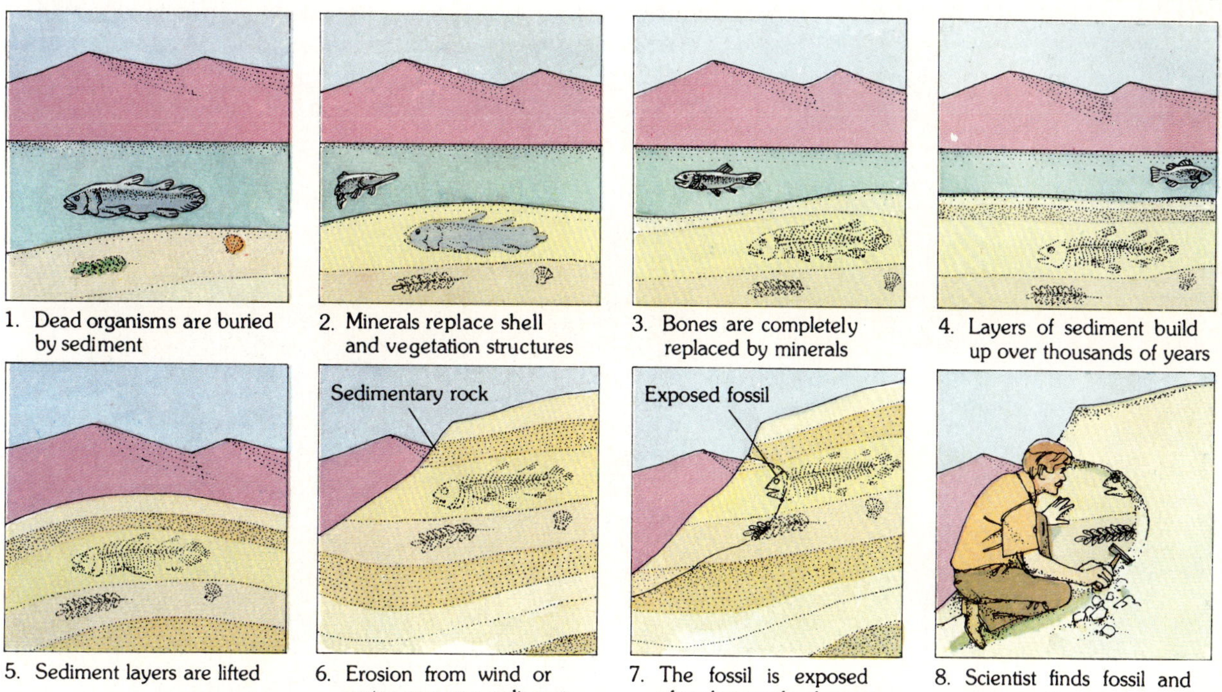

1. Dead organisms are buried by sediment
2. Minerals replace shell and vegetation structures
3. Bones are completely replaced by minerals
4. Layers of sediment build up over thousands of years
5. Sediment layers are lifted
6. Erosion from wind or water removes sediment
7. The fossil is exposed after thousands of years
8. Scientist finds fossil and removes it

Radioactive Carbon Dating

More recently, the discovery of natural radioactivity has provided another method of measuring the ages of rocks and fossils. Some substances that occur in nature are **radioactive.** This means that their atoms give off high-energy particles and become changed into atoms with different characteristics. For example, potassium-40 becomes argon-40; uranium-238 changes to lead-206; and carbon-14 (radioactive) becomes nitrogen-14 (non-radioactive).

Challenge!
Read about radioactive isotopes. What is an isotope? How do atoms of one isotope of an element differ from atoms of another isotope of the same element? What makes some isotopes radioactive?

Now, it turns out that it takes a certain amount of time for one-half of any amount of a particular radioactive substance to change in this way. This period of time is called the **half-life** of that substance. The half-life of uranium-238 is 4,510,000,000 years. However, the half-life of carbon-14 is only 5,568 years. Because of its short half-life, carbon-14 has been widely used for measuring ages of up to about 40,000 or 50,000 years. This method of dating fossils is called **radioactive carbon dating.** Uranium, potassium, and rubidium are used for dating materials thought to be older than 50,000 years.

How do scientists measure age using radioactive substances? Let's use carbon as an example. Both carbon-14 and carbon-12 occur in carbon dioxide (CO_2) in the air. Carbon dioxide is taken up by living things and becomes incorporated into their cells. So, if a fossil has the same carbon-14 to carbon-12 ratio as the carbon dioxide of the air, it must have died recently. On the other hand, if only one-half of the expected carbon-14 remained, the fossil would be judged to be about 5,568 years old. Remember, the half-life of carbon is 5,568 years. So it takes that long for one-half of the carbon-14 present at the time the organism died to decay. Similarly, if only one-fourth of the expected carbon-14 remained, the fossil would be about 11,136 years old (5,568 + 5,568). The diagram below shows how this works.

Determining fossil age with radioactive carbon 14

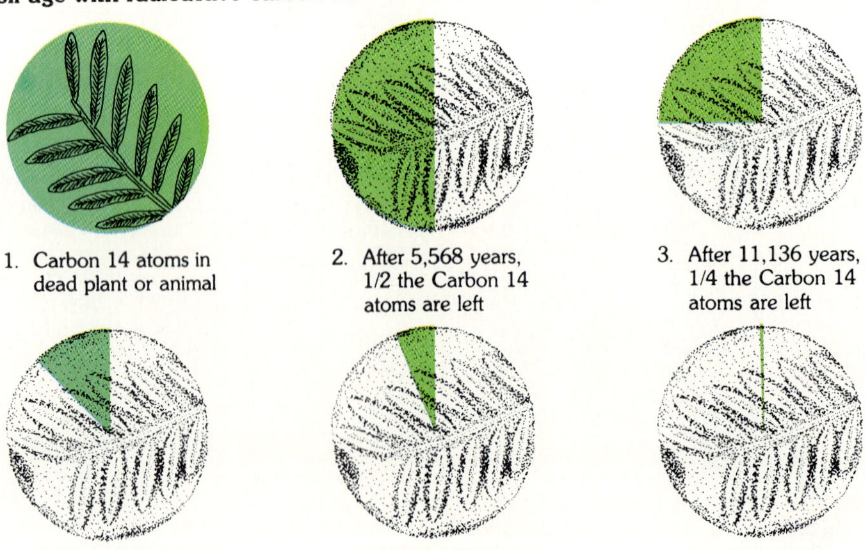

1. Carbon 14 atoms in dead plant or animal
2. After 5,568 years, 1/2 the Carbon 14 atoms are left
3. After 11,136 years, 1/4 the Carbon 14 atoms are left
4. After 16,704 years, 1/8 of the Carbon 14 atoms are left
5. After 22,272 years, 1/16 of the Carbon 14 atoms are left
6. After 60,000 years, about 1/1,000 of the Carbon 14 atoms are left

Types of Fossils

Here you can see some examples of different kinds of fossils that scientists have discovered. You can also see the estimated age of each fossil. The dates were determined by the sedimentation and radioactive dating techniques you have been reading about.

Petrified wood
255 million years old

Fern leaves
360 million years old

Sea lilies
465 million years old

Trilobites
280 million years old

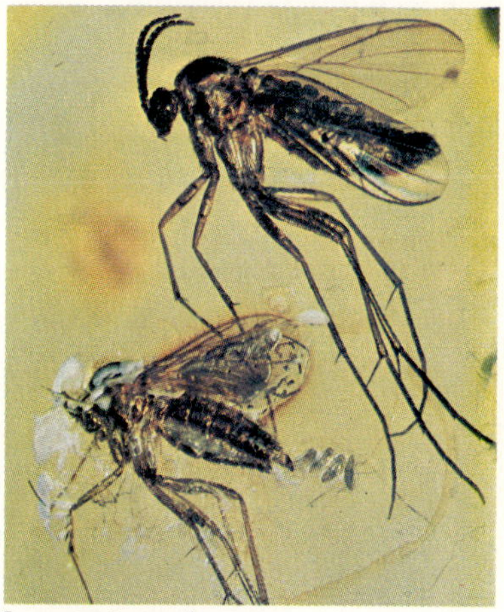

Insects in amber
35 million years old

Review It

1. What is a fossil?
2. How are fossils formed? In what kind of rock are fossils usually found?
3. What is meant by the half-life of a radioactive substance?
4. How do scientists use half-life to measure the age of fossils?

8–2 The Fossil Record

You have learned that layers of sedimentary rock and their fossils can be compared. This provides information about which fossils are older and which ones are younger. You have also learned that radioactive elements like carbon-14 can be used to estimate actual ages of fossils. In these ways, scientists have been able to prepare a standard worldwide record of sedimentary rock layers and their fossils. This is called the **fossil record.**

The table at the right is an outline of the *geological* time record. Notice that the time intervals are described as eras, periods, and epochs. The longest time intervals are known as **eras.** Beginning with the Paleozoic era, these are subdivided into **periods.** The most recent era, the Cenozoic, has periods subdivided into **epochs.** The fossil record is an important part of the geological time record.

There are several important things to remember about the fossil record.

1. It does not exist in a complete form at any one place. Rather, the record comes from many rock layers found at many separate locations.
2. It is not just guesswork, but the result of careful observations, including comparisons of rock layers and fossils and measurements of age.
3. The oldest parts have only fossils of very simple organisms. Recent portions have fossils of more and more complex organisms.
4. Similar fossils are found in rock layers of similar age. Where age differences are big, differences among fossils are big too. This suggests that evolution is ordinarily a gradual process.
5. Many fossils represent species that do not exist today. They have become extinct. Other species have undergone little change during long intervals of time. Some of these "living fossils" are pictured below.

"Living Fossils"

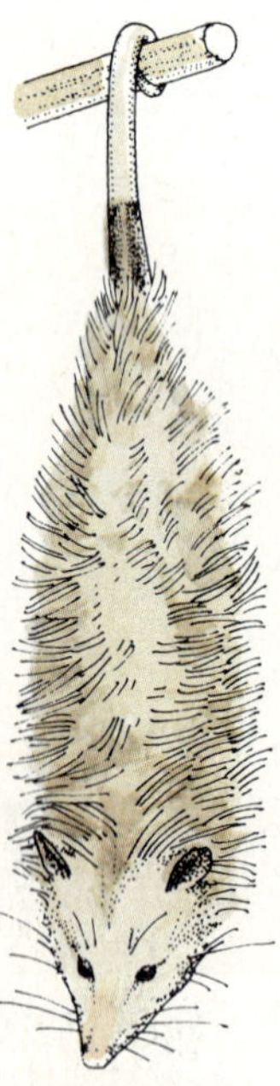
Opossum

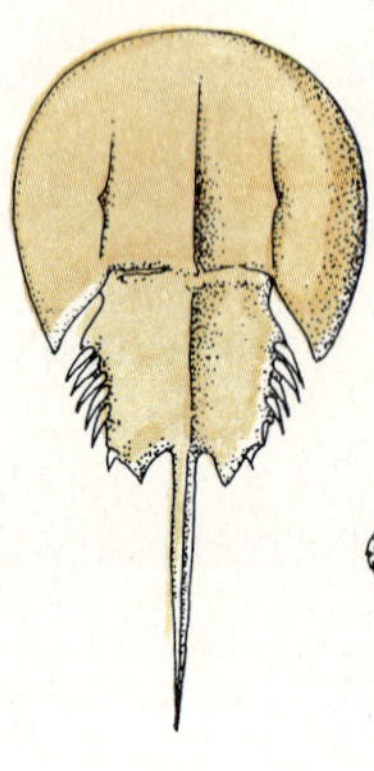
Horseshoe crab

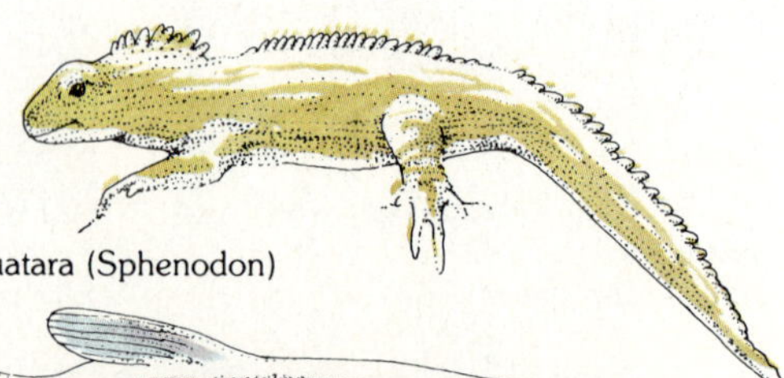
Tuatara (Sphenodon)

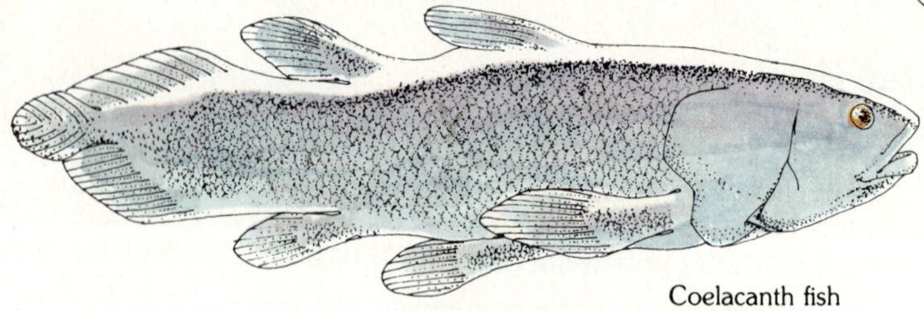
Coelacanth fish

The Fossil Record

Era	Period	Epoch	Outstanding events	Millions of years ago
Cenozoic	Quaternary	Recent	Major human civilizations	Less than 1
		Pleistocene	*Homo sapiens*	2?
	Tertiary	Pliocene	Later hominids	6
		Miocene	Increase in mammal populations	22
		Oligocene	Early hominids; grasses, grazing mammals	36
		Eocene	Primitive horses	58
		Paleocene	Mammals; dinosaurs become extinct	63
Mesozoic	Cretaceous		Flowering plants	145
	Jurassic		Birds; dinosaurs climax	210
	Triassic		Early mammals; conifers, cycads, dinosaurs	255
Paleozoic	Permian		Mammallike reptiles	280
	Pennsylvanian		Insects, amphibians, reptiles, coal forests	320
	Mississippian		Club mosses Horsetails	360
	Devonian		Amphibians	415
	Silurian		Land plants and animals	465
	Ordovician		Early fishes	520
	Cambrian		Marine animals	580
Pre-cambrian			Early marine animals	1 billion
			Green algae	2 billion
			Bacteria, blue-green algae	3 billion

Visualizing the Geologic Time Record

The amount of time represented by the fossil record is so enormous that it is hard to visualize. Looking below at the diagram, all the oceans of the world could be used to represent the estimated 4.5 billion years since the origin of the earth.

A single beach from the many kilometers of beaches surrounding all of these oceans could represent the estimated time humans have been on earth. Finally, a single shell on this beach would represent the time since the earliest known civilizations in Asia.

Visualizing the Time Record

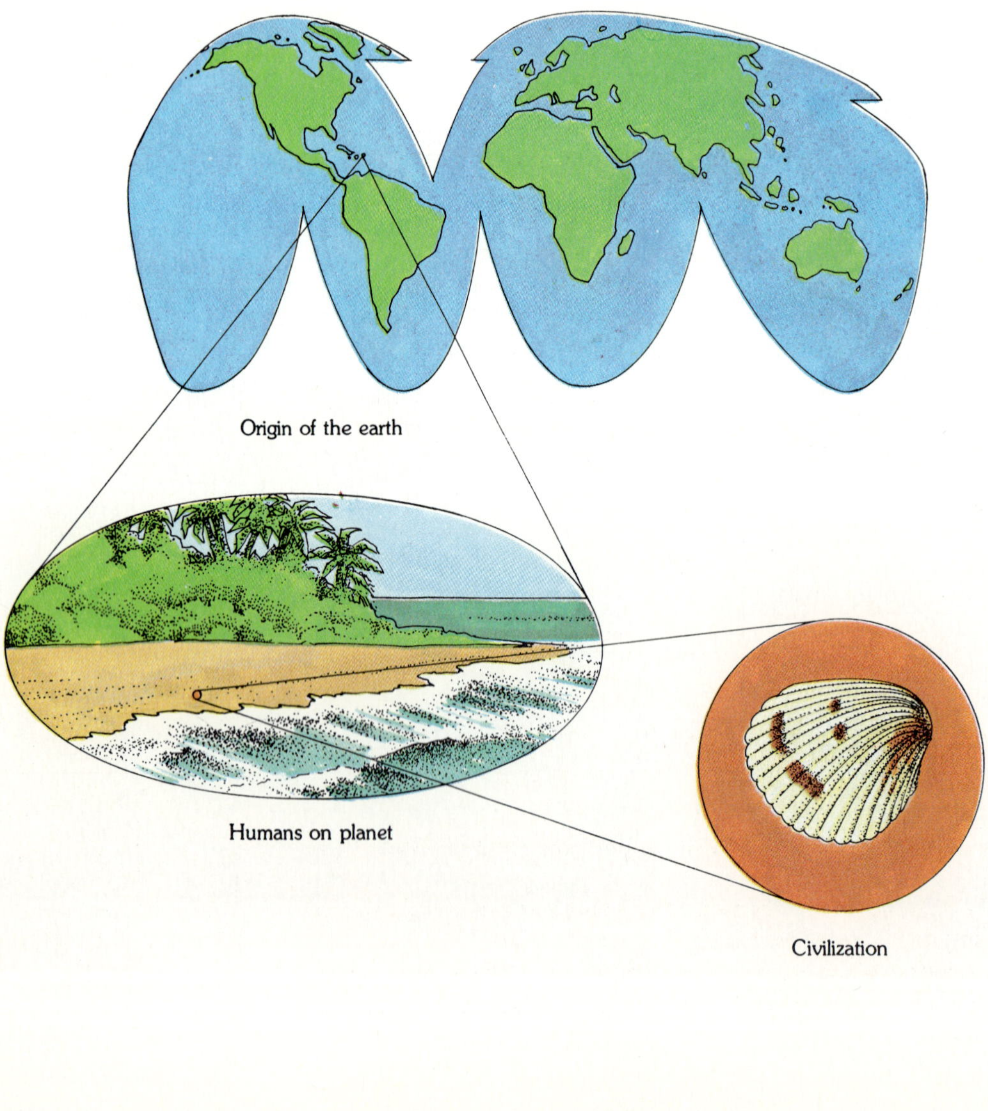

Relationships Among Life Forms

Now suppose the fossils described in the fossil record were arranged according to age and to similarity of structure. An interesting pattern appears. The oldest forms are always simple. And the many small changes through time add up to more and more differences and complexity. The overall pattern resembles a tree. The oldest, simplest forms make up the trunk of the tree. The more recent forms make up the branches. You can see the relationship pictured below.

The actual record of rocks and fossils is not perfect. There are gaps in it, meaning that there are layers of rock in which fossils of some kinds of organisms have not been found. When the fossils occur again in a more recent layer, they may be quite different than in the earlier layers. At these points, the fossil record does not provide evidence of gradual evolution of that species. However, new evidence is continually filling these gaps. Scientists generally think the gaps represent incomplete evidence, rather than special exceptions to the pattern of the fossil record.

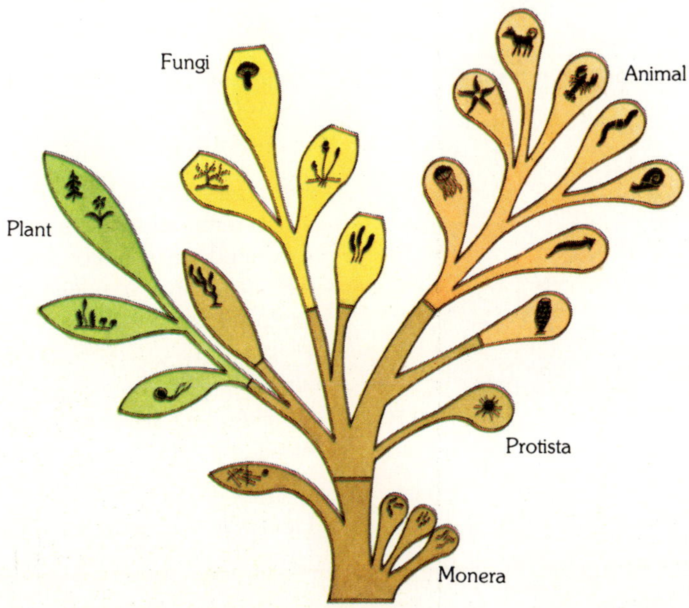

"Tree of Life": Relationships among organisms through time

Review It

1. What are the names of the eras of the geologic record? What is the estimated length of each?
2. List three kinds of observations and measurements that are used in putting together the fossil record.
3. Describe the overall pattern of the "tree of life." Where are species simplest? Most complex? Where do the most differences among species exist?

Activity

Making Fossils

Materials
Leaf
Dead insect
Corn syrup or balsam
Glass slide
Petroleum jelly
Glass dish
Plaster of Paris
Cardboard strip
Eyedropper

A. Procedure
Small insects and spiders sometimes get stuck in a sticky substance called **resin.** Resin comes from the bark of certain trees. After a long time, the resin hardens. Insects trapped in the resin are preserved as fossils. Parts of plants may also be preserved in the same way.

1. Get a small, dead insect, some corn syrup, and a glass slide.
2. Place the insect on the slide. Drop a small amount of syrup on the insect.
3. Allow the syrup to dry. Add more layers of syrup, if necessary, until the insect is completely covered. You now have an insect "fossil," similar to those formed millions of years ago.

Analysis
1. What could a scientist learn from this kind of fossil?
2. What other kinds of organisms could be preserved in this way? Which ones could not? Explain your answer.
3. What substances, besides resin, could form whole-organism fossils, such as those made by resin?

Insect fossil trapped in resin

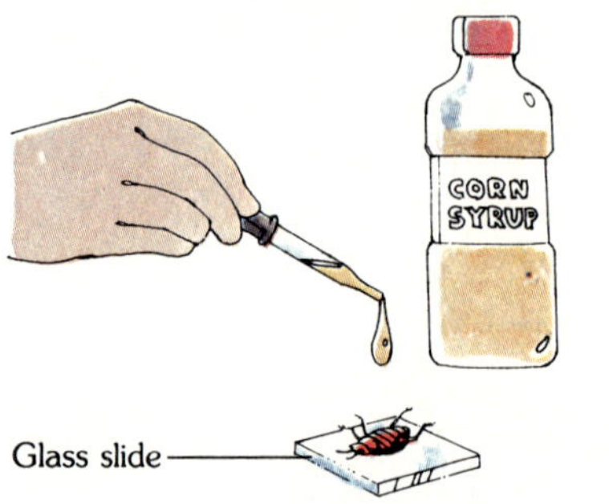

Fossil leaf impression

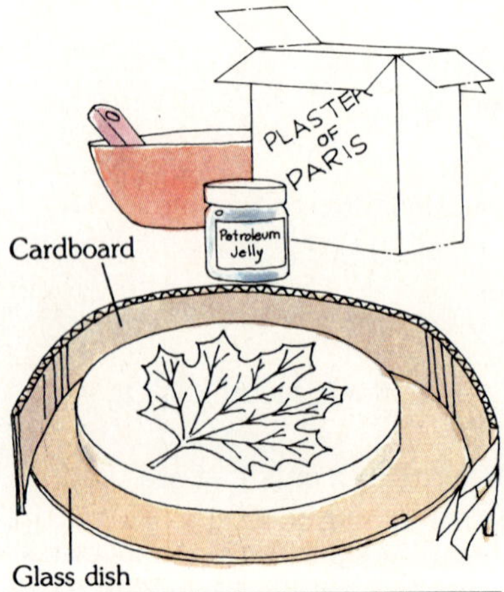

B. Procedure
Many plants became fossils when they were buried in material at the bottom of lakes, rivers, or swamps. The soft parts of the plant, such as leaves, slowly decayed. But as the mud changed to rock, a flat *print* of the leaves was left in the rock.

1. Smear petroleum jelly on both sides of a leaf. Put the leaf in a glass dish.
2. Place a strip of cardboard around the leaf. Now pour plaster of Paris over the leaf.
3. After the plaster has hardened, remove the leaf. The hardened material you have left is a "print" of a leaf, similar to a fossil print.

Analysis
1. What could a scientist learn about a leaf from the imprint left in mud?
2. How is a mud-print fossil different from one made in resin?
3. Tracks of animals, trails of worms, and burrows of animals can be preserved as prints. What could a scientist learn about the habits of such animals from studying prints?

8–3 Evidence from Observing Living Things

The observations and conclusions of Charles Darwin on the five-year voyage of the ship *Beagle* stimulated much scientific thought about how species adapt to their environment and survive. Although Darwin's work has made the trip famous in history, the main purpose of the voyage was to continue the survey and mapping of South America. Darwin was allowed to come along as the ship's naturalist. He gathered together the scientific equipment that he would need to make his collections and fit it into the small space given to him at the stern (back) of the ship. He would receive no pay.

The year was 1831. The ninety-foot ship—about the size of two classrooms—was packed full of provisions for the long journey. In his quarters, Darwin had to remove one of the drawers from some cabinets in order to have a place for his feet when he slept. He was seasick for most of the voyage, and developed a lifelong loathing for sea travel. But he was fascinated with what he observed on the land at each of the stops along the ship's route, and he recorded these observations with great care and detail.

Darwin became the main contributor to the early theory of evolution. When he published his book, *On The Origin Of Species,* in 1859, two kinds of evidence influenced his thinking. One was the physical differences he observed among some closely related species of organisms. The other was his observation that different species have similar body parts. His observations of 14 species of finches in the Galápagos and Cocos islands are especially well known.

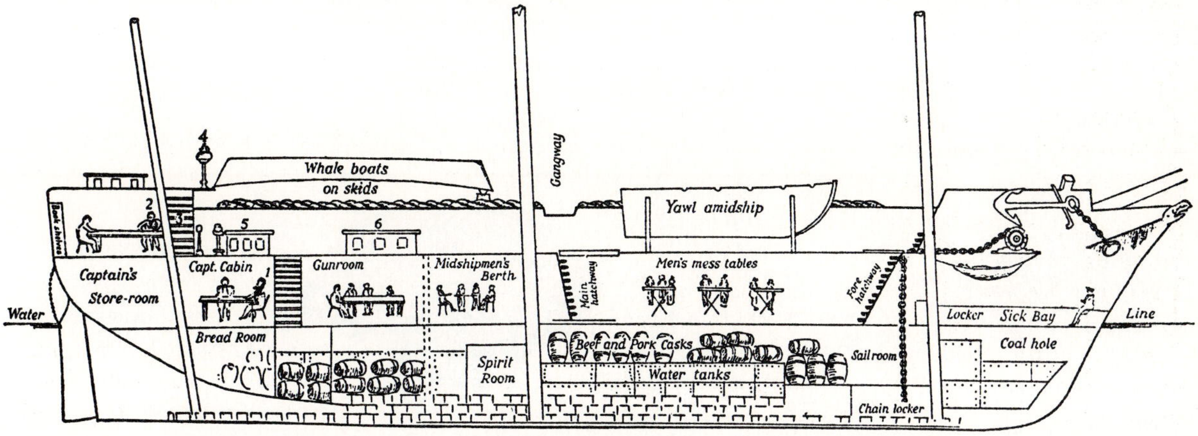

H.M.S. Beagle 1832

1 Mr. Darwin's seat in Captain's Cabin 2 Mr. Darwin's seat in Poop Cabin 3 Mr. Darwin's drawers in Poop Cabin 4 Azimuth Compass 5 Captain's skylight 6 Gunroom skylight

Darwin's Finches

Overall, the 14 species of finches that Darwin observed were very similar. But, as you can see, there were distinct differences in their beak structure. The differences were related to differences in diet and in location of each bird population. Such differences among populations are exactly what one would expect if the changes arose through evolution. For example, the differences in beak structure could easily have been the result of many small genetic changes. The changes might have occurred in separate finch populations over a long period of time.

Of course, the observations of beak structure do not *prove* evolution. The function of a theory in science is to explain as many observations as possible. However, in this case, the theory of evolution seems to be a reasonable explanation for Darwin's observations.

Darwin's Finches

Shape of Bill	Type of Bill, Habits	Shape of Bill	Type of Bill, Habits
 Warbler finch	Slender, pointed bill Feeds on small insects; lives in trees	 Large cactus ground finch	Heavy, pointed bill; suited for crushing Feeds on small cactus seeds; lives on ground
 Vegetarian finch	Thick, grasping bill Feeds on fruits, seeds, and leaves; lives in trees	 Medium ground finch	Very powerful, heavy bill; suited for crushing Feeds on large seeds and nuts; lives on ground
 Woodpecker finch	Thick, pointed bill Uses cactus spines or twigs as tools to probe for insects; lives in trees	 Large ground finch	Very thick, curved bill; parrotlike Feeds on buds and fruit; lives on ground

Evidence from Similarities Among Organisms: Homologous Structures

The second source of evidence that influenced Darwin's thinking was his observation that different species had similar body parts. The study and comparison of the body parts of different species is called **comparative anatomy.**

A study of the limbs of different vertebrates, for example, shows interesting similarities. Below you can see the forelimbs of several different animals. The skeletons of the limbs are quite similar, even though the limbs have different functions. The forelimbs of the bird are used for flight. However, the forelimbs of the whale are used for swimming. And human forelimbs are used for manipulating tools.

Organs and body parts in different species that have a similar structure, but different functions are called **homologous** (hō mol'ə-gəs). How do such homologies occur in the first place? Darwin supplied an answer to this question. He suggested that different organisms possess homologous organs because they are descended from a common ancestor. For example, a "bird like reptile" might have been the early ancestor of the reptiles, birds, and mammals that exist today. The existence of a common ancestor would explain the similarities that can be observed among the forelimbs of reptiles, birds, and mammals.

But again, homology does not *prove* evolution. Nobody has actually seen such gradual changes occurring over millions of years. Nobody has actually seen a reptile like ancestor slowly develop into a bird on the one hand, and a whale or a human on the other. But homology does *suggest* evolution.

Homologous Structures

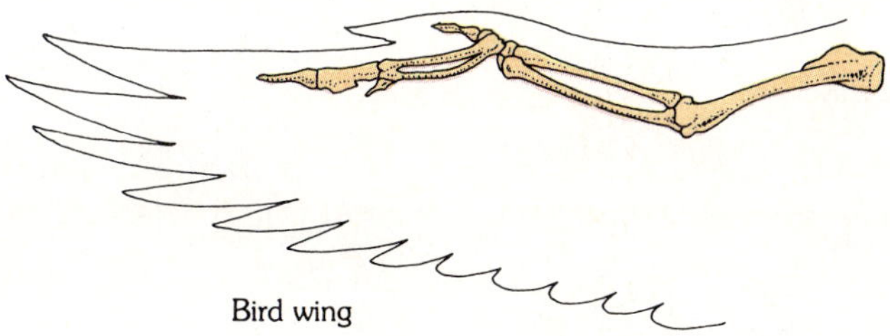

Bird wing

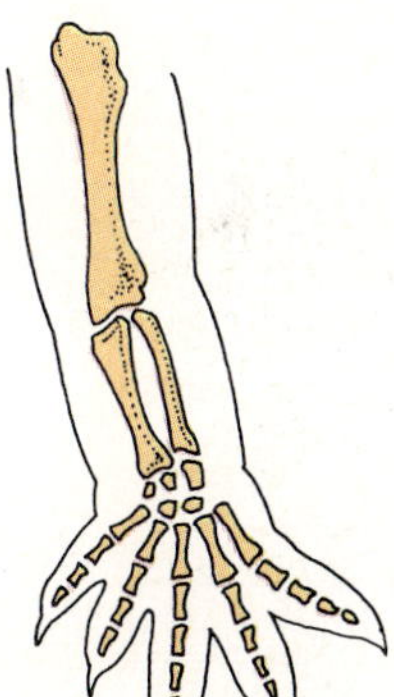

Alligator forelimb

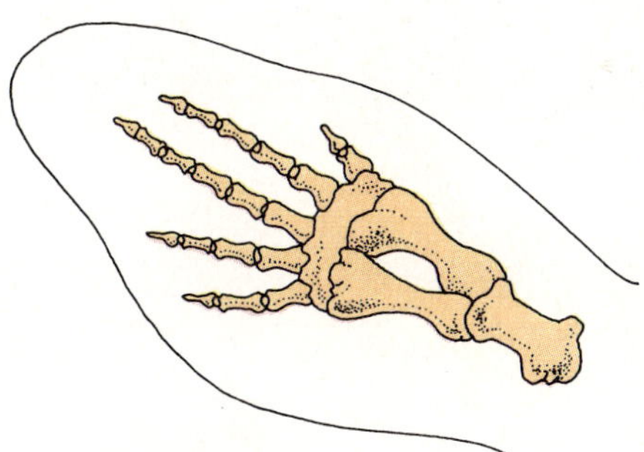

Whale flipper

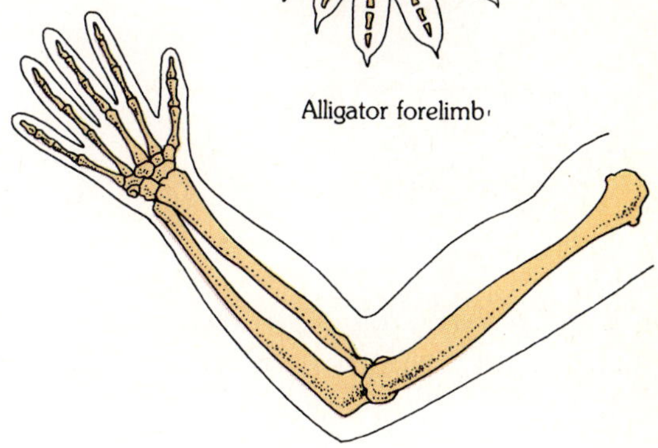

Human arm

Analogous Structures

Animal or plant structures that are similar are not necessarily homologous. For example, a butterfly wing and a bat wing may, at first glance, look similar. The wings certainly have similar functions. However, upon closer examination, differences do appear. The bat wing you see below has an *internal* skeleton of bone. The skeleton, in turn, is covered by layers of muscle and skin that are many cells in thickness. The butterfly wing, however, has a delicate *external* skeleton. The "skeleton" is made up of wing veins. The wing itself consists of a membrane made up of a thin layer, only two cells in thickness. Organs or body parts that have a similar function, but that differ in structure are called **analogous** (ə nal′ə gəs). Analogous structures probably indicate that the organisms did not originate from a common ancestor.

Analogous Structures

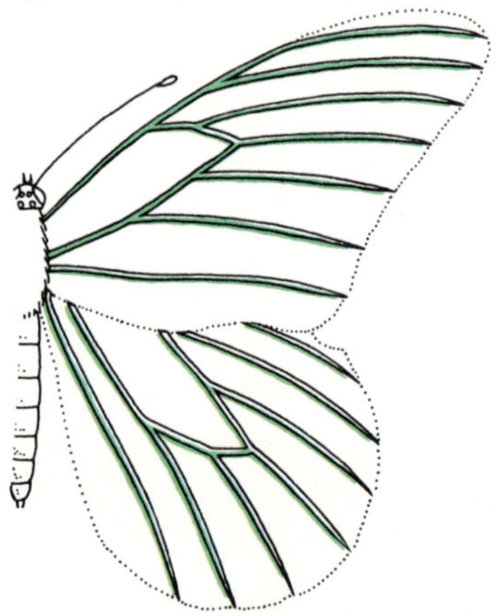

Butterfly wing

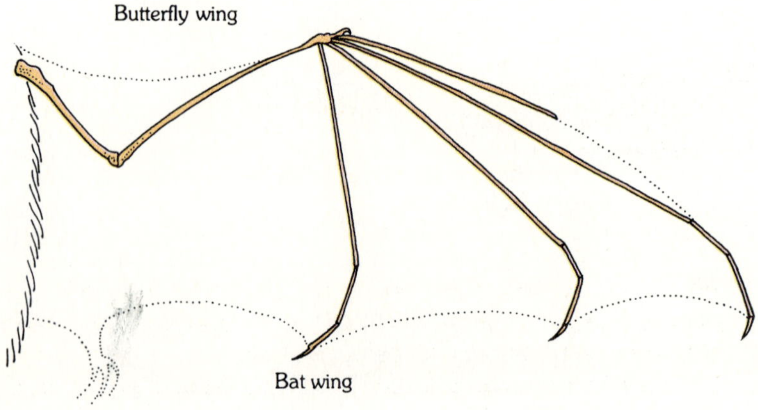

Bat wing

Analogous Eye Structures

The image-forming eye of a squid and a human also appear to be very similar. Each of the two eyes has an iris to regulate the amount of light entering the lens. Each eyeball is filled with fluid. And each uses a lens to focus a sharp image on a light-sensitive screen.

Based on these observations, you might conclude that a human and a squid are closely related. However, if you studied the details of how each organism develops and how each eye develops, you would find many differences. Tissue structure and organization of the eyes are very different. In fact, the eye of a squid and the eye of a human are analogous, *not* homologous. You can study these differences at the right.

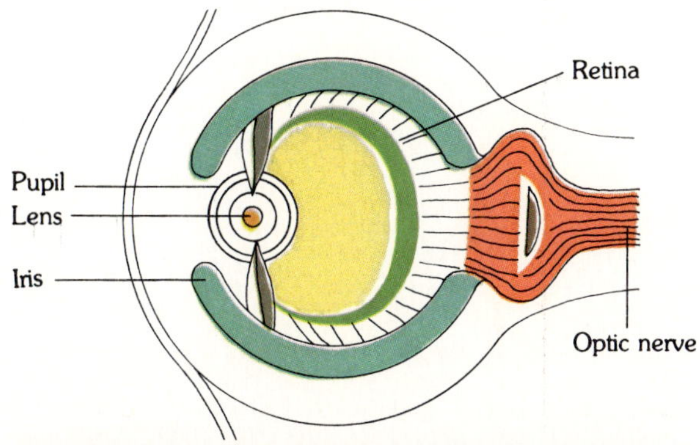

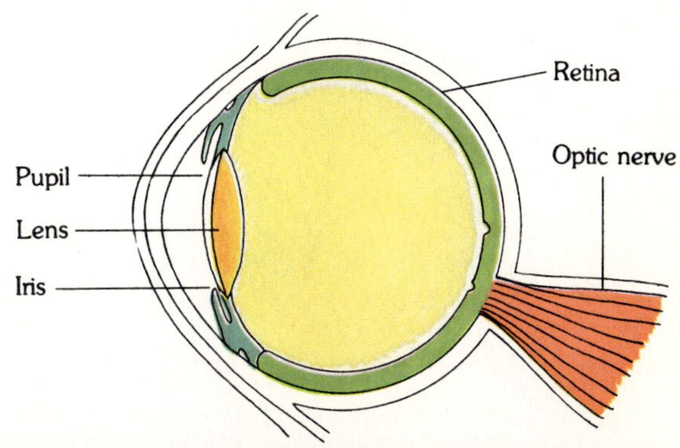

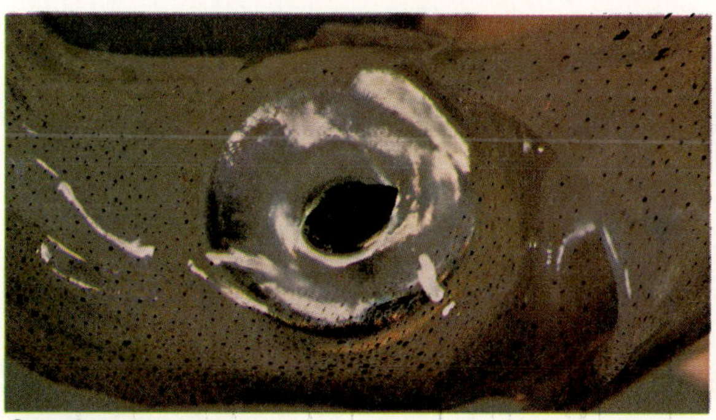

Squid eye structure

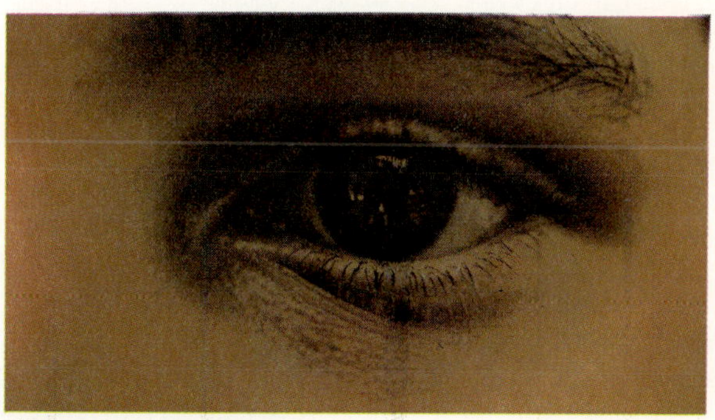

Human eye structure

Review It

1. What are some differences among the 14 species of finches of the Galápagos and Cocos islands that Darwin observed? How are these differences evidence of change by evolution?
2. What are homologous structures? How are homologous structures evidence of evolution?
3. What are analogous structures? How do analogous structures differ from homologous structures?

8–4 Evidence from Observing Embryonic Development of Organisms

Scientists have often compared the embryos of different kinds of vertebrates. For example, they might study embryos of fish, salamanders, tortoises, chickens, rabbits, and humans. In such studies, they find an interesting pattern. All of these organisms are similar to each other early in their embryonic development. They become progressively different from each other as development proceeds. The drawings below show you this pattern for a variety of vertebrate animals.

Embryonic Development of Five Animals

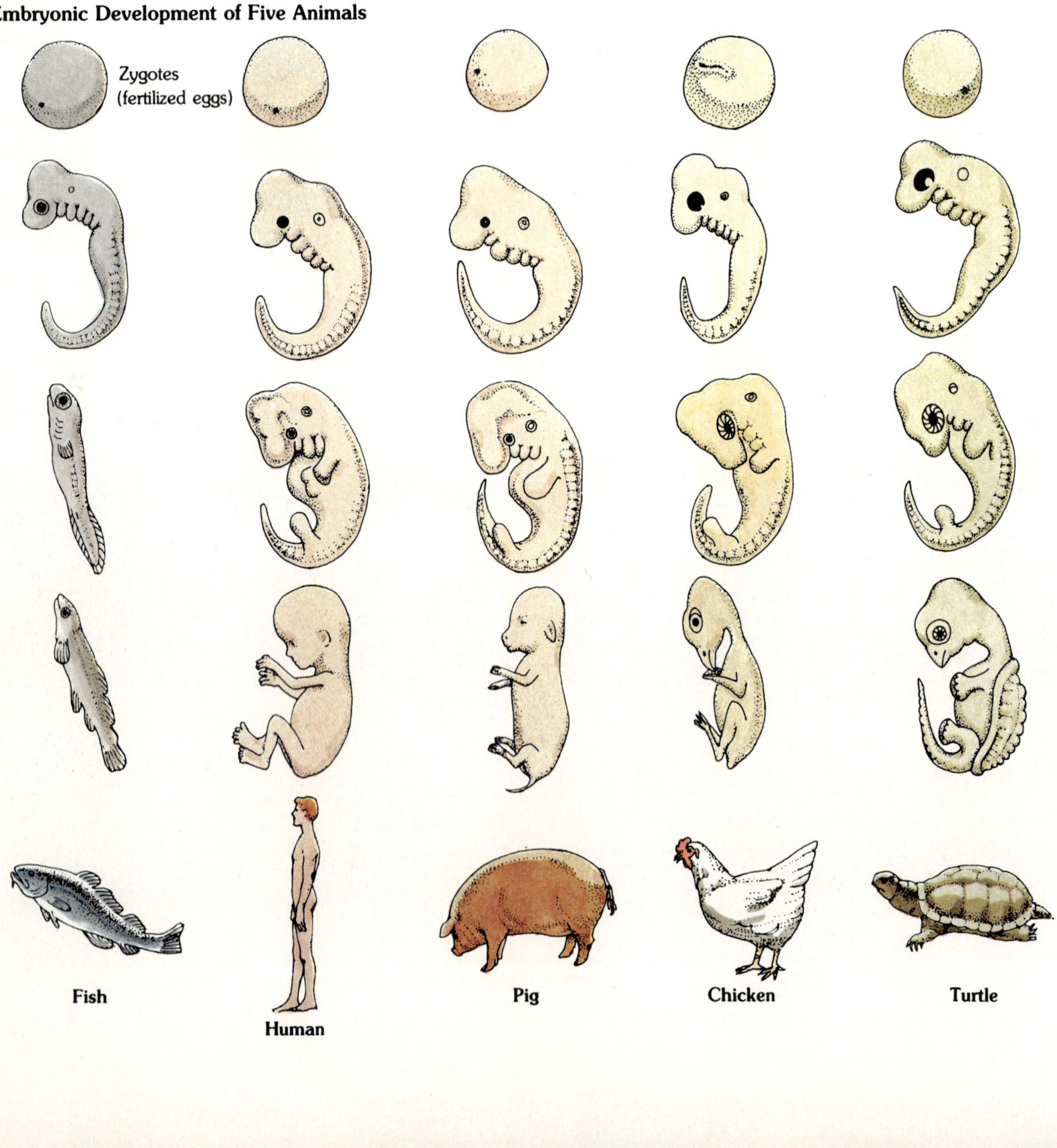

Development of Body Organs

Often, the internal organs of vertebrates also resemble each other during embryonic development. For example, the kidneys of vertebrates are of three major types. The drawings below show you the basic structure of each type.

Early in embryonic development, most vertebrates (except possibly mammals), have Type A kidneys. This type remains throughout adulthood in hagfish and some true (bony) fish. However, during the development of most bony fish and amphibians, the Type A kidney is replaced by a Type B kidney. Fish and amphibians retain the Type B kidney into adulthood. The embryos of reptiles, birds, and mammals also have Type A and B kidneys during embryonic development. But the Type B kidney is finally replaced by the Type C kidney. The kidneys of adult reptiles, birds, and mammals remain Type C.

The important thing to remember here is that adult organisms with Type C kidneys also had Type A and B kidneys sometime during their embryonic development. The same pattern of similarities is true for other organs and organ systems.

In summary, the genetic differences which make one group of organisms different from another show up more and more as the organism develops. Genetic differences seem to take effect late in the embryonic development of each organism. This pattern suggests that differences among species have evolved in many steps from a common ancestor.

However, here is an important warning. The pattern of similarity in embryonic development does not mean that a bird embryo goes through a fish and a reptile embryo stage. It also does not mean that birds have evolved first from fish and then from reptiles. But it does suggest that very young bird embryos are similar in some ways to very young embryos of related organisms. Genetic differences that show up during the development of different species and groups are collections of such genetic changes. Biologists think these changes have been accumulating for thousands, even millions, of years.

Kidney Types

Type A
Kidneys of Lungfish

Type B
Kidneys of Frog

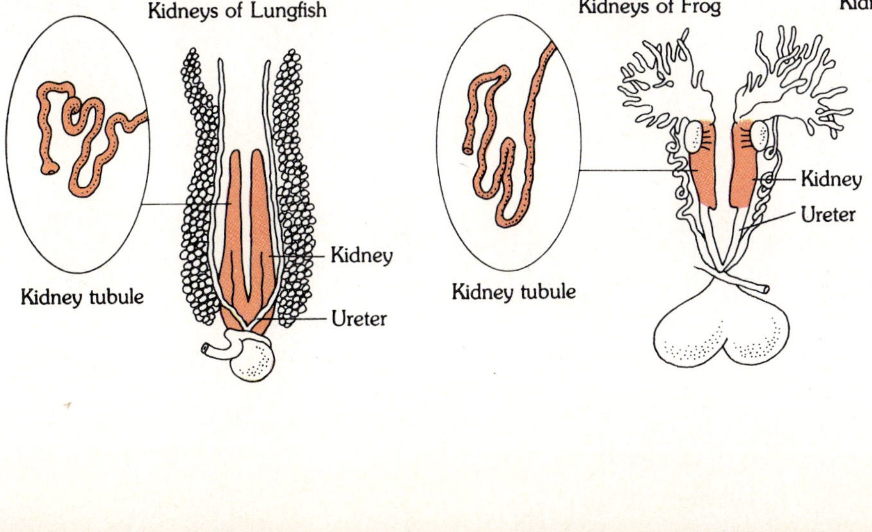

Type C
Kidneys of Pigeon

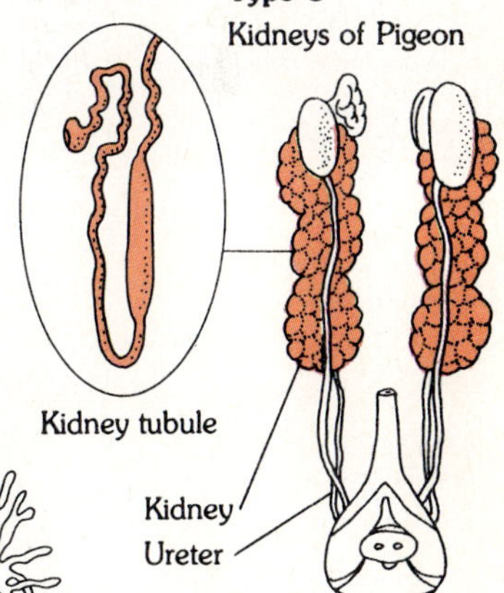

Evidence from Vestigial Structures

Cave-dwelling lizards have vestigial eyes

Cormorant with vestigial wings

A careful study of an animal may reveal certain body parts that do not seem to perform any function. But a similar animal may have a homologous part that does have an important function. The parts of an organism that do not seem to have a function are called **vestigial structures** (ve stij'ē əl).

The flightless cormorant of the Galápagos Islands has unusual vestigial structures. The wings of this bird are small and poorly developed. They cannot be used for flight, but they may help a little for balance. It is thought that the ability to fly has not been helpful to the survival of this bird. So the ability to fly has gradually been lost. The wings of the flightless cormorant are vestigial.

Some cave-dwelling fish and salamanders have eyes. But the organisms cannot see with these eyes. In other species, the eyes are essentially missing. There is no advantage in having vision in entirely dark caves. So the eyes have become vestigial structures.

Numerous vestigial structures also occur in humans. Probably the best known one is the appendix. In some mammals, this organ is much larger than in humans and plays an important part in digestion. However, it has no such function in humans. Some other vestigial parts of humans that have been identified are (1) muscles to move the ears, (2) body hair (which does not keep us warm), and (3) caudal vertebrae. (These are homologous to the tails of vertebrate animals.)

Again, vestigial structures do not prove evolution. When you observe such structures, you are not observing evolution. But vestigial structures do provide strong evidence for evolution. They provide evidence that gradual changes in traits have occurred through time. They also provide evidence that organisms with vestigial structures have common ancestry with organisms that have functional homologous structures.

Humans have some vestigial structures

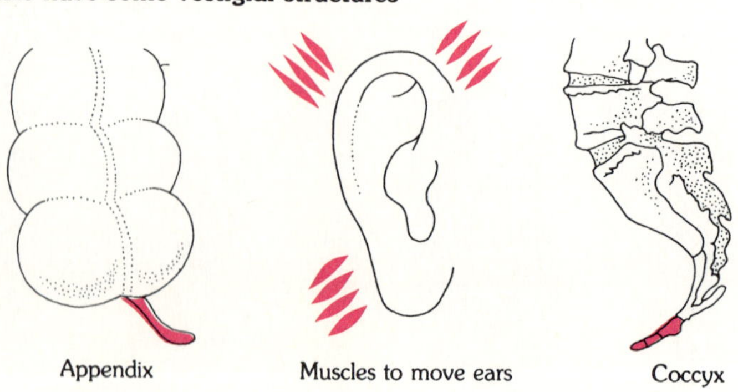

Evidence from Similarities Based on Biochemical Tests

Some fairly recent techniques in chemical analysis have brought out more evidence of evolution. Remember that observed traits, such as eye color, are programmed by an individual organism's DNA. The sequence of bases in the DNA determines the sequence of amino acids in the polypeptide chains of protein molecules. If two groups of organisms, such as bullfrogs and leopard frogs, are very similar, it is because their DNA structures are very similar.

Many chemical analyses and comparisons of proteins and DNA have been made. In these studies, similar organisms are found to have similar DNA and proteins. Different organisms have different DNA. For example, when species from different orders of mammals are compared, the differences in their DNA structures is greater than when species within the same genus are compared.

Studies of blood serum are often used to help determine chemical relationships among organisms. A researcher may inject blood serum from a rabbit into a guinea pig. The guinea pig's circulatory system builds antibodies against the "foreign" proteins of the rabbit's blood. If the guinea pig blood serum is then mixed with the rabbit blood serum, a precipitate forms and settles to the bottom. The precipitate is the result of a chemical reaction between the two blood types.

Now, if the sensitized guinea pig blood is mixed with blood from a *pig,* a precipitate also forms. But not as much precipitate forms. And if the sensitized guinea pig blood is mixed with *rat* blood, almost no precipitate forms. This indicates that the blood of a pig is more closely related to the blood of a rabbit than it is to the blood of a rat. You can see how this works by studying the diagram.

Because of studies like the one above, it was decided that rabbits should no longer be classified with the rats, as rodents. Now, a separate order called *Lagomorpha* was named for the rabbits. Therefore, observations in **comparative biochemistry** tell us how similar the DNA and proteins are between species or groups that scientists wish to compare. This is a basic measure of genetic similarity. The genetic differences are interpreted as indications of how much evolution has taken place.

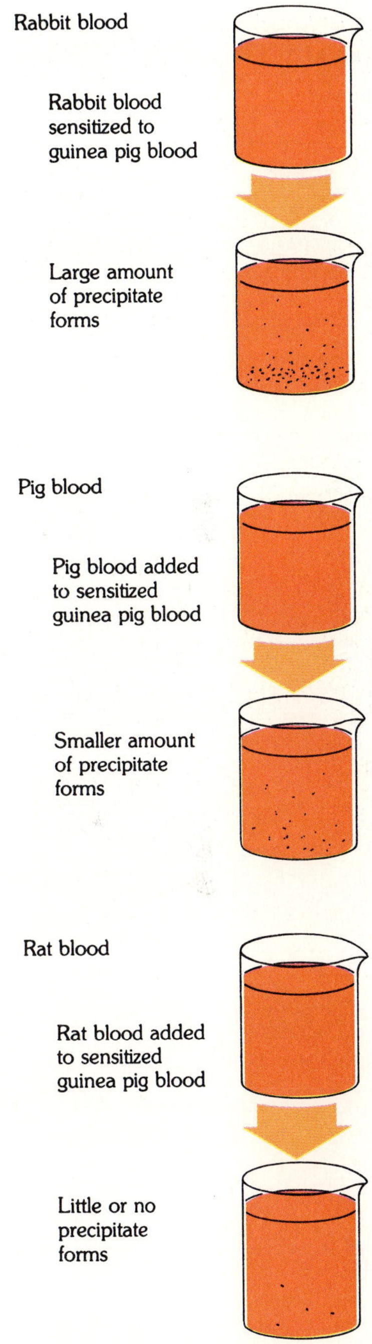

Biochemical test used to determine similarities among blood samples of rabbits, pigs, and rats

Evidence from Race Circles

Recall that males and females of the same species can breed and bear fertile offspring. But, members of separate species cannot breed and produce fertile offspring. The result is that the genetic make-up of two different species is kept separate.

However, when the geographical distribution of a species is studied, scientists often find a number of subgroups of species called **races.** They differ in various ways. But sometimes you can tell races apart by color, body shape, or body size. The members of the races can interbreed, however. Therefore, they are members of the same species.

Studies of races and their distributions have produced very interesting results. The photos on the right show seven races of the salamander *Ensatina eschscholtzi.* The distribution of these salamander species in California is also shown. At five of the borders, where pairs of races overlap, the races can interbreed. However, the overlapping races *klauberi* and *eschscholtzi* interbreed very little, if at all. All of these races are considered to be members of the same species. However, it could also be argued that *klauberi* and *eschscholtzi* are members of different species. Remember, these two races cannot interbreed.

A similar example has been studied in the frog family. The common leopard frog *(Rana pipiens)* occurs from Mexico to Canada and through much of the United States. Generally, these frogs can interbreed. However, if a frog from the Mexican population is crossed with one from the Canadian population, the young often do not survive. The distribution here is not a circle, as in the California salamander. Otherwise, the examples are similar.

In both the case of the salamander and the frog, it can be inferred that **speciation** (spē shē ā′shən) is taking place. That is, it looks as if one species is in the process of becoming two species. We are observing a pattern of similarities and differences that can be explained with the theory of evolution.

Review It

1. At what stage of development are embryos of vertebrates most similar? When are they most diverse? How is this pattern evidence of change by evolution?
2. What is a vestigial structure? What are some examples? How are vestigial structures evidence of change by evolution?
3. How does a study comparing DNA (or a protein) of two groups of organisms provide evidence about evolution?
4. What is speciation? How is a "race circle" evidence of speciation?

A "Race Circle"

8–5
Direct Observations

Indirect evidence based on the fossil record, the development of organisms, and other observations seem to show that changes in populations have occurred through time. These studies have been complemented by observations of actual changes in populations that have taken place within recorded history. Such examples of "microevolution" indicate that changes in the genetic make-up of a population do take place.

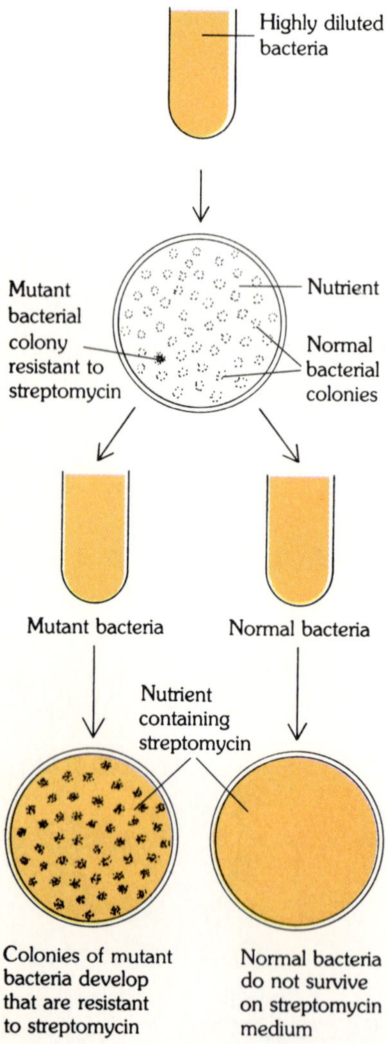

Developing a strain of bacteria resistant to streptomycin

Observations Based on Experiments

Direct observations of changes in the genetic make-up of a population have been made in the laboratory. Changes in the gene make-up, or traits, of a population are transmitted in reproduction. So species with the shortest life span give us the best chance to observe evolution in progress. Since bacteria and insects have very short life spans, they are favorites for experimental studies of evolution. Many generations can be observed in the space of a few days or weeks.

Bacteria and Antibiotics

Antibiotic drugs have been a great advance in medicine. Such drugs kill certain microbes when applied in small doses. However, the efficiency of the drug is reduced when strains of microbes become resistant to the antibiotics. Such resistant bacteria can be developed in the laboratory.

A concentration of the antibiotic *streptomycin,* as low as 25 milligrams per liter, stops the growth of the colon bacteria *E. coli.* However, if several billion of these bacteria are placed on a growth medium containing streptomycin, one or several bacterial cells will survive. These cells will grow and divide and eventually form large colonies. Strains of bacteria permanently resistant to high doses of streptomycin may be developed from such a colony.

During such experiments, a kind of "microevolution" seems to be happening. A major change in the gene make-up of the population occurs. Bacteria with genes for the trait "resistance to streptomycin" survive and form new populations. Those without the trait appear to die out. You can see the set up for such an experiment to the left.

Insects and DDT

Many insecticides, or poisons that kill insects, have also been discovered. A very powerful insecticide is DDT. Houseflies were once controlled by DDT sprays. Yet, houseflies from certain locations became relatively resistant to this poison. In many places DDT could no longer be relied upon to control the flies. Insects, it seems, also possess traits for resistance to certain chemical poisons.

Strains of such DDT-resistant flies have been developed in laboratory experiments. A normal, sensitive strain of houseflies can be exposed to a dose of DDT that kills about 90 percent of the individuals. Now the survivors are used as parents of the next generation. And this generation is also exposed to DDT. After three generations of such selection, a significant resistance to DDT appears in the housefly population.

The emergence of strains of insect pests resistant to various insecticides has become a major problem in recent years to those concerned with pest control. Resistance to DDT has been observed in several species of mosquitoes, body lice, cockroaches, and bedbugs. So here again, we have some examples of what appears to be "microevolution" occurring in insect populations. The changes are due to gradual changes in the gene make-up of the population over a period of time.

Developing a strain of houseflies resistant to DDT

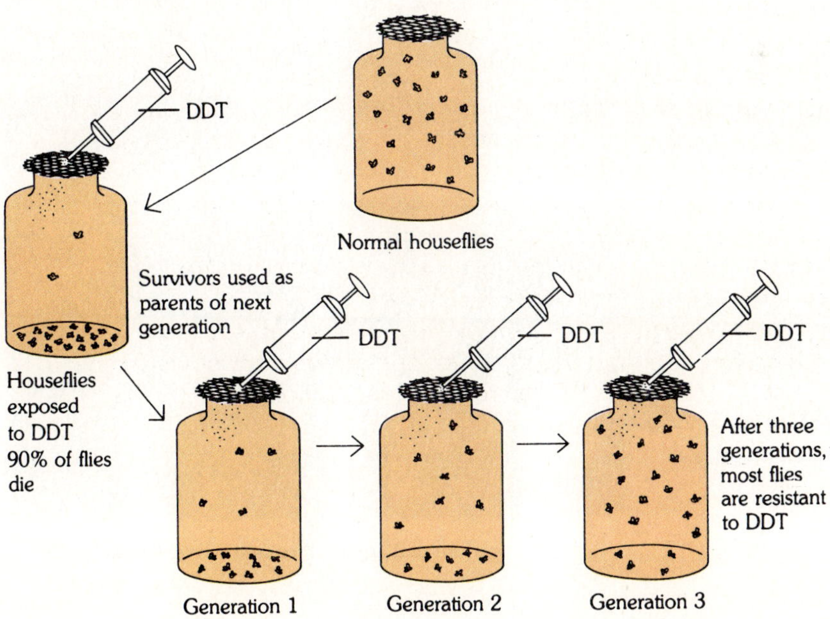

Observations in Nature

The peppered moth *(Biston betularia)* has been studied carefully in its natural setting. It is common in wooded areas of England. Before 1840, most members of the species were gray with dark markings on the wings. Black moths occurred, but were very rare. However, in 1850, the black moths began to increase in number. About 98 percent of the peppered moths were completely black by 1895. The light gray, speckled form had become rare. This change took place in about 50 moth generations.

About the time all this was happening, England was becoming increasingly industrialized. Trees, leaves, and rocks were turning black from the soot in industrial areas. It was suggested that the darker color was the result of a change in the moth's environment. In unpolluted environments, tree bark was light and lichens were abundant. The light-colored moths with speckles were almost completely concealed. However, the black moths stood out and were easily spotted by birds. In industrial areas, where tree trunks were blackened by soot, the situation was reversed. The dark moths were hidden on the tree trunks. But light-colored moths made easy prey.

Moth Coloration in England

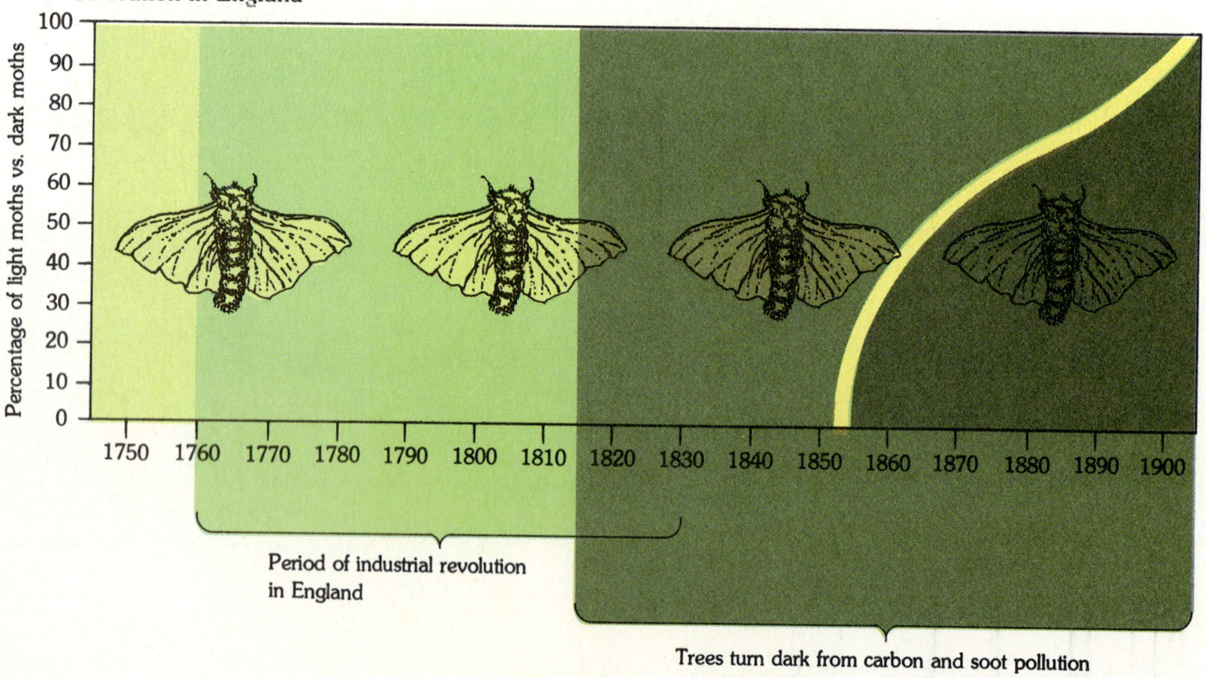

Moth Population Shifts from Light to Dark Coloration

Over a period of time, the moth population changed from light- to dark-colored moths. Genetic studies show the dark color of peppered moths is produced by the dominant allele of a single gene. In polluted environments, bird predators greatly reduce the number of moths carrying the allele for light, speckled color. As a result, the dark moths become more numerous. At mating time, the dark moths have a better chance of passing on their alleles to the next generation. Over time, the frequency of the allele for dark coloration becomes very high, and most moths are dark colored.

The rapid shift toward darker coloration of the peppered moth is called *industrial melanism.* At least 150 other species of moths in Europe and the United States have exhibited industrial melanism. In unpolluted areas, these species are light. But in polluted areas with darkened tree trunks, they are dark.

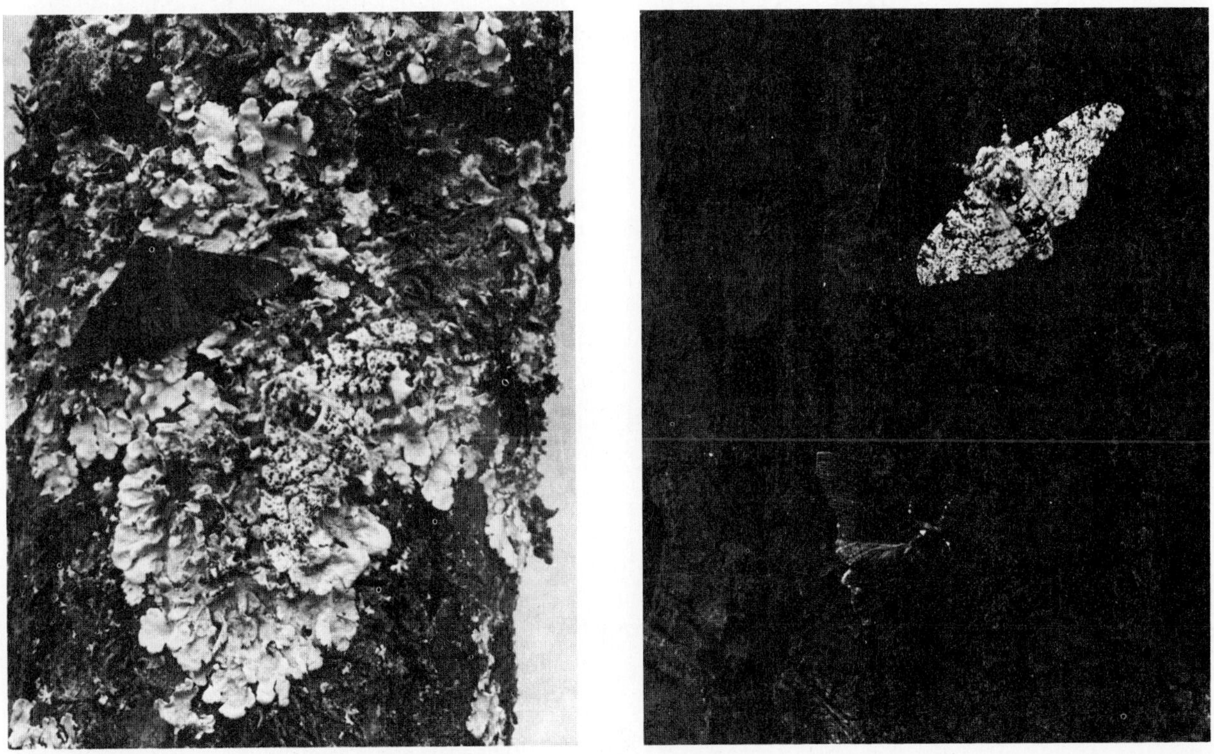

Peppered moth (Biston betularia)

Other Examples of Observed Evolution

Many similar examples of evolution are known. For instance, house sparrows were introduced to North America from Europe in 1852. Descendants of these birds now inhabit the entire North American continent. Environmental conditions vary over wide extremes within the range of these birds. In 1964, birds from the limits of the range were compared. Results seemed to indicate that evolutionary changes have taken place in different directions in various populations.

For example, birds in the rainy northwestern part of Canada are much more darkly pigmented than those from the dry southwestern United States. Birds in Louisiana and Mexico have obvious yellow coloration on the underside of the body. This color is missing from sparrows from other regions. Birds in coastal California are the lightest in weight (27 grams). The heaviest live in the northern parts of Canada (31 grams). Thus, various populations of the North American house sparrow seem to have evolved in different directions, depending on environmental factors in different parts of the continent.

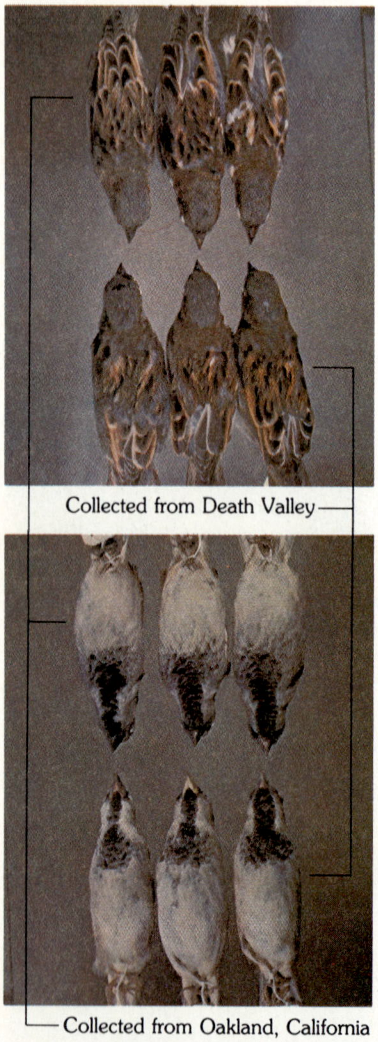

Color variations among house sparrows

Distribution of body sizes of house sparrows in North America

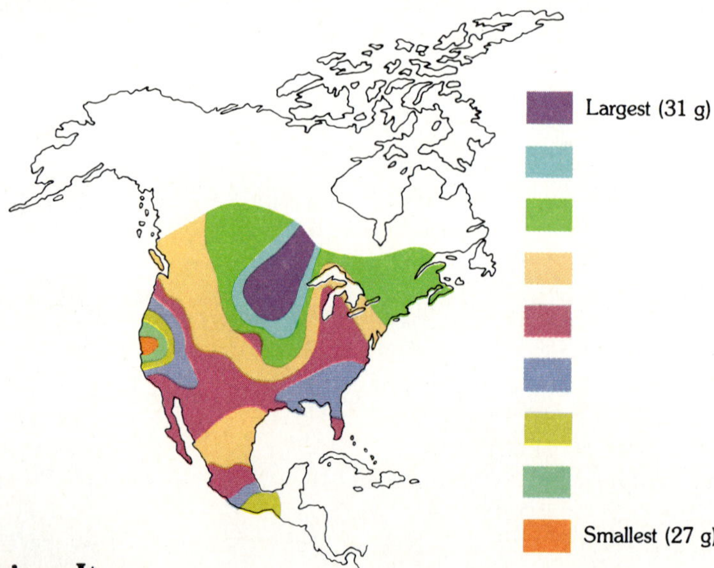

Review It

1. Give at least two examples of observed evolution. Exactly what genetic trait was observed to change in each case?
2. Explain why evolution is more likely to be observed over a short period of time in bacteria and insects, than in mice or rabbits.

Breakthroughs in Biology

New Fossil Finds Raise New Questions about the Ancestors of Birds.

Archaeopteryx (är'kē op'tər iks), "ancient wing," lived about 130 million years ago. Fossil remains of six of these animals were found in the nineteenth century. It had a long jointed tail, cone-shaped teeth, and many other features in common with reptiles. If it had been identified on the basis of its bone structure alone, *Archaeopteryx* surely would have been called a reptile. However, by a rare and fortunate chance, the fossil imprint of feathers on its reptilelike forelimb and long tail were also preserved. Thus, *Archaeopteryx* was seen to be almost exactly intermediate between reptiles and birds. For many biologists, this ancient animal seemed the perfect answer to the question "Who were the ancestors of birds?"

In 1977, Dr. James Jensen of Brigham Young University made a discovery in eastern Colorado that calls into question the role of *Archaeopteryx* as the ancestor of birds. Dr. Jensen found a fossil thighbone (femur) of what appears to be a bird that is as old as *Archaeopteryx.* More recently he has located another fossil femur just meters away from the site of the first one. In this new fossil find, the head of the femur, the upper end that fits into the hip socket, appears to be small and birdlike. This indicates that the animal possessing it was probably a good flyer. By contrast, the femur of *Archaeopteryx* has a well-developed head and is more like that of a ground animal.

If these new finds become generally accepted as fossils of flying birds that lived at the same time as *Archaeopteryx,* the question of "Who were the ancestors of birds?" will have to be raised again.

Analysis

1. What additional fossil evidence might scientists need to support a decision about the place of *Archaeopteryx* in the fossil record?
2. Try to find the latest information on Dr. Jensen's new fossil find.

Chapter Summary

The theory of evolution may be the most important theory in biology because it explains so many observations. Evolution is a change in the gene frequencies of a population across successive generations. Most evidence of evolution is indirect. But some direct observations also exist.

Differences among fossils provide indirect evidence of evolution. Patterns of similarities and differences can be observed throughout the fossil record. Comparative anatomy of living things also provides indirect evidence. Body parts of different organisms which are similar in structure and in embryonic origin are called homologous structures. Often, a study of homologous structures is combined with a study of geographical distribution for useful evidence of evolution.

Comparative biochemistry is used to compare the genetic make-up of organisms and populations. The similarities and differences scientists find are evidence that evolution might have taken place.

Direct observations of evolution are also possible. Recent experimentation with bacteria and insects has produced changes in the genetic make-up of the populations. For example, bacteria with increased resistance to antibiotics have been produced. Hundreds of cases of industrial melanism have also been observed for various species of moths.

All observations of evolution have been based on small-scale changes. However, large-scale evolutionary changes involve enormous periods of time. And according to the theory of evolution, the large-scale changes are made up of the many small genetic changes.

Questions to Think About

1. Is it correct to say that an *organism* evolves? Why or why not?
2. If you found a fossil with 1/8 of the original carbon-14 left, about how old would you estimate that fossil to be?
3. Fossils provide strong evidence of evolution. Yet, fossils are not proof or observations of evolution. What is the difference between evidence and observations in this case?
4. Let us imagine for the moment that all vertebrates begin development as tiny versions of the adult form (microscopic fish, birds, etc.). This is not the case, but if it were, would this evidence support the theory of evolution or conflict with it? Defend your position.
5. Try to think of at least one possible explanation for vestigial structures that does not involve evolution. Do you think your explanation is reasonable? Do you think it is likely? Why or why not?

Extra Research

1. Look up other homologous structures in comparative anatomy books. Look for patterns of similarity and differences.
2. Prepare a report on human hereditary diseases. Try to find out which ones are caused by changes in DNA base sequences. Find out what protein changes are caused by the mutations.
3. Do additional research on fossil formation. What are some processes involved? What different ways do fossils form?
4. Look up the fossil evidence on the evolution of the horse. Develop a chart showing the major changes.
5. Look up the term *dendrochronology*. What does it add to your understanding of measuring the age of fossils?

Chapter Test

Part I

Copy the number of each test item. If the item is *true*, write *T* after the number; if the item is *false*, write *F*.

1. The eras of the geological time record are all equally long. (192)
2. There are some gaps in the fossil record. (195)
3. Only the skeletons and shells of animals have been preserved from the past. (188)
4. Similarities among organisms was an important part of Darwin's evidence on evolution. (197)
5. Differences among organisms was an important part of Darwin's evidence on evolution. (197)
6. During development, birds pass through a fish stage and a reptile stage. (203)
7. Races of the same species are usually considered to be interfertile. (206)
8. Homologous structures have similar embryonic development. (203)
9. In circles of races, we are observing speciation taking place. (206)
10. A homologous structure can also be vestigial. (204)
11. Comparative biochemistry can be used to help classify organisms. (205)
12. Darwin's finches with their beak structures were observations of physical differences among closely related species. (198)
13. Insect resistance to DDT is an example of an individual organism's ability to change its genetic makeup. (209)

Part II

Copy the number of each sentence below. After each number, write the word or words that complete the sentence correctly.

14. Two ways that fossil age can be measured are _____ and _____. (189–190)
15. The pattern in the fossil record indicates that organisms have developed from simple to _____ over a long period of time. (195)
16. Similarities in the embryonic development of organisms suggests that they have developed from a common _____. (203)
17. _____ is used to study chemical relationships in the DNA and protein structure of organisms. (205)
18. One example of a change that has been observed in a population through time is _____. (209)
19. A rapid change in the coloration of a moth population that is influenced by an industrial environment is an example of _____ _____. (211)
20. Fossils are mainly found in _____ rock. (189)

Part III

Copy each numbered item from List A. After each item, write the letter and words from List B that best describe the item.

List A

21. half-life (190)
22. homologous structures (199)
23. vestigial structure (204)
24. epoch (192)
25. analogous structure (200)

List B

a. subdivided periods in geological record
b. length of time it takes for half the atoms of a radioactive substance to disintegrate
c. similar structure and origin
d. similar in origin, but functionless
e. similar in function, but not origin

Chapter 9

Mechanisms of Change

Some parts of evolution can be observed. Other parts can be inferred. Quite a few ways that gene frequencies change are now known. If we can explain how gene frequencies change, we can explain how evolution occurs.

How did Darwin explain evolution? How do we explain it today? In what ways is our explanation more complete than his? How do we explain the formation of separate species from a common ancestor? You will find answers to these and other questions in this chapter.

Chapter Objectives

1. Describe the difference between Lamarckism and Darwinism.
2. Define artificial selection and natural selection.
3. Describe the part mutations play in evolution.
4. Summarize how natural selection works.
5. Describe at least five ways populations become genetically isolated from each other.
6. Define adaptation.
7. Explain speciation using isolation of populations.
8. Compare Darwin's theory of evolution with the modern theory of evolution.
9. Describe the difference between convergent and divergent evolution.

This angler fish lives at depths where it is almost totally dark. It has a luminous lure that dangles from its jaw and attracts its prey. The luminous lure is an example of the angler fish's adaptation to its dark environment.

9–1
Early Attempts to Explain the Fossil Record

Jean Baptiste de Lamarck (1744–1829) was a well-known French naturalist. During his time, evidence for **evolution** (ev'ə lü'shən) was not available from genetics, biochemistry, and physiology. He and many others thought evolution could be explained by inheritance of **acquired characteristics.** Lamarck said that at one time an animal with a short neck stretched its neck trying to reach leaves on trees. In order to reach higher leaves, the animal would stretch farther and eventually develop into the animal we know as a giraffe. According to Lamarck's theory, its offspring would inherit the long neck. This means that evolution would take place by acquired characteristics. Inherited traits would be the result of accidents and also of the way ancestors used their body parts.

Evidence does not support Lamarck's theory of evolution. Special use of a body part does not change the genetic make-up of that part. Acquired characteristics are *not* inherited.

The evolution of giraffes: Lamarck's hypothesis

1. Ancestral giraffes had short necks. These giraffes stretched their necks in order to reach tree leaves.
2. The offspring of short-necked giraffes had longer necks than the parents. These giraffes, in turn, stretched their necks to reach tree leaves.
3. After many generations of neck stretching, the long-necked giraffes developed.

Darwin's Hypothesis

Darwin developed a different theory of evolution. According to his theory, ancestral giraffes had differing neck lengths. Because of this, longer-necked giraffes were able to reach leaves better than shorter-necked ones. More food was available to these giraffes so they were more likely to survive and reproduce. Longer necks were inherited by the offspring, and eventually the giraffe population had only long-necked giraffes.

The evolution of giraffes: Darwin's hypothesis

1. Ancestral giraffes had necks of various lengths. The various neck lengths were inherited.
2. Giraffes with longer necks could more easily reach tree leaves. Longer-necked offspring had a better chance of surviving than those with shorter necks.
3. The trait for long necks was passed on to each generation. After many generations, only long-necked giraffes survived.

Artificial Selection

Darwin's theory was based on many observations. He had traveled over much of the world and observed a wide variety of animal life. Darwin also experimented by breeding pigeons. He observed the big differences between wild pigeons and fancy domestic varieties developed by careful selection and breeding for desired traits. Darwin was impressed that such a great variation in breeds could be produced through selective breeding. Such selection for desired traits is called **artificial selection.** It has been practiced by humans for thousands of years. The traits of domesticated animals and cultivated plants have been accumulated in this way.

Breeds of domestic pigeons

Variations in a species due to artificial selection

Natural Selection

Darwin said that, in nature, species change through **natural selection.** By this he meant that conditions in nature determine which organisms survive and reproduce. Extreme cold, shortage of food, and natural enemies may cause some organisms to die before they reproduce. A selection process takes place, but the selection is not made by humans. It occurs rather through environmental factors acting upon variation in the population.

Darwin said that evolution works in the following way:

1. Populations tend to overproduce. They produce more offspring than are likely to survive.
2. The members of a population are not identical. They show a variation from one to the other, and these variations are inherited.
3. In nature, members of a population must struggle to survive, and not all will survive.
4. Since members of a population are not identical, some are likely to be better suited to survive than others. The members that are better suited to the environment will survive and have offspring.

Darwin did not know how differences among members of a population arise or how differences are inherited. His theory just stated how natural selection operates upon these variations. Darwin's theory of evolution explained many observations, but it was an incomplete explanation.

Review It

1. According to Lamarck, how would a population of giraffes inherit long necks?
2. According to Charles Darwin, how would a population of giraffes inherit long necks?
3. What is artificial selection? How do changes in genetic make-up come about through artificial selection?
4. What is natural selection? How do changes in genetic make-up come about through natural selection?

Pioneers in Biology

Charles Robert Darwin: Pioneer in developing the theory of evolution.

When he was 21, Charles Darwin had a chance to travel around the world. The voyage changed his entire life. It has also changed the thinking of many people since then.

Charles Darwin was born on February 12, 1809, in Shrewsbury, England. As a young man, he enjoyed long walks and insect-and plant-collecting excursions in the country. Charles began his career education at Edinburgh University with the intention of entering his father's profession of medicine.

After two years, and a great dislike for anatomy, Charles changed his career plans to the ministry, entering Christ's College, Cambridge, in 1827. While pursuing his studies, Darwin's love of nature, of collecting, and the study of geology, zoology, and botany persisted. At Cambridge, he became friends with Professor Henslow, a botanist. It was Henslow who suggested Darwin's name to the British Admiralty as the one best qualified to be the ship's naturalist during their upcoming voyage to continue surveying South America.

On December 27, 1831, the journey on the 10-gun brig, H.M.S. *Beagle,* began. For five years, Darwin made observations and collections, and took detailed notes in geology, botany,

and zoology. These observations and notes formed the basis for much of the writing that he did later in his life. He was especially impressed with what he saw on the Galápagos Islands. He found many species of insects, birds, and plants which did not occur anywhere else on earth.

When Darwin returned home in 1836, he set about organizing his materials. During the busy years that followed he wrote his journal, published books and papers, and continued his work as a naturalist. From 1838–1841, he was secretary of the Geological Society.

In 1842, continued ill health forced Darwin to leave London and settle for a quieter life in the country. During that year, he wrote his first rough draft of his theory on the origin of species. In 1844, the original work was expanded. He continued to increase his reading and study of history, horticulture, animal and plant breeding, and geology. He corresponded with other scientists. He carried out numerous cross-breeding experiments with pigeons, and studied the ways that plant seeds can be carried from place to place.

In 1856, Darwin's friends urged him to write out his results. By 1858 he had completed about ten chapters of what was promising to be a detailed explanation of his theory. But in that same year another naturalist, A. R. Wallace, sent Darwin an essay that proposed a theory of species origin that was the same as his own. To be fair to both parties, Wallace's essay, and a letter that Darwin had written to a scientist friend in 1857 describing his theory were published together in the Linnaean Society's Journal in 1858.

Darwin next organized his notes into a more concise work than he had originally planned, and on November 12, 1859, published *On the Origin of Species by Natural Selection, or the Preservation of Favored Races in the Struggle for Life.*

There has been much debate over Darwin's theory, and the impact of his work was far greater than he had imagined. He had only thought to show that there were two sides to the question of change in species.

Much of Darwin's final years were spent defending and clarifying his theory. But he continued to observe, to experiment, and to write, mainly about plants. He died on April 19, 1882.

9–2
Modern Theory of Evolution

Peach

Nectarine

White and red carnations

Variations produced by gene mutations

The major ideas in Charles Darwin's theory of evolution by means of natural selection are accepted by most biologists today. Some of the observations discussed in Chapter 8 were not available to Darwin, yet they do support his theory. Also, no major pattern of scientific evidence that conflicts with his theory has turned up.

Sources of Variation

Charles Darwin could not explain where differences among individuals of a population arise. As you read in Chapter 7, we now know that genes are the units of hereditary information. Gene mutations are a source of variation among members of a population.

Sometimes a gene mutation occurs during the growth of an organism. The effects of such a case were observed in 1741. One branch of a peach tree produced smooth-skinned peaches which we now call nectarines. The difference between peaches and nectarines is caused by one pair of genes. The change originally took place by gene mutation. Many other varieties of food and ornamental plants have originated in similar ways.

If gene mutations are transmitted from one generation to the next, they become part of the diverse genetic information of a population. The results are often damaging to the individual who inherits them. This is because they are changes to a genetic make-up that has been refined by natural selection for a long time. But sometimes the results are neutral or beneficial. Also, the environment may change so that a variation that was once a disadvantage becomes an advantage. This was apparently the case with the dark mutant peppered moth discussed in Chapter 8. You will remember that the dark peppered moths began to survive in areas where tree bark turned black from soot. Their dark color became an advantage because birds could not see them. The dark coloration was protection from the birds.

A second source of variation is chromosomal accidents, or **chromosomal mutations.** In this case, chromosomes or sections of chromosomes are lost, added, or rearranged. Chromosome losses usually have severe effects since genes are lost. The genetic change is often not transmitted to the next generation since the carrier is not likely to survive and reproduce.

Polyploidy

Flowering plants normally possess two sets of chromosomes per cell. Such cells are termed **diploid.** But sometimes whole sets of chromosomes may be added by chromosome mutation. Cells with three or more sets of chromosomes are called **polyploid.** Polyploidy is fairly common in plants, but very rare among animals. In humans, polyploidy usually results in death before birth.

In plants, polyploidy has probably been important as a method of genetic change. In chrysanthemums, the diploid number is 18, and species are known that have 18, 36, 54, 72, and 90 chromosomes. Species of roses are known with chromosome numbers of 14, 21, 28, 35, 42, and 56. Some polyploids, depending on how they originate, are sterile and thus do not pass on their genetic make-up to offspring. They can be propagated by cuttings, of course. Other polyploids are fertile, however, and may have been important in evolution.

Diploid lily

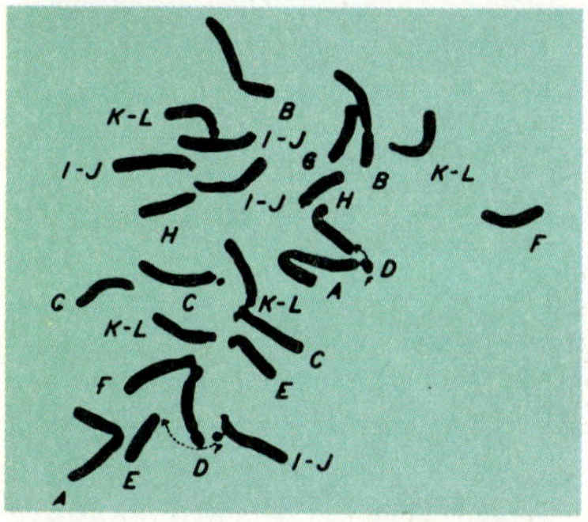

Chromosomes from a diploid lily

Distribution of Variation

Genetic variations in a population are introduced by gene mutations or by chromosome mutations. If a polyploid from a chromosome mutation is fertile but does not interbreed with other species, we have an "instant" new species. As the population grows, its members will have certain special characteristics in common. Sometimes a new variation comes from a gene mutation. The new variation is distributed through a population by independent assortment of chromosomes at meiosis, and by random fertilization.

Migration

The movement of individuals into or out of a population is called **migration** (mī grā'shən). Migration is another way that variations originate within a population and are distributed among populations. If an individual leaves a population, its genetic make-up is lost from that population. Similarly, if an individual enters a population, its genetic make-up is added to the population.

Tetraploid lily

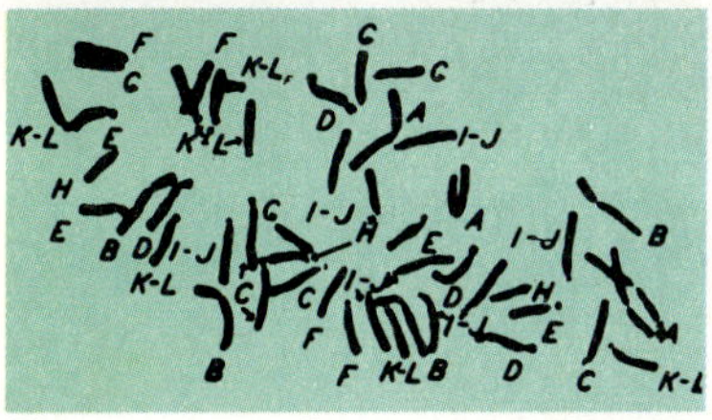

Chromosomes from a tetraploid lily

How Natural Selection Works

In most populations, individuals are not identical. Organisms with certain genetic make-ups are likely to survive and have fertile offspring more often than others. When this happens, a change of gene frequencies in the population takes place. Two conditions are necessary if genes are to stay in a population: (1) individuals carrying the genes must survive; and (2) they must reproduce.

In many birds, the male of the species is more colorful than the female. This is an advantage in reproduction, because in birds females often choose the mating partner. The eye-catching coloration of the male mallard duck is an advantage because females select these males for mating. Other examples are the bright coloration of male pintail ducks and the male lyre bird. Sometimes the bright coloration of male birds may also be a disadvantage. It may make them more visible to their enemies. In these cases, the reproductive advantage must be greater than the disadvantage of being seen by enemies.

Coloration of female birds affects the population differently. There is little or no reproductive advantage in females being recognized for mating, since females choose the mating partner. But there is a definite advantage to not being seen by enemies during nesting and caring for the young. The dull coloration of female birds is an advantage which allows them to blend in with the environment. The females survive longer than the males and are able to care for the young.

Male sage grouse

Spring mating dance in sage grouse

Differences in Survival Rates

Natural selection by differences in survival rates is probably shown best by the studies of the peppered moth in different woodlands. Professor Kettlewell of Oxford University in England has done research on these moths in the sooty, industrialized Birmingham area and in the unpolluted area of Dorset. He marked and released dark and light moths in both areas. Later, both kinds of moths were collected from both woodlands. The table in the margin shows the results. More light moths survived in the light woods and more dark moths survived in the sooty woods. Observations in the study showed that birds were able to see and capture more light moths in the sooty woods. And the birds captured and ate more dark moths in the light woods. In each case, the more visible moths were eaten more.

The moth is normally a light, grayish color. The dark form is a mutant. Once the dark variation appeared in the soot-covered woods, its success at survival enabled it to reproduce and transmit the genes for dark color to its offspring. In the woods with light, lichen-covered tree trunks, the moth population kept the original light color, which was successful in that area.

Differences in survival rates also explain how populations of insects become resistant to pesticides. Insects that have a genetic make-up that enables them to survive the insecticide will transmit that genetic make-up to the next generation. Those that die from the insecticide will not. Resistance of microorganisms to antibiotics is explained the same way.

Death rate of peppered moths released in different woodlands

Type of woodland	Number eaten by birds	
	Dark	Light
Grayish	164	26
Sooty	15	43

Locations of light and dark peppered moth

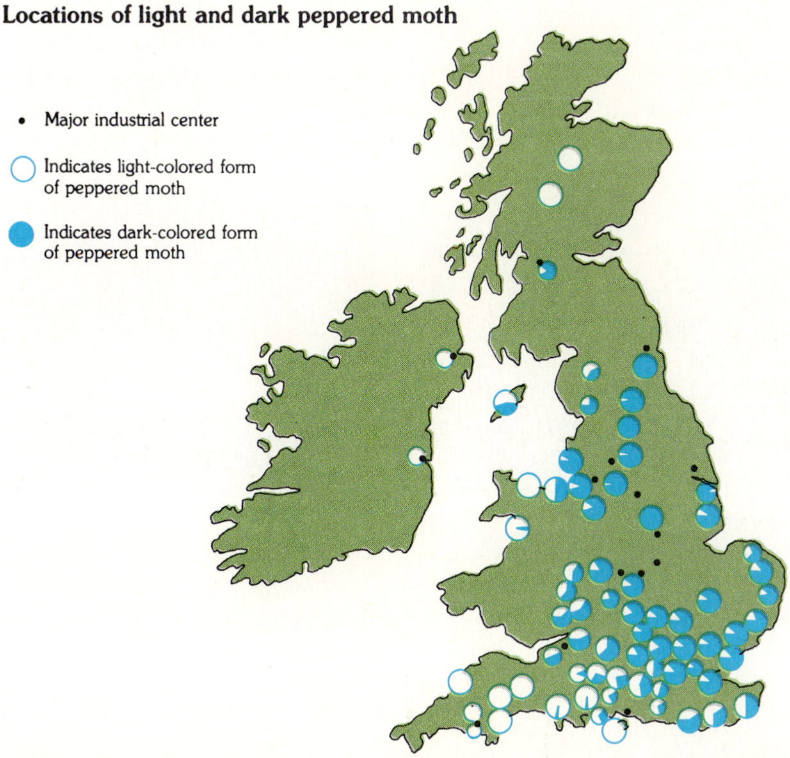

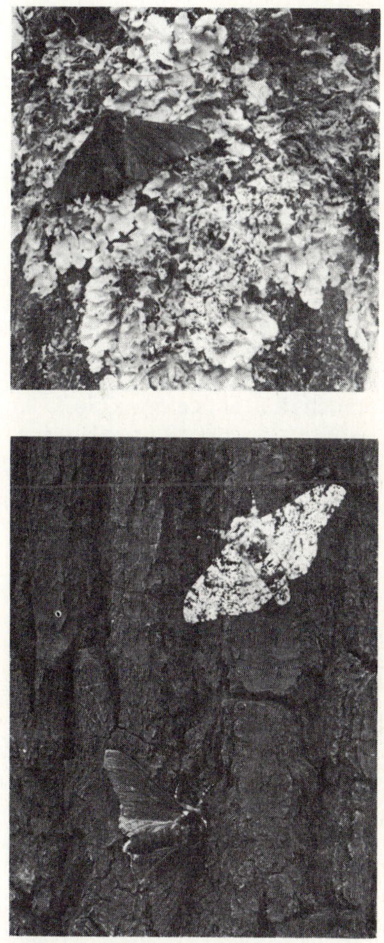

Light and dark peppered moth in both environments

Adaptation

An important result of natural selection is **adaptation** (ad′ap tā′shən). Through adaptation, a population becomes better suited to survive in its environment through changes in genetic make-up. When the environment changes, the direction of adaptation changes too. Adaptations may be expressed as changes in body structure, body function, coloration, or behavior. You can see some examples of adaptation below.

Structure and Coloration

Walking stick
The walking stick resembles a twig. When the insect rests on a branch, it is almost invisible to its enemies.

Stone plant
The stone plants above grow in stonelike shapes and even have the color of stones. They are protected because hungry animals may not identify them as food.

Body Functions

Venus's-flytrap
The Venus's-flytrap lives in nitrogen-poor bogs. To supplement its diet, the plant uses a special traplike structure to capture and digest insects.

Fennec
Ear size is often related to climate. The fennec of North Africa has large ears, which help to get rid of body heat.

Behavior and Coloration

Caligo butterfly
The Caligo butterfly has markings that resemble the head of an owl. When perched on a twig, the butterfly can often startle and frighten off potential enemies.

Bittern
The bittern conceals itself from enemies by standing absolutely still among the reeds. It points its long, slender bill upward. In this way, the bird's appearance blends with the water reeds.

Isolation of Populations

When populations of the same species are in contact with one another, interbreeding occurs. In the process, genes are exchanged by the populations. But populations often become separated from one another. They may be separated by barriers such as mountain ranges, rivers, highways, an oil pipeline, or a canyon. Such isolation is called **geographic isolation.** When this happens, the flow of genes between the populations stops. Without the exchange of genes, the populations tend to become more and more different from each other. The three main reasons that their differences increase are: (1) the different environments will cause different adaptations among the populations; (2) the populations may have slightly different gene frequencies when the isolation takes place and the differences are able to develop; and (3) the populations may have different mutations.

Abert squirrel

Kaibab squirrel

The Abert squirrel and the Kaibab squirrel of the Grand Canyon area illustrate geographic isolation. The two squirrel populations are thought to have the same ancestry but seldom interbreed now because of the Grand Canyon.

Another example of geographic isolation is the 14 species of finches on the Galápagos Islands and Cocos Islands. You read about Darwin's observations of these finches in Chapter 8.

Speciation

If populations stay geographically isolated for very long, they may become very different from one another. Eventually, they cannot interbreed even if they are brought back together. Such isolation is called **genetic isolation.** When this happens, separate species have been formed. The process of forming a new species is called **speciation** (spē'shē ā'shən). Recall that speciation occurs by polyploidy too.

Genetic isolation happens in many ways. Some of them are:

1. Each population is so adapted to its own environment that it cannot survive in the environment of the other.
2. The body structure of the individuals may have become so different that they cannot interbreed.
3. Two populations may have become adapted to different habitats and seldom encounter members of the other population.
4. Behavior may have changed so much that mating is prevented.
5. The populations may no longer mate during the same season.
6. Fertilization may not take place even though mating occurs.
7. Hybrid offspring from two different populations may die during development or be sterile.

Two races of squirrels found on either side of the Grand Canyon

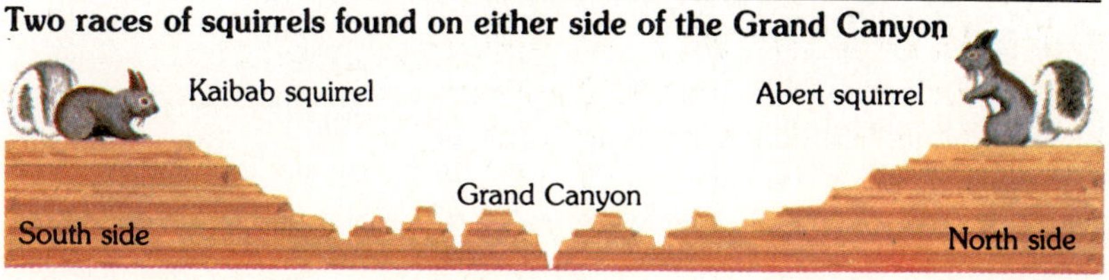

Examples of Genetic Isolation

A few examples will show how genetic isolation can happen. A certain hybrid flax seed never germinates in nature. But if you take off the seedcoat before planting the seed, it germinates into a perfectly healthy plant. This is a case where a healthy hybrid offspring (seed) is produced, but it cannot develop beyond the seed stage on its own. Since the hybrids cannot develop and reproduce, the two parent populations are genetically isolated.

The distributions of leopard frogs and salamanders you read about in Chapter 8 provide examples of genetic isolation taking place. Leopard frogs *(Rana pipiens)* occur north into Canada and south to Mexico. Populations that overlap are interbred through the entire range. So the frogs must all be members of the same species. Experiments have been done in which frogs from many populations are crossed. The findings were interesting. When there was a large north-south distance between the frog populations crossed, many offspring died or were not normal. In some cases almost all the offspring died.

Examples of possible speciation among leopard frogs

In this example, some of the evidence suggests that the frogs are all members of the same species. Other evidence implies that the frogs at extreme ends of the range are not members of the same species.

Divergent Evolution and Convergent Evolution

Geographic isolation can lead to genetic isolation. Isolated populations adapt to their own environments and increased variation comes about. Many biologists think that this process accounts for most of the species of plants and animals we observe today. The overall process of producing more and more different forms from one ancestral form is **divergent evolution.** By this process, many different species can arise from a common ancestor. The differences among the species come from the adaptations of each species to its own environment.

For an example, think again about the finches of the Galápagos Islands. All 14 species are quite different from any other birds in the world, though they are fairly similar to one another. So it is thought that they have evolved from a single ancestor species. The divergence from a single ancestor species to the 14 species today illustrates divergent evolution.

In some cases, species with very different ancestry may come to occupy similar environments. It is reasonable to think that the common environment may exert similar selection pressures on the two species, and they may come to resemble each other. Such resemblance between two unrelated species that is due to the similarity of their environments is called **convergent evolution.** For example, sharks and whales look fairly similar. Yet, a detailed study of the anatomy and embryology of each shows that they are members of different groups. The similarities are superficial. The whales are warm-blooded like other mammals and have lungs rather than gills. The whale's flippers are similar in structure to appendages of mammals that live on land and are not like the fins of sharks or fish.

Whale

Shark

Examples of unrelated organisms that have come to resemble each other

Different species of Galápagos finches developed from a common ancestor

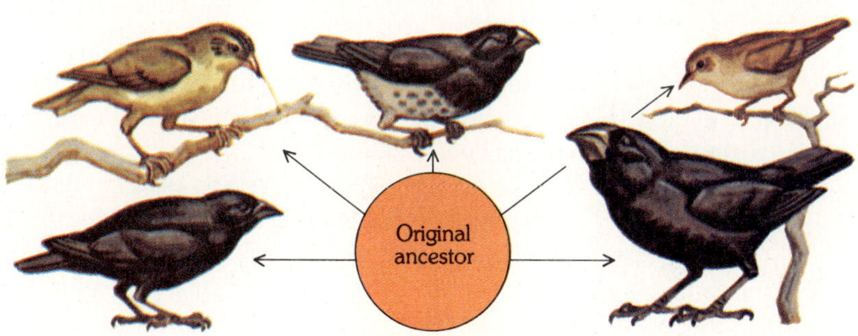

Review It

1. What are three ways in which variations originate in populations?
2. Give two examples that show how natural selection operates.
3. Define adaptation.
4. How can populations become genetically isolated?
5. Contrast divergent evolution and convergent evolution.

Chapter Summary

Lamarck tried to explain change and diversity of forms in the fossil record. He suggested that acquired characteristics could be inherited. Evidence from genetics does not support this theory. Darwin thought that change and diversity could be explained by natural selection. His theory was that environmental factors like climate, food supplies, and natural enemies determine which organisms survive and reproduce. The characteristics of those organisms that survive would be transmitted to the next generation.

The modern theory of evolution is more complete than Darwin's. Scientists have found that variations within populations arise through gene mutation, chromosome mutations, and migration. Gene mutations occur when there is a change in the base sequence of DNA.

Biologists think that geographic isolation and adaptation account for most of the species of plants and animals that we observe today. The process of producing many different forms from a single ancestral form is divergent evolution. However, sometimes two populations that are basically very different in structure but live in similar environments resemble each other. Thus a whale, which is a mammal, may resemble a shark. Their resemblance is due to convergent evolution.

Questions to Think About

1. Which population do you think would have the better chance to survive in nature: one that had evolved through artificial selection or one that had evolved through natural selection? Defend your position.
2. In looking back, it may be particularly significant that Charles Darwin's observations included the Galápagos Islands. List two or three features of the Galápagos Islands that provided very important evidence for his theory of evolution by natural selection.
3. One of Darwin's problems was that he could not explain to his critics how information about traits could be transmitted from one generation to the next. This question can now be answered in great detail. Using your knowledge from Chapters 5 and 7, answer this question in terms of cell division, chromosomes, genes, and DNA.
4. Explain why two populations that are geographically isolated for a long period of time tend to become genetically isolated.
5. Look at the list of mechanisms shown below. Based only on the information in your text, which of the mechanisms have been *observed* to occur and which ones have been *inferred* from other observations? Defend your position on each.

   (a) artificial selection
   (b) gene mutations
   (c) chromosome mutations
   (d) natural selection through survival rate differences
   (e) adaptation
   (f) speciation by geographic isolation
   (g) speciation by polyploidy
   (h) convergent evolution

Extra Research

1. Do library research on the history of domestication of animals and cultivation of plants. What were some of the earliest domesticated animals and cultivated plants?
2. Talk to an agricultural extension agent in your area about artificial selection. Ask if there are researchers who work in selective breeding for certain desired traits in a particular animal or plant species.
3. Prepare a report on the *green revolution*. What does it have to do with natural selection and artificial selection?
4. Do some reading on evidence of early humans in the fossil record. In what part of the world did some of the very early human populations live?

Chapter Test

Part I

Copy the number of each test item. If the item is *true*, write *T* after the number. If the item is *false*, write *F*.

1. According to Lamarck, evolution would take place by the accumulation of acquired characteristics. (218)
2. According to Darwin, evolution would take place through natural selection. (220)
3. Darwin said that variations in a population arise through mutations. (220)
4. Polyploidy is a gene mutation. (223)
5. Adaptation causes natural selection. (226)
6. If two populations are geographically isolated in nature, there will be little or no gene flow between them. (227)
7. Geographical isolation is the same as speciation. (227)
8. Adaptation sometimes causes genetic isolation. (228)
9. Genetic isolation may occur even though interbreeding occurs. (228)
10. The finches of the Galápagos Islands are an example of convergent evolution. (229)

Part II

Copy the number of each sentence below. After each number, write the word or words that complete the sentence correctly.

11. Selection of organisms by humans for breeding for desired traits is called _____ _____. (220)
12. Which one of the following ideas does *not* fit with the other three: genetic isolation; species formation; convergent evolution; speciation. _____ (227)
13. Which of the following ideas does *not* fit with the other three: variation; gene mutations; chromosome mutations; adaptation. _____ (222)
14. The condition where there are three or more sets of chromosomes per cell is called _____. (223)
15. The superficial similarities between whales and sharks illustrate the process of _____ _____. (229)
16. Dark peppered moths evolved in sooty woods because they had higher _____ rates than light peppered moths. (225)
17. According to the modern theory of evolution, changes in _____ and _____ may alter populations. (222)
18. _____ is the movement of individuals into and out of a population. (223)
19. An animal's ability to change its outer coloration to blend with the environment is an example of an _____. (226)
20. When several species of animals appear to have had a common ancestor, _____ _____ may have occurred. (229)

Part III

Copy each item from List A. After each item, write the letter and words from List B that best describe the item.

List A

21. speciation (227)
22. adaptation (226)
23. acquired characteristics (218)
24. natural selection (220)
25. geographic isolation (227)

List B

a. condition in which no gene flow occurs between members of two populations
b. traits from body cell changes that were once thought to be inherited.
c. process by which new species develop
d. genetic changes that enable a species to adjust to altered conditions in the environment
e. process by which favorable genetic variations accumulate in a population over generations

Focus on Careers

Paleontologist

The work of paleontologists has fascinated most people at one time or another. Perhaps you have spent happy hours reading about stegosaurs and pterodactyls. You may have seen one of these animals assembled from fossilized remains. Maybe you have read some theories about why and how dinosaurs became extinct. If you have done any of these things, you have some idea about the work of a paleontologist. A paleontologist is a research scientist who collects and studies the fossils of plants and animals in order to trace the development of past forms of life.

A paleontologist's work is divided into three phases. The first phase involves locating and obtaining specimens. How does a paleontologist know where to dig? He or she begins by knowing where *not* to dig. Certain rocks, granite for example, never contain fossils. Lava rarely contains fossils, and if it does have fossils they are poorly preserved. Fossils are found mainly in areas where rocks have formed by sedimentation.

After fossils are located and carefully removed, a paleontologist performs various laboratory tests on them. He or she classifies them according to their botanical or zoological family and tries to determine their probable age.

The third phase of a paleontologist's work involves writing reports summarizing his or her findings. These reports are used as evidence to support or disprove various theories concerning the history of the earth or as aids in locating natural resources, such as petroleum.

Paleontology is a branch of geology—the study of the composition, structure, and history of the earth's crust. A person interested in this field should first earn a bachelor's degree in geology, taking courses in invertebrate and vertebrate paleontology. This is necessary for entry-level jobs. An advanced degree in paleontology is necessary for research, supervisory, and teaching jobs. Many paleontologists also specialize: a paleobotanist studies mainly plant fossils; a micropaleontologist studies fossilized microorganisms.

A paleontologist's work may be physically demanding. Doing field work often means spending considerable time working in remote areas in uncomfortable climates. Then too, many hours are spent analyzing specimens in the lab and preparing reports. However, a paleontologist can find great excitement and satisfaction in contributing to our knowledge of Earth and its history.

Career Information

Biochemist: American Society of Biological Chemists, 9659 Rockville Pike, Bethesda, MD 20014.

Museum Technician: American Association of Conservators and Restorers, 1250 East Ridgewood Avenue, Ridgewood, NJ 07450.

Paleontologist: American Geological Institute, 5205 Leesburg Pike, Falls Church, VA 22041.

Physical Anthropologist: American Anthropological Association, 1703 New Hampshire Avenue NW, Washington, DC 20009.

Biochemist

Duties: Studies chemical composition and behavior of living things. A biochemist conducts research in a laboratory and may investigate the effects of food, hormones, or drugs on various organisms. For example, a biochemist might study how the body forms a certain hormone and then try to discover how the hormone might be manufactured in a laboratory. Biochemists are involved in research efforts to find cures for cancer and heart disease. They also work in areas of environmental protection.

Training: Requires a Ph.D. degree.

Opportunities: Favorable through the 1980s.

Museum Technician

Duties: Prepares and arranges specimens for museum exhibits. The technician might clean fossils and then brush on preservatives, such as resin. Or, he or she might help construct skeletal mounts of fossil animals using welding and carpentry tools. The technician also maintains museum files; cleans, catalogs, labels, and stores specimens; and helps install exhibit materials.

Training: Requires a bachelor's or advanced degree in anthropology with specialized study in botany, zoology, or geology.

Opportunities: A limited number of jobs are available in museums.

Physical Anthropologist

Duties: Investigates the physical characteristics of different races or groups of people to learn how they have been influenced by heredity and environment. A physical anthropologist may identify human skeletal remains or teach a chimpanzee to communicate with sign language.

Training: Requires a Ph.D. degree. Students study four areas: cultural anthropology, archaeology, linguistics, and physical anthropology before specializing. There is extensive training in human anatomy, biology, genetics, and the study of primates. Training should include archaeological field work which may mean traveling to remote areas and working in uncomfortable climates.

Opportunities: Most anthropologists teach in colleges and universities or work in museums.

Bibliography: Unit Four

Bates, Marston. *Man in Nature.* Prentice-Hall, 1964. Deals with mankind as an animal, our ancestors, the varieties of humans, problems of domestication, and the environment.

Colbert, E. H. *Dinosaurs: Their Discovery and Their World.* Dutton, 1961. Information about ancient reptiles.

DeBeer, Sir Gavin. *Atlas of Evolution.* Thomas Nelson and Son, 1964. Beautifully illustrated volume covering the plant and animal kingdoms, including humans.

Fenton, Carroll. *Tales Told by Fossils.* Doubleday, 1966. Discusses fossil evidence of evolution.

Goodall, Vanne. *The Quest for Man.* Praeger, 1975. Contains lavish illustrations and essays dealing with varied aspects of human evolution. Intriguing information for a general audience.

Hogarth, P., and Salomon, J. J. *Prehistory.* Dell, 1962. Story of early humans with many pictures.

Howell, F. Clark. *Early Man.* Time Inc., 1965. Well-illustrated volume depicting human evolution.

Hutchinson, Peter. *Evolution Explained.* David and Charles, 1974. An introduction to modern evolutionary theory written for readers with little background in biology. Well illustrated and contains a selection of classic studies as major sources of evidence.

Huxley, Julian. *Evolution in Action.* Harper and Row, 1953. Explains in easily understood terms the pattern of genetic structure and how natural selection works.

Huxley, Julian. *The Wonderful World of Life.* Doubleday and Co., Inc., 1969. Presents evidence for evolution. Good illustrations of organisms.

Karp, Walter. *Charles Darwin and the Origin of Species.* Harper and Row, 1968. Interesting biography of Darwin. Includes many photographs.

Lehrman, Robert. *The Long Road to Man.* Basic Books, Inc., 1961. Traces the development of animals from the invertebrates to the mammals. Also discusses human evolution and the major geologic eras.

Montagu, Ashley. *The Human Revolution.* World Publishing Co., 1965. Account of evolution of humans from primate beginnings.

Moore, Ruth. *Charles Darwin.* Alfred A. Knopf, 1966. Biography of Darwin.

Moore, Ruth. *Evolution.* Time Inc., 1964. Beautifully illustrated volume that discusses many aspects of evolution, including adaptation, natural selection, and development of mutations.

Moore, Ruth. *Man, Time, and Fossils.* Alfred A. Knopf, 1965. An interesting account of contributions of famous scientists to evolution theory. Includes chapters on Darwin, Lamarck, DeVries, Leakey, and others.

Morris, Desmond. *The Naked Ape.* Dell, 1967. Interesting comparisons of social behavior and evolution of body structure between humans and other primates.

Simpson, George Gaylord. *The Geography of Evolution.* Chilton Books, 1965. Collection of essays on the origin and development of animal life, especially the relationship of that development to the geography of our planet.

Stebbins, G. L. *The Processes of Organic Evolution.* Prentice-Hall, Inc., 1966. Up-to-date account of why evolution takes place.

Wallace, B., and Srb, A. M. *Adaptation.* Prentice-Hall, 1964. A clear account of the meaning of adaptation and its relation to genetics and evolution.

White, Edmund, and others. *The First Men.* Time Inc., 1973. A picture essay of the transformation of primate ancestors to humans. Significant fossil discoveries and work of numerous scientists are described.

White, J. F., ed. *Study of the Earth.* Prentice-Hall, Inc., 1962. Includes discussions of fossils and how age of rocks can be determined.

Articles

Cusack, Michael. "What Are We? Why Are We?" *Science World,* March 13, 1973.

Leakey, Richard. "Skull 1470." *National Geographic,* June, 1973.

Schiller, Ronald. "New Findings on the Origin of Man." *Reader's Digest,* August, 1973.

UNIT FIVE

MONERANS, PROTISTS, AND FUNGI

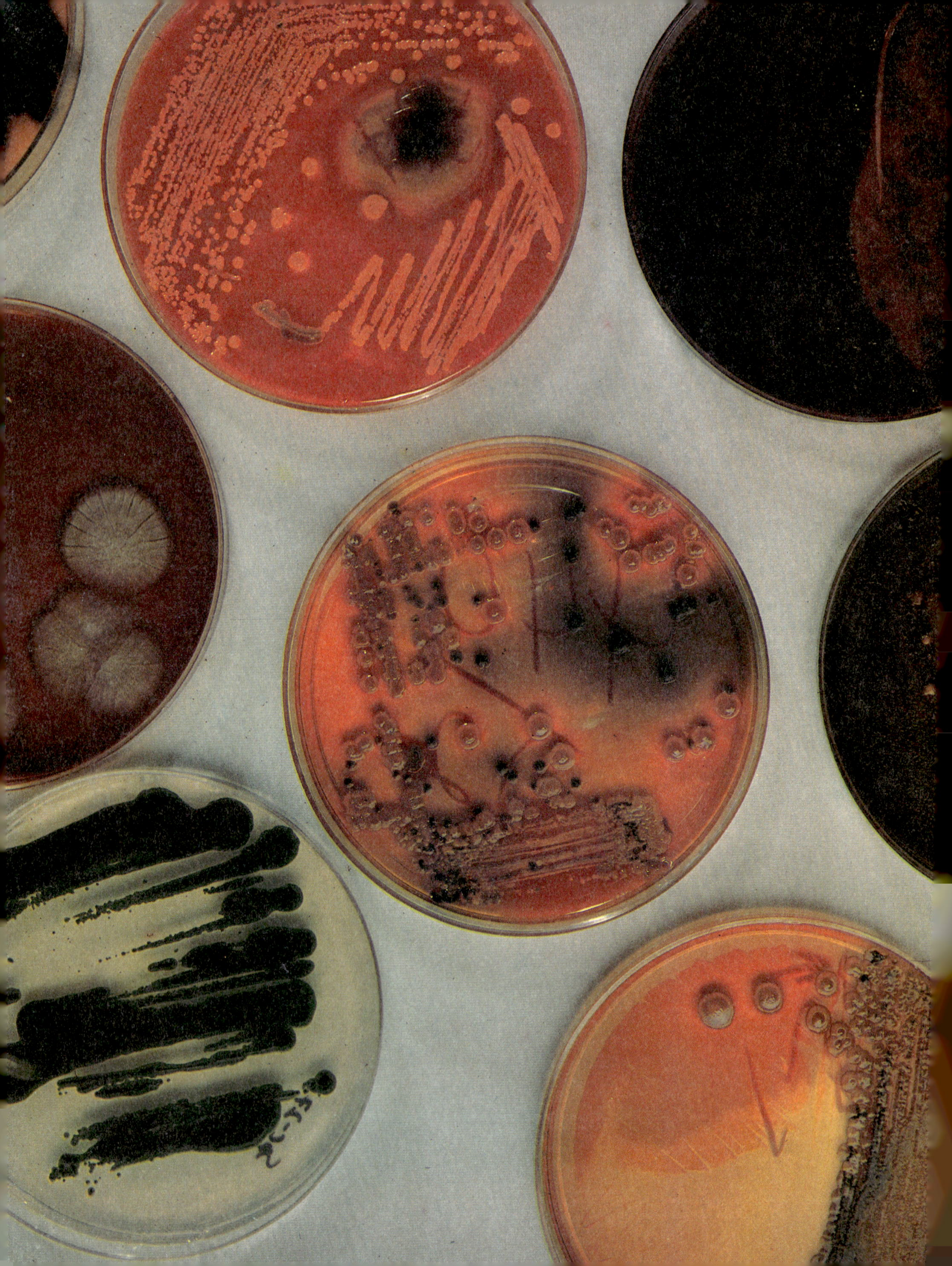

Chapter 10

The Monerans

Monerans are small and, generally, one-celled organisms. They are probably the most numerous kind of organism on Earth. And they are very widespread, occurring just about everywhere. Monerans can be found on virtually any surface—including your skin and the inside of your intestines.

In addition to being so numerous and so widespread, monerans have a very long history. They appear in the fossil record long before any other kind of organism. As a result, they are of particular interest, for they are thought to have the closest resemblance to the earliest forms of life on Earth.

In this chapter you will have a chance to find out why monerans are so numerous and so widespread as you learn about their structure and activities. You will also find out why life on Earth cannot exist without them.

In addition, part of this chapter is devoted to the viruses. At one time, viruses were thought to be relatives of the monerans. You will find out more about *that* idea as you learn about the structure, the activities, and the importance of the viruses.

Chapter Objectives

1. Describe the structure of the moneran cell.
2. Name and differentiate four kinds of moneran nutrition.
3. Describe the sexual and asexual reproduction of monerans.
4. List ways monerans affect you and other living things.
5. Describe the structure of viruses.
6. Describe the reproduction of viruses.
7. List ways viruses affect you and other living things.

Petri dishes are used to grow many types of bacteria.

10–1
Kingdom Monera

The **Kingdom Monera** (mə nir′ə) includes the blue-green bacteria (blue-green algae) as well as all other kinds of bacteria. So learning about the monerans is actually learning about bacteria. All the various kinds of bacteria are grouped together in a single kingdom because their cells are all quite plain, lacking many structures found in the cells of other organisms.

The Moneran Cell

Bacteria, or monerans, are among the smallest of all cellular organisms. If 5,000 average-sized bacterial cells are placed end to end, they would form a line about one centimeter long. An individual moneran, such as the one shown below, can be seen only with the use of a microscope. The microscopic view shows a *prokaryotic cell.* All monerans, or all bacteria, are **prokaryotes** (prō kar′ē ōtz).

Did You Know?
Biologists believe that the lack of specialized organelles and the plain design of the prokaryotic cell is evidence that prokaryotes, or monerans, have the closest resemblance to the first forms of life on Earth.

Most prokaryotic cells have a definite cell wall, a cell membrane, and a grainy-looking cytoplasm filled with ribosomes and a few clumps of genetic material. Many common cell structures, such as the membrane-bound organelles, are absent. For instance, no membrane encloses the genetic material to form a distinct nucleus. Instead, the single strand of DNA, called the *bacterial chromosome,* is in direct contact with the cytoplasm. No Golgi bodies and no endoplasmic reticulum are seen. Even the energy-releasing mitochondria are absent. Their job is done by enzymes found along the cell membrane.

The layers around the cell—the membrane and the wall—are protective. Without the cell wall, water pressure inside the cell would burst the cell membrane. The strength of the wall is not provided by cellulose but by another complex organic substance called **peptidoglycan** (pep tī′do glī′kan). This substance is found only in prokaryotes. Underlying the cell wall is the cell membrane, which regulates the passage of materials in and out of the cytoplasm. Only dissolved materials are able to enter and leave. Digestion occurs outside the cell. There is no phagocytosis, and no food vacuoles are formed. The cytoplasm is similar to that found in other cells, but it does not undergo the streaming that is normal in other cells.

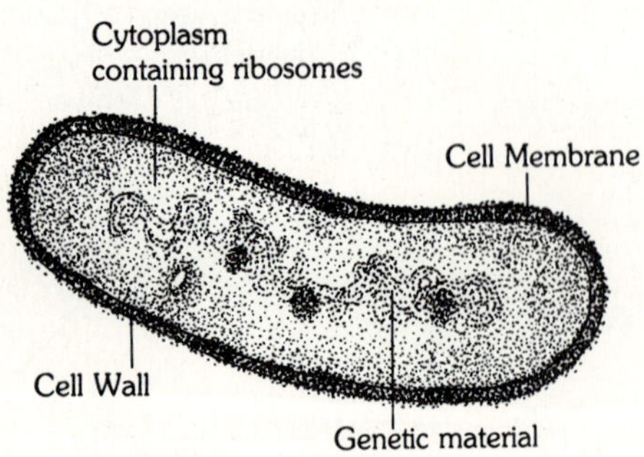

Structures of moneran cell

Dividing moneran cell

Other Cell Structures

In addition to the cell wall, which itself may be many layered, some prokaryotes secrete one or more layers of slimy materials. This coating is called the *sheath* or the *capsule.* The slime may enable the cell to move by gliding, although how this is done is unknown. Also, the slime may provide additional protection against unfavorable conditions. Or it may hold a number of cells together in a chain or a clump.

Some prokaryotes have elastic, threadlike structures called **fibrils** (fī'brəlz). Fibrils aid in movement. Some fibrils are found inside the cell. Spirochetes (spiral bacteria) use internal fibrils to move by twisting. Other structures called **flagella** (flə jel'ə) may be outside the cell.

Photosynthetic prokaryotes have special, unenclosed photosynthetic membranes. The blue-green bacteria—once known as blue-green algae—carry the green pigment *chlorophyll* and the blue pigment *phycocyanin* on their photosynthetic membranes. The structure of a blue-green bacterium is shown below.

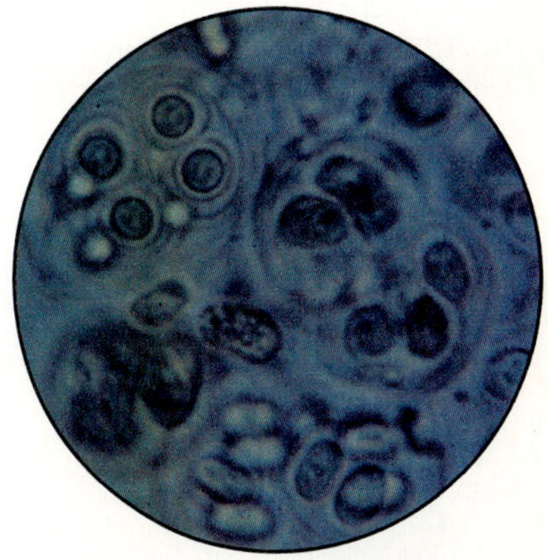

Blue-green bacteria in sheath

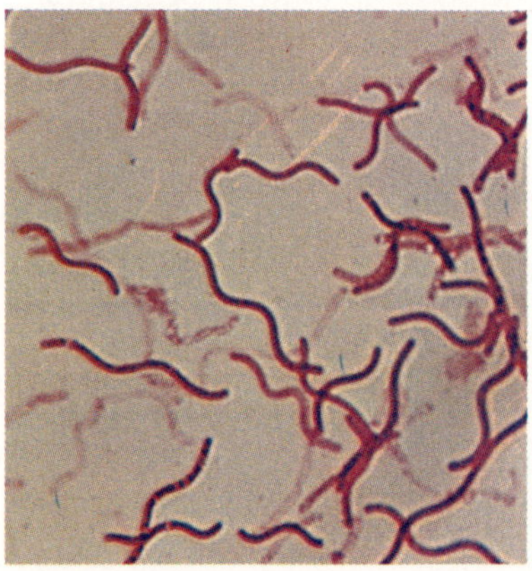

Spiral bacteria

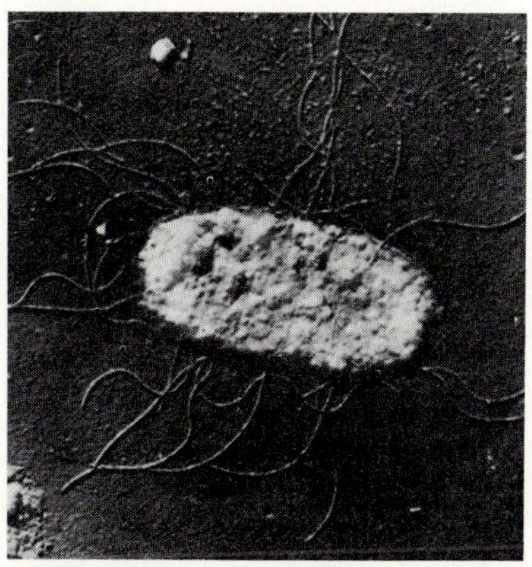

Flagellated bacterium

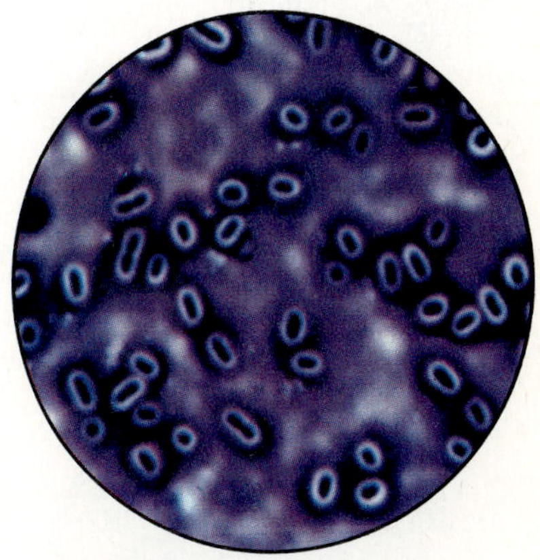

Encapsulated bacteria

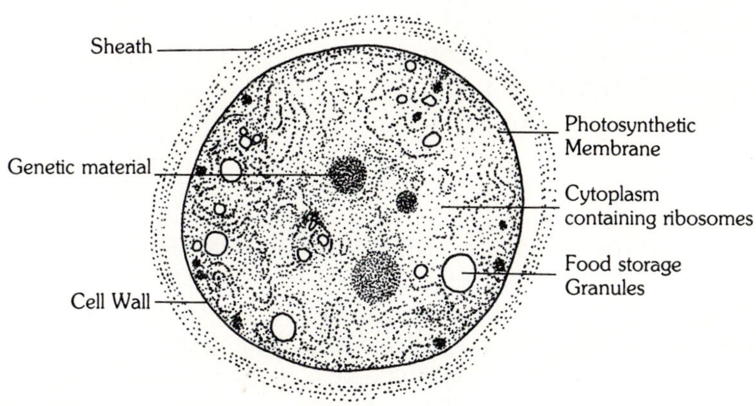

Structures of blue-green bacterium

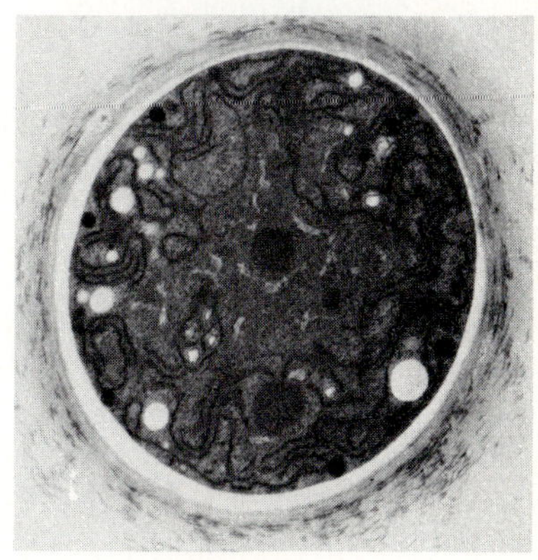

Blue-green bacterium

Did You Know?
Anaerobic monerans are thought to be the descendants of monerans that lived before the earth's atmosphere contained any oxygen.

Purple sulfur bacterium

Nitrifying bacterium

Did You Know?
Considering the widespread occurrence of monerans on Earth and the fact that monerans can live where no other organisms can, space scientists thought it might be possible for bacteria to be present in outer space. So they kept returning astronauts in isolation to check for possible bacterial infections.

Activities of the Moneran Cell

Two prokaryotes may look like each other. But their cell activities, or metabolism, may be completely different. Diversity among the prokaryotes, or monerans, is not so much in structure as it is in metabolism.

One example of metabolic diversity among monerans is in their need for oxygen. Some monerans require oxygen to live. These monerans are the **aerobes** (er′ōbz). Other monerans, the **anaerobes** (a ner′ōbz), are poisoned by oxygen and can live only in its absence. Still other monerans can live either with or without oxygen.

Another example of metabolic diversity among monerans is in their nutrition, or their need for raw materials and energy. In raw material needs, monerans may be either **autotrophs** (ô′tə trofs) or **heterotrophs** (het′ər ə trofs). A heterotroph requires organic compounds. And an autotroph requires simple inorganic compounds. In energy needs, monerans may be either **phototrophs** (fō′tə trofs) or **chemotrophs** (kē′mə trofs). A phototroph is photosynthetic, obtaining energy from light. And a chemotroph is chemosynthetic, obtaining energy from chemical bonds. Examples of the various kinds of nutrition are shown in the table below.

Kind of Nutrition	Energy	Raw Materials
Phototrophic autotrophs blue-green bacteria purple sulfur bacteria plantlike protists all plants	light	inorganic compounds
Phototrophic heterotrophs purple nonsulfur bacteria	light	organic compounds
Chemotrophic autotrophs iron bacteria nitrifying bacteria	chemical bonds	inorganic compounds
Chemotrophic heterotrophs the majority of bacteria animallike protists all fungi all animals	chemical bonds	organic compounds

The widespread occurrence of monerans is largely a result of their metabolic diversity. Monerans can live in places and under conditions where other organisms are unable to survive. For example, monerans are found in the cold ices of Antarctica, in the boiling hot springs of Yellowstone Park, and in the dark depths of the ocean. While some monerans survive without oxygen, others survive where conditions are very salty, or very acidic, or very alkaline.

Monerans as Decomposers

As *decomposers,* monerans affect not only humans but all life on Earth. Life on Earth depends on a steady supply of energy and on a recycling of essential raw materials. The sun provides the steady supply of energy. And the metabolism of the decomposers—many of which are monerans—provides the recycling of sulfur, oxygen, carbon, nitrogen, and other elements essential for life. Monerans, because of their great metabolic diversity, are able to carry on chemical changes that no other kinds of organisms are able to carry on.

Consider what happens when a tree falls or when a fish or some other organism dies or leaves behind solid waste materials. Decomposers, in the course of carrying out their metabolic activities, use the materials left by the fish or the tree. Those materials provide raw materials, or food, for the decomposers. For example, the solid organic waste material in sewage provides raw materials for the bacteria that live in sewage-treatment plants. As a matter of fact, the metabolism of those bacteria provides the most efficient means of purifying human sewage.

The metabolism of decomposers changes complex organic compounds into simple inorganic compounds. In that way, useless materials are changed into a useful form. Because there are only limited amounts of raw materials on Earth, all the materials needed by living organisms would eventually be bound up and unavailable. Life would cease to exist. But because of the metabolism of monerans and the other decomposers, raw materials bound up can be returned to the environment for use by the next generation of organisms.

The metabolism of monerans is also important in other far-reaching processes. For example, only monerans can convert nitrogen in the atmosphere into a usable chemical compound. That process is called **nitrogen fixation.** Without nitrogen fixation, plants would be unable to grow. And without plants, animals would have no food. Nitrogen fixation is carried out by various kinds of bacteria including the blue-green.

The blue-green bacteria (blue-green algae), which are photosynthetic, are also thought to be largely responsible for the development of the oxygen-rich atmosphere of Earth. Without the past activity of blue-green bacteria, the earth's atmosphere would not have become suitable for oxygen-breathing organisms.

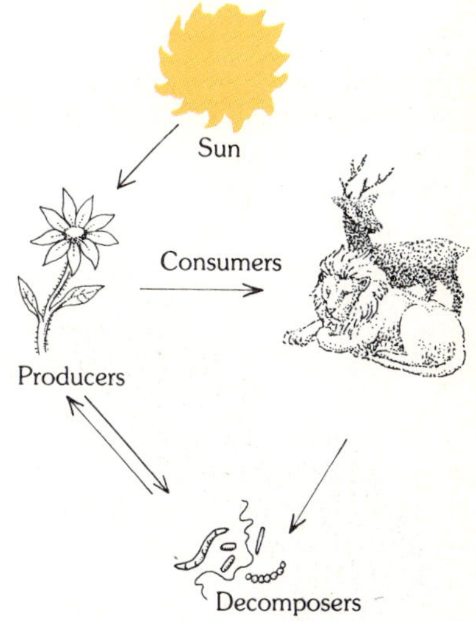

The relation of food producers, food consumers, and decomposers. Arrows show the flow of energy and materials.

Did You Know?
Introducing blue-green bacteria (blue-green algae) into rice paddies to fix nitrogen can increase the rice crop as much as 20 percent.

Monerans in Manufacturing

The metabolism of monerans also plays an important role in the manufacture of many products useful to humans. Those products include food, drugs, and chemicals. For example, bacterial enzymes help improve the flavor of coffee, cocoa, and tobacco. Swiss, cheddar, and provolone cheese owe their flavor to the action of bacteria on milk. Vinegar, buttermilk, and yogurt are the result of bacterial metabolism. Lactic acid, acetic acid, and various enzymes and vitamins are also produced by bacterial metabolism.

Challenge!
Try making yogurt from milk. You will need a bacterial "starter" culture and other ingredients as well as the proper directions. Make observations with and without a microscope. What changes do you observe? Find out how the bacteria take part in those changes.

Classification of Monerans

Did You Know?
When biologists began to use a new dating method—called RNA sequencing—to study monerans, they made some discoveries that startled them. They found that one group of anaerobic monerans, the methanogens, have very unique RNA molecules. Now some biologists want to group these monerans apart from other monerans. Methanogens are anaerobes found in the intestine of animals, in sewage, and in hot springs. They are called methanogens because they produce methane gas.

Biologists have not yet agreed among themselves about how to classify the monerans. There are actually a variety of classification schemes. Some schemes are based on structural traits. For example, a moneran can be classified as photosynthetic or nonphotosynthetic depending on whether it has photosynthetic membranes. Other schemes are based on metabolic traits. You saw, for example, that monerans can be classified by nutrition as autotrophs or heterotrophs, or by their use of oxygen as aerobes or anerobes, and so on.

One of the simplest ways to classify monerans is by shape. Moneran cells occur in many shapes, but the three most common are the spherical, or **coccus** (kok′əs), shape; the rod, or **bacillus** (bə-sil′əs), shape; and the spiral, or **spirillum** (spī ril′əm), shape.

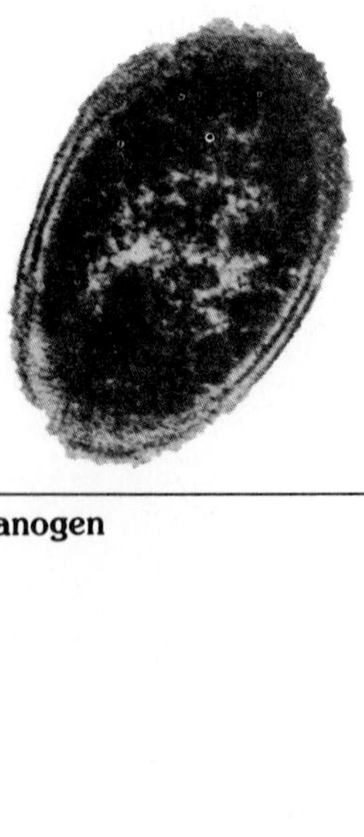

Methanogen

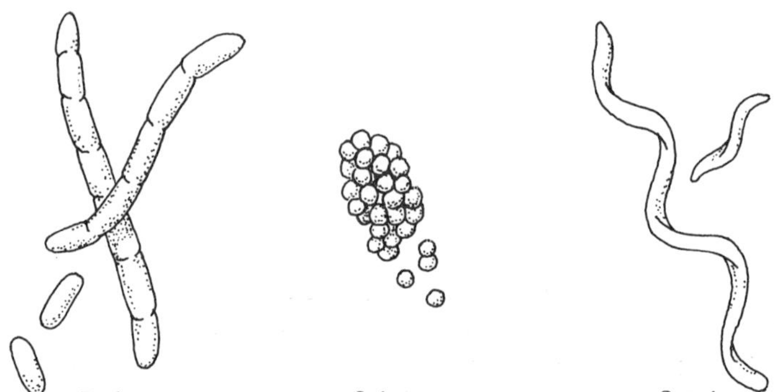

Moneran cells have three basic shapes and may occur singly or in clusters.

Bacteriologists have also classified monerans as gram-positive or gram-negative depending on the way they take the gram stain. Gram staining relates to a structural trait: the cell wall. Gram-negative monerans have a single, thick cell wall and are colorless after staining because they do not retain the initial stain. Gram-positive monerans have a many-layered cell wall and are blue after staining because they do retain the initial stain. However, that method of classification allows no place for the **mycoplasmas** (mī′kō plaz′məz), which lack walls.

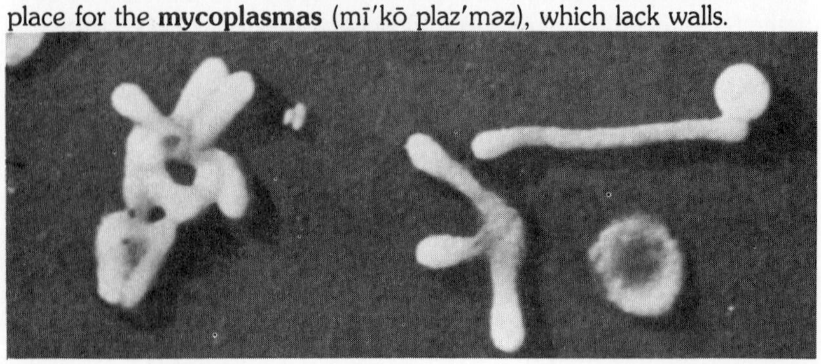

The mycoplasmas shown here cause pneumonia. The shapes of their cells vary because they have no cell wall.

Monerans and Symbiosis

Symbiosis (sim'bē ō'sis) is the relationship between two different kinds of organisms living in close association. Just about every kind of organism on Earth lives in symbiosis with one or more kinds of moneran. This symbiosis may be either **mutualism** or **parasitism.**

Mutualism

In mutualism, both of the partners of symbiosis benefit. An example of mutualism is the relationship between you and some of the bacteria in your intestine. They are provided with a comfortable home and the organic materials and energy they need whereas they produce vitamins you are unable to produce yourself.

Another example of mutualism is between cows and some of their intestinal bacteria. Those bacteria are able to break down, or digest, cellulose. Without those bacteria, a cow would be unable to digest its food.

Parasitism

In parasitism, only one organism, known as the *parasite,* benefits from the association. The other partner, known as the *host,* is harmed. Parasitic bacteria use organic material from their host as a source of energy and raw materials. Sometimes parasites seriously damage the host, causing disease. Such parasites are called *pathogens.* Pathogenic bacteria may damage the tissues of their host as they grow and reproduce. Or they may produce materials as they carry on metabolism that are toxic to their host.

Pathogenic bacteria cause hundreds of diseases in humans as well as in other organisms. These diseases include tuberculosis, tetanus, diphtheria, pneumonia, meningitis, leprosy, syphilis, typhoid fever, cholera, dysentery, gonorrhea, legionnaire's disease, and plague.

Pathogenic bacteria can reach you in various ways. They may move from one person to another by direct contact. Or they may be spread indirectly through water, dust, food, feces, insects, or other animals. Pathogenic bacteria enter the body through a body opening or any break in the skin.

The human body has several built-in defenses against pathogenic bacteria. Your skin and mucous membranes can stop the bacteria *before* they enter the body. Or once *inside* the body, special blood cells called *phagocytes* can attack and engulf those bacteria. Or special substances called *antibodies* can inactivate the bacteria. In addition, bacterial infections can be stopped by antibacterial drugs, such as penicillin.

Did You Know?
Penicillin and other antibacterial drugs are toxic to bacteria because they interfere with the making of peptidoglycan in the cell wall of bacteria.

In the past, in spite of the various defenses of the human body, pathogenic bacteria were a major factor in maintaining a relatively constant level of human population. Today, largely because of our ability to control pathogenic bacteria, we have a high level of general health, long life-spans, and expanding human populations.

Growth and Reproduction

When environmental conditions are favorable to growth, monerans can reproduce rapidly, forming a large group, or *colony*. Such a colony may contain millions of cells and can be seen easily without magnification.

Blooms, or population explosions of algae in lakes, are often blue-green bacteria (blue-green algae). Blue-green bacteria can also be seen in moist places—on river banks, in stream beds, tidal flats, or even on tree trunks. Have you ever lost your balance on a slimy rock? If you looked very closely to see what made you slip, what you probably saw was a colony of blue-green bacteria.

Moneran colonies are often cultivated in the laboratory. The color, texture, and growth patterns of pure cultures, grown as colonies, can be used as aids in identification.

Cell Division

Many monerans reproduce by a very simple kind of cell division called **fission** (fish'ən). Unlike the mitotic division common in other cells, there are no special structures involved. In fission, the single strand of DNA doubles. Then the cell divides in half, taking identical genetic material to each new cell. The two offspring are identical to the parent. Because no transfer of genetic material occurs, fission is an *asexual* process of reproduction. Under favorable conditions, fission can occur every 20 minutes. The abundance of monerans is largely a result of their ability to reproduce rapidly.

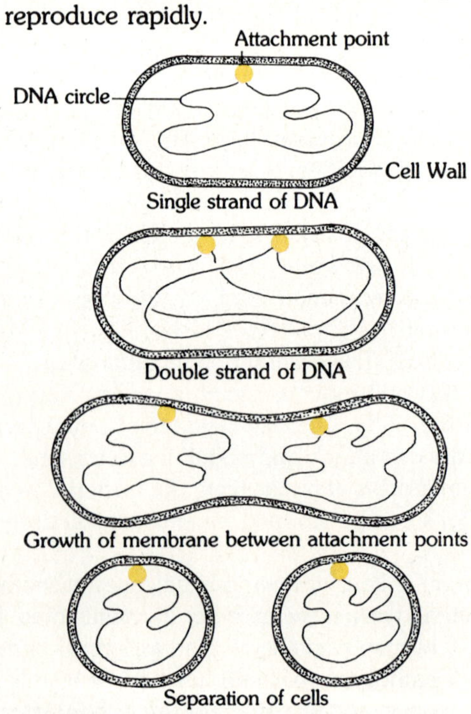

Fission of a Moneran Cell

Did You Know?
A single bacterium can produce as many as 250,000 offspring in just six hours. But as the size of the population grows, food becomes scarce and waste products become toxic. As a result, the size of the population levels off. Otherwise, a single bacterium could produce a 2,000-ton population in just one day.

Resting Cells

When environmental conditions become unfavorable to growth and reproduction, a moneran may then form an inactive cell called a *resting cell.* A resting cell is dormant: it does not reproduce, grow, take in food, or show any other sign of life. A resting cell may withstand boiling and freezing temperatures, lack of moisture, damaging radiation, and toxic chemicals. When conditions become favorable, a resting cell may germinate to form a normal, active cell.

The resting cell shown here is an **endospore** (en'dō spôr). It is formed inside a bacterial cell. During formation, a thick, almost indestructible wall encloses all of the genetic material and part of the cytoplasm. The rest of the cell is left to disintegrate. Endospores look glassy under a microscope.

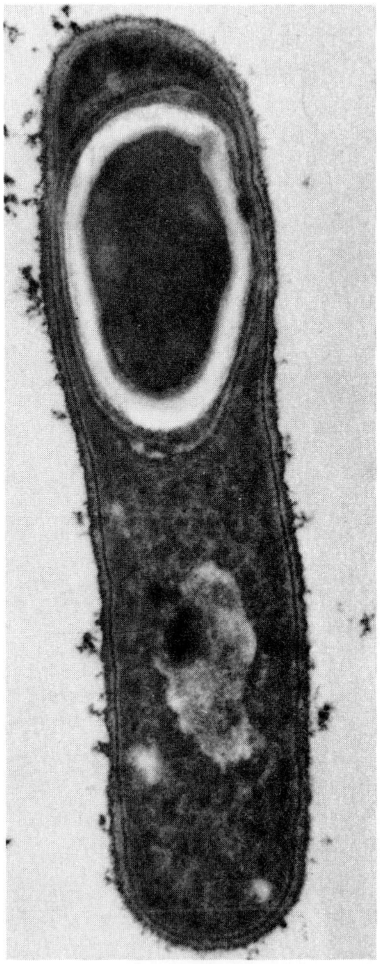

Endospore inside a moneran cell

Sexual Reproduction in Monerans

Sexual reproduction involves the transfer of genetic material. It is known to occur infrequently in some monerans. It may occur in various ways, one of which is **conjugation** (kon'jə gā'shən). Conjugation involves a cell-to-cell contact by a connecting bridge. One cell acts as a *donor* and one as a *recipient* as DNA passes through the bridge. After some genetic material is transferred, the two monerans separate. Because genetic material is transferred, the offspring are different from the conjugating parents.

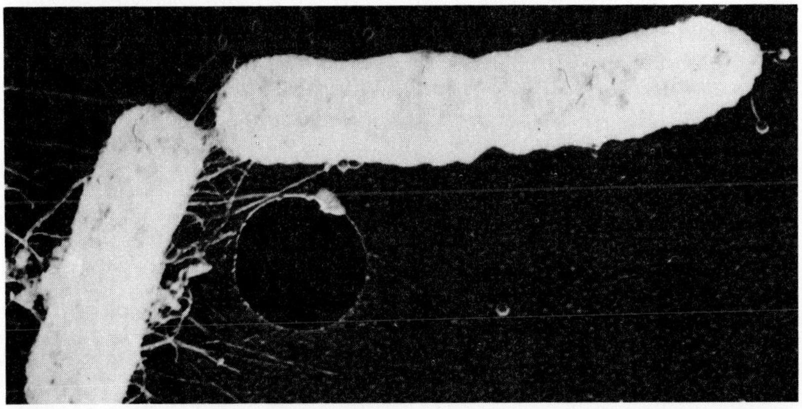

Conjugating bacteria

Review It

1. Make a drawing to show the structure of a typical prokaryotic cell.
2. How do monerans reproduce? Describe several ways.
3. Compare the nutrition of blue-green bacteria (blue-green algae) to the nutrition of bacteria living in your intestine.
4. What would our world be like if monerans became extinct?
5. How can you explain the abundance and the widespread occurrence of the monerans?

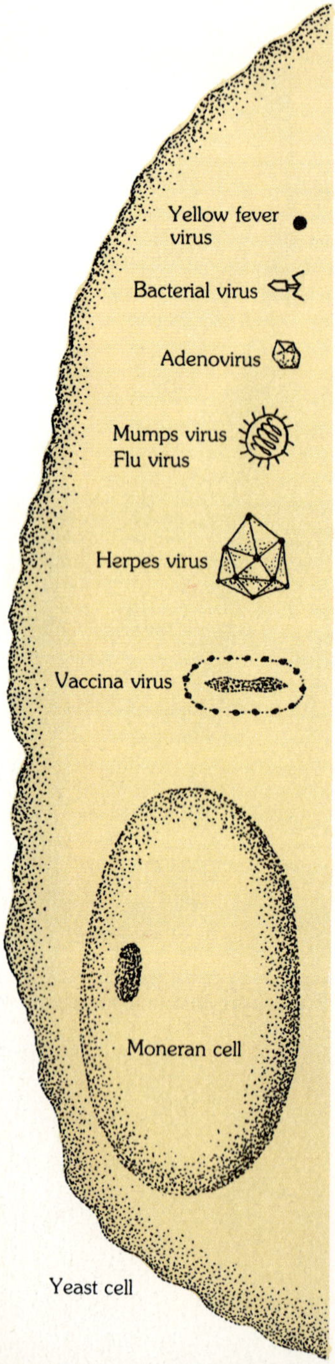

Relative sizes of cells and viruses

10–2
Viruses

A **virus** is essentially a piece of nucleic acid wrapped in a protective coat of protein. The nucleic acid of the virus is able to enter, or infect, a living cell. And it is able to make copies of itself, or reproduce, by taking over the machinery of the host cell. The nucleic acid of a virus can direct the cell to stop its usual activity and, instead, to make and assemble the parts of new viruses. So inside a living cell, a virus displays one of the traits of all living things: the ability to reproduce. But outside a living cell, a virus is lifeless. It neither carries on metabolism nor grows.

A virus is *not* a cell. It is thought of as a molecular structure for several reasons.

1. A virus, unlike a cell, never grows, respires, or carries on metabolism.
2. A virus, unlike a cell, does not arise *directly* from a previously existing similar structure.
3. A virus contains only one kind of nucleic acid—either RNA or DNA. A cell contains both RNA and DNA.
4. A virus carries only a few enzymes. A cell carries many enzymes.
5. A virus, unlike a cell, can be *crystallized.*

Although biologists agree that viruses are not cellular organisms, there is no agreement about whether viruses are organisms of some other kind or whether viruses are alive. An agreement would be easier to reach if more were known about the origin of viruses. Although various ideas about the origin of viruses have been proposed, no one has yet been proven. One idea is that viruses began as cellular organisms but lost their cell structures as they became parasitic. Another idea is that viruses are the noncellular ancestors of cellular organisms. A third idea is that viruses are pieces of nucleic acid detached from cells.

The Size of Viruses

Each kind of virus has a definite size. In general, a virus is far smaller than the smallest known cellular organism. A virus can pass through an opening that holds back a bacterium. In fact, a virus is too small to be seen with an ordinary microscope. To be seen, viruses must be magnified tens—or even hundreds—of thousands of times. Such magnification is possible only with an electron microscope. The diagram shows the relative sizes of a common bacterium and some viruses.

The Shape of Viruses

In addition to a definite size, each kind of virus also has its own definite shape. The shape depends on the arrangement of the protein molecules that cover the nucleic acid. A virus may have one of three basic shapes. It may be many-sided, somewhat like a crystal. It may be spirallike, or **helical** (hel′ə kəl). Or it may be a combination of helical and many-sided.

The three basic shapes are shown in the examples below. The adenovirus, which causes respiratory infections in various animals, has a many-sided shape. Each of its twenty sides, or faces, is triangular. The virus of mosaic disease in tobacco plants has a helical shape. It looks like a long, slender cylinder. But when examined closely, the molecular units of the mosaic virus appear to be wound in a spiral. The tadpolelike T_4 virus found in bacteria has a combination of shapes. The head is many-sided, whereas the tail is helical.

Adenovirus

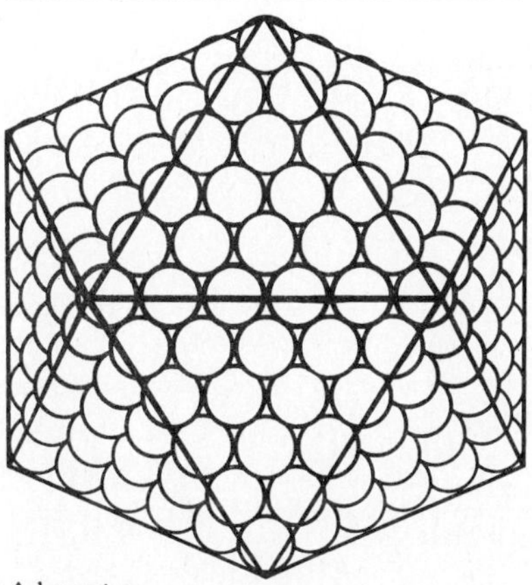

Adenovirus

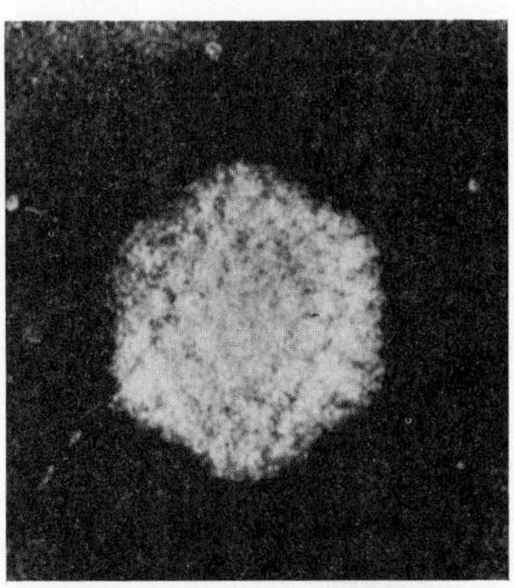

A closeup of adenovirus

Bacterial Virus

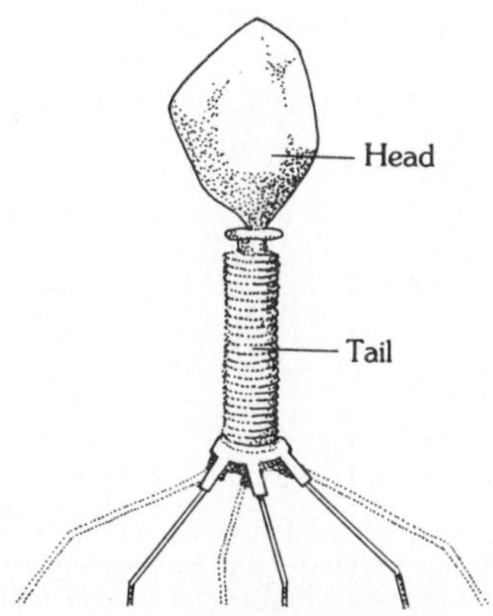

A model of a bacterial virus

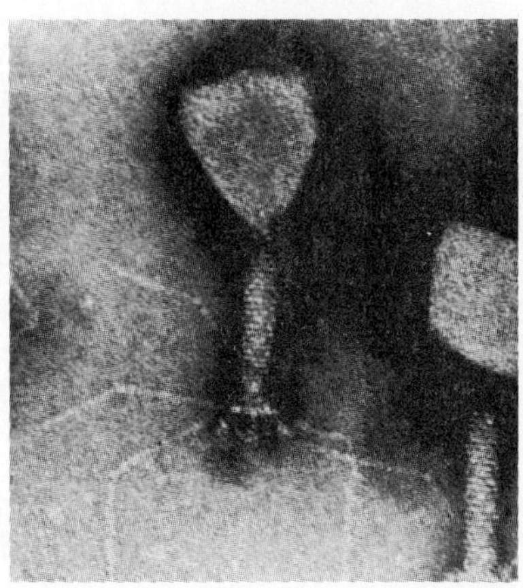

A closeup of a bacterial virus

Tobacco Mosaic Virus

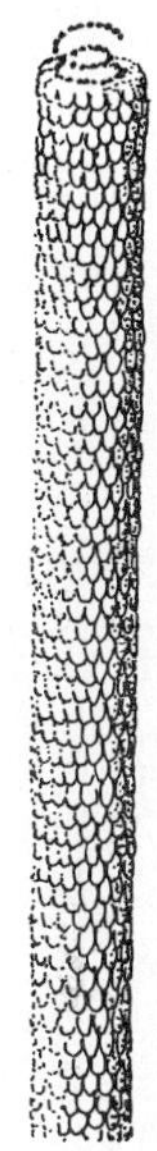

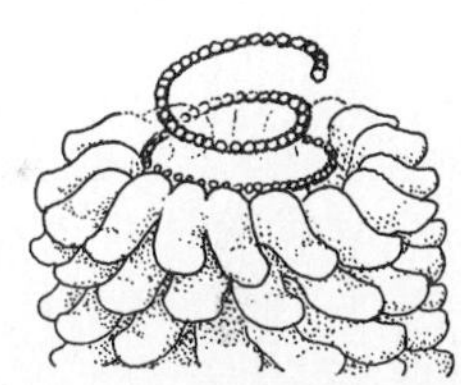

Spiral structure of tobacco mosaic virus

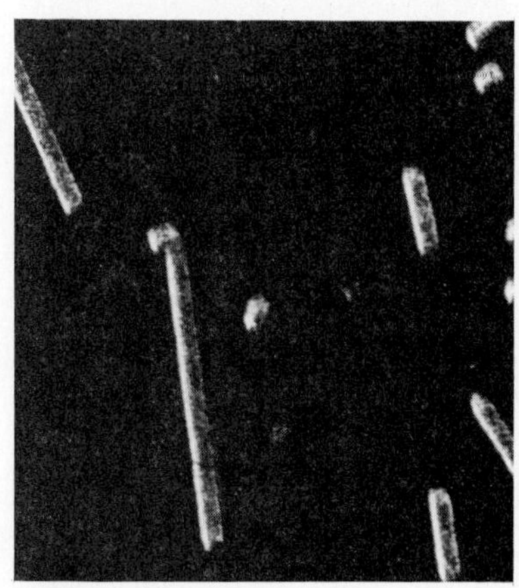

Tobacco mosaic virus

Virus Infections

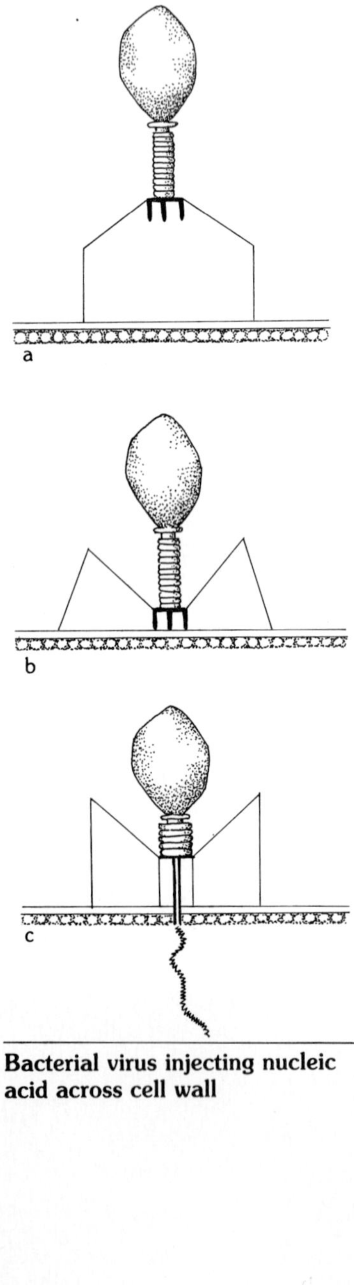

Bacterial virus injecting nucleic acid across cell wall

A lifeless virus existing outside of a host is unable to move by itself. Instead, it may be carried from one organism to another by direct contact. Leaves rubbing together or roots touching are examples of direct contact between organisms. Or, the virus may be carried indirectly by way of water, air, dust, food, or feces.

Once in contact with a new host, the virus must then attach itself. This depends on the outer protein coat of the virus. Generally, the outer coverings of both cell and virus must "fit" to provide a place of attachment. Because of this requirement, a bacterial virus can attach only to certain bacterial cells. Similarly, an animal virus can attach only to cells of certain animals, and a plant virus can attach only to cells of certain plants.

After attachment, the virus is then able to enter, or infect, the cell. The nucleic acid of bacterial and plant viruses must be injected across a cell wall. In animals, which have no cell wall, viruses are carried inside the cell membrane by phagocytosis.

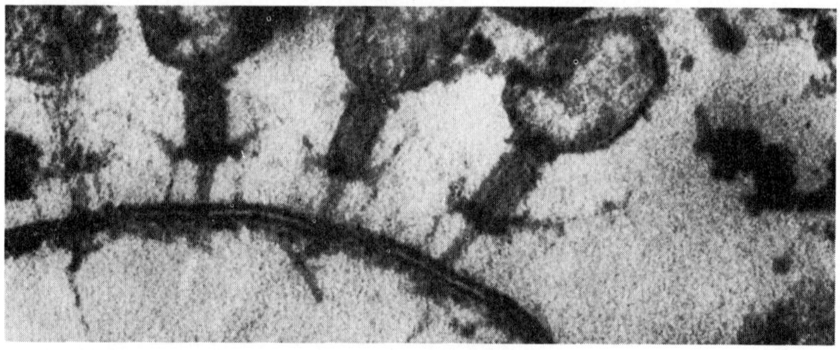

Photograph of bacterial virus injecting nucleic acid

Defenses Against Virus Infections

Humans have several lines of defense against viruses. These defenses must be overcome before an infection can occur. One of the defenses involves **phagocytes.** These cells are able to engulf and destroy a virus before the virus can cause an infection.

A second line of defense involves **immune responses.** An immune response may occur naturally or may be acquired through the use of vaccines. For example, smallpox, polio, measles, and influenza are virus infections that are controlled by vaccines. The immune response involves production of substances called antibodies. Antibodies inactivate a virus *before* the virus enters a cell.

A third line of defense involves an "interfering" substance. In humans, that substance is a naturally occurring protein called **interferon** (in'tər fir'on). Interferon is effective against many different viruses instead of against just one kind of virus. Interferon operates *inside* the cell by interfering with virus reproduction without interfering with the normal processes of the cell. Researchers are currently trying to find drugs that stimulate production of interferon inside the human body.

The Effects of Virus Infections

Once established inside a cell, the infecting virus may have one or more of several possible effects.

1. The cell may be unchanged. The virus exists inside the cell as an unseen, or **latent,** infection. This infection may never appear, or it may appear at some later time. Cold sores are examples of latent infections that have become visible.
2. The cell is changed in some way. For example, it may be changed into a *tumor* cell. Sometimes these tumor cells are **benign** (bi-nīn'), or harmless. Warts are examples of benign tumors. At other times, the tumor cells divide very rapidly becoming **malignant** (mə-lig'nant), or harmful. *Cancer* is the development of a malignant tumor. Various kinds of cancers in both plants and animals are caused by viruses.
3. The cell is killed. This occurs because of rapid reproduction and release of new viruses. Such viruses and infections are called **virulent** (vir'yə lənt). In the case of a many-celled organism—a plant or an animal—virulent infections damage tissue, causing disease. In plants, virulent infections include mosaics and stunts. In humans, virulent infections include polio, measles, mumps, rabies, influenza, smallpox, yellow fever, hepatitis, and the common cold.

Did You Know?
Viruses may cause reactions that produce the tissue damage known as allergy.

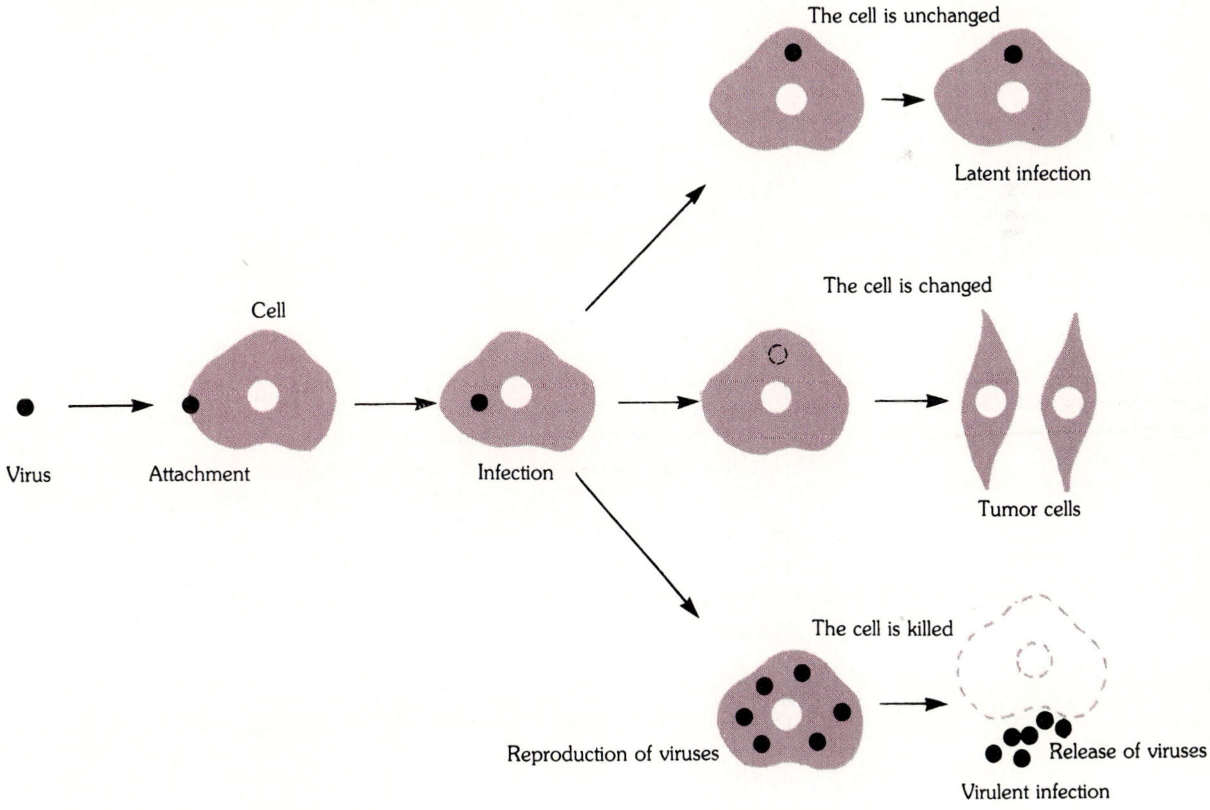

The Effects of Virus Infections

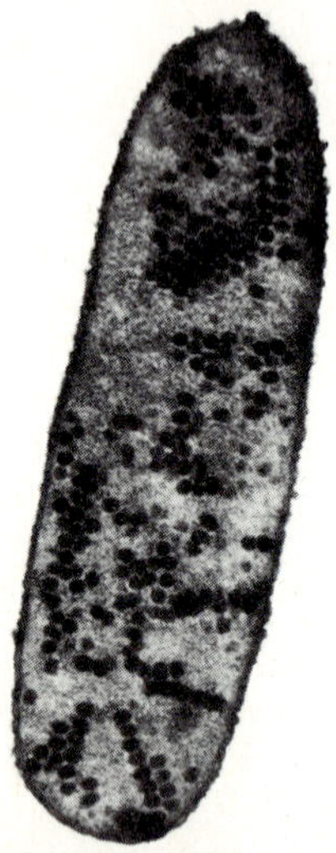

Bacterium infected by viruses

Reproduction of Viruses

The exact details of virus reproduction vary depending on the specific cell involved. For example, the reproduction of an animal virus is somewhat different than the reproduction of a bacterial virus. Nevertheless, virus reproduction generally has the same basic steps. Those steps include attaching and entering, making of protein and nucleic acid units, putting the units together, and releasing new viruses.

One of the best known examples of virus reproduction is that of bacterial viruses, or *bacteriophages.* The virus attaches itself to a bacterial cell wall and injects its nucleic acid across the cell wall into the cell. The protein coat of the virus remains outside the cell. The nucleic acid of the virus takes over and directs the making and assembling of new nucleic acid and new protein units. After 50 to 200 new viruses are produced, the bacterial cell dissolves, or *lyses,* releasing the new viruses. This entire process takes less than one hour. It results in the death of the bacterium.

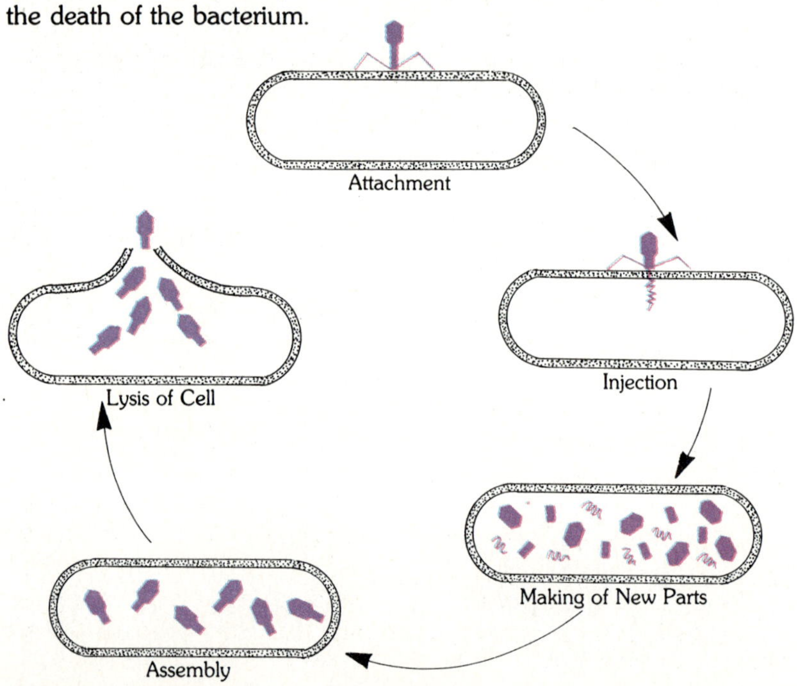

Reproduction of Virulent Bacterial Viruses

Review It

1. Make a drawing to illustrate the three main shapes of viruses.
2. What are some of the things that might happen when a virus enters a living cell?
3. Describe your own experience with viruses. Compare your experiences with experiences of others in your class.
4. How are viruses like cellular organisms? How are they different?

Issues in Biology

Should chimpanzees be imported to test hepatitis *B* vaccine?

Hepatitis *B* is a debilitating liver disease caused by a virus. In 1978 there were about 15,000 reported cases in the United States hospitalized with the disease. Since many cases go undetected or unreported, it was estimated that there were ten times that many, or 150,000 cases of hepatitis *B* in the United States alone. According to the center for disease control, hepatitis *B* is fatal in about one percent of hospitalized cases. In the U.S. there are an estimated one million recovered patients who are carriers of the hepatitis virus. Many of these people develop cirrhosis of the liver, a major cause of death in modern society.

A vaccine has been developed against the virus that causes hepatitis *B*. Chimpanzees are used to test the safety of this vaccine for use on humans, as the test must be carried out on a primate. Currently, about four chimpanzees are used to test each vaccine lot. They are given a multiple of the human dose. A chimpanzee can be used only once for this test. They are watched for several months, and their antibody levels tested. After the safety tests, they can be used in other research. For the vaccine to be available in large quantities, many wild chimpanzees will have to be captured. They cannot be bred quickly in a laboratory as can rats.

There are an estimated 50,000 chimpanzees in the wild. Private and public agencies have taken stands against the use of large numbers of chimpanzees that are needed for testing the vaccine. People in favor of using chimpanzees for these tests point out that hundreds of them are killed annually in their native land by the farmers whose crops they eat.

There are many questions about the methods of capture as well as how the captured breeding animals are housed and treated. In the minds of many people, the intelligence and social structure of the higher primates put them in a different category than traditional laboratory animals.

Analysis

1. Would you vote for a permit to import chimpanzees for laboratory use?
2. What standards would you follow for keeping chimpanzees in a laboratory?

Chapter Summary

The Kingdom Monera is made up of all the various kinds of bacteria. All monerans have a prokaryotic cell. The prokaryotic cell is quite plain, lacking membrane-bound organelles.

Although monerans are small, they are able to exist by the millions. Their abundance is largely a result of their rapid rate of reproduction.

The diversity of monerans is more in metabolic traits than in structural traits. The widespread occurrence of monerans is largely a result of their metabolic diversity.

Monerans are able to take part in many important processes. They are particularly important as decomposers, as nitrogen fixers, as producers of oxygen, and as pathogens.

A virus is a nucleic acid molecule protected by a coat of protein. Viruses display one trait of living organisms: the ability to reproduce. They can make copies of themselves by taking over the machinery of the cell they invade. As a result, viruses cause many diseases.

Questions to Think About

1. Suppose there were no limits on the growth of bacteria. How would that affect you?
2. Why are pathogenic bacteria that form tough, resistant resting cells more feared than pathogenic bacteria that form none?
3. Of the four kinds of moneran nutrition, which might have occurred first?
4. Penicillin kills harmful bacteria. What might it do to the beneficial bacteria in your intestine?
5. Explain how an atom that was once a part of some dinosaur could now be a part of you?
6. Why must people be concerned about the fate of the blue-green bacteria (blue-green algae)? of monerans in general?
7. Suppose a virus infection were treated with a drug that harmed the virus. How would you expect the cells to be affected?
8. How does the idea that viruses originated as pieces of nucleic acid relate to the fact that viruses infect only certain kinds of cells?
9. Suppose one kind of virus was found to reproduce outside of cells. How would that affect current ideas about viruses?

Extra Research

1. In addition to conjugation, a few bacteria reproduce sexually by transformation and transduction. Find out about those processes and make a drawing to illustrate each one.
2. Make drawings to show the various kinds of monerans and viruses.
3. Find out about the rickettsias. What roles have these monerans played in history?
4. What is a lysozyme? Where is it found? What roles does it play in defense against disease?
5. What are the most recent "breakthroughs" in the study of virus disease?
6. Oncogenic viruses are cancer-causing. Find out what is currently known about these viruses.
7. Bacterial viruses are being avidly studied by molecular biologists. Find out why.
8. Find out how bacteria and bacterial viruses have been used in genetic research. What are the possible advantages and disadvantages of such research?
9. Blooms of blue-green bacteria (blue-green algae) are often the result of environmental pollution. Find out exactly why such blooms occur, why those blooms cause problems, and what can be done to prevent them.

Algal Bloom

Chapter Test

Part I

Copy the number of each test item. If the item is true, write *T* after the number; if the item is false, write *F*.

1. Viruses are cells. (246)
2. Bacteria have enclosed genetic material. (238)
3. Blue-green bacteria contain chlorophyll. (239)
4. Bacteriophages invade animal cells. (250)
5. Blue-green bacteria require light for energy. (240)
6. Interferon inactivates viruses. (248)
7. Prokaryotic cells have ribosomes. (238)
8. Peptidoglycan is found in all cells. (238)
9. Viruses are the same size as bacteria. (246)
10. Viruses do not infect plants. (248)
11. A latent virus infection is never seen. (249)
12. Some monerans are parasites. (243)
13. A virus may contain both DNA and RNA. (246)
14. Viruses exist in colonies. (246)
15. Some monerans have no cell wall. (242)

Part II

Copy the number of each sentence below. After each number, write the word or words that complete the sentence correctly.

16. Organisms that live without oxygen are ____. (240)
17. The raw materials of autotrophs are ____ compounds. (240)
18. A cell that lacks membrane-bound organelles is a ____. (238)
19. The relationship between organisms that live in close association is called ____. (243)
20. The material in the capsule or sheath of a moneran is ____. (239)

Part III

Copy each item from List A. After each item, write the letter and the words from List B that best describe it.

List A

21. conjugation (245)
22. flagella (239)
23. colony (244)
24. fission (244)
25. ability to be crystallized (246)

List B

a. a way for bacteria to exchange genetic material
b. asexual reproduction of monerans
c. whiplike structures used for movement
d. trait of viruses
e. many of the same kind of organism living together

Chapter 11

The Protists

The protists, like the monerans, are one-celled organisms. But protists are quite different than monerans in several ways. For one, the protists are "newer" than the monerans. According to the fossil record, protists have been living on Earth for only about 1.5 billion years whereas monerans have been around for probably about 3.5 billion years. For another, the cells of protists are generally larger and more complex than the cells of monerans.

There is a great diversity among protists just as among monerans. But unlike the monerans, protists are generally similar to one another in metabolic activities. Diversity among protists is more a diversity of structure than a diversity of metabolism. A huge array of variations of structure exists among protists. Even the most active imagination could not dream up all their bizarre shapes and forms. And when compared to the cells of other organisms, the cells of protists are among the most elaborate.

This chapter will give you a brief glimpse of the great structural diversity found in the protist kingdom as you meet some animallike protists and some plantlike protists. You will find out not only about their structure and activities but also about how protists affect the survival of other living organisms.

Chapter Objectives

1. Describe the structure of a typical protistan cell.
2. Discuss two hypotheses for the origin of the protistan cell.
3. Name and describe the three main kinds of plantlike protists.
4. Name and describe the four main kinds of animallike protists.
5. Describe at least two ways protists affect other organisms.

A paramecium after ingesting stained yeast cells.

11-1
The Protistan Cell

The protistan cell, unlike the moneran cell, is made up of compartments. These compartments, or *organelles,* are formed by internal membranes. For example, the protistan cell has a membrane-bound nucleus containing two or more chromosomes and connected in some way with the endoplasmic reticulum. The protistan cell has one or more mitochondria and one or more Golgi bodies. It may also contain other organelles such as flagella, lysosomes, and contractile vacuoles. And the cytoplasm circulates, or *streams.* In short, the protistan cell is a *eukaryotic* cell, the kind of cell you studied in Unit Two. The eukaryotic cell of a nonphotosynthetic protist is shown below.

The trait of possessing special structures for different tasks is called division of labor, or **specialization** (spesh′ə lə zā′shən). In both nonbiological and biological systems, specialization is generally thought of as providing an efficient way to use available resources. In the case of protists, the specialization involves the cell organelles. Each of the separate, membrane-bound organelles is specialized to carry out its own particular task. For example, the mitochondrion specializes in respiration. The chloroplast, if present, specializes in photosynthesis. And the nucleus specializes in the transfer of information.

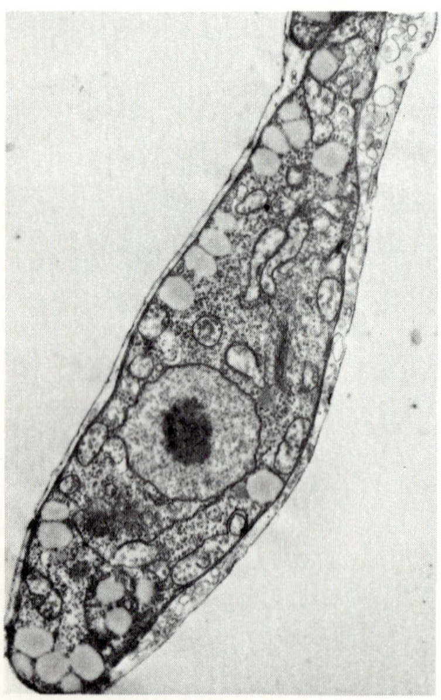

Nonphotosynthetic protist

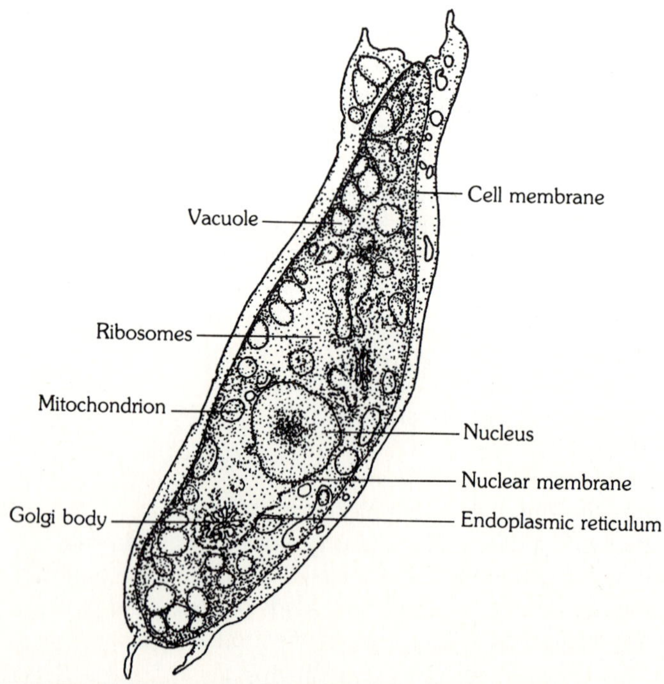

Structures of nonphotosynthetic protist

The Origin of the Protistan Cell

The emergence of the protistan cell may be considered one of the most important single events in the history of life on Earth. Many biologists have tried to find an explanation for how this event occurred. Among the various hypotheses proposed, the two most discussed at this time are the hypothesis involving symbiosis and the hypothesis involving infolding of the cell membrane.

In both hypotheses, the ancestral protist was relatively large and its cell was amebalike, or **ameboid** (ə mē′boid). This means that its outer layer was flexible enough to change shape as cytoplasm pushed against it. This flexibility would permit the formation of vacuoles and the ingestion of smaller organisms. Perhaps this ancestral protist was similar to a *Mycoplasma* bacterium, which does have a flexible outer layer.

In the *symbiosis hypothesis,* several kinds of organisms lived together for mutual benefit and eventually lost their ability to live apart. In this hypothesis, the chloroplasts were originally blue-green bacteria (blue-green algae), the mitochondria were originally aerobic bacteria, and the flagella were originally spiral bacteria.

In the *infolding hypothesis,* the outer layer developed deep folds that eventually broke off and began to exist as separate structures, or organelles. Diagrams of both hypotheses are shown below.

Various bits of evidence can be found to support each hypothesis. It may be that some organelles are the result of symbiosis while others are the result of infolding.

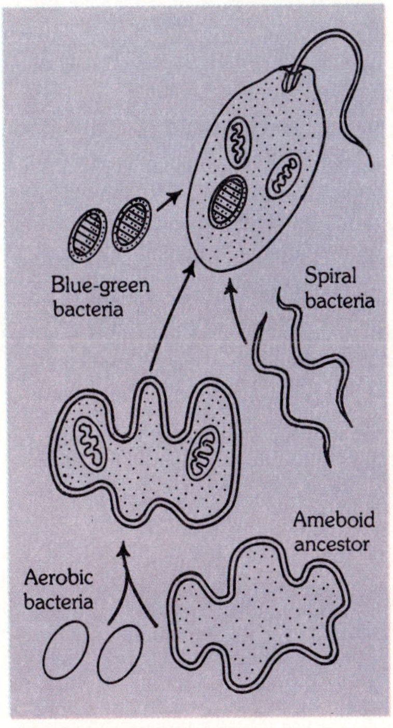

Symbiosis Hypothesis

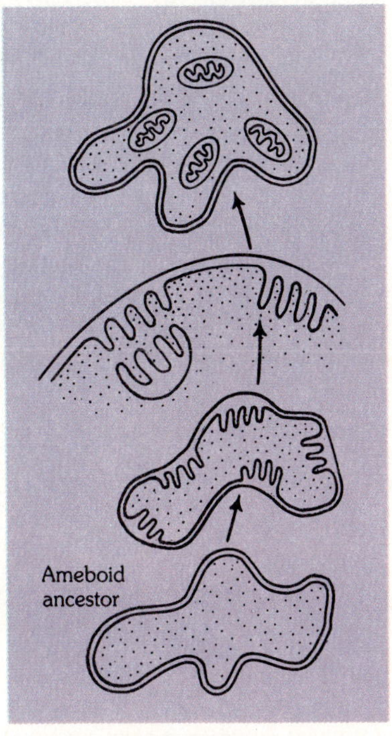

Infolding Hypothesis

Did You Know?

An Australian termite is the host to not just one but a series of symbionts. Without the smooth-swimming, flagella-covered protist that lives in its digestive system, the termite would be unable to break down the cellulose it eats. And without the spiral bacteria that act as flagella, the protist would be unable to swim. Further, the bacteria that live in the cytoplasm of the protist are believed to be responsible for the actual breakdown of cellulose.

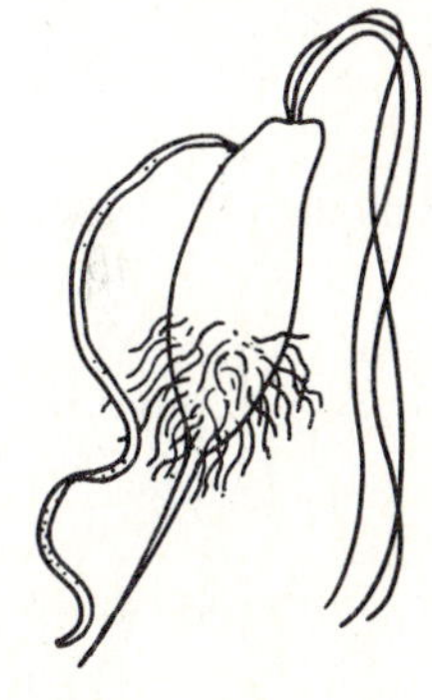

Symbionts of termite

Challenge!
Look for Euglena *in freshwater pools and ponds—particularly where the water hardly moves and where the water is covered with a greenish scum.*

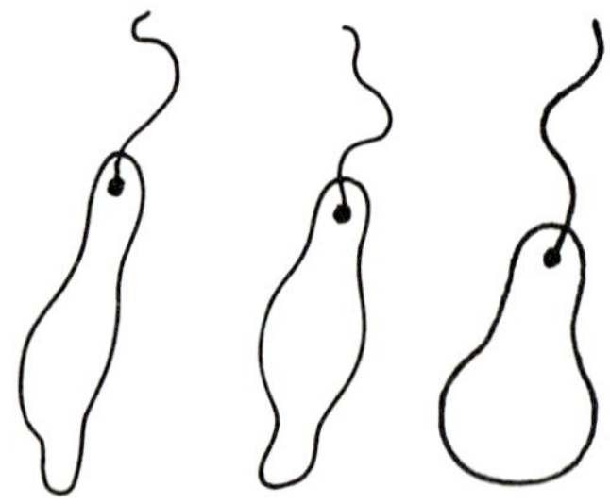

Ameboid motion of *Euglena*

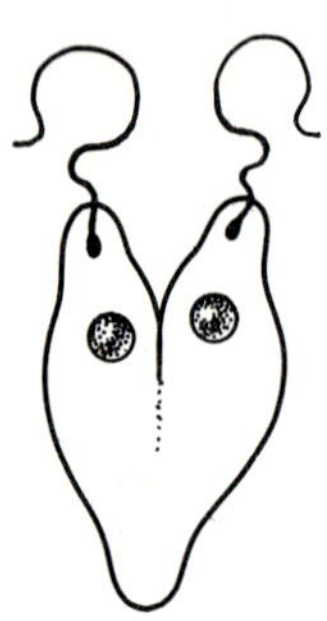

Fission of *Euglena*

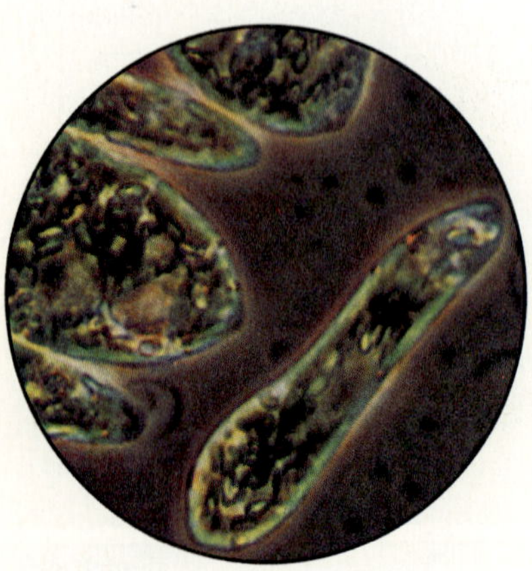

Euglena

Euglena, a Representative Protist

As a one-celled eukaryote, *Euglena* is obviously a member of the kingdom Protista. But before the five kingdom scheme was invented and biologists recognized only the plant kingdom and the animal kingdom, *Euglena* was the object of a classification "tug-of-war." *Euglena* was claimed by some for the animal kingdom because of its animallike traits. It has a flexible outer covering and no cell wall. Also, it has special organelles for motion. At the same time, it was claimed by others for the plant kingdom because it contains the green pigment chlorophyll.

Because of its combination of plantlike and animallike traits, *Euglena* is quite versatile in the way it lives. Ordinarily, *Euglena* uses light and simple inorganic materials in photosynthesis. In other words, it is a phototrophic autotroph. But in the absence of light, *Euglena* becomes a chemotrophic heterotroph. This means it takes in rather than makes organic materials. This versatility enables *Euglena* to survive in a variety of environments. And because of this versatility, some biologists think *Euglena's* ancestors may have existed before any other kind of protist.

If you were to look at *Euglena* under a microscope, the first structures you would probably notice are the green chloroplasts and the centrally located, well-formed nucleus. Another noticeable structure is the flasklike cavity called the reservoir. Near the reservoir may be one or more vacuoles that collect and remove excess water; one or more whiplike flagella, which are used both for swimming and for sensing; and a red **eyespot,** which is thought by some to be sensitive to light. In addition, the cytoplasm of *Euglena* contains food-storage granules, mitochondria, and an endoplasmic reticulum.

Euglena ordinarily swims, using its flagella. But when there is not enough water for swimming, *Euglena* is able to move by a creeping motion. This motion is possible because the outer covering is flexible enough to change shape when cytoplasm pushes against it.

Euglena reproduces asexually. The nucleus of *Euglena* divides by mitosis. After mitotic division, the rest of the cell divides along its length. When conditions are unfavorable to growth and reproduction, *Euglena* may form a thickly coated resting cell.

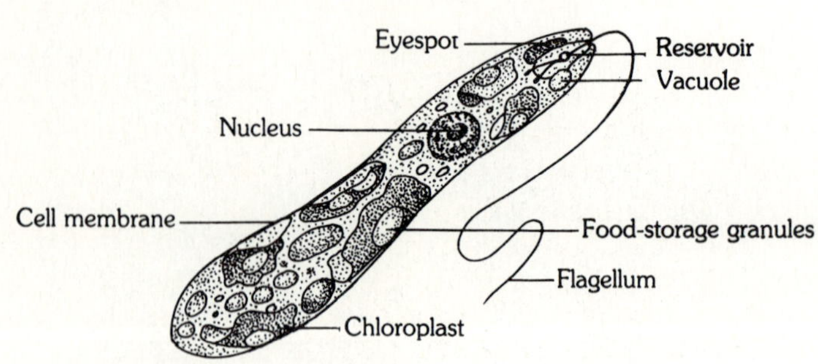

Structure of *Euglena*

Activity

Euglena in Action

Materials

Culture of *Euglena*
Microscope
Eyedropper
Glass slides
Cover slips
Black construction paper
Scissors
Methyl cellulose or fibers
Ice
Aluminum foil

Procedure

1. Put a drop of the *Euglena* culture on a glass slide and cover it with a cover slip.
2. Make a second slide the same way. Put black construction paper over half of the cover slip. Then place the entire slide in bright light for 20–30 minutes. The light should not be so strong that the slide dries out.
3. Examine the *Euglena* culture on the first slide. Use low power first. Look for chloroplasts, the nucleus, the eyespot, and various other organelles. Make a full-page, labeled drawing of the *Euglena* to show what you see.
4. Adjust the light with the mirror to observe a flagellum on a single *Euglena*. If the *Euglena* moves too fast, slow it down with methyl cellulose or fibers. Notice whether *Euglena* is pushed or pulled by its flagellum. Make drawings to show what you see.
5. After 20–30 minutes, remove the second slide from bright light. Compare the distribution of the *Euglena* on the uncovered and covered part of the slide. Record your observations.
6. Cut a strip of aluminum foil so it is about 30 cm (1 ft.) long and, in its middle, has a hole the size of the hole in the microscope stage. Place the strip on the microscope stage with the slide as shown.

7. Heat one end of the strip and/or cool the other end. Do not allow the slide to dry out. Observe the distribution of *Euglena*. Record your observations.
8. If the slide should start to dry out, the *Euglena* would then be unable to move by swimming. Instead, it would use a creeping motion. Before adding more water, look for this motion.

Analysis

1. Of the traits you observed, which of *Euglena's* traits are plantlike? Which are animal-like?
2. What is the advantage of being able to respond to changes in heat and temperature?
3. Do you think *Euglena* would prefer a cool, bright place; a cool, dark place; a warm, bright place; or a warm, dark place? Design an experiment to test your hypothesis.

Challenge!

Euglena *also responds to different colors of light and to different chemicals such as vinegar and salt. Describe a way to demonstrate one or more of these responses.*

Plantlike Protists

Challenge!
Obtain a sample of fresh water or salt water. Look at a drop or two under a microscope. Try to identify as many different kinds of protists as possible.

All of the plantlike protists can be called **algae** (al′jē). *Algae* is a rather informal term that can be used to refer to almost all chlorophyll-bearing organisms found living in water. Generally, algae have different reproductive processes than most plants. While many of the various algae are members of the protist kingdom, others are members of the plant kingdom. And the blue-green "algae" (blue-green bacteria) are members of the moneran kingdom.

There are three main groups of plantlike protists: the **euglenophytes** (yü glē′nə fīts), the **chrysophytes** (kris′ə fīts), and the **pyrrophytes** (pir′ə fīts). The euglenophytes include euglenids such as *Euglena*. The chrysophytes include diatoms and golden algae. The pyrrophytes include the dinoflagellates. The three kinds of plantlike protists are differentiated by two traits: photosynthetic pigments and food-storage materials. These and other traits are shown in the table below.

The Importance of Plantlike Protists

Plantlike protists are abundant in both fresh water and salt water. A liter of salt water taken near the ocean surface may contain as many as a quarter of a million of these protists. But in spite of their abundance, these protists usually go unnoticed by most people because of their size. Nevertheless, they are very important. The earth would be unable to support life without them.

As photosynthetic organisms, the plantlike protists use light and simple inorganic materials to make their food and to produce oxygen. These plantlike protists along with other kinds of algae maintain over half of the earth's oxygen supply. More importantly, the plantlike protists are indispensable as primary producers. **Primary producers** are the main source of food—either directly or indirectly—for all animals of the sea and shore.

Plantlike Protists

	Euglenophytes	Chrysophytes	Pyrrophytes
Examples	euglenids	diatoms golden algae	dinoflagellates
Pigment	chlorophyll a chlorophyll b	chlorophyll a chlorophyll c	chlorophyll a chlorophyll c
Food storage	fat, paramylum	oil, leucosin	oil, starch
Special material in cell wall	no cell wall	silica	cellulose
Flagella	1–3	2	2

Diatoms, Jewels of the Sea

Diatoms are probably the most numerous of all plantlike protists. Because of this abundance, they are one of the most important primary producers of the sea.

The cell wall of a diatom contains silica and is formed in two halves. The rim of one half tightly overlaps the rim of the other half, like a pillbox. The *silica,* which is a hard, inorganic substance, forms a skeleton, or shell, that remains long after the organic part of the diatom has been decomposed. Diatom shells occur in a seemingly endless variety of intricate designs.

Diatom shells piling up at the bottom of oceans and lakes over millions of years formed deposits called **diatomaceous earth** (dī'ə tə-mā'shəs). These deposits may be a thousand meters thick with as many as five million diatom shells in each cubic centimeter.

Challenge!
Obtain a sample of diatomaceous earth. Examine it under a microscope. Make a slide collection of as many diatom shells as possible.

Diatom

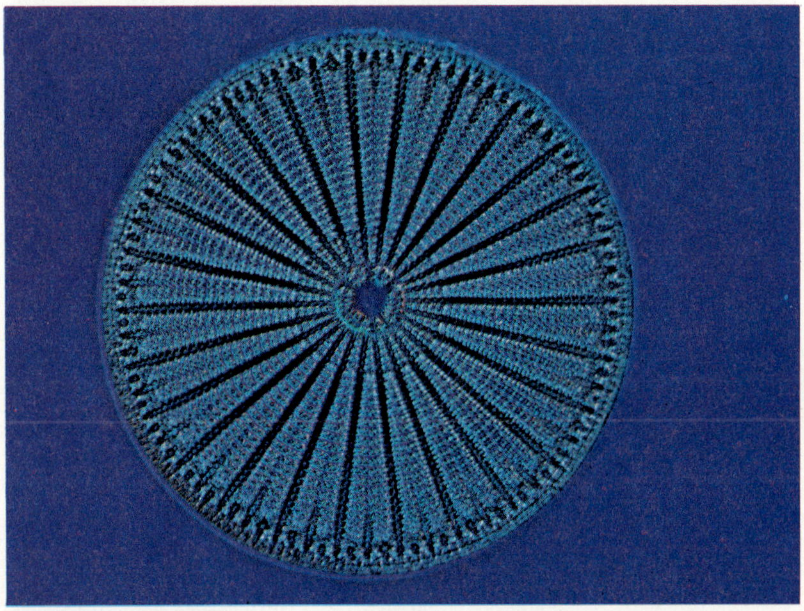

Did You Know?
Population explosions of certain species of the dinoflagellates Gymnodinium *and* Gonyaulax *cause special blooms called "red tides." The rusty color of the water is a result of the yellow-orange color of their photosynthetic pigments. Red tides are toxic to various kinds of sea life. The toxin can cause a fatal paralysis in humans who eat contaminated oysters and clams.*

Dinoflagellates

Dinoflagellates are second only to the diatoms in abundance and importance as primary producers. Along with other pyrrophytes, many dinoflagellates are **bioluminescent** (bī'ō lü mə nes'nt). This means they can change chemical energy into light. This bioluminescence can be seen at night from a ship passing through waters inhabited by dinoflagellates. The disturbed water—as in a ship's wake—seems to dance with tiny flashes of light.

The *dino* in dinoflagellate means "whirling." A dinoflagellate appears to whirl, or spin, as it moves. It has two flagella housed in two grooves. One groove extends the length of the cell while the other encircles the midsection. With the exception of the grooves, a dinoflagellate is covered by armorlike, cellulose plates.

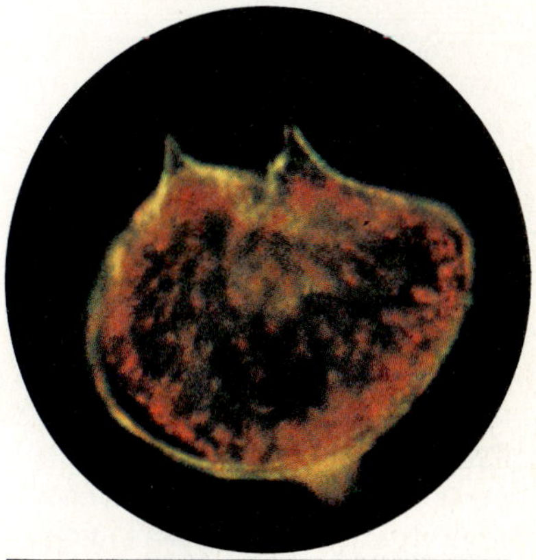

Dinoflagellate

Did You Know?
Nine pairs and two single microtubules run from the base to tip of both cilia and flagella. This is true for the cilia and flagella of all eukaryotic organisms. Two central microtubules are surrounded by nine pairs of microtubules. This is called the "9 plus 2" pattern.

The flagella of prokaryotes are thinner and are made of a single fibril.

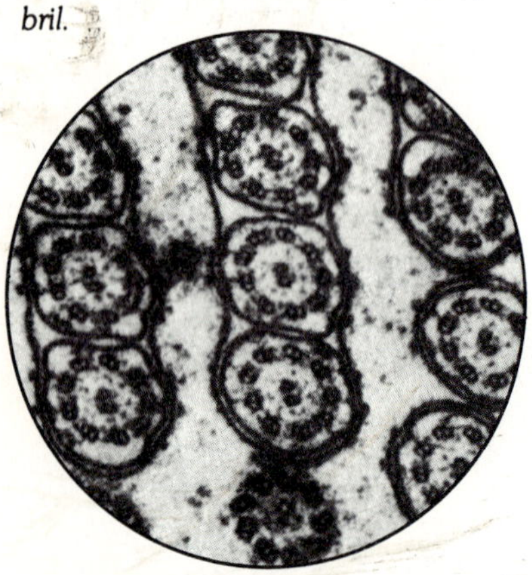

The 9 plus 2 pattern

Animallike Protists: The Protozoa

The animallike protists are often called **protozoans** (prō'tə zō'ənz), which means "first animals." All are chemotrophic heterotrophs, unable to make their own food. Once they find food, they may either *engulf* it or *absorb* it. Engulfed food is digested in the cell, whereas absorbed food must be in a predigested form.

Protozoans are classified by their organelles of motion into four main groups: the **zooflagellates** (zō'ə flaj'ə lāts), the **sarcodines** (sär'kō dīnz), the **ciliates** (sil'ē its), and the **sporozoans** (spôr'ə-zō'ənz). The zooflagellates, which often resemble a nongreen *Euglena,* use flagella for motion. The sarcodines use fingerlike projections of cytoplasm called **pseudopods** (sü'də podz), or false feet. The ciliates use short flagellalike organelles called **cilia** (sil'ē ə). The sporozoans have no special organelle for motion. They move by gliding. These and other traits are shown in the table below.

Protozoans normally reproduce asexually by *fission.* They may also reproduce sexually. This generally involves the fusion of a pair of gametes. This fusion forms a *zygote.* Reproduction may often be followed by the formation of a resting cell.

Protozoans are widely distributed, found in all moist environments: soil, salt water, and fresh water. Saltwater protozoans are important links in the food chains of the ocean. These protozoans feed on small organisms and, in turn, are fed on by larger organisms. Also, many protozoans are important as decomposers, breaking down complex organic materials into simpler materials. And some protozoans—particularly the sporozoans—are important as *pathogens.*

	Zooflagellates	Sarcodines	Ciliates	Sporozoans
	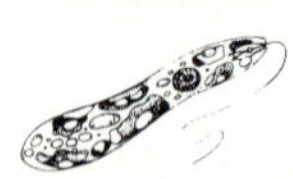	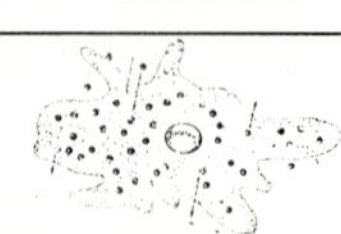	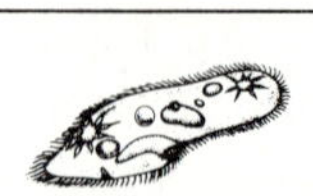	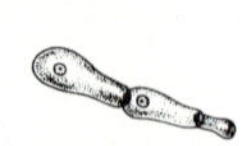
Examples	*Peranema* *Trichomonas* *Trypanosoma* *Trichonympha*	*Ameba* *Foraminifera* *Radiolaria* *Heliozoa*	*Paramecium* *Balantidium*	*Plasmodium*
Motion	by flagella	by pseudopods	by cilia	glides
Feeding	absorbs	engulfs	engulfs	absorbs
Parasitic	a few	a few	a few	all
Skeleton or shell	none	some lime some silica	none	none

Challenge!
Look at finely divided natural chalk under a microscope. Make drawings to record your observations.

Ameba: A Famous Animallike Protist

A look at a living *Ameba* under a microscope would show constant changes of shape. As the *Ameba* moves, it slowly pushes out its pseudopods. These disappear as the *Ameba* pulls itself around them.

By using its pseudopods in ameboid motion, the *Ameba* can engulf food, such as bacteria or other protists. And it can form food vacuoles by phagocytosis. In addition to its importance in feeding, ameboid motion also enables *Ameba* to respond to various stimuli such as light, chemicals, and contact.

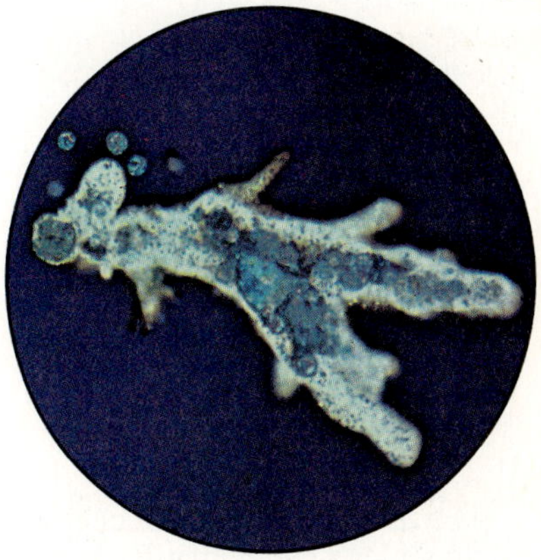

Ameba

Foraminiferans and Radiolarians: The Rock-formers

Both foraminiferans and radiolarians are related to the *Ameba* and have pseudopods. But, unlike the *Ameba,* they have hard external skeletons. Their pseudopods extend through holes in the skeleton. Accumulations of foraminiferan shells, which are made of calcium carbonate, form the mineral *chalk* and the rock *limestone.* And accumulations of radiolarian shells, which are made of silica, form silica-containing rock.

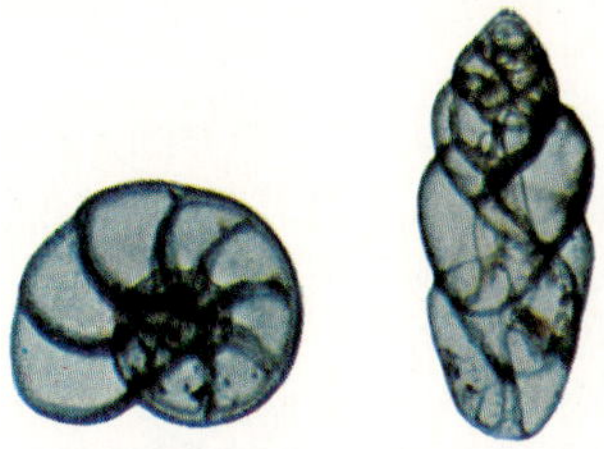

Foraminiferans

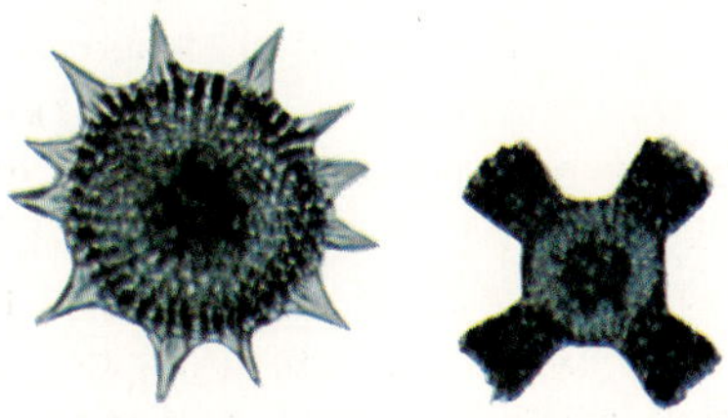

Radiolarians

Paramecium: A Common Animallike Protist

The *Paramecium* is both common and abundant in fresh water. Unlike the *Ameba,* it has a permanent shape. That shape is slipperlike. The *Paramecium* swims rapidly using the cilia that cover its body. Cilia operate like paddles moving the *Paramecium* backward as well as forward.

Cilia also direct food into the gullet. From the gullet, food enters a food vacuole where enzymes break down, or digest, the food. Digested wastes are expelled from the food vacuole through an **anal pore.** And just as in other freshwater protists, excess water is expelled by a contractile vacuole.

Paramecium has a small and a large nucleus. The small **micronucleus** controls reproduction, and the large **macronucleus** controls respiration and other cell activities. *Paramecium* generally reproduces asexually but occasionally it may reproduce sexually by **conjugation.** During conjugation, a pair of *Paramecia* join to exchange the genetic material of their micronuclei.

Paramecium has an interesting defense mechanism. When threatened, *Paramecium* shoots dartlike projectiles called trichocysts at its enemy.

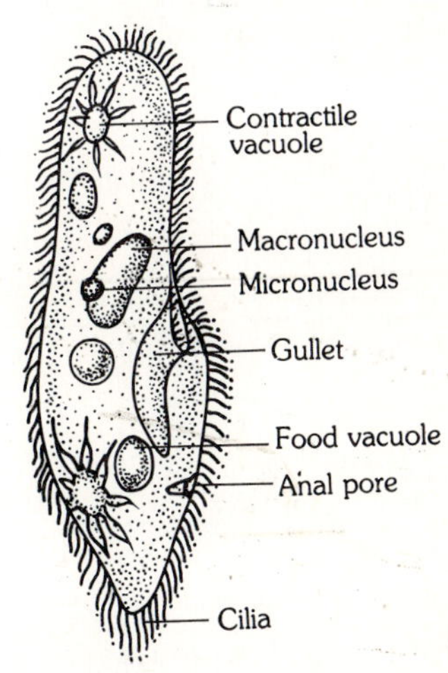

Paramecium

Pathogenic Protozoans

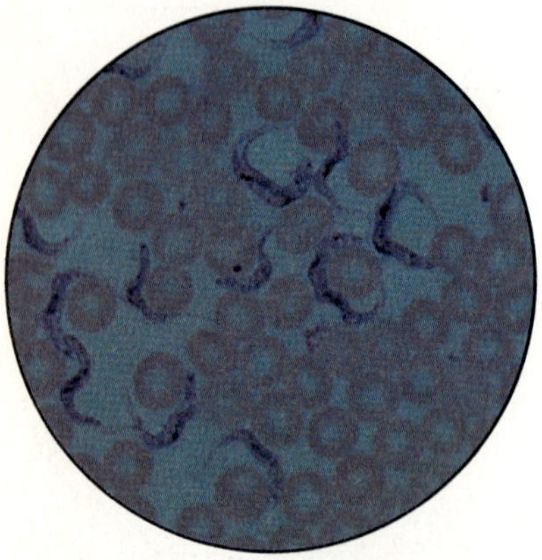

Human blood containing *Trypanosoma*

Pathogenic protists are second only to pathogenic monerans in the amount of misery they cause in human and other hosts. Among protozoans pathogenic to humans are the sporozoan *Plasmodium,* which causes malaria; the sarcodine *Entamoeba,* which causes amebic dysentery; the ciliate *Balantidium,* which causes balantidial dysentery; and the zooflagellate *Trypanosoma,* which causes African sleeping sickness.

Biologists have long studied the complex life cycles of various species of *Plasmodium* to find a way to control malaria. A *Plasmodium* never occurs outside of its two hosts: mosquitoes and humans. In the female *Anopheles* mosquito, it completes a stage of *sexual* reproduction involving gametes and zygotes. In the human, it completes several multiple *asexual* mitotic divisions in liver and red blood cells. The chills and fever of malaria, which occur at regular intervals, coincide with the release of *Plasmodium* cells from red blood cells.

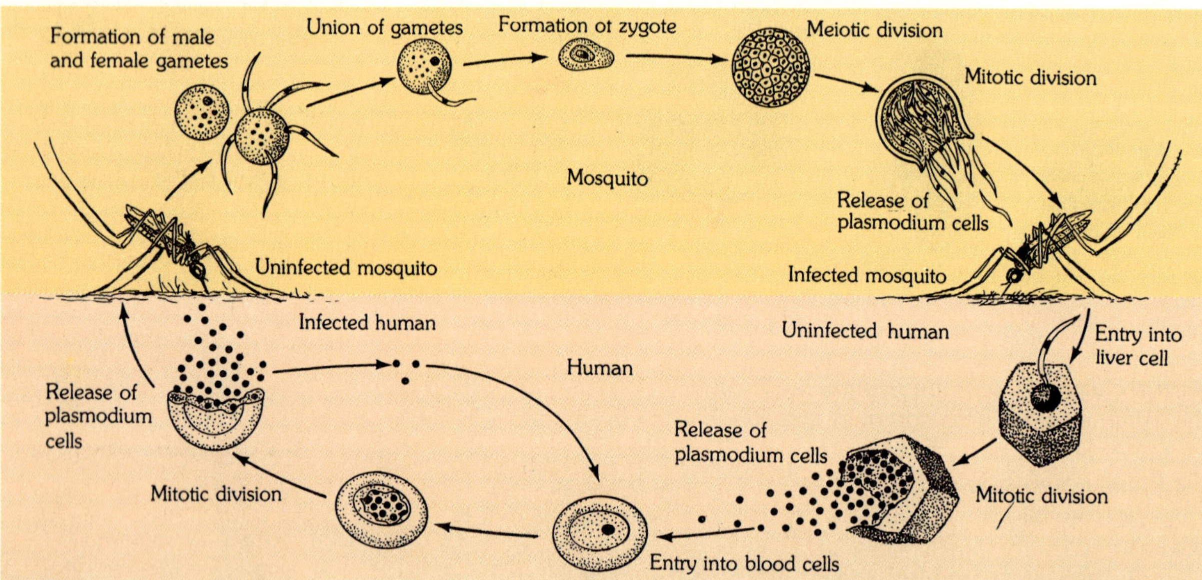

Life cycle of *Plasmodium*

Review It

1. Draw and label a protistan cell.
2. What are some of the ways the protistan cell might have originated?
3. What are the four main kinds of animallike protists? How are they different from one another?
4. What are the three main kinds of plantlike protists? How are they different from one another?
5. Discuss several ways protists affect your life.

Breakthroughs in Biology

Bacteria, viruses, protozoa, and fungi can be used to get rid of crop pests without harming the environment or contaminating the food chain. These microscopic organisms are living pesticides known as microbials.

Since the 1940s the agriculture industry has used chemicals to poison crop pests. That method has failed in many areas because the chemicals proved to be harmful to the environment, and the pests developed a resistance to those pesticides. With the safety of the environment and the health of humans in mind, the U.S. Department of Agriculture is putting more emphasis on developing microbials and on helping industry to market them.

One bacterial agent, *Bacillus thuringiensis,* has been used since the late 1950s to control several types of leaf-eating caterpillars that destroy crops and forests. But the problems and expense involved in developing, producing, and storing some other live pest fighters had delayed their widespread use.

In January of 1978, the Agricultural Research Service of the U.S. Department of Agriculture had 40 scientists working on programs to develop effective pesticides from live agents. One of their more promising microbials is a virus that is now being used to protect cotton and tobacco crops. Another is a protozoan that kills grasshoppers.

Tussock Moth

A virus that kills the Douglas fir tussock moth and one that kills the gypsy moth have been developed by the U.S. Forest Service and registered by the Environmental Protection Agency. Both moths defoliate trees.

The Canadian Forest Service, Forest Pest Management Institute is developing viral agents against the European pine sawfly, the Douglas fir tussock moth, and the Eastern spruce budworm—all of which have destroyed millions of acres of trees in Canada and the U.S.

The bacterium, *Bacillus popilliae,* kills Japanese beetles and is being used to protect home lawns and grasses on the east coast, but it is still very expensive to produce.

Researchers are also working on a bacterium to kill certain mosquito species, a fungus to master beetle grubs and other mosquitoes and aphids, and viruses to stifle insects that damage important fruit and vegetable crops.

Analysis

1. Write to the U.S. Department of Agriculture and find out if other microbials have been developed.
2. Find out about the Environmental Protection Agency guidelines on safety tests for microbials.

Chapter Summary

Although there are thousands of different protists, all are one-celled and eukaryotic. Protist cells may be quite elaborate and may have a variety of specialized organelles.

Protists are either plantlike or animallike depending on whether they have chlorophyll. Plantlike protists include euglenophytes, chrysophytes, and pyrrophytes. The major groups are distinguished by their photosynthetic pigments and their food storage materials. They are important as primary producers of food as well as oxygen producers.

The animallike protists, or protozoans, include zooflagellates, sarcodines, ciliates, and sporozoans. The major groups are distinguished by their organelles for motion. They are important as links on many foods chains, as decomposers, and as pathogens.

Questions to Think About

1. Compare the traits of plantlike and animallike protists. Consider nutrition, feeding, and other metabolic processes as well as structures. Describe similarities and differences.
2. How does the outer layer of a protist that engulfs food compare to the outer layer of a protist that absorbs food?
3. What are two requirements for ameboid motion? Why is ameboid motion generally seen only in eukaryotic organisms?
4. The term *protozoan* means "first animal." The term *protophyte* means "first plant." What does the term *protist* mean? Is this a good name for one-celled eukaryotic organisms?
5. Instead of protists being considered simple, they are considered very complex. Discuss this idea.
6. How can the similarities in the structure of cilia and flagella be explained?
7. Compare a typical moneran cell and a typical protistan cell. What are their similarities and their differences? Consider size, type of cell, organelles, metabolism, and reproduction.

Extra Research

1. Certain membrane-bound organelles have their own DNA. How might this occurrence be explained?
2. How might plantlike protists be used to benefit humans? List various possibilities.
3. What causes "red tides," the population explosions of reddish dinoflagellates? What can be done to prevent or control red tides?
4. What are the uses of diatomaceous earth?
5. *Paramecium* has spearlike organelles called "trichocysts." What is their job? How do they operate?
6. Dinoflagellates are found as symbionts in coral reefs. What are the benefits?
7. What evidence supports the various hypotheses about the origin of the protistan cell?
8. Flagellated, ameboid protists are considered least complex whereas ciliated, stiff-walled protists are considered the most complex of protists. Why?
9. Invent a new protist. Describe its environment, its structure, and its life style. What are its needs? How are they obtained?
10. Which organisms provide the most oxygen to the atmosphere?
11. In what eukaryotic organisms do cilia occur? What jobs do these cilia perform?

Chapter Test

Part I

Copy the number of each test item. If the item is true, write *T* after the number; if the item is false, write *F* after the number.

1. *Euglena* is both plantlike and animallike. (258)
2. Protist cells are simpler than moneran cells. (256)
3. Conjugation is a kind of asexual reproduction. (263)
4. Diatoms and dinoflagellates are found in the ocean. (261)
5. All algae are protists. (260)
6. The *Ameba* has a red eyespot. (258)
7. Zooflagellates move by using pseudopods. (262)
8. Ciliates move by using cilia. (262)
9. Sporozoans have no organelle for motion. (262)
10. Sarcodines move by using flagella. (262)

Part II

Copy the number of each sentence below. After each number, write the word or words that complete the sentence correctly.

11. Membrane-enclosed compartments inside a eukaryotic cell are _____. (256)
12. *Plasmodium* is the cause of the disease _____. (264)
13. When an organism gives off light, it is _____. (261)
14. A _____ is a plantlike protist covered with cellulose plates. (261)
15. Limestone and chalk are formed of deposits of _____. (263)

Part III

Copy each item from List A. After each item, write the letter and words from List B that best describe it.

List A

16. chlorophyll (258)
17. specialization (256)
18. ameboid (257)
19. animallike protist (262)
20. form silica-containing rock (263)
21. diatomaceous earth (261)
22. *Paramecium* (263)
23. symbiosis hypothesis (257)
24. diatom (261)
25. pseudopod (262)

List B

a. collection of radiolarians
b. two silica shell halves
c. found in plantlike protist
d. has both a small and large nucleus
e. flexible enough to change shape
f. protozoan
g. describes origin of cell organelles
h. collection of diatom shells
i. false foot
j. division of labor

Chapter 12

The Fungi

The fungi make up a kingdom that includes molds, mildews, morels, mushrooms, yeasts, rusts, smuts, and blights. Although these organisms often go unnoticed, they actually can be found thriving just about everywhere—in your backyard, in the air you breathe, in your food, and even in you.

In this chapter, you will find out why fungi often go unnoticed—even though many fungi are *macroscopic,* that is, large enough to be seen without a microscope. You will study some common fungi and find out how they grow, feed, and reproduce. You will have a chance to compare several different kinds of fungi and see how fungi are classified. And you will learn why fungi are important—to you and to all other living things.

Chapter Objectives

1. Describe the distinguishing traits—including structure, growth, feeding, and reproduction—of a common fungus.
2. Differentiate three main groups of fungi.
3. Discuss the importance of fungi as decomposers, as symbionts, as food producers, as food, and as agents of disease.

Mushrooms are part of the fungi kingdom.

12–1
The Life of a Fungus

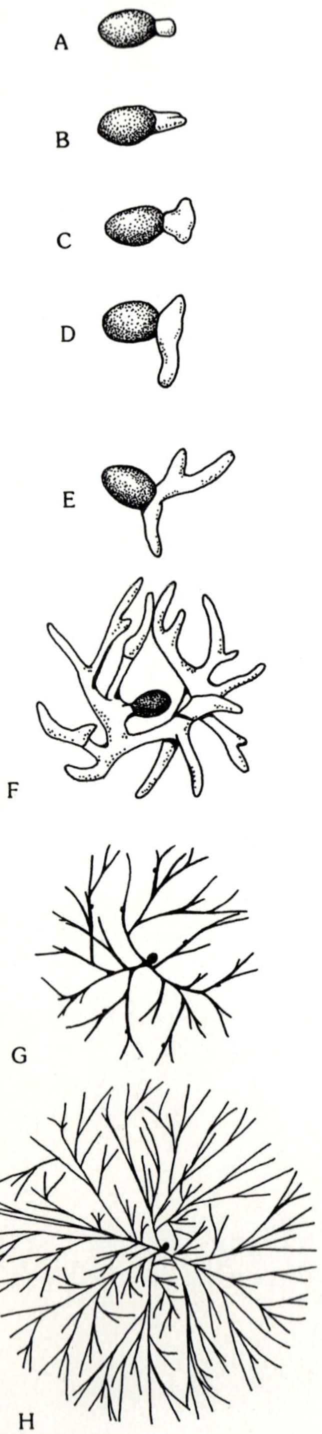

Growth from spore to mycelium

The life of a **fungus** (fung′gəs) can be thought of as beginning when a fungus germinates from a reproductive cell called a **spore** (spôr). Much like a resting cell, a spore is often covered by a thick, resistant outer layer to protect it from unfavorable conditions. A few fungi (fun′jī) continue life in the form of a single cell. Such fungi are called **yeasts.** But most fungi develop larger, filamentous, or threadlike, structures.

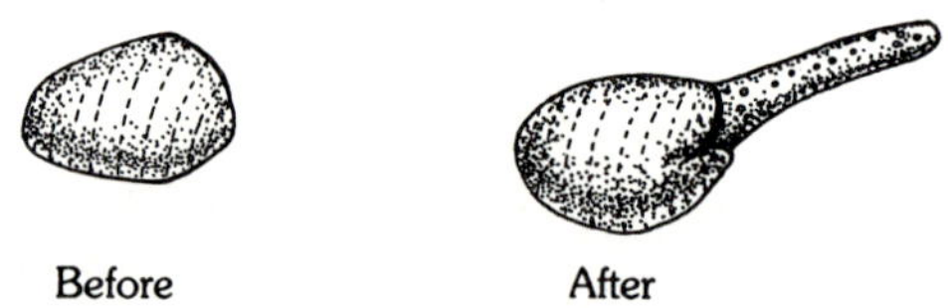

Germinating spore

Filamentous Fungi

Almost everyone is familiar with at least some filamentous fungi. Mushrooms, for example, are commonly seen on lawns and in the forest. Molds are well-known kitchen pests. When a filamentous fungus, such as a mushroom, begins to grow, it sends out a tiny, tubular filament from the spore. This filament is called a **hypha** (hī′fə).

The hyphae of fungi are **eukaryotic** (yü′kar ē ot′ik), containing membrane-bound organelles. The rigid outer covering of a hypha may contain the tough substance **chitin** (kīt′n) or cellulose. Inside a hypha, the cytoplasm is able to move freely. This is possible because either there are no crosswalls to separate the hypha into compartments, or the hypha has only partial crosswalls.

As a hypha grows, it branches out, twisting and turning. Eventually a tangled mat of hyphae is formed. This mat is called a **mycelium** (mī sē′ lē əm).

The mycelium may be green, blue, pink, black, yellow, or orange depending on the pigments present. However, none of these pigments is chlorophyll. So a fungus cannot make its own food. Instead, it must take in food. To do this, the hyphae secrete digestive enzymes into the food source—the soil, the rotting log, the carpet of needles or leaves—where the fungus is growing. The food is digested externally and then absorbed through the walls of the hyphae.

A mycelium of a mushroom may live for a long time. As it uses up its food supply, growth occurs at the tips of the hyphae. The mycelium grows outward in an ever widening circle much like pond ripples.

The Growth of a Mushroom

Although most people are familiar with mushrooms, few people realize that a mushroom is only part of the entire fungus organism. The mycelium that is specialized for feeding and growing remains hidden *below* the surface in soil or in whatever material nourishes the fungus. The only part that can be seen *above* the surface is the mycelium specialized for reproduction. This reproductive mycelium is the mushroom. Mushrooms appear only when conditions are favorable for spore production. And they are generally rather short-lived.

Mushrooms occur above the most recent growth of the hidden mycelium. A large mycelium produces many mushrooms. Such a mycelium may produce a circle of mushrooms called a *fairy ring.*

A mushroom begins to develop as a knot of hyphae covered by a thin membrane. With continued growth, the membrane breaks, revealing a distinct cap and stem. The cap covers a series of paper-thin tissue layers. These layers are called **gills.** The gills radiate like spokes of a wheel around the stem. The surface of the gills are covered by countless spore-producing hyphae. These special, club-shaped hyphae are called **basidia** (bə sid′ē ə). Each basidium produces four spores.

The spore-producing capacity of a single mushroom is amazing. Half a million spores a minute may be produced for several days. A spore is carried by air currents and, if it lands in a suitable place, may germinate and eventually form a new mycelium.

Growth of reproductive mycelium

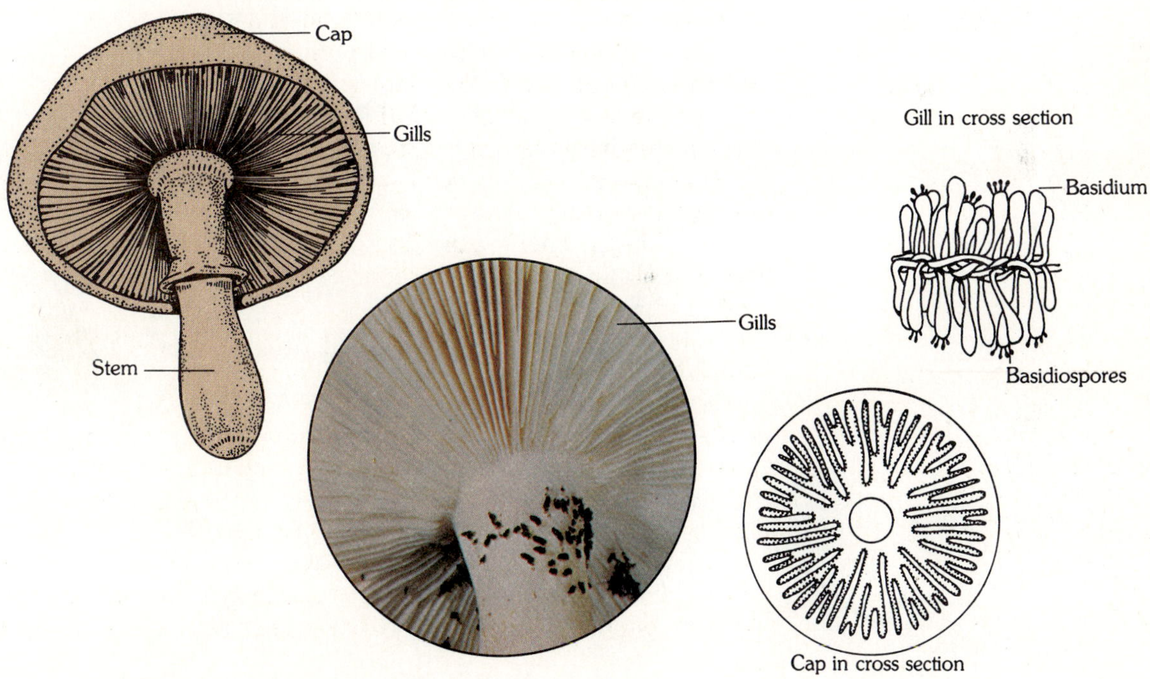

Reproductive structures of mushroom

Body Plan and Classification of Fungi

Both plants and fungi have part of their bodies below and part above the surface. But unlike a plant, a fungus has both the growing and the absorbing part below the surface. The vegetative mycelium, which does the growing and absorbing, lives embedded in the materials it feeds on. As a result, it is not readily seen. For that reason and also because the reproductive mycelia of fungi show a wide diversity of structures, classification of fungi within the fungi kingdom is based largely on reproductive traits. The chart you see as you turn the page shows three main groups of fungi and important traits of each.

Haustoria in host cells

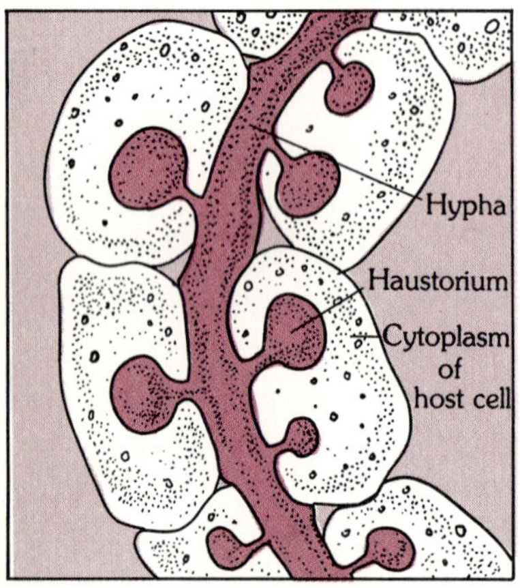

Other Specialized Hyphae

In addition to specialized reproductive structures, fungi may also have specialized vegetative structures. For example, certain parasitic fungi have hyphae designed to penetrate the cells of their hosts. These specialized hyphae are called **haustoria** (hô stôr'ē ə).

A few fungi have hyphae that are specialized to capture living animals. For example, the fungus shown below has hyphae that develop ring-shaped branches. When a roundworm crawls through the ring, the ring swells, trapping the victim. Haustoria from the ring quickly penetrate the living tissues of the victim to secrete digestive enzyrnes and to absorb the digested tissues. Other predatory fungi have sticky knobs that hold the victim.

Fungus capturing roundworm

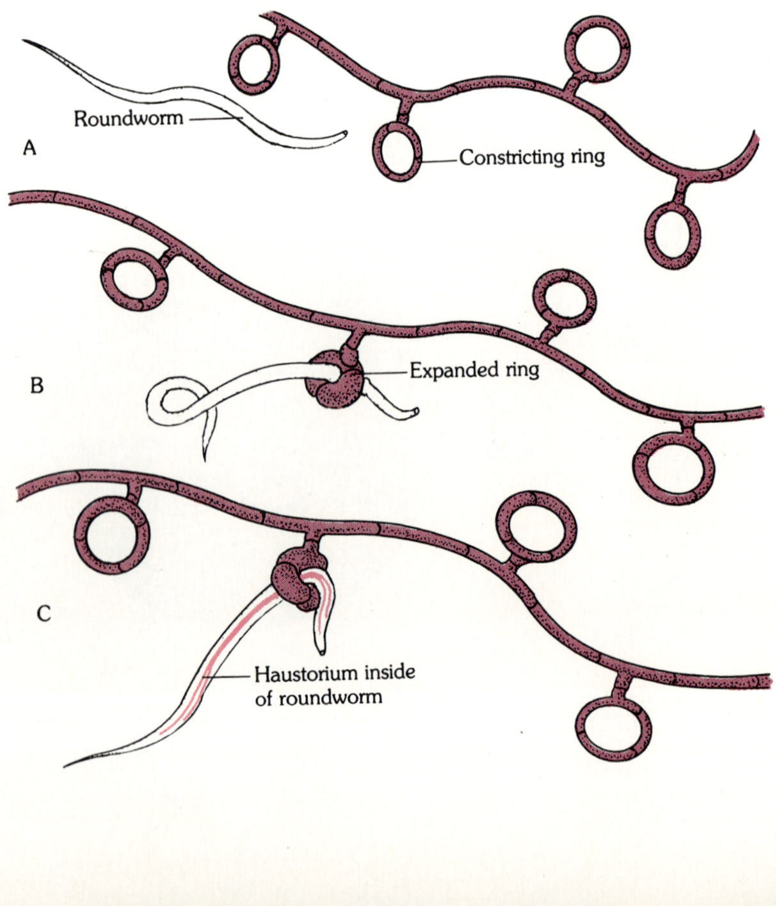

One-celled Fungi: The Yeasts

Certain fungi develop as a filamentous form under some conditions and as a one-celled form, or yeast, under other conditions. For example, a species of *Mucor* develops as a filamentous form in the presence of oxygen and as a yeast in the absence of oxygen. Still other fungi exist only as yeasts.

Yeasts occur abundantly both in soil and in association with living plants and animals. They are particularly abundant where there is a high content of sugar, such as in the nectar of flowers or on the surfaces of ripening fruits. A yeast cell absorbs the sugar and then breaks the sugar down to obtain needed energy and raw materials.

The method used to break the sugar down depends on whether the yeast is operating in the presence or in the absence of oxygen. In the presence of oxygen, sugar is broken down so that the final waste products are carbon dioxide and water. This metabolic process is *respiration.* In the absence of oxygen, sugar is broken down so that the final waste products are carbon dioxide and alcohol. This metabolic process is *fermentation.* In the next lesson you will find out more about yeast fermentation and why yeast fermentation is important to humans.

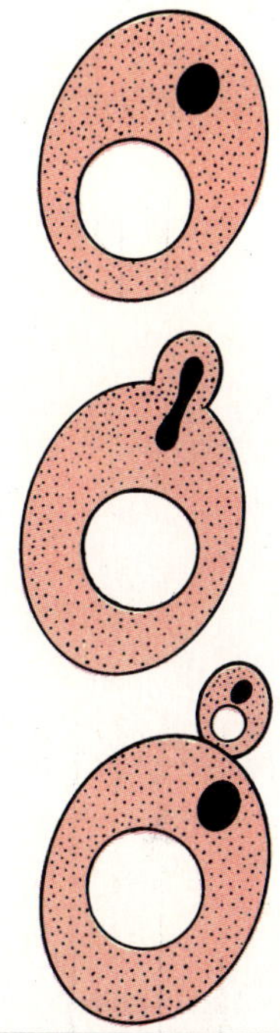

Budding of yeast cell

The Life Cycle of a Yeast

Yeast cells exist both as *haploid* cells with unpaired chromosomes (n) and as *diploid* cells with paired chromosomes (2n). Both the diploid and the haploid cells can reproduce asexually by **budding.** Sexual reproduction involves fusion of two haploid cells to form a zygote. Eventually, a diploid cell undergoes meiosis, and new haploid cells are formed.

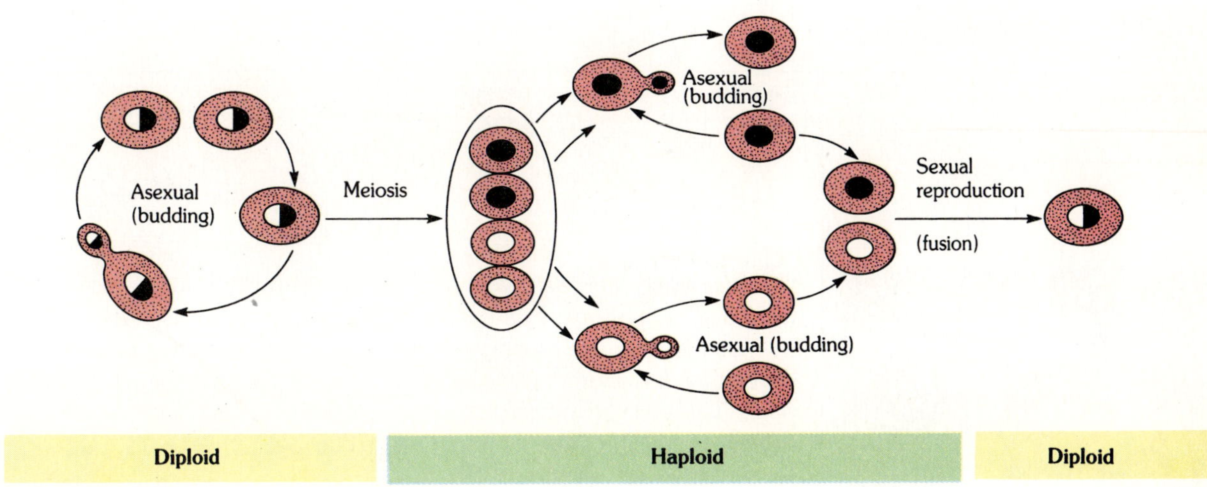

Life cycle of yeast

Name of group	Vegetative Structures	Asexual Reproductive Structures
Phycomycetes algalike fungi example: *Rhizopus,* bread mold a common algalike fungus 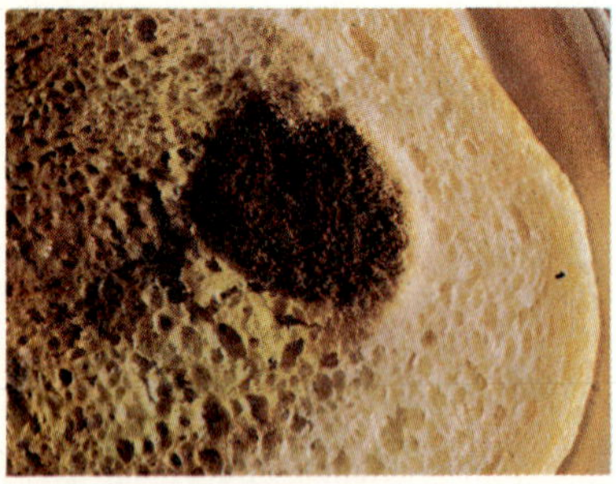	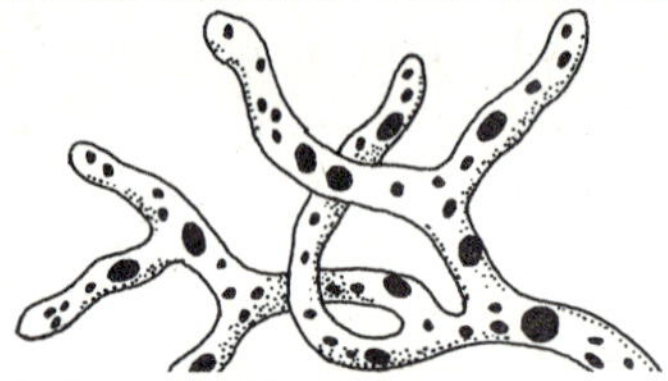 A few algalike fungi exist as yeasts but most are filamentous. In the filamentous forms, the vegetative hyphae are separated from the reproductive hyphae by crosswalls, or **septa** (sep′tə). The vegetative hyphae, lack crosswalls. Their cytoplasm is able to move freely through the filaments.	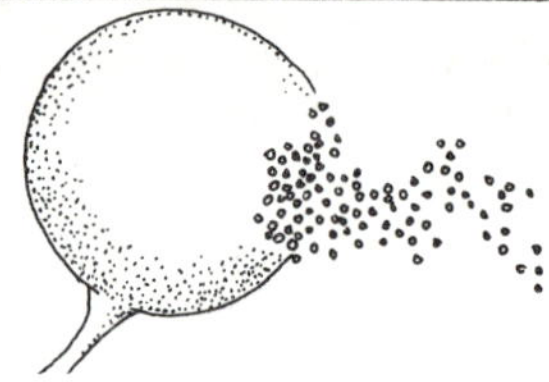 Asexual spores of an algalike fungus, such as a bread mold, develop within a **sporangium** (spə ran′jē əm). A sporangium is a spore case found at the tip of a spore-producing hypha.
Ascomycetes sac fungi example: *Peziza,* a cup fungus a common sac fungus  Courtesy Carolina Biological Supply Company	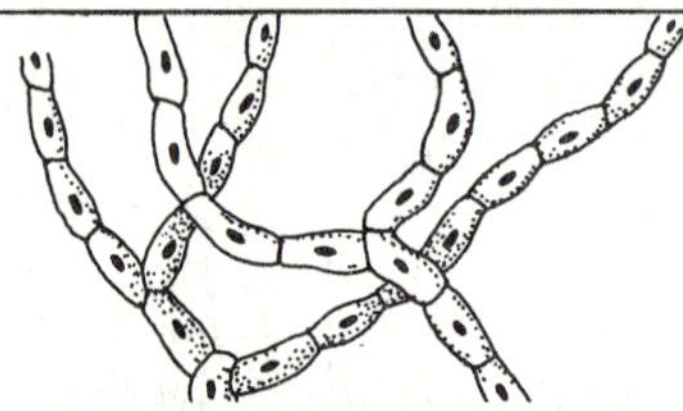 A few sac fungi exist as yeasts, but most are filamentous. In the filamentous forms, the vegetative hyphae are separated from the reproductive hyphae by septa. Also, the vegetative hyphae have crosswalls with openings. As a result, cytoplasm is still able to move freely.	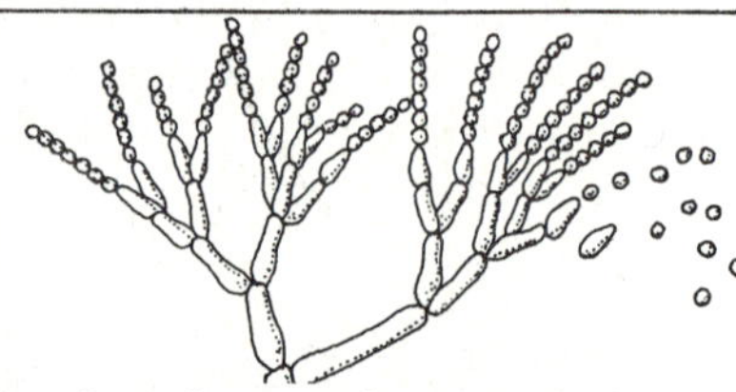 In a few sac fungi, including the morels, asexual spores are unknown. When asexual spores do occur, they generally develop in the open. The specialized hypha that produces those spores is the **conidiophore** (kə nid′ē ə-fôr). The spores are **conidia** (kə-nid′ē ə).
Basidiomycetes club fungi example: *Agaricus campestris* an edible club fungus 	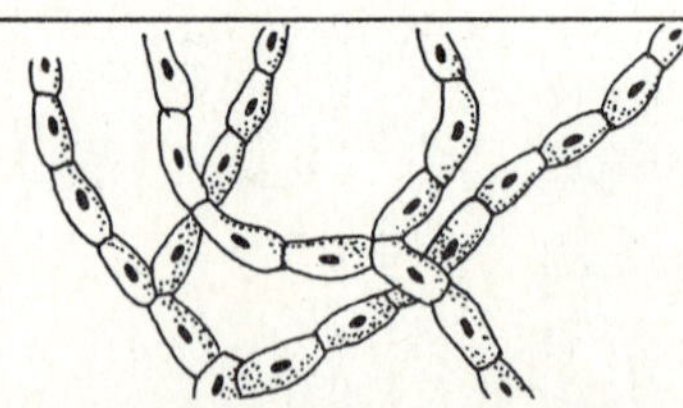 A few club fungi are yeasts but most are filamentous. In the filamentous forms, septa separate vegetative and reproductive hyphae. The vegetative hyphae are septate much like those of the sac fungi. However, the septa are believed to regulate the flow of cytoplasm.	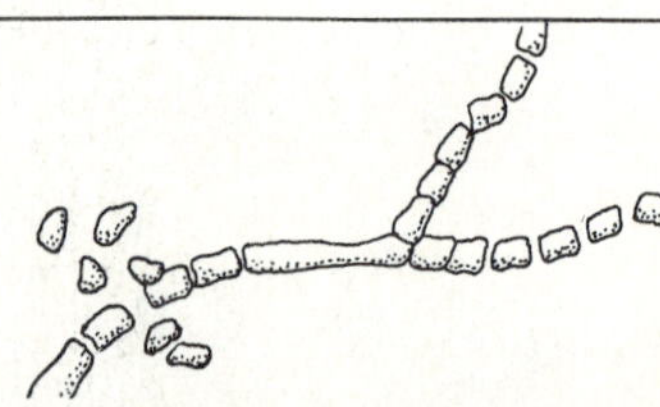 In many club fungi, including the mushrooms, asexual spores are unknown. When asexual spores do occur, they are formed as buds, as fragments, or as conidia.

Sexual Reproductive Structures

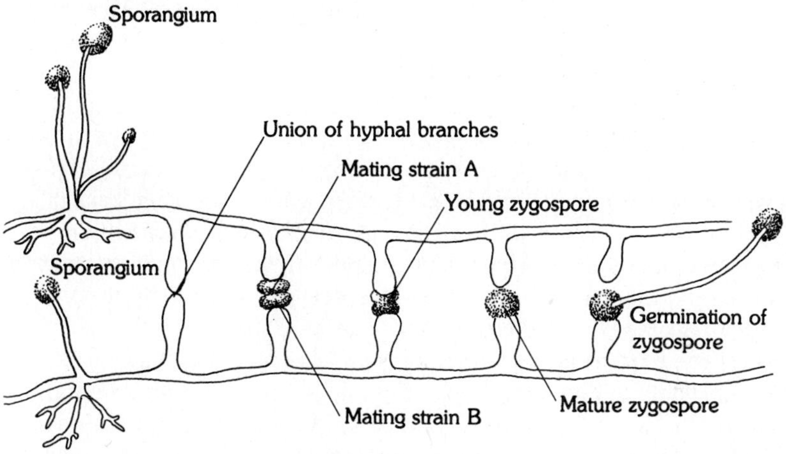

In bread mold, sexual reproduction requires two mating types, or strains. When the two strains come together, a zygote called a **zygospore** (zī′gə spôr) is formed. This results in a diploid nucleus, which contains paired chromosomes. After a period of rest, the diploid nucleus undergoes meiosis, producing haploid spores. These spores may germinate to form a new mycelium.

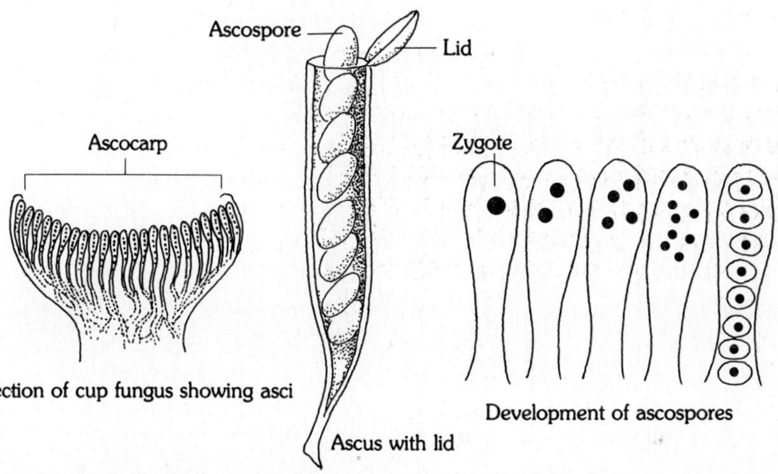

Section of cup fungus showing asci

Ascus with lid

Development of ascospores

In sac fungi, the fruiting mycelium is called an **ascocarp** (as′kə-kärp). Fusion of two nuclei forms a diploid zygote. Then division by meiosis and mitosis results in eight haploid **ascospores** (as′kə spôrz). These spores are contained in the saclike **ascus** (as′kəs). Numerous asci line the ascocarp. A cup fungus, such as *Peziza*, has a lid at the top of each ascus.

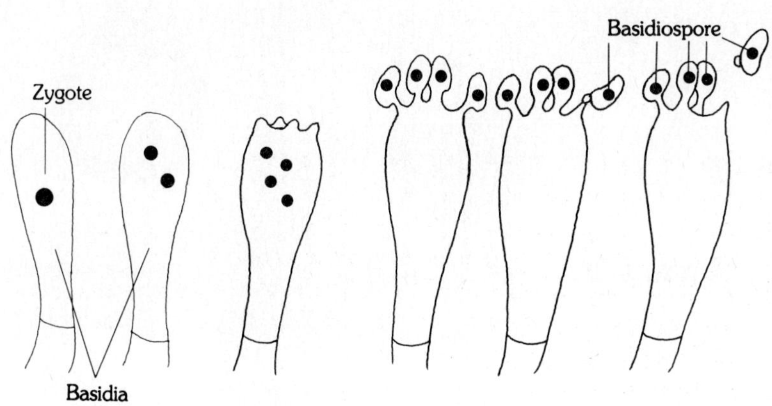

In club fungi, the fruiting mycelium is called a **basidiocarp** (bə sid′ē ō kärp), a structure more commonly known as a mushroom. The basidiocarp protects the club-shaped zygotes, or **basidia.** When the diploid zygote undergoes meiosis, four haploid basidiospores are formed.

Challenge!
Obtain an actively growing plasmodium of a slime mold. Observe its cytoplasmic streaming under a microscope. Dry it to encourage the formation of fruiting bodies. Describe what you see.

The Slime Molds

Slime molds are often studied with the fungi because they resemble the fungi. However, the resemblance does not occur in all four stages of their life cycle. Of the four stages, two are inactive, involving reproduction. And two are active, involving feeding and moving.

The first stage in the life of a slime mold is a spore. The spore is an inactive, reproductive cell. When the spore germinates, it produces the second stage: a single-celled, active form. This form may be flagellated, ameboid, or both. Like all other slime-mold forms, it lacks chlorophyll and, as a result, is heterotrophic and cannot make its own food. Because it lacks cell walls, it is able to feed by engulfing bacteria, yeasts, and other small organisms. However, it may also feed by absorbing nonliving organic materials.

The other active form occurs when the single-celled forms swarm together, becoming macroscopic. This stage is called a **plasmodium** (plaz mō′dē əm). The plasmodium can often be found in cool, moist, shady places and under fallen leaves or rotting logs. The plasmodium may be cellular or filamentous. If cellular, its cytoplasm is restricted by crosswalls to the cells. If filamentous, its cytoplasm is not restricted by crosswalls and may be seen streaming inside the tubular filaments. In either case, the plasmodium acts like a single organism. The fourth and last stage is a fruiting, or reproductive, form. The plasmodium stops moving and feeding to form spore cases. These upright structures produce and release spores, completing the cycle.

Life cycle of slime mold

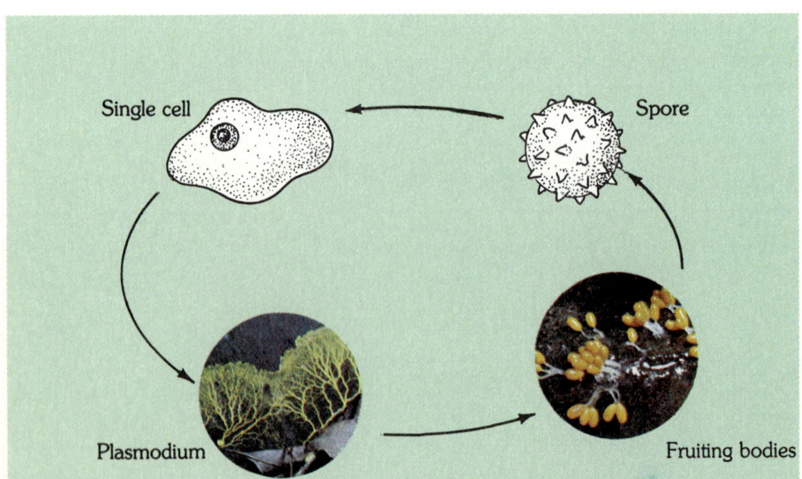

Review It

1. Describe the structure, growth, and feeding of a filamentous fungus.
2. Compare filamentous and one-celled fungi.
3. Compare the reproductive structures of club and sac fungi.
4. Compare filamentous fungi to slime molds.

Activity

Investigating Mushrooms

Materials
Several fresh mushrooms
Razor
Hand lens
Microscope slide
Cover slip
Compound microscope
White paper card
Scissors
Glass or beaker

Procedure A

1. Obtain and examine a single mushroom. Locate the stem, the cap, and the gills. Make a drawing to show your observations. Estimate and record the number of gills under the cap.
2. Using a razorblade, cut a thin slice along the stem of a mushroom. Examine the slice with a hand lens or microscope, estimating the number of hyphae that make up the stem. Make a drawing to illustrate your observations. Record your estimate.
3. Place a single gill on a microscope slide and cover with a cover slip. Examine the edges of the gill under low power. Look for the club-shaped, spore-producing hyphae. Make a drawing to show your observations. Estimate and record the number of spore-producing hyphae on a single gill. Estimate and record the number of spores produced by each.

Procedure B

1. Obtain another mushroom and a paper card. Using a scissors, cut a hole in the card so that the hole is large enough for the stem of the mushroom. Slip the stem through the hole so that the gills rest against the card. Place the mushroom and card on a glass as shown. Be sure the stem is in water. Allow the mushroom to stand undisturbed for at least 24 hours.
2. Carefully remove the mushroom from the card and observe the print made by the falling spores. Describe the print and record your observations.

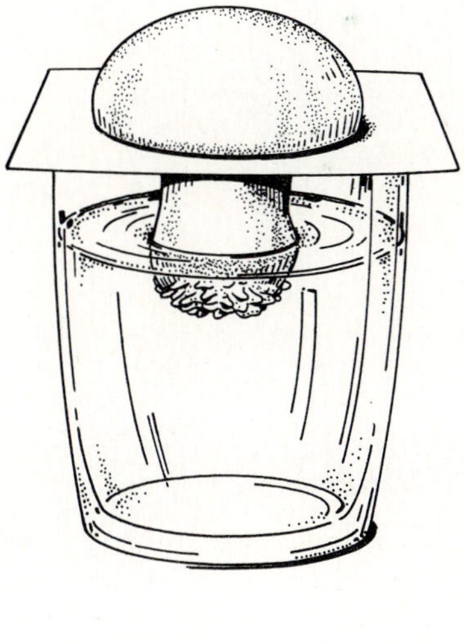

3. Scrape some of the spores from the print. Place them in a drop of water on a microscope slide. Cover the slide with a cover slip. Examine the spores using both the high and the low power. Make a drawing to illustrate your observations. Estimate the number of spores on your slide and on the entire spore print. Record your estimates.

Analysis

1. Use your estimates from Procedure A to estimate the total number of spores produced by a single mushroom.
2. Use your estimates from Procedure B to estimate the total number of spores released by a mushroom over a 24-hour period.
3. Compare the two estimates. If they are not alike, try to explain why that is so.
4. Considering the large number of spores produced and spores released, try to explain why the earth is not covered by mushrooms.

Challenge!
Collect any mushrooms that grow where you live. Make spore prints of those mushrooms and examine their spores under a microscope. How do the prints and spores compare to those you observed in this activity?

12–2
Fungi as Decomposers

Shelf fungi

Jelly fungi

Fungi are adapted to live very close to their food sources. In fact, many fungi are embedded in their food sources. Fungi obtain nutrients to absorb by secreting digestive enzymes into the food source. The enzymes break down, or digest, the food, changing complex organic materials into simpler organic materials. This process, carried on by unseen vegetative mycelia, goes on all the time. The breakdown, or digestion, of organic materials can also be called *decomposition*. And any organism that causes decomposition can be called a *decomposer*.

Fungi are important as decomposers. They are responsible for the decomposition of a wide variety of organic materials. Almost any organic material can provide food for fungi. In soil, fungi work along with the monerans and the protists to decompose the wastes and remains of plants and animals. Fungi can also decompose fabric, leather, fur, wood, food, paint, and even some plastics. While the decomposition of organic materials valued by humans is indeed unfortunate, decomposition of organic materials in the soil is very fortunate. For without this process, life on Earth would eventually become impossible.

Without decomposition, the organic materials in the wastes and remains of plants and animals would eventually be completely tied up and useless to living plants and animals. But when these wastes and remains are decomposed, the organic materials are returned to the soil. In that way, it is possible to use these organic materials again. One scientist estimated that a layer of organic debris about 12 miles thick would now cover the earth if decomposition had never occurred.

***Polyporus*, a sulfur mushroom, grown on oak tree**

Fairy ring

Aspergillus on shoe leather

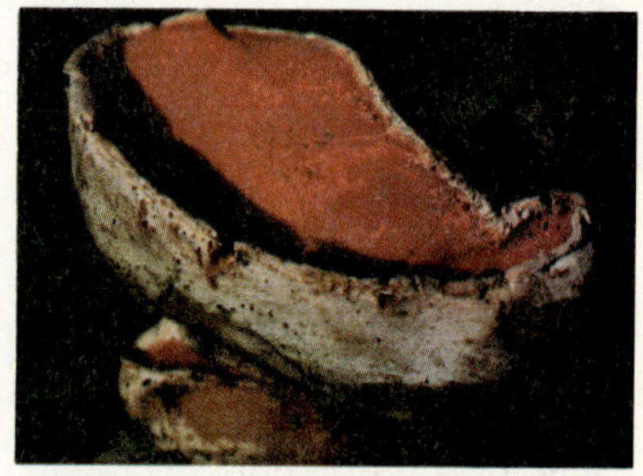

Scarlet cup

Polyporus, varnish-shellac fungus

Bird's nest fungus

Fungi as Symbionts

Fungi often benefit from close associations with other kinds of organisms. These symbiotic relationships may be either *parasitic* or *mutualistic*. Two of the most important kinds of mutualistic relationships of the fungi are those with plant roots and those with algae. Information about parasitic relationships of fungi appears on the following pages.

Mycorrhizae

A **mycorrhiza** (mī′kə rī′zə) is a mutualistic relationship between a plant root and a soil fungus. The soil fungus surrounds the root of a green plant and the hyphae penetrate the root tissues. The green plant, a photosynthetic organism, provides organic nutrients to the soil fungus. In turn, the soil fungus provides the root with a greater surface area for the absorption of water and minerals from the soil.

Fossils over 300 million years old show evidence of mycorrhizae, indicating that the mycorrhiza relationship has existed for a long time. But because mycorrhizae occur in a hidden world beneath the surface of the soil, people were often unaware of their existence. Now, mycorrhizae are believed to be very common, occurring in almost every kind of plant. In some cases, a plant will grow in the absence of soil fungi, but the growth is poor. For example, orchids, cranberries, and various forest trees do not thrive in the absence of soil fungi. Some plants require a particular kind of soil fungus, while others form mycorrhizae with a variety of soil fungi.

Cross section of mycorrhiza

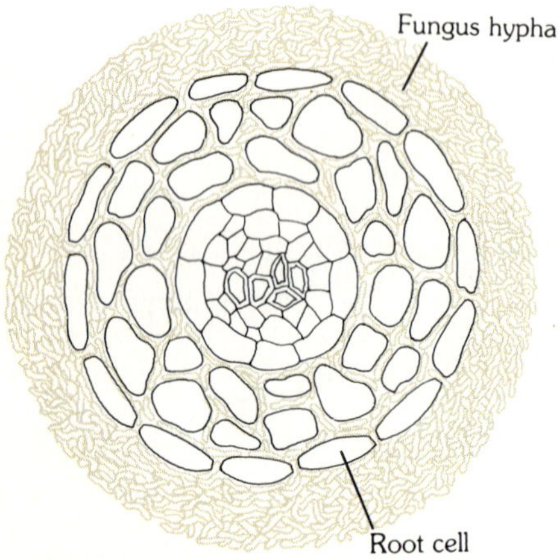

Mycorrhizae: symbiotic associations of fungus and roots

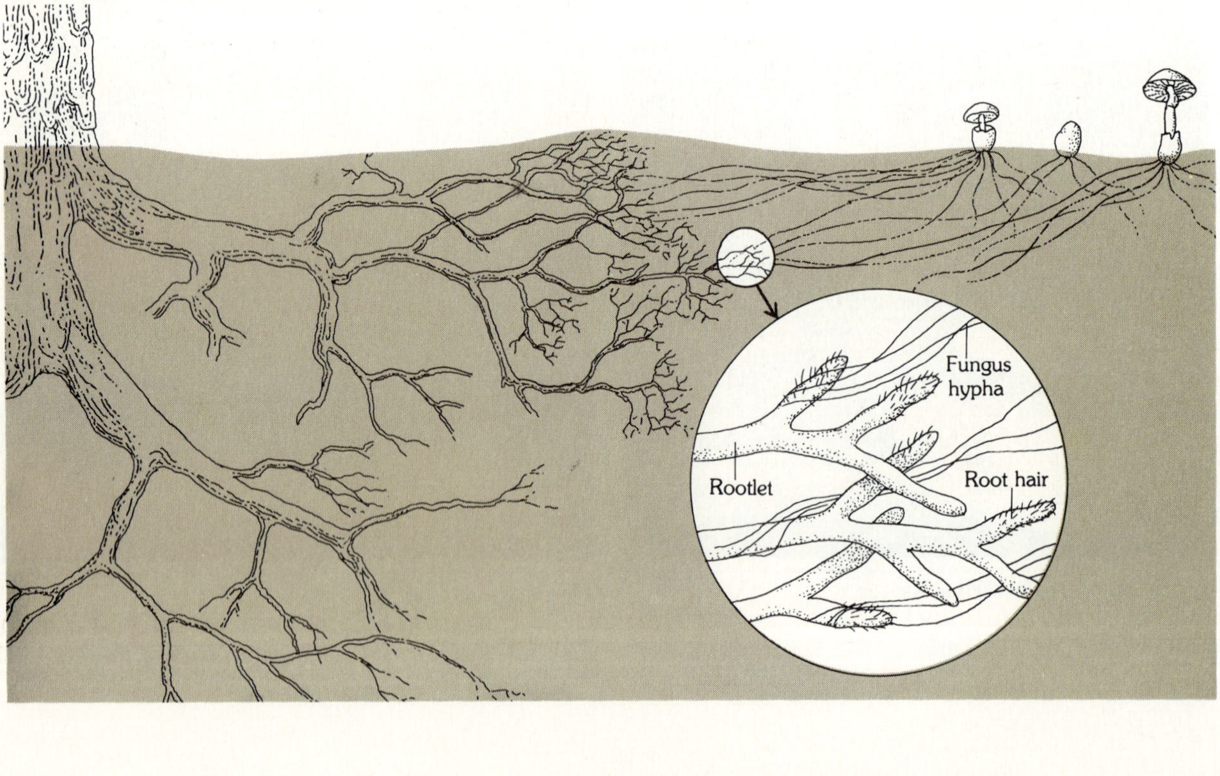

Lichens

A **lichen** (lī'kən) is a mutualistic relationship between a fungus and an alga. The fungus is generally an ascomycete, or sac fungus. And the alga is generally one of the green algae. The algae cells are dispersed among the hyphae of the fungus, which is usually the more bulky and conspicuous part of the lichen. The photosynthetic algae cells provide nutrients for the fungus. The fungus, in turn, protects the algae cells from dryness. It may also provide inorganic nutrients. New lichens are generally formed by fragments from an existing lichen.

Lichens grow in a variety of forms. They may be crustlike, leaflike, or shrublike. The larger lichens, such as reindeer moss, are often prevalent enough in arctic and subarctic regions to provide food for grazing caribou herds.

Lichens can grow on bare surfaces, such as rock, and in other places that do not permit the survival of any other kind of organism. Lichens can also survive harsh conditions: extreme heat, extreme cold, and dryness. A dry lichen can exist for months without water. However, lichens cannot survive where the air is polluted. Because of their sensitivity to air pollutants, they can be used as pollution indicators.

In plant communities, lichens are usually considered pioneers. The chemical action of lichen acids on rocks and the organic materials from generation after generation of lichens contribute to the formation of soil. Eventually larger plants of the community succeed these pioneers.

Cross section of lichen

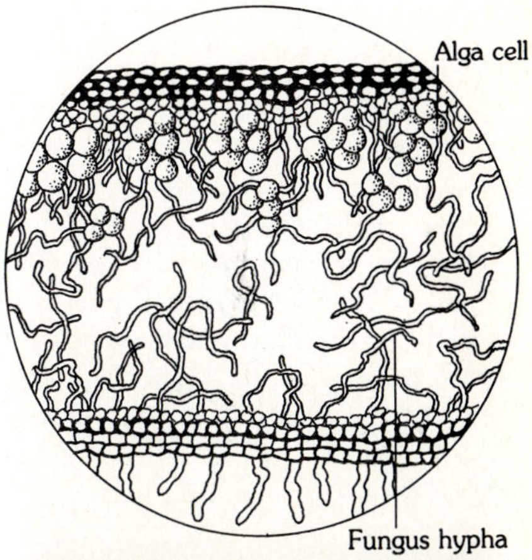

Foliose lichen

Shrublike lichen

Leaflike lichen

Fungi in Industry

The metabolic activities of the fungi play an important role in the making of many foods, beverages, medicines, and chemicals useful to humans. Some of the manufacturing processes employing fungi have been in use for centuries while others are, by comparison, very recent. Of the fungi in industrial use, the most common are the *Saccharomyces* yeasts and the blue-green molds. Both of those are classified as ascomycetes, or sac fungi.

Yeasts in Industry

A poet once described happiness as "a loaf of bread, a jug of wine, and thou." Such happiness would be impossible without the metabolic activities of the *Saccharomyces* yeasts.

Challenge!
Prepare a culture of active yeast in grape juice or in a sugar solution. Observe the living cells and their buds under a microscope. Try to identify the waste products of their metabolism.

Bread gets its light, fluffy texture from the carbon dioxide produced when *Saccharomyces* ferments the carbohydrates in bread dough. The rising of bread occurs as the carbon dioxide is produced. Similarly, wine gets its alcoholic content when the same yeast ferments the carbohydrates in grapes. The same yeasts used in breadmaking and winemaking are also used in brewing beer. Without this yeast fermentation, there would be no bread, no wine, and no beer. Because of its use in these processes, *Saccharomyces* yeasts are often considered among the most important of all domesticated organisms.

Blue-green Molds in Industry

The most important blue-green molds in industry belong to the genus *Aspergillus* and the genus *Penicillium*. Some species of *Penicillium* produce an antibiotic substance called **penicillin** (pen′ə sil′ən). Penicillin acts as a chemical weapon, protecting the fungus from attack by some bacteria. It either stops or inhibits the growth of the bacteria. Penicillin is important as a medicine because it has the same effect in the human body as it has in the fungus that produces it.

Blue cheeses

Other species of *Penicillium* are important in the production of blue cheeses. For example, the metabolic activity of *Penicillium roquefortii* grown on milk in the caves of Roquefort, France, produces Roquefort cheese. Other cheeses—including Camembert, Gorgonzola, and Stilton—also owe their unique colors, flavors, and textures to the blue-green molds.

Aspergillus is important in the making of citric acid. Citric acid is the material that gives lemons their sour taste. Many lemon-flavored soft drinks, candies, and some medicines owe their sour taste not to lemons but to the citric acid produced by *Aspergillus*. *Aspergillus* is also used in making soy sauce from soybean meal.

Fungi as Food

Among the edible fungi are yeasts, truffles, morels, puffballs, and mushrooms. Yeasts are used in dried form as a protein, vitamin, or amino acid supplement. All the other edible fungi are eaten for their unique flavors.

The favorite fungus of gourmets is the truffle, which grows underground near tree roots as a mycorrhiza. Truffles take about six years to mature. They are harvested by trained pigs who sniff them out. As yet, it is not possible to cultivate truffles commercially.

Another gourmet favorite is the morel, especially the species of *Morchella.* Unlike the truffles, morels can be cultivated commercially.

Puffballs are edible when firm and white. Gourmets use giant puffballs for puffball steaks, breaded puffball cutlets, and creamed puffball purée.

Morel

Puffball

Shaggy mane

Poisonous Mushrooms

Some mushrooms are good to eat, but most are not. If a poisonous mushroom is eaten, it can produce severe nausea and even death. Although there are more than fifty species of edible wild mushrooms in the United States, only a few—such as the shaggy mane—are easy to identify. Many different mushrooms look alike. And even experienced mushroom hunters can be fooled. So when it comes to eating wild mushrooms, when in doubt, *don't.*

Amanita mushrooms such as fly amanita and destroying angel are all similar in appearance. And all are deadly poisonous. They are found in open woods and clearings in summer and fall. Most of the deaths from mushroom poisoning are caused by these Amanita mushrooms.

Fly amanita

Destroying angel

Powdery mildew

Corn infected with smut

Leaf infected with rust

Cedar apple rust

Fungi as Plant Parasites

Parasitic fungi cause some of the most destructive plant diseases, and these diseases can have widespread effects on humans. For example, a parasitic fungus wiped out the potato crops of Ireland during the 1840s. At that time, the Irish people depended upon potatoes as their main food item. As a result of the fungus, more than a million Irish people died in one of the most terrible famines in recorded history.

Fungal diseases are generally easy to recognize on infected plants because the spores or the hyphae eventually become apparent. For example, mildews eventually appear as white, fuzzy masses. And rusts form patches of yellow, orange, or black.

Powdery Mildews

Powdery mildews are parasites of many familiar green plants. These include lilacs, grapes, roses, apples, and wheat. The mycelium of powdery mildew forms on the surface of the leaves and then sends down short hyphae into the leaf tissues to absorb nutrients. If you have a garden, it is quite likely you have powdery mildew spreading its hyphae over your phlox, dahlias, and zinnias.

Rusts and Smuts

Both rust and smut are parasites that can reproduce rapidly, affecting large areas in a short time. The name *rust* comes from the color of the rust spores on the host. The name *smut* comes from the dark, dusty, often smelly spores found within the host. Smuts attack cereal grains such as wheat, corn, barley, and rye. The smut enters a seedling. The hyphae grow up the stem as the plant grows. Eventually the kernels of the grain are replaced with the foul-smelling smut balls.

A rust infects more than one host plant. Wheat rust, for example, infects two host plants: wheat and barberry. White pine blister rust infects the white pine and either gooseberry or currant. Wheat rust has probably caused greater losses in the United States than any other fungus. In 1916 it ruined nearly 300 million bushels of wheat. But a much greater disaster occurred during the Depression of the 1930s. Farmers watched green fields turn red-brown as rust appeared. Nearly half of the world's farmland is used to grow wheat. Wheat rust ruins a tenth of that. Losses are in the hundreds of millions of bushels each year.

The Life Cycle of Wheat Rust

In the fall, infected wheat produces thick-walled, black resting spores called **teliospores** (tē'lē ə spôrz). These spores spend the winter in soil or in wheat stubble. In early spring, the teliospores germinate to form basidiospores. The basidiospores, carried by the wind, are able to infect barberry plants. Infected barberry plants develop yellow-red patches on their leaves. A sexual union of hyphae in the barberry plant produces **aeciospores** (ē'sē ə spôrz). Aeciospores, carried by the wind, are able to infect wheat plants. Infected wheat plants develop red-orange patches. **Urediospores** (yü rē'dē ə spôrz) are released from those patches to infect other wheat plants. In fall, the cycle begins again as teliospores are formed.

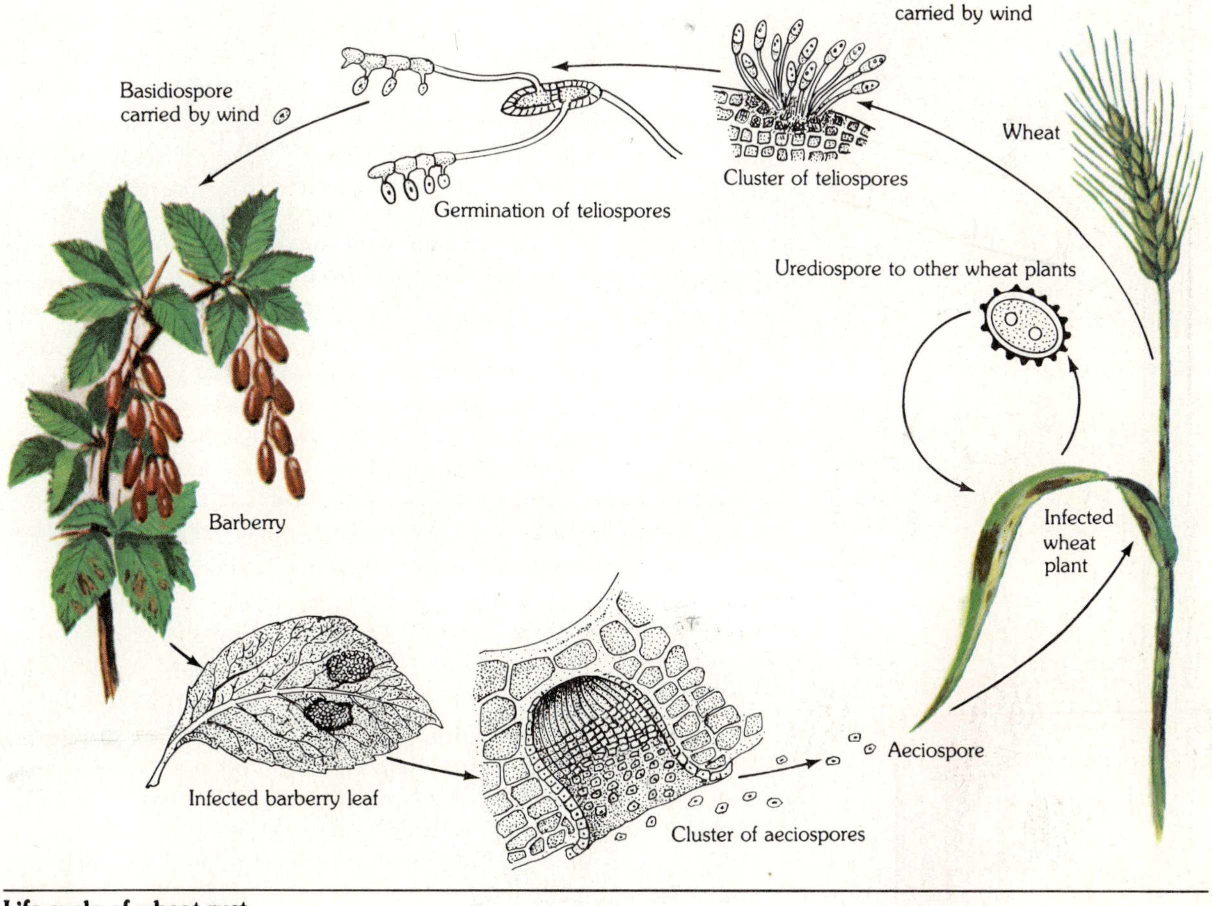

Life cycle of wheat rust

Fungi as Human Parasites

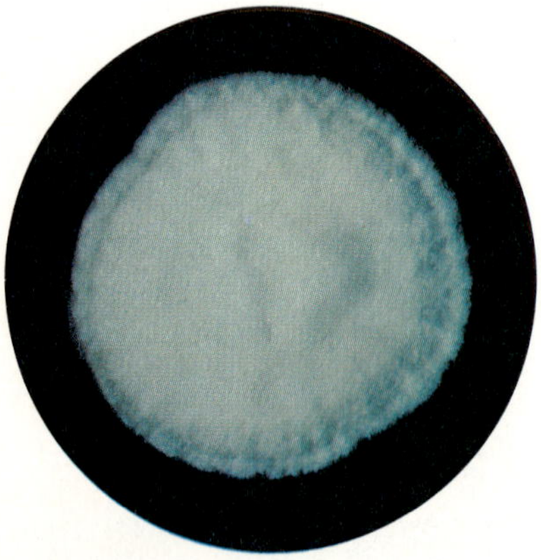

Ringworm

Although plants are far more commonly attacked by fungi, humans are not immune. A number of fungi can live both on and in the human body. Susceptibility to fungal infections may be related to a deficiency of the specific antibodies in blood that prevent the growth of fungi.

Fungus infections are often difficult to treat. For one thing, drugs strong enough to harm the fungus are strong enough to harm the infected person. Also, fungus spores act as resting cells and are able to survive inside of the human body for long periods. So a fungus disease that appears to be cured may often reappear.

Fungi that grow on skin and scalp are called skin plants, or **dermatophytes** (dėr′mə tə fīts). Dermatophytes cause the itching and inflammation of athlete's foot and ringworm. Athlete's foot is very common. It is estimated to infect more than 50 percent of men in the United States at least once in their lives. The same fungus causes ringworm on the scalp. Actually, no worm is involved. Instead, the fungus mycelium grows outward in all directions forming a ring or a circle on the scalp. Ringworm seems more common in children than in adults and more common in boys than in girls. It occasionally appears as an epidemic in schools, where it is spread from one child to another. Fungicides and antibiotics are used to treat ringworm, but symptoms may last for months.

Fungus spores breathed into the lungs can cause *Valley Fever (coccidiomycosis)* and *histoplasmosis*. These lung diseases are often confused with tuberculosis on X-ray pictures. Spores of both of those disease-causing fungi are found in the soil.

Cryptococcosis is a yeast infection that can occur in almost any organ of the body. But it is most frequent in membranes of the spinal cord. The yeasts causing the disease are carried by pigeons, so this disease is more frequent in cities than in rural areas.

Review It

1. Describe the role of fungi as decomposers. In what ways are decomposers helpful? How are they harmful?
2. Describe the role of fungi in mycorrhizae and in lichens.
3. Describe the role of fungi in causing diseases.
4. Make a list of the ways fungi affect other living things. Mention both good and bad effects.

Pioneers in Biology

Alexander Fleming: Pioneer in antibiotic treatment of disease.

In 1928, Alexander Fleming was studying staphylococcus bacteria. He kept several specimens of the germ in culture dishes for microscope work. One dish became contaminated by a mold, as was common under non-sterile laboratory conditions. Usually, contaminated cultures were thrown away. But Fleming noticed that the harmful staphylococcus bacteria were disappearing around this mold. It was a common mold found on old bread and cheese, *Penicillium notatum*. This chance contamination led to the discovery of penicillin.

Alexander Fleming was born August 6, 1881, on a farm in Scotland. Though school often ended by age 12 for farm boys, he continued his education to age 16. After working for four years he enrolled in St. Mary's Medical School and remained as laboratory bacteriologist under Sir Almroth Wright, the leader in immunizing with killed vaccines rather than live ones. In the early 1900s it was known that the chemical antiseptics, the external disinfectants, were too strong for internal use. They damaged the body's defense system. Vaccines were used to fight disease.

With the Royal Army Medical Corps during World War I, Fleming saw the terrible infections that developed from battle wounds. Vaccines could prevent epidemics such as typhus, but

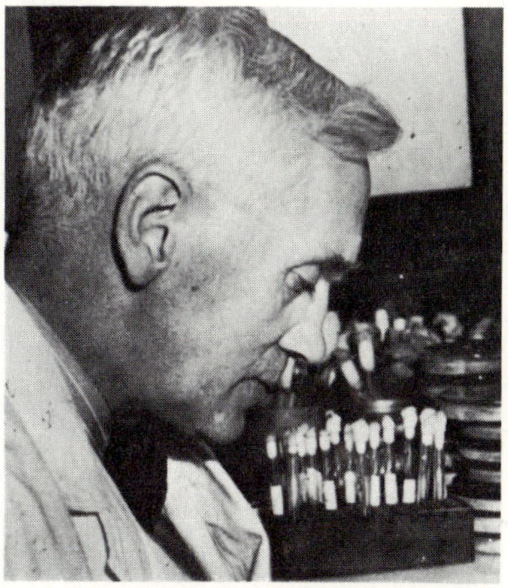

could do nothing against infection. After the war, Fleming dedicated himself to finding a way to help the body's leucocytes do their work.

Experimenting in 1922, Fleming found that tears could destroy powerful bacteria. He called the antibiotic substance, found also in mucus, *lysozyme* (lī'sə zīm). This natural antiseptic was harmless to body cells. Fleming's experience with this first antibiotic prepared him for recognizing the significance of what he would see in the culture of staphylococcus six years later.

After Fleming discovered the antibacterial action of *Penicillium* in 1928, he produced a pure culture of the mold. Then he experimented with it on other bacteria. Some were unaffected. Others, like staphylococcus, were stopped by it. Next, he produced the antibacterial substance free of the mold by planting it on a meat broth and filtering the resultant growth. He also found that penicillin did not harm the body's leucocytes.

But there were problems. In the crude form, penicillin was unstable. Fleming lacked both the chemistry training and the equipment to produce a concentrated form. He published his results in 1928, but the "wonder drug" would not become publicly available until the 1940s.

In 1938, Dr. E. B. Chain read Fleming's original paper on penicillin and recognized its importance. He and Dr. Howard Florey were part of the team of Oxford scientists who began working on *Penicillium* and succeeded in isolating its antibacterial substance. Next, they built up a supply of penicillin. During the year that followed, over 100 patients were successfully treated. Problems of producing enough units of penicillin were immense. But after the proof of successful treatments, a laboratory in the United States took responsibility for mass-producing the substance. By 1943, penicillin was ready for the Allied troops in World War II.

The success of penicillin, the strongest weapon ever used against bacterial disease, exceeded even Fleming's hopes. In 1945 Fleming, Florey, and Chain were awarded the Nobel Prize in Medicine. Fleming was overwhelmed with the acclaim and awards. He remarked, "Nature makes penicillin; I just found it."

Alexander Fleming died in London on March 11, 1955.

Chapter Summary

At one time, fungi were classified as plants. A fungus is rooted in one place, like a plant, and its cytoplasm is enclosed by a rigid wall. But classification of fungi as plants occurred before the five-kingdom classification scheme was devised. Today, the fungi are classified as a separate kingdom, quite distinct from the other four kingdoms.

Unlike a moneran, a fungus is eukaryotic. Its cytoplasm contains distinct membrane-bound organelles including membrane-bound nuclei. Unlike protists, the unit of structure in a fungus is generally not a cell but a tubular filament. And unlike protists, most fungi are best adapted for life on land rather than life in the water. Unlike a plant, a fungus cannot make its own food because it contains no chlorophyll. Instead, a fungus is a chemotrophic heterotroph, like an animal. But unlike an animal, which generally ingests food and digests the food internally, a fungus digests its food externally and absorbs predigested food.

Fungi are of great importance both biologically and economically. On the negative side, parasitic fungi, which feed on living organic materials, cause disease. Also, decomposing fungi destroy food, lumber, fabric, paper, and leather. Yet, on the positive side, without the breakdown of organic materials by fungi, the bodies and wastes of organisms would litter the earth and lock up useful materials. Thus, fungi are important biologically as decomposers. Fungi are also important biologically as symbionts with roots and with algae. And economically, fungi are important in the production of medicines, foods, and industrial chemicals.

Questions to Think About

1. The vegetative body, or mycelium, of a fungus may be present but unnoticed. Why?
2. Why do fungi sometimes grow in circular patterns?
3. Compare slime molds to fungi. How are they similar? How are they different? Consider size, structure, nutrition, feeding, and reproduction. Repeat the comparison for monerans, protists, plants, and animals.
4. How would life on Earth be different without the decomposing fungi? Without the parasitic fungi? Without other symbiotic fungi?
5. Life on Earth would be impossible without the fungi. Explain that statement.
6. Suppose a new fungicide were developed that wiped out the entire kingdom of fungi. How would you be affected?

Extra Research

1. Fuel produced by yeast fermentation may someday replace or supplement petroleum. Find out what materials may be fermented and what machinery is needed.
2. What is zymase? Why is it important?
3. Fungi that grow in the presence of petroleum deposits act as Geiger counters for oil prospectors. Find out more.
4. There are many pros and cons about the use of hallucinogens obtained from fungi. Find out more.
5. Interview a physician about the prevention and treatment of various fungal diseases in humans.
6. What could be done to a material to protect it against attack by fungi?
7. Make a list of various plant diseases caused by fungi. Choose one. Describe its life cycle and what is being done to control this fungus.
8. What might be done to control rusts and smuts?
9. What should one do after accidentally eating a poisonous mushroom?

Chapter Test

Part I

Copy the number of each test item. If the item is *true*, write *T* after the number; if the item is *false*, write *F*.

1. Fungi contain small amounts of chlorophyll. (270)
2. The life cycle of a yeast includes both sexual and asexual stages. (273)
3. An ascospore is formed by mitosis and meiosis. (275)
4. The cytoplasm of fungi contains membrane-bound organelles. (270)
5. Fungi digest food outside of their hyphae. (270)

Part II

Copy the number of each sentence below. After each number, write the word (or words) that completes the sentence correctly.

6. Fungi that cause disease are _____. (284)
7. Fungi that break down organic materials are _____. (278)
8. The process of _____ is used in making beer, wine, and bread. (282)
9. The zygote of black bread mold is a _____. (275)
10. The plasmodium is a macroscopic stage in the life of a _____. (276)
11. The thin layers of tissue found under the cap of a mushroom are _____.(271)
12. Yeasts reproduce asexually by _____.(273)
13. The spore case of a bread mold is a _____.(274)

Part III

Copy each numbered item from List A. After each item, write the letter and words from List B that best describe the item from List A.

List A

14. tubular filament (270)
15. body of a fungus (270)
16. symbiotic alga and fungus (280)
17. symbiotic root and fungus (280)
18. septa (274)
19. *Penicillium* (282)
20. spore (270)
21. one-celled fungus (270)
22. conidia (274)
23. basidiospore (275)
24. fairy ring (271)
25. haustorium (272)

List B

a. spore of a mushroom
b. lichen
c. crosswalls
d. mycorrhiza
e. reproductive cell
f. yeast
g. asexual spores
h. hypha
i. mycelium
j. produces antibiotic
k. feeding hypha of parasitic fungus
l. circle of mushrooms

Focus on Careers

Microbiologist

"I would encourage anyone who is interested in biology, chemistry, and math and likes working in a lab to consider studying microbiology," says Frank, who has been an industrial microbiologist for fifteen years. "Almost all my work is done in a lab equipped with the latest scientific instruments. I work in a manufacturing plant where the primary operation is the conversion of starch—from corn—to sugar. You may know that bacteria are present anytime grain and water come together. It's my job to isolate and identify any harmful bacteria. This is done by studying samples in question under a microscope. Actually, all the organisms I investigate can be seen only with a microscope. That fact frustrates some people, but I enjoy studying the little universe I find under my microscope.

"A microbiologist also has many opportunities for employment. This is partly because there are so many different microorganisms, or microbes—both helpful and harmful. Microbes are used in the production of cheese, yogurt, alcoholic beverages, antibiotics, vitamins, and cosmetics. This means work for industrial microbiologists, who experiment to find the most efficient ways to use these helpful microbes. An industrial microbiologist also investigates ways of inhibiting the growth of harmful microbes in these processes.

"A medical microbiologist studies the relationship between microbes and human diseases. He or she may study only bacteria and be called a bacteriologist or only viruses and be called a virologist. A medical microbiologist makes laboratory cultures of disease-causing microbes and identifies types and characteristics under a microscope. He or she may investigate methods to prevent, treat, and cure diseases by doing tests on lab animals.

"There are many new ideas in the field of microbiology. Microbiologists are studying ways of producing inexpensive foods by fermentation and ways of recycling animal wastes into energy. It's an exciting field—one I'm glad I planned for."

You might plan for a career in microbiology if you have scientific aptitude, adequate vision, and the dexterity needed to work with small instruments. You will need a bachelor's degree in microbiology, biochemistry, or food science. Such a program will include courses in math, chemistry, statistics, genetics, biophysics, and general microbiology. Microbiologists work in hospitals, food and drug industries, manufacturing, and in agricultural agencies.

Mushroom Farm Worker
Duties: Plants, cultivates, and harvests mushrooms in caves, mines, or specially constructed dark buildings. A worker mixes manure to be used as growing medium, plants mushroom spawn, keeps beds properly watered, and checks periodically for damaged or defective mushrooms. He or she harvests mushrooms when they are ready.

Training: Requires on-the-job training.

Opportunities: With experience, the person may eventually own or manage a mushroom farm.

Career Information
Microbiologist: American Society of Microbiology, 1919 I Street NW, Washington, DC 20006.

Mushroom Farm Worker: Mycology Society of America, University of Michigan, Ann Arbor, MI 48104.

Pharmacist: American Pharmaceutical Association, 2215 Constitution Avenue NW, Washington, DC 20037.

Sewage Plant Operator: Water Pollution Control Federation, 2626 Pennsylvania Avenue NW, Washington, DC 20037.

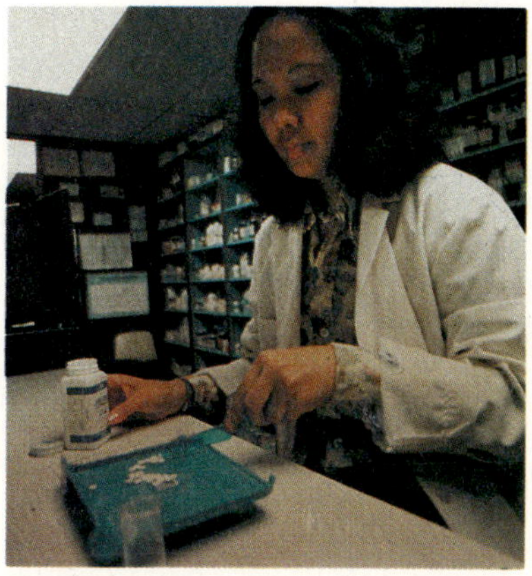

Pharmacist
Duties: Dispenses drugs and medicines prescribed by physicians and dentists. A pharmacist has knowledge of the chemical properties of drugs and the effects they have on the body. He or she is able to advise consumers on the use of hundreds of medicines—both prescription and nonprescription products. A pharmacist may test drugs for purity and strength.

Training: Requires five years of college-level study with an emphasis on science and math courses, including many hours of laboratory work. All states require pharmacists to pass a licensing exam.

Opportunities: Outlook is excellent through the 1980s. Pharmacists may find jobs in drugstores, hospitals, and drug manufacturing laboratories.

Sewage Plant Operator
Duties: Controls the equipment in waste-water treatment plants that removes harmful domestic and industrial wastes from water. The operator maintains the pumps, pipes, and valves that connect the collection system to the site of treatment. He or she adds chemicals to remove pollutants, takes samples of the water for laboratory analysis, and tests and adjusts the level of chlorine in the water. Occasionally, operators work under emergency conditions, as when a heavy rain causes vast amounts of waste water to flow through the sewers possibly exceeding the plant's treatment capacity.

Training: Requires on-the-job training. To get the better paying jobs, two-year programs in waste-water technology are available.

Opportunities: Extremely favorable.

Bibliography: Unit Five

Anderson, Dean A. *Introduction to Microbiology.* C. V. Mosby Co., 1973. Comprehensive textbook covering historical development of microbial knowledge, classification, and identification of microorganisms, practical applications of microbiology, and relationship of microorganisms to disease.

Anderson, M. D. *Through the Microscope.* Natural History Press, 1965. Well-illustrated volume describing importance and historical information about viruses, fungi, and microbes.

Aylesworth, Thomas. *The World of Microbes.* Franklin Watts, Inc., 1975. Beautiful illustrations of protists. Discusses the importance of microbes in disease and in the food industry.

Christensen, Clyde M. *The Molds and Man,* 2nd ed. University of Minnesota Press, 1965. A general account of the fungi and their impact on humans.

Christensen, Clyde M. *Molds, Mushrooms, and Mycotoxins.* University of Minnesota Press, 1975. A reference book that explores the lives of fungi and their relation to humans. Focuses on harmful effects that certain molds, mushrooms, and other fungi have on humans.

Dubos, Rene. *The Unseen World.* Oxford University Press, 1962. Discusses general structure of viruses and bacteria and their practical applications in medicine and the food industry.

Grant, Madeleine. *Wonder World of Microbes,* 2nd ed. McGraw-Hill, 1964. Discusses viruses, bacteria, and their importance in human life.

Kavaler, Lucy. *Mushrooms, Molds, and Miracles.* The John Day Co., 1965. Discusses the many roles of fungi in food production, their effects on human health, their effects on crops, and more.

Kavaler, Lucy. *The Wonders of Fungi.* The John Day Co., 1964. The story of fungi as both destroyer and saver of humans, from prehistoric days to the present-day age of wonder drugs.

Lapedes, Daniel N. *Helpful Microorganisms.* World Publishing Co., 1968. Interesting discussion of the role of microorganisms in our lives.

Lindemann, Edward. *Water Animals for Your Microscope.* Macmillan Co., 1967. Guide to identifying and examining the variety of animals found in ponds, rivers, lakes, and oceans.

Prescott, Gerald W. *The Diatoms.* Coward, McCann, and Geoghegan, Inc., 1977. Discusses structure, uses, and reproduction of diatoms and algae.

Schneider, Leo. *Microbes in Your Life.* Harcourt Brace Jovanovich, Inc., 1966. Background information and techniques used in microbiology.

Stanley, Wendell, and Valens, Evans. *Viruses and the Nature of Life.* E. P. Dutton and Co., 1961. Clear account of the structure and physiology of viruses.

Walter, W. G., and others. *Introduction to Microbiology.* D. Van Nostrand, 1973. Basic information on different categories of microorganisms and techniques used in the study of microbiology.

Wheeler, Margaret, and Volk, Wesley. *Basic Microbiology.* J. B. Lippincott Co., 1964. Describes categories of microorganisms and practical application of knowledge about microorganisms.

The World of Microbes, Doubleday and Co., Inc., 1965. Picture-essay encyclopedia volume that discusses bacteria and their activities and viruses.

Wyss, Orville, and others, *Elementary Microbiology.* John Wiley and Sons, Inc., 1963. Includes chapters on bacteria, molds, protists, and viruses.

Articles

Zahl, Paul A. "Bizarre World of the Fungi." *National Geographic,* October, 1965.

UNIT SIX

THE PLANTS

Chapter 13

The Plant Kingdom

The first forms of life on Earth existed only in water. When life first appeared on Earth, the land was rocky and without soil. Eventually, simple water-dwelling plants developed traits that enabled them to become less and less dependent on water. About 400 million years ago, land plants began to appear in the fossil record. Today, land plants exist in great variety and abundance, covering much of the once-barren land with a thick carpet of green.

In this chapter, you will look at each of the main groups of plants. You will learn about important traits of each group. And you will find out how those traits made plants less dependent on water and more able to exist successfully on land.

Chapter Objectives
1. Name and differentiate the algae.
2. Describe some of the traits developed by simple water-dwelling plants to solve the problems of life on land.
3. Differentiate the two main groups of land plants: the nonvascular and the vascular plants.
4. Discuss the importance of a diploid, sporophyte plant in the life cycle of a vascular plant.
5. Differentiate the two main groups of vascular plants: the seedless plants and the seed-bearing plants.
6. Discuss the importance of vascular tissue.
7. Differentiate the two main groups of seed-bearing plants: the gymnosperms and the angiosperms.
8. Discuss the importance of the seed and of the flower.
9. Differentiate the two main kinds of flowering plants: the dicots and the monocots.

This scene in Cook's Forest in Pennsylvania includes many plants common to the plant kingdom.

13–1
The Plant Kingdom

The plant kingdom includes the flowering plants, the conifers, the ferns, the mosses, and certain kinds of algae. While plants exist in a wide variety, ranging from simple to complex, they do share certain common traits. In general, they are multicellular and photosynthetic. Their cells are walled and eukaryotic, containing one or more of the various forms of chlorophyll and generally storing food as starch. They are generally attached and nonmotile, that is, unable to move. The more complex plants have developed highly specialized vegetative and reproductive structures. Those structures serve for attachment, support, transport of materials, photosynthesis, and protection.

The simplest of the plants live surrounded by water. The surrounding water satisfies many of their needs. It keeps the cells from drying out. It supports the plant and brings nutrients. And it provides for the dispersal of spores and the meeting of gametes.

The move from water to land offered many advantages, including an abundance of nutrients, light, and space. But life on land is life outside of and away from water. Life on land involves various water-related problems, such as the danger of drying out. To live on land, a plant has to find a way to obtain water and to transport that water from where it is found to where it is needed. A land plant must protect itself from drying and find ways to carry on reproduction without water. It has to develop a support system and to find ways to survive extremes of temperature, light, moisture, and air motion that occur outside of water. The development of new traits and specialized structures from the simpler to the more complex plants is best understood as a way to solve the problems of living on land.

Red algae, brown algae, and green algae are simple, water-dwelling plants. They are differentiated mainly by their pigments. A discussion of these plants follows here, and their traits are summarized in the table. The more complex land plants are discussed later.

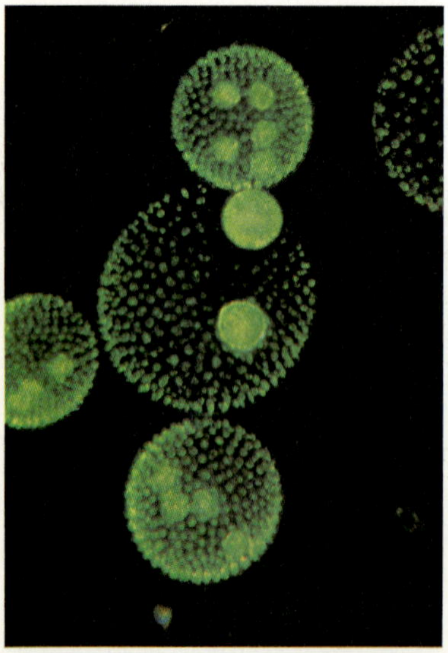

Volvox, a colonial green algae

The Green Algae

Green algae are of particular interest in biology because the plants that first invaded land are thought to be closely related to the green algae. Unlike the red and the brown algae, green algae share certain traits with land plants. Those traits include materials of their pigments, their stored food, and their cell walls.

Green algae display a wide variety of structure and function, ranging from simple to complex. By comparing simple green algae with more complex members, it is possible to see a few of the new traits that developed in plants as they began to invade land. For instance, green algae range from motile and unattached to nonmotile and attached. Also, green algae range from small and one-celled to large and multicellular. In *multicellularity,* cells become interdependent and develop specialized tissues and organs.

Red Algae and Brown Algae

Both red algae and brown algae—commonly called *seaweeds*—are generally large, multicellular plants of the oceans. Both contain green chlorophyll pigments, but, in red algae, the greens are masked by the red *phycoerythrin* (fi′kō ə rith′rin); while in brown algae, the greens are masked by the brown *fucoxanthin* (fyü′kō zan′thin). Both red and brown algae have rootlike, stemlike, and leaflike parts, but brown algae are usually larger and sturdier than red algae. Giant kelps, which are brown algae, may grow to lengths of over 30 meters. These algae can withstand exposure to air and to wave action. Other brown algae are also found in shallow water of the intertidal zones or in floating beds as in the Sargasso Sea. Red algae generally occur in the protection of deep water. Various red and brown algae are eaten as vegetables and are a good source of minerals, particularly iodine. Red algae are important as a source of agar, the jelly used to make a growth medium for bacteria and fungi.

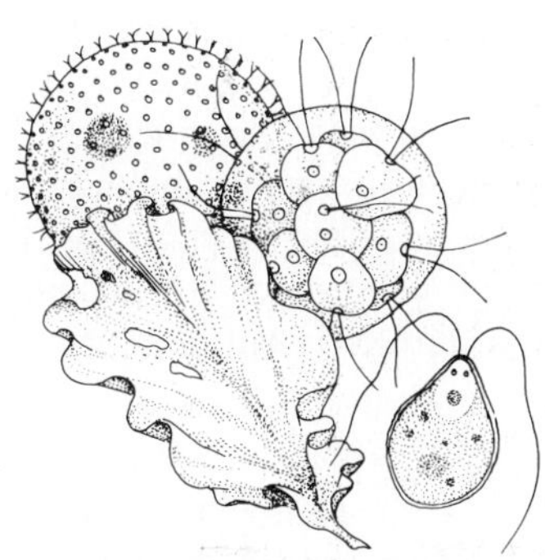

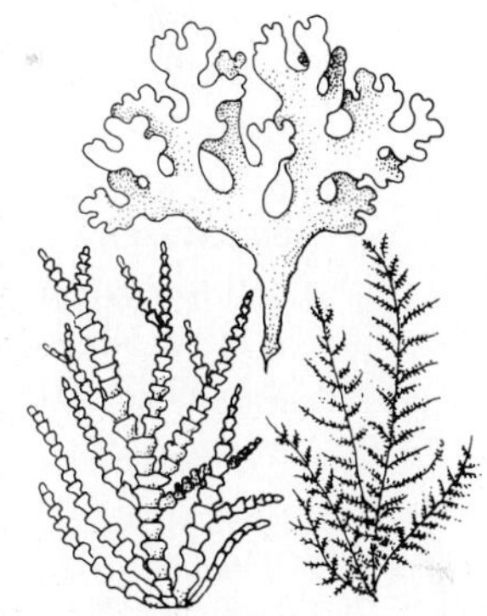

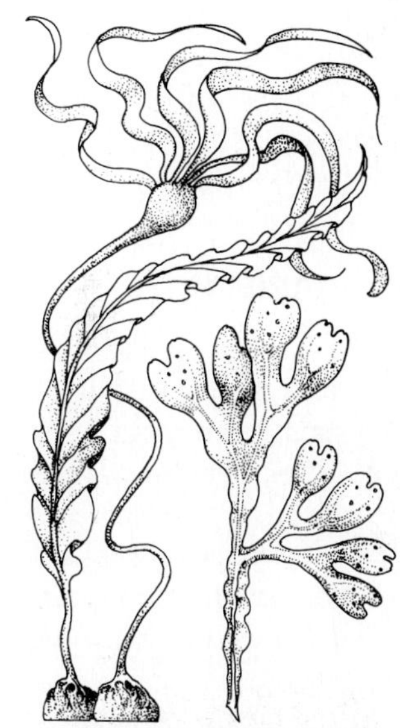

	Green Algae (Chlorophyta)	Red Algae (Rhodophyta)	Brown Algae (Phaeophyta)
Names of Examples	*Chlamydomonas* *Ulva* *Volvox* *Oedogonium* *Spirogyra*	*Chondrus crispus* *Porphyra* *Gelidium* *Plumaria*	*Laminaria*, a kelp *Fucus*, a rockweed *Sargassum*, a rockweed *Nereocystis*, giant kelp
Number of Species	7,000	4,000	1,500
Pigments	chlorophyll a and b	chlorophyll a and d phycoerythrin	chlorophyll a and c fucoxanthin
Food Storage	starch	floridian starch	laminarin
Cell Wall	cellulose	cellulose pectic materials	cellulose alginic acid
Distribution	mostly in fresh water a few in seawater	mostly warm seawater a few in fresh water	mostly cool seawater
Form	one-celled to multicellular	multicellular	multicellular

Reproduction in Green Algae: from Simple to Complex
Comparing the reproduction of the one-celled *Chlamydomonas* to that of the multicellular *Ulva,* suggests some of the new traits that developed as plants began to invade land. Both *Chlamydomonas* and *Ulva* reproduce both sexually and asexually. Asexual reproduction maintains constancy of traits within a population, while sexual reproduction produces variation. All sexual life cycles have the same pattern. All involve a *diploid* stage formed by fusion of *haploid* gametes. And all involve the reduction of the diploid stage by *meiosis.* However, the details of this pattern may vary.

Chlamydomonas is diploid only in the zygote. *Ulva,* on the other hand, has alternating multicellular diploid and haploid stages. The haploid is called the **gametophyte** (gə mē′tə fīt) **generation;** the diploid is called the **sporophyte** (spôr′ə fīt) **generation.** Occurrence of the sporophyte is thought to be the result of postponing meiosis. The diploid stage is important because specialized tissues and organs seem to occur only in diploid organisms.

The gametes of *Chlamydomonas* are generally alike, that is, **isogametes** (ī′sō gə mētz). But the gametes of some species of *Ulva* may be different in size, that is, **anisogametes** (an′ī sə gam′ets). The smaller gamete is the *sperm* and the larger gamete is the *egg.* The egg is important because it can store food for the developing plant.

Another important new trait of green algae, although not seen in these life cycles, is the development of structures that protect the egg and zygote.

Review It

1. How are the red, the brown, and the green algae different from each other?
2. What new traits were developed to help solve the problems of living on land?
3. Compare reproduction in *Ulva* with that in *Chlamydomonas.*

Sexual Life Cycle of *Ulva*

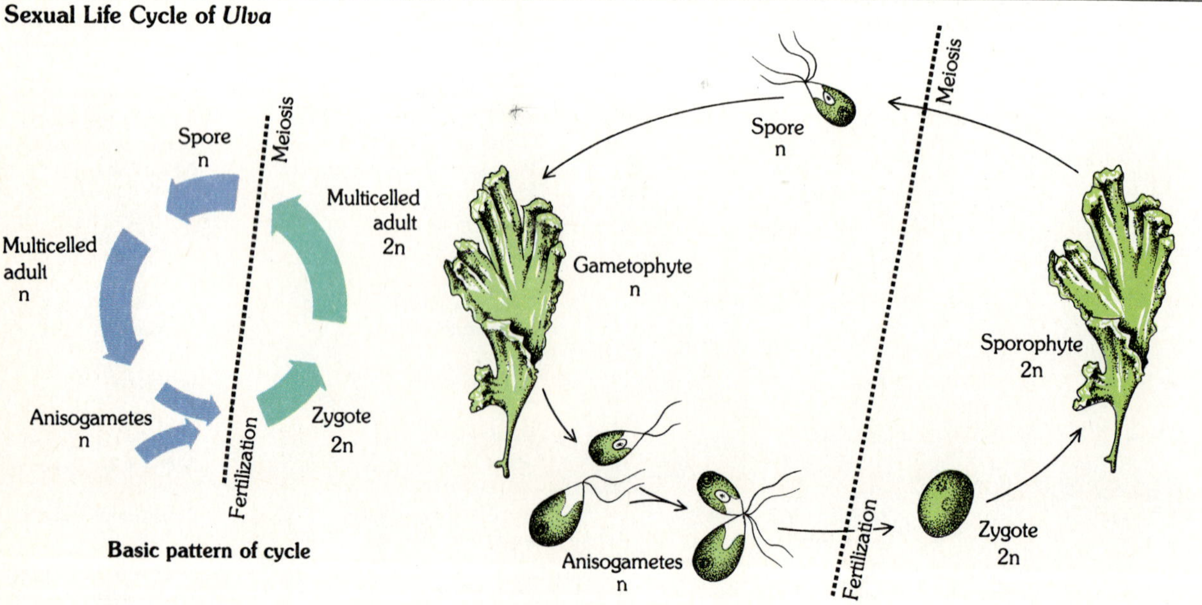

Activity

Observing Reproduction in *Chlamydomonas*

Materials

Two cultures of *Chlamydomonas*, different mating strains
Slides
Medicine droppers, sterile
Dissecting needle
Compound microscope
Lugol's iodine solution

Procedure

1. With a medicine dropper, put a drop of one of the cultures on a slide. Cover the slide with a cover slip. Observe *Chlamydomonas* first under low power and then under high power. Make a drawing, labeling the flagella, cell wall, chloroplasts, nucleus, and any other structures you observe.
2. Using a medicine dropper, put a drop from each culture about an inch apart on another slide. Do not cover the drops. Examine both drops under the low power of the microscope. Look for differences. Record your observations.
3. Using a dissecting needle, mix the two drops. Observe under the microscope. Mating will occur within five minutes. After five minutes, stain the slide with iodine solution. Add a cover slip and examine the zygotes under high power. Make drawings to record your observations.

Analysis

1. Describe the mating process in *Chlamydomonas.*
2. If you were to collect some *Chlamydomonas* in the field, how could you find out if they were different mating strains?

Challenge!

Put both mating strains on an agar plate. Store in the light for one week and in the dark for two weeks. Then transfer some of the culture to a fresh agar plate. Invert the fresh plate over chloroform. Store in the light for two days. Then examine the plate for germinating zygotes. Add water to permit motility.

Sexual Life Cycle of *Chlamydomonas*

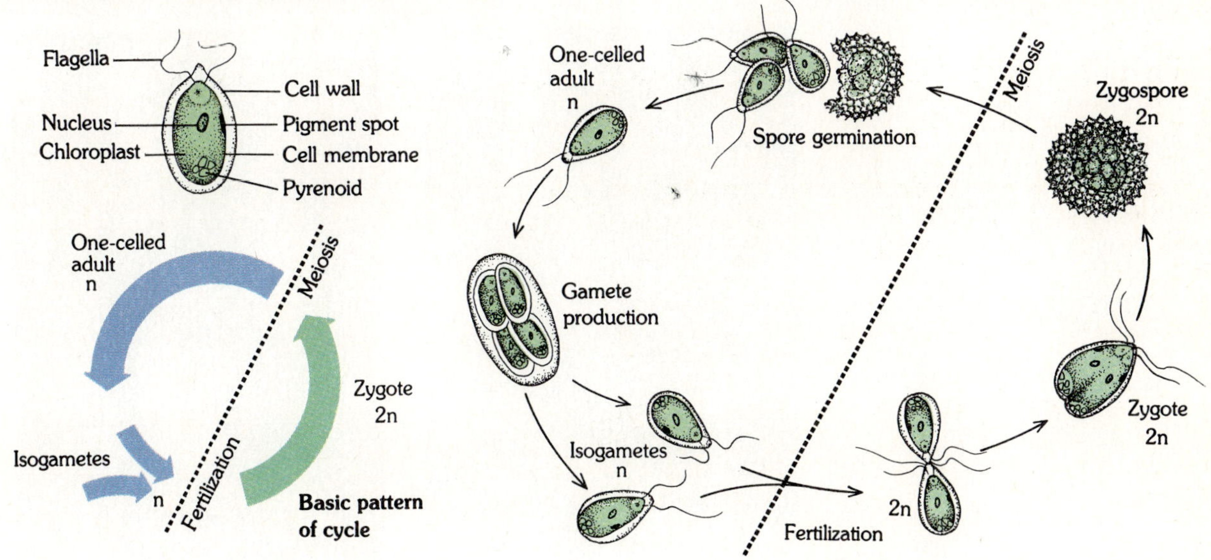

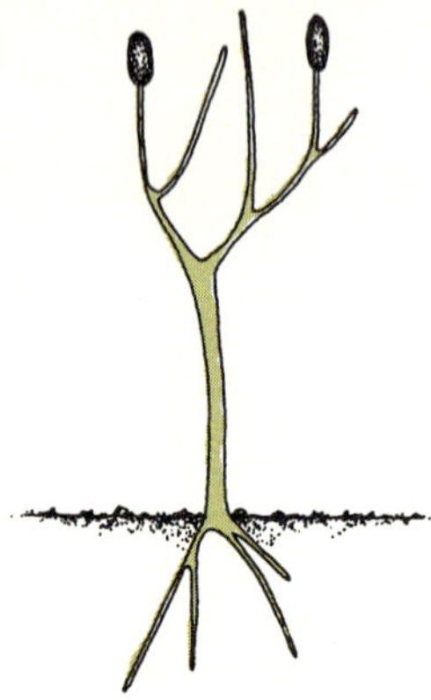

Body plan of land plant

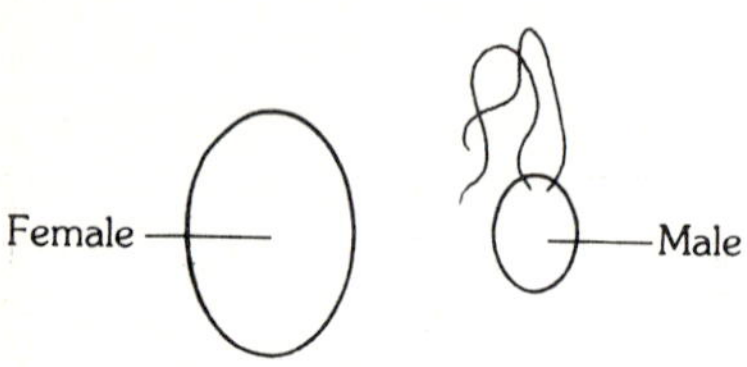

Oogametes

13-2
Land Plants

The land plants include the mosses, liverworts, ferns, conifers, and flowering plants. Although as a group they are quite diverse, they do share a few common traits. For instance, the multicellular, vegetative body of land plants generally has the same basic plan. It is made up of an aboveground, photosynthesizing part and an underground, absorbing part.

Their reproductive structures—many of which are multicellular—include an egg-producing organ that serves to protect not only the egg but also the zygote and the developing young plant.

Another common trait is **oogamy** (ō og′ə mē). In oogamy, the female gamete (egg) is not flagellated, and it is much larger than the male gamete (sperm). The storage of excess food accounts for the large egg size. And because the egg is retained in a protective structure to await the sperm, it is nonmotile. This is another way to avoid drying out on land.

Another common trait can be seen in the sexual life cycle of all land plants. All have alternation of a sporophyte (diploid) generation with a gametophyte (haploid) generation. However, those two generations are rarely alike in either form, size, or length of life.

There are two main groups of land plants: those that have and those that do *not* have an internal "plumbing" system. This system, the **vascular** (vas′kyə lər) **system,** provides for support and for transport of water and nutrients. It is made up of specialized tissues that tend to develop only in diploid plants. Land plants with this system are called the **vascular plants;** those without, the **nonvascular plants.** Information about the simpler, nonvascular plants, such as the mosses and liverworts, follows here. The more complex, vascular plants are discussed later.

***Marchantia*, a common liverwort**

***Polytrichum*, a common moss**

Nonvascular Plants: the Bryophytes

The bryophytes include the mosses and liverworts. They grow in moist, often shady places. Liverworts have flat bodies, while mosses grow upright, gaining support by growing close together.

In the life cycle of a **bryophyte** (brī′ə fīt), the haploid gametophyte is larger and longer-lived than the dependent, diploid sporophyte. The diagram below shows the gametophytes and the attached sporophytes of both mosses and liverworts.

In liverworts, the sporophyte develops from a diploid zygote on the underside of the egg-producing stalk of the female gametophyte. The sporophyte of a liverwort lasts only a few days.

In mosses, the sporophyte is the single stalk that grows from the female gametophyte. The sporophyte wilts after it releases its spores, while the gametophyte continues to make food and grow.

Vascular tissues in bryophytes have little chance to develop because the diploid sporophyte is so short-lived. And without vascular tissue for support, bryophytes must remain small. In the competition for sunlight, they are overshadowed by the larger-bodied plants.

Without vascular tissue to transport water through their bodies, bryophytes must remain close to water. Water is transported through their bodies by diffusion. Also, without water, the sperm is unable to swim to the egg.

According to the fossil record, bryophytes have changed very little over millions of years. Nevertheless, they have been important in making land a suitable environment for other plants. They help to make soil by breaking down rocks and by adding their remains to the soil. Also, they can absorb and hold large amounts of water and, as a result, are important in preventing soil erosion and flooding.

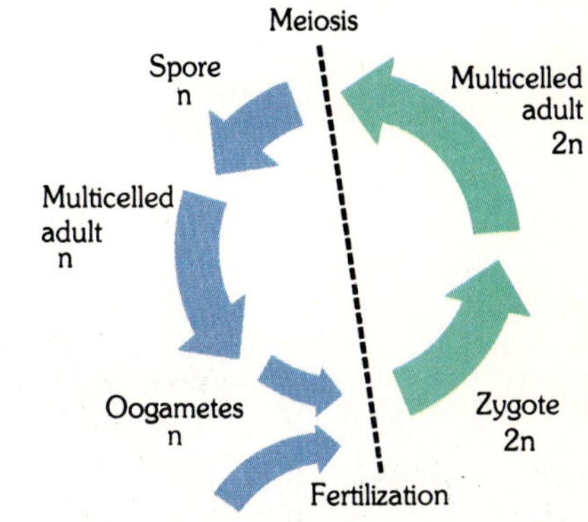

Sexual life cycle of bryophyte, basic pattern

Did You Know?
Thick mats of Sphagnum, *a moss that grows in bogs, are eventually compressed into a coallike material, called peat, that can be burned for fuel.*

Liverwort

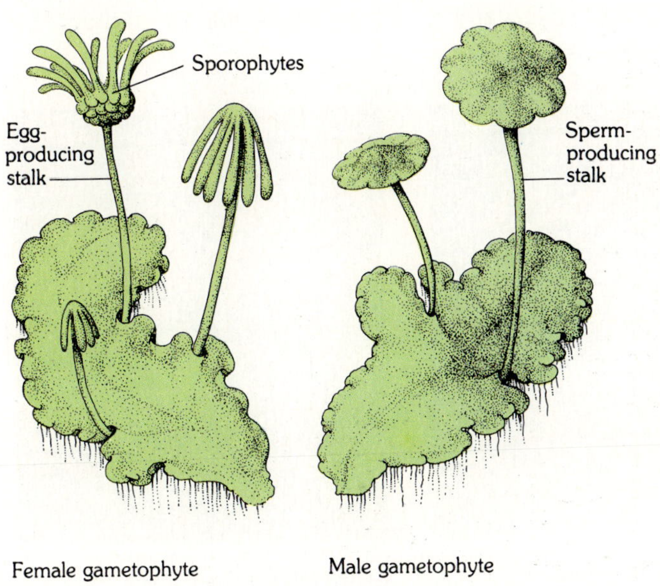

Moss

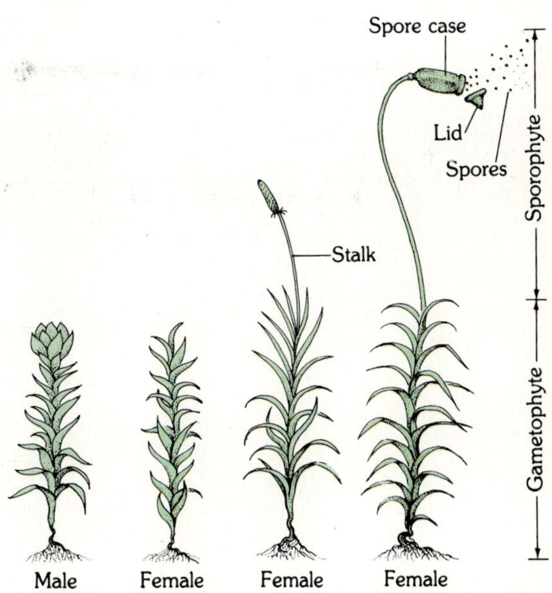

Vascular Plants

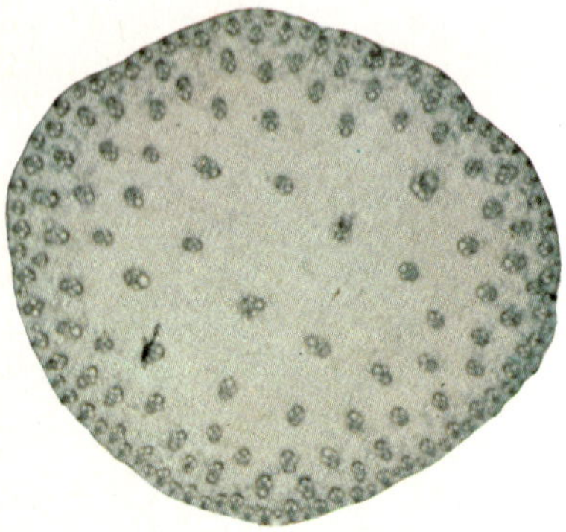

Courtesy Carolina Biological Supply Company

Vascular tissues in stem

Vascular plants include the ferns, the conifers, and the flowering plants. As a land plant, a vascular plant has alternating sporophyte and gametophyte generations in its sexual life cycle. But, as a vascular plant, the sporophyte is conspicuous, free-living and independent. In contrast, the gametophyte is small, dependent, and short-lived.

It is in the free-living, independent sporophyte that the specialized tissues, systems, and organs so characteristic of the vascular plant are found. These specialized traits help the vascular plant survive with greater ease on land. Two of the more important of these traits are the waterproof layer on the aboveground parts and the vascular tissue. Vascular tissue forms a continuous, internal system running through the stem between leaves and roots. Because stems, roots, and leaves of vascular plants contain vascular tissue, they are considered true organs.

Vascular tissue also provides support. Vascular plants—trees, for example—attain their large size and long life because of those supporting tissues.

There are two main kinds of vascular plants: those that do and those that do *not* bear seeds. A discussion of the simpler, seedless plants follows here. The more complex, seed-bearing plants are discussed later.

Seedless Plants

About 300 million years ago, large seedless plants flourished in swamplands and formed large forests. Today, the thick and compressed layers of those plants form coal deposits. Most seedless plants are now extinct. The few that remain include the ferns, the horsetails *(Equisetum)*, the ground pines *(Selaginella)*, and the club mosses *(Lycopodium)*. With the exception of a few tropical tree ferns, all of the living seedless plants grow close to the ground. They are important primarily as ground covers and as pioneer plants, following lichens and mosses into bare areas.

***Equisetum*, horsetails**

***Lycopodium*, club mosses**

***Selaginella*, ground pines**

The Fern: a Common Seedless Plant

Ferns, the most common of all seedless plants, are widely distributed in moist places from the poles to the equator. The vegetative body of the fern sporophyte is made up of an underground stem from which roots grow downward and from which large leaves, or *fronds,* grow upward. Coiled leaf buds, called *fiddleheads,* unroll as the leaf develops.

The fern sporophyte, a diploid organism, produces haploid spores by meiosis in spore cases. Large groups of spore cases form rust-colored spots that can be plainly seen on the underside of a frond. When the spores are mature, they are released into the air. If the spore lands in a suitable place, it germinates and forms a small, heart-shaped, haploid gametophyte. The gametophyte develops both egg-producing and sperm-producing structures on its underside. Sperm are flagellated and require water—from rain or dew—to swim to the nonmotile egg. Fertilization of the egg by the sperm forms a diploid zygote. The zygote divides by mitosis, forming the diploid sporophyte. The young sporophyte remains attached to the parent gametophyte for just a short time.

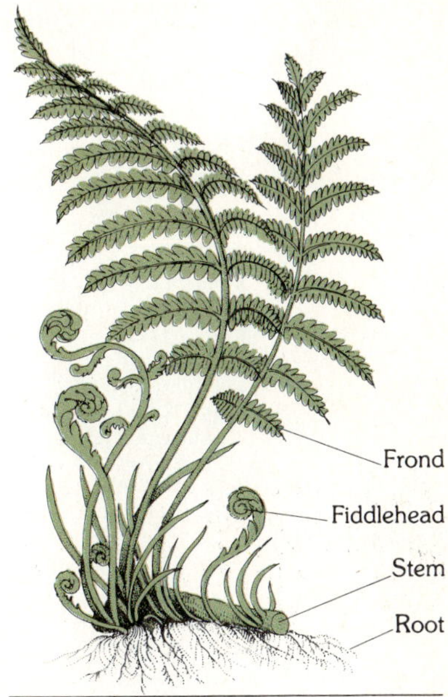

Fern sporophyte

Sexual Life Cycle of a Fern

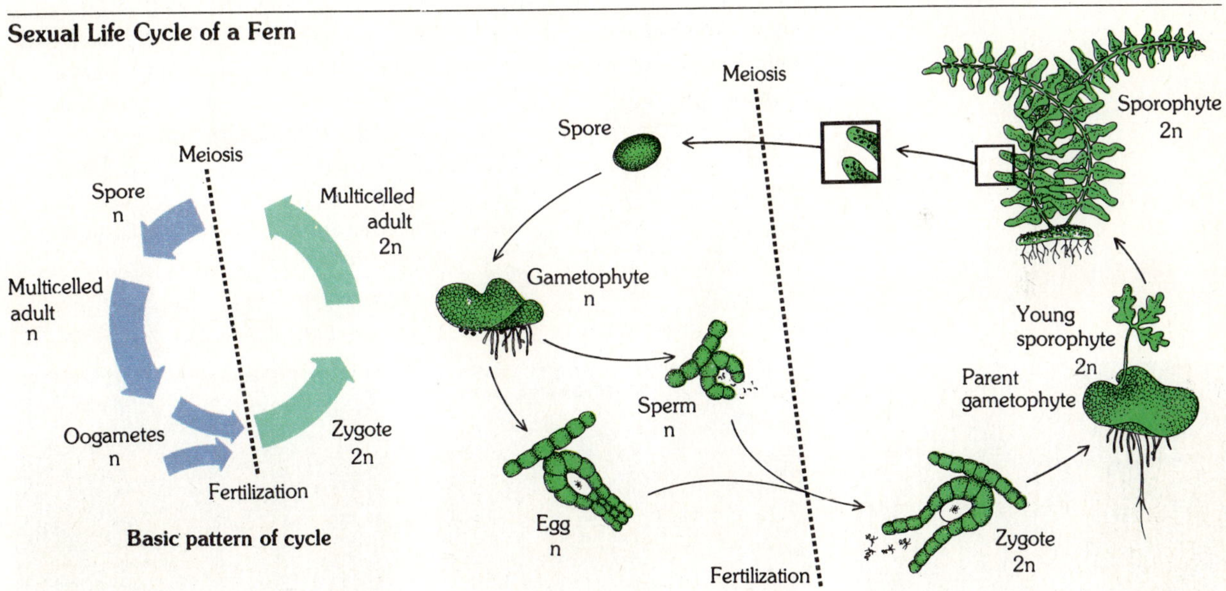

Review It

1. How do vascular and nonvascular plants differ?
2. Compare the diploid sporophyte to the haploid gametophyte of the vascular plant. Why might this difference occur?
3. What are some of the advantages of vascular tissue?

13–3
The Seed-bearing Plants

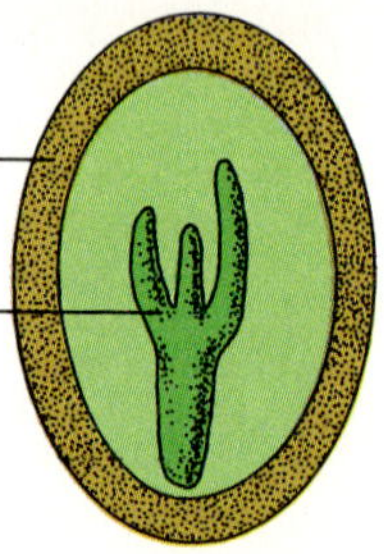

Structure of seed

The seed-bearing plants include the conifers and the flowering plants. Their abundance and widespread occurrence when compared to the seedless plants is largely a result of two reproductive traits involving **pollen** and **seeds.**

Seed-bearing plants have separate male and female gametophytes. The female gametophyte is produced in an ovule while the male gametophyte is produced in a grain of pollen. A pollen grain grows a tube to bring the sperm to the egg. After fertilization, the zygote develops inside a seed.

A seed is a complex structure made of tissues from three different generations. It includes tissues from the parent sporophyte, the entire gametophyte, and the young sporophyte seedling. Tissues from the parent sporophyte form the seed coat while the gametophyte furnishes a food supply. The "carry-along" food supply as well as the protective coating permit the seedling to remain dormant until conditions favor growth.

The two main kinds of seed-bearing plants are those that do and those that do *not* form their seeds in flowers and enclose them in fruits. **Angiosperms** (an′jē ə spėrmz′) are the plants with the enclosed seed. Angiosperms are also called *anthophytes,* or flowering plants. **Gymnosperms** (jim′nə spėrmz′) are the plants with the exposed, or "naked", seed. This does not mean they *lack a seed coat.* It does mean they *lack an enclosing fruit.* The gymnosperms include the conifers, cycads, and ginkgoes. A discussion of the simpler group, the gymnosperms, follows here. The more complex, flowering plants are discussed later.

Sago palm, a cycad

Ginkgo, the maidenhair tree

Leaves of the ginkgo

The Gymnosperms

In the past, both ginkgoes and cycads were widespread. They formed forests at a time when dinosaurs roamed the earth and, perhaps, provided food for the dinosaurs. But today many cycads are extinct and only one species of ginkgo remains. The conifers, or conebearers, are by far the most common of the gymnosperms, although even they are not as widespread as they once were. The conifers of today are important as a source of lumber and paper pulp.

Conifers generally have rounded, spiny leaves called *needles.* Conifers include both woody tree and shrub forms of pine, spruce, hemlock, cedar, juniper, yew, and sequoia. The sequoia are among the oldest and largest of all living things.

Conifers

Reproduction in Gymnosperms

Gymnosperms generally produce both *seed cones* and *pollen cones.* Of the two, the seed cones are larger. In cycads and ginkgoes, seed cones and pollen cones are found on separate plants. But in the conifers, both seed and pollen cones are usually found on the same plant.

In pines, pollination usually occurs in the spring. The pollen cones ripen quickly, releasing clouds of pollen grains to the wind. Seed cones either on the same tree or on a different tree may trap the floating pollen. Only a few of all the pollen grains released actually reach a seed cone. Fertilization may not occur until a year after pollination. After fertilization, the ovules develop into seeds. In the autumn, when the seed cones mature and the cone scales open, winged seeds are released to the wind. Most of the cones you find on the ground are seed cones that have already emptied.

Pollen Cone and Seed Cone

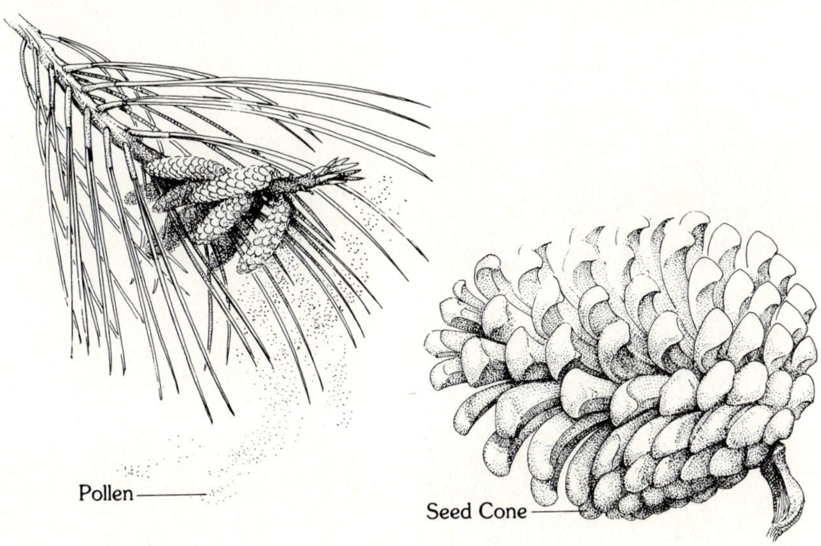

The Flowering Plants

Flowering plants provide humans with all of their cereal grains: wheat, corn, rice, oats, and barley. Without these grains, the rise of human civilization might have been impossible. Flowering plants also provide fruits and vegetables such as potatoes, tomatoes, apples, grapes, spinach, beans, oranges, peaches, onions, celery, and carrots. In addition to food, flowering plants provide various useful materials such as wood and fibers.

The flowering plants are the most common of all land plants. They are the plants you are most likely to see around you—on the lawn, in the field, in the florist shop, among the produce at the supermarket. Flowering plants are the most abundant and widespread. They are found in all climates: hot, moist, cold, and dry. Flowering plants are the most varied. They display great diversity both in vegetative and in reproductive structures. There are more species of flowering plants than all the other kinds of plants put together. In short, flowering plants are the most dominant and successful of all plants. This success is largely a result of a unique organ, the flower.

Flowering plants produce seeds deep within their flowers and protect those seeds by enclosing them in fruits. The diagram below shows the relation of the seed to the seed-bearing structure in both gymnosperms and angiosperms. By looking at the diagram, you can also see some of the important parts of the flower. The flower is a highly specialized reproductive organ. Seeds are produced in the ovary of a flower. A fruit is the ripened ovary of a flower. Fruits protect the seed and aid in seed dispersal. You will find out more about these reproductive structures and processes in Chapter 16.

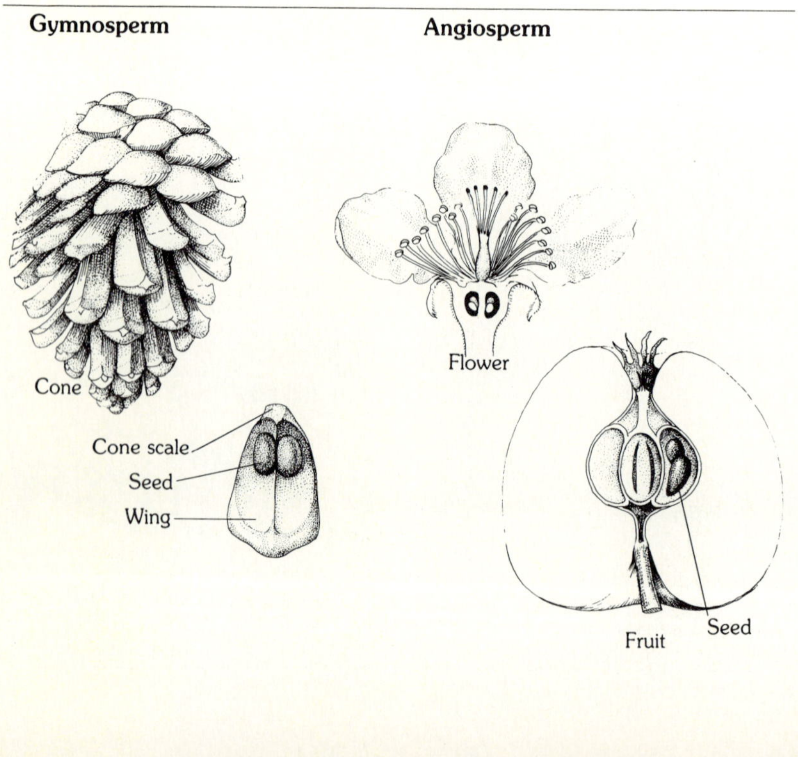

Classification of Flowering Plants

There are two main groups of flowering plants: the **monocotyledons** (mon′ ə kot′l ēd′ nz) and the **dicotyledons** (dī kot′l ēd′nz). A *cotyledon* is a seed leaf. The monocotyledons, or *monocots,* have one seed leaf in their seeds. And the dicotyledons, or *dicots,* have two seed leaves in their seeds. The seed leaves store food for the young sporophyte and become the first leaves when the seed germinates.

There are also other, more obvious differences in these two groups. For example, they vary in the number of their flower parts. Monocots usually have flower parts of three or multiples of three. Dicots usually have flower parts of four and five or multiples of four and five.

If you tried to classify a flowering plant when it is without flowers or seeds, you could look at its vegetative structures. For example, the leaves of monocots have veins that run parallel to each other, while the leaves of dicots have an interconnecting network of veins. If you were to slice across the stem, you would see that the vascular bundles are scattered in monocot stems, while in dicot stems the vascular bundles are arranged in a ring. Also, the roots vary. Monocots have many nonwoody, fibrous roots. Dicots may have a large, woody taproot.

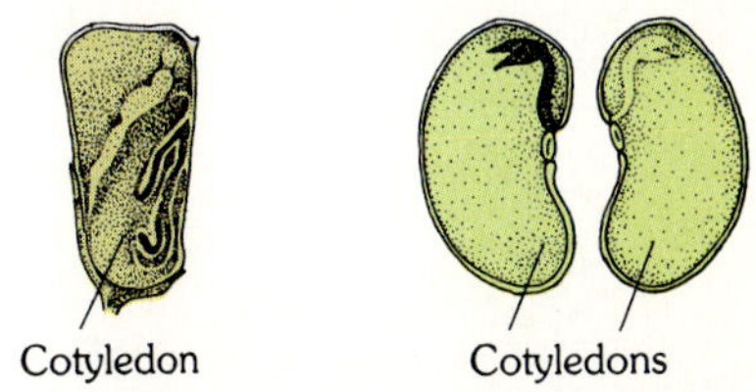

Monocot seed **Dicot seed**

Monocot

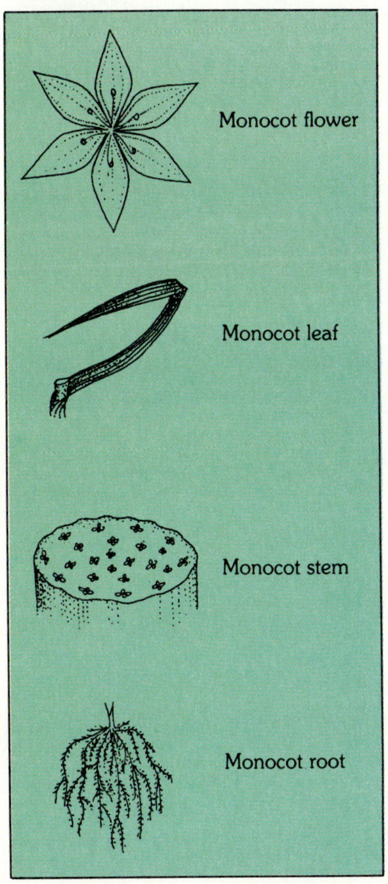

Dicot

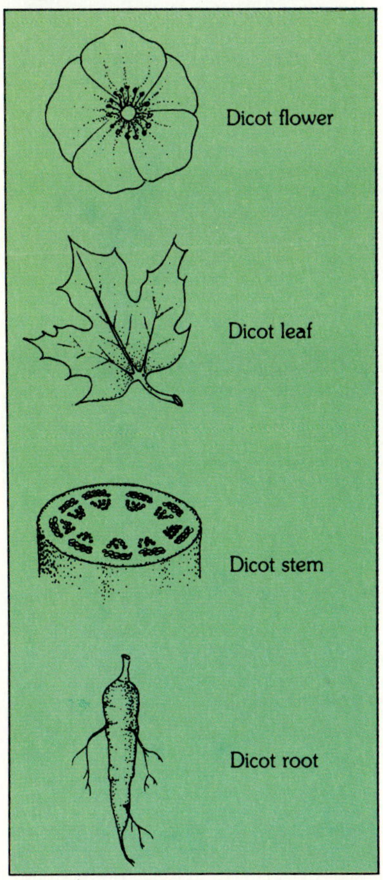

Common Dicots and Monocots

Dicots include both woody and nonwoody plants. Nonwoody plants are **herbaceous** (hėr′bā′ shəs), that is, they have soft, fleshy stems. Woody dicots are useful as a source of lumber and paper pulp, while herbaceous dicots are important as a source of food, fibers, drugs, and various oils.

With the exception of the conifers and a few monocots, most large, woody plants are dicots. For example, oak, elm, beech, maple, willow, and birch trees are all woody dicots. Unlike the conifers, the dicot trees are broad-leaved and most shed their leaves once a year. Many of the fruits and vegetables eaten by humans are from herbaceous dicots. They include beets, lettuce, tomatoes, and beans. Other herbaceous dicots are the rose, buttercup, aster, holly, and cactus.

Monocots are generally herbaceous. The exceptions include the woody palms, bamboos, pineapples, and bananas. Common herbaceous monocots include the cattail, amaryllis, lily, orchid, and grass. The grasses include all of the cultivated cereal grains such as wheat, rice, corn, oats, and barley.

Oak, a woody dicot

Roses, herbaceous dicots

Grass, a monocot

Banana, a monocot

Firewheel, herbaceous dicots

Review It

1. Name some examples of gymnosperms and explain how they are different from angiosperms.
2. What is the advantage of the seed? of the flower?
3. Give five different ways monocots differ from dicots.

Issues in Biology

Genetic variability: can a safe amount of seed resources be maintained while satisfying the world's high production needs?

Workers covering corn ears to prevent cross-pollination

Hybrid seeds developed for maximum yield, ease of harvesting, and disease resistance have contributed to the high crop production in the United States. Large areas in other countries that once supported a variety of plant species are being cleared and the native plants are being replaced with these high-yield hybrid strains. Are there reasons for maintaining the native plants? There is disagreement as to what the world's future needs will be.

In many parts of the world people do not have enough food to eat. Increasing crop yields in these countries is a way to feed the people and to stimulate economic growth for the country. When populations can develop their own food resources, they won't need to rely as much on importing necessary supplies.

Some scientists concerned about genetic variability point out that the diverse native varieties can become extinct through overgrazing and replacement by introduced seeds. Each variety of plant species has unique genetic material (germplasm) to contribute. Overdependence upon one variety has proven to be unwise.

In the 18th century, a South American plant, the potato, was introduced and growing well in Irish soil. The population of the country had tripled by 1830. But in 1845 and again in 1846, the Irish potato crop was destroyed by the fungus *Phytophthora infestans.* The Irish economy was dependent upon that one crop, so its failure meant disaster. During the next few years, almost half of the eight million people in Ireland either died or left the country.

In the United States, in the late 1960s, one strain of hybrid seed corn was so successful that few seed companies maintained other strains. Twenty percent of the U.S. corn crop was wiped out by a disease in 1970 and it took two years before the disease resistant seed was again available in enough quantity to meet the demands.

Plant strains that are resistant to various fungi, viruses, bacteria, and insects can only be developed by careful genetic breeding, requiring a supply of DNA from the different varieties of a given plant species. Collection and storage programs such as plant gene banks freeze thousands of seeds for future use. But some botanists point out that a power crisis would damage the viability of those seeds. Future disease/plant relationships are not known because life forms are constantly changing through mutation and adaptation.

Analysis

1. Can you think of ways to provide for both high production and genetic variability?
2. Contact a seed company and find out how many varieties of corn, wheat, and soybeans are now grown in the United States.

Chapter Summary

As plants invaded land, they developed many new traits to enable them to solve problems of life away from water. Some developed large, multicellular bodies. This was a prerequisite for the development of tissues, organs, and systems. Also, meiosis was postponed, resulting in a macroscopic, multicellular, diploid organism called the sporophyte. In the life cycle of all land plants, a sporophyte alternates with a gametophyte, which is a multicellular, haploid organism.

Nonvascular plants, such as mosses and liverworts, are characterized by reproductive structures that protect the egg, the zygote, and the developing young plant. Seedless vascular plants, such as ferns, are characterized by vascular tissue that provides support and transport of water and nutrients. Vascular tissues, which enable land plants to attain their large size, are found only in sporophytes. This results in the dominance of the sporophyte in the life cycle of all vascular plants.

Seeds, which are formed in both gymnosperms and flowering plants, protect the young plant. The flowering plants further protect the young plant by forming seeds deep inside of flowers and enclosing these seeds in fruits. In flowering plants, the gametophyte generation occurs only in the flower and the seed and is quite reduced in size. The greatest diversity of plants occurs among the flowering plants. Almost all of the more important plants in your life belong to this group.

Questions to Think About

1. How are plants different from organisms of other kingdoms?
2. What evidence exists that ancient green algae were ancestors of the land plants?
3. List the traits that helped plants exist on dry land. Describe the importance of each.
4. Nonvascular plants have showed little change over millions of years. Why?
5. Land plants need to protect their various reproductive cells from drying out. How is this done in mosses, ferns, and seed-bearing plants?
6. Why are ferns considered better suited than liverworts for life on dry land?
7. The ancient forests of ferns, horsetails, and club mosses have all disappeared. What are some possible reasons for their disappearance?
8. Compare sexual reproduction in mosses and in ferns.
9. Discuss alternation of generations in algae, bryophytes, and vascular plants.
10. Ferns do not normally grow in dry places even though they have vascular tissues. Why?
11. Compare ferns and seed plants
12. Why can vascular plants grow so tall?
13. Fossils of oak and cedar trees are not found in beds of coal. Why?

Extra Research

1. Paleobiologists use information from the study of fossils to describe ancient forests. Find out more about how this can be done.
2. The *psilophytes* are a rare group of seedless vascular plants. Why are these plants of interest to biologists?
3. What might be involved in attracting sperm to the eggs of liverworts and mosses?
4. Many plants called mosses are not really mosses at all. Irish moss is an alga, reindeer moss is a lichen, club moss is a vascular plant, and Spanish moss is a flowering plant. Why are these not true mosses? Find out how these plants were named.
5. Make a list of the various flowering plants used for food. Identify the edible structures as a flower, fruit, seed, leaf, root, or stem.
6. Find out some of the ways *Sphagnum*, or peat, moss is used.
7. Can the trees of today form the fossil fuels of tomorrow? Explain your answer.
8. Find out why the leaves of conifers are rounded and spiny rather than flat.
9. Find out more about the cereal grains. Where is each of them grown? When were they first cultivated by humans?
10. Make a collection of either moss spores or fern spores or both. Try to grow plants from the spores you collect.

Chapter Test

Part I

Copy the number of each test item. If the item is true, write *T* after the number; if the item is false, write *F*.

1. Green algae reproduce sexually and asexually. (298)
2. All green algae are one celled. (296)
3. The gametes of green algae may be alike or different. (298)
4. The horsetails *(Equisetum)* have jointed, pipe-like stems. (302)
5. The gametophyte generation of mosses is far smaller than the sporophyte generation. (301)
6. A fern is a nonvascular plant. (302)
7. A club moss is a vascular plant. (302)
8. The main task of the flower and the seed is protection. (306)
9. The seeds of gymnosperms are protected by fruits. (304)

Part II

Copy the number of each sentence below. After each number, write the word (or words) that completes the sentence correctly.

10. Ancestors of the first land plants are thought to be ____. (296)
11. All plants contain the green pigment ____. (297)
12. A fern leaf is called a ____. (303)
13. Mosses and liverworts are ____ plants. (301)
14. Monocots have ____ roots. (307)
15. Another name for flowering plants is ____. (304)
16. A dicot seed has ____ cotyledons. (307)

Part III

Copy each numbered item from List A. After each item, write the letter and words from List B that best describe it.

List A

17. phycoerythrin (297)
18. fucoxanthin (297)
19. starch (296)
20. sporophyte generation (298)
21. gametophyte generation (298)
22. vascular tissue (302)
23. monocot stem (307)
24. dicot flower (307)
25. monocot leaf (307)

List B

a. scattered bundles of vascular tissue
b. pigment of red algae
c. haploid, multicellular plant
d. five petals
e. pigment of brown algae
f. diploid, multicellular plant
g. parallel veins
h. food storage material of plants
i. provides transport and support

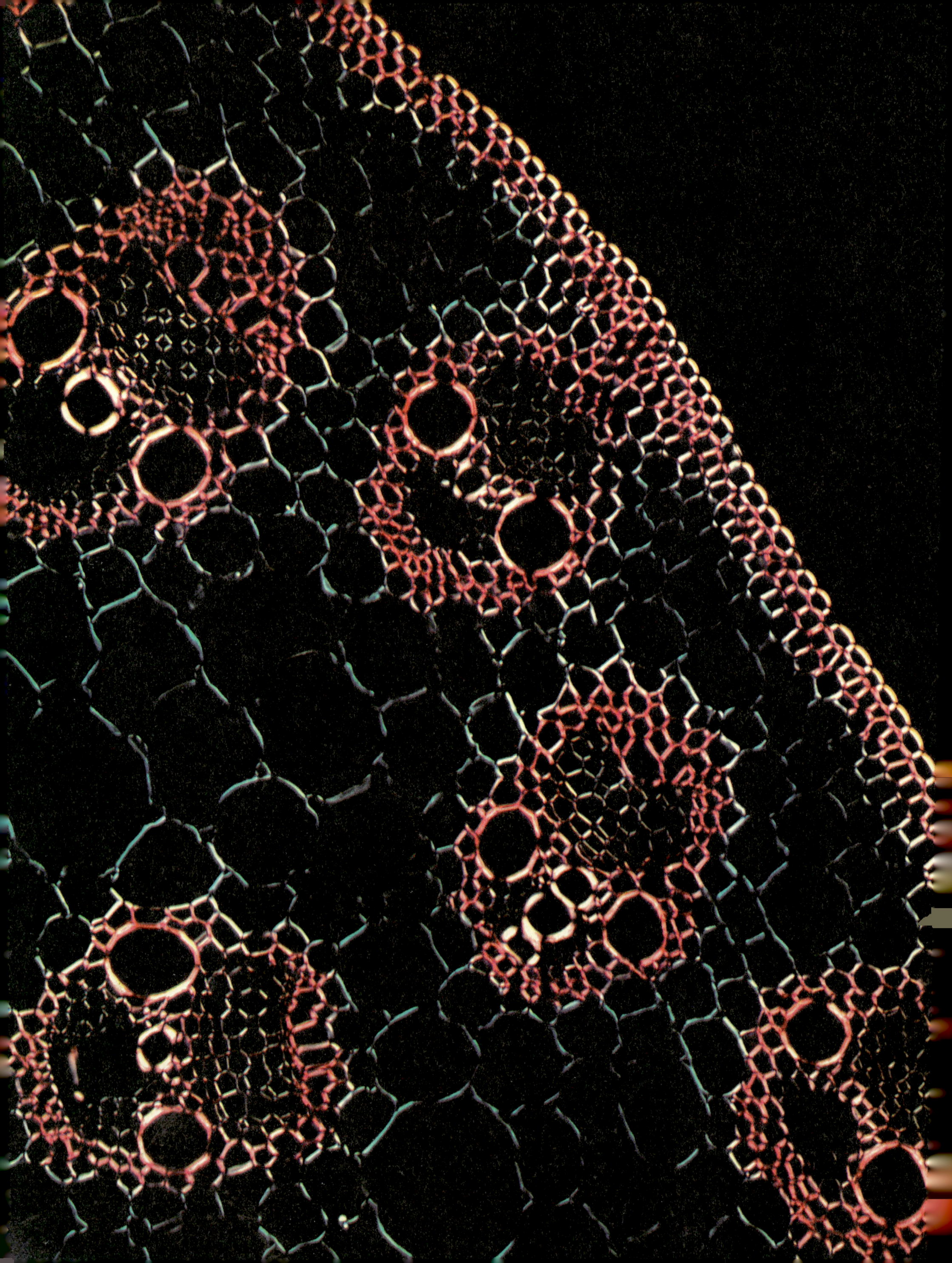

Chapter 14

Roots and Stems

Members of the plant kingdom range from simple aquatic plants, such as algae, to complex land plants, such as flowering plants and conifers. The more complex plants have problems that water plants do not have. Roots and stems help solve some of these problems.

The requirements of a land plant are simple, consisting mainly of getting air, sunshine, water, and minerals. Getting minerals from the soil requires an organ that pushes down into the soil to absorb the water and minerals. This organ, the *root,* also attaches the plant firmly in the ground and stores food. The root is usually the first organ to develop as a seed germinates.

As the root develops, the plant needs another organ, the *stem,* for support and for attachment of leaves. The stem develops above the root. Different tissues develop in the stem to give rigidity to the plant for support and protection, to conduct food and water between the root and the leaves, and to provide storage of food and growth of cells. Besides the root and stem, plants develop leaves and certain reproductive parts, such as flowers and seeds. Leaves, flowers, and seeds are discussed in later chapters.

Chapter Objectives

1. Describe various means of vegetative propagation.
2. Distinguish between monocot and dicot stems.
3. Identify several kinds of roots.
4. Describe the effects of various plant hormones.
5. Define tropism and describe various kinds of tropisms.

A corn stem seen through a microscope shows the vascular bundles clearly.

14–1
Roots

The **root** is the simplest organ of a plant body. It serves as an anchor, as you know if you have ever tried to pull weeds from a lawn or garden. The root also performs the vital function of absorbing water and minerals from soil and transporting them to the rest of the plant.

A third function of the root is the storage of food. Green plants often produce more food than they can use. This excess food is most often stored in roots in the form of sugars or starches. Beets, sweet potatoes, carrots, and radishes are a few examples of roots that store food. Many of these *fleshy root plants* are grown commercially for food for humans.

Dandelion

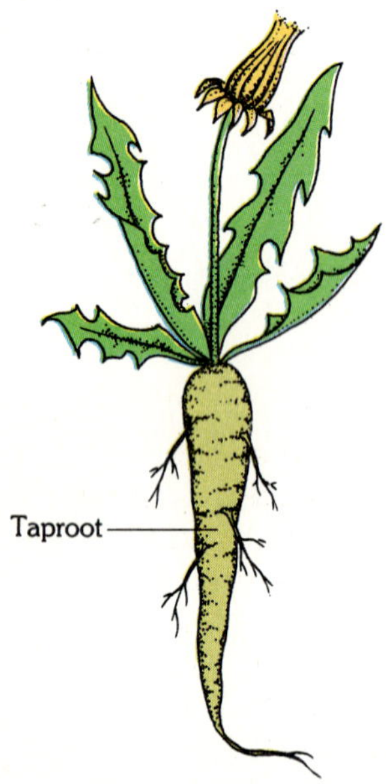

Root Systems
Anchorage in the soil is achieved in two main ways. One is the **taproot system** and the other is the **fibrous root system.**

Taproot System
The first root to appear when a seed germinates is the **primary root.** This root grows down into the soil and may become strong and thick as it grows. This type of root is the **taproot.** Taproots are also able to reach deep sources of water. Many plants in dry areas have very long taproots and are thereby able to survive during the hot, dry weather. Some trees, such as the live oak, can grow in very dry areas.

After a short time, the taproot gives off little side branches. These branches are called **secondary roots.** The secondary roots help hold the plant more firmly in the ground and increase the root surface available for water absorption. The taproot and its smaller secondary roots make up the taproot system.

Grass

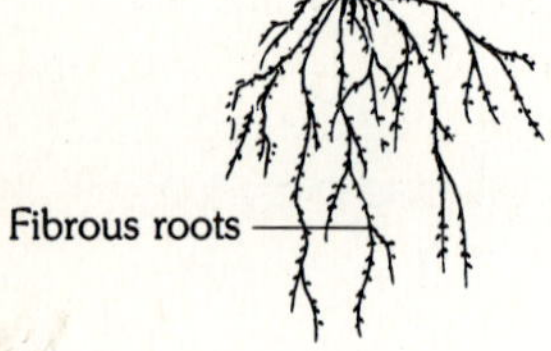

The Fibrous Root System
In plants such as grass and corn, the primary root is soon joined by numerous roots that become long and thin and spread out all through the soil. These roots, called **fibrous roots,** make up the fibrous root system. The branching, spreading network of roots grows rapidly at a shallow depth over a fairly large area. Fibrous roots are efficient absorbers of water and dissolved minerals. They are also excellent soil binders and, as a result, are effective in the prevention of soil erosion. All grasses have fibrous roots.

Fibrous roots may serve to store food just as taproots do. Every sweet potato we eat is one of several roots of a sweet potato plant. They are called **fascicled roots** because they resemble a bundle.

The Root Tip and Its Growth

Roots burrow their way into the ground while growing. If you were to examine the root of a young plant under a microscope, you would see that the root seems to be fuzzy. This fuzziness is caused by fine thread-like growths called **root hairs.** These root hairs are extensions of the outer layer of cells.

If you were to cut a very thin slice lengthwise through a young carrot root and look at it under a microscope, you would see three main regions of root growth. Those are the region of cell division, the region of elongation, and the region of maturation.

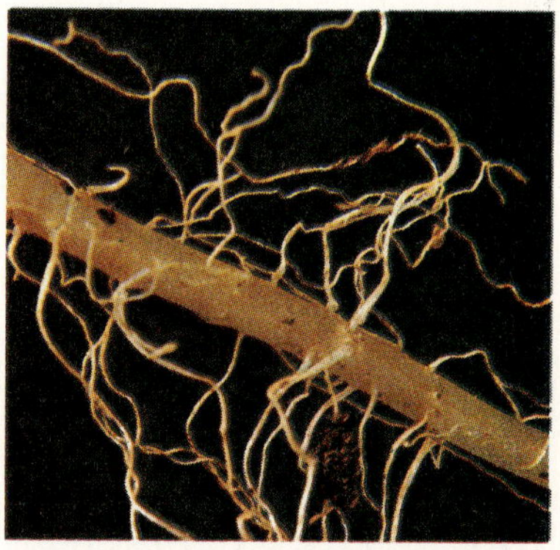

Root hairs

The Region of Cell Division

The **root cap** is the region at the tip of the root that covers the dividing tissue called **apical meristem** (ap′ə kəl mer′ə stem). The root cap protects the apical meristem. All cell division goes on in the apical meristem. New cells are added both downward and upward. This process is called *apical growth.* Cells of the protective root cap are worn away as the root grows and pushes through abrasive soil. However, the cells are continually being replaced by new cells from the apical meristem.

The Region of Elongation

Above the apical meristem, cells stop dividing. Instead the newly developed cells grow longer. As the cells lengthen, they push the root tip forward, driving it farther into the ground. The lengthening of the cells in this region is responsible for all of the lengthening, or **primary growth,** of the root.

Longitudinal section of root

The Region of Maturation

In the region behind the region of the elongation, the fully grown cells differentiate, that is, they change so that they can carry on special tasks. These cells form the various tissues of the root. Some cells become the tissues of the water-conducting **xylem** (zī′lem) or of the food-conducting **phloem** (flō′em). Other cells develop a thin wall that is only one cell-layer deep, called the **epidermis** (ep′ə dėr′mis). The epidermis covers the outside of the root. As epidermal cells develop, they project part of themselves into the soil, becoming the root hairs. Root hairs are densest and appear fuzziest near the root tip. The root hairs are very important. They absorb water from the soil by osmosis. The millions of root hairs enlarge the absorbing surface of the root. The root can thereby absorb more water and minerals and get a better grip in the soil.

Cross section of dicot root

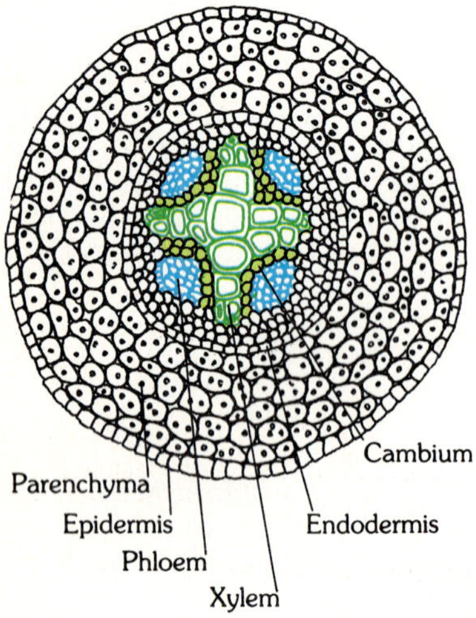

Primary Root Tissues

The primary tissues of the root include the apical meristem and the three tissues produced by the apical meristem. Those three are the protective epidermal tissues; the conductive vascular tissues; and **parenchyma** (pə reng'kə mə), sometimes called ground tissue. By looking at the cross section of the root shown here, you can see how those tissues are arranged.

The entire outer surface of the root is protected by a single layer of protective, or epidermal cells. Similar cells protect the entire surface of the plant. The job of the epidermis of the stems and leaves is to prevent water loss. But the epidermis of the roots must also provide a surface for absorbing water and its dissolved minerals. Root hairs provide that surface.

The vascular tissues are grouped together at the center of the root. These tissues extend the entire length of the plant, from the roots to the leaves by way of the stem. Recall that *xylem* and *phloem* are the two kinds of vascular tissue. The xylem cells are thick-walled and are fused end-to-end. These cells do not live very long and the ends dissolve, leaving a set of continuous hollow tubes through which water and dissolved minerals move upward from the root.

Food manufactured in the leaves moves down to the roots through the phloem. Phloem cells are thin-walled and retain their protoplasm. Only their nuclei die. Unlike the xylem cells, phloem cells do not form continuous tubes. The walls that separate the cells are perforated so that they are sievelike. Food moves downward or upward through the phloem in a continuous stream, depending on the season. In the cross section shown, the xylem has the appearance of a thick "plus" sign or four-pointed star. The phloem is located between the arms of this xylem plus sign.

The thin-walled **pericycle** (per'ə sī kəl) cells are the one-layered cells that surround the xylem and phloem. The pericycle together with the vascular tissues make up the *vascular cylinder.* This cylinder provides rigidity for the root. *Secondary,* or branch, roots can develop from pericycle tissues.

Cortex (kôr'teks) is made up of unspecialized parenchyma and is a food-storage tissue. Cells of cortex are loosely packed; therefore, water and minerals can easily flow around them from the root hairs to the xylem.

Water must pass through the walls of the **endodermis** (en'dō-der'mis), which is the layer of thick, waxy cells that surrounds the vascular cylinder. Water cannot easily flow between these cells.

Activity

Observing Primary Growth in Roots

Materials

5 soaked bean seeds
5 soaked corn seeds
Moist paper towel
Container
String or elastic band
Metric ruler
Thread or marker
Waterproof ink

Procedure

1. Roll the soaked seeds into a moist paper towel so that they are about 5 centimeters from each other and 5 centimeters from one edge of the towel. Fasten the roll with a string or elastic band to keep the seeds in place. Stand the seed roll on end with the seeds at the top. Place the seed roll in a container that has some water at the bottom. Place the container in a warm, dark place and observe every other day.
2. When the roots are about 3 centimeters long, mark them as shown. Open the seed roll and remove one seedling. Cover the other seedlings with a moist towel to prevent drying. Blot the seedling to remove excess moisture. Place the seedling to be marked on a metric ruler.

3. Use a fine-point waterproof marker or a thread painted with waterproof ink. Be sure to wipe the excess ink from the thread to prevent smudging. Quickly make fine lines on the root about 1 mm apart, starting at the root tip.
4. Make a sketch of one marked bean seedling and one marked corn seedling.
5. Roll the marked seedlings in the moist paper towel as before. Again, place the seed roll in the container and place the container in a warm, dark place.
6. Make a chart that you can use to record your measurements.
7. Sketch and measure the seedlings after one day and after two days.

Analysis

1. Are the spaces between the marks still 1 mm apart after one day? after two days? Explain.
2. Where did most growth occur?
3. Compare the growth of the corn seedlings and the bean seedlings. How are their growths different? How are they alike?

Challenge!

Grow radish seeds on moist blotter paper in a covered shallow glass container. Keep the container closed and the blotter moist. Use a hand lens or a microscope to observe the root hairs. Where are most of the root hairs? Which root hairs are the longest?

Spanish moss

Corn plant

Aerial roots

Secondary Root Tissues

All tissues produced by meristem tissues, other than the apical meristem, are called **secondary tissues.** You have already read about the apical meristem of the root tip. In dicots, which are seed plants with two seed leaves, another meristem tissue is found. This tissue is a **lateral meristem,** so called because it produces an increase in thickness rather than in length. Cells of the lateral meristem, also known as **vascular cambium** (kam′bē əm), appear as the root grows older. They are located between the xylem and the phloem. Cells of the vascular cambium are forever young. They never lose their ability to divide. They live as long as the root lives. After the vascular cambium forms in a root, that part of the root can no longer easily absorb water and minerals from the soil. It can, however, store nutrients and conduct them up and down the vascular cylinder.

Other Kinds of Roots

In addition to taproots and fibrous roots, some plants may develop roots that grow out from various parts of the stems or even the leaves. These roots are called **adventitious roots** (ad′ven tish′əs). When a begonia cutting is put in water or soil, the roots formed are adventitious roots. These roots show the ability of plants to regrow new parts.

Not all roots are underground. Corn plants develop **prop roots** aboveground as well as underground fibrous roots. The prop roots grow out from the stem at a point several inches aboveground. They extend into the soil and help support the cornstalk.

Not all roots anchor in soil. Floating plants, such as water lilies, have **aquatic roots** with no root hairs. Others roots hang freely in the air. The roots of some orchids that grow clinging to trees are called **aerial roots.** These roots are spongy in texture, which enables them to absorb moisture from the air. Spanish moss, which may be seen draping trees in the southern part of the United States, also has aerial roots. Some vines grow clusters of **climbing roots,** which are adventitious roots that grow from their stems. These roots help the vine cling to walls, trees, or any other surfaces on which it climbs.

Review It

1. Why are roots important?
2. Compare fibrous roots and taproots.
3. Describe the growth regions of a root tip.
4. Describe the main tissues of a root.

14–2 Stems

We have just been learning about the part of the plant that develops under the ground. Now we will learn about the part of the plant that develops above the ground. This part is called a **shoot.** The shoot grows into a **stem.** The stem conducts the raw materials from the root to the leaves, which manufacture food to be stored in the roots also. The stem carries water and minerals to the leaves, which manufacture the plant's food, and to the flowers which enable the plant to reproduce itself. Stems also serve to support the aboveground part of the plant.

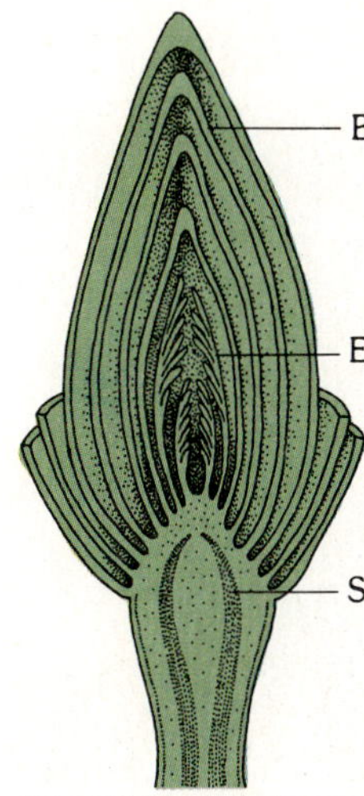

Longitudinal section of bud

Structure of Stems

Examination of a woody twig after all of its leaves have fallen will give you a great deal of information about stems. At the tip of a twig you will notice a swelling, or **bud.** All parts of a plant aboveground start as buds. A bud is a knob of cells that divide and develop into different tissues. These tissues make up the structures of the stems, leaves, and flowers. The bud has a protective outer covering called the **bud scales.** If you were to peel away the bud scales, you would find a soft mass of green tissue. This tissue is called the **shoot tip,** or **shoot apex.** A longitudinal section of the shoot tip shows tiny layers that will grow into leaves or flowers.

Not all buds are the same. The bud at the shoot tip is called the **terminal bud.** It usually develops first. Buds growing along the sides of the twig are called **lateral buds.** Branches, as well as leaves and flowers, develop from lateral buds. Each **node** is a region whose cells may give rise to a bud, leaf, flower, or branch. The section of the stem between nodes are called **internodes.**

Look at the twig and you can see various scars. When a terminal bud opens, the bud scales drop off. They leave a ring of marks around the twig. By measuring the length of the stem between rings, you can tell how much the twig grew in length each growing season.

Leaf scars occur at the nodes where leaves were once attached. These may be shaped somewhat like half-moons. If you look closely at leaf scars, you will see tiny spots called **bundle scars.** These scars are the ends of vascular tissues that carried water and minerals to the leafstalk from the stem.

In many places on the stem of a twig, the bark forms small, round openings. These little holes are called **lenticels** (len'tə selz'). Lenticels allow the exchange of air and water vapor between the atmosphere and the tissues of the stem.

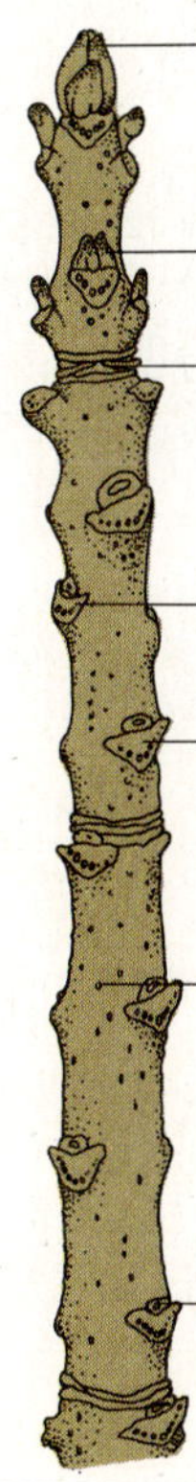

Twig

Excurrent growth

Deliquescent growth

Growth Patterns

The shoot tip is a region of dividing cells called *the apical meristem,* the same name given to dividing tissue in the root tip. When the growing season begins, the cells of the apical meristem begin rapid cell division. The new cells differentiate into stem tissues. All growth in length, or **primary growth,** depends on the activity of the apical meristem.

Terminal Bud

In some plants, the terminal bud dominates growth. In others, the lateral buds dominate. These two kinds of growth dominance produce characteristic plant shapes.

When the terminal bud dominates growth, it gives rise to a kind of growth called **excurrent growth.** The main stem, or trunk, grows much faster than the branches. This produces a plant with a long, large, straight central trunk with small lateral branches. Many coniferous trees exhibit excurrent growth. Their tall, straight trunks make them valuable for lumber.

When lateral buds dominate growth, the trunk divides into several large branches that spread out from a common center. This kind of growth is called **deliquescent growth** (del′ə kwes′nt). Many dicot trees exhibit deliquescent growth. They make good shade trees because they spread outward as they grow.

Internal Structure of Stems

There are two main kinds of plant stems: **woody stems** and **nonwoody stems.** Examples of woody stems include those of vines, shrubs, conifers, and dicot trees. Examples of nonwoody, or herbaceous, stems include those of young seedlings, the sunflower, the tomato, the buttercup, corn, and most annuals. In comparison to woody stems, herbaceous stems are usually thin, weak, soft, green, and short-lived. All of these traits occur because vascular cambium is absent. Without this tissue, there is no **secondary growth,** or growth in thickness. And there is no secondary xylem, which is more commonly known as **wood.**

Woody Stems

Woody stems are generally strong and hard. They may last for many years and may become quite thick. In trees, woody stems become very large and strong enough to support many tons.

In its first year of growth, the woody stem resembles the herbaceous stem. As the cells of the apical meristem at the shoot tip divide and mature, they become the primary tissues of the stem: epidermis, ground tissue, primary xylem, and primary phloem.

But primary growth alone will not give the woody stem its strength. The strength, hardness, and size of a woody stem are a result of secondary growth. The vascular cambium produces the tissues that make the stem grow thicker. The cells that develop on the outer side of the cambium become secondary phloem, and on the inner side secondary xylem develops. Secondary xylem cells develop thick, stiff cell walls. These form the tissue known as wood. Wood gives strength to the stem.

As growth progresses from year to year, many changes occur. No new ground tissue is added to the stem. In fact, the primary tissues may be crowded out by secondary tissues.

A layer of parenchyma cells just beneath the epidermis develops into the **cork cambium.** The cork cambium produces the waterproof cork cells, which gradually harden to form a protective layer around the stem. The epidermis is replaced by the cork.

Cork is much thicker and tougher than epidermis. It protects the stem from injury by insects, blows, fire, and temperature extremes. Because cork is subjected to all of these hard knocks on its outer surface, it has to renew itself. As the stem grows in diameter, the cork layer cracks and is renewed.

The outermost tissue on a tree trunk consists of cork, cork cambium, cortex, and phloem. This entire outer layer is called **bark.** Under the bark and forming the bulk of the stem is secondary xylem, or wood. Between the bark and wood is a thin layer of vascular cambium. If the bark is stripped off and the phloem is destroyed, the tree will die because food cannot pass between the leaves and roots.

You can find the vascular cambium in a stem by simply peeling back the bark. The cambium is where the bark and wood separate. Tissues near the vascular cambium are often green in color and quite moist. The moisture is water, dissolved minerals, and food being moved between leaves and roots.

In the trunk of an old tree, the wood at the center is darker than the outer wood. This older, darker wood is the **heartwood.** The dark color is caused by an accumulation of waste products—resins, gums, and tannins. These materials clog the cells of the heartwood so that no liquids can move up or down.

The light-colored wood surrounding the heartwood is called **sapwood.** This wood conducts water and dissolved minerals between roots and leaves. With each year of growth, some sapwood is changed into heartwood.

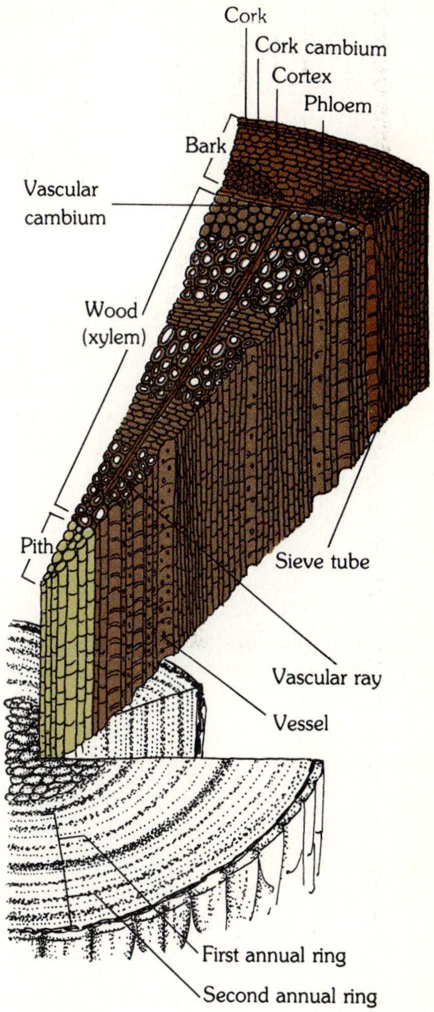

Tree trunk

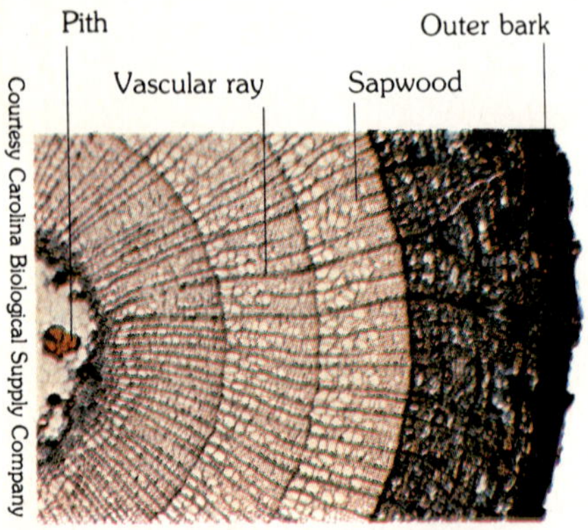

Cross-section of a woody stem

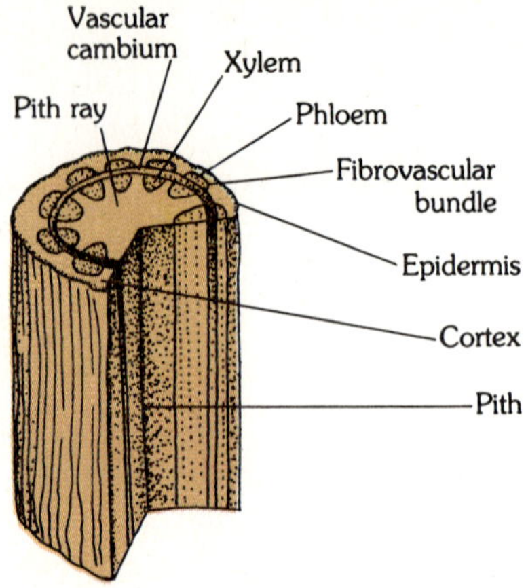

Dicot stem (sunflower)

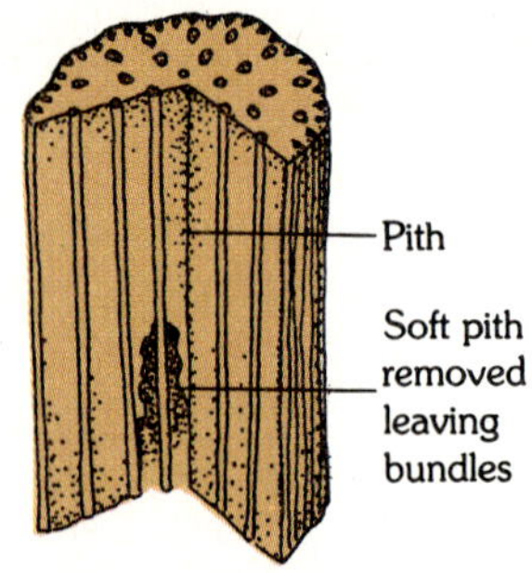

Monocot stem (corn)

Vascular Rays

In some woody stems, you can see lines extending from the bark through the heartwood. These are **vascular rays.** They are produced by activity of the vascular cambium, and they carry materials from the xylem and phloem to the center of the trunk.

Another feature to observe in the cross section of a tree trunk is that of the annual rings. Annual rings result from unequal growth of secondary xylem in spring and summer. Cells of secondary xylem produced in spring are large with thin walls. These produce *spring wood.* Cells produced in summer are small with thick walls. These produce *summer wood.* Summer wood appears darker and denser than spring wood. As a result, distinct rings are formed by the summer wood of one season and the spring wood of the next. These rings can be counted to find the age of the tree.

Herbaceous Dicot

The stem of a sunflower is an example of a herbaceous dicot stem. A layer of epidermis makes up the outer protective tissue. Ground tissue makes up the bulk of the stem. Ground tissue cells that are close to the epidermis may contain chloroplasts and carry on photosynthesis. If so, the ground tissue is green in color; otherwise it is white.

If you look at the cross section of the sunflower stem, you can see that the vascular tissues are arranged in a ring in the sunflower stem and embedded in ground tissue. Ground tissue inside this ring is called **pith.** Outside this ring it is called cortex. Ground tissue usually contains stored food. A closer look at the vascular tissues in the sunflower stem shows that the xylem and phloem are separated by the vascular cambium.

Monocot Stems

The corn stem shown in cross section is a good example of a monocot stem. By looking at the two cross sections you can see how they are alike and how they are different. Both types of stems have protective, ground, and vascular tissues. These tissues were formed during primary growth, but the tissues are arranged differently in the two stems. Instead of being arranged in a ring, the vascular tissues, called bundles, in the corn stem are scattered throughout the stem. There is no vascular cambium in monocot stems. When the stem reaches full seasonal growth, it is about as thick as it will ever be. It may grow taller but not thicker. All ground tissue of a monocot stem is called pith.

Review It

1. Why is the stem important?
2. How is a monocot stem similar to a herbaceous dicot stem? How is it different?
3. Describe how the stem of a woody plant changes as it grows.

14–3
Plant Growth and Response

If two identical plants are placed in environments that differ in light, water, temperature, and available minerals, the two plants may become quite different in size, form, color, and various other traits. These differences may appear because of the activity of plant hormones.

Plant Hormones

Hormones are substances produced in one part of an organism that influence growth, cell division, and other traits in another part of the organism. They occur in both plants and animals. Plant hormones are typically active in very small quantities, but play a major role in regulating the growth and change in position of roots, stems, leaves, and flowers. They should not be thought of as substances that act only as stimulators of growth. In some cases they are inhibitors. They are best thought of as *chemical messengers*. The best known plant hormones are **auxins** (ək′sənz), **gibberellins** (jib′ə rel′ənz), and **cytokinins** (sī tə-kin′inz).

The effect of root-forming hormones on plant cutting

Auxins

Charles Darwin and his son Francis are thought to be the first to observe the effects of auxins. They reported their discoveries in *The Power of Movement in Plants*, published in 1881. The Darwins had noticed that grass seedlings had a tendency to bend toward light. They noticed that the bending took place in the zone of cell elongation, below the tip of the seedling. They tried covering the tip of the growing shoots with a small metal cap and then exposing the shoots to light coming from the side. They found the usual bending did not occur. Then they found that normal bending does occur if a lightproof collar was placed *below* the growing tips. From these experiments the Darwins concluded that something produced in the growing tip was transmitted downward in the stems and somehow caused the bending.

Scientists have now isolated the auxin responsible for the plant growth characteristics demonstrated in the Darwins' experiments. It is called *indolacetic acid* (IAA). This substance is produced in all growing parts of plant organs, but especially in shoot tips. Auxin stimulates cell division and elongation of cells, which result in the growth of stems.

Apical buds, those at the very tip of a plant stem, are stimulated to grow by auxins. But lateral buds, those on the sides of stems, are inhibited in their growth by auxins. When a plant's apical bud is removed, the lateral buds begin to grow, and a bushier plant is produced. If you then apply auxin to the cut where the apical bud was removed, growth of the lateral buds is retarded.

Auxin is produced in fruit seeds and helps ripen the fruit. Auxin may also prevent the fruit from falling too soon. When the seeds are mature, they stop making auxin. The dwindling auxin supply in the fruit triggers the fall of fruit from the tree. To control this, commercial growers spray fruit with auxin solution.

Normal and gibberellin-treated plants

Seedless Fruits

It is also common practice to produce certain seedless fruits by using auxins. Ordinarily a fruit will not develop without pollination and fertilization. These processes are necessary to produce seeds, and auxins from seeds are necessary for a fruit to develop. But if auxins are applied to the unfertilized flowers of certain plants—for example, watermelons, cucumbers, and tomatoes—seedless fruit will develop.

Synthetic auxins have been developed. The most commonly available one is 2,4-D. It is used as a weedkiller because it kills broad-leaved plants when applied to their leaves and buds. Since the concentration of 2,4-D that kills dandelions, thistles, and other broad-leaved plants does not affect grass, this synthetic auxin is commonly applied to lawns.

Auxins function by causing cells to elongate. They do this by softening the cell walls, thus making them more plastic. When walls of cells soften, their internal water pressure, or **turgor,** pushes outward and expands the volume of the cells. As volume expands, more water enters the cells, keeping them *turgid.* After a certain amount of expansion, the cell walls once again become rigid. The result is larger cells.

Gibberellins

Gibberellins are another type of naturally occurring plant hormone. Gibberellins were accidentally discovered when scientists studied a problem encountered by Japanese rice farmers. Rice plants infected by a fungus grew gigantically tall for a while; then, fell over and died. Japanese botanists found that a substance produced by the fungus, which is called *Gibberella,* caused the peculiar growth. The substance was named "gibberellin," after the fungus. Since then it has been found that many plants produce and respond to gibberellins.

Some plants grow to giant sizes when treated with gibberellin. This hormone also stimulates flowering of some plants and early germination of certain seeds.

Cytokinins

Cytokinins, a name that means "cell division," are plant growth hormones chemically similar to nucleic acids. They have the ability to stimulate mitosis in some plant tissues. These hormones also appear to be important in the development of seed embryos and in promoting leaf and lateral bud growth. The "bushiness" of a plant may depend on the delicate balance between auxins and cytokinins.

Tropisms

Tropisms (trō'piz əmz) are plant responses to external stimuli. Environmental factors, such as light, gravity, touch, water, and chemicals, may cause tropic responses. **Positive tropisms** are movements or growth *toward* a stimulus. **Negative tropisms** are movements or growth *away* from a stimulus.

Phototropism

Phototropism

The tendency of seedlings to grow toward light, as observed by the Darwins, is a **phototropism.** This type of positive response is controlled by auxins. There are two different theories used by scientists to explain how auxins cause phototropisms. The *auxin migration theory* postulates that light causes auxins to migrate from the light side to the dark side of an unevenly lighted growing tip. The cells on the dark side then have more auxin and grow more rapidly than those on the light side. This causes the plant to bend toward the light.

The *auxin oxidase theory* explains phototropisms differently. This theory agrees with the auxin migration theory in assuming that auxin is produced by stem tips from which it diffuses to all parts of a plant where growth is stimulated. However, the auxin oxidase theory makes the assumption that an enzyme, auxin oxidase, is present in all plant tissues and continually destroys auxins. Proponents of this theory think that the destructive action of auxin oxidase is accelerated by light energy. Thus, the side of a plant stem exposed to more intense light will have a lower concentration of auxin than the side exposed to less intense light. Thus, the shaded cells will grow more, and the stem will bend toward the light.

Which theory best describes what actually causes stem-bending? Only further research can answer the question.

Roots that grow aboveground show the opposite reaction to light—they are *negatively phototropic.* How can this be explained? Root cells apparently behave just the opposite to stem cells in their reaction to high concentrations of auxin: their growth is inhibited. The auxin migration theory hypothesizes that auxin migrates away from light in roots, just as it does in stems. The greater auxin concentration in shaded areas retards growth and results in root growth away from light. The auxin oxidase theory would explain this phenomenon in this way: more auxin is destroyed on the lighted sides of roots and thereby reduces auxin concentration on that side to a level that stimulates growth. Thus, roots grow away from light.

Other Tropisms

Imagine a seedling grown in a glass tube, as shown in the picture, with no light present to influence direction of growth. The stem will turn upward, and the root will turn downward.

This response of the plant to gravity is called a **geotropism.** Stems are negatively geotropic, and roots are positively geotropic. This response is also controlled by auxin. If you place a seedling on its side, auxins are pulled to the lower sides of stems and roots by gravity. This causes *more* growth on the lower side of stems and *less* growth on the lower sides of roots. As a result of this unequal growth, stems grow upward and roots grow downward. When shoots and roots grow vertically, auxin concentration is the same on both sides, and growth continues in a vertical direction.

Growth, particularly in roots, also appears to be affected by the presence of water and minerals. Early observers thought that roots grew toward water and minerals. They named these growth habits **hydrotropism** and **chemotropism.** Scientists no longer think these are really tropisms. Roots simply grow better where water is present in soil than where water is scarce. More roots develop when water and/or minerals are present to support growth. Growth *to* water and minerals is probably just a matter of chance.

Did You Know?

Several plant scientists have exposed plants to magnetism. They have found that during germination and seedling development, the growing parts of some plants bend in response to a magnetic field and some seedlings are stimulated to grow at a significantly faster rate if exposed to magnetism.

The Russian researchers Krylov and Tarakanova claimed in 1960 that wheat and oat seedlings grew two to five times as fast if the germinating seeds were kept in constant alignment with a magnetic field. Krylov called this growth behavior magnetotropism.

During the same year, a London University researcher, L. J. Audua, exposed cress and oat seedlings to a strong magnetic field. He found that growing plants consistently bent from stronger to weaker parts of magnetic fields.

Seedling's response to gravity

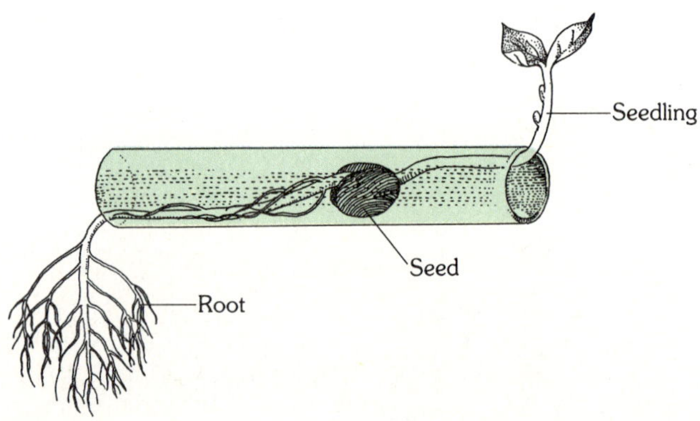

Review It

1. What are some of the ways in which auxins might most directly affect your life?
2. What are the effects of phototropism and geotropism on roots?
3. How do phototropism and geotropism bring about their effects on roots?
4. What is the difference between the auxin oxidase theory and the auxin migration theory?

Breakthroughs in Biology

New germ plasm for the corn plant

Corn, *Zea mays,* is the most important grain crop in the western hemisphere. In 1978, farmers in the United States alone harvested over six billion bushels of corn to be used mainly as a livestock feed crop. The price of such commodities as beef, hogs, chickens, milk, and eggs is determined largely by corn prices.

Domestic corn has been bred for large ears, well-developed cobs, and high yields. Some hybrid corn varieties are resistant to certain diseases. But since conditions that determine pest infestation and disease change constantly, plant breeders need new genetic material to develop hardy new varieties.

Domestic hybrid corn cannot survive on its own because its kernels are permanently attached to a solid cob. People must remove the kernels from the cob and plant them. *Zea mays* is an annual grass and must be planted each year.

In 1977, a University of Guadalajara botany student, Rafael Guzmán, made an important discovery. In a tiny, remote mountain region of Jalisco, Mexico, he found what may be the most primitive member of the *Zea* genus, the new species *Zea diploperennis.* It was the first discovery of a wild grass with the same number of chromosomes as *Zea mays.* It crosses with corn and produces fertile seeds.

***Zea mays* and *Zea diploperennis* ears with separated husks**

Zea diploperennis does not look like domestic corn. Each plant has 5 to 100 small ears and about 10 triangular seeds on each ear. A perennial grass, it springs up every year.

Zea diploperennis can offer these characteristics for crossbreeding with *Zea mays:*

1. It thrives in a cool, damp climate and is resistant to several types of fungus diseases.
2. It grows at 6,000 to 10,000 feet, so it has adapted to climates that would allow winter growth in the southern United States.
3. It is a perennial grass, so a cross may be developed that needs to be planted only about once every few years, saving the farmers millions of dollars in plowing and sowing costs.

Botanists from the University of Wisconsin who first identified this new species, Dr. Hugh Iltis and John Doebley, are experimenting with seeds from *Zea diploperennis.* With Guzmán, they re-explored the beautiful and diverse Sierra de Manantlán and found, near the original site, three other small areas where the plant is growing. It was fortunate that the areas were discovered as they are unprotected and could easily be destroyed by grazing and plowing. Iltis and Doebley have sent about 70 packets of seeds to plant breeders all over the world for experimentation.

Analysis

1. What is the significance of the discovery of *Zea diploperennis?*
2. Wild plants are often destroyed because they are not recognized as important. What could be done to avoid losing vital germ plasm?

14–4
Vegetative Reproduction

Potato tuber

Onion bulb

Iris rhizome

When a farmer decides to grow potatoes, he doesn't plant potato seeds. Instead, he cuts up potatoes so that each piece has at least one "eye." The eyes are the dark spots found in small depressions on the surface of the potato. The pieces of potato are planted, and the eyes develop into leafy stems. Roots develop from one end of the stem, and starch from the potato piece is used as food to nourish this developing young plant. Eventually, an entire, new potato plant develops from the original potato eye.

Note that *no flower or seed* is involved in the formation of the new potato plant. The new plant simply develops from a piece of the original plant. The new plant is genetically identical to the original plant that produced the cut-up potato. Since the new potato plant was produced from vegetative, rather than reproductive, parts of the original plant, this form of reproduction is called **vegetative.** All of the new potato plants that are developed in this way from a single potato plant constitute a **clone** (klōn). The process is sometimes called **cloning.**

There are several naturally occurring means of vegetative reproduction. These involve specialized stems, roots, and leaves. Plant growers have developed several techniques for artificial vegetative reproduction. The term **propagation** is often used in place of reproduction.

Natural Propagation by Roots

Some plants have thick, fleshy roots. These roots are filled with stored food. If these roots are separated, the pieces can form new plants. Each of the fleshy dahlia rootlets has the capacity to form a new plant. Each new shoot from the sweet potato root can form a new set of roots and may grow as a separate plant.

Natural Propagation by Stems

A number of plants can reproduce by developing special stems. Some of these stems grow aboveground, whereas others develop below ground. Potatoes are a type of underground storage stem called **tubers.** "Eyes" of potatoes are actually buds that can grow into complete potato plants. Some plants reproduce by developing bulbs or corms. **Bulbs** are short, flattened underground stems encased in thick storage leaves. **Corms** are much like bulbs, except they are surrounded by thin, dry, scaly leaves. Onion bulbs produce new bulbs at the bases of the fleshy leaves. Gladiolus corms produce new corms at the bases of the dry leaves.

Creeping buttercups send out stems, called **runners,** that grow along the ground. Buds along these runners develop roots and quickly develop into new plants that are able to live by themselves. Some plants, such as the iris, have thick underground stems called **rhizomes.** Rhizomes have buds that can form entire, new iris plants.

Propagation by Layering
If you bend a rose bush branch and bury part of it, a buried bud can produce a new rose plant. This new plant can be separated from the parent and it will grow by itself. This reproductive process is called **layering.**

Propagation by Cuttings
Plant growers use pieces of stems, leaves, and roots to form new plants. These **cuttings** usually include stems with buds or entire leaves. When cuttings are planted in wet sand or soil, roots form, followed by stems. Many woody plants, such as ivy and willow, can be propagated by stem cuttings.

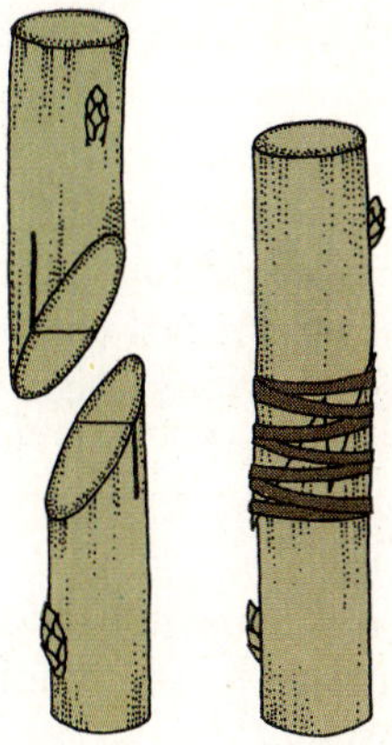

Propagation by Grafting
Grafting involves cutting a twig, stem piece, or bud from one plant and attaching it to the slashed stem of another plant. Grafting can only be done with woody plants. There are several methods of grafting, but it is essential in all of the methods that the cambium layer of the piece being grafted be placed in contact with the cambium layer of the stem to which it is being attached. This is important because the cambium layers must grow together and form water- and food-conducting vessels across the grafted tissues.

Grafting is frequently used to combine a plant that has a particularly strong root stock with another plant that has desirable upper stem characteristics. For example, one variety of apple tree may have very strong roots but produce inferior apples. Another variety may have weak roots but produce superior apples. By joining the quality stem from one to the strong roots of the other, a very fine apple tree can be produced.

Review It

1. Describe some examples of natural propagation by plants.
2. Explain some methods of artificial propagation of plants.
3. What is a clone?

Chapter Summary

Roots absorb water and minerals from the soil and deliver these raw materials to the stem. Stems are the bridges between the food-manufacturing leaves and the raw-material-supplying roots. Stems and roots contain bundles made up of xylem and phloem tissues. Through these tissues, water and minerals flow upward and food flows downward. In dicot plants, a tissue called cambium is produced in stems and roots. In plants that live for more than one year, the protective outer bark forms from the cambium.

Plants manufacture hormones. The best known are the auxins, gibberellins, and cytokinins. Auxins stimulate cell elongation, apical bud growth, and the development and ripening processes in fruits. Gibberellins stimulate growth, flowering, and seed germination. Cytokinins stimulate cell division and promote lateral bud and leaf growth.

Plants respond to external stimuli. These responses are called tropisms. Phototropism is the response of plants to light. Geotropism is the response of plants to gravity.

Many plants that produce seeds are not grown from these seeds. Instead, they are reproduced vegetatively from the roots, stems, or leaves of the parent plant. These new plants are genetically identical to the parent plant. The parent plant and the plants reproduced from it form a group called a clone. Vegetative reproduction may occur naturally or may be caused by plant growers. In either case it is called propagation. Through vegetative reproduction, desirable genetic traits can be passed along from generation to generation without the risk of being lost through crossbreeding.

Extra Research

1. Make a collection of photographs or drawings of trees that exhibit excurrent and deliquescent growth patterns. Do this for deciduous trees after the leaves have fallen in late autumn or in winter, or in early spring before new growth appears. Identify each tree and label each picture with the name of the tree.
2. Make a collection of plant stems and try to raise new plants from these stems by vegetative reproduction. Keep records of the kinds of stems you collected and of how long it takes for new growth to appear from each stem.
3. Examine the twig of a tree in the winter. How old is the twig? (Remember that every time a terminal bud grows and sheds its bud scales a ring is formed around the twig.) Measure and record by years the distances between these bud scars. Did the length of the twig increase an equal amount each year? See if you can obtain a record of weather in your area going back at least as far as the age of the twig. Were the growing seasons in any of these years very wet or very dry? very hot or very cool? Did unusual weather in any one year influence growth of the twig in that year?

Questions to Think About

1. What kinds of problems do you think land plants have that water-dwelling plants do not have? Explain your answer.
2. Why is it advantageous to a plant for the roots to develop first when the seed germinates?
3. What are the advantages to a plant of a taproot system? of a fibrous root system?
4. What are the advantages to people of plants with taproot systems? of plants with fibrous root systems?
5. How do root hairs benefit a plant?
6. Generally, if a plant is pulled from the ground, the plant will die, even if it is replanted. Why do you think this is so?
7. If a tree is *girdled,* that is, if a strip of bark is removed all the way around the trunk, the tree will die. Why do you think this happens?
8. What are auxins? What effects do they have on plants?
9. Gibberellin greatly stimulates plant growth and development. How could the effect of gibberellin benefit farmers?
10. What are the advantages of vegetative reproduction?

Chapter Test

Part I

Copy the number of each test item. If the item is *true*, write *T* after the number; if the item is *false*, write *F*.

1. The root of a plant manufactures food for the plant. (314)
2. Plants with taproot systems are likely to be found in hot, dry climates. (314)
3. Cell division in the apical meristem causes the plant to increase in thickness. (320)
4. Herbaceous stems are usually thin, weak, and short-lived. (320)
5. Rings in the cross section of a tree trunk can provide information about climate changes over periods of years. (322)
6. Auxins stimulate apical bud growth. (323)
7. Geotropism is defined as the response of plants to light. (326)
8. A plant produced by grafting will be genetically identical to the plant upon which it was grafted. (329)
9. The main purpose of the potato is to serve as food for people. (328)
10. In monocot stems, the vascular bundles are scattered throughout the stem. (322)

Part II

Copy the number of each sentence below. After each number, write the word (or words) that completes the sentence correctly.

11. _____ absorb water from the soil. (314)
12. Tissue at root or stem tips where cell division occurs is called _____. (315)
13. The two kinds of vascular tissue are _____ and _____. (316)
14. The growth pattern in plants with dominant lateral buds is called _____ growth. (320)
15. The bulk of a stem is composed of _____ tissue. (321)
16. Growth of lateral buds is promoted by hormones called _____. (324)
17. The response of a plant to light is called _____. (325)
18. The potato is an example of an underground storage stem known as _____. (328)
19. A parent plant together with all of the plants that have developed from it vegetatively form a group known as _____. (328)
20. _____ are hormones that help ripen fruit. (323)

Part III

Copy each item from List A. After each item, write the letter and the words from List B that best describe the item.

List A

21. fibrous roots (314)
22. tuber (328)
23. root hairs (315)
24. shoot (319)
25. propagation (328)

List B

a. extensions of epidermal cells
b. part of a plant that develops aboveground
c. a fleshy, food-storing, underground stem
d. spread out in a network through the soil
e. vegetative reproduction

Chapter 15

The Leaf

Leaves are strikingly beautiful when they turn bright red or yellow in autumn. Not only are leaves important in our lives because they are beautiful but they are important because of the functions they perform. In higher plants the process of photosynthesis occurs mainly in leaves. Some of the energy captured by photosynthesis is used by plants through respiration, but much of the energy becomes available to humans and animals that eat plants. Leaves are important in controlling water loss and the movement of water and minerals in plants. Leaves may play a part in controlling the time of flowering.

Where do photosynthesis and respiration occur? How do leaves control water movement in plants? How do leaves turn color? What makes leaves fall from trees? These are some of the questions that will be answered in this chapter.

Chapter Objectives

1. Describe the internal and external structure of some common leaves.
2. Explain how plants absorb and transport water.
3. Describe how plants control water loss from leaves.
4. Present theories of how food and water are transported in the phloem.
5. Relate the processes of photosynthesis and respiration to leaf structure.
6. Explain some of the effects of light and temperature on plant growth and development.

The color of the leaves, as well as the hoar frost on the leaves, of this pin oak tree indicate cold weather. Leaf coloration is due to a decrease in light.

15–1
Leaf Structure

Pitcher plant

Leaf with plantlets

Leaves are familiar parts of plants. They vary in shape and size and in the way they are arranged on a plant. Many plants can be identified just by studying their leaves. Whether a leaf is needle-shaped as in pine trees, or broad and flat as in maples, all green leaves have an important function. They are food-manufacturing factories, the centers of photosynthesis. The structure of the leaf makes possible the capture of the sun's energy and the building of complex molecules from simpler raw materials. Sugar, proteins, fats, and starches can be produced by plants through the photosynthetic action of leaves. Because of this, green plants are the basic source of food for most heterotrophs.

Specialized Leaves

The structure of some leaves is such that they can carry out special functions other than or in addition to photosynthesis. Some plants have thick fleshy leaves which store water or food. Plants which store water in their leaves are called **succulents** (suk′yə lənts). An onion bulb is made up of food storage leaves, which are attached to a very small stem. When we eat onion bulbs, we are really eating leaves.

Some of the most fascinating leaves are those which are modified to catch insects. Some, such as Venus's-flytrap, have leaves which are hinged along the midrib and close over the insect. Others, such as the pitcher plant, have pitcher-shaped leaves which are usually filled with water and digestive enzymes. The inside of the pitcher is lined with stiff downward-pointing hairs, which allow insects to crawl down into the pitcher but do not allow them to crawl out again.

Propagation by Leaves

The leaves of a few plants will produce small plantlets which have small leaves and roots. These tiny plantlets separate from the leaves and fall on the ground where they may take root and form new plants. Plant growers use pieces of leaves to form new plants. These *cuttings* usually include entire leaves. Cuttings are planted in wet sand or soil, and roots form, followed by stems. African violets and geraniums can be grown from cut leaves.

Dicot Leaves

The leaves of most familiar dicot plants are similar. They have two main parts, the **petiole** (pet'ē ōl) and the blade. The petiole, often called the "stem" of the leaf, attaches the leaf to a branch. The flattened blade is the main body of the leaf. If you examine a variety of dicot leaves, you will find that many of them have one main continuation of the petiole that runs through the blade. This is called the *midrib.* Other leaves have several large ribs that branch out through the blade from the petiole. These ribs, in addition to giving support to the leaf, also act as veins. The veins contain **xylem** (zī'lem), which supplies all cells of the leaf with water and minerals, and **phloem** (flō'əm) to transport manufactured foods out of the leaf to other parts of the plant body.

Monocot Leaves

Most monocot leaves do not have a petiole. The blades of many monocot leaves narrow toward the base of the leaf and often form a sheath around the stem to which the leaf is attached. Instead of having a branching network of veins as in dicots, most monocots have leaf veins which run more nearly parallel to one another from leaf tip to leaf base. Even though there is a difference in the appearance of the leaves of monocots and dicots, the basic function of leaves is the same for both plant types.

Simple and Compound Leaves

If the leaf of a plant consists of a single blade, it is called a **simple** leaf although the edge may be indented in various ways. If a leaf blade is divided into two or more parts, the leaf is **compound.** Each of these parts of a compound leaf is called a **leaflet.** It may be difficult to tell a leaf from a leaflet. However, a bud will form at the base of a leaf where the petiole joins the stem, but a bud will never form at the base of a leaflet.

Compound leaves grow in two distinct patterns: **pinnate** (pin'āt) and **palmate** (pal'māt). Pinnately compound leaves have leaflets attached along a single midrib. Leaves of ash trees are a good example. Palmately compound leaves have leaflets radiating from a central point, like a fan. Leaves of the horse chestnut tree grow in the palmately compound pattern.

Simple dicot leaf

Simple monocot leaf

Palmately compound dicot leaf

Pinnately compound dicot leaf

Inside a Leaf

If you examine a leaf without a microscope, you would be likely to think the green chlorophyll pigment is spread evenly throughout the leaf tissue. You might also assume that the cells making up the leaf would all be much alike. However, if a very thin slice of a leaf is examined under a microscope, many distinct structures become visible. You would find the chlorophyll concentrated in the middle cell layers. The upper and lower layers of the leaf usually do not have **chloroplasts** (klôr′ə plasts) except in certain specialized cells called **guard cells.** The upper and lower layers, each of which is only a single cell in thickness, are called the **epidermis** (ep′ə dėr′mis). Sandwiched between the upper epidermis and the lower epidermis are layers that make up the **mesophyll** (mes′ə fil). This is where photosynthesis takes place. Vascular tissue runs throughout the mesophyll. Air spaces that connect to openings in the epidermis are also present, thus enabling gases to move in and out from the atmosphere.

The Leaf Epidermis

The cells of the leaf epidermis may have different shapes in different types of leaves. Sometimes the cells are interlocking like the pieces of a picture puzzle and sometimes they look more like bricks. When viewed from the side they almost always look like cubes. Some epidermal cells develop hairlike growths on the outer surface. The effect of thousands of these tiny hairs is to give the leaf a fuzzy texture. In some plant species epidermal hairs secrete a substance that makes the leaves feel sticky. This secretion can be protective to the plant because it can be irritating or poisonous to humans or grazing animals.

One of the most important structures of the leaf epidermis is an opening called the **stoma** (stō′mə). It controls the exchange of gases that are involved in photosynthesis and respiration. It also regulates the loss of water from the leaf.

In some plants, stomata are found only in the lower epidermis. In other plants, stomata are found on both upper and lower leaf surfaces.

Leaf cross section

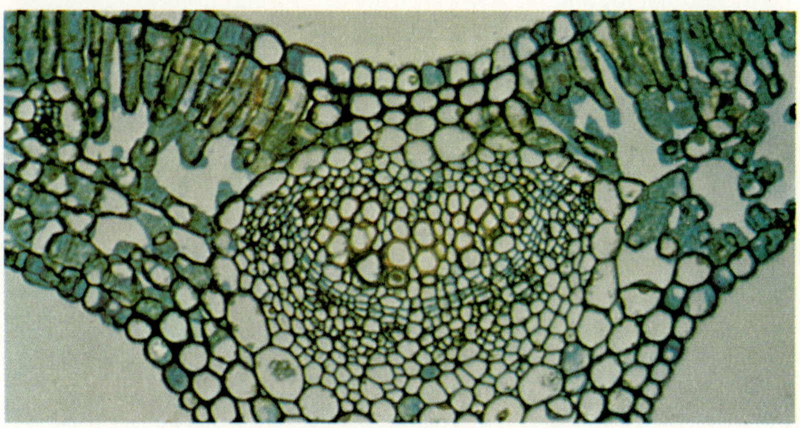

Courtesy Carolina Biological Supply Company

The Mesophyll

The mesophyll of leaves is the tissue sandwiched between the upper and lower epidermis. Most leaves have two different layers of mesophyll cells: a **palisade** (pal′ə sād′) layer and a spongy layer. The palisade layer is found directly under the upper epidermis. This layer is ordinarily one or two cells thick. The cells are elongated and arranged in orderly rows like bricks standing on end. Each cell contains chloroplasts, sometimes as many as 75. This layer is a major site for photosynthesis. The chloroplasts of palisade cells stream around in cycles, allowing for equal exposure of all chloroplasts to light energy. This streaming movement is known as **cyclosis** (sī klō′sis).

Beneath the palisade layer, cells of the spongy layer are irregular in shape and loosely packed. There are large air spaces between cells. These cells have fewer chloroplasts than the palisade cells. The air spaces of this layer are all connected to stomata.

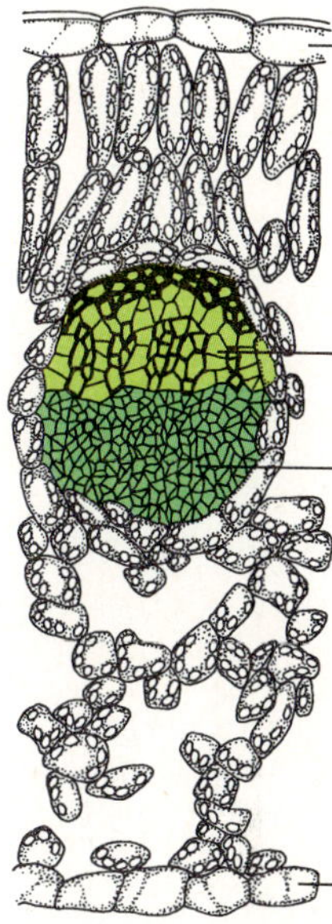

Leaf vein cross section

Leaf Veins

Veins run through the mesophyll of leaves. These consist of one or more **fibrovascular** (fī′brō vas′kyə lər) **bundles.** Each fibrovascular bundle contains xylem, phloem, and support fibers. Water and minerals are brought up from the roots through the stem and leaf petiole by xylem. Phloem conducts manufactured foods out of the leaf to other parts of the plant. Tough fibers are also present, providing rigidity. You can easily see fibrovascular bundles when you eat celery. They are the long, tough "strings" you notice in the celery stalk. The stalk is the petiole of the celery leaf.

Leaf cross section

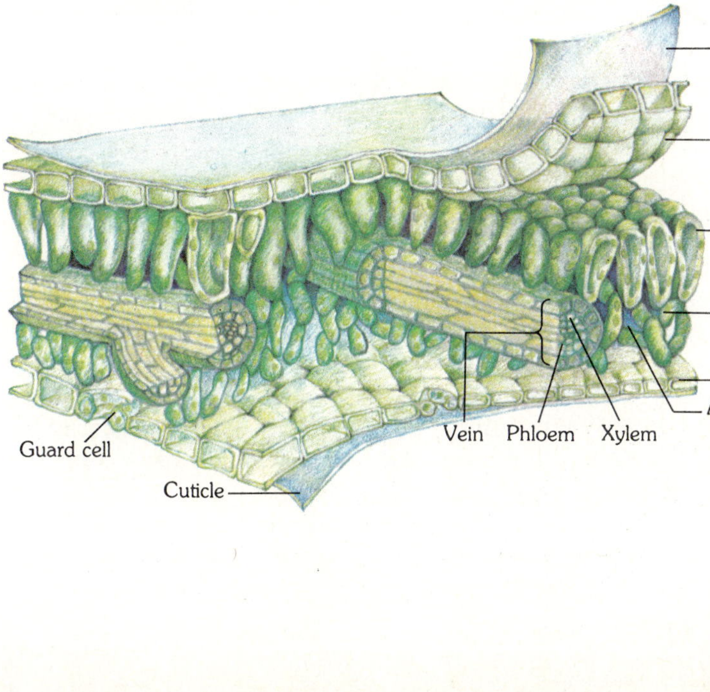

Energy Transfer in Green Leaves

A very important function of green leaves is the capture of energy from the sun. This process, called photosynthesis, was described in Chapter 4. Once the sun's energy has been captured by the molecule-building process of photosynthesis, the energy must be released in order to do work. This process of energy release is called respiration. Respiration occurs in nearly all living cells, including the cells of the leaf. The energy released by respiration is used in the leaf to accomplish things such as growth of cells, addition of cells for reproduction, streaming of cytoplasm, and active transport across the cell membrane. The process of cellular respiration and the use of energy for things such as active transport were also presented in Chapter 4.

Energy Transfer in Green Leaves

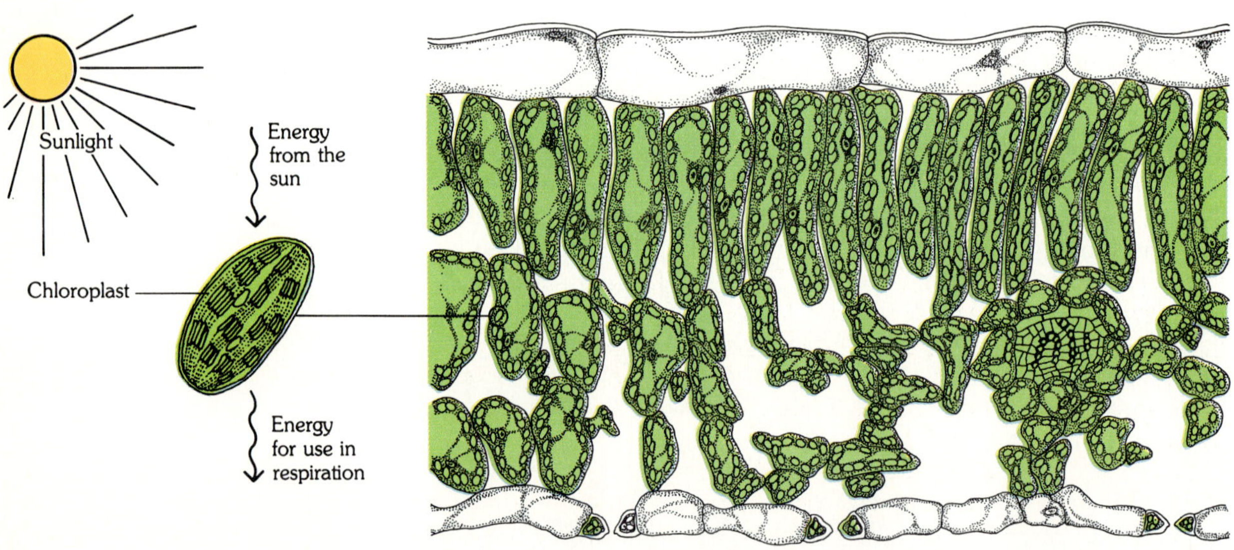

You should remember that photosynthesis occurs only in those cells where chlorophyll is present, but respiration occurs in most living cells. Those plant cells which are not carrying on photosynthesis must have energy brought to them. This is done by means of transporting glucose molecules through the phloem. These glucose molecules are then broken down in the process of respiration so that energy is released. If animals eat the green leaves or other plant parts that contain glucose, the animal has access to the energy from the sun. The glucose then can be broken down by respiration in the animal cells so that there is a source of energy for cells to grow, reproduce, and carry on other functions.

Gas Exchange in Green Leaves

In order to accomplish energy transfer, molecules must be built and broken down. This building and breaking down requires the use of oxygen and the release of carbon dioxide. These gases move to and from the cells of the leaf through the spaces of the mesophyll and through the stomata. The diagram shows the flow of gases. It must be remembered that respiration occurs during the day as well as at night, but in bright light more carbon dioxide is used by photosynthesis than is given off by respiration.

Light

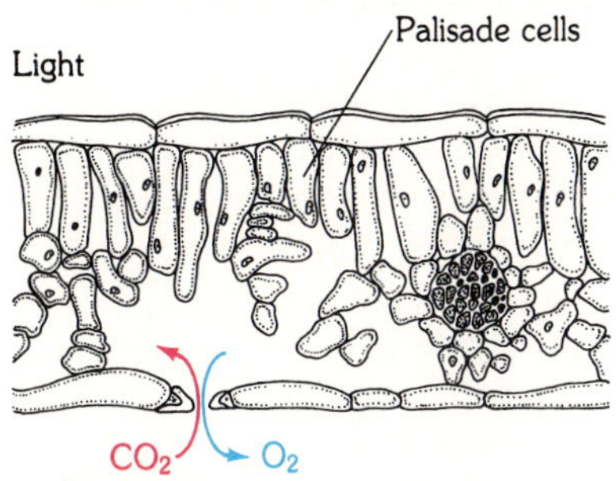

Photosynthesis

Light and dark

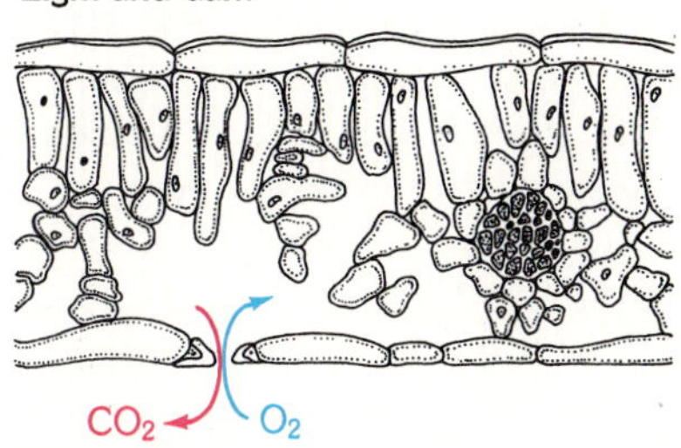

Respiration

Molecules of Energy Transfer

Oxygen and carbon dioxide are only a few of the molecules involved in the transfer of energy in green leaves. There are numerous molecules within the leaf cells, particularly in the chloroplasts and mitochondria. Some molecules are moved through the leaf and into and out of the leaf. Water, necessary for photosynthesis and other leaf cell functions, is brought into the leaf primarily through the veins. Much more water than is used in photosynthesis passes through the leaf and exits through the stomata. Water is also a by-product of respiration. Some of this water is reused in the leaf and some passes out through the stomata. Glucose is one of the molecules produced in photosynthesis. It is significant because it holds energy captured from the sun. Glucose molecules are moved from cell to cell and the energy they hold is moved with them.

Light

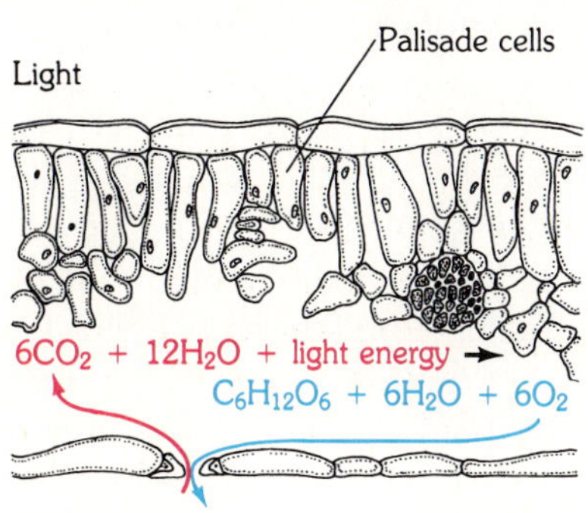

Photosynthesis

Light and dark

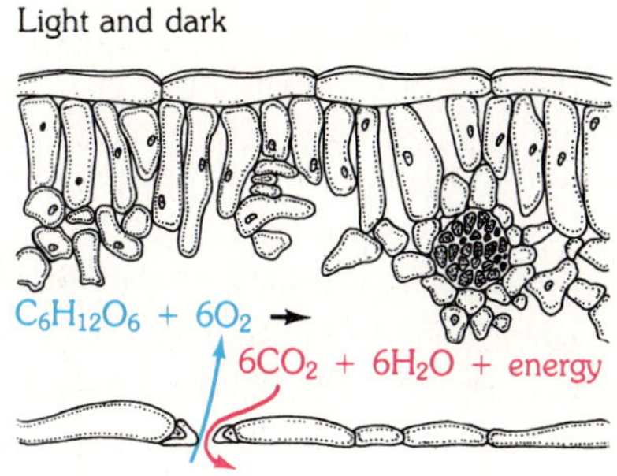

Respiration

Stomata

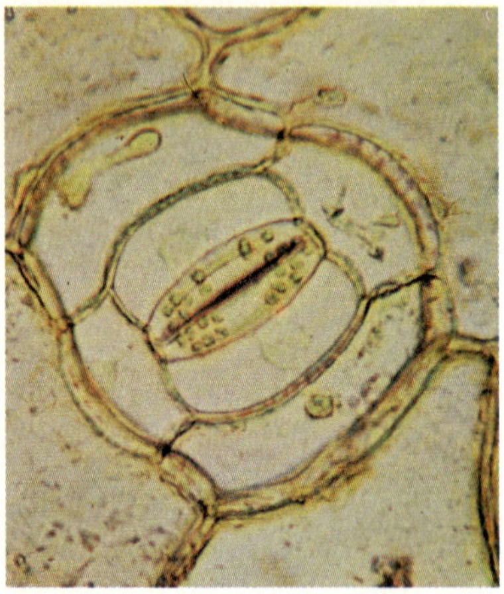

Stoma

Stomata are formed by guard cells. When a guard cell absorbs water from the surrounding epidermal cells, the shape of the guard cells changes. This change occurs because the wall away from the stoma is thin, but the wall next to the stoma is thick. The thin outer wall expands but the thick inner wall does not and the inner wall is pulled into a crescent shape.

There are a number of factors which play a part in controlling the amount of water taken into and held by a guard cell. One of these factors is the amount of sugar present in the guard cell. As sunlight strikes the guard cell, photosynthesis produces sugar in the guard cell. As the concentration of sugar in the guard cell increases, more water will move into the guard cell by osmosis. As water moves in, the cell swells and the stoma opens.

Stomata Close at Night and Open during the Day

At night or in the dark, when photosynthesis has stopped, no sugar is being produced. Sugar which may still be present from photosynthesis is changed to starch by an enzyme. There is another enzyme present which does the opposite of the first enzyme. It changes starch into sugar. This enzyme does not function at night. It does not function because: (1) There is an accumulation of carbon dioxide from cellular respiration. (2) The carbon dioxide dissolves in water and forms carbonic acid. (3) The carbonic acid keeps the enzyme from functioning. This enzyme starts functioning again when sunlight strikes the leaf because: (1) Light begins photosynthesis. (2) Photosynthesis uses CO_2. Therefore CO_2 can not dissolve in water and form carbonic acid. (3) The lack of acid allows the enzyme to work. (4) The enzyme changes starch to sugar. (5) Sugar build-up causes water to enter guard cells by osmosis. (6) Water causes the guard cells to open the stomata. As a result of all this, the stomata are open during the day and closed at night. The opening and closing of stomata is *directly* caused by water pressure in the guard cells but is indirectly caused by the amount of carbon dioxide present.

The regulation of the opening and closing of stomata is vital to the life of plants. In addition to allowing exchange of gases between plant cells and the atmosphere, the amount of water loss from the plant is regulated.

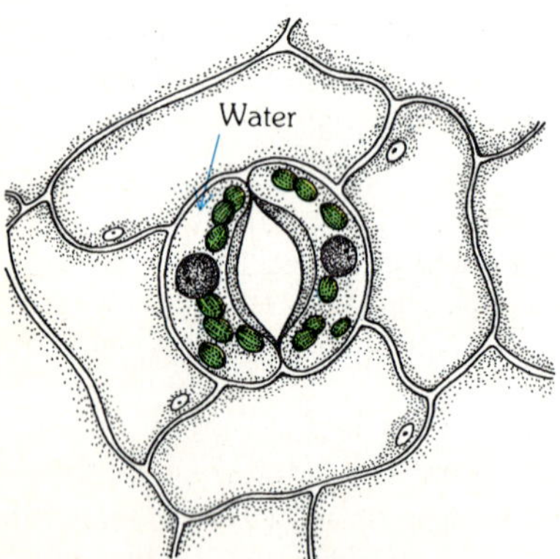

Stoma open

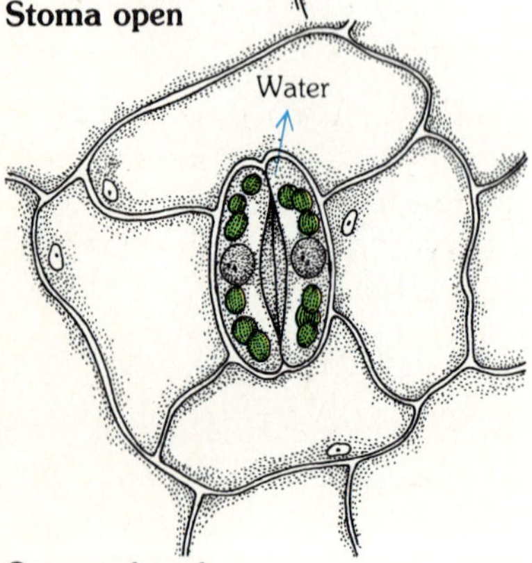

Stoma closed

Review It

1. Distinguish between pinnate and palmate leaves.
2. Describe the structure and function of various leaf tissues.
3. Explain the operation of leaf stomata. What factors are involved in their opening and closing?
4. Describe several specialized leaf adaptations.

Activity

How Many Stomata Are on a Leaf?

Materials
Several kinds of fresh leaves
Microscope
Slide
Cover slip
Medicine dropper
Razor blade
Forceps

Procedure

1. Holding the under surface of a leaf toward you, tear the leaf downward on an angle. This tearing action should reveal portions of the lower epidermis. The epidermis will appear as a narrow, colorless zone along the torn edge of a leaf.
2. Quickly cut off a small piece of this colorless border and put it in a drop of water on a microscope slide. Cover with a cover slip.
3. Under low power of the microscope, locate a stoma. Then while observing the stoma make a labeled sketch of it and a few of the cells around it.

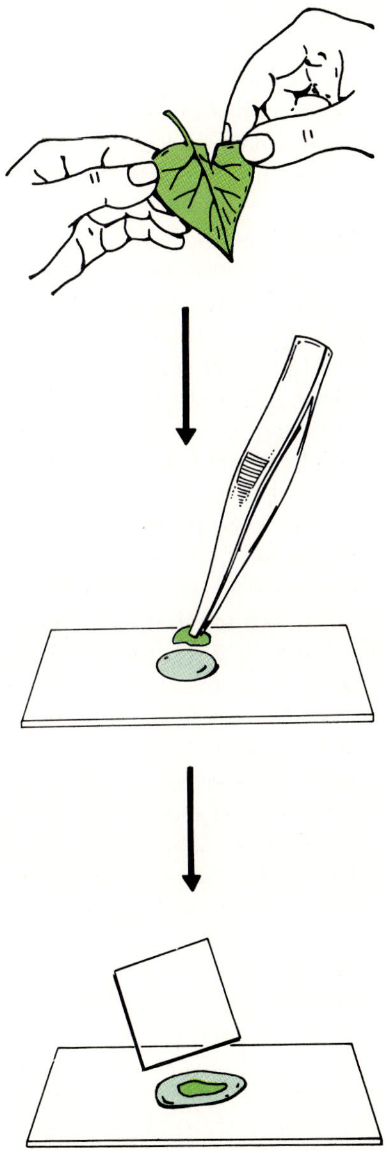

4. Find the average number of stomata observed in ten different high power fields. Record this number.
5. Now make a slide of a piece of upper epidermis from the same leaf. Find the average number of stomata.
6. Examine several kinds of leaves and find the average number of stomata.
7. Find the average number of stomata on the upper and also lower surface of each kind of leaf.

Analysis

1. What do your data suggest about the distribution of stomata on the leaf surfaces?
2. On the basis of your data, what assumption can you make about gas exchange between leaves and the atmosphere?
3. Can you suggest any reason why one leaf surface might be more effective in controlling gas exchange than another leaf?

15–2
Transport Through Leaves, Stems, and Roots

Living things that are small and composed of a few cells do not need a special system to transport materials. Materials such as water, minerals, and food move from cell to cell by processes such as osmosis and active transport. You will recall learning these processes in Chapter 4. Plants that have leaves, stems, and roots do have need of a special transport system. The function of roots, stems, leaves, and flowers are all dependent on one another and require materials that come from other parts of the plant. In order to move these materials long distances a transport system is necessary. This system is called a *vascular system.* The structure of the vascular system in the roots and stems was described in Chapter 14. The structure of this system in the leaf was just described in this chapter. Now you will see how these structures all function together to transport materials in larger plants.

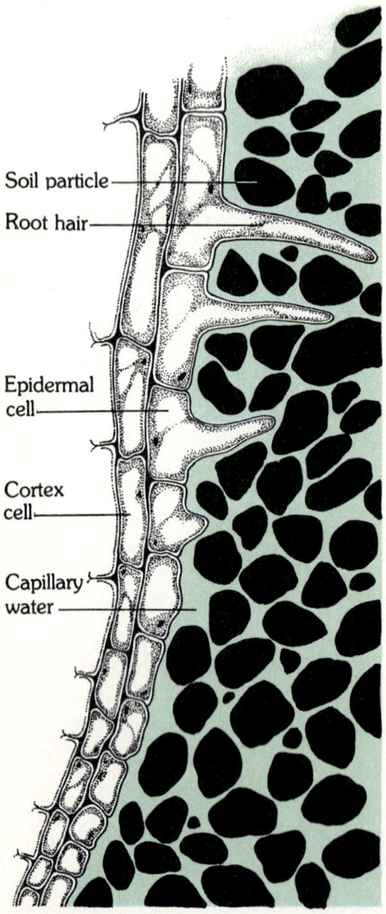

Capillary water and root hairs

Absorption and Transport of Water

The absorption and transport of water is not difficult for a single-celled organism living surrounded by water. These single cells obtain water through their cell membranes by *osmosis.* You may recall from Chapter 4 that osmosis is the process in which water diffuses through a selectively permeable membrane from an area of greater water concentration to an area of lower water concentration. Cell membranes are selectively permeable membranes.

Very few of the cells of a multicellular plant can depend upon osmosis to provide water directly from outside the plant. Millions of cells are located in the upper regions of the plant, perhaps many meters away from the soil where water is obtained. In a large tree, this distance from soil to tip may be as much as 100 or more meters.

How is water moved through such large plants? In order to find the answer, let's start in the soil and trace the water's pathway through a vascular plant. Water is available in soil in tiny spaces among soil particles. This water is called *capillary water.* The water tends to stay in these tiny spaces because of a molecular attraction between the particles of water and the particles of soil. When a lot of water is available, such as after a rainstorm, water tends to fill all of the small spaces between soil particles until it evaporates or is used by plants. To absorb this water, root hairs of plants grow into the spaces between soil particles. Root hairs are long extensions of epidermal cells, and almost all water absorption is accomplished through them. The root hairs are so numerous on most plants that the absorptive surface is 5 to 20 times greater than the surface area of the large roots visible to the naked eye.

Water Transport in Roots and Leaves

There are two very important processes that cause water movement through plants. These are osmosis and transpiration. In osmosis, diffusion of water takes place from outside of root hair cells, through the cell membranes, and into the cytoplasm. This occurs as long as there are more salts, sugars, and other dissolved substances in the cytoplasm of root hair cells than are in solution in surrounding soil water. As the water moves into these cells, the cells tend to develop greater proportions of cytoplasmic water than neighboring cells further inside the root tissue. Thus, water tends to move by osmosis from outer root cells to inner root cells. Water is passed by osmosis from cell to cell and finally enters the xylem.

During daylight hours, the stomata are open in the plant's leaves. Evaporation of water takes place rapidly through these openings in the process called transpiration. A continuous column of water exists from the roots, through the xylem channels of the roots, stem and leaves and out to the stomatal openings of the leaves. As water evaporates, a tension develops on the column of water in the xylem. As a result, water is moved upward continuously to replace water lost by evaporation through the leaf openings. Through this process, enormous amounts of water move through large trees every day.

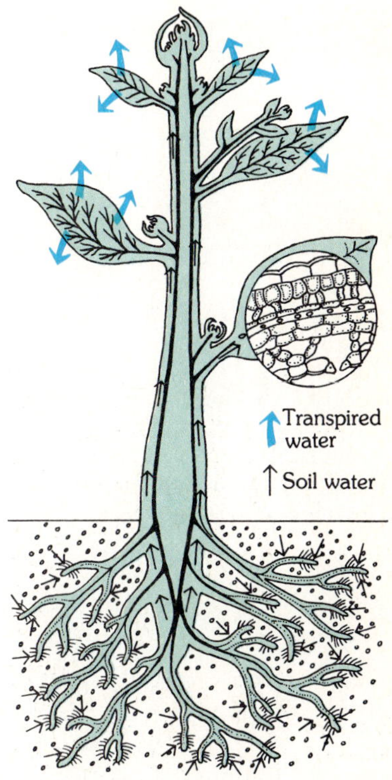

Translocation

Theories of Water Movement Through Plants

How tons of water can move from tiny root hairs in the soil through many meters of woody tissue, finally evaporating through leaves far from the ground has long been a question of great fascination to biologists. This process of movement of quantities of water and other materials through plant tissues is called **translocation.**

Unlike animals, which are equipped with pumping mechanisms to move fluids throughout their tissues, plants must rely on nonliving xylem tissues to translocate liquids. An early theory, considered in about the year 1700, was that xylem tissues somehow pumped or pushed water upward. This idea was discarded when it was found that some totally dead stems could still conduct water.

In about 1725, Stephen Hales conducted a series of experiments with water movement through plants. Hales noticed that the water moved in very narrow tubes in the xylem tissue, and suggested that the movement might be due to *capillary action.* This idea is best described by a simple demonstration: If you take a very fine glass tube and stand it on end in a container of water, you will observe something interesting. The water in the tube will rise to a higher level than the water in the container. If you try a variety of tubes of different diameters, you will find that the narrower the space inside the tube, the higher will be the water level. The liquid is thought to rise in narrow tubes due to an attractive force, or capillarity, between the molecules of the liquid and the molecules of the glass inside the tube. This attraction is greater than the attraction of one liquid molecule for another liquid molecule. As a result one water molecule will "jump" above another to reach the glass.

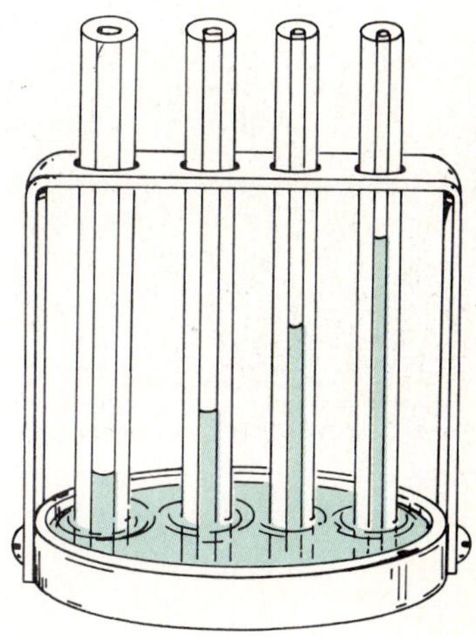

Capillary tubes of differing diameters.

Root Pressure

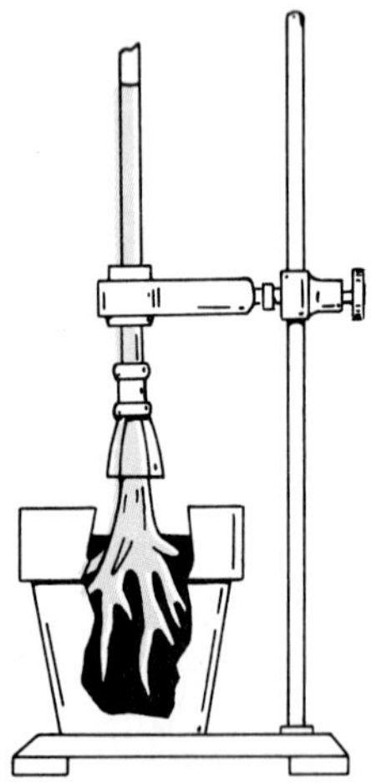

Root pressure demonstration

Although capillarity could account for a rise of water of many centimeters in a plant stem, it could not begin to accomplish water movement to great heights in large plants. Even the narrowest tube imaginable could not transport water to the upper leaves of even medium-sized plants through capillarity alone. Stephen Hales apparently realized this because he also investigated another possible explanation of the upward rise of water in plants. He observed that water and dissolved materials oozed out of the cut stems of plants left attached to their roots. When a glass tube was attached to the cut stem, Hales found that water rose a considerable distance up the tube. He hypothesized that the rise was due to **root pressure** in which roots of a well-watered plant somehow push liquids upwards in xylem channels. Biologists now believe root pressure results from the osmotic absorption of water by the root hairs and the resulting pressure in root cortex cells. This pressure has been measured and it has been estimated it could not account for a water rise of more than 30 meters. So there must be still other forces contributing to water conduction in very tall trees.

Transpiration-Cohesion

The explanation of water translocation that is most widely accepted by scientists at present is the *transpiration-cohesion theory*. The main assumption of this theory is that water exists in a continuous column from the leaves to the root tips of a tree. As water is lost through transpiration in the leaves, a force acts on the column of water in the plant. This pulling force causes water to move up the xylem from the roots.

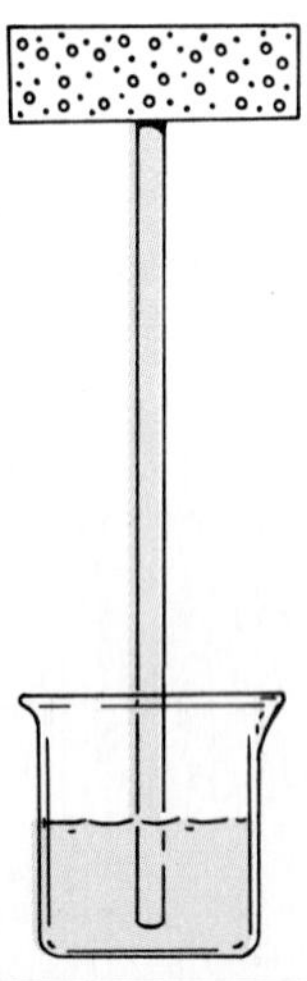

Transpiration—cohesion theory model

The transpiration-cohesion theory can be demonstrated with a sponge, a container of water, and a long glass tube, as shown in the diagram. The tube is filled with water so an unbroken column of water exists between the water in the container and the wet sponge. As water evaporates from the sponge, more water moves up the glass tube to take its place.

Why doesn't the water column break? Apparently, molecules of water have very strong attractive forces for each other. These are known as **cohesive forces.** These forces have been measured and found more than sufficient to account for the cohesion of the water molecules in columns as tall as those in the tallest trees. Wind and heat from the sun cause leaves to lose water by evaporation from the top of the column. This causes the water in the xylem to be placed under tension and it is pulled upward in an unbroken column. More water is then brought to the xylem by root cells. Root pressure undoubtedly contributes to the process, as does capillarity to a small degree. But scientists generally believe the transpiration-cohesion forces to be the major factors in water rise in large plants.

Problems of Water Balance in Plants

Plants require a constant supply of water to maintain healthy, functioning cells. Cells are composed largely of water, and all important physiological functions—respiration, digestion, synthesis of chemical compounds—require water. Water, in the vascular system, transports minerals from the roots to all parts of the plant. In addition, herbaceous plants depend upon the maintenance of internal water pressure as a means of support. As osmosis brings water into cells, a pressure is built up internally. This pressure, called *turgor,* keeps the cell membrane tightly stretched. Since plant cells are surrounded by cell walls, there is a limit to the size a cell can grow as turgor pressure increases. Eventually this pressure forces the cell membrane and underlying cytoplasm firmly against the inside of the cell wall, keeping the cell very stiff. As long as sufficient water is available, turgor pressure remains high, and soft plant tissues such as leaves and flower petals remain firm.

Normal turgor pressure

Low turgor pressure

Remember that leaf stomata are generally open in the daytime and closed at night. When stomata are open, tremendous amounts of water are continuously lost to the atmosphere. In addition, some water is lost by diffusion through the surfaces of leaves and stems. Woody stems often have small openings called **lenticels** (len′tə sels′) through which water can be lost.

Herbaceous plants have their leaves and stems covered by a waxy layer of **cutin** (kyüt′n). This greatly retards water loss but does not completely stop it. The thickness of the cutin varies considerably among plant species. In general, the cutin is thicker in leaves of plants exposed to intense sunlight or dry environments and thinner in plants growing in shade or moist environments.

When plants lose water through leaves and stems at a faster rate than they can absorb it from soil, they may undergo temporary wilting. Plant cells lose their turgidity and the plant tissues appear limp. This condition can be overcome if more water is made available to the plant or when stomata close at night and water builds up in plant tissues. If there is a very severe lack of water in a plant's environment or if its roots are damaged, permanent wilting may take place. This situation cannot be reversed by making water available or by slowing the rate of transpiration. Permanently wilted plants usually die.

Guttation

Some plants growing in moist soil surrounded by humid air will form droplets of water along the edges of their leaves. This loss of water droplets is known as **guttation** (gə tā′shən). The process occurs when conditions favor water absorption by a plant but do not favor transpiration. In other words, when water uptake is greater than water loss, water pressure increases in xylem ducts. Excess water is allowed to escape through pores at the ends of leaf veins. Scientists believe guttation prevents internal damage to plants that might occur due to excessive water pressure.

Cactus

Plant Adaptations for Water Balance

Plants have adapted to very difficult environments. Deserts, for example, receive almost no rainfall, yet a few plants manage to live in them. These plants get water by having very long roots that can reach water deep underground or by having large, spreading root systems that can rapidly absorb large quantities of water when it does rain. To prevent evaporation of water, desert plants often have very small leaves. The leaves are covered by thick layers of cutin. Leaves may also be rolled up. This helps cut down the surface area of the leaf over which wind blows and thus retards water loss. Most cactus plants have no leaves at all. The entire plant visible above ground is a fleshy stem that makes food and stores quantities of water.

Translocation of Foods in Phloem

Foods produced by photosynthesis are moved from leaf cells through sieve tubes of phloem tissue to other parts of the plant. The sieve tubes form a network of ducts composed of living cells. If the phloem tissue is killed, all translocation stops. The movement is many times more rapid than could be accomplished through diffusion alone. Frequently the movement is in a direction not possible to explain by the laws of diffusion, since sugars are moved from areas of less concentration to storage areas of far greater concentration. What makes the process even more puzzling is the observation that phloem transport can move either upward or downward in a plant. Research results indicate phloem transport is a process of active transport and requires energy. Exactly how this system operates in transporting sugars is still an open question.

The *protoplasmic streaming theory* is one attempt by biologists to explain phloem transport. This theory suggests that materials are transported by rapidly circulating cytoplasm in the sieve tube cells. You may recall this process, called *cyclosis,* from your earlier study of living cells. The theory is that materials are rapidly carried by this process from one end of a cell to another where they are passed along cytoplasmic strands from one sieve cell to the next. This theory can account for the speed of transport through phloem vessels, the movement of materials in opposite directions, and the requirements of energy and oxygen. Research is currently being done to test the theory.

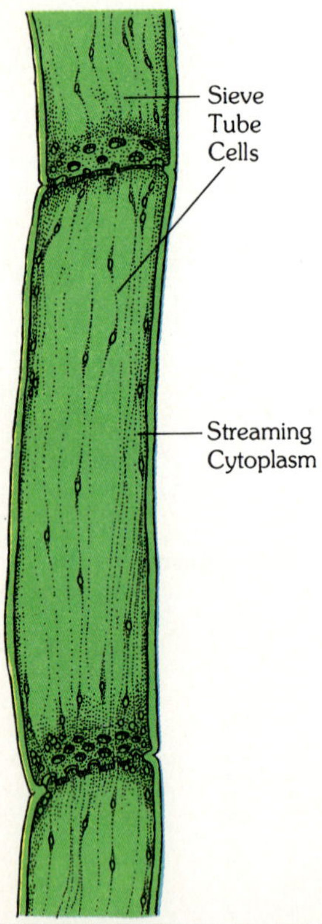

Protoplasmic streaming theory of phloem transport

Review It

1. Describe the process that maintains turgidity of plant cells.
2. Explain the transpiration-cohesion theory.
3. Describe the movement of materials through the phloem.
4. Describe the process of guttation. What are some possible advantages of this process to plants?
5. List some adaptations of various plants to problems of water balance.

15–3
Plant Growth and Response

Leaves play an important part in growth and responses of the entire plant. Leaves are often the first parts to be affected by environmental factors such as light, temperature, and touch. These factors have direct effect on the leaves and through the leaves affect the entire plant.

Challenge!
Set up an experiment to find why plants which are grown in the dark tend to have **longer** *stems. Garden beans make a good experimental plant, but you may wish to try other plants as well. Try some plants in total darkness, partial darkness, and light from only one direction. If you have trouble interpreting your data, see the section on plant growth and response in Chapter 14.*

Influence of Light

There are many plant growth processes controlled by the amount and duration of light. Within certain limits, the rate of photosynthesis increases as the total amount of available light increases. This means that the amount of light available for absorption by green plants has an effect on the total food supply of the world. Remember, animals ultimately depend upon green plants for food.

Light is necessary even for the formation of chlorophyll in plants. The tissues of plants grown in darkness are said to be **etiolated** (ē′tē-ə lā′təd). They are yellow or white in color, since chlorophyll does not develop. Photosynthesis, of course, cannot take place in etiolated plants. Plants grown in darkness grow longer stems. But the growth is weak and spindly, and leaves are smaller and often fewer in number.

If you compare leaves from the same plant that are exposed to sunlight with those that are shaded, definite differences can be seen. The surfaces of shaded leaves tend to be larger, while light-exposed leaves tend to be thicker. This is because light-exposed leaves may develop more than one layer of palisade cells (which carry on most of the photosynthesis) and thicker vascular tissues. Also, *more* sugar is normally found in the sunlit leaves. How can you explain the presence of more sugar? Usually less water is found in sunlit leaves. Can you think of any reason for this situation?

Plant grown in sunlight and plant grown in darkness

Blooming caused by lengthening the period of light

Photoperiodism

Many plants respond to the length or light and dark periods of their environment. Variations of light and dark periods can affect various growth patterns. These include flowering, seed germination, stem length, and leaf shedding.

The response of plants to varying periods of light is called **photoperiodism** (fō'tō pir'ə diz əm). The length of days in the earth's temperate zones varies during different times of the year. In the north temperate zone, June 21 has the longest amount of daylight and December 21 has the shortest. Many plants seem to be triggered to flower by very precise combinations of daylight and darkness. The Maryland mammoth tobacco plant, for example, is triggered to flower when days are very short. It is called a *short-day plant.* Short-day plants flower only when the day is less than a certain length. If days are longer than this critical length, the plants produce only vegetative growth (stems, roots, and leaves). Some short-day plants are cocklebur, Biloxi soybean, tulips, and goldenrod. Some short-day plants flower during short days in spring, others during short days in fall. A few flower in both seasons.

Spinach, sugar beets, iris, and radishes are *long-day plants.* They flower only after a certain critical day length has been exceeded. Other plants are not affected by photoperiods at all. They simply flower after a period of growth. Among these *neutral-day plants* are tomatoes, many varieties of peas, zinnias, and marigolds.

Experiments with short-day plants indicate it is a long night rather than a short day which causes the flowering. However, the terms *short day* and *long day* continue to be used.

Horticultural Use of Phototropism

Horticulturists have found they can artificially regulate light and dark periods in greenhouses. Thus they are able to make plants bloom or bear fruit during months when this would not happen under natural conditions. Long-day plants can be brought into bloom during the short days of winter by adding hours of artificial light in the morning or evening. Short-day plants can be brought into bloom during the summer, when days are long by keeping them in the dark in the morning or evening.

A good deal of research has been devoted to photoperiodism. So far scientists have learned that plant leaves seem to be the part of the plant that is sensitive to day length. Many plant scientists believe a flower-producing hormone, *florigen,* to be involved in the photoperiodic response. Florigen is thought to be formed in the leaves under certain conditions of day length. It is then transported to the growing plant shoots where it causes flowers to be formed.

Some types of seeds have been found to require definite photoperiods to trigger their germination, even when all temperature, moisture, and other factors are satisfactory. In some other seeds, light has been found to prevent germination.

Influence of Temperature

Rates of photosynthesis, transpiration, translocation, and respiration of plants are all affected by temperature. Growth stops or slows down when temperature is too high or too low. Each species of plant has an *optimum temperature range* in which it grows best.

In some plants, temperature plays a controlling role in the development of flowers, fruits, and seeds. These plants grow only vegetatively if kept at a high temperature. They need to be exposed to cold periods before they will flower. Cold treatment, necessary for flowering in these plants, is called **vernalization** (ver′nal i zā′shən). Celery, winter rye, and henbane are plants requiring vernalization. Annual plants are normally vernalized as seedlings. Winter wheat, for instance, is sown in the fall and the following winter provides the required cold treatment. Biennial and perennial plants are vernalized during adult growth. Beets and celery grow during a first season and are then vernalized during winter. In their second season, they flower.

Leaf Coloration

In colder parts of the north temperate zone, hardwood tree leaves "turn" in the fall to bright yellows, reds, and oranges. This, too, is a temperature-controlled event. As fall temperatures get cooler, chlorophyll production stops and what chlorophyll is left is destroyed by sunlight. Yellow and orange pigments are present in the chloroplasts of the leaves at all times, but chlorophyll masks them. When chlorophyll disappears, the yellow *xanthophyll* and the orange *carotene* become visible. The red color appearing in some leaves is the pigment *anthocyanin.* Its formation in the vacuoles of leaf cells is stimulated by cold temperatures.

Falling Leaves

Low fall temperatures also help trigger the fall of leaves from the trees. As temperatures drop, so does the production of the plant hormone, IAA (indoleacetic acid). When the amount of IAA produced by the leaf is lower than the amount in the stem, the cementing material between the cells of the **abcission** (ab sizh′ən) **layer** dissolves. This causes the cells to die. An abscission layer of dead cells forms at the base of each leaf petiole. Eventually the leaves are attached only by their fibrovascular bundles, and the action of wind or rain causes the leaves to break away from these weak attachments.

Abcission layer

Oxalis during the day

Oxalis at night

Nastic Movements

Tropisms, which were described in Chapter 14, are plant movements accomplished through unequal growth of different parts of a plant due to environmental stimuli. Tropisms are directional movements: growth is either toward or away from a stimulus.

There are other plant movements that are not directional. Many flowers, for example, open and close during a 24-hour cycle. Some flowers and leaves seem to have a built-in rhythm of movement that takes place even when kept in complete darkness and at constant temperature. Most people are familiar with morning glory flowers that open during the day and close up at night. These are movements called **nastic movements.** These movements are controlled by turgor pressure in certain cells. They do not require growth of the plant. When a flower opens in the morning, turgor pressure is increased in the cells at the base of the flower petals, causing each petal to move. At dusk, these cells lose turgor; the petals droop and close. Nastic movements can occur quickly, often happening within a few minutes. They can occur over and over again. A good example of this is the sensitive plant *Mimosa pudica.* Its leaves react to touch by immediately seeming to wilt. The compound leaflets of the sensitive plant are supported by swellings at their points of attachments to petioles. A slight touch or heat causes water to leave the cells of the leaflet-base swellings, causing the leaflets to fold together. There is also a swelling at the base of the petiole. When the cells in this swelling lose turgor, the entire leaf will drop down from an erect position. This happens in a matter of seconds. The sensitive plant is unusual in that the stimulus picked up in one leaf may be transmitted through the cytoplasm to other leaves and cause those leaves to droop. Within ten minutes to a few hours, turgor returns to the cells controlling the petiole and leaflets, and they return to normal positions.

Nastic movements are the result of turgor changes in cells, while tropisms are brought about by growth of cells. Nastic movements are reversible while tropisms are permanent.

Did You Know?

Plants can respond to touch, but the response is not always immediate. Pea tendrils, the wiry organs that hold pea plants to vertical supports, coil when they are stroked. M. J. Jaffee, an Ohio University researcher, has demonstrated that sensory information can be stored for hours before the plant responds.

Apparently the pea tendrils need light in order to respond to stimulation. Tendrils stroked in the dark did not coil, but as soon as they were illuminated, even hours later, they began to spiral.

Review It

1. Make a list of the ways in which light affects plants. In how many of these do leaves play a part?
2. Describe several ways in which temperature affects the growth and response of plants.
3. Compare tropisms and nastic movements. What does osmosis have to do with nastic movements?
4. Give examples of different ways in which leaves function in reproduction.

Pioneers in Biology

Percy Lavon Julian:
Pioneer in chemical research with soybeans.

Fire can be hazardous at any time. But during wartime, when combustibles are everywhere, the threat of fire is ominous. What does fire have to do with a bean? One of the soya products that Percy Julian developed was used to extinguish oil and gasoline fires. Called "bean soup" by the Navy during World War II, it was credited with saving thousands of lives and many warships.

Percy Julian was born April 11, 1899, in Montgomery, Alabama. His grandfather had been a slave, later owning his own farm. Percy's father worked as a railway mail clerk and had strict ideas about education. Percy and his five brothers and sisters graduated from De Pauw University, Indiana, and all went on to receive advanced degrees.

At De Pauw, Julian majored in organic chemistry, receiving his B.A. in 1920. He was a member of Sigma Xi, the science research honor society. The nine years following graduation were spent in teaching and research. He taught first at Fisk University in Nashville, then studied biophysics at Harvard. He taught at West Virginia State College, and while there, in a sparsely equipped laboratory, he did research. Julian trained by repeating the experiments of the famous Viennese scientist Ernst Späth, who was an expert in building artificial compounds to

copy natural ones. In 1929, with help from a wealthy Harvard classmate, he got financial backing to go to Vienna and study with Späth himself. He was taken into Späth's home and treated as one of the family.

After receiving his Ph.D. from the University of Vienna, Dr. Julian returned to teach at Harvard, and later at De Pauw. After three years of painstaking research, Julian succeeded in synthesizing physostigmine, a chemical that scientists had been working on for over ten years. His synthetic drug, used to treat the eye disease glaucoma, became available to the public. Dr. Percy Julian's fame as a chemist was assured.

In 1936, Julian was called by the Glidden Company. They were looking for the best chemist to head their soya products division and develop new uses for the soybean. Julian became director of research at Glidden, a position he held for 17 years. He resigned in 1953 to form the pharmaceutical business, Julian Laboratories.

Julian developed over 100 products from the soybean, including the chemical fire extinguisher used by the Navy. Out of soya sterols he synthesized the female and male hormones, progesterone and testosterone, making them available at low cost; and amino acids for use in cases of infection and emergency therapy. But he is best known for his Compound S—synthetic cortisone.

Synthesis of cortisone made it available for the thousands who suffer from rheumatoid arthritis and certain muscular diseases. The natural hormone was available in such a limited supply that treatment had been impossible.

Throughout his career, Percy Julian retained a love of farming, and he gardened for relaxation. Late in his life, he cared for a garden that included 10,000 tulips.

Dr. Julian died of cancer at age 76. He had attained his goal of making life a little easier for those who came after him.

Chapter Summary

Leaves are very prominent organs of vascular plants. The structure of leaves varies so much that many different plants can be identified by their leaves without using other organs for identification. All leaves are alike in that they have vascular tissues for conducting water, minerals, and glucose. The structures of some leaves are modified to store water, or food, or even to catch insects. However, the internal structure of most leaves is arranged primarily to carry on photosynthesis. The internal layers, called mesophyll, are the photosynthetic layers. The epidermal cells do not have chloroplasts except in the guard cells that surround the openings called stomata. These openings, which are controlled by the guard cells, regulate the exchange of gases used in photosynthesis and respiration. The stomata also control water loss from the leaf. The amount of water lost from the leaf controls the water flow through the entire plant. The water movement up through plants is thought to be accomplished mainly by transpiration pull, cohesion and root pressure.

Many leaves are broad and flat in structure so as to better catch the light necessary for photosynthesis. Light reception by leaves not only affects the process of photosynthesis but also many other processes and structures of the plant. For example, plants grown in bright light tend to have thick leaves with each leaf having a smaller surface area. Also, the time of flowering and coloration of leaves in fall is controlled in many species by the light that strikes the leaves.

As leaves grow, they tend to move either toward or away from certain stimuli. This response is a tropism. Some leaves respond to stimuli in other ways. The response of *Mimosa* to touch is a good example of this type of movement. It is called a nastic movement.

Leaves are not usually thought of as reproductive organs, but in some cases plantlets form along the edges of leaves. These plantlets drop to the ground and form new plants. Also, leaves are sometimes used for propagation by cuttings. African violets are often propagated in this way.

Questions to Think About

1. What is the function of each of the following in photosynthesis: sunlight, water, chlorophyll, oxygen, carbon dioxide?
2. What are some of the factors involved in the rise of water in trees?
3. Some vascular plants do not carry on photosynthesis. Predict some differences in the leaves, stems, and roots of these plants as compared to photosynthetic vascular plants.
4. What would happen to the rate of water movement through a plant on a rainy day as compared to a hot, sunny day?
5. Suppose that you discover a previously unknown plant species. This plant seems to have a movement in response to wind. How could you determine if this movement is a tropism or a nastic movement?
6. Suppose that all animal life suddenly disappeared from Earth, but plant life remained. Would plants eventually use up all the carbon dioxide and die out? Why?

Extra Research

1. Go to a supermarket and make a list of foods that are actually plant leaves, stems, modified stems, and roots.
2. Use a microscope to compare cross sections of gymnosperm leaves and angiosperm leaves. What similarities and differences can you find?
3. Do a study of insect-catching plants. Find out about sundew, Venus's-flytrap, pitcher plants, and bladderwort.
4. Do some library research on transpiration. Then try to design a way to measure the rate of transpiration in a potted plant. Compare the transpiration rates under various conditions.
5. Do an experiment to determine the effects of different colors of light on growth of a plant species. One way to do this is to grow plants in cardboard boxes with light entering only through windows in the boxes. Cover the windows with different colored cellophane.
6. What might be some advantages of photoperiodism or vernalization to a plant species?
7. Do some library research to find out about the factors involved in leaf abscission.

Chapter Test

Part I

Copy the number of each test item. If the item is *true,* write *T* after the number; if the item is *false,* write *F.*

1. Evaporation of water from tree leaves is a factor helping move water up from the tree's roots. (333)
2. Translocation means the movement of a plant from one place to another. (343)
3. Capillary rise of water in narrow tubes is caused by atmospheric pressure. (343)
4. Root pressure is the tendency of liquids to move downward toward the roots of plants. (344)
5. Guttation is another name for transpiration. (345)
6. Transpiration is the process that moves carbon dioxide into a leaf. (344)
7. Monocot leaves have netted venation patterns. (335)
8. Mesophyll is a photosynthetic tissue. (336)
9. Cyclosis is the process of water movement through xylem tissue. (346)
10. Leaves have veins. (337)

Part II

Copy the number of each sentence below. After each number, write the word (or words) that completes the sentence correctly.

11. Leaves carry on a number of functions, but the one that is probably most important is ____. (333)
12. A leaf blade that is divided into two or more parts is called a ____ leaf blade. (335)
13. The opening and closing of stomata is controlled by cells called ____. (340)
14. The only epidermal cells that contain chloroplasts are the ____. (336)
15. The ____ layer of cells is the layer that carries on the greatest amount of photosynthesis. (336)
16. Plants grown in darkness are yellow or white and this condition is called ____. (347)
17. ____ is the response of plants to the length of the period of light received by plants. (348)
18. The ____ color in many fall leaves is due to the presence of anthocyanin. (349)
19. Plant movements which are caused by unequal growth because of a stimulus are called ____. (350)
20. Some leaves carry on ____ by forming plantlets along the edge of the leaf. (334)

Part III

Copy each numbered item from List A. After each item, write the letter and words from List B that best describe it.

List A

21. petiole (335)
22. nastic (350)
23. stoma (340)
24. guard cell (340)
25. tropism (350)

List B

a. opening in a leaf
b. irreversible change in a plant
c. regulates size of opening in a leaf
d. attaches a leaf to a branch
e. reversible change in a plant

Chapter 16

Flowers

Flowers provide us with a great deal of beauty. But flowers do not exist primarily to delight the eyes—and the noses—of people. They are a necessary part of the perpetuation of plant species, the reproductive organs of the flowering plants.

How are seeds formed in flowers? How are fruits developed? These and other questions will be considered as we explore reproduction in flowering plants.

Chapter Objectives

1. Describe the parts of a flower.
2. Explain the process of fertilization of flowering plants.
3. Compare the structures found in monocot seeds with those of dicot seeds.
4. List the factors necessary for seed germination.

An amaryllis flower always has six stamens. The other pronglike part seen among the stamens is the pistil.

16–1
The Flower

Did You Know?
The plants that produce grains, from which the bread and cereals you eat are made, are flowering plants. All of the grains belong to the grass family. Members of this family have tiny flowers that you can see with the help of a magnifying glass.

Wheat

Flowers are specialized sexual reproduction organs. Great variations in size, shape, and color can be observed in flowers. Some flowers, such as lilies and roses, are bright and showy; other are small and inconspicuous. Sexual reproduction allows new combinations of genes to be formed. Offspring have genetic combinations different from those of the parents. This allows for variation within a species and gradual adaptations to a changing environment.

Structure of Flowers

There is no such thing as a ''typical'' flower; variation in appearance is vast. But certain parts are common to most flowers.

Many flower parts are only accessories to the main job of seed production. The **sepals,** (sē′pəlz), which are modified leaves, form an outer ring, the **calyx** (kā′liks), around the flower bud. The calyx forms a protective outer coat for a flower bud while it is developing. As the flower opens, the calyx opens and the sepals bend back. In most flowers, the sepals are green. But some flowers, such as the tulip and certain lilies, have colored sepals in addition to colored petals. The thickened stalk at the base of the calyx is called the **receptacle** (ri sep′tə-kəl). The stem that attaches the flower to the main plant body is called the **pedicel** (ped′ə səl). The brightly colored **petals** are not essential for forming seeds. But petals can be a means of attracting insects or other animals to assist with pollination or seed dispersal. All of the petals of a flower together make up the **corolla** (kə rol′ə)

The essential seed-forming organs are found inside the protective ring formed by the petals. The **stamens** (stā′mənz) are the flower's male reproductive organs bearing **pollen** (pol′ən). A single stamen consists of a **filament** (fil′ə mənt) supporting an **anther** (an′thər) at its top. The anthers produce pollen.

The flower's female reproductive organ is called a **pistil** (pis′tl); it is often shaped like a vase and has a swollen base called an **ovary** (ō′vər ē). Seeds are formed in **ovules** (ō vyülz) inside the ovary. The upper part of the ovary narrows to a stalk called a **style.** The tip of the style, called the **stigma** (stig′mə), has a sticky surface. Study the table to help learn about the function and appearance of flower parts.

Challenge!
Many flowers are thought to be capable of movement. They are said to turn toward the sun, open and close corollas, and droop at night. Design a way to get evidence that flowers do move. You might consider photography as a possible research tool.

Catalog of Flower Parts

Flower part	How it looks	What it does
pedicel	green stem	supports flower
receptacle	green stem swelling	supports pistil, calyx, and petals
calyx	green or colored ring of leaflike sepals around petals	protects and supports flower bud
petals	brightly colored; leaflike	protect sex organs; attract insects or birds
stamens		
anther	knoblike tip	forms pollen
filament	thin stalk	supports anther
pollen	dustlike powder	contains sperm nuclei
pistil		
ovary	swollen base of pistil	contains ovules
ovules	cavities inside ovary	contains egg nuclei
style	stalk of the pistil	supports stigma
stigma	sticky top of pistil	traps pollen

Strawberry flower

Cross section of perfect flower

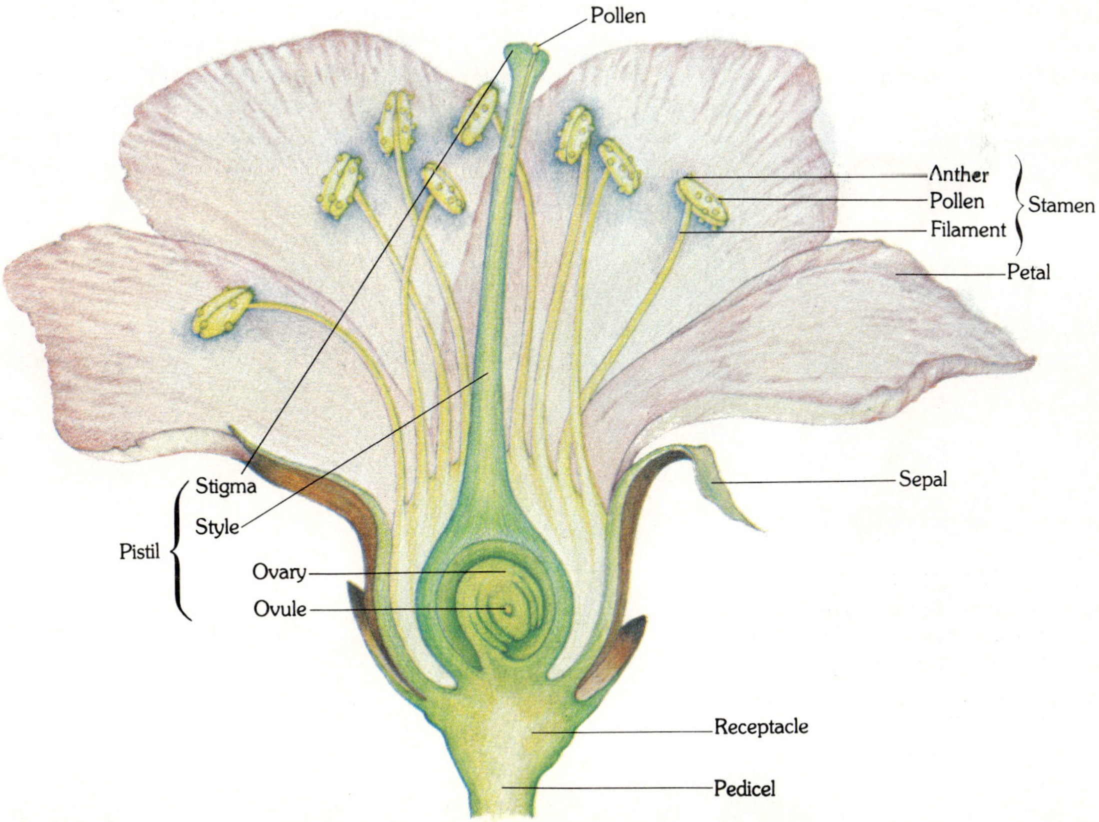

Kinds of Flowers

Everyone is aware that flowers vary tremendously in color, size, and shape. Other important variations occur in the reproductive parts. Flowers that have both pistils and stamens are called **perfect flowers** by botanists. Flowers that have stamens but lack a pistil are called **staminate** (stəm′ə nāt) **flowers.** Those having a pistil but no stamens are called **pistillate** (pis′tl āt) **flowers.** Some plant species have male plants with staminate flowers and female plants with pistillate flowers. Many plant species produce both staminate and pistillate flowers on the same plant body. Corn plants, walnut trees, and oak trees are of this type.

You may recall the two main groups of angiosperms, monocots and dicots. Do you remember the differences in their stem tissues and leaf venation? Their flowers also differ. Monocots usually have flower parts in threes or multiples of three. Dicot flower parts are usually in multiples of four or five.

Monocot Flowers

Dicot Flowers

Pollination

If a flowering plant is to form seeds and fruits, pollen grains must be transported from an anther to a stigma. This is called **pollination** (pol ə nā shən). Then a sperm nucleus from the pollen grain must fuse with the egg cell nucleus inside an ovule. This is called **fertilization** (fêr'tl ə zā'shən). Some species of flowering plants accomplish this by **self-pollination:** pollen is moved from the anthers to the stigma of the same flower or from the anthers of one flower to the stigma of other flowers of the same plant.

Self-pollination can take place in many ways. Wind or insects can simply dislodge pollen from the anthers of a flower and cause it to settle on the stigma of the same flower. Other plants, such as violets, have flowers that never open enough for pollen to leave. The pollen stays inside the closed flower and fertilizes the ovules of that flower.

Other species of flowering plants depend on **cross-pollination.** Pollen is moved from the anthers of one plant to the stigma of another flower of the same species. Cross-pollination is more common than self-pollination. Cross-pollination relies on a number of outside agents to move the pollen from one flower to another: wind, insects, water, bats, birds, and people can be pollinating agents.

Many flowers have a sweet nectar that attracts insects and birds. The pollen of some flowers is used as food by some wasps, bees, and moth larvae. In moving from one flower to another in search of nectar or pollen, these insects transport pollen from the anthers of one flower to the stigmas of other flowers of the same species. Insect-pollinated flowers are often brightly colored and sweet smelling. These characteristics may assist insects in locating them.

Many trees, such as birch, oak, and beech, produce clouds of pollen that drift in wind currents. Corn plants have exceedingly long stamens that grow high on the stalk. Pollen is carried by wind from staminate flowers to pistillate flowers that grow on other corn plants. Plants that are pollinated by wind have flowers that lack aroma and nectar. They often do not even have petals.

Some aquatic plants are pollinated when pollen is carried by water currents from one plant to another. These plants, like those pollinated by wind, produce abundant pollen. Their stigmas are long and feathery, thus increasing their chances of "capturing" drifting pollen.

People often intentionally pollinate plants, thus taking the place of natural pollinators. For instance, date palms are often artificially pollinated. Strands of staminate flowers are placed inside clusters of pistillate flowers. The clusters are then tied together. Cotton that has been impregnated by pollen may also be placed within the pistillate flower clusters.

Insect pollination

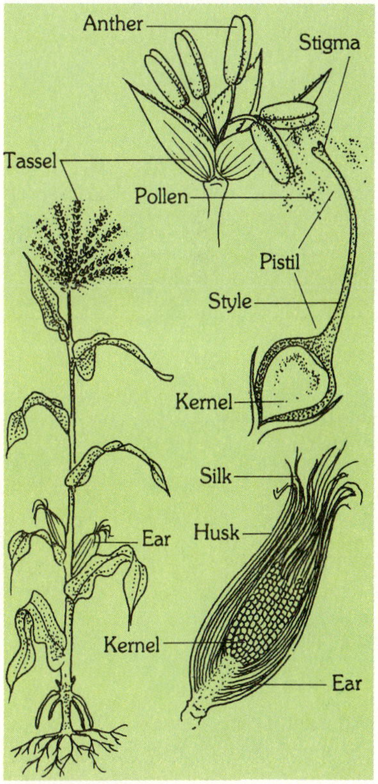

Corn plant

Challenge!
Collect and examine as many different kinds of pollen grains as you can. You can get pollen of wild flowers, garden flowers, flowers in florist shops. You can also collect pollen on slides that you lightly coat with vaseline and place outdoors on a windowsill or roof. You can examine the pollen grains with a microscope. Loose pollen can be placed in drops of water before being placed on the slide.

Pollen grains

Did You Know?
For many years, it was thought that bees used only the natural sugar of nectar from flowers as food. It is now known that bees depend on pollen as a source of protein.

Pollen Formation

Examine the cross section of an anther. Notice the four chambers inside the anther. These chambers are known as **pollen sacs.** These pollen sacs are full of cells called **microspore mother cells.** While the anther is growing, each microspore mother cell divides by *meiosis* (mī-ō'sis), forming a group of four cells called **microspores** (mī'krə spôrz). These microspores develop into pollen grains. A great number of pollen grains is formed in each sac. The pollen of many species of plants has its own characteristic shape.

The nucleus of each pollen grain then divides by *mitosis,* so that each grain contains two nuclei: a tube nucleus and a generative nucleus. The function of each nucleus is described below. Each nucleus has one-half the number of chromosomes found in root, stem, or leaf cells of the plant. This reduction in the number of chromosomes is accomplished by the meiotic division of microspore mother cells inside the pollen sacs.

Egg Cell Formation

Examine the cross section of an ovary. There may be one or several chambers inside an ovary. Each healthy ovule produces an egg cell. If fertilization occurs, this egg cell develops into an embryo and a seed is formed. The ovary itself develops into a fruit to enclose and protect the seed or seeds.

The ovule starts development as a tiny knob on the inside wall of the ovary. It begins to grow and, at the same time, develops a short stalk that holds it away from the ovary wall. Protective walls develop around the ovule until only one tiny opening to the inside of the ovule remains. This opening is called the **micropyle** (mī'krə pīl').

Inside the ovule, a highly important cell forms. This is the **megaspore** (meg'ə spôr) **mother cell.** This mother cell undergoes meiotic division to form four **haploid megaspores** (hap'loid meg'ə spôrz), each of which has only one half of the number of chromosomes found in other cells of the plant. Three of the four megaspores die; the remaining one enlarges and forms a cell called the **embryo sac** (em'brē-ō sak), the female gametophyte. As the embryo sac develops, the megaspore nucleus divides by **mitosis** (mī tō'sis) into two haploid nuclei. Each nucleus undergoes two more mitotic divisions, which result in a total of eight haploid nuclei in a single cell. One of these nuclei is an egg nucleus that takes a position near the micropyle to await fertilization. Two others form **polar nuclei** in the center of the embryo sac. After fertilization, the polar nuclei will form food storage tissue for the developing seed. Three of the eight haploid nuclei migrate to the end of the embryo sac opposite the micropyle and disintegrate. Two others lodge near the egg nucleus. Their function is not known. The egg nucleus is surrounded by a thin membrane and may now be called the egg cell.

All of the activities described above only prepare the plant for reproduction. When these activities have taken place, the egg cell nucleus is ready for fertilization by a sperm nucleus from the pollen grain.

Sexual Life Cycle of Flowering Plant: Gamete formation to Fertilization

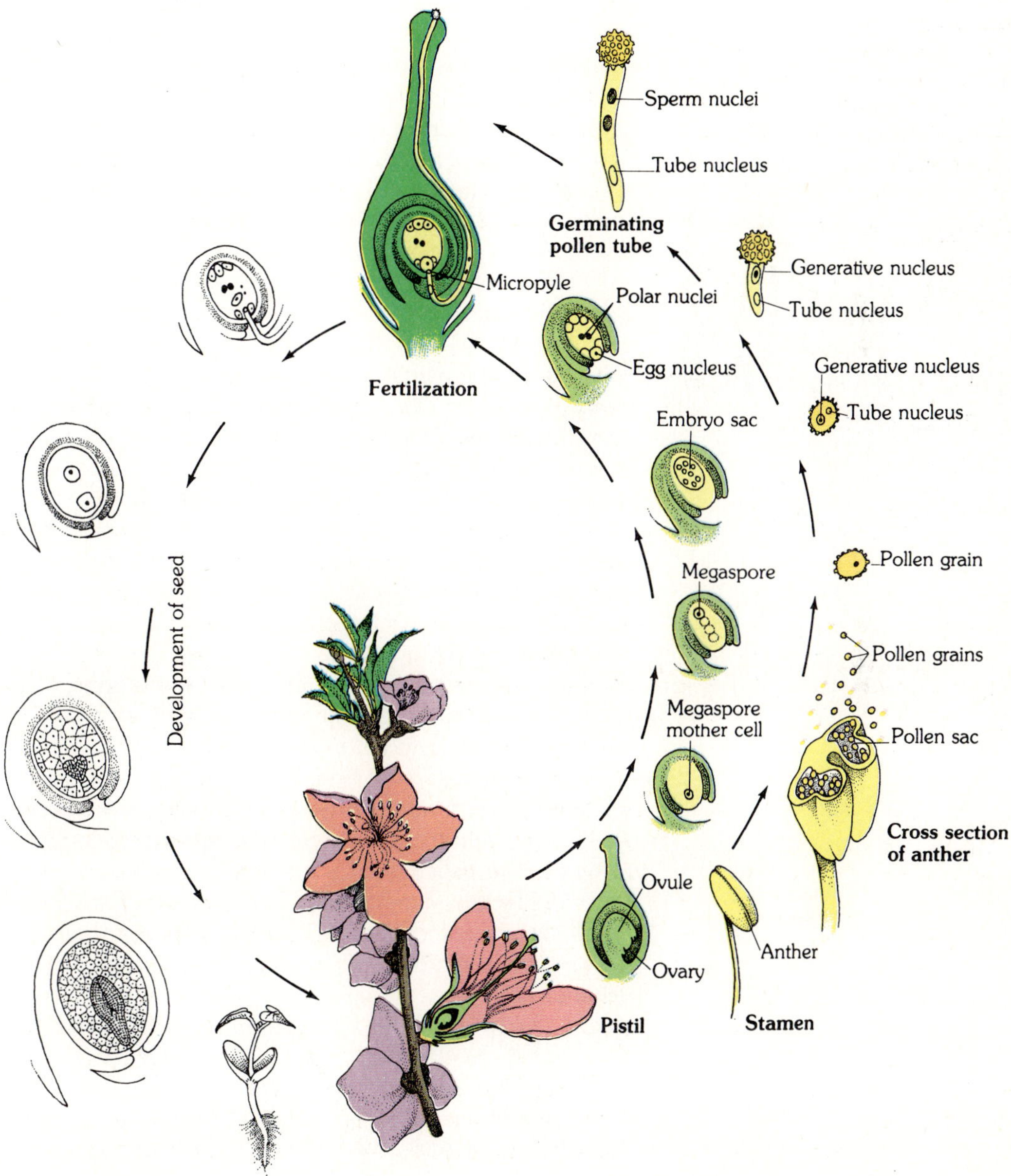

Fertilization

Did You Know?
As a pollen grain germinates and the pollen tube grows, active protoplasmic streaming can be seen in the growing end of the tube. Following behind the region of tube growth is a region filled with dense clusters of materials that probably serve to nourish tube growth. They diminish as the tube grows.

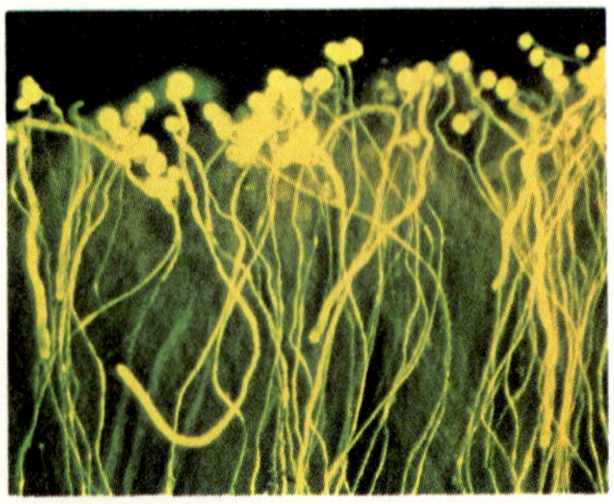

Pollen germination

Stigmas secrete a sticky fluid that becomes a trap for pollen grains. Any pollen grain that touches this substance is glued in place on the stigma. Certain substances in the sticky coating stimulate **germination** (jėr′mə nā′shən) of pollen grains.

When a pollen grain germinates, it produces a slender, thin-walled **pollen tube,** the male gametophyte. As the pollen tube grows, it pushes and digests its way through the soft tissues of the stigma, style, and ovary. The tube finally penetrates the *micropyle* of the ovule and enters the embryo sac.

As the pollen tube develops, the tube and **generative nuclei** (jen′ə rā′tiv nü′klē ī) of the pollen grain move down into it. The tube nucleus moves along with the growing tip of the tube and apparently is involved in the tube's growth. During its journey down the tube, the haploid generative nucleus divides mitotically into two sperm nuclei, which are also haploid. When the pollen tube penetrates the embryo sac, the end of the tube opens and the two sperm nuclei enter the sac. Each sperm nucleus has a special function. One unites with the egg to form a **zygote** (zī′gōt), the fertilized egg. The other unites with the two polar nuclei at the embryo sac's center. The combination of one sperm nucleus and two polar nuclei forms the **endosperm** (en′dō-spėrm) **nucleus.** This nucleus divides repeatedly to form the stored food tissue of the developing seed. This food tissue is called **endosperm.**

Flowering plants are the only organisms known that require fertilization by *two* sperm nuclei for successful reproduction. This process is known as **double** fertilization.

Alternation of Generations

Flowering plants exhibit **alternation of generations,** or stages in their life cycles, just as certain algae, mosses, and ferns do. The **sporophyte generation** (spôr′ə fīt jen′ə rā′shən) involves that part of the plant that we see: the stem, leaves, roots, and flowers of the plant. In the sporophyte generation, megaspores and microspores are produced. These spores grow into the **gametophyte generation** (gə mē′tō fīt jen′ə-rā′shən). The pollen grain is the male gametophyte, and the embryo sac is the female gametophyte. The male and female **gametes** (gam′ētz), or reproductive cells, unite to form a zygote that grows into the embryo in the seed. The embryo is the next young sporophyte generation.

Review It

1. Make a list of the various parts of a flower. What are the functions of each of the parts?
2. Describe the production of a pollen grain in an anther. What develops inside a pollen grain?
3. Describe the events that take place inside an ovule before the fertilization of an egg nucleus.
4. Describe the processes of pollination and fertilization in seed plants.

Sexual Life Cycle of Flowering Plant: Fertilization to Mature Sporophyte

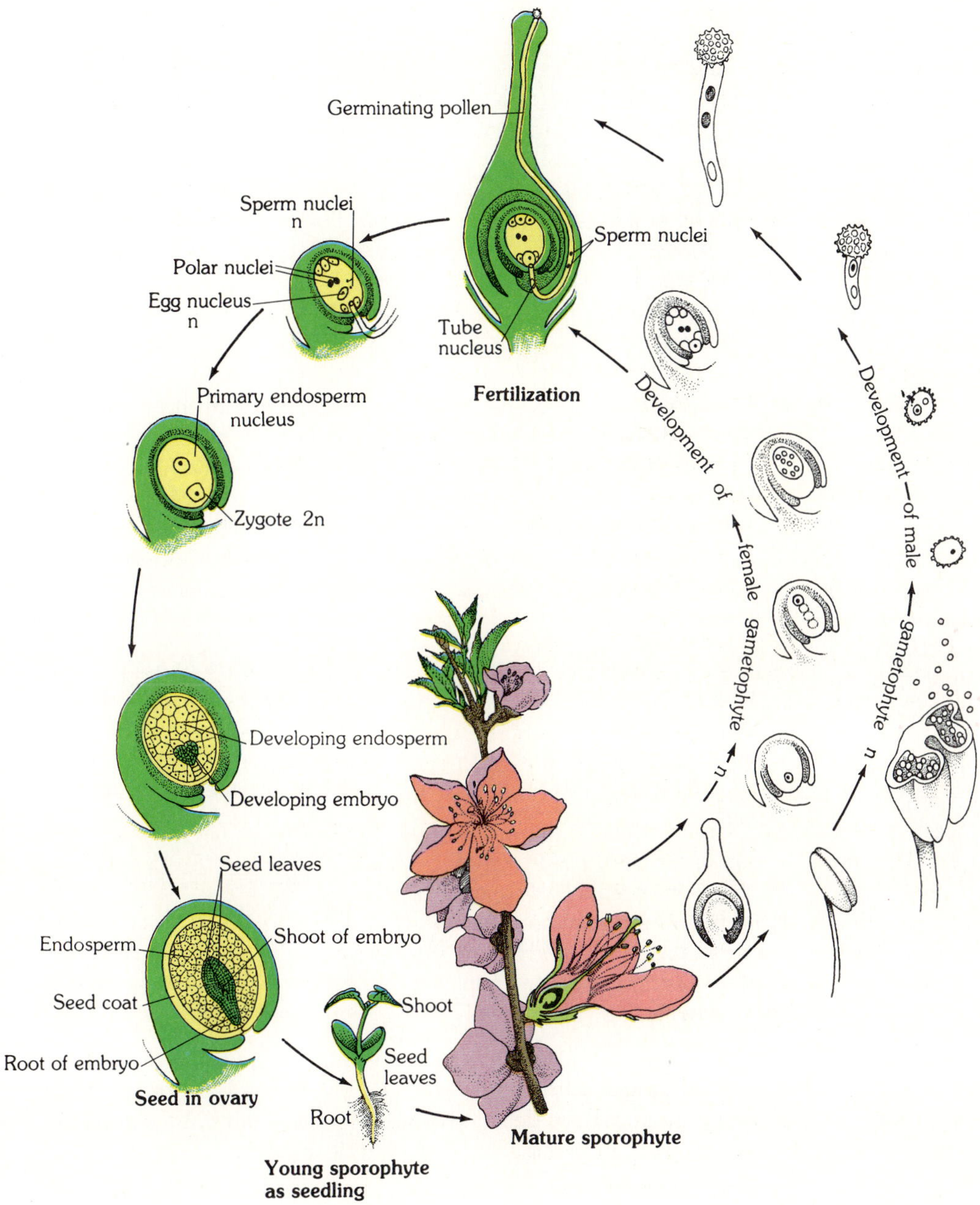

Issues in Biology

Soil resources and land use: how will people resolve the conflict?

For a long time the world population grew very slowly. High birth rates balanced by high death rates kept population figures under one billion until about 1830. About 200 years ago, this balance was altered as improved health practices lowered the death rates. It took only 100 years, from 1830 to 1930, to add the second billion people. In the next 30 years, the world population was three billion. The fourth billion arrived in 15 years. By 1985, an estimated five billion people will be on the earth.

The planet's soil resources have not increased. Land cleared of trees and bushes whose roots held soil firm is more vulnerable to erosion by water and wind. When erosion leaches valuable minerals out of the soil, it is no longer able to support food crops. The conversion of the Great Plains area in the U.S. from a grass land to a dust bowl in the 1930s is an example.

Increasing population growth affects our soil resources in the following ways:

1. Potential farmland is covered as more surface area is used for expanding cities, housing, waste disposal, and for energy and transportation system needs.
2. Good croplands become scarce and marginal areas such as tropical jungles are converted to permanent agriculture. Tropical soil, not chemically suited for crop farming, soon becomes infertile.
3. Pressures to produce large crops encourages agricultural practices that do not guard against hazards such as erosion. The end result is lower yield.

But Dr. William H. Brown of the Hudson Institute's *Prospects for Mankind* program contends that the ability to produce food will far outstrip the earth's population growth. He gives the following reasons for his belief.

1. Modern irrigation methods and replacement of depleted soil minerals by fertilization can restore depleted land.
2. By the end of this century, production of food in growth chambers—greenhouses with a totally controlled environment—could increase yields by 100 times those of conventional agricultural methods.
3. In countries with a scarcity of food, the scarcity is due to problems of planning and management practices rather than to a lack of resources.

Analysis

1. Should steps be taken to protect soil resources?
2. Why can people be fed more economically on grain than on meat?
3. What were some problems of the Aswan Dam project for irrigation and flood control in SE Egypt?

16–2
Fruits and Seeds

When a sperm nucleus fuses with an egg nucleus in a flower, a number of changes are triggered. As one or more seeds begin to develop, hormones are produced. These hormones cause the petals and stamens of the flower to wither. However, the ovary begins to enlarge.

Challenge!
How are seedless fruits produced? Do some library research to find out. Try horticultural magazines or books as a source of information.

Fruits

The walls of the ovary develop to form a fruit. In some plants, the receptacle or calyx fuses with the ovary and becomes part of the fruit.

A fruit, then, is an enlarged and ripened ovary. Depending on the type of plant, the walls of the mature ovary may be soft, juicy, and fleshy or dry and hard. The familiar tomato, peach, and apple have fleshy outer walls. Corn grains, acorns, and bean pods are also fruits. The seeds within a fruit are matured, fertilized ovules.

Seeds

After fertilization, the ovule changes rapidly. It greatly increases in size due to growth of its embryo and endosperm. The embryos of some seeds resemble miniature plants; others appear simply as lumpy tissue.

A seed develops one or two **cotyledons** (kot′l ēd′nz). Cotyledons are "seed leaves," which often serve as reservoirs for stored food. In some seeds, for example, bean seeds, the embryo absorbs the food stored in the endosperm before the seed matures. In these cases, the cotyledons hold the food reserve and form the bulk of the seeds. In other seeds—a grain of corn, for example—the endosperm is not absorbed and the cotyledon remains small. Endosperm forms the bulk of the corn seed.

In addition to one or two cotyledons, a seed embryo consists of an **epicotyl** (ep′ə kot′l), a **hypocotyl** (hī′pə kot′l), and a **radicle** (rad′ə kəl). The epicotyl is the embryonic shoot above the point of attachment to a cotyledon. The hypocotyl is the embryonic stem located below the cotyledon attachment. The radicle is the embryonic root.

As an ovule matures into a seed, the walls of the ovule are transformed into the **seed coat,** which protects the seed. Seed coats usually are tough and water resistant. They protect the seed against injury and dehydration.

A monocot seed: the corn seed

A grain of corn is a good example of a monocot seed. If you examine it, you will be able to observe all of the monocot seed characteristics. Look first at the outside surface of the grain. On one side you will see an oval shape that is lighter in color than the rest of the seed. This is the embryo. You are looking at it through the transparent seed coat. At one end of the grain, below the embryo, you can find the point where the seed was attached to the cob. On the opposite end, above the embryo, you will find a small indentation. This is the **silk scar,** where the style was once attached.

The seed coat is easily identified. The white tissue filling more than half the seed is endosperm. In sweet corn it contains starch and sugar. "Field" corn, usually fed to livestock, stores only starch in its endosperm and so does not have a sweet flavor.

Corn seed: a complete fruit

The lower part of the corn seed contains tissue that is the embryo. The largest, uppermost part is the cotyledon. Just below the cotyledon is a rodlike shape, which makes up the epicotyl and hypocotyl. As a corn seed develops, the epicotyl forms the shoot, and the hypocotyl develops into the lower part of the stem. Below the hypocotyl is the embryonic root tissue, called the radicle.

Each corn seed is actually a complete fruit. The outer covering of the grain is made up of the tough ovary wall fused to the seed coat. The outer wall corresponds to the fleshy part of a peach or the entire pod from a bean plant.

A dicot seed: the bean seed

Beans are dicot seeds. Examination of the outside of a bean seed will reveal many structures that are different from those of a grain of corn. The kidney-shaped bean has a very smooth outer coat called a **testa** (tes'tə). The concave side of the bean has a light-colored mark called the **hilum** (hī'ləm) where the bean was attached to the pod. Near the hilum is a tiny indentation, the micropyle. This is where the pollen tube penetrated the ovule during fertilization.

A bean seed, soaked overnight, can easily be separated into two parts. When separated, the seed will look like the one shown in the illustration.

The two "halves" of the seed are the cotyledons. They are quite fleshy and are different in appearance from the single cotyledon of the corn grain. The rest of the embryo is sandwiched between the two cotyledons. The upper epicotyl is easily identified by the tiny leaves. The hypocotyl, or lower stem, and radicle, or root can be seen attached to the epicotyl.

The stored food of the bean is concentrated in the cotyledons. Endosperm was absorbed during the growth of the embryo.

When a bean seed germinates, an *apical bud* starts to grow between the leaves of the epicotyl. This develops into the terminal bud of the shoot. The cotyledons provide food for initial growth. After the plant has established photosynthesizing leaves, the cotyledons wither and drop off the seedling.

Comparison of Representative Monocot and Dicot Seeds

Corn seed (monocot)	Bean seed (dicot)
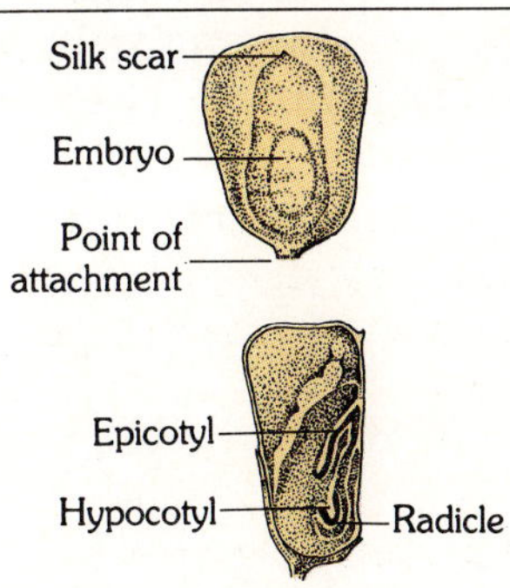	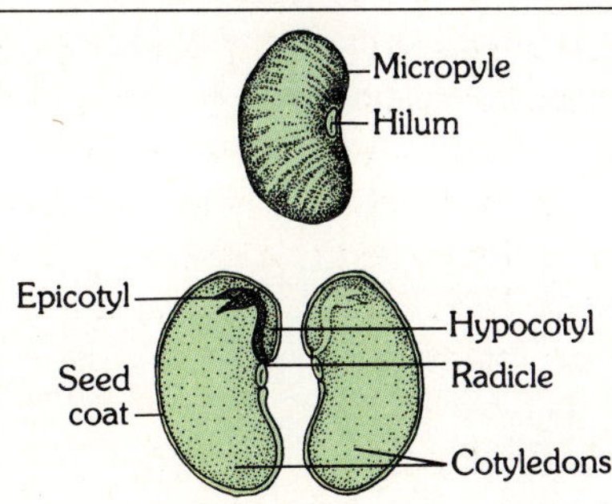
One cotyledon	Two cotyledons
large endosperm at maturity	no endosperm at maturity
small, round epicotyl with rolled leaves	large epicotyl with two folded leaves
outer cover is fruit coat; seed coat lies underneath	outer cover is seed coat; pod is the fruit coat
external silk scar shows where style was attached	external hilum and micropyle visible

Did You Know?

Some aquatic plants produce seeds with very hard seed coats that are completely waterproof. A valve in the hilum of these seeds closes when there is moisture outside the seed, and opens when it is dry. When the valve is open, air can enter. As a result, these seeds get drier and drier, and their dormancy deepens. These seeds never germinate unless the seed coat is mechanically ruptured so that water may enter.

Seed Dormancy

After dispersal from their parent plants, many seeds undergo a period of dormancy. This resting period may last for a few weeks, for a winter, or for many years. Some seeds will break dormancy only after **vernalization** (vėr′nl ī zā′shən), or exposure to a period of cold temperatures. Other seeds will germinate only after exposure to specific periods of light and darkness.

Annual plants, which live for one season only, depend entirely on their dormant seeds for survival from one year to the next.

Germination

The awakening and sprouting of a seed is called *germination.* Germination will not take place unless certain conditions are present.

1. Many seeds require periods of dormancy from several weeks to several months before germination will take place. Other seeds do not require a dormant period.
2. If germination inhibitors are present, they must be overcome. The inhibitors must be inactivated by special weather conditions or by being washed or dissolved away.
3. Resistance of the seed coat to penetration by water and/or oxygen must be overcome. The seed coat may only require soaking in water, or it may have to be cracked or broken. In some seeds the seed coat is eroded away by such abrasive materials as sand.
4. Sufficient water must be available.
5. Sufficient oxygen must be available.
6. The temperature must be within certain favorable limits.

The first phase of germination involves the absorption of sufficient water by the seed. The enzymes of the seed embryo cannot begin to operate until sufficient water is available. These enzymes start to change the stored food of the seed into chemical forms that the embryo can use for growth. During this water absorption phase, most seeds swell, and their seed coats soften.

Bean seedling development

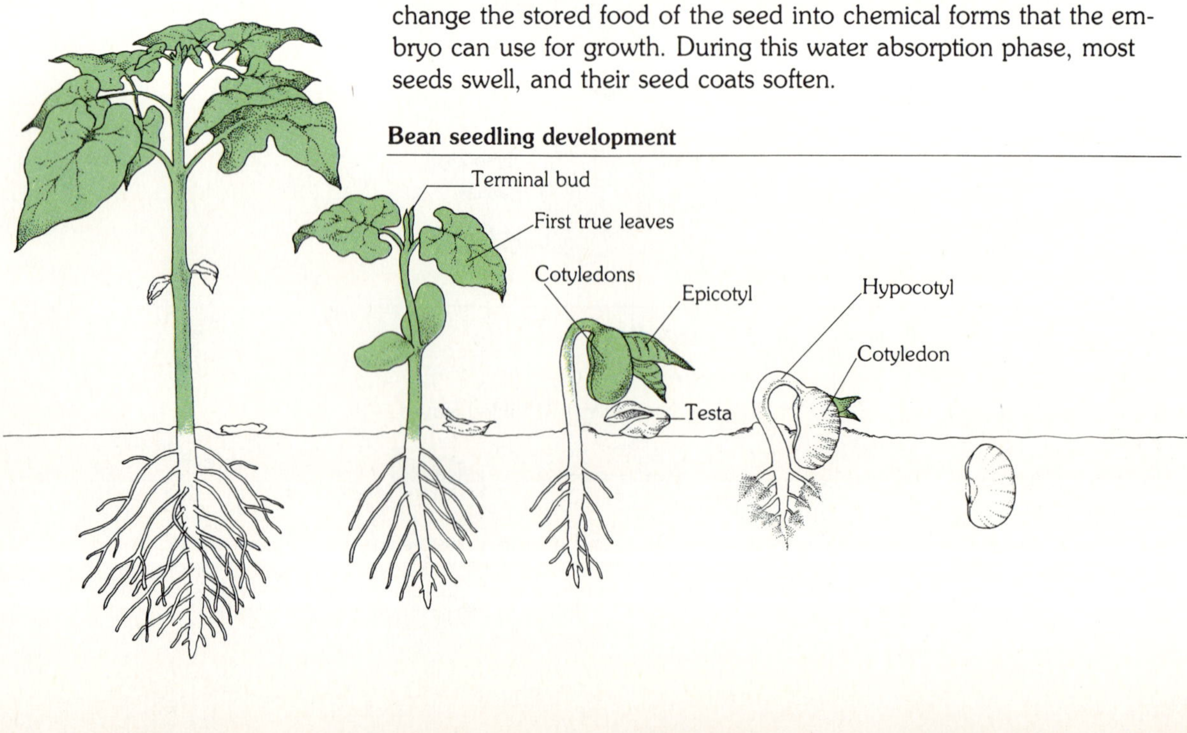

The Role of Water and Oxygen in Early Growth

When enough water is present inside a seed, rapid cell division begins in the embryo. A supply of oxygen is also required because respiration of the developing cells is an oxygen-consuming process. For this reason, seeds germinate best in well-aerated soil. They usually will not germinate if they are entirely submerged in water, because the available oxygen is insufficient to support respiration.

Many *enzymes* are involved in germination. An important enzyme is **amylase** (am′ə lās), which converts stored starch into usable sugar. Enzymes can operate only within a narrow range of temperatures. That is why certain temperatures are required for germination of seeds.

To germinate, a bean seed must absorb water. The seed swells, and a split develops in the seed coat at the micropyle. The radicle grows downward through the split seed coat to become the primary root. Fine root hairs develop near the growing tip of the new root.

The hypocotyl lengthens and arches, lifting the seed out of the soil. The seed coat drops off. The cotyledons separate, and two small leaves appear between them. As these leaves develop, a terminal bud emerges between them to produce the main stem. The cotyledons shrivel as their stored food supply is used by the seedling.

Water absorption in a corn seed is through the point where it was attached to the cob. The radicle grows and breaks through the water-softened seed coat. It grows downward to form the primary root. The epicotyl breaks through the seed coat and grows upward. The leaves of the epicotyl remain tightly rolled in the seed's single **coleoptile** (kō′lē op′təl), a protective sheath enclosing the leaves. After the coleoptile breaks the soil's surface, the leaves unroll and grow upward. A stem grows from between these leaves to form the cornstalk. Unlike the bean seed, the corn seed remains underground. Nourishment for the seedling comes primarily from the endosperm. A cotyledon absorbs the food from the endosperm and transfers it to the growing part of the embryo.

Corn seedling development

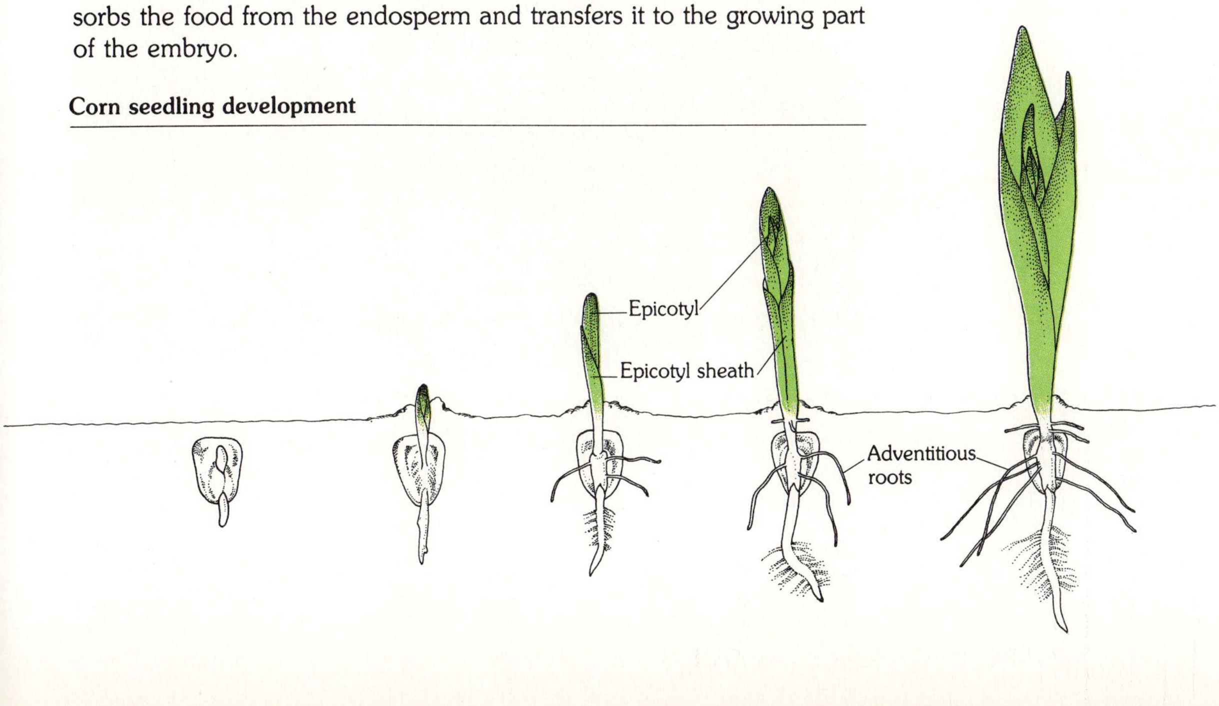

Jewelweed

Burdock adhering to fur

Seed Dispersal

If all seeds were simply to drop to the ground and germinate immediately below their parent plants, difficulties would arise. Developing seedlings would have to compete with the parent plant and one another for light, water, and minerals. Few would survive; those that did would probably be undernourished.

It is important for seeds to be dispersed or carried away from parent plants to new growth sites. A number of interesting adaptations ensure dispersal by a variety of agents. These dispersal agents include mechanical propulsion, wind, water, and animals.

Mechanical propulsion—Some plants have pods that "explode" when mature, causing seeds to be thrown some distance away from the parent plant. Jewelweed distributes its seeds in this way.

Water—Many seeds float and can be carried over long distances by water. The water iris, for instance, has seeds enclosed by a waterproof, waxy coating. They are carried for long distances in streams before the waxy coating wears away so that water can penetrate the seed and trigger germination.

Animals—Many fruits are equipped with barbs so that they cling to the fur of animals. As the animals travel through bushes and grass, the seeds are brushed off. Other fruits, such as cherries, berries, and tomatoes are attractive to animals as food. Animals eat the fruit, often swallowing the seeds. Undigested seeds are deposited along with the animals' excreta in places far from the parent plant. Birds and bats are particularly active in this form of seed dispersal.

Wind—Wind disperses the seeds of a great number of plants. Dandelions and milkweed have fluffy, parachute-like structures attached to their seeds. Maple and ash trees produce winged seeds that have been given the name *samara.*

Dandelion flower and seed head

Dandelion seed

Review It

1. How do botanists and nonbotanists differ in their definition of a fruit?
2. Compare the structures of bean and corn seeds.
3. Compare the development of bean and corn seedlings.
4. Describe four different means of seed dispersal.

Activity

Light and Seed Germination

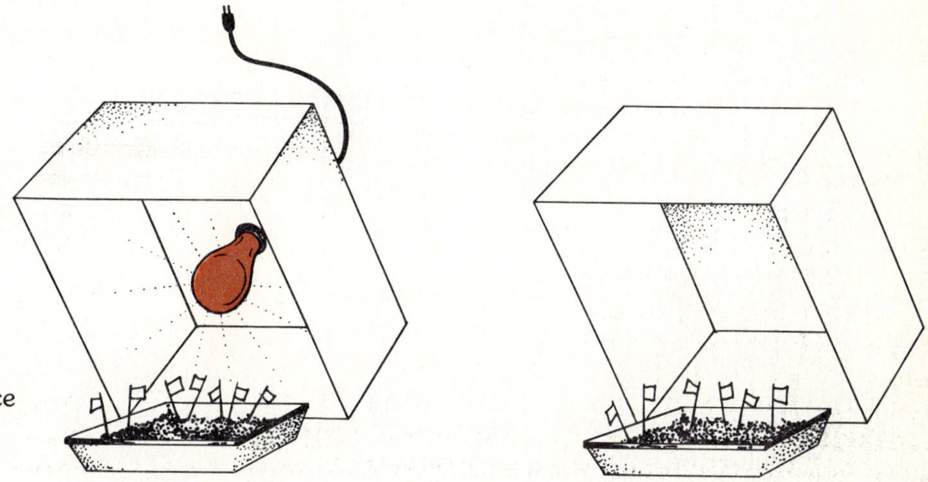

Materials

Two planting trays
Potting soil or vermiculite
Plant labels
Darkroom safety light
Two boxes to cover trays
Corn and bean seeds

Procedure

1. Fill each planting tray with soil.
2. Scatter seeds over the surface of the soil in each tray.
3. Scatter different kinds of seeds in different locations. Label the location of each kind of seed. Use a piece of paper attached to a stick.
4. After the seeds are planted, cover each tray with a light-tight box.
5. Water the seeds often enough so that they will remain moist.
6. Keep one tray covered by the light-tight box all of the time except when watering the seeds.
7. Remove the light-tight box from the other tray to expose the seeds to red light once a day for about one minute. Be careful not to use the red light when watering seeds.
8. Record your results.

Analysis

1. How many seeds germinated in the tray that was exposed to red light?
2. How many seeds germinated in the tray that was *not* exposed to red light?
3. What conclusions can you draw from your results?

Challenge!

Do an investigation to find out if water enters bean seeds through the hilum and/or the micropyle. Place dabs of waterproof glue over the micropyle of some seeds, over the hilum of others, and over both the micropyle and hilum of a third group of seeds. As a control, have one group on which no glue has been placed. Place all of the seeds on damp paper towels and observe germination or the lack of germination.

Chapter Summary

The reproduction of the angiosperm plants is carried on primarily by special organs called flowers. The flower parts, which are thought to be modified leaves, are highly varied in structure. The flower is part of a sporophyte generation and both megaspores and microspores are produced within flower parts. Megaspores develop into gametophyte embryo sacs, and microspores develop into gametophyte pollen grains.

Pollen develops in flowers in structures called anthers. Sperm nuclei from pollen fertilize egg nuclei developed in floral structures called ovules. Mature fertilized ovules are commonly called seeds. Fruits are formed from ovaries that surround ovules.

There are two types of seeds—those from monocotyledonous plants and those from dicotyledons. These two seed types have many features in common, but they differ in means of food storage and in number of cotyledons. Monocot and dicot embryos have different growth patterns.

Questions to Think About

1. What advantages might a period of dormancy have for seeds? In what climate would you expect seeds to have dormant periods?
2. If a species of plant could reproduce *only* vegetatively, how might this affect the future of the species?
3. Certain chemicals, such as ethylene gas or thiourea, have been found to be effective in breaking dormancy of some seeds. Try to propose some hypotheses to explain how these chemicals may work to overcome dormancy.
4. What might the advantage be to a species to have cross-pollination rather than self-pollination?

Extra Research

1. Find as many examples of wild fruits as you can. How are the fruits related to methods of seed dispersal?
2. Plants and their parts are often used as inspiration for artwork and decorative designs. Find as many examples of this kind of art as you can. Look both at present-day designs and designs of earlier societies.
3. Dig up some clover plants, roots and all. Examine the roots for swellings or nodules. Find out what the nodules are. How do they function in the life of the plant?
4. Try to stimulate pollen grains to germinate and to form pollen tubes. Dampen paper towels with various concentrations of sugar solutions. Then place the pollen on the towels in petri dishes or glass jars. You will need to use a microscope to see the pollen tubes, if any form.
5. Collect some garden flowers, or ask a florist if you can have some wilted or discarded flowers. Dissect the flowers, and try to identify their anatomical structures.
6. What features of various flowers are probably adaptations for attracting pollinating insects?
7. In what respects is an apple like an acorn? Explain.
8. List as many ways as you can to *prevent* seed germination.

Chapter Test

Part I

Copy the number of each test item. If the item is true, write *T* after the number; if the item is false, write *F*.

1. A pedicel is another name for a flower petal. (357)
2. A calyx is a structure formed from many sepals. (356)
3. Flowers that produce only male sex cells are called perfect flowers. (358)
4. Pollen grains always contain a single nucleus. (360)
5. Polar nuclei form pollen grains. (360)
6. Corn seeds have no cotyledons. (366)
7. Another name for a seed coat is *testa*. (367)
8. A samara is a structure used by some trees for dispersal of pollen. (370)
9. Seeds can germinate without oxygen if sufficient water is available. (368)
10. Some seeds are prevented from germinating by chemical growth inhibitors. (368)

Part II.

Copy the number of each sentence below. After each number, write the word or words that complete the sentence correctly.

11. Nourishment for development of a corn seedling comes from _____. (369)
12. The upper stem of a seedling is called a(n) _____. (366)
13. Factors necessary for germination of seeds generally include _____. (368)
14. Treatment by cold temperatures to break seed dormancy is called _____. (368)
15. A micropyle can be observed on a _____ seed. (360)
16. Fruits are actually ripened _____. (365)
17. Fertilization in angiosperm flowers is called _____ fertilization. (362)
18. The part of a flower that traps pollen is called the _____. (357)
19. Brightly colored, sweet-smelling flowers are frequently pollinated by _____. (356)
20. A megaspore mother cell would most likely be found in the _____ of a flower. (360)

Part III

Copy each numbered item from List A. After each item, write the letter and words from List B that best describe it.

List A

21. germination (362)
22. micropyle (360)
23. radicle (366)
24. vernalization (368)
25. pollination (359)

List B

a. transfer of pollen from anther to stigma
b. exposure to a period of cold temperatures
c. sprouting of a seed
d. embryonic root
e. opening through which the pollen tube penetrates the ovule

Focus on Careers

Nursery Owner/Landscape Architect

"I enjoy being creative—working out designs, seeing them executed, watching plants grow. Working all day behind a desk is not for me. I love the outdoors and this job gives me the chance to be there."

Ronald is part owner and manager of a nursery in California. He designs planting arrangements for home patios and yards as well as for industrial sites. This can include planting lawns, shrubs, trees, and flowers. He uses both hand and power-operated equipment such as mowers, trimmers, and sprayers. His management responsibilities at the nursery include placing orders for stock, supervising employees, and attending sales shows around the country to learn about new ideas and products.

Landscaping requires an ability to work well with customers. The customer usually provides a description of the area to be landscaped and information about drainage, amount of sunlight available, and how the area is to be used. The landscape architect then makes a detailed plan and presents it to the customer for approval. The landscape architect then sees that the plan is carried out properly.

While a person with a farm background might have one basic qualification for the job—experience with the outdoors—

Ronald did not grow up on a farm. But his love for the outdoors led him to enter college with a forestry career in mind. He became interested in horticulture, especially as it relates to landscaping. (Horticulturists grow plants for beauty and environmental uses.)

During summers and after college Ronald worked as a landscape laborer with a local nursery. A few years later, he decided to start his own business.

A four-year course in horticulture with landscaping emphasis is helpful in getting work as a landscaper. Some junior colleges and vocational-technical schools have programs that prepare a person for certain aspects of the landscape trade. However, a self-employed landscaper is both a designer and a businessperson and therefore needs courses in communication, economics, business, and bookkeeping.

At present, opportunities in landscaping are good. People want attractive surroundings full of living trees, bushes, and plants. And this provides opportunities for the landscape architect and nursery owner.

Career Information

Floral Designer: Society of American Florists and Ornamental Horticulturists, 901 N. Washington Street, Alexandria, VA 22314.

Landscape Architect: American Society of Landscape Architecture, Inc., 1750 Old Meadow Road, McLean, VA 22101.

Produce Clerk: National Association of Retail Grocers, 360 N. Michigan Avenue, Chicago, IL 60694.

Tree Surgeon: Institute of Agricultural Technology, Michigan State University, East Lansing, MI 48823.

Floral Designer

Duties: Assembles cut flowers and greenery into arrangements to be sold as gifts and decorations. This requires knowledge of the physical properties of flowers, including which flowers remain fresh when cut and which can be successfully dyed different colors. A floral designer must have a good eye for colors and a sense of proportion and design.

Training: Requires at least six months of on-the-job training in a floral shop. Trainee begins by copying simple arrangements, preparing flowers for storage, and delivering flowers. Courses in floral design and management are offered at junior colleges. Courses vary in length from six weeks to two years.

Opportunities: Job opportunities for floral designers are expected to increase for the foreseeable future.

Produce Clerk

Duties: Maintains a supply of healthy-looking fruits and vegetables in a grocery store. The clerk knows how to display produce attractively, how to date and rotate produce to reduce spoilage, and how to keep greens crisp.

Training: Requires a high-school diploma and six months of on-the-job training.

Opportunities: With experience and ability, a clerk can advance to produce department manager.

Tree Surgeon

Duties: Prunes and treats trees and shrubs to improve appearance and health. For example, the surgeon tries to prevent deterioration by scraping decayed matter from holes in trees and then filling them with cement. He or she may apply fertilizers using hand or machine sprayers. A tree surgeon must be healthy and willing to work irregular hours and in all kinds of weather. Work is physically demanding.

Training: Requires two years of study in an arboriculture technician program.

Opportunities: Homeowners, local governments, businesses, and utility companies hire tree surgeons. Work-related experience is helpful in getting a job.

Bibliography: Unit Six

Anderson, A. W. *How We Got Our Flowers.* Dover Publications, Inc., 1966. Fascinating history of common flowers of America.

Bold, Harold C. *The Plant Kingdom,* 3rd ed. Prentice-Hall, Inc., 1970. Describes structure, function, and reproduction of plants.

Devlin, Robert M. *Plant Physiology.* Reinhold Publishing Corp., 1966. Includes recent findings in plant physiology.

DeWit, H. C. D. *Plants of the World.* E. P. Dutton and Co., Inc., 1967. Illustrated volume of representative species of plants.

Dowden, Anne. *Look at a Flower.* Thomas Crowell Co., 1963. Sketches and descriptions of plant structures, seed development, and techniques of pollination.

Duddington, C. L. *Evolution and Design in the Plant Kingdom.* Thomas Crowell Co., 1969. Describes adaptations of plants and their means of survival.

Edlin, Herbert. *Atlas of Plant Life.* John Day Co., 1973. Presents a review of wild and cultivated plants around the world.

Emboden, William A. *Bizarre Plants.* Macmillan Publishing Co., 1974. Interesting account of superstitions and legends of plants that are monstrous, mythical, and magical.

Everett, Thomas H. *Living Trees of the World.* Doubleday and Co., Inc. (no date). Beautifully illustrated volume describing characteristics, distribution, and origin of a wide variety of trees. Describes their many uses too.

Frisch, Rose E. *Plants That Feed the World.* D. Van Nostrand Co., 1967. Informative account of the flowering plants that feed the world.

Hutchins, Ross E. *Plants Without Leaves.* Dodd, Mead, and Co., 1966. Description of uses and odd ways of survival of leafless plants.

Kavaler, Lucy. *The Wonders of Algae.* The John Day Co., 1961. Discusses the importance of algae as a food source.

Milne, Lorus Johnson. *Because of a Flower.* Atheneum Press, 1975. Discusses various plants that provide centers for small communities of animal life attracted by their flowers, fruits, and seeds.

Northen, Henry, and Northen, Rebecca. *Ingenious Kingdom.* Prentice-Hall, Inc., 1970. Examines the reasons behind longevity of individual plants and other secrets of the plant world.

Perl, Philip, and others. *Ferns.* Time Inc., 1977. Lavish illustrations and information about proper care for ferns.

Poling, James. *Leaves: Their Amazing Lives and Strange Behavior.* Holt, Rinehart, and Winston, 1971. A study of structure and behavior of leaves, their evolution and specialization.

Riedman, Sarah. *Trees Alive.* Shepard Co., 1974. Useful reference for structure and function of tree parts. Also poses the problem of depletion of forest resources by humans.

Tribe, Ian. *The Plant Kingdom.* Grosset and Dunlap, 1970. Illustrates and describes characteristics of all major plant groups.

Wendt, Frits, and the Editors of *Life Magazine. The Plants.* Time Inc., 1963. Beautiful picture-essay book covering interesting aspects of plants.

Vallin, Jean. *The Plant World.* Sterling Publishing Co., Inc., 1967. Interesting discussions of common plants, their structures, pollination, propagation, fruits, and seeds.

Wohlrabe, Raymond A. *Exploring the World of Leaves.* Thomas Crowell Co., 1976. Discusses types of leaves, why leaves turn color, physiology of flowering plants, gymnosperms, and ferns.

Articles

Galston, Arthur W. "The Language of the Leaves." *Natural History,* January, 1973.

Goldstein, Nancy. "Potassium—Plant Traffic Cop." *Science World,* December 1, 1969.

Spencer, Patricia W. "The Turning of the Leaves." *Natural History,* October, 1973.

Zahl, Paul A. "Where Would We Be Without Algae?" *National Geographic,* March, 1974.

UNIT SEVEN

THE INVERTEBRATES

Chapter 17

Sponges to Starfish

The animals you will find out about in this chapter are *invertebrates,* or animals without backbones. You can find almost all of these animals living near the edge of the sea. However, you may also find some in your backyard, or in nearby rivers, streams, ponds, and lakes. And, although invertebrates are not easily exhibited, you may also see some of them in a zoo or an aquarium. But regardless of where they are seen, invertebrates always attract a great deal of attention because of the variety of their shapes, colors, and life styles. You will have a chance to observe many such traits as you study this chapter.

Objectives

1. Differentiate between invertebrates and vertebrates.
2. List the general characteristics of the phylum Porifera and explain how sponges are classified.
3. Compare the feeding behavior of the coelenterates to that of the sponges.
4. Describe the different ways a flatworm reproduces.
5. Explain the differences between animals with bilateral symmetry and those which have radial symmetry.
6. Differentiate between parasitic invertebrates and free-living invertebrates and give two examples of each.
7. Describe the various systems of a segmented worm.
8. List ways that mollusks are important to humans.
9. Describe the water-vascular system of an echinoderm.
10. For each phylum, describe and identify general characteristics, body plan, activities, and reproduction.

This piling in the ocean off Massachusetts supports many types of sea life, as well as a pier. Some of the invertebrates seen here are anemones, starfish, sponges, and barnacles.

17-1
The Invertebrate Phyla

Organisms of the animal kingdom are far different from organisms in the plant, fungi, protist, or moneran kingdoms. Unlike the one-celled protists and monerans, animals are made up of many cells. They can be described as *multicellular.* Animals also have a definite size and shape. And they have far different feeding habits than the fungi and the plants. Animals do not absorb food as do fungi. And they cannot make their own food as do plants. Instead, they are *heterotrophs.* Animals depend—either directly or indirectly—on plants for food. And animals very often must move around from one place to another to find food.

In this unit and the next, you will study the animal kingdom. This unit is devoted to **invertebrates,** or animals without backbones. And the next unit is devoted to **vertebrates,** or animals with backbones. Long ago, the presence or absence of a backbone was considered an important trait and was used as a basis for classifying animals. Today that trait is no longer considered so important. Nevertheless, the terms *vertebrate* and *invertebrate* are convenient and still commonly used by biologists.

In this chapter you will study seven phyla of invertebrate animals. All seven phyla were chosen for a reason. Some were chosen because they contain members that are familiar to you. Others were chosen because they are useful in illustrating biological principles. Still others were chosen because they are important to humans. You can see representative members of each phylum below and on the next page.

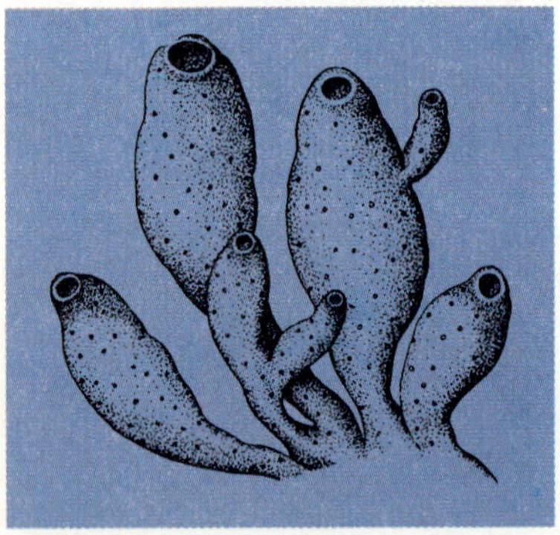

Sponges
Members of the **phylum Porifera** (pə rif'ər ə) are very simple animals. Nevertheless, sponges are quite well suited to their environment. They have been in existence for hundreds of millions of years.

Coelenterates
A representative member of the **phylum Coelenterata** (si len'tə-rā'tə) is the *Hydra.* A *Hydra* looks something like a piece of string with a frayed end.

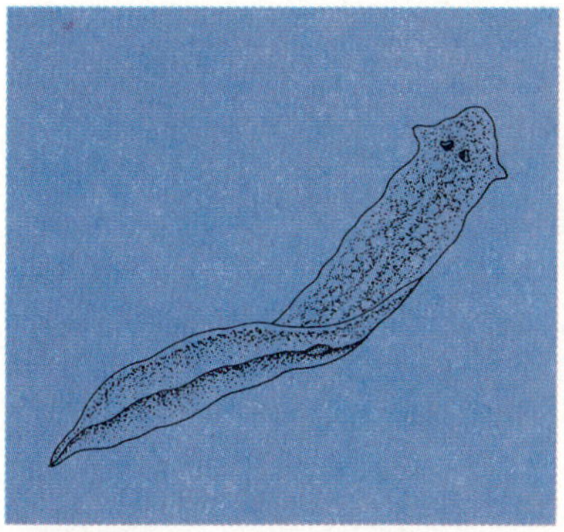

Flatworms
Some members of **phylum Platyhelminthes** (plat′ə hel-min′thēz) are free living. Others are parasitic. You will study the common, free-living planarians and the parasitic flukes and tape-worms

Roundworms
Members of **phylum Nematoda** (nem′ə tō′də) are common as parasites. They are important to humans as a cause of disease in both plants and animals. You will study the trichina worm and the hookworm.

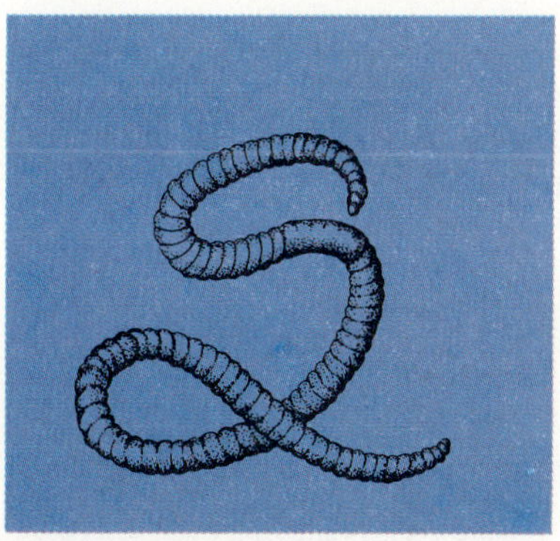

Segmented Worms
The most familiar member of **phylum Annelida** (ə nel′ə də) is the earthworm. Members of this phylum have bodies divided into segments by rings on the outside and by partitions on the inside.

Mollusks
Members of the **phylum Mollusca** (mə lus′ kə) have soft bodies, generally covered by a hard shell. Of all the phyla described in this chapter, mollusks are the most abundant in numbers of species. You will study the clam.

Echinoderms
Members of the **phylum Echinodermata** (i kī′nə dėr′mə tə) are thought to be the most closely related to the vertebrates. The member of this phylum that you will study is the starfish.

Sponge Colonies

The Sponges: Phylum Porifera

People once thought sponges were plants because a sponge does not move. Sponges remain permanently attached in one place, often living with other sponges in groups or colonies. Most sponges are found on the ocean floor in shallow water. And a few are found in fresh water. Some sponges are green, orange, red, or purple, whereas other sponges are dull in color. Sponges range in diameter from 1 to 200 centimeters. They are often slimy and may have a disagreeable odor.

Body Structure and Organization of Cells

The body of a simple sponge looks like a thin-walled sac. This sac has one large opening, or **osculum** (os'kyə ləm), at the top. And it has many small openings, or **pores,** along its sides. The osculum and pores lead to a large internal cavity. The outside of a sponge is covered by protective **epidermal** cells. The inside cavity is covered by **collar cells.** Each collar cell has at least one whiplike flagellum and a circle of filaments. Between the two layers of cells is a middle, jellylike layer that contains various other kinds of cells.

Biologists disagree about whether or not the two layers of cells in the sponge are tissues. For example, a sponge may be divided into separate cells by squeezing it through a cheesecloth. The separated cells survive alone. Yet, when they touch, the cells adhere to one another. Eventually, large groups of cells collect and a new sponge, identical to the original, is formed.

Skeleton

A sponge may be flat, round, cuplike, fanlike, or vaselike. The shape of a sponge is formed by its skeleton. The skeleton is made up of tiny **spicules** (spik'yülz), or of a fibrous organic material called **spongin,** or of a combination of these. Spicules are found in the jellylike layer. They are made of hard materials such as calcium carbonate or silica. Sponges are classified by the kind of spicules they contain. You can see a spicule at the right.

Challenge!
Look at pieces of genuine and artificial sponges under a microscope. What differences do you see? Find out which sponge can hold more water.

Feeding

Water and food are drawn into a sponge through its pores. The flagella of the collar cells beat back and forth rapidly, creating currents. The filaments on each collar cell strain the incoming water. Food particles, such as bacteria and diatoms, are trapped by the filaments and then ingested by the collar cells. Unused materials leave through the osculum.

Food moves from the collar cells directly into the ameboid cells, or **amebocytes** (ə mē'bə sīts), that wander through the jellylike layer. The ameboid cells deliver food to the epidermal cells.

Reproduction

A sponge may reproduce both asexually and sexually. Asexual reproduction may occur when pieces of sponge break off. Each piece grows back, or **regenerates,** into an entirely new sponge. Asexual reproduction also occurs by **budding.** Once a bud is formed, it may break off, attach to a new place, and develop into an adult sponge. Or the bud may remain attached to the parent and so become a part of a sponge colony.

In sexual reproduction, the ameboid cells develop into eggs and sperm. The sperm are released by one sponge and enter another sponge through pores. The sperm may fertilize an egg and, if so, a **larva** develops. A larva is an immature form that does *not* resemble the adult. The larva swims away from the parent, attaches in a new place, and grows into an adult.

Sponge Colony

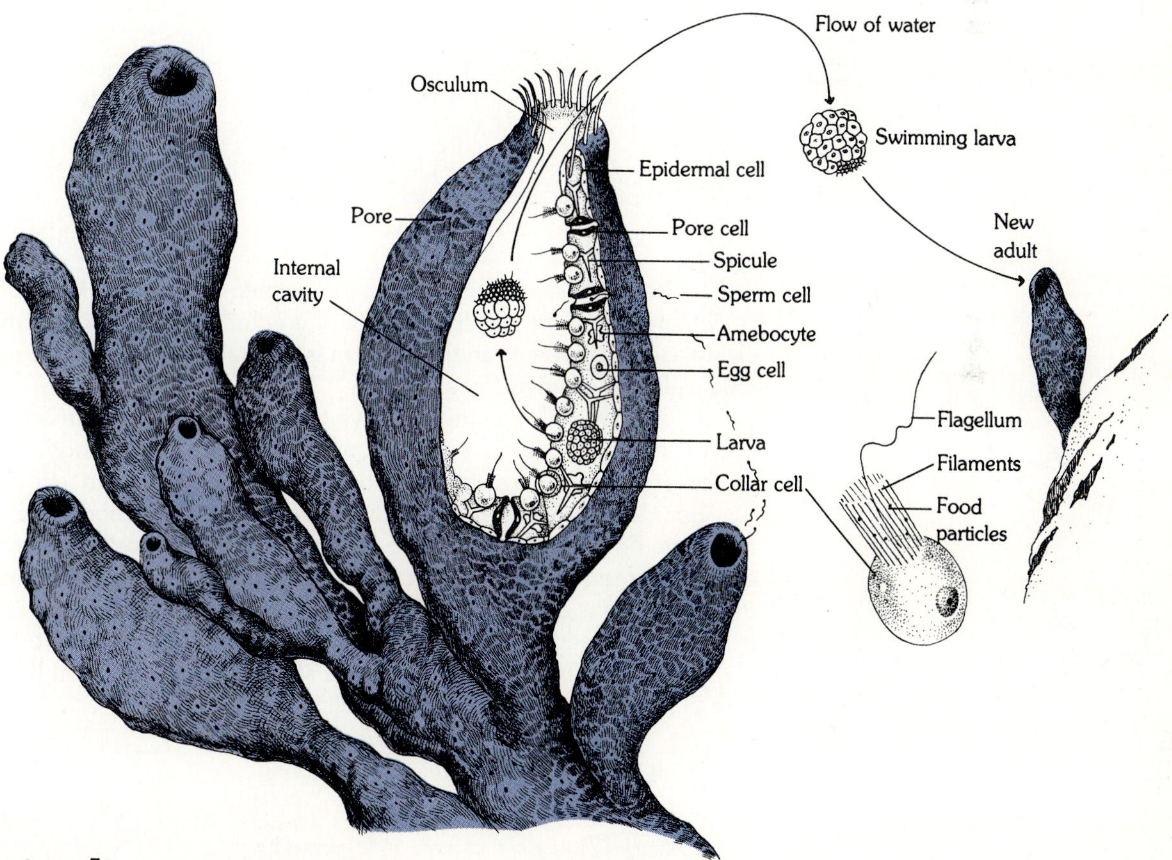

Review It

1. Why are animals considered to be heterotrophs?
2. Explain the difference between invertebrates and vertebrates.
3. Describe the structure of a sponge.

17-2
The Coelenterates: Phylum Coelenterata

The coelenterates include the jellyfish, hydroids, sea anemones, and corals. Most of the nearly 9,000 species of coelenterates live in warm seawater. But a few species live in fresh water. And one of the most spectacular jellyfish, *Cyanea artica,* lives in the cold water of northern oceans. This jellyfish measures as much as 4 meters across with tentacles 30 meters long.

Sea anemones, with their colorful bodies and petallike tentacles, are often called flowers of the sea. Their relatives, the corals, live together by the trillions in colonies. Many ocean islands and reefs are made of the calcium carbonate skeletons of coral.

Structure

Coelenterates have two body forms: **medusa** (mə dü′sə) and **polyp** (pol′ip). The life cycle of many coelenterates include both of these forms. The medusa, a bell-shaped form, swims close to the surface of the water with its mouth facing down. The polyp, a tube-shaped form, is attached at its base with its mouth facing up.

Both the polyp and medusa have a single, saclike digestive cavity called the **coelenteron** (si len′tə ron). Food enters and waste materials leave the coelenteron through its single opening. One or more rings of **tentacles** are located around the mouth. A tentacle is a long, slender appendage used to move, touch, and hold.

Medusa

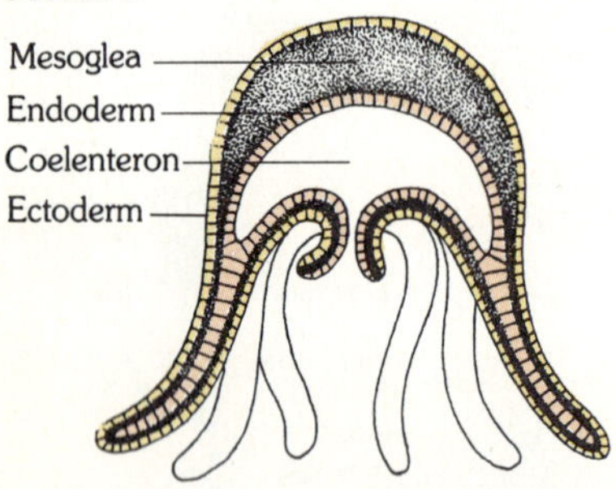

Polyp

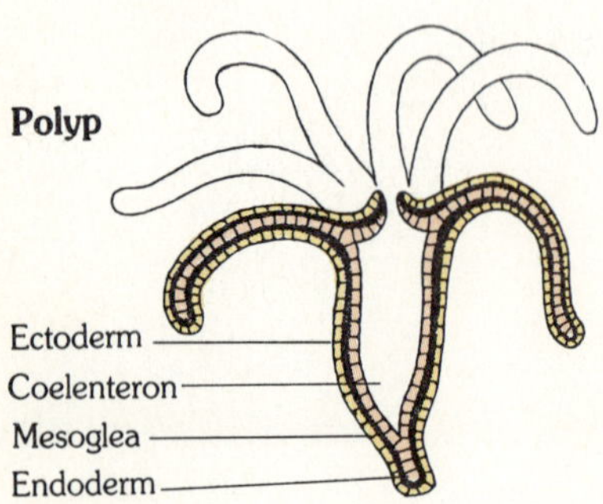

Tissues

Unlike the sponges, coelenterates have definite tissues. The body of coelenterates has two layers of tissue: the **ectoderm** and **endoderm.** Between the ectoderm and endoderm is a thin layer of a jellylike material called **mesoglea** (mes′ə glē′ə).

Endoderm cells line the digestive cavity. Some endoderm cells are specialized gland cells that secrete enzymes to break down food. Other specialized endoderm cells ingest food.

Ectoderm cells cover and protect the outside of the body and the tentacles. Ectoderm cells may be specialized as muscle cells, nerve cells, or **thread cells.** Each thread cell contains a **nematocyst** (nem′ə-tə sist), which is a bulb containing a coiled thread. When a thread cell is touched, it seems to explode. A thread pops out of the bulb to tangle the victim. The thread may also be sticky or contain poison barbs. The poison seems to paralyze the victim.

Sting Cell Before and After Release of Thread

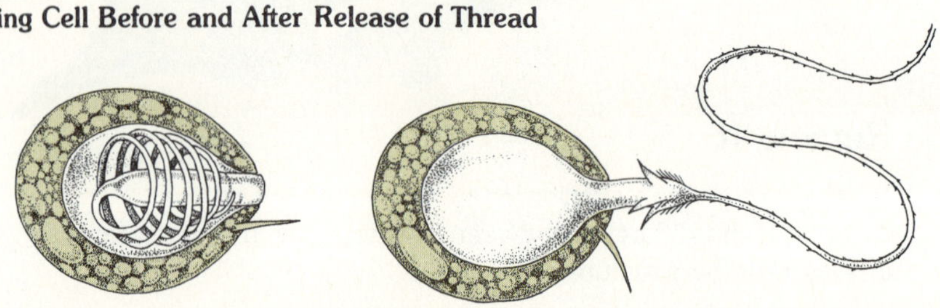

Radial Symmetry

Both the polyp and medusa forms have **radial symmetry.** This means that if you look at the mouth, the body parts are arranged like spokes of a wheel. If an organism with radial symmetry is cut in half any place through the mouth, the two halves will look exactly the same.

Radial Symmetry

Reproduction

The medusa form swims around and reproduces *sexually*. The medusa of a jellyfish, for example, produces eggs and sperm. Generally, the eggs can be fertilized only by sperm from a separate medusa. After fertilization, a swimming larva develops, attaches to the bottom, and changes into a polyp. The polyp form grows and reproduces *asexually* by budding or fission. Fission produces a platelike medusa. The medusa eventually detaches and swims away, completing the life cycle.

Life Cycle of Jellyfish

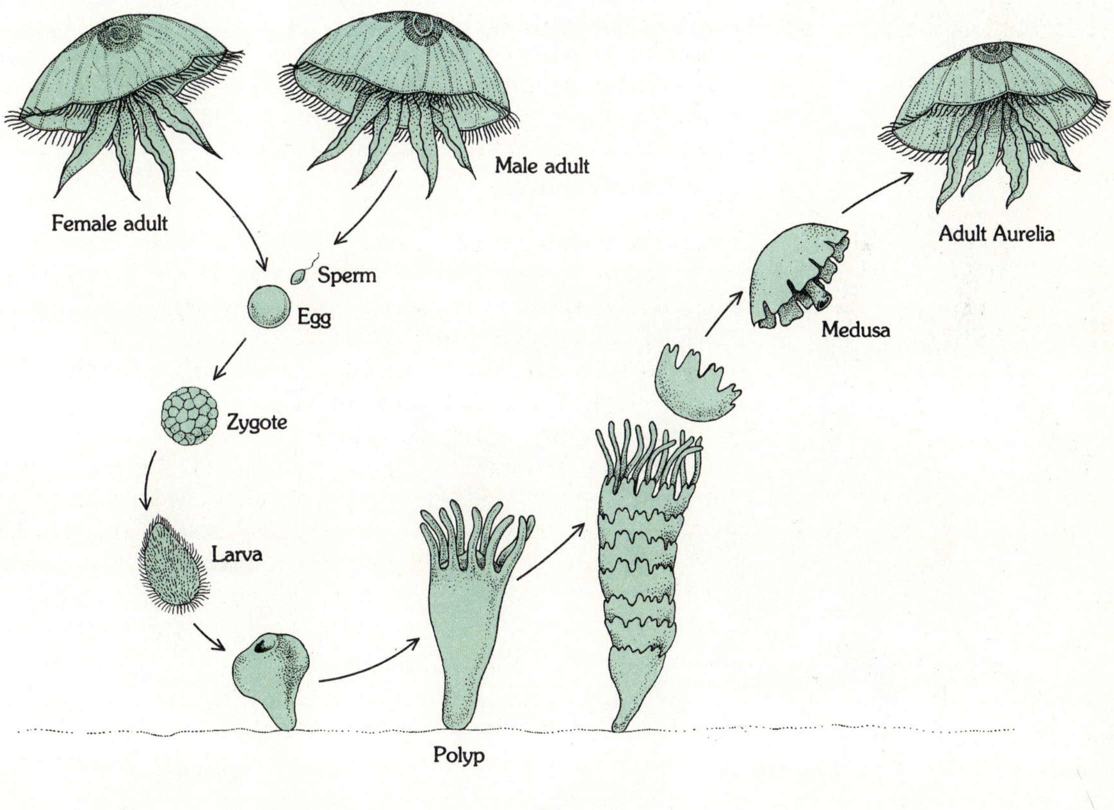

Sommersault of the Hydra

Hydra

Hydra is a saclike polyp that lives in fresh water. Unlike many other coelenterates, it has no medusa form. Depending on the species, *Hydra* may be from 1 to 30 millimeters long. To the unaided eye, *Hydra*, looks like a short piece of string with one end frayed. Actually, the frayed end is made up of tentacles, which form a ring around the mouth.

Hydra generally stays in one place. It attaches itself to underwater plants and debris by secreting a gluelike material at the base of its body. But *Hydra* can also move by swimming, by gliding on its base, or even by somersaulting.

Feeding

Hydra captures food by discharging the threads of several nematocysts at the body of its victim, or prey. The threads entangle and a poison from the nematocysts paralyzes the prey. The tentacles then push the prey into the mouth. Prey are often much larger than the *Hydra*. So after feeding, *Hydra* appears to be like a stocking that is stuffed with a softball. When food enters the body cavity, gland cells begin to secrete enzymes that digest the food. Other cells in the body cavity ingest the food.

Body Plan of the Hydra

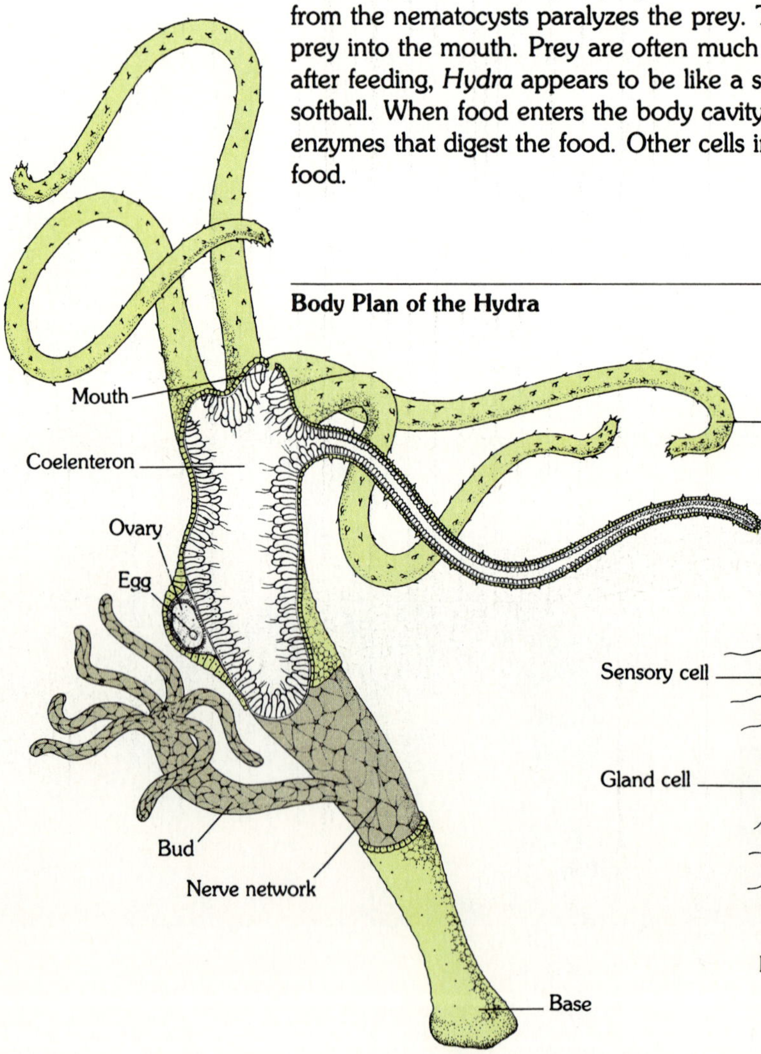

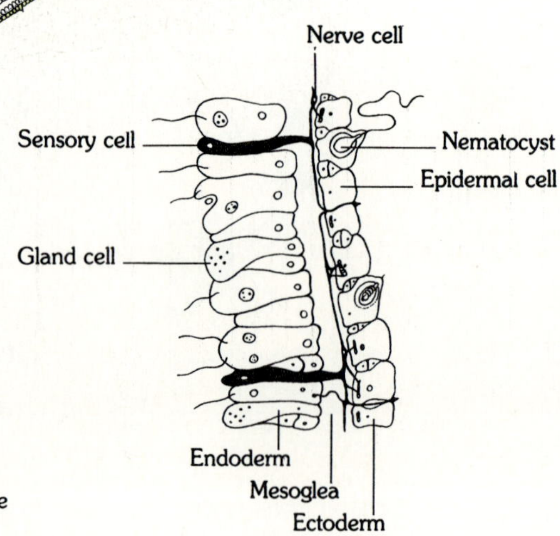

Nerve Network

The action of feeding requires *Hydra* to sense the presence of prey, to sting and grasp it, and to coordinate the movement of nearly all parts of the body in forcing the meal through the mouth. To carry out such complex behavior, *Hydra* uses its **nerve network.** The nerve network is made up of a series of sensory cells that are interconnected. These cells receive information about, and react to, changes in the environment. The nerve network then directs the muscle cells to respond. The nerve network can be thought of as a simple nervous system.

Respiration and Excretion

Hydra does not have even the beginning of a respiratory or an excretory system. *Hydra* obtains oxygen by diffusion through cells at its surface. Carbon dioxide wastes are removed by the same process. Undigested food materials pass out through the mouth.

Reproduction

Hydra reproduces both sexually and asexually. The most common kind of asexual reproduction is budding. Buds form in the midregion of the body as outgrowths of the body wall. Several buds may form at one time. After tentacles and mouth appear on the bud, the base of the bud pinches off from the parent. Male *Hydra* buds to form other males. And female *Hydra* buds to form other females.

Hydra also has the ability to reproduce asexually by regeneration. It regenerates whole bodies from pieces that have been cut off. Because of this ability, *Hydra* gets its name from Hydra, the many-headed monster of Greek mythology, who grew two heads to replace each one that was cut off.

When conditions are unfavorable, *Hydra* develops sex organs and reproduces sexually. When sperm formed in the testes of the male are mature, they are released. The sperm swim to the ovaries of a female *Hydra,* where one sperm cell fertilizes an egg. After the egg is fertilized, it divides to form an **embryo.** The embryo becomes covered by a protective casing. After the casing forms, the embryo drops off the parent and lies dormant—generally until the next spring. Then the embryo develops into an adult *Hydra.*

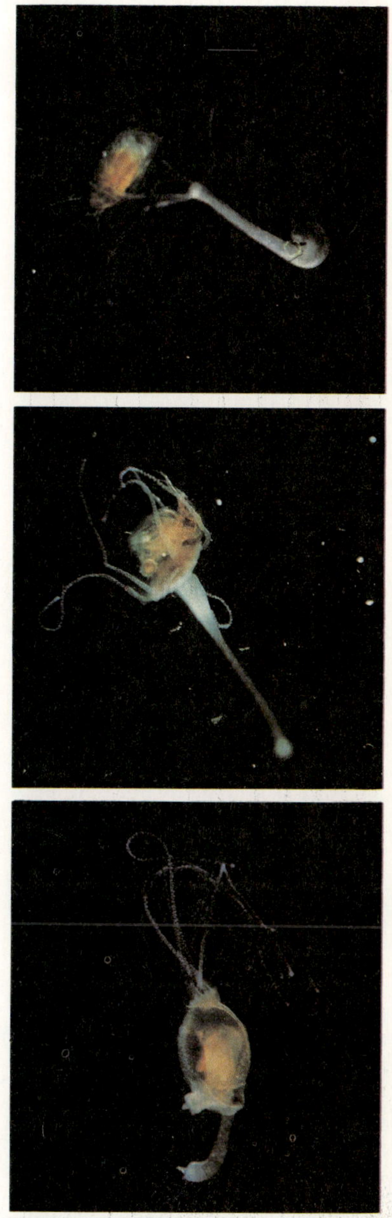

Hydra Feeding on Daphnia

Review It

1. Compare the feeding ways of the coelenterates with those of the sponges.
2. List some general traits of the coelenterates.
3. How is a polyp different from a medusa?
4. How do you know if an organism has radial symmetry?
5. How does *Hydra* take in and digest food?
6. Describe two methods of asexual reproduction in *Hydra.*

Activity

Observing *Hydra*

Materials

Culture of *Hydra*
Daphnia or other small crustaceans
Watch glass
Dissecting needle
Dissecting microscope
Medicine dropper

Procedures

1. With the medicine dropper, take a *Hydra* from the *Hydra* culture and put it onto your watch glass. Observe *Hydra* carefully for awhile to see how it adjusts to the new environment. Record your observations.
2. Study and sketch the structure of the *Hydra*. Label base, mouth, and tentacles.
3. After *Hydra* has fully extended, tap the edge of the dish sharply with your finger. Make a drawing to show what you see.
4. Use the medicine dropper to direct a stream of water at the *Hydra*. Record your observations.
5. Touch the tip of a tentacle with the dissecting needle. Repeat the procedure, touching the base of a tentacle and then the body. Record your observations.

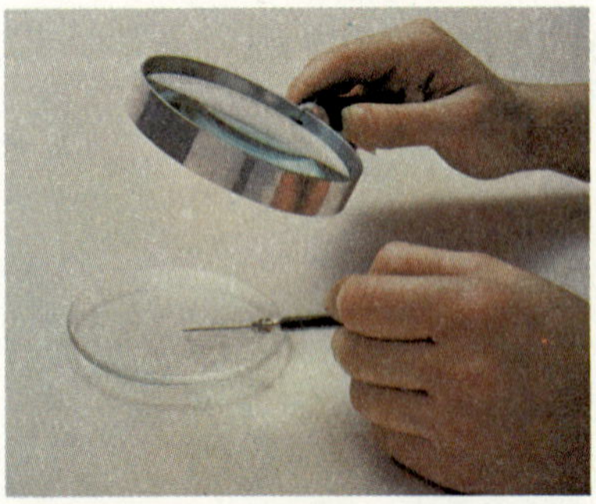

6. If there are any *Hydra* left, put a new *Hydra* onto your watch glass. Add several *Daphnia* and watch the *Hydra* closely. If the *Hydra* eats a *Daphnia*, observe what happens to the mouth and tentacles. Record your observations.

Analysis

1. What response did *Hydra* make to a sharp blow? to a stream of water? to touch? What do these responses tell you about the nerve network of *Hydra?*
2. How do the tentacles help *Hydra* eat?

Challenge!

Place several dead Daphnia *onto a watch glass with* Hydra. *Observe with a dissecting microscope. Explain the reaction* Hydra *has to* Daphnia.

17-3
The Flatworms: Phylum Platyhelminthes

Whereas some sponges and most coelenterates have radial symmetry, flatworms have **bilateral symmetry.** Organisms with bilateral symmetry can be divided into two halves that are mirror images of each other. Organisms with bilateral symmetry also have a head, or **anterior,** end. And they have a tail, or **posterior,** end. The back of such organisms is called the **dorsal** side, and the belly is called the **ventral** side. You can see the bilateral symmetry of the planarian below.

Bilateral Symmetry of Planaria

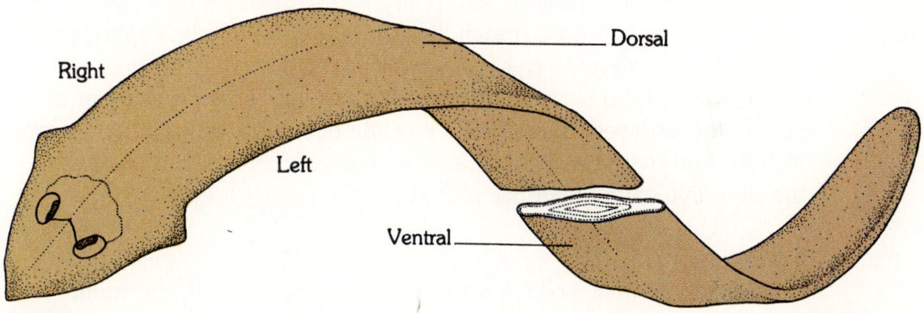

Tissue Structure

The tissue structure of flatworms is also different from that of the sponges and the coelenterates. In addition to having layers of ectoderm and endoderm, flatworms have a middle tissue layer called **mesoderm.** The mesoderm makes up most of the body and forms true **organs.** Organs are structures made up of two or more tissues. The arrangement of the three body tissues of the planarian is shown below. When viewing the tissues, imagine that a slice has been made through the center of the body of the planarian.

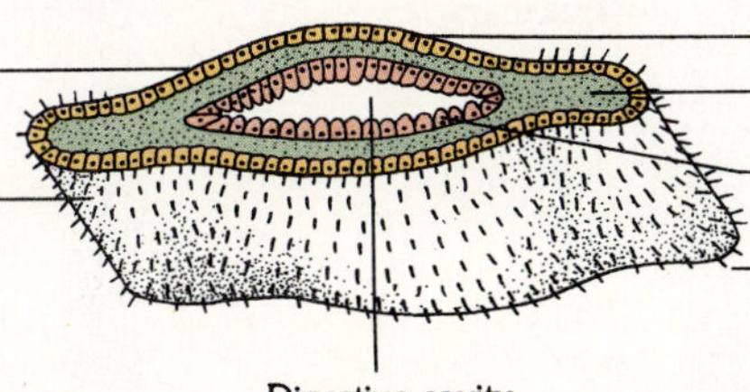

Planarians

Planarian

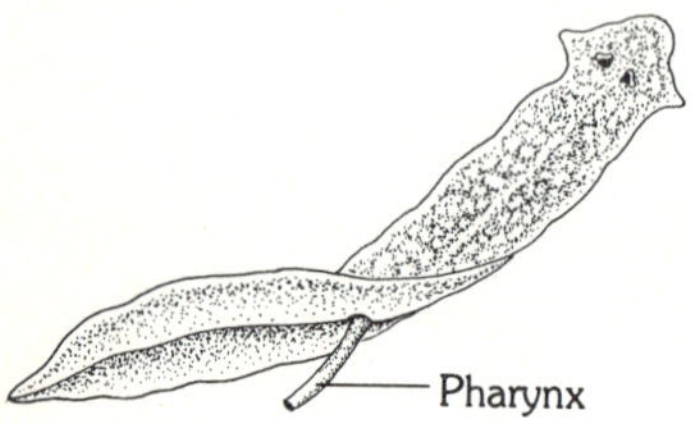

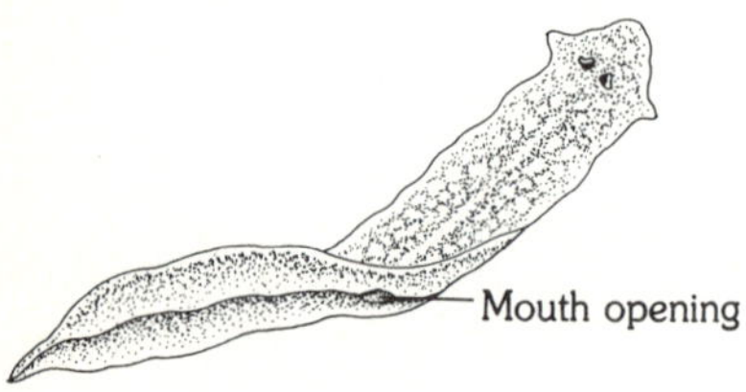

Many of the traits of the flatworm phylum can be seen in free-living planarians. Planarians live in fresh water and eat small organisms. Glands on the ventral side of the planarian secrete a slime. Planarians glide over the slime, propelled by cilia also located on the ventral side. Also, planarians use muscles that contract in wavelike motions. Planarians appear to roll in a spiral pattern as they move.

Feeding and Digestion

The mouth of a planarian can be seen on the ventral side of the body near the posterior end. When a planarian feeds, it thrusts out a movable tube, or **pharynx** (far'ingks). The pharynx moves in and out to draw food into the mouth. The pharynx also secretes enzymes to soften food so it can be drawn into the pharynx. In some species of planarian, the pharynx may be used to swallow the entire prey. From the pharynx, food passes into a digestive cavity where it is either digested or absorbed. The digestive cavity has three main branches, one in front and two along the sides. The many branches and pockets of the digestive cavity take the place of a circulatory system.

Excretion

As in coelenterates, solid wastes are eliminated through the mouth. There is no opening at the posterior end for elimination of solid wastes. Liquid wastes are removed by light-bulb shaped organs that connect by tubes and pores to the outside of the body. There are no respiratory organs to exchange gases. Instead, gases are exchanged by diffusion through the cells of the body surface.

Digestive System

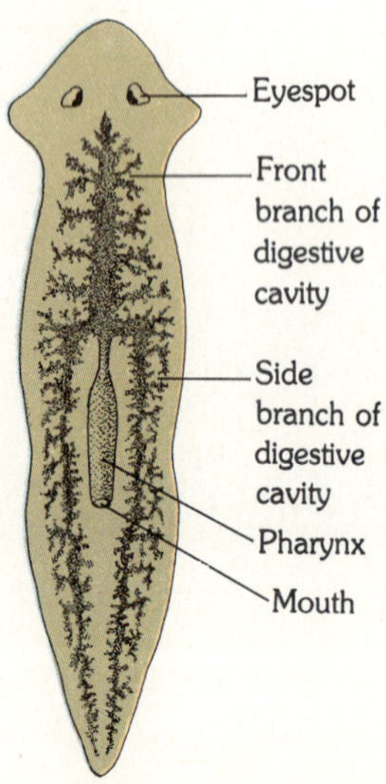

Excretory System

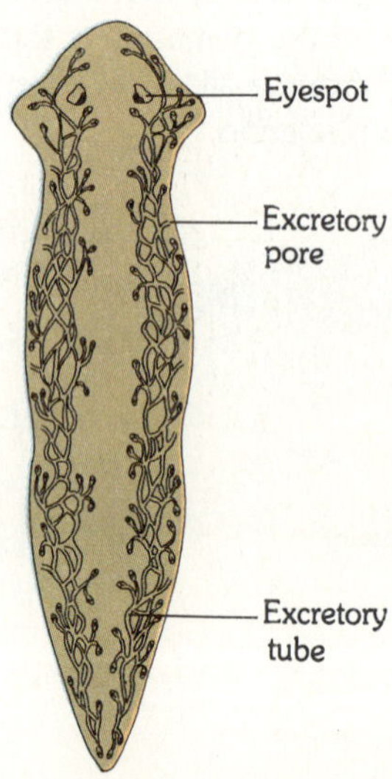

Nervous System

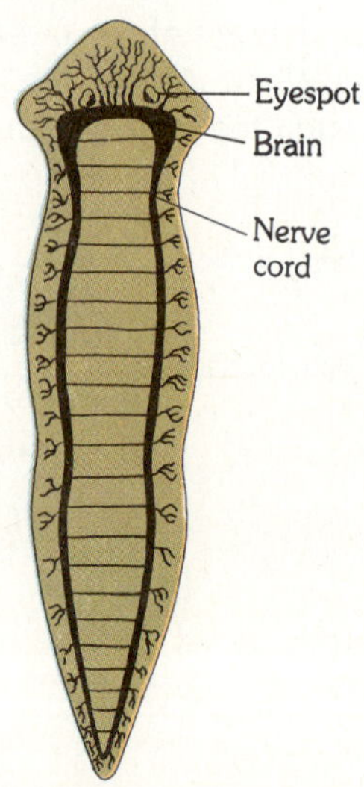

Nervous System

On the dorsal side of the arrow-shaped head of a planarian are two **eyespots.** These eyespots cannot see shapes. They can only detect light. In other words, planarians can see only as well as you can with your eyes shut. Beneath the eyes is a two-lobed mass of nerve tissue. This tissue acts as a **brain,** or communication center, for the entire organism. Running from this brain are two sets of nerves, one on each side of the body, to keep the brain informed of events throughout the body.

Reproduction

Planarians reproduce both sexually and asexually. For sexual reproduction, planarians have both male and female sex organs in the same individual. However, planarians do not fertilize themselves. Fertilization requires two separate planarians. After fertilization, a capsule is formed with ten eggs or less. The capsule is fastened to an object in the water. The eggs hatch two or three weeks later.

In asexual reproduction, the posterior part of the planarian begins to act as if it were separate from the rest of the body. Eventually, the posterior part breaks off. Then, both anterior and posterior parts regenerate the missing parts.

If the head of a planarian is split and the two halves are kept apart for awhile, a two-headed planarian will develop. It will eventually split into two planarians. A planarian can be cut into as many as six parts. Each part will regenerate missing parts to form a new planarian.

Asexual Reproduction

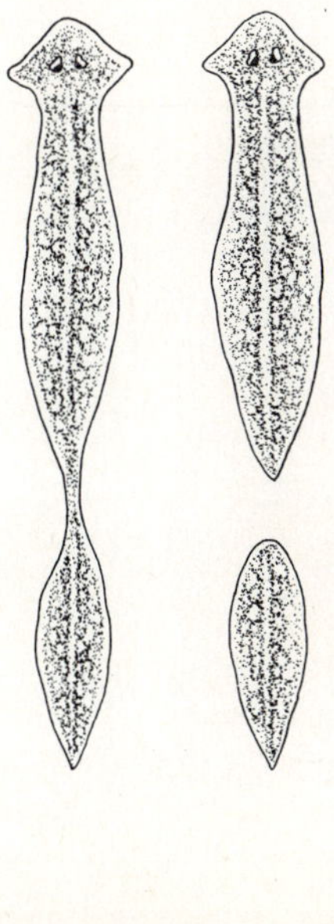

Regeneration after Head is Split

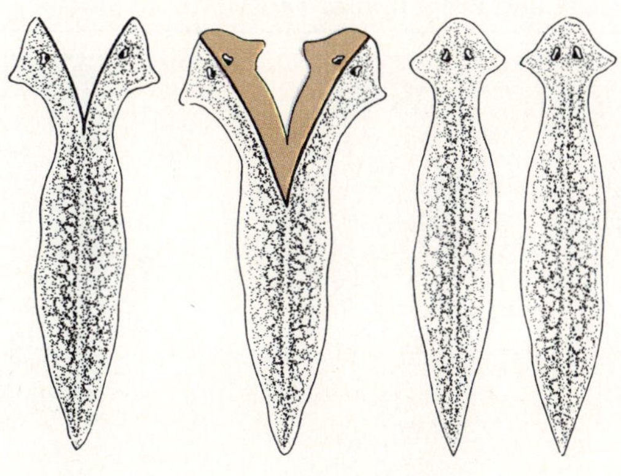

Regeneration from Widthwise Cuts

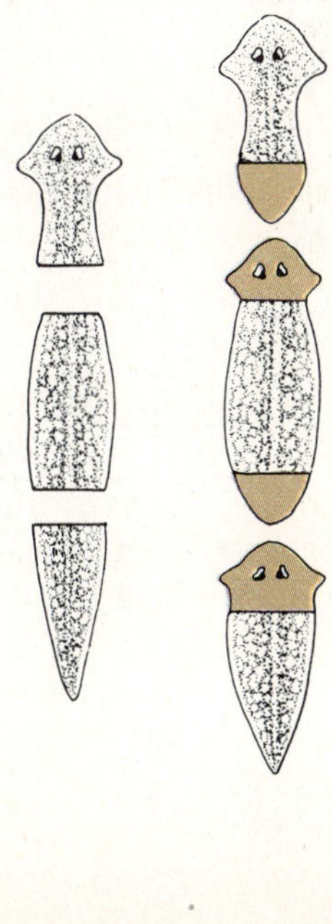

Parasitic Flatworms

Tapeworm

Two of the three main groups of the flatworm phylum are entirely parasitic. These include the flukes and the tapeworms. Flukes look somewhat like planarians. However, the mouth of a fluke is located at the anterior end. Also the fluke has one or more suckers on its ventral side, so it can cling to its host. A fluke has a tough outer layer that will not dissolve in the digestive juices of the host. The life cycle of a fluke involves changing into various forms as the fluke moves from one host to another. Humans living in areas where public health standards are low may suffer the effects of blood and liver flukes.

Tapeworms

Tapeworms are long, ribbonlike flatworms. Adult tapeworms live in the intestines of nearly every kind of vertebrate, including humans. The head of an adult tapeworm is made up of hooks and suckers by which it attaches itself to the wall of the host's intestine. A tapeworm feeds by merely absorbing food that has already been digested. The body of a tapeworm is made up of sections called **proglottids** (prō glot′idz). A tapeworm grows by adding new proglottids in the region behind the head. The older and larger proglottids are located near the posterior end. The sections are made up mostly of reproductive organs, both male and female in the same proglottid. The sections fill with larvae, which develop from self-fertilized eggs. The filled proglottids break off from the tapeworm. When the separated proglottids pass out of the host's body with solid waste, the larvae may be taken in by a second host, who keeps the cycle going. Usually there are two or more hosts in a complete life cycle.

Biologists often try to explain why so many flatworms are parasitic. One explanation suggests that flatworms could not compete successfully with the more specialized organisms. Living in and on the bodies of other organisms has enabled the parasitic flatworms to live in a less competitive environment.

Head and Proglottid of Tapeworm

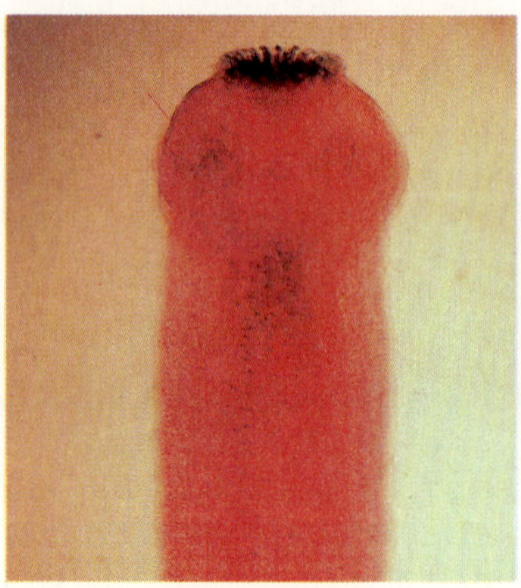

Courtesy Carolina Biological Supply Company

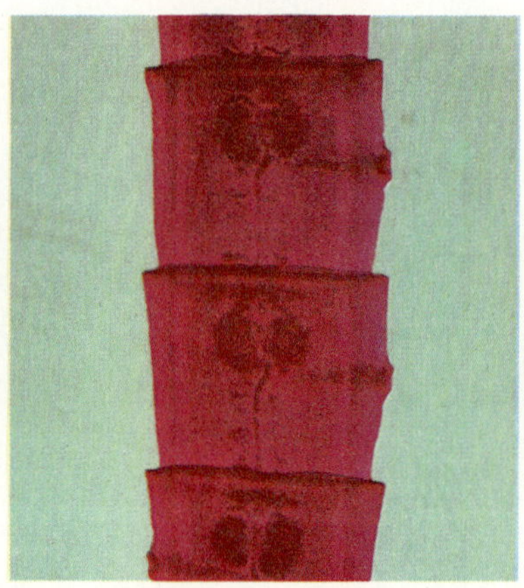

Courtesy Carolina Biological Supply Company

Life Cycle of a Tapeworm

Humans cannot drink the water from the pure-looking streams and lakes of Isle Royal National Park in Lake Superior. Because of a tapeworm, the water of Isle Royal must be boiled before use.

Both the wolves and moose of Isle Royal serve as hosts for this tapeworm. Humans can also serve as hosts. The mature proglottids and the eggs they contain pass out with the wolves' solid wastes into the water or onto plants. The moose may drink the water or eat the plants and, at the same time, take in the eggs. The larvae hatch from the eggs in the stomach of the moose. They eventually move through the circulatory system to the lungs where they form cysts. If the wolves eat the moose, they also eat the larvae. A larva develops into an adult tapeworm in the intestine of the wolf. You can see how this cycle works by studying the diagram below.

Life Cycle of the Tapeworm on Isle Royale

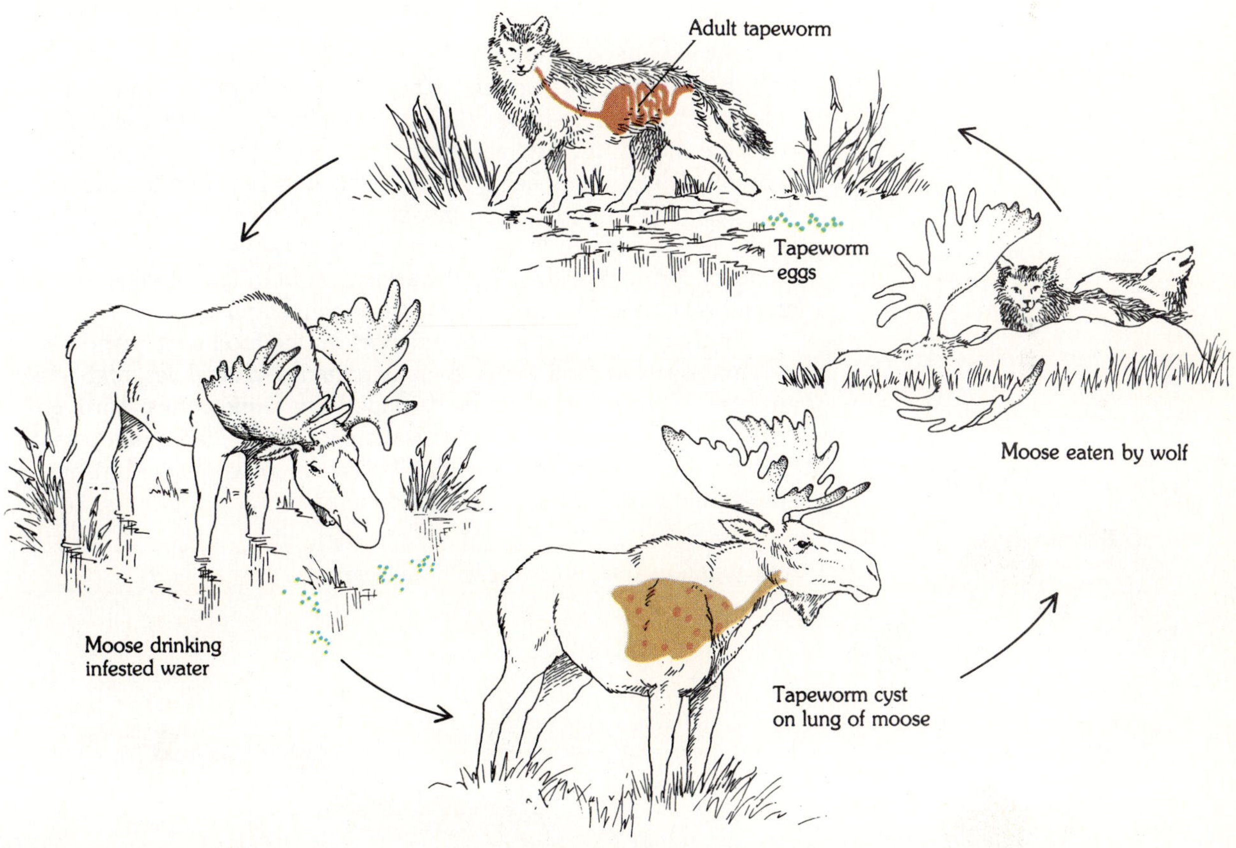

The Roundworms: Phylum Nematoda

Roundworm

If you want to discover, describe, and name a new species of animal, you should study the roundworms. New species of roundworms are being discovered at the rate of about one per day. At present, about 10,000 species of roundworms are known. And biologists think there may be as many as 500,000 species. Roundworms seem to occur everywhere on earth, from polar regions to hot springs and from desert sands to ocean floors. Many of them are parasitic in plants and animals. Humans serve as hosts for at least 50 species of parasitic roundworms.

The number of individual roundworms is also staggering: 90,000 were found inside *one* rotting apple, and 10 million were found in one square meter of pasture land. One biologist observed that if all the earth—except the roundworms—were suddenly to become invisible, the soil, water, plants, animals, highways, and cities could be seen in outline by the presence and position of the visible roundworms.

Structure

Roundworms have a very simple body plan: a tube-within-a-tube. The inner tube is the digestive tube. This tube runs from a mouth at the anterior end to an opening called the **anus** at the posterior end. Beyond the mouth, the first part of the digestive tube is the muscular pharynx. The pharynx helps the worm suck its prey into the mouth. After being taken in, the food passes into the intestine where it is digested and absorbed. A roundworm must move its entire body to mix the food during digestion. Solid wastes pass from the intestine into the rectum and, then, out of the body through the anus.

Roundworms have a harmful effect on the food supply and health of humans. Basic food crops, such as potatoes, pineapples, and celery, are continuously threatened by damage from roundworms. Throughout the world, people suffer the weakening effect of large numbers of parasitic roundworms living in their bodies. Hookworms and trichina are examples of such parasitic roundworms.

Tube-Within-a-Tube Structure of Roundworm

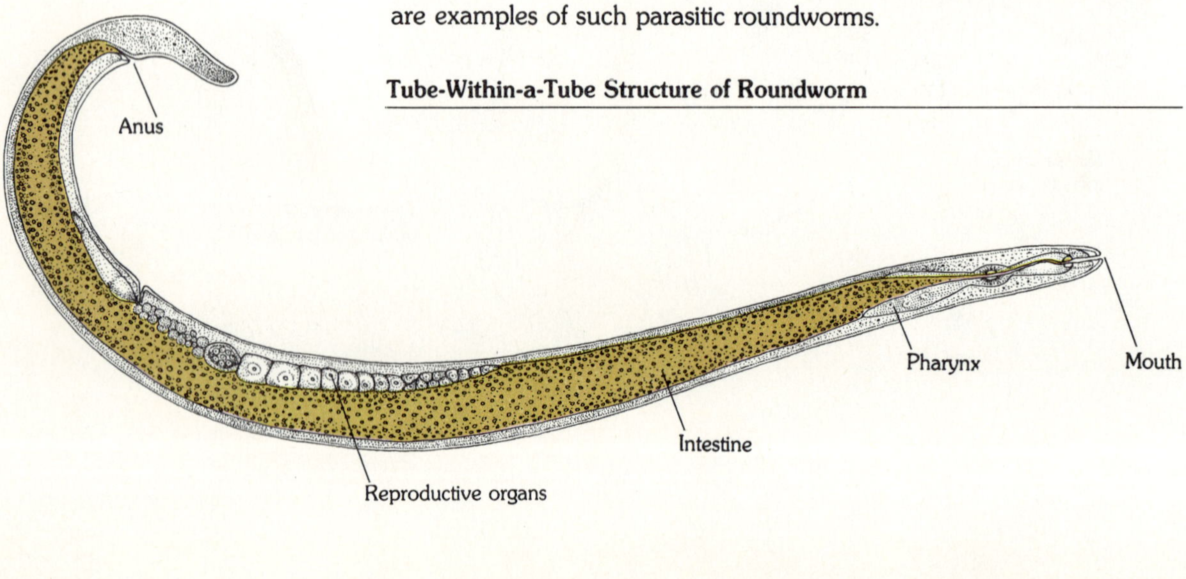

Hookworms

A hookworm can enter the human body through any body part that touches the ground—for example, the feet. The larvae bore through the skin and enter the bloodstream. They pass through the heart on their way to the lungs. They grow larger. Eventually, the larvae move up the windpipe to the throat. They are swallowed and go to the intestine. They live there as adults and reproduce. The eggs pass out with the solid waste. Wherever such wastes are deposited, the eggs hatch into larvae.

Adult hookworms live on blood. Because hookworms take part of the host's blood, the cells of the host's body are deprived of oxygen. This makes the host feel weak and tired. To prevent the spread of hookworms, human wastes should be disposed of properly. And people should wear shoes outside so the hookworm cannot reach their skin.

Trichina

Trichina is a parasitic roundworm that appears in humans who have eaten partially cooked or uncooked pork. One ounce of infected pork may contain as many as 100,000 trichina roundworms. Each roundworm can produce as many as 1,500 larvae.

Trichina roundworms live mainly in pigs, dogs, and rats. As adults, trichina roundworms do no damage to the intestine where they live. But as larvae, they damage the tissues of their host. The diaphragm, ribs, tongue, and eyes are tissues most often damaged as the larvae move through the body looking for muscle to live in. Humans who eat the uncooked muscle of the host can become infected with the worm larvae. Each larva may develop into an adult in the human intestine, and the cycle is started once more.

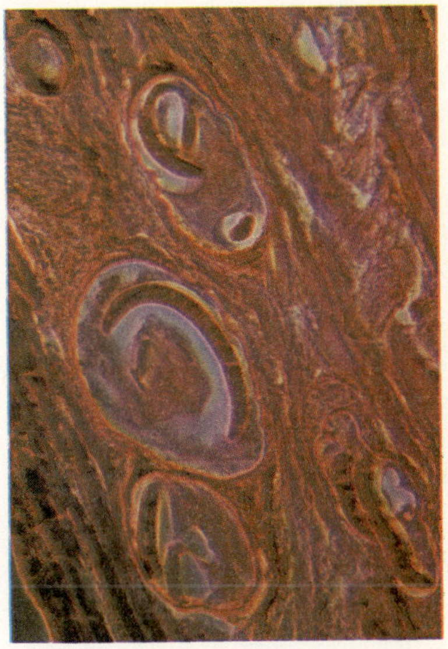

Larvae of Trichina Worm in Muscle

Review It

1. Compare and contrast bilateral and radial symmetry.
2. To what part of an animal do each of these terms refer: *dorsal, ventral, anterior, posterior?*
3. How are flatworms different from roundworms?
4. Describe the body plan of a typical flatworm and a typical roundworm.

17-4
The Segmented Worms: Phylum Annelida

The most familiar members of this phylum are the earthworm and the leech. In addition, there are many other kinds of segmented worms, or **annelids.** Most annelids live in the ocean. But a few live in fresh water or on land. They may vary in size from microscopic to very large. The Australian earthworm, for example, may grow to a length of more than two meters.

The name *annelid* means "ringed" and refers to the series of segments that make up the body of the worm. Because of this *segmentation,* some species resemble beaded bracelets. Partitions divide the body on the inside so that the segmentation occurs on the inside as well as on the outside. Each segment contains muscles, digestive tube, nerve cord, blood vessels, and excretory organs.

Segmentation is an advantage because there can be division of labor among segments. Segments can specialize in different tasks. For example, several segments might pump blood and contain a heart. Other segments might contain a brain and direct the activities of the other segments.

Overall, annelids are more complex than flatworms and roundworms. Specialized muscles in the body wall and digestive tract allow an annelid to move in many ways. And a much larger nervous system makes an annelid able to learn more than a simpler invertebrate animal is able to learn.

Structure

Annelids have a tube-within-a-tube body plan. The inner tube is the digestive tube. Surrounding the digestive tube is a closed circulatory system. The blood stays in tubes, or vessels, as it travels throughout the body.

In annelids and in all the animals that you will study later, the space between the body wall and the digestive cavity is the **coelom** (sē'ləm). The coelom is generally a fluid-filled cavity lined with mesoderm. Recall that, in flatworms, this space is solid mesoderm. And in roundworms the space between the body wall and the digestive cavity is an open cavity that contains organs.

Flatworm

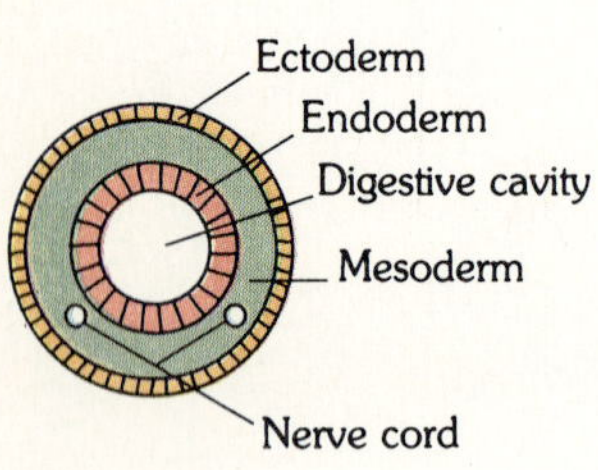

Roundworm

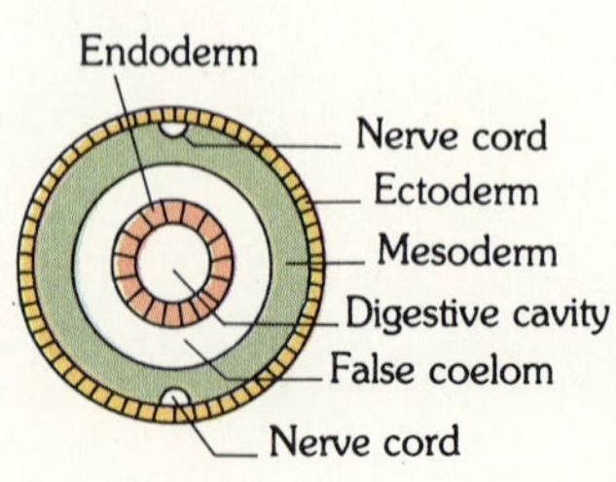

Segmented Worm

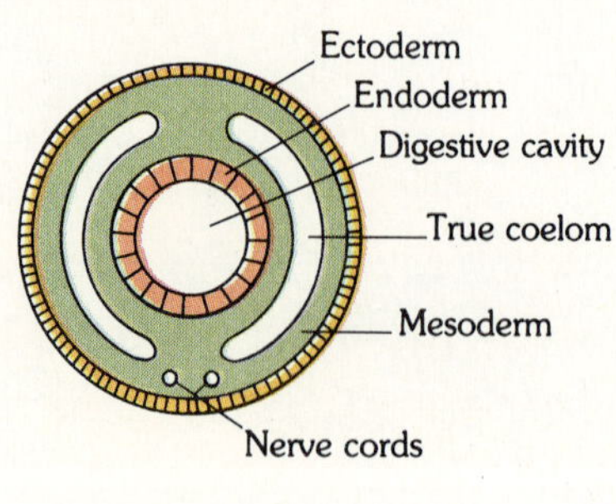

The Earthworm

Most of the traits of annelids can be observed by studying the earthworm. You can see the external structure of an earthworm below. Note that a transparent **cuticle** is secreted by the outer layer of the body wall. This keeps the skin moist. A moist skin surface is necessary for gas exchange to occur through the skin during respiration. A slimy **mucus** secreted by skin cells also helps keep the skin moist. And it helps with locomotion as well.

At the sides of the body on the ventral surface of each segment are four pairs of short, stubby bristles, or **setae** (sē'tē). The setae provide traction for movement. Earthworms also use the setae to hold them in their burrows when something is trying to remove them.

The anterior end of the body is darker and more pointed than the posterior end. Segments are usually numbered from the anterior end, starting with the mouth as segment one. A saddle, or **clitellum** (kli-tel'əm), which has a role in reproduction, is located from segment 32 to segment 37.

Did You Know?
When heavy rains soak the earth, oxygen cannot enter the soil. A low oxygen level in soil forces earthworms to the surface. Once above the surface of the soil, the skin of an earthworm can lose its moisture if the air becomes dry or if exposed to heat. So the earthworm may be unable to respire even when there is plenty of oxygen. And it may die from lack of oxygen.

External Features of Earthworm

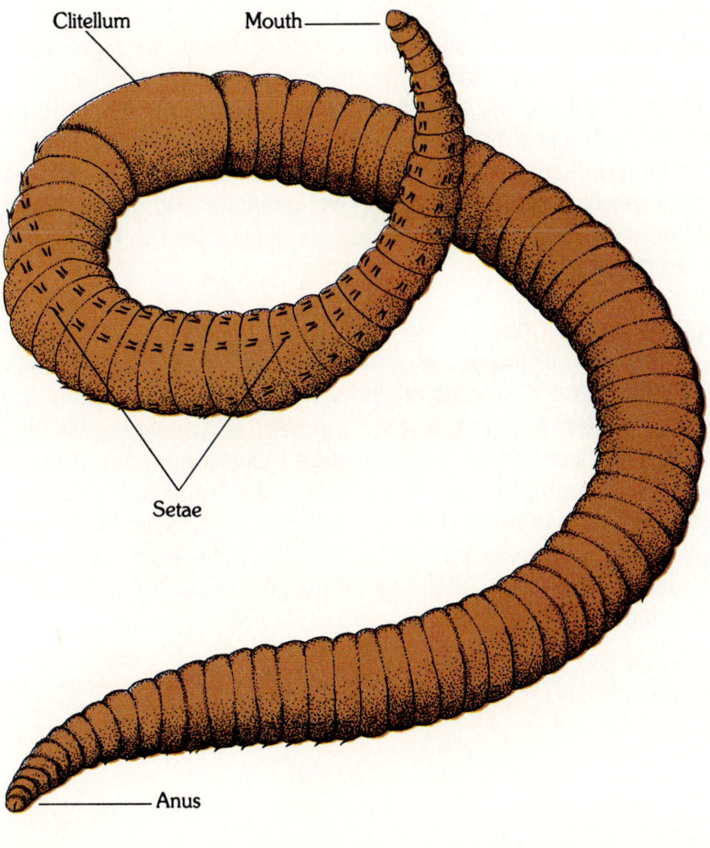

Feeding and Digestion

The mouth of the earthworm is partially covered by an upper lip, or **prostomium** (prō stō′mē əm), and opens into a muscular pharynx. Soil, which contains organic materials and rock particles, enters the mouth. From the mouth, the soil is sucked into the digestive tract by the pharynx. From the pharynx, soil enters a narrow **esophagus** and then a **crop,** where it is stored briefly. Then the soil goes into a muscular **gizzard** where it is ground up. The hard particles in soil help grind up the organic particles of food. From the gizzard, food enters the **intestine** and is broken down by bacteria and enzymes. The blood in the walls of the intestines absorbs the food. Undigested materials in the soil pass out of the body through the anus in the form of **castings.** Piles of castings may be found on the surface of the soil. The organic material in the castings adds to the soil fertility. And the feeding action of earthworms keeps the soil loose by providing spaces for air and water.

Excretion

Excretion in the earthworm is handled by organs called **nephridia** (ni-frid′ē ə). Each segment of the earthworm has one pair of nephridia. The nephridia collect waste materials from blood and from fluids in the coelom. The wastes are eliminated through pores on the ventral side.

Circulation

The dorsal blood vessel is found on the dorsal side of the digestive tract. This vessel collects blood from the body and contracts to force blood toward the head. At segments 7 to 11, blood enters one of the five pairs of muscular tubes called "hearts" or **aortic arches.** The blood is pumped from the aortic arches to a ventral blood vessel. In the ventral vessel, blood flows toward the posterior end. It also flows to the skin to exchange carbon dioxide for oxygen and to the intestine to pick up digested food.

Challenge!
Earthworms often burrow deeply into the ground—especially during cold weather. Dig into the ground to find out how deep the earthworms have burrowed where you live.

Nervous System

The brain of the earthworm lies above the anterior end of the pharynx. A ventral nerve cord runs on the ventral side of the intestine from the brain to the last segment. In each segment, the nerve cord is enlarged to form a small mass of nerve tissue called a **ganglion.** Short nerves from each ganglion run to muscles and sensory cells.

Although the earthworm has no eyes or ears, it responds to both light and sound. However, it is not sensitive to red light. People who fish with earthworms take advantage of this. They use a red flashlight at night when they collect their worms.

Structure of Earthworm

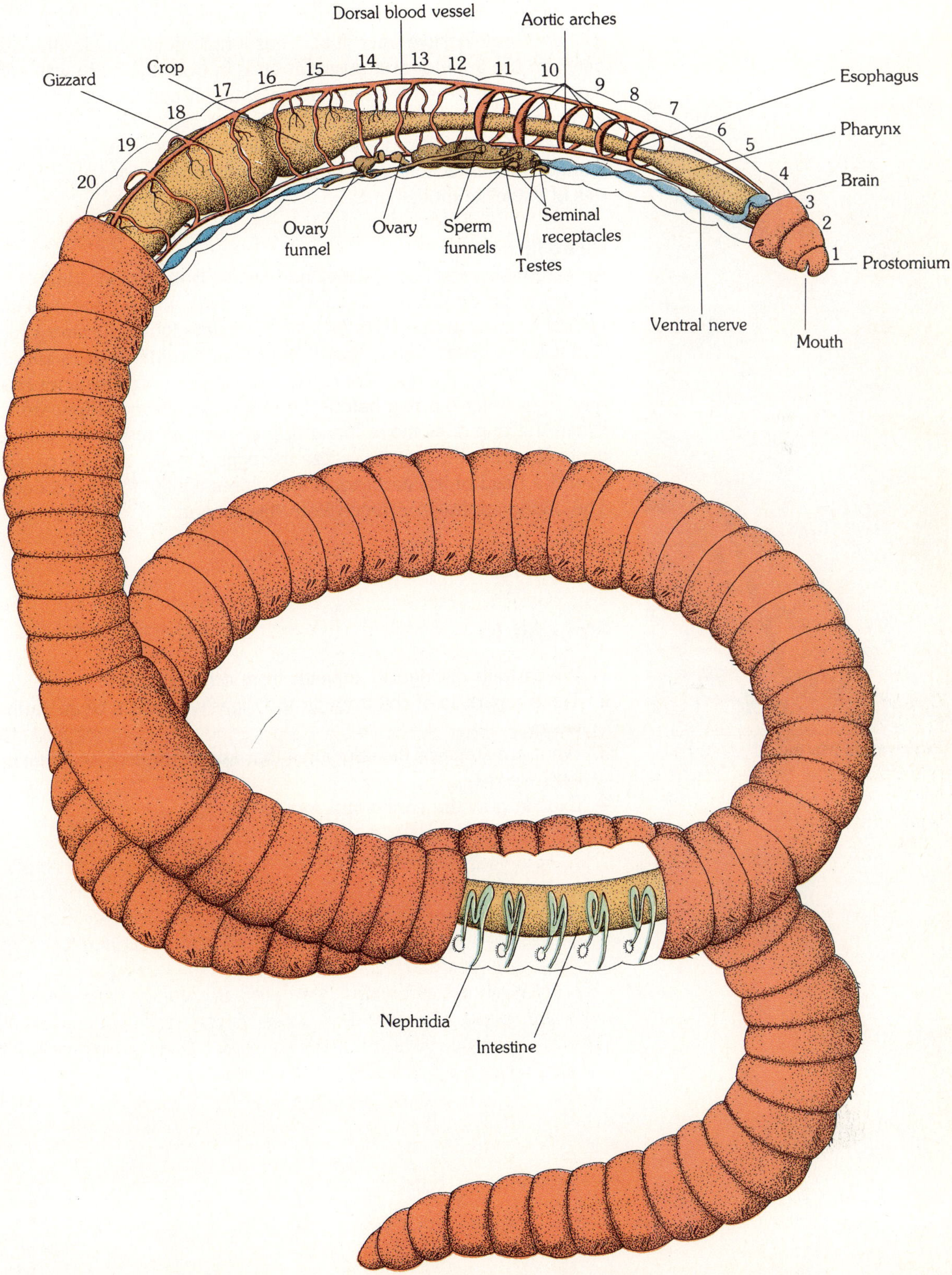

Locomotion

The earthworm has circular muscles that wrap around both the body wall and the digestive tract. It also has longitudinal muscles that run between the anterior and posterior ends in both the body wall and the digestive tract. When the longitudinal muscles contract, they carry the earthworm forward by shortening its body length. When the circular muscles contract, bulges are formed along the length of the earthworm. The interaction of these two sets of muscles can produce a series of wavelike movements.

Earthworms Mating

Reproduction

An earthworm has both ovaries and testes. But two individuals are necessary to produce fertilized eggs. When earthworms mate, two individuals come together. This usually occurs at night, following a rain. The two worms exchange sperm from their seminal vesicles and then separate. When the eggs are ready, the clitellum secretes a mucous ring. Eggs enter the ring before the ring begins to move forward. When the ring does move forward, the sperm stored in the seminal receptacles enter the ring to fertilize the eggs. Finally, the ring passes over the head of the earthworm and closes tightly, forming a **cocoon.** The young earthworms hatch from this cocoon.

Review It

1. Which traits distinguish annelids from other animals?
2. Trace a particle of soil through the digestive system of an earthworm.
3. What are some of the structures that appear in each segment of an earthworm?
4. Explain how the earthworm moves.

17-5
The Mollusks: Phylum Mollusca

Mollusks are abundant throughout the oceans from the mud and rocks of tidal zones to deep ocean waters. They are also found in fresh water and in moist areas on land. Mollusks are quite important to humans as a source of protein food. They make up such dishes as oysters on the half shell, clam chowder, abalone steak, escargot, and squid foo yung.

Mollusks are the second largest phylum of animals in number of species. The three main groups are the **bivalves,** the **gastropods,** and the **cephalopods.** Bivalves, or two-shelled mollusks, include the clam and the oyster. Gastropods, or stomach-footed mollusks, include the snail and the slug. Cephalopods, or head-footed mollusks, include the octopus, the squid, and the chambered nautilus.

Structure

Although mollusks appear to be quite different from one another, all have the same basic body plan. All mollusks have soft bodies that are covered by a fold of tissue called the **mantle.** And all mollusks have a muscular foot that may be used for locomotion and capturing prey. The foot may also contain sensory organs.

All mollusks except the bivalves have a tongue-shaped scraper in their mouth. The scraper is used to loosen and tear food and to push food into the digestive tract.

A shell secreted by the mantle may be present. The shell is important in providing a retreat in case of danger and in preventing loss of moisture. The mollusk shell is made of calcium carbonate, a hard material. As a result, mollusk shells are easily preserved as fossils. The fossil record contains many thousands of extinct mollusk species. Mollusk shells are used by humans to make buttons and jewelry.

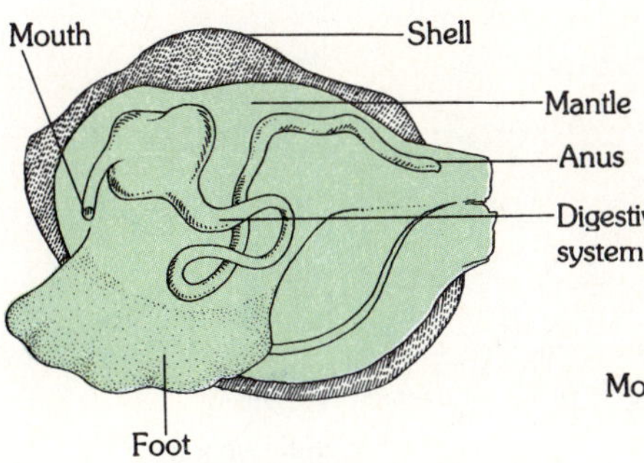

Bivalve

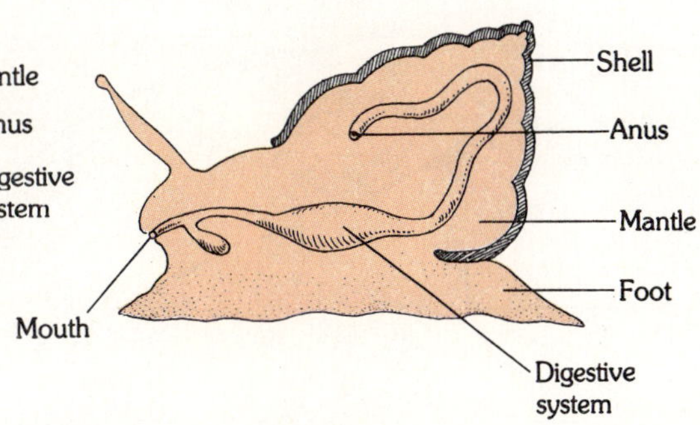

Gastropod

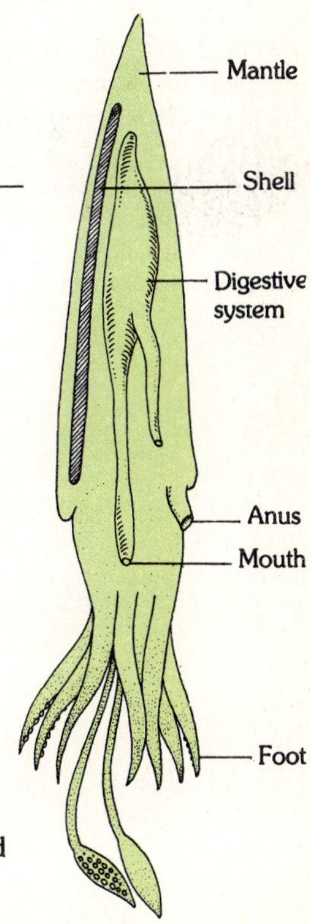

Cephalopod

Clam

Bivalves, the Two-shelled Mollusks

The bivalves include the clam, the scallop, and the oyster. All three bivalves are found in salt water. The clam, our representative bivalve, is also found in fresh water.

Structure

The two shells, or **valves,** of the clam are held together and hinged by two large muscles on the dorsal side. The shells open on the ventral side to allow the hatchet-shaped foot to move in and out. Near the posterior end are two openings, or siphons, which allow water to flow in and out of the clam. Water flows into the **incurrent siphon** and out of the **excurrent siphon.**

When the shells are completely open, you can see the mantle, which covers the foot and other organs. Under the mantle are the **gills,** which are respiratory organs. Water that flows through these sievelike gills provides oxygen for respiration. The space between the mantle and the other body parts is the **mantle cavity.**

The Shell

One of the tissues of the mantle continuously secretes the layers of calcium carbonate that form the shell. A clam shell has concentric lines. These are lines of shell growth, similar to tree rings.

The inner surface of the shell, a layer known as *mother-of-pearl,* appears shiny and colorful. Mother-of-pearl was once a major raw material in the making of buttons.

Nervous System

The clam has a small, simple nervous system made up of three ganglia connected by nerve cords. The clam also has sensory cells that respond to touch, light, and certain chemicals.

Internal Structure of Clam

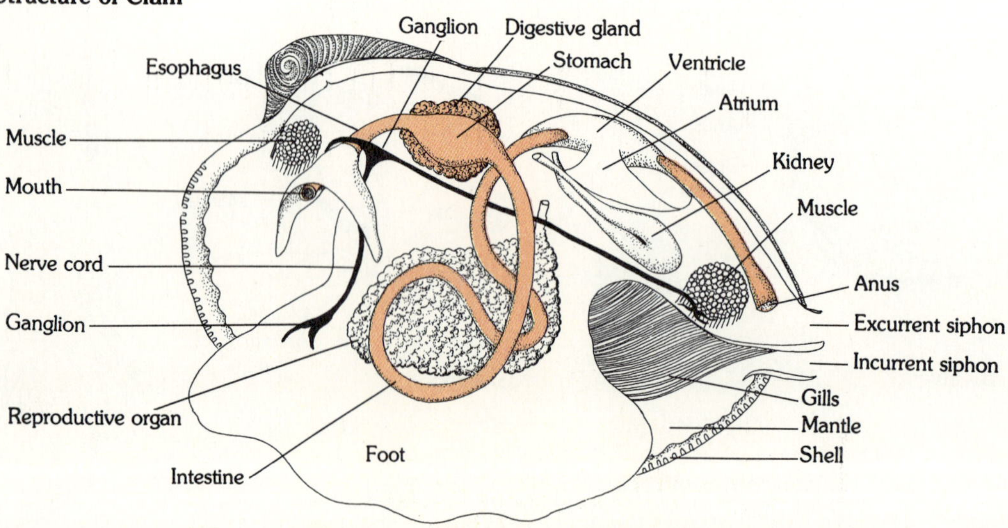

Locomotion and Feeding

When the clam moves, the foot is thrust out to burrow into the sand. Once burrowed, the foot fills with blood, forming an anchor. Then the muscles of the foot contract, pulling the body of the clam forward.

Because the clam moves so slowly, it is not able to chase prey. Instead, the clam strains food particles from the water that flows through the gills. Cilia on the gills sweep food forward along the gills to the opening of the mouth. From the mouth, food passes into the esophagus, the stomach, and the intestine. Undigested food and waste pass from the anus into the mantle cavity. From there, waste is carried out of the clam by the flow of water through the excurrent siphon.

Did You Know?
In some bivalves, when a foreign object, such as a grain of sand, becomes lodged between the mantle and the shell, the mantle secretes calcium cabonate around it, and a pearl forms. Cultured pearls are formed when a mother-of-pearl bead is placed between the mantle and the shell of an oyster. Large offshore areas are devoted to farming oysters to raise cultured pearls.

Circulation

The circulatory system is made up of the heart and blood vessels. The heart has three chambers: a **left atrium,** a **right atrium,** and a **central ventricle.** Each atrium collects blood coming from the gills. The ventricle pumps blood through arteries to the body. Many of the arteries lead to open body cavities called **blood sinuses.** Oxygen is absorbed directly by cells surrounding these sinuses. From the sinuses, blood flows through the gills and then back to the heart.

Reproduction

The clam reproduces sexually with the help of its siphons. The male sheds sperm into the water that flows out through his excurrent siphon. Fertilization usually occurs in the body of the female. The sperm may be taken into a female clam by the way of her incurrent siphon. A large number of larvae develop from the fertilized eggs. These larvae leave the female clam through her excurrent siphon. Most of the larvae are eaten by other organisms. The few that survive attach themselves to fish. During growth, the young clam is parasitic, depending on its host for food. When development is complete, the adult clam sinks to the bottom and becomes a free-living individual.

Circulatory System of Clam

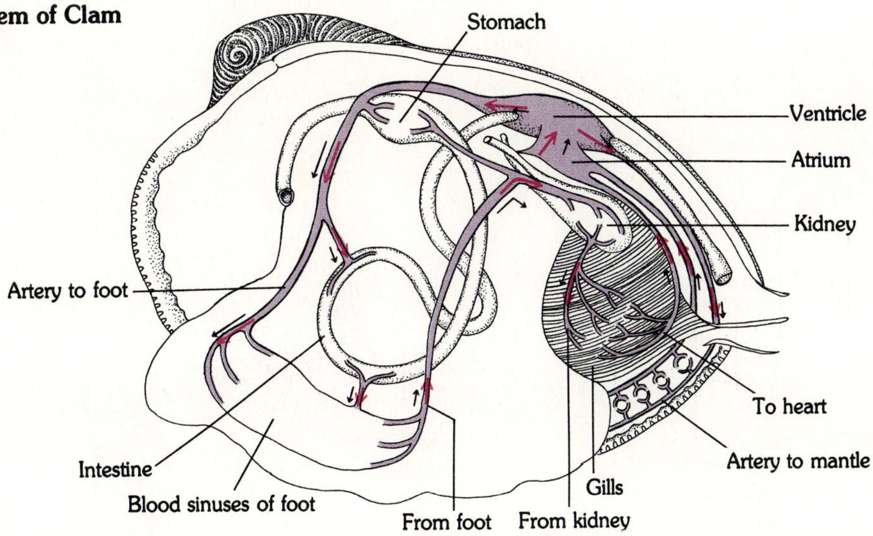

Cephalopods, the Head-footed Mollusks

Octopus

Cephalopods include the octopus, cuttlefish, squid, and chambered nautilus. The giant squid is the largest of all invertebrate animals. A giant squid may be 15 meters in length, 6 meters around, and 2,000 kilograms in weight. Sightings of these squids by sailors may have led to wild tales about ship-eating sea monsters.

The foot of an octopus has eight arms, or tentacles. The foot of a squid has ten tentacles. The heads of these active mollusks are large and well developed. The head includes a brain, complex eyes remarkably similar to our own, and two powerful jaws.

The mantle of a cephalopod is usually a thick, muscular cape with a funnel projecting out of it. When a cephalopod moves rapidly, it contracts the mantle, vigorously forcing the water out of the funnel. By this method, a cephalopod can jet propel itself at great speed in a direction opposite to that at which the funnel is pointed. When attacked or threatened, some cephalopods are able to release a cloud of inky fluid that confuses and may temporarily numb the senses of its enemy.

Nervous System of Octopus

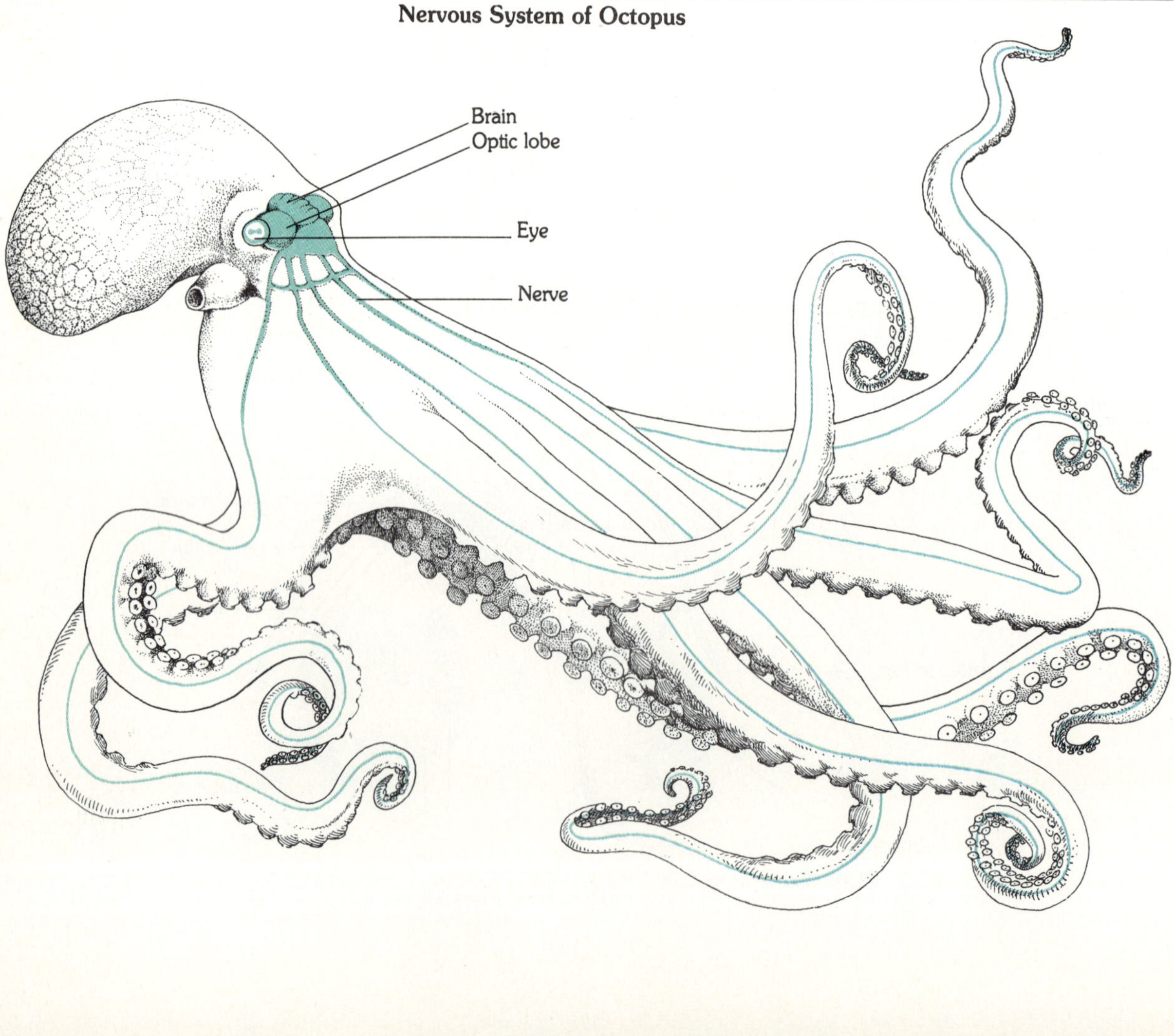

Gastropods, the Stomach-footed Mollusks

Gastropods, the largest group of mollusks, include the snail, the slug, the periwinkle, and the abalone. Gastropods have a single shell or no shell at all. For instance, the snail, the commonest of all gastropods, has a twisted hump of a body protected under one twisted shell. The limpet has a dunce-hat shell. The abalone has a flat shell. The slug has tiny shell remnants buried in its mantle. And the nudibranch has no shell.

A gastropod usually has a broad, muscular foot on its ventral side, explaining the name *stomach-footed.* The foot of a land snail is an organ for creeping. It has numerous strands of muscle running in all directions allowing movement in every direction. The snail is richly supplied with mucous glands that lubricate the sole of the foot. This lubrication makes moving easier. Snails move along by short, rhythmic muscular waves, leaving a trail of fluid.

The head of a snail is at the anterior end of the foot. The head has a pair of touch-sensing tentacles with eyes at their tips. Inside the mouth opening is the scraper, which loosens and cuts up food. The ingested food follows a long curving path through the esophagus to the complex stomach. Food is ground up in the stomach and mixed with the secretions of the large digestive gland. From the stomach, food passes into the intestine. Solid wastes are mixed with mucus before leaving through the anus.

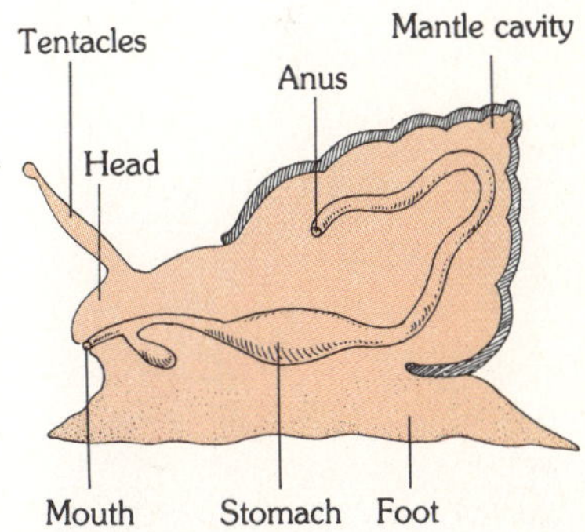

Body Plan of Land Snail

Respiration in snails takes place in the mantle cavity. Oxygen in water and carbon dioxide in blood are exchanged by diffusion through gills. However, land-dwelling snails do not have gills. The space the gills would normally occupy is instead filled with blood vessels. Carbon dioxide and oxygen are exchanged across the walls of those blood vessels.

Review It

1. Describe the digestive system of a clam.
2. How does a squid move?
3. Compare the respiratory systems of a clam and a land snail.

Starfish

17-6
The Echinoderms: Phylum Echinodermata

There are around 6,000 species of echinoderms, which includes starfish, sea urchins, brittle stars, sea cucumbers, and crinoids. Echinoderms live only in the oceans. Echinoderms, also called *spiny-skinned animals,* are covered with a skeleton of plates and spines made up of calcium carbonate. The skeleton lies just beneath a layer of skin cells. Adult echinoderms have radial symmetry with arms usually appearing in multiples of five. Echinoderms are not segmented. And they do not have a head with a central nervous system. Individuals are usually either male or female and have simple reproductive systems. Many species are able to regenerate lost or injured parts.

Structure
A unique trait of echinoderms is the **water-vascular system** that they use for locomotion. Water enters this system through a small, sieve-like opening called the sieve plate. Then the water moves into the **ring canal** around the central disk. Leading off the ring canal are **radial canals,** one down the center of each arm. Off these canals are many elastic, water-filled **tube feet.** The tube feet look like hollow cylinders with suckers on the end. An echinoderm moves by extending an arm until the tube feet strike an object and become attached by suction. Once an arm is attached, the water is removed from its tube feet. The tube feet are then shortened, causing the body to move forward. This method of locomotion is slow, but it allows echinoderms to twist and turn their bodies. The starfish is an echinoderm that has this system of locomotion.

Structure of Starfish

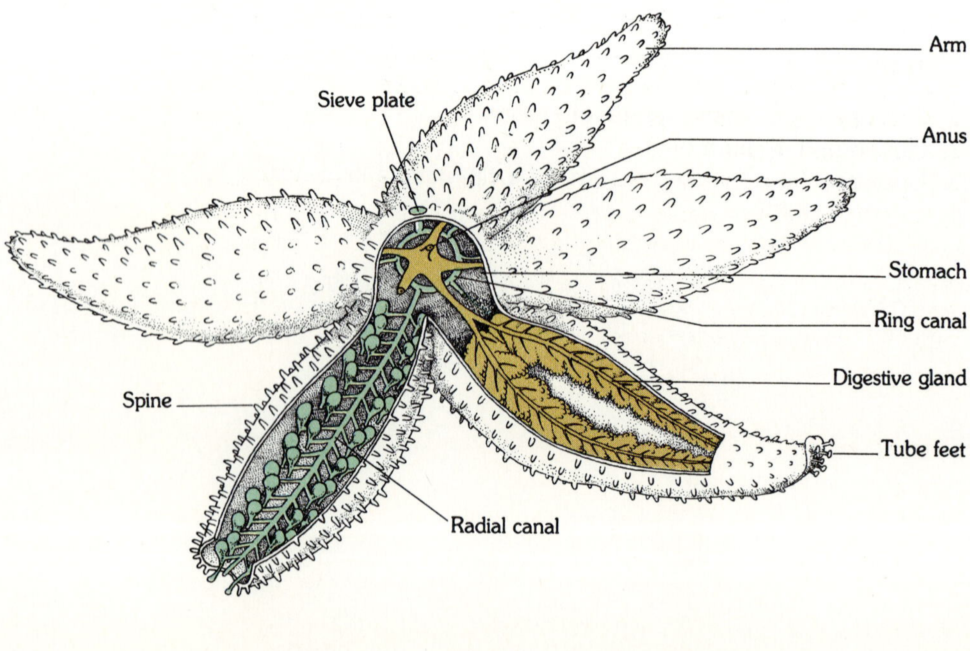

Starfish

Starfish feeding on mussel

The body of a starfish has five or more arms, or rays, attached to a central disk. The entire dorsal side is protected by skeletal plates and rounded spines. The sieve-plate, the opening to the water-vascular system, is located on the dorsal side of the central disk between two arms. Less easily seen on the dorsal side is the anus, where wastes are eliminated. The ventral surface of each arm is grooved to provide a channel for the tube feet. At the tip of each arm is a light-sensitive eyespot.

Feeding and Digestion

A mouth, located in the center of the ventral surface, leads into a large stomach. The stomach may be turned inside out through the mouth so a starfish can digest its food outside the body. Starfish feed on many kinds of animals, including clams and oysters. For example, the arms of the starfish will press on a clam shell, enabling the tube feet to form suction cups. Then, in this humped position, the starfish will pull. It will use different tube feet and change them periodically to outlast the clam. The clam uses its two large muscles to keep its shell closed. In some instances, the clam will tire, and the shell will open wide. But even an opening as small as 0.1 millimeter is large enough for the starfish to force its stomach inside the clam. The inside-out stomach then secretes powerful digestive enzymes. These enzymes turn the soft body of the clam into a kind of thick soup that can be easily slurped up by the starfish.

Economic Importance of Starfish

Starfish have been known to eat as many as ten oysters a day. Tremendous numbers of starfish appear periodically in the oyster beds of the Northeast. An oyster company in Rhode Island once removed 10 million starfish from its oyster beds. Originally, oyster farmers would cut the starfish into pieces and throw the pieces back into the water. But this method only increases the number of starfish. Starfish are able to regenerate. So each piece the fishermen threw back into the water developed into a new starfish.

The two most common methods now used to remove starfish are by mopping and by lime deposits. In mopping, the mop is a horizontal iron bar with a large amount of rope attached. Starfish are tangled in the rope as the mop is dragged across the oyster bed. In the lime-deposit method, lime is poured onto the oyster beds. The lime kills the starfish, but does no harm to the oysters or any other animal with protective scales or shells.

Other Echinoderms

Sea Urchin

Sand Dollar

Sea Cucumber

Crinoid

Sea urchins have oval-shaped skeletons with long, movable spines on the top of the body. They resemble colorful pincushions. Sea urchins are active and feed on both living and dead plant and animal materials. With its unusual chewing structure, known as Aristotle's lantern, a sea urchin can chew through coral and other hard or rocky materials to make a place to live. Sea urchins are related to sand dollars, which have round, flat skeletons.

Adult sea cucumbers vary in size and shape from small, long wormlike creatures to large, spherical ones. The body is soft, leathery, and generally dull in color. A mouth at one end is surrounded by tentacles. Sea cucumbers are generally sluggish. They feed on materials at the bottom of the ocean. Sea cucumbers use their ability to regenerate parts as a defense mechanism. If disturbed, the sea cucumber will split open and expel part of its internal organs. These secrete a fluid that becomes stringy and tough to enmesh the enemy. The sea cucumber then regenerates these organs.

Brittle stars resemble starfish at first glance, but they have no tube feet. Their arms move like snakes, enabling them to move rapidly to catch prey. The arms are loosely attached and can break off easily. So the brittle star can leave an arm behind to escape an enemy. A new arm will be regenerated shortly. Some species of brittle stars have branching arms.

Crinoids include the sea lilies and feather stars. Sea lilies resemble tall flowers. They are attached by stalks to the bottom in deep ocean water. Feather stars move about by swimming and crawling. According to the fossil record, crinoids were very abundant about 350 million years ago.

Brittle Star

Brittle Star

Review It

1. How are echinoderms different from other animals?
2. How does the water-vascular system of starfish work?
3. How is a starfish able to open a clam shell?

Pioneers in Biology

Libbie Henrietta Hyman: Pioneer in invertebrate zoology.

Dr. Libbie Hyman spent the last 36 years of her life, from age 43 to 79, writing six of the most comprehensive volumes ever published in English on the invertebrates. She started with the protozoa, and worked up to the mollusks before she died. This invaluable resource details invertebrate anatomy, embryology, physiology, and ecology. The volumes are illustrated with Dr. Hyman's detailed pen drawings. There are about one million classified invertebrates, so the task she set for herself was enormous.

Hyman was born on December 6, 1888, in Des Moines, Iowa. Her father was a tailor who later opened a clothing store in Fort Dodge, Iowa, where Libbie spent her childhood. Mr. Hyman's business was not financially successful so there was little money for the family. When Libbie completed high school, with the highest grades in her class, she did not plan to attend college.

Hyman was fascinated with nature, enjoying the flowers, plants, and insects that were so plentiful in the woods around Fort Dodge. After high-school graduation, she returned for an extra year in order to take all of the science courses she had missed. She was not aware of the opportunities for further education at various universities in the country. With no plans for

the future, she went to work at a factory, pasting labels on boxes.

But something happened to change the direction of her life. One of Libbie's former teachers discovered that she did not have money to pursue her interests through education. She arranged for Libbie to receive a scholarship to the University of Chicago. Libbie worked hard to keep her scholarship, and earned both her B.A. and Ph.D. degrees. She stayed at the University of Chicago until 1931, first as laboratory assistant, and later as research assistant to Dr. Charles Manning Child.

Dr. Child was a teacher who believed in studying an organism as a whole, and not just as a series of parts. He taught her that the whole being, whether insect, worm, or human, was something special by itself. Though this philosophy was not popular at the time, it struck Hyman as correct. She grew in her field, becoming a respected scientist. While at the University of Chicago, in 1919 and 1922, Dr. Hyman wrote *A Laboratory Manual for Elementary Zoology* and *A Laboratory Manual for Comparative Vertebrate Anatomy,* books that are still used today. But her strongest interests were centered on invertebrate study. In 1931, Dr. Hyman went to New York to begin her life work.

Hyman lived near the American Museum of Natural History in New York, where she could study the many original materials written on the invertebrates in Italian, German, and French. In 1937, the museum appointed her research associate and made a laboratory available for her use. There, her usual workday lasted from 8:30 in the morning to 7:30 at night. When she took time to relax, she enjoyed gardening, listening to music, or reading detective stories.

In 1955 Dr. Hyman was awarded the Daniel Giraud Elliot Medal of the National Academy of Sciences for Volumes II and III of *The Invertebrates.* In 1960 she was honored with the Gold Medal of the Linnaean Society of London. The Gold Medal is the Society's highest award and a mark of international distinction in the field of biological research. Dr. Hyman was the first woman zoologist, and the third American, to receive the award.

For the final ten years of Dr. Hyman's life she suffered from Parkinson's disease. She continued to work, much of the time confined to a wheelchair. Dr. Libbie Hyman died in 1969 at the age of 80.

Chapter Summary

All animals, the invertebrates as well as the vertebrates, are unable to make their own food. Instead, they are heterotrophs. That is, they must find outside sources of the energy and materials they need for survival. Very often the job of getting food requires structures specialized for movement and/or capture of prey.

The invertebrates are a very large group of animals making up more than 90 percent of all known animal species. The trait they all share is the absence of a backbone. Then, additional traits are used to subdivide this large group. By looking at the table below, you can see the main groups of invertebrates studied in this chapter. These groups are arranged in a sequence, according to the degree of structural complexity and specialization each group possesses. Sponges, for example, are the least complex. So they are listed first as the simplest of all invertebrates. Each succeeding group represents some major advance in structural complexity and specialization.

Questions to Think About

1. Compare and contrast the following: one-celled and multicelled organisms, animals and plants, vertebrates and invertebrates, animals with radial and those with bilateral symmetry.
2. The body structures of multicelled animals are said to be specialized. What does that mean?
3. Why are earthworms more active at night than during the day?
4. How do attached animals, such as sponges and crinoids, move to other places?
5. How have invertebrates affected *your* life?

Extra Research

1. Expand the chart shown on this page. Add columns for new traits such as larval forms, systems, and life cycles. Add columns for one or more of the animal phyla *not* described in this chapter.
2. Make a poster showing one or more of the invertebrates described in this chapter.
3. Plan and prepare a meal using invertebrates in place of vertebrate animals.
4. In addition to parasitism, invertebrates display various other symbiotic relationships. Find out about some of those relationships.

Phylum	Symmetry	Tissue	Body cavity	Digestive system	Special traits
Sponges	radial	none	none	none	
Coelenterates	radial	2 layers	none	one opening	nematocysts
Flatworms	bilateral	3 layers	none	one opening	
Roundworms	bilateral	3 layers	false coelom	mouth, anus	
Annelids	bilateral	3 layers	true coelom	mouth, anus	segmentation
Mollusks	bilateral	3 layers	true coelom	mouth, anus	radula
Echinoderms	radial	3 layers	true coelom	mouth, anus	water-vascular system

Chapter Test

Part I

Copy the number of each test item. If the item is *true*, write *T* after the number; if the item is *false*, write *F*.

1. Coelenterates have a unique water-vascular system. (406)
2. Parasitic roundworms harm their hosts. (394)
3. Some mollusks have no shell. (401)
4. Flatworms are in phylum Platyhelminthes. (389)
5. Mollusk shells are secreted by the mantle. (401)
6. Sponges can reproduce sexually by producing buds. (383)
7. Some nematodes are parasites inside the human body. (395)
8. The two body forms of coelenterates are the medusa and the polyp. (384)
9. Adult echinoderms have radial symmetry. (406)
10. The crinoids are an extinct group of echinoderms. (408)

Part II

Copy the number of each sentence below. After each number, write the word (or words) that completes the sentence correctly.

11. The muscular organ of mollusks used for locomotion is called the ____. (403)
12. A fluid-filled cavity lined with mesoderm is called a ____. (396)
13. The starfish can turn its ____ inside out to digest food. (407)
14. Planarians live in what kind of environment? ____ (390)
15. Food is carried to epidermal cells of the sponge by ____. (382)
16. ____ is the ability of organisms to regrow lost or injured parts of the body. (383)
17. The simplest kinds of worms, with both a mouth and an anus as part of their digestive tract, are the ____. (394)
18. A pearl is made when a particle is trapped between the ____ and ____ of an oyster. (403)
19. ____ is an example of a gastropod. (405)
20. Eggs of the earthworm are fertilized in a ____. (400)

Part III

Copy each numbered item from List A. After each item, write the letter and words from List B that best describe it.

List A

21. invertebrates (379)
22. nematocyst (384)
23. spicules (382)
24. posterior (389)
25. annelid (396)

List B

a. animals without backbones
b. tail end
c. segmented worm
d. calcium carbonate and silica
e. bulb containing a thread

Chapter 18

Arthropods

The phylum Arthropoda is the largest phylum in the animal kingdom. There are over one million species, with many still unknown and unclassified. Some of the animals in this phylum are crabs, ticks, dragonflies, ants, barnacles, centipedes, and spiders. Much of the success of the arthropods can be traced to specific body features and to their ability to adapt to different environments. Members of this phylum live in the oceans, in fresh waters, on land, in the air, and inside other organisms. No other group of animals affects humans in more ways than do arthropods. They eat a variety of foods and are themselves food sources for many organisms. Although certain species can be harmful, humans would have difficulty surviving without arthropods.

Chapter Objectives

1. Describe the distinguishing traits of the phylum Arthropoda.
2. Describe the advantages and disadvantages of an exoskeleton.
3. Name the major classes of arthropods and give an example of each.
4. Identify and state the function of the parts in the crayfish, spider, and the grasshopper.
5. Explain the processes of incomplete and complete metamorphosis and give examples of each.
6. Describe the social structure of the bee.

The leaflike legs and coloring of this tropical insect provide it with good camouflage to protect it from its enemies.

18–1
The Phylum Arthropoda

Hermit crab

Krill

Green bush cricket molting

The phylum **Arthropoda** (är throp'ə də) is the largest of all the phyla. In fact, there are more members in this phylum than in all the phyla combined. Estimates of well over a million species of arthropods have been given. Some estimates are as high as 10 million species, the majority of these not yet classified. Among invertebrates, the arthropods are the most highly developed and specialized. A vast diversity among members of this phylum exists, but all arthropods exhibit several characteristics common to every member of this phylum.

The Exoskeleton

All arthropods have a jointed skeleton on the exterior of their bodies. This is known as an **exoskeleton** (ek'sō skel'ə tən). It is secreted from the epidermal layer of tissue. Most of the exoskeleton is composed of a substance called **chitin** (kīt'n). Chitin is a very strong but flexible material. The exoskeleton provides an armor of protection for its wearer. It is also waterproof. But there are some disadvantages to having an exoskeleton.

As the arthropod grows larger, its exoskeleton does not. To accommodate the increase in size, the organism must undergo the process of **molting** (mōlt'ing). During molting the animal sheds its protective armor. Arthropods are very vulnerable at this time, and most will seek shelter during this period. The frequency of molting and the length of time needed to form a new exoskeleton varies among species.

Another disadvantage of an exoskeleton is its limitation of size. Arthropods with exoskeletons will never grow to be very large. Imagine an arthropod the size of a horse. The weight of the exoskeleton would be enormous, and the muscle structure needed to carry such a heavy skeleton would in itself be prohibitive.

The muscles of arthropods are attached to the exoskeleton. The system of muscles runs in all directions. This feature along with the jointed skeleton allows for numerous intricate movements. Most of the muscle is striated, consisting of parallel muscle fibers or strands.

Scorpion

Body Features

Several body features also distinguish the members of the phylum Arthropoda. When looking at the exterior of any member of the arthropods, you will notice that it has *bilateral symmetry.* The body is also segmented, or divided into sections. The three major regions of segmentation are the *head, thorax,* and *abdomen.* Some of the segments, especially in the more advanced arthropods, are fused together, or united. There are numerous segmented appendages. These appendages contain many joints and are often highly specialized. Some examples of modifications are claws, pincers, ovipositors, gills, jaws, fangs, paddles, and antennae. You can see some of these examples below.

Internally, the arthropods have a tube-within-a-tube structure consisting of a mouth, digestive tract, and anus. They also have an open circulatory system with a tubular heart. This means that the blood is pumped out of the heart toward the head region and flows toward the posterior through open spaces or **sinuses** (sī′nəs əz). The blood circulates food and picks up wastes as it passes directly over the body tissues. Other than the tubular heart with an occasional artery extension, there are no organs or vessels in the circulatory system to contain the blood.

The nervous system of the arthropods consists of a double chain of ganglia on the ventral side. There are two dorsal ganglia connected to the double chain which may be slightly larger. These ganglia act like a brain to regulate some behavior, although many activities are controlled by the segments. Many arthropods can still perform functions, such as movement, after the brain has been removed.

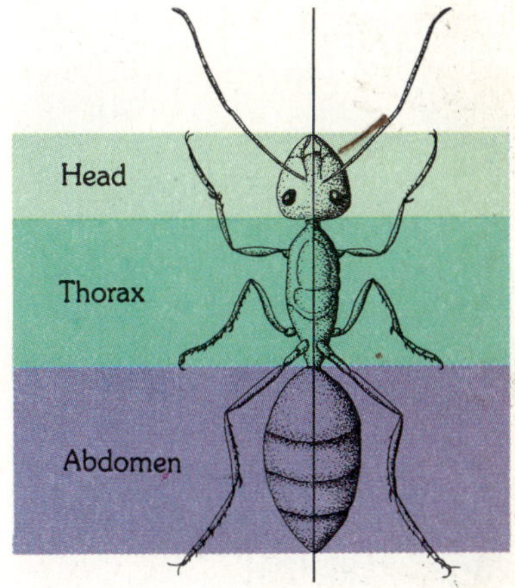

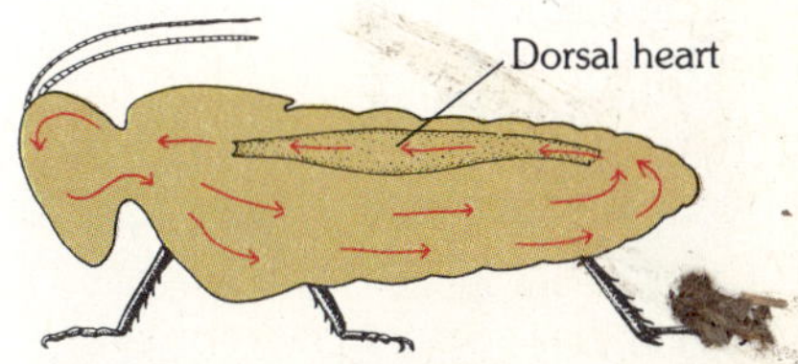

Open circulatory system

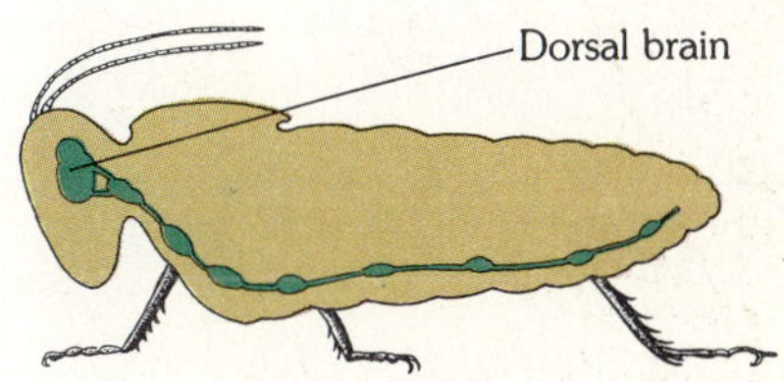

Ventral nervous system

Pincers

Paddles

Ovipositor

Claws

The Success of the Arthropods

As a phylum, the arthropods are considered to be one of the most successful groups of species. Two reasons for their success are the development of the exoskeleton and the adaptations of their appendages. The exoskeleton allows for greater mobility, support, and protection. The appendages have been developed into highly specialized extensions, including sophisticated sense organs such as compound eyes.

There are three major classes of arthropods that will be discussed in this chapter. Focus will be placed on one particular member in each class. These classes are: Crustacea, Arachnida, and Insecta. Two smaller classes, Chilopoda and Diplopoda, will also be discussed.

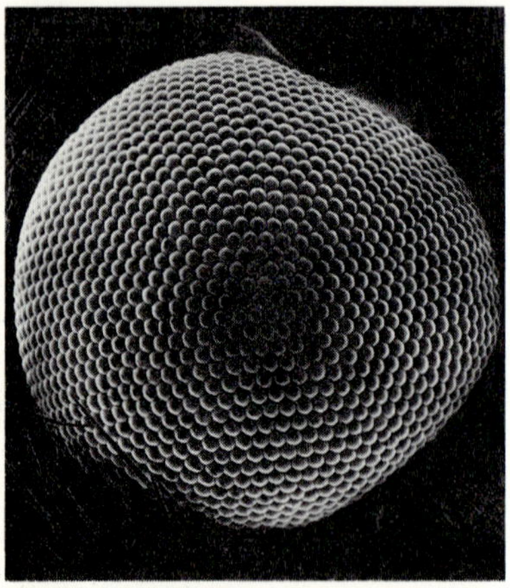

Compound eye

Class Insecta

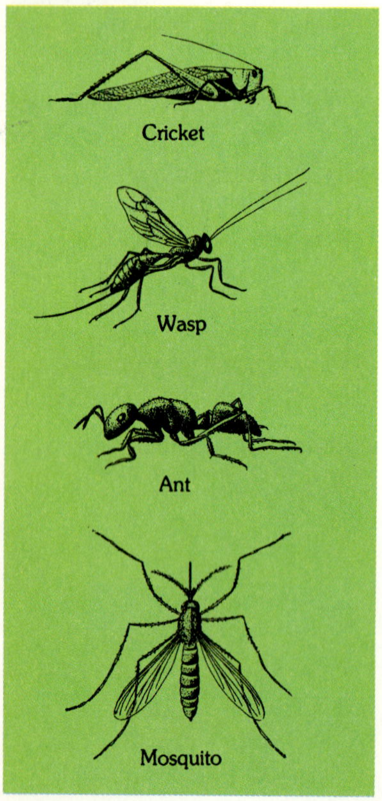

Class Crustacea

Class Arachnida

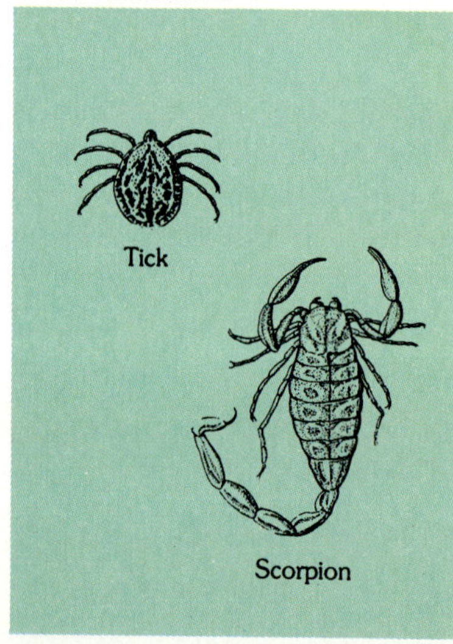

Classes Chilopoda and Diplopoda

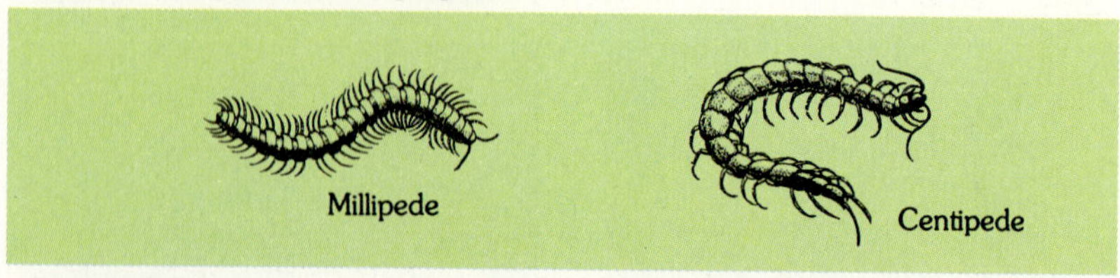

Review It

1. What are the advantages and disadvantages of an exoskeleton?
2. What other body features characterize the arthropods?
3. Explain the workings of an open circulatory system.

18–2
The Class Crustacea

The class **Crustacea** (krus'tā'shə) contains many familiar animals. Members include shrimp, crabs, lobsters, barnacles, crayfish, and water fleas. Most crustaceans are aquatic, living in marine or freshwater environments. A few, like land crabs and sowbugs, are terrestrial, living in damp environments on land.

Crustaceans are characterized by two major body regions and at least five pairs of legs. The unique trait of these animals is the diverse forms of paired appendages.

Humans regard many crustaceans as an excellent food source. Shrimp, crab, and lobster are held high on the list of delicacies and play a significant economic role in commercial businesses.

A great deal of money is also spent to correct vast amounts of damage that may be caused by crustaceans. Barnacles can hinder the operations of a ship and its equipment by attaching themselves to underwater surfaces. Other crustaceans, like the crayfish, dig burrows several feet deep into stream banks which can weaken an embankment to the point where it will collapse. This may contribute to flooding if these banks are part of a levee or dike system.

Let us examine more closely one member of the class Crustacea that is fairly common and clearly exhibits the characteristics of its class. As previously mentioned, the crayfish is an aquatic organism that lives in fresh water. It prefers the muddy bottoms of streams and ponds, where it is a scavenger. It feeds on insect larvae, worms, and dead matter. It will also consume members of its own species. In turn, the crayfish provides food for fish, birds, and mammals, including humans.

Prawn

Barnacles on a whale

Crayfish

The Crayfish

The External Structure

The external structure of the crayfish is like other members of its class. There are two major body regions, the **cephalothorax** (sef′ə lō-thôr′aks) and the **abdomen.** The cephalothorax is the fused segments of the head and thorax regions. It is covered by a piece of exoskeleton called the **carapace** (kar′ə pās).

The most noticeable feature of the crayfish is its numerous paired appendages. At the head region are five pairs of appendages. These include a pair of **antennules** (an ten′yülz), used as organs of balance and hearing. The second pair of appendages are long **antennae** (an-ten′ē). These are organs of touch, taste, and smell. The jaws, or **mandibles** (man′də bəlz), are the third pair. The mandibles of a crayfish crush the food, moving from side to side, unlike our own jaws, which move up and down. The fourth and fifth pairs of appendages on the head are called the **maxillae** (mak sil′ē). They hold food, tear it up, and pass it to the mouth.

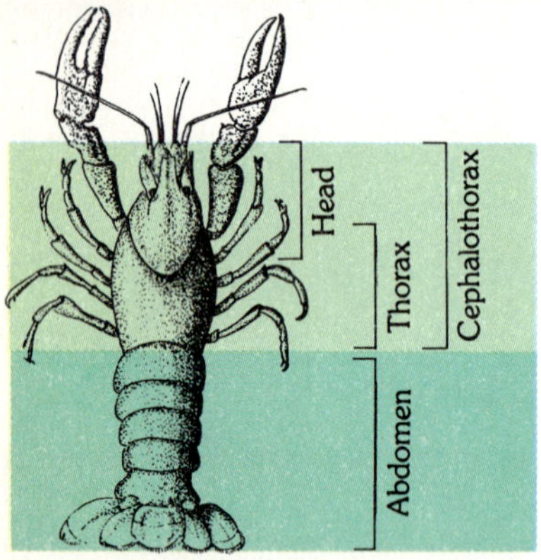

Dorsal view of crayfish

The thorax region of the crayfish has even more pairs of appendages. The first three pairs are called the **maxillipeds** (mak sil′ə pedz′). They hold the food. The next pair are the claws, or **chelipeds** (kē′lə-pedz). They are used in prey capture and in defense. In back of the claws are four pairs of **walking legs.** There are seven joints per leg. These joints allow the crayfish to walk forward, backward, and sideways. Should an appendage be lost in a battle, the crayfish will replace that appendage during the next molting process.

The *abdomen* is the posterior region. It, too, has several pairs of appendages. Unlike the cephalothorax, the abdomen is clearly divided into seven segments. The first five segments have paired appendages called **swimmerets** (swim′ə rətz). They function in swimming and in the reproductive process. The first pair of swimmerets in the male crayfish are slightly larger and aid in transfer of sperm to the female.

External structure of the crayfish

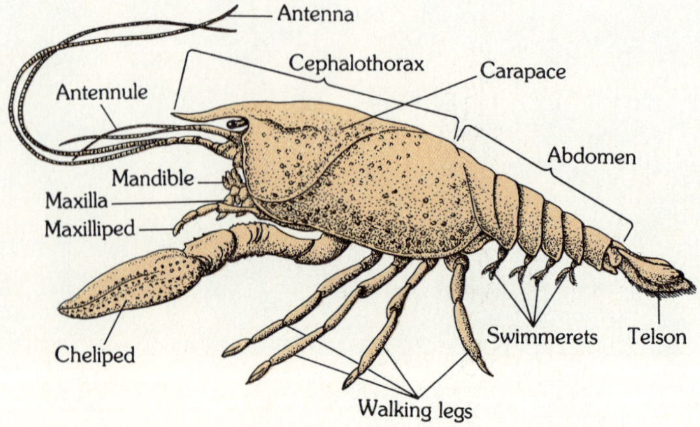

The Telson

The sixth segment of the abdomen is modified into a flipper, and the seventh segment is the tail, called the **telson** (tel′sən). Strong muscles flip the telson swiftly in a ventral direction forcing water forward. This causes the crayfish to be propelled backward. The action causes quite a stir on the bottom of the stream, and the crayfish is usually lost to its predators in the murky cloud.

The Digestive System

After the food is sufficiently torn and shredded, it passes into the mouth, down a short esophagus to the stomach. Rows of teeth made of chitin line the stomach to further grind the food. After being mixed with digestive juices, the nutrient material is strained and passed into the digestive glands where it is absorbed. Material that is not digestible is passed on to an intestine and eliminated through the anus.

The Circulatory System

The circulatory system of the crayfish is like that of other arthropods. It is an open system consisting of a thin, dorsal, tubular heart, extended by a few arteries. The heart pumps the blood forward to the head. The blood then trickles toward the posterior and delivers nutrients and picks up wastes as it flows over the tissues. Sinuses and spaces collect and return the blood to the heart to be recirculated.

The Nervous System

The crayfish possesses a ventral nerve cord with numerous ganglia that control many activities throughout its body. A dorsal brain, consisting of two enlarged ganglia, is located just beneath the **compound eyes.** The compound eyes have many lenses, although the sight of the crayfish is thought to be limited. Hearing is also limited. But the senses of touch, taste, and smell are all highly developed in the crayfish. The antennules, antennae, and maxillipeds are appendages specialized for these senses.

Internal structure of crayfish

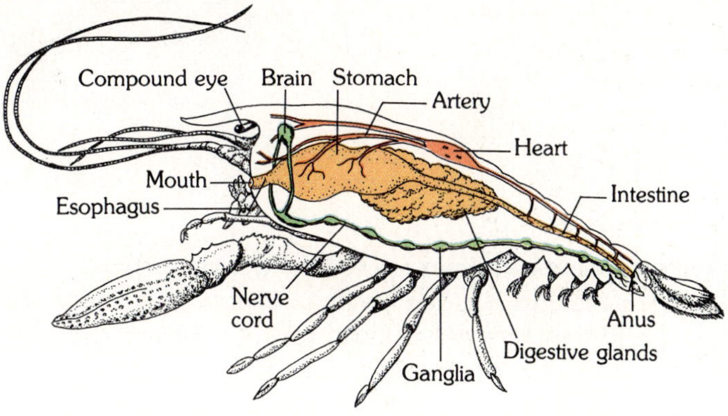

Location of gills in a crayfish

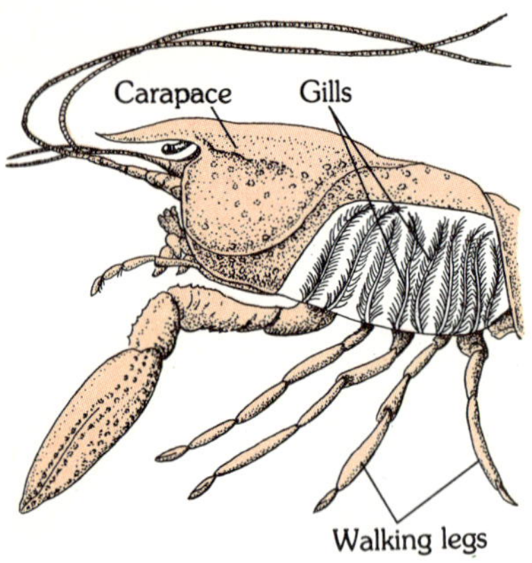

Cross section of crayfish

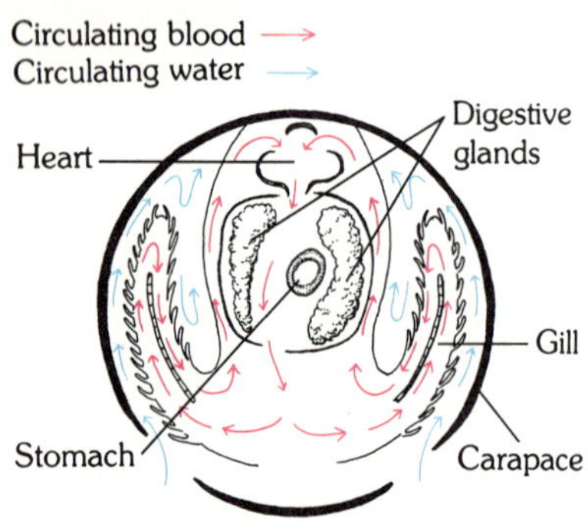

The Respiratory System

Since crayfish are freshwater, aquatic organisms, they must obtain their oxygen from the watery environment. Specialized structures called *gills* (gilz) provide the surface for gas exchange. The featherlike gills are located underneath the carapace and are attached to the legs of the crayfish. The surrounding waters are continually being circulated up under the carapace by the movements of the second maxillae and the legs, supplying an ample amount of oxygen. A constant flow of blood to the gills picks up the oxygen by diffusion and releases carbon dioxide to the water as a waste.

The Excretory System

Waste materials from metabolic processes are removed from the tissues by the blood. Wastes other than carbon dioxide, such as those containing nitrogen, are collected by the blood and delivered to organs called the **green glands.** These are located in the head in front of the stomach. Waste material from the green glands is excreted to the environment through a pore at the base of each antenna.

The Reproductive System

The crayfish, like many organisms, reproduces sexually. Sperm are transferred from the male to the female and stored in small receptacles on the ventral side. As the eggs are laid, they are fertilized by the sperm. The eggs are covered by a secretion that causes them to stick to the female's swimmerets. She carries the eggs until they hatch and develop enough to be on their own.

The excretory and reproductive systems of the male crayfish

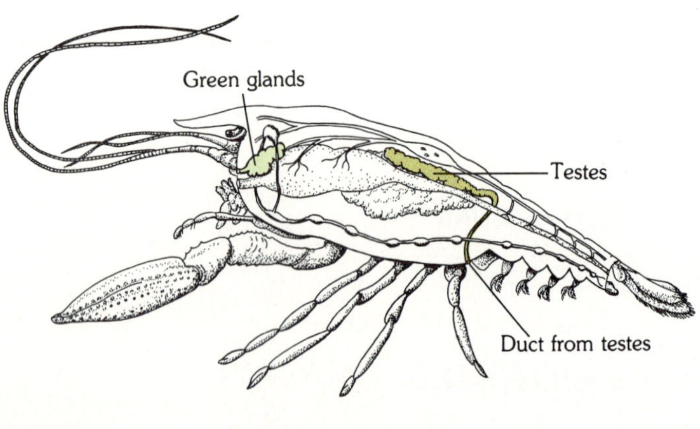

Review It

1. Of what importance to humans are the crustaceans?
2. What body features help the crayfish adapt to its environment?
3. What special structure allows the crayfish to obtain oxygen? Explain how this structure functions.

Activity

Observing Land Crustaceans

Purpose
To discover how a land-dwelling crustacean has adapted to living on land.

Materials
15 live sowbugs
2 stacking culture dishes
Blotters or paper towels
Black, lightproof paper
Pencil

A. Procedure
Sowbugs are small land crustaceans. They are about one centimeter long and have flattened, oval bodies.

1. Take a culture dish and line one half with a damp blotter or paper towel. Line the other half with a dry blotter or paper towel.
2. Put about five sowbugs in the dish.
3. After 10 minutes, record how many animals are on each side of the dish.

Analysis
1. In what areas of the culture dish did you find the most sowbugs?
2. What does this tell you about their reaction to moisture?

Sowbugs

B. Procedure
1. Line the bottom of another culture dish with a damp blotter or paper towel.
2. Place five or more sowbugs in the dish. Cover one half of the dish with black, lightproof paper.
3. Place the dish in a well-lighted area.
4. After 10 minutes, record how many animals are on each side of the dish.

Analysis
1. In what area of the dish did you find the most sowbugs?
2. Do you think sowbugs would be more active during the day or during the night? Why?
3. Why is moisture essential to the sowbug?

C. Procedure
1. Take the Petri dish with the moist paper towels and place it in front of you.
2. With the eraser end of a pencil, gently prod one of the animals.
3. Try gently prodding the other sowbugs.

Analysis
1. Was the sowbug sensitive to touch?
2. Explain how this reaction is a defense against attack.
3. How long did it take before the sowbug returned to its normal position?

18–3
The Class Arachnida

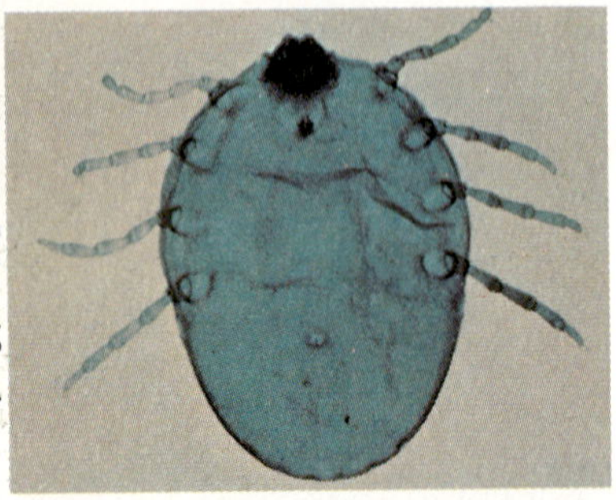

Tick—arachnid

Louse—insect

The class **Arachnida** (ə rak′nə də) also contains many members familiar to humans. Most, such as spiders, scorpions, ticks, and mites, are regarded with distaste. Several groups, such as the ticks, are dangerous because they carry disease and are parasitic. Others, such as the spider and scorpion, can deliver a formidable sting or bite.

Members of the arachnid class exhibit much diversity in form, but all possess several common characteristics. For example, arachnids have no antennae and possess only simple eyes, sensitive only to changes of light intensity. Their bodies are divided into two regions, the cephalothorax and the abdomen. The cephalothorax has six pairs of appendages. The first pair are fangs or pincers and are used for capturing prey. The second pair of appendages vary in appearance and function. Walking legs compose the last four pairs of appendages. This trait is used to quickly distinguish arachnids from insects, which possess only three pairs of walking legs.

The abdomen of the arachnids may appear either long and segmented or short and rounded. The abdomen contains the openings to the respiratory organs located within. Organs of respiration are either **book lungs** or **tracheae** (trā′kē ē′). Book lungs are air-filled sacs that have several leaflike folds. The lungs receive air through openings on the abdomen. Leaves of the lungs are held open by bars so that air can circulate freely. This allows a large surface area of the lungs to be exposed to air. Oxygen can diffuse quickly into the blood that flows through the lungs.

The tracheae are air tubes that end in small fluid-filled **tubules** (tü′byülz). Like book lungs, they receive air through slitlike openings on the abdomen. Once the air reaches the tubules, the oxygen goes directly into the body tissues. Both of these types of respiratory systems are unique to the phylum Arthropoda.

Book lungs

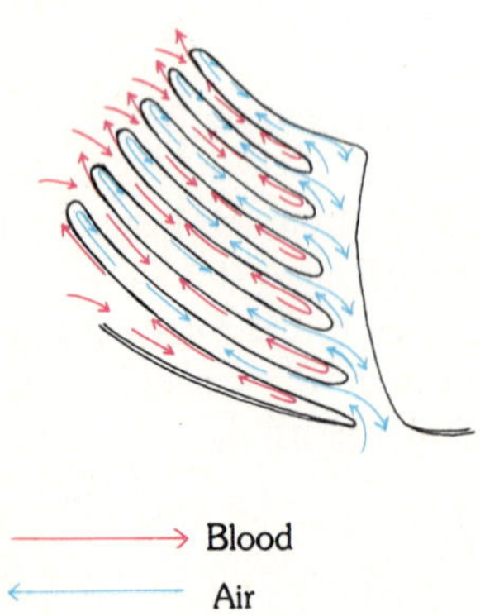

Location of respiratory organs in a spider

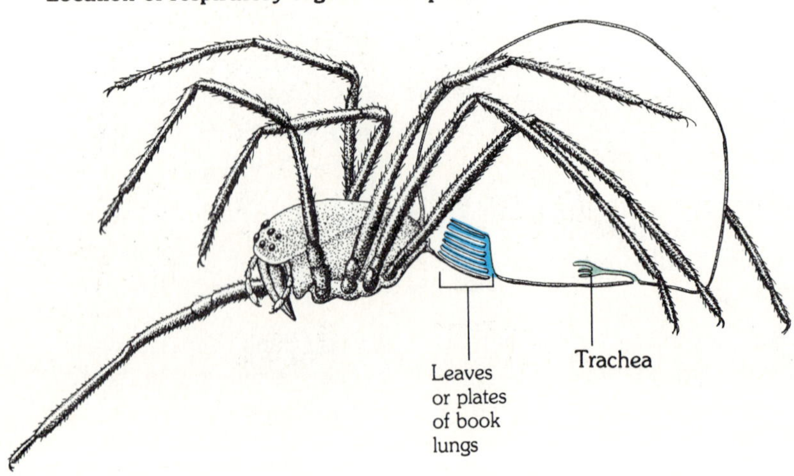

The Spider

Spiders are the largest order of arachnids. There are numerous species of spiders, each with large populations. Although you may consider them to be enemies, spiders are truly beneficial to humans. Spiders destroy many, many insect pests that would otherwise cause extensive damage to crops each year.

A spider's anatomy is typical of its class and phylum. Its body is divided into two segments: the cephalothorax and the abdomen. Appendages on the cephalothorax are paired and specialized. The first pair contains poison glands. These are called the **chelicera** (kə lis'ər ə) and act as fangs to inject the poison. The second pair of appendages, the **pedipalps** (ped'ə palps'), are sensory organs and aid in holding and tearing the prey. The male spider also uses these appendages to transfer sperm to the female. There are usually eight simple eyes located on the cephalothorax, and there are four pairs of walking legs.

The spider, like all arachnids, lives on a completely liquid diet. Digestive enzymes must be injected into the prey along with the poison. These enzymes predigest the food to a liquid state. The liquid can then be pumped into the spider by muscular branches of the stomach. Digestion is then completed internally with the aid of digestive glands.

The spider's book lungs are located on the abdomen. There are several slitlike openings on the ventral side that allow sufficient oxygen to enter. Air is drawn in and expelled by muscular action. The book lungs have leaflike plates of chitin. These leaves provide ample surface area for gas exchange.

Female lynx spider

Fisher spider on egg case

Wolf spider

Female spider—internal and external structure

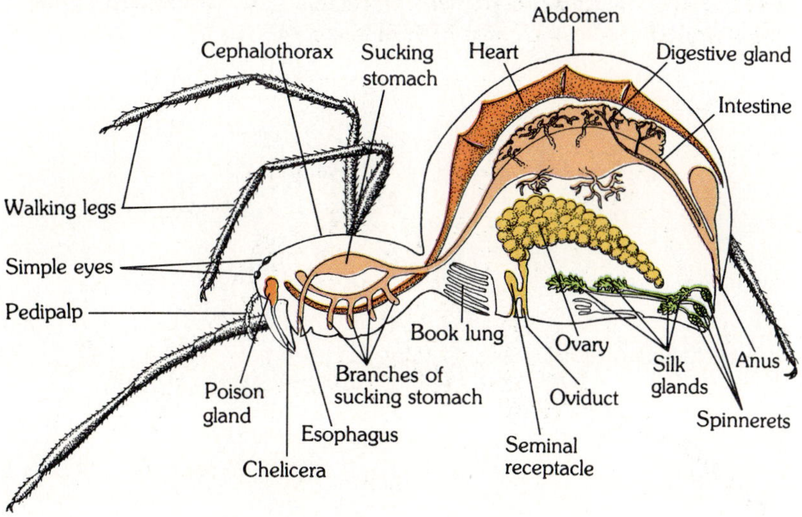

Spider's Silk: A Special Adaptation

The abdomen of the spider also has appendages known as **spinnerets** (spin′ə rets′). These complex spinning organs serve as outlets for the silk-gland ducts. In most spiders, there are six spinnerets, and each spinneret contains many duct openings. The spider's silk is released as liquid protein. It hardens as it is pulled out of the spinnerets. The several kinds of silk glands produce different types of silk threads for various functions. Most spiders use silk threads as a *dragline.* This is the line of silk that is left behind as the spider moves from place to place. It acts as a safety line similar to that used by mountain climbers. Some spiders use silk threads to build webs for capturing food. It has been found that web-spinning is not a learned behavior, but that the spider inherits the knowledge as instinct. Spiders also use silk threads to make protective cocoons for their eggs and to distribute newly hatched spiders on air currents.

Did You Know?

The web of a spider is like a telegraph line. When the web shakes, sensory hairs on the spider's legs let the spider know how big the prey is. If the vibrations are too weak, the spider may ignore them. If the vibrations are violent, the spider may cut the prey loose so that it will not destroy the whole web, or the spider may hide until the intruder has left. Only a medium-sized vibration means food to the spider. When one is felt, the spider rushes to wrap the prey in silk before it can escape.

Fisher spider with prey (fish) in web

Classes Chilopoda and Diplopoda

Both the smaller classes of **Chilopoda** (kī lop'ə də) and **Diplopoda** (dip lop'ə də) are multisegmented organisms. Chilopoda, or centipedes, and Diplopoda, or millipedes, live in moist environments. They may be found under rocks, leaves, and logs.

Centipedes have long, flat bodies. The head has a pair of antennae, mandibles, two pair of maxillae, and two groups of simple eyes. The first segment has poisonous claws to capture and kill prey. Centipedes will feed on earthworms, insects, and other small arthropods. Some larger varieties that live in the tropics will hunt frogs and small snakes. Sometimes their bite can be poisonous to humans.

Except for the last two segments, every other remaining segment has one pair of legs. It is this feature that lends the centipede its name of "one hundred feet." Most species do not usually have that many legs. Thirty-five pairs of legs are average for the class. Centipedes can move rather quickly.

Millipedes have long, heavy rounded bodies. They are primarily vegetation eaters. The head region bears antennae, mandibles, maxillae and, in some species, simple eye clusters.

The major difference between the centipedes and the millipedes is in the number of legs per segment. Whereas centipedes have only one pair per segment, the millipedes have two pairs per segment. Although the name means "one thousand feet," they do not normally possess that many. Their legs are weak and the millipede moves very slowly. When disturbed, a millipede will roll itself into a ball for protection.

Two species of centipedes

Courtesy Carolina Biological Supply Company

Millipede

Millepede rolled into a ball

Review It

1. What physical traits distinguish arachnids from other arthropods?
2. How is the spider adapted for its role in the environment?
3. What physical traits do centipedes and millipedes have in common? In what ways do they differ?

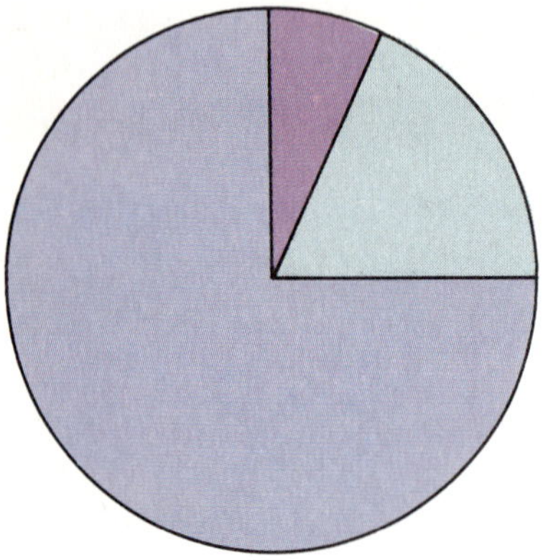

18–4 The Class Insecta

The most familiar group of arthropods belongs to the class **Insecta.** This is the largest class of arthropods, both in the number of different species and the number of individual animals. Scientists have identified more than 700,000 individual species of insects, but they believe many more remain to be identified.

Insects are the most successful group of arthropods. One reason for their success is their high rate of reproduction. *Drosophila melanogaster,* commonly called the fruit fly, can reproduce a new generation every twenty days. Insects have also adapted to almost every environmental condition. Various species of insects can be found in tropical jungles, in cold polar regions, on mountains, and in underground caves. The only place where few insects are found is in the oceans.

The ability to eat a variety of foods is another reason why insects are such a successful group. Some insects live on liquids, such as plant nectar or blood from animals. Others eat fresh or decaying vegetation, and some are meat eaters. Still others eat substances such as manure, hair, glue, and oil.

Insects and humans affect each other in many ways. Since both are very successful species, they sometimes affect each other in negative ways. Insects are considered by most people to be annoying. They can bite or sting and can infect humans and other mammals with disease. Insects also destroy ten to twenty percent of our crops each year. On the other hand, insects pollinate many of the plants we need for food, clothing, shelter, and other resources. They provide us with useful products, such as honey, and are themselves a source of food for fish, birds, and other animals.

Common Characteristics of Insects

Variations among insects are numerous, but a list of the general traits of an adult insect applies to most species.

1. Insects have three clearly defined body regions: head, thorax, and abdomen.
2. The head has a pair of antennae; mouth parts; and compound and simple eyes.
3. The thorax has three pairs of legs.
4. If the insect has wings, these are on the thorax.
5. The abdomen usually does not have any appendages.
6. Insects breathe through tracheae which are similar to those of some arachnids.

The Grasshopper

The grasshopper is a typical example of the class. It is a good insect to study because it is large and found in great numbers. It feeds on vegetation. In fact, at times, grasshoppers cause extreme losses of crops due to their enormous appetites. It is considered a harmful pest by many farmers.

The major external features of the body can be easily examined with the aid of a magnifying glass. The grasshopper, like all insects, has three major regions: the head, the thorax, and the abdomen.

Grasshopper

The Head

The head of the grasshopper is made up of six fused segments. On the frontal side of the head are a pair of jointed antennae. The antennae are sensory organs used for feeling and smelling. Two compound eyes are located on the sides of the head behind the antennae. Each eye is made up of many six-sided lenses. Light enters each lens, therefore the object the grasshopper sees is divided into several parts. The illustration shows how all of these parts combine to make the whole picture. Grasshoppers also have three simple eyes called **ocelli** (ō-sel'ī). They are located between the antennae and the compound eyes. The ocelli are sensitive only to light and dark.

The mouth parts, unlike human mouth parts, are located outside of the mouth. The frontal region of the mouth is covered by an upper lip called the **labrum** (lā'brəm). Behind the labrum are the rough-edged crushing jaws, or *mandibles*. These are lined with chitinous teeth. They move sideways to cut and tear as well as chew food. Behind the mandibles are the maxillae. They assist in holding the food and pushing it into the mouth. The second pair of maxillae are joined together to form the lower lip, or **labium** (lā'bē əm). The labium covers the lower mouth region. Both pairs of maxillae bear sensory **palps** which aid in sorting out the edible particles of food.

Vision of a grasshopper

Grasshopper head structure

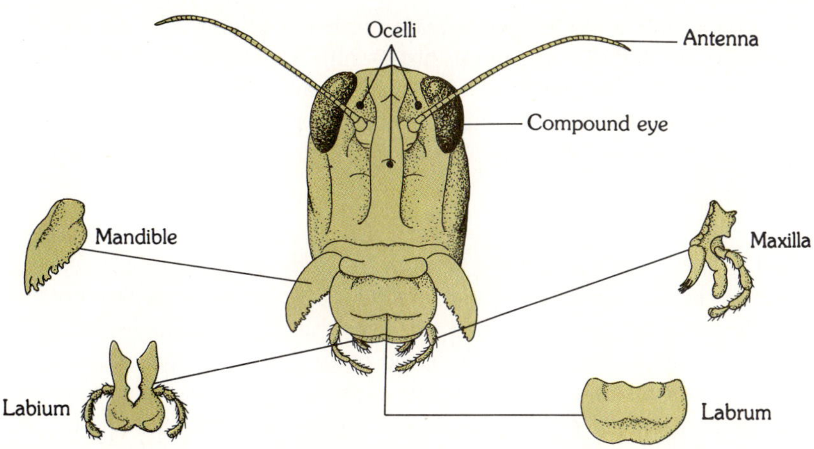

The Thorax

The thorax consists of three segments: the prothorax, the mesothorax, and the metathorax. Tough and flexible tissues attach the head to the prothorax. The head moves freely, even though there is no neck. One pair of legs is attached to each segment of the thorax. The first two pairs of legs are adapted for walking. The last pair of legs is larger and more muscular than the other pairs. These are the legs used for jumping. Each leg is made up of five parts. Pads and claws on the **tarsi** (tär′sī), or feet, help the grasshopper climb and cling to vegetation. In some species of grasshoppers, sound-sensitive organs are located on the **tibias** (tib′ē əz), the long joints next to the tarsi.

The mesothorax and metathorax have one pair of wings each. Each wing is made up of two chitinous flaps of the body wall. Veins running through the wings give them strength. The first pair of wings is not used for flying. They are protective covers for the thin hind wings, which unfold like a fan before flight. The second pair of wings is made to flap by the alternate contraction of horizontal and vertical muscles. Other muscles change the wing angle, giving the grasshopper more speed and the ability to change direction.

The external features

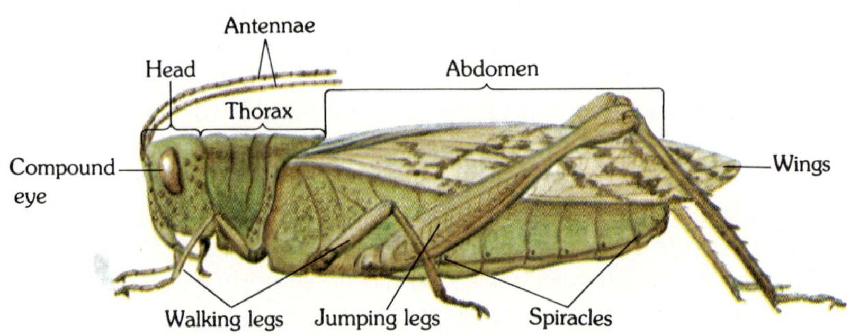

The Abdomen

The abdomen is made up of eleven flexible segments. The sensory organs for hearing, the **tympana** (tim′pə nə), are located on each side of the first abdominal segment. The tympana are similar to our eardrums. The tympana are thin, flat membranes that vibrate when sound waves hit them.

Grasshopper's respiratory system

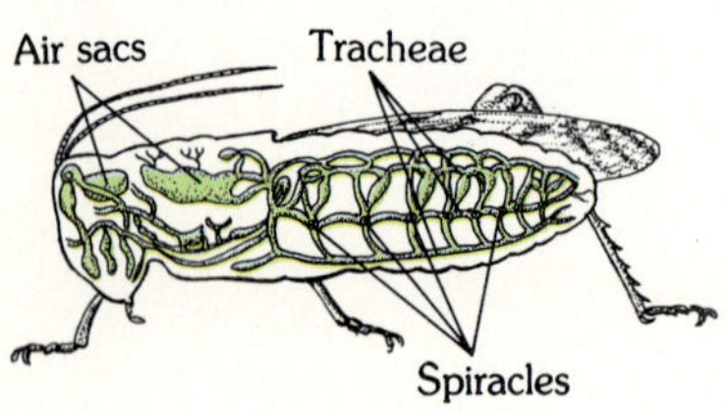

The Respiratory System

The grasshopper breathes through pairs of tiny holes called **spiracles** (spī′rə kəlz). Located on the sides of the abdomen and thorax, each spiracle leads into an air tube or *trachea*. The tracheae divide into smaller and smaller tubes. The smallest tubes, the **tracheoles** (trā′kē-ōlz), branch out to all the cells of the body. These tubules carry oxygen to the cells and remove carbon dioxide. The movements of the wings and abdomen pump air in and out of the spiracles.

The Circulatory System

The circulatory system of the grasshopper, like that of other arthropods, is an open system. The system is made up of one dorsal vessel containing a tubular heart and an **aorta** (ā ôr′tə). The heart forces blood to flow forward and out of the aorta. The blood flows into spaces among the tissues of the head. The blood then trickles back through the blood spaces in the thorax and abdomen, bathing the tissues of the whole body. It then returns to the heart. During the course of its movement, the blood carries food to the body cells and picks up waste materials.

The Digestive and Excretory Systems

The digestive system of the grasshopper begins at the mouth. Juices from the salivary glands are secreted into the mouth to moisten the food. The food can then flow easily through the esophagus into the **crop.** The crop is a large thin-walled sac that is mainly used to store food until the grasshopper is ready to digest it. After leaving the crop, food passes into a muscular *gizzard.* The gizzard contains teeth made of chitin, which grind the food before it passes into the stomach.

The stomach is the main organ of digestion and absorption. At the anterior end of the stomach, are six pairs of pouches called **gastric ceca** (gas′trik sē′kə). The gastric ceca secrete digestive juices containing enzymes. Digested food is absorbed into the blood from the stomach. Undigested food passes through the stomach and into the intestine, or hindgut. The intestine is made up of the large intestine, the *colon,* and the *rectum.* Waste materials leave the grasshopper's body through the anus, which is at the end of the rectum.

Between the stomach and intestine are numerous **Malpighian** (mal pig′ē ən) **tubules.** Waste materials containing nitrogen are collected by the blood from the body cells and absorbed by the Malpighian tubules. These wastes are converted to **uric** (yůr′ik) **acid crystals.** Uric acid crystals are then emptied into the intestine. The waste material is then discharged through the anus. This process conserves water and allows the grasshopper to live in very dry lands.

Circulatory, digestive, and excretory systems

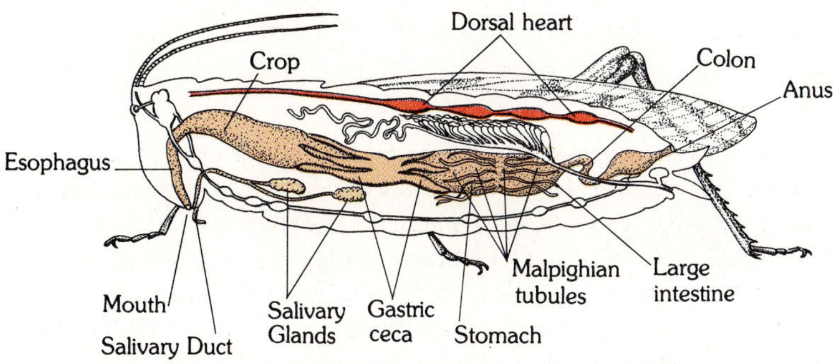

Did You Know?
Insects have many protective devices to prevent the total destruction of their species. One interesting device is camouflage. This is a simple way of hiding from an enemy or victim by blending in with the background. In the picture below, the ambush bug blends with the nearby flowers.

Ambush bug with fly

The Nervous System

The nervous system of the grasshopper is a double cord of ganglia running along the ventral side of its body. Throughout the grasshopper's body are areas of fused ganglia that act as nerve centers for messages from the head, thorax, or abdomen. The fused ganglia in the head are larger than those of the thorax and abdomen. This type of brain receives messages from the sense organs on the head and responds to those messages. The brain does not control all muscular activity but acts as a coordinator. For example, even with its brain removed the grasshopper can still walk, jump, or fly. However, these movements are irregular and undirected.

The Reproductive System

Grasshoppers reproduce sexually. This means that a female egg cell must unite with a male sperm cell in order for a new grasshopper to develop. The male deposits the sperm into a special storage sac in the female. Mature eggs are fertilized as they pass the sperm storage sac and enter the **ovipositors** (ō′və poz′ə tərz). The ovipositors are organs at the end of the abdomen used to lay the eggs. The female digs a small hole in the ground with the ovipositor and deposits the fertilized eggs. The young grasshoppers hatch the following spring.

Nervous and female reproductive systems

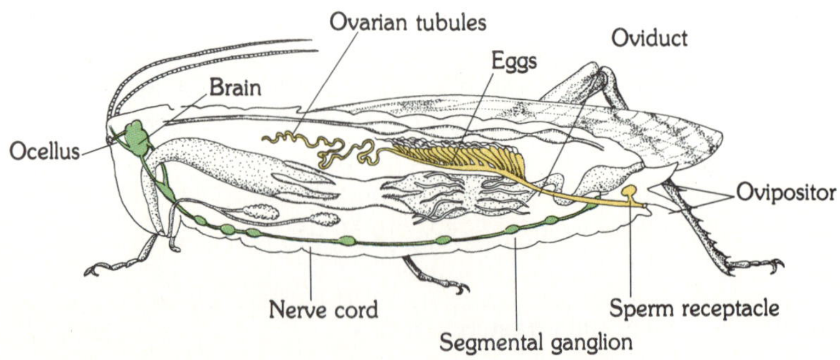

Development of the Insect

Insects spend more of their lives in developmental stages than in the adult life. The method of development from egg to adult varies. A few wingless insects, such as springtails and silverfish, merely become larger as they grow older. However, most insects go through major body changes called **metamorphosis** (met′ə môr′fə sis) as they pass through stages in their life cycles. Two kinds of metamorphoses will be identified: incomplete and complete.

Incomplete Metamorphosis

Grasshoppers develop by means of **incomplete metamorphosis.** It occurs in three stages: the *egg,* the *nymph* (nimf), and the *adult.* When the grasshopper's egg hatches, the young nymph emerges. The nymph closely resembles the adult but lacks certain features, such as wings. A series of molts (usually five) takes place. The nymph will increasingly resemble the adult with each successive molt. Other organisms with incomplete metamorphosis are the aphid and the cricket.

Incomplete metamorphosis

Complete metamorphosis

Egg

Larva

Complete Metamorphosis

Butterflies and moths develop by means of **complete metamorphosis.** There are four distinct stages in complete metamorphosis. These are: the *egg,* the *larva* (lär′və), the *pupa* (püy′pə), and the *adult.*

In this type of life cycle, the young butterfly has no resemblance to the adult. When the egg hatches, the larva emerges. A butterfly larva is called a *caterpillar.* Beetle larvae are called *grubs,* and fly larvae are called *maggots.* The larva does not usually inhabit the same area or eat the same food as the adult. This is an advantage, because it eliminates any competition between larvae and adults of the same species.

During the larval state, the caterpillar has an enormous appetite and with each molt becomes larger and heavier. Once growth is completed, the larva enters the pupa stage. In the butterfly, the caterpillar hangs upside down and forms a protective covering called a *chrysalis.* The insect is quite helpless at this stage. Within the chrysalis, hormones control major body changes and restructure much of the organism. In the final molt, the chrysalis splits open and the adult butterfly emerges.

Pupa

Pupa

Adult

Review It

1. What are some reasons for the success of insects?
2. What traits distinguish the insect from other arthropods?
3. Suppose you observe a grasshopper eating a blade of grass. What process will that blade of grass go through before it is completely digested?
4. Describe the processes of incomplete and complete metamorphosis. What is the advantage of metamorphosis?

18–5
Insect Behavior

Everything an organism does is called its behavior. What it eats, how it protects itself, and how it reproduces are part of an organism's behavior pattern. Behavior may be simple or complex, learned or inherited. Whatever the case, behavior patterns determine how well an organism can survive in various environments.

Among the insects, behavior is almost entirely inherited. Collecting food, eating, and laying eggs are simple activities performed by insects as a result of this instinct. On a more complex level these same activities are carried out in the colonies of the social insects. Social insects include bees, wasps, termites, and ants. They are called social insects because they live together as cooperating members of a colony. Within the colony there is a division of labor. Each member of the colony has a specific job and is physically adapted to perform it.

Drone bee

The Honeybee

The honeybee colony is an example of complex insect behavior. The colony is made up of one queen, several hundred drones, and several thousand workers. The queen is the central figure in the colony. She develops from a fertilized egg that has been laid in a royal cell. When the larva hatches, it is fed special food called **royal jelly.** This ensures the development of a fertile queen. The queen has a large body which is adapted for reproduction. Her function in the colony is the production of eggs.

The drones, or male bees, develop from unfertilized eggs. They are bigger than the workers, but not as large as the queen. Their only function is to mate with the queen. The drones have thick bodies, large eyes, and strong wings. Their mouth parts are not developed enough to gather food. They are fed by the workers. When fall arrives and the supply of food gets low, the workers no longer take care of the drones. The drones die of starvation or the workers sting them to death.

Queen bee

Worker bee

Workers tending cells in hive

The Worker Bee

The workers are female bees. They are not able to reproduce, because their ovipositor has been modified into a stinger. The workers make up the majority of the colony. Their lives are fairly short due to the amount of work they do. The workers are responsible for feeding and caring for the queen, drones, and larvae. They collect pollen and nectar which are used for food. They may even fan their wings to cool the air in the hive. The workers build the hive as well as protect it and keep it clean. They use wax to build a series of combs divided into many cells. Some cells are for the developing bees and others are filled with stored food.

The body of the worker bee is modified to carry out all of its jobs. The mouth parts are adapted for lapping nectar. Their last four abdominal segments contain glands that secrete wax for building the comb, and their legs are modified for collecting pollen.

Worker bees sucking honey

The Life Cycle of the Colony

The queen bee may live from three to five years. Since the hive can only have one queen at a time, the queen secretes an inhibiting substance called **queen substance** that prevents the workers from rearing new queens. The queen secretes enough queen substance for each worker as long as she is able to lay eggs and the hive is not overcrowded. When overcrowding does occur, or the queen is no longer productive, she cannot produce enough queen substance for each worker. The workers then begin to prepare the hive for swarming, the process by which the hive is divided.

First, the workers build a group of drone cells. The queen lays unfertilized eggs in these cells and the drones develop. When the mature drones are four days old, they begin leaving the hive on mating flights. They join groups of drones from other nearby hives in search of a virgin queen to mate with.

After building the drone cells, the workers then build royal cells in the lower portion of the comb. The queen lays one egg in each royal cell. When the eggs hatch, the larvae are fed royal jelly by the workers. It only takes sixteen days for a new queen bee to mature. While this is taking place, the original queen begins to lose her popularity. Eventually she is forced from the hive and flies off with a group of workers to start a new colony. This is the *prime swarm* and it takes place before the new queens have matured.

After the prime swarm, the first young queen emerges from her royal cell. She checks the hive for other mature queens. If another queen is present, a fight occurs between the two. Eventually one queen is either killed or forced out of the hive. The queen that remains is encouraged by the actions of the workers to leave the hive for her mating flight. She makes up to twelve or more flights, mating with a different drone each time. The queen now has enough sperm to last for her lifetime. She will either lead another swarm from the hive, allowing another young queen to emerge and take over, or remain in the hive and destroy the other young queens.

Developing larvae

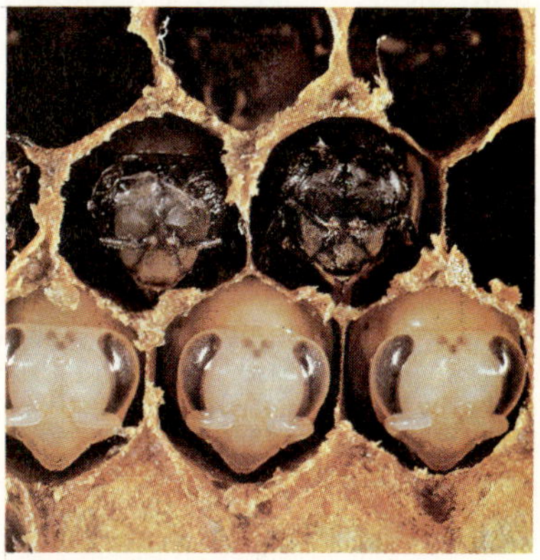

Maturation of pupae

Swarming

Challenge!
Some biologists do not agree with Von Frisch's analysis of the bee dances. They feel that the bee dances are not the only factors used by the honeybee to determine the direction of food. Two biologists who have done other research are D. L. Johnson and A. M. Wenner. Look up their research and make a report on their views.

Variations of the waggle dance

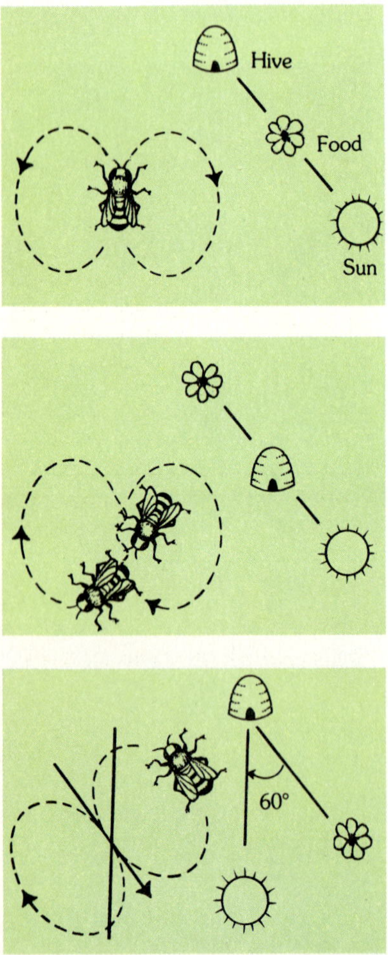

How Honeybees Communicate

Communication plays an important role in the success of a society—human as well as insect. Karl von Frisch and other biologists have done extensive studies of honeybee communication. Von Frisch observed that honeybees communicate with each other through a series of bee dances. When the worker bees find food, they return to the hive and tell the other worker bees about the food. Through their dances, the workers are able to communicate what kind of food is available, how much is available, how far away it is, and in what direction.

If the food is within 100 meters, the worker bee does a round dance on the vertical side of the hive. In the round dance, the bee makes a complete circle, first clockwise, then counterclockwise. This is repeated several times. The other bees gather around and some begin to imitate the dance. Then they fly off in all directions to find the food.

Food that is farther away than 100 meters leads to a different type of dance. This is the waggle dance. It is also performed on the vertical side of the hive. The waggle dance has four steps:

1. The bee runs a short distance in a straight line.
2. She makes a 360-degree turn to the left.
3. She runs straight ahead again.
4. She makes a 360-degree turn to the right.

This figure-eight pattern is repeated several times. During the pattern, the bee continually wags her abdomen. The faster the waggle dance is done, the farther away the food is from the hive.

The direction of the food is revealed by the angle the worker bee makes her straight run. This gives the direction of the food in relation to the sun's position. In the first picture, the bee is heading straight up. This means the food is located in a direct line toward the sun. In the second picture, the bee is heading straight down. This means that the food is in a direct line away from the sun. The third picture shows the bee heading 60 degrees to the left of the vertical. Therefore, the food can be found in a direction 60 degrees to the left of the sun's position.

As with the round dance, the other bees watch and join in the waggle dance. They then fly off in the correct direction to the food.

Review It

1. Name the different types of members found in a honeybee colony and describe their function.
2. How do honeybees communicate with one another?
3. What advantages are there in this type of social behavior?

Breakthroughs in Biology

Pheromones

Hormones have been a part of the biologist's research for a long time. These chemical substances are released within the body by special glands. They do not leave the body of the organism that secretes them. They remain inside, directing or influencing the development and behavior of the organism.

Much is being learned about a similar group of chemical substances called **pheromones** (fer′ə mōnz). Pheromones are different in that they are released from the body of the organism and influence the development or behavior of other organisms of the same species. Pheromones seem to be a major means of animal communication. They play a part in mating, growth, and relaying other information.

Synthetic pheromones of the female gypsy moth are placed in a box to capture the male gypsy moth. This is one of the methods being used to control the spread of the destructive insect.

Pheromone involvement in mating is evident among the gypsy moths. Sexually mature female moths secrete a certain pheromone during mating season. This pheromone is detected on the antennae of male gypsy moths. The pheromone can be detected as far away as two kilometers. Once the male detects the pheromone, he flies to the female to mate. Humans use traps baited with the female pheromone to catch the male moths in an attempt to control the population of this pest.

Some pheromones act as inhibitors of development. This is evident in the termite colony. The termite colony is made up of a queen, a king, workers, and soldiers. The queen and king secrete pheromones that prevent the development of another queen or king. If one of them dies, the "anti-queen" or "anti-king" pheromone is no longer present. Several nymphs are then stimulated to develop into queens or kings.

Pheromones are also useful in food gathering. In the ant colony, the worker ants are responsible for collecting food. When the worker finds food, she returns to the nest leaving a trail of pheromone behind her. Other workers then follow the trail back to the food. The pheromone evaporates in several minutes, thus eliminating the confusion of old trails and preventing too large a following to the food. If the food source is rich, returning ants may lay down their own trails and attract more workers.

Analysis

1. In what ways are pheromones different from hormones?
2. How do pheromones affect the population of the termite colony?
3. What role do pheromones play in food gathering?

Chapter Summary

The phylum Arthropoda contains thousands of species that are highly diversified. Specific body features such as the exoskeleton and specialized paired appendages have helped to make the members of this phylum extremely successful. They are the most numerous of all living organisms.

The major classes of arthropods are the crustaceans, the arachnids, and the insects. Two minor classes include the centipedes and the millipedes. Each class has specific characteristics which adapt it to the environment it lives in.

Occasionally species of arthropods are organized into a social structure. The honeybee is one example. It utilizes a division of labor and an elaborate communication system to successfully survive in its habitat.

Questions to Think About

1. What would happen to an arthropod if it were unable to molt?
2. How do arthropods ensure the survival of their species?
3. What do you think is the significance of a harmless fly resembling a stinging bee?
4. How are insect societies and human societies similar? In what ways are they different?
5. Humans and insects are the dominant animals on Earth. How would you explain the fact that there is only one species of humans and more than 700,000 species of insects?

Extra Research

1. Set up a laboratory situation in which you can keep a living aquatic arthropod, such as a crayfish or water strider.
2. Study the tarsi of houseflies and work out an explanation for how they can walk on ceilings.
3. Read about plants that eat insects. Find out how different species feed on insects. How important are insects to their diets?
4. Examine crabs, shrimp, and lobsters at a fishmarket or in the seafood section of a grocery store. Make a list of the structures you are able to identify. Make sketches of parts you cannot identify and look them up later.
5. Make a poster illustrating the life cycle of an arthropod.
6. Toxicology is the study of poisonous substances. Write a report about the toxins of centipedes, spiders, and bees.
7. Look up the research of Carroll Williams and others on the hormonal control of metamorphosis in the Cecropia moth. Did he find a "fountain of youth" for these insects?
8. Make a report on research being done on pheromones and humans. Do humans produce and react to pheromones?

Chapter Test

Part I

Copy the number of each test item. If the item is true, write *T* after the number. If it is false, write *F*.

1. Arthropods have an exoskeleton made of chitin. (414)
2. All arthropods have eight legs. (422)
3. Arthropods have jointed feet. (415)
4. A crayfish has a ventral heart. (419)
5. Green glands are used for respiration. (420)
6. Millipedes have one pair of legs per segment. (425)
7. An ovipositor is used for laying eggs. (430)
8. Some bees communicate with each other about food by means of dances. (436)
9. Arthropods have an open circulatory system. (419)
10. Spiders belong to the class Arachnida. (422)

Part II

Copy the number of each sentence below. After each number, write the word or words that complete the sentence correctly.

11. The type of arrangement in which the body consists of two similar sides is called ____ ____. (415)
12. Lobsters and crayfish are members of the class called ____. (417)
13. A disadvantage of an external skeleton is that it limits the animal's ____. (414)
14. The appendages to which eggs are attached in a female crayfish are called ____. (420)
15. The tympana of a grasshopper are used as organs of ____. (428)
16. The type of metamorphosis that a butterfly shows is called ____ metamorphosis. (432)
17. An insect has three body regions: head, thorax, and ____. (426)
18. A caterpillar represents the ____ stage in the life cycle of a moth. (432)
19. The only kind of bee in a hive that is able to lay eggs is the ____. (433)

Part III

Copy each numbered item from List A. After each item, write the letter and words from List B that best describe the item.

List A

20. mandibles (418)
21. pheromone (437)
22. crayfish (418)
23. molting (414)
24. tracheae (422)
25. insect (426)

List B

a. has a cephalothorax
b. used for respiration
c. jaws
d. chemical released from the body that influences other organisms
e. shedding of the exoskeleton
f. has 6 legs

Focus on Careers

Marine Life Technician

"My job helps plan the future," says Angela, a marine life technician who works for a marine biologist. "We're finding answers to some of the world's biggest problems in the oceans: new food and energy sources, medicines, new methods of weather control.

"Of course, my day-to-day work isn't all that earthshaking," Angela says lightly. "Today in the lab, for example, I started counting the number of plankton in each of fifty samples of ocean water taken from different depths. It will take days, but we'll get an idea of how much food is available for marine animals in this area." (Plankton are tiny organisms that are a vital food source for most marine animals.)

Marine life technicians, although all concerned with life in the oceans, may be involved in a wide variety of jobs. Many, like Angela, work in laboratories where they assist marine biologists in studying ocean life and the environmental conditions that affect it.

A technician may test samples of seawater periodically to determine temperature, saltiness, and signs of pollution. He or she may work in a lab that specializes in a certain form of life. In an oyster lab, for example, one technician might study oyster larvae to discover whether oysters could be grown commercially in laboratories. In similar laboratories experiments with sponges, sea cucumbers, starfish, and other species have produced antibiotics, fungicides, and cancer-fighting agents. A marine life technician uses a variety of equipment, including microscopes, pH meters, spectrophotometers, and other analytic instruments. Some use photomicrography and computer analysis in their work.

Not all jobs are in laboratories. A marine life technician might be a specialist on a fishing boat or on a fish-processing ship. He or she might gather and process life forms for use in research or as biological specimens.

Two years of study at a junior college will prepare a person to become a marine life technician. The course of study emphasizes biology, chemistry, zoology, and marine botany. The student spends much time doing laboratory work and is expected to spend some part of the study on a research ship.

There are many opportunities for marine life technicians. And, since the ocean still harbors many unknowns, the job of a marine life technician is likely to be exciting and worthwhile.

Career Information

Beekeeper: U.S. Department of Agriculture, Personnel Division, Federal Center Building, Hyattsville, MD 20782.

Biological Photographer: Biological Photographic Association, Box 2603, W. Durham Sta., Durham, N.C. 27705

Marine Life Technician: American Society for Oceanography, 1730 M Street NW, Suite 412, Washington, DC 20036.

Parasitologist: American Society of Parasitologists, Southwestern Medical School, Dallas, TX 75235.

Beekeeper
Duties: Maintains beehives for the commercial production of honey and for the fertilization and pollination of orchards, clover, and other crops. Beekeepers use special equipment to handle bees and remove honey and beeswax from hives.

Training: Requires on-the-job experience working with bees and beekeeping equipment. Commercial beekeepers are required to have special training available at vocational schools or agricultural colleges.

Opportunities: Beekeeping is a profitable sideline for farmers and for people living in large cities and small towns, where regulations permit. There are limited opportunities for full-time beekeepers.

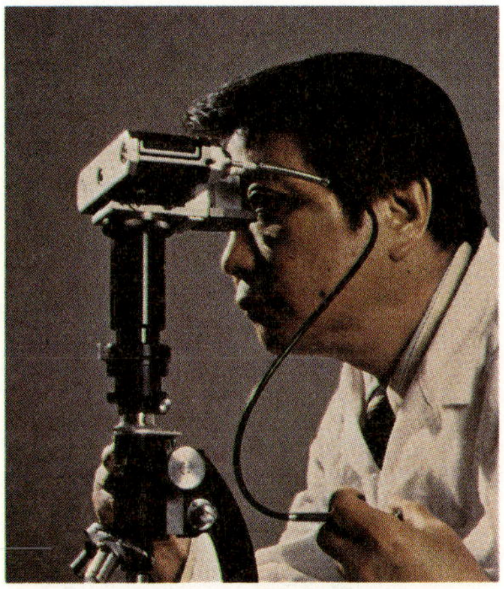

Biological Photographer
Duties: Photographs scientific subjects to illustrate books, magazines, research articles, and teaching activities. Subjects might include a new surgical technique or a virus colony. The photographer uses his or her knowledge of scientific procedures as well as photographic techniques to plan photos and set up required equipment such as microscopes, telescopes, and ultraviolet lighting. The photographer may process film and paper to make slides, prints, and motion picture film. He or she often designs equipment for special jobs.

Training: Requires background in science, photographic training (available in colleges and art schools), and on-the-job training.

Opportunities: Job prospects for photographers with scientific or medical background are very good. These people play an important part in research and education.

Parasitologist
Duties: Studies tapeworms, roundworms, mites, ticks, and other parasites that attack and cause diseases in humans and other mammals. (Parasites are organisms that exist in or on a living host. Many are pathogens; that is, they cause disease. Syphilis, tetanus, ringworm, malaria, and amebic dysentery are some diseases caused by parasites.) Working in a laboratory, a parasitologist investigates the characteristics, habits, and life cycles of various parasites, perhaps specializing in a particular kind. He or she may develop methods or drugs to combat parasites or to treat diseases caused by them.

Training: Requires an advanced degree in parasitology. People with good science background and laboratory experience may be hired to work with a parasitologist.

Opportunities: Jobs are available in private and federal research laboratories.

Bibliography: Unit Seven

Angel, Martin, and Angel, Heather. *All Color Book of Ocean Life.* Octopus Books, 1975. Beautiful photographs and descriptions of common and unusual inhabitants of the ocean.

Bellamy, David. *The Life-Giving Sea.* Crown Publishers, Inc., 1975. An integrated view of the sea, its origins, and the make-up of its inhabitants.

Breland, Osmond. *Animal Life and Lore.* Harper and Row, Inc., 1963. Fascinating and improbable facts about the ways of invertebrates, fish, mammals, and other vertebrates.

Buchsbaum, Ralph, and Milne, Lorus. *The Lower Animals: Living Invertebrates of the World.* Doubleday and Co., Inc., 1960. A complete discussion of representative invertebrates with accompanying photographs.

Callahan, Phillip. *Insect Behavior.* Four Winds Press, 1970. Account of many aspects of insect behavior with a section on possible projects and experiments.

Dethier, Vincent. *To Know a Fly.* Holden-Day, Inc., 1962. Witty and humorous presentation of physiology of flies. Extremely entertaining.

Dozier, Thomas. *Dangerous Sea Creatures.* Time-Life Films, Inc., 1976. Beautifully illustrated volume on poisonous and venomous invertebrates and vertebrates.

Farb, Peter, and the editors of *Life Magazine. The Insects.* Time Inc., 1962. Beautiful picture-essay book that highlights the history of a wide range of insects.

Grzimek, H. C. Bernard, ed. *The Lower Animals.* Van Nostrand Reinhold Co., 1974. Comprehensive encyclopedia of lower vertebrates, from protozoans to arthropods.

Hoyt, Murray. *The World of Bees.* Coward McCann, Inc., 1965. Describes all aspects of honeybee life.

The Illustrated Encyclopedia of the Animal Kingdom. The Danbury Press, 1971. Twenty-volume pictured encyclopedia covering invertebrates, vertebrates, and endangered species.

Jenkins, Marie M. *The Curious Mollusks.* Holiday House, 1972. Discusses all aspects of mollusk life.

Mason, Herbert. *The Fantastic World of Ants.* David McKay Co., Inc., 1974. Story of ant societies.

Newman, L. H. *Man and Insects.* Doubleday and Co., Inc., 1967. Basic book on study of insects, their role in ecology, and their effects on humans.

Schisgall, Oscar. *That Remarkable Creature, the Snail.* Julian Messner, 1970. Easy-to-read introduction to freshwater, saltwater, and land snails.

Shuttlesworth, Dorothy. *The Story of Spiders.* Garden City Books, 1959. Fascinating information and pictures of a wide variety of spiders.

Snow, Keith. *The Arachnids.* Columbia University Press, 1970. Valuable reference text dealing with spiders and their relatives.

Tweedie, Michael. *All Color Book of Insects.* Octopus Books, 1973. Lavish illustrations of insects along with brief information about their characteristics and behavior.

Articles

Cooke, John A. L. "Unveiling the Black Widow." *Natural History,* February, 1973.

"Following the Ladybug Home." *National Geographic,* April, 1970.

Gilbert, Voss L. "Shy Monster, the Octopus." *National Geographic,* December, 1971.

Herrnkind, William F. "Strange March of the Spiny Lobster." *National Geographic,* June, 1975.

"New Tricks Outwit Our Insect Enemies." *National Geographic,* September, 1972.

Wellington, William G. "Tents and Tactics of Caterpillars." *Natural History,* January, 1974.

UNIT EIGHT

THE VERTEBRATES

Chapter 19

Fish, Amphibians, and Reptiles

You probably don't think you have much in common with sea squirts, lampreys, or sharks. Yet biologists place you in the same phylum with these organisms. You are a *chordate.* So are fish, frogs, snakes, and birds. In addition to being a chordate, you are a *vertebrate,* which simply means you have a backbone. This is another characteristic you share with birds, reptiles, amphibians, and fish.

Although vertebrates are much less numerous than insects, they have successfully invaded most parts of the world—land, water, and air. This chapter will introduce you to some of your fellow chordates, particularly fish, amphibians, and reptiles.

Chapter Objectives

1. Describe the traits that characterize the chordates.
2. Name the classes of chordates and briefly describe a representative of each class.
3. Compare the three classes of fish.
4. Explain how amphibians and reptiles are adapted for life on land.

This mother iguana and her baby are a type of American lizard. They grow four to six feet long and live in tropical trees.

19–1
Phylum Chordata

The most diversified and complex phylum on Earth today is the phylum **Chordata** (kôr'dä'tə). Chordates, the members of phylum Chordata, are found in the air, in the water, and on land. They range in size from the smallest fish to the largest whale.

At some time in their life cycle, all chordates possess three characteristics not found among the animals you have studied so far. The first characteristic is a dorsal **notochord** (nō'tə kôrd), a stiff, solid cord that runs down the back of the body. The second characteristic is a dorsal *hollow nerve tube.* The third characteristic is the presence of *pharyngeal* (fə rin'jē əl) *gill slits,* openings found in the throat region. In all but the most primitive chordates, the notochord is not apparent except in very early stages of development. In land-dwelling chordates, such as humans, the gill slits also are not apparent except in the early stages of development.

The phylum Chordata is usually divided into three subphyla: **Urochordata** (yür'ō kôr dä tə), **Cephalochordata** (sef'ə lō kôr dä tə), and **Vertebrata** (vėr tə brä'tə).The urochordates and the cephalochordates never develop true backbones. For this reason, they are considered to be more primitive than the vertebrates. Only the vertebrates, which include humans, develop true backbones. It is thought that the vertebrates developed from an ancestor like a urochordate.

Urochordata (tunicate)

Cephalochordata (lancelet)

Vertebrata (alligator)

Subphylum Urochordata

If you live near the seashore, you may have stepped on the most well-known urochordate: the **tunicate** (tü'nə kit), or sea squirt. Urochordata is sometimes called *Tunicata,* after the tunicate.

The adult tunicate is a baglike animal that lives attached to rocks along the seashore. The tunicate feeds on tiny food particles in the water. One opening in the body, called the *incurrent siphon,* takes in water. The tunicate filters out the food particles and expels the water through another opening called the *excurrent siphon.*

Close examination of the adult tunicate shows only one chordate feature: gill slits. You must look at the larval form to see the other two characteristics: the notochord and the tubular nerve cord.

The larval tunicate is a tadpolelike organism. It swims actively, feeding as it goes. As the tunicate larva matures, it sheds or partially absorbs its tail and finds a rock upon which to attach itself.

Urochordates have little importance to humans except as a reminder of what an ancient precursor of Vertebrata may have looked like.

Tunicate: adult form

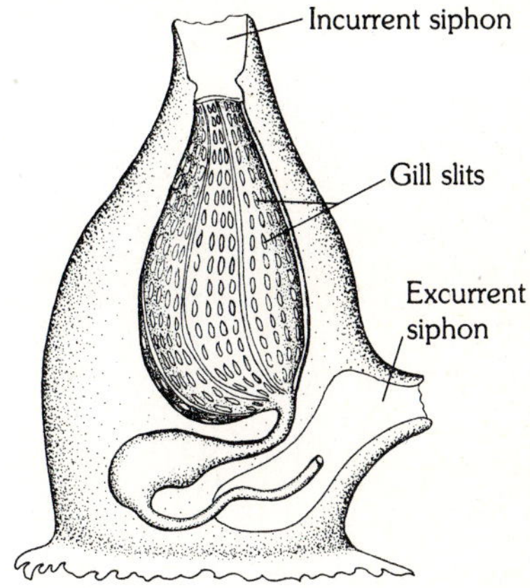

Subphylum Cephalochordata

The members of this subphylum more closely resemble the vertebrates with which you are familiar. The cephalochordates are active, fishlike animals. The most common example is the **lancelet** (lans'lit), or **amphioxus** (am'fē ok'səs). The Chinese use the lancelet for food, but you probably have never seen one. Lancelets live mostly in the waters off the coast of China.

The lancelet shows all three chordate characteristics: notochord, tubular nerve cord, and pharyngeal gill slits. The slender lancelet also has an eyespot, segmented muscles that enable it to swim, and folds of skin extending along both sides of its body. The folds work the way the fins of a fish do.

Tunicate: larval form

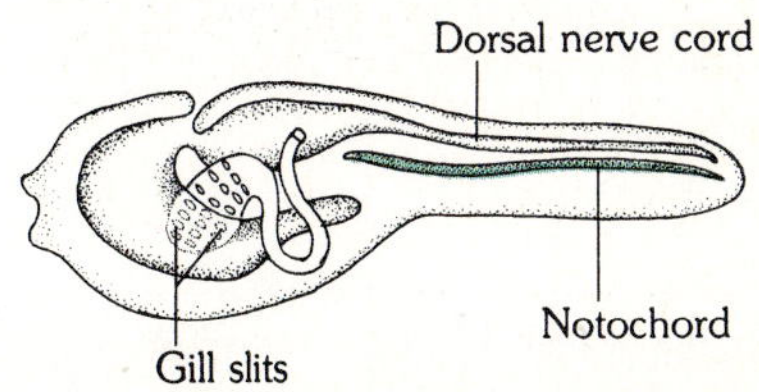

Adult lancelet

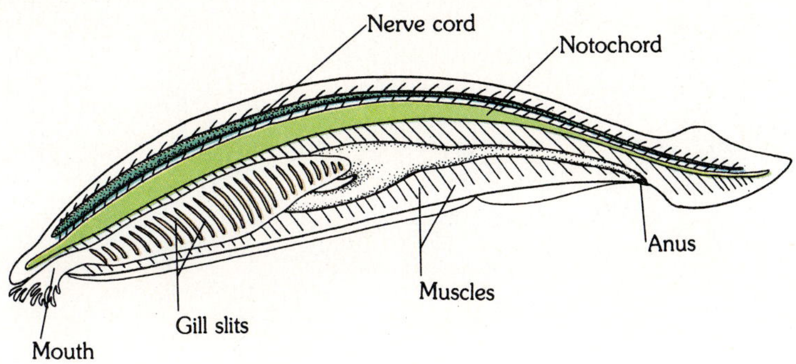

Subphylum Vertebrata

Certainly the most familiar and widespread chordates are the vertebrates, animals with backbones. You can find vertebrates almost everywhere you look. Beneath the ground are moles and gophers. In the sky are bats and birds. In the waters of the world are fish, whales, dolphins, and porpoises. On land are snakes, lizards, frogs, turtles, dogs, deer, and cats. Altogether, there are about 54,000 vertebrate species.

Though there is a wide variety of forms, vertebrates share six characteristics:

1. an internal skeleton
2. a backbone
3. two pairs of appendages (fins or arms and legs)
4. a closed circulatory system
5. a heart with two or more chambers
6. a large, central cavity containing vital organs

Living vertebrates are divided into seven subgroups, or classes: (1) *Agnatha,* (2) *Chondrichthyes,* (3) *Osteichthyes,* (4) *Amphibia,* (5) *Reptilia,* (6) *Aves,* and (7) *Mammalia.* A representative of each class is shown below. An eighth class, *Placodermi,* consists of extinct species of fish with jaws and bony scales. In this chapter, you will learn the first five living classes of vertebrates. Aves and Mammalia will be introduced in the chapter that follows.

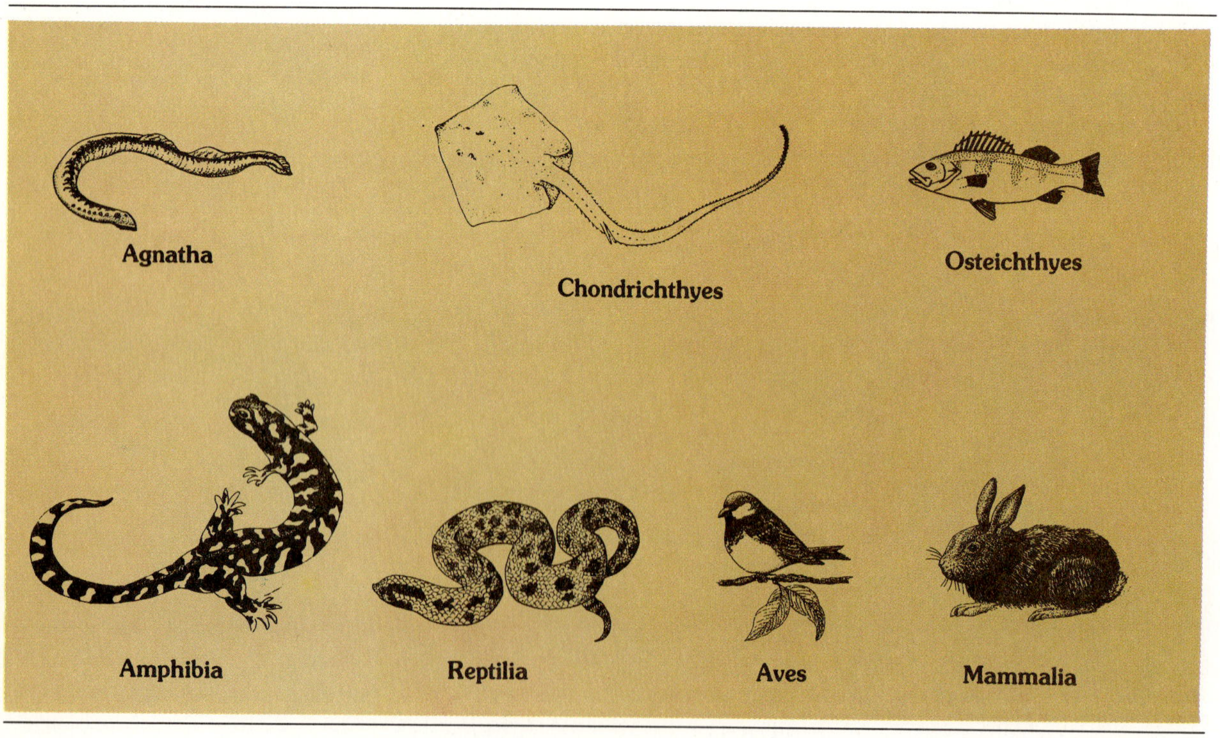

Ectotherms and Endotherms

It is important to note that the five classes discussed in this chapter all share a common trait. All of these animals have a variable body temperature. The temperature of the organism changes with the temperature of its surroundings. Commonly, these animals are referred to as "cold-blooded." The more scientific term for this is **ectothermic** (ek'tō thėr'mik). Humans and all other mammals and birds are **endothermic** (en'dō thėr'mik), or "warm-blooded." This means that the internal body temperature remains constant, regardless of external temperature changes.

Whether an organism is endothermic or ectothermic is important. This feature plays an important role in controlling how active an organism is in its surroundings. Many ectotherms must regulate their activities according to the external temperature. As the temperature decreases, so does an organism's *metabolic rate,* or the speed at which the chemical and physical changes occur within an organism. Modifications in behavior have evolved to help an ectotherm cope with fluctuations in external temperatures. Some of these behaviors include basking in the sun and hibernation.

Garter snakes hibernating in mud

Lizard basking in the sun

Review It

1. What three characteristics are always found in chordates during some stage of their life cycle?
2. What are six basic traits of all vertebrates?
3. Why are ectotherms (e.g. lizards) less active in cooler temperatures? In what ways can ectotherms modify their activities so that they can inhabit colder regions?

19-2
The Fish: Agnatha, Chondrichthyes, and Osteichthyes

Adult lamprey (right)

Lamprey mouth

The fossilized remains of class **Agnatha** (ag nə′ thə), or jawless fish, have been found in rocks that are as old as 450 million years. The fossils reveal that extinct jawless fish were covered by thick, bony plates. The only known living representatives of the class Agnatha are the **lamprey** (lam′prē) and the *hagfish*. The lamprey and the hagfish are slippery, scaleless, eellike animals. They may reach a length of up to one meter. Living agnaths show five distinctive features:

1. The notochord exists in the adult form.
2. The skeleton is all cartilage.
3. Blood is pumped through a two-chambered heart.
4. A single nostril connects with an organ of smell.
5. The mouth is round and jawless.

The Lamprey
Most species of lamprey are parasitic. The parasitic species feed by attaching themselves to other fish. Their mouths act like suction cups. Horny teeth inside the mouth enable the lamprey to tear a hole in the host. The lamprey then feeds on the blood and body fluids of the host. The lamprey attack need not be fatal to the victim. Often fish are caught with circular scars—evidence of surviving a lamprey attack.

Lamprey attached to fish

Life Cycle of the Lamprey

Sea lampreys swim up rivers to shed eggs, or **spawn** (spôn). The lampreys make a protective hole for the eggs by removing stones from the river bottom with their mouths. The female sheds about 65,000 eggs. The male fertilizes the eggs as they are deposited in the hole. After about 20 days, immature lampreys, or larvae, hatch from the eggs. Each larva is about one centimeter long.

The larvae migrate downstream to live in mud beds. The larvae lack mouth suckers. They are *filter-feeders:* they strain small plants and animals from the water. After about four years, the larvae mature into adults.

The adult sea lampreys leave the mud beds and return to the sea for a year. The lampreys then return to the rivers to spawn and die, completing their life cycle.

Did You Know?

During the last 50 years, one species of sea lamprey found new feeding grounds in the Great Lakes. The sea lamprey first invaded Lake Ontario through the St. Lawrence Seaway. Niagara Falls prevented the lamprey from going farther for over a hundred years. When the Welland Canal was completed, the lamprey gained access to Lakes Michigan, Huron, and Erie.

Lake trout were the lamprey's major food source because trout swim in the colder depths that the lamprey prefers. The lamprey entered the upper Great Lakes in such numbers that they almost destroyed the trout population. The fishing industry was severely hurt. The only successful method of reducing the lamprey threat has been the use of selective poison on the larval beds of the lamprey.

Class Chondrichthyes

Chondrichthyes (kon drik'thē ēz'), such as sharks and rays, have skeletons made entirely of *cartilage.* Cartilage is a strong, flexible protein secreted by cells. The shape of your ears and nose depends on the cartilage inside.

Chondrichthyes show two great advances over the lamprey and the hagfish: biting jaws and paired fins. Biting jaws enable Chondrichthyes to eat a greater variety of food. Paired fins make Chondrichthyes good swimmers. In addition to a skeleton made of cartilage, biting jaws, and paired fins, living Chondrichthyes show six distinctive features:

1. A reduced notochord persists in adult forms.
2. Blood is pumped through a two-chambered heart.
3. Two nostrils lead to sensitive organs of smell.
4. The mouth is on the ventral, or under, side of the body.
5. Fertilization occurs inside the body of the female.
6. Scales cover the body.

Paired fins in the shark

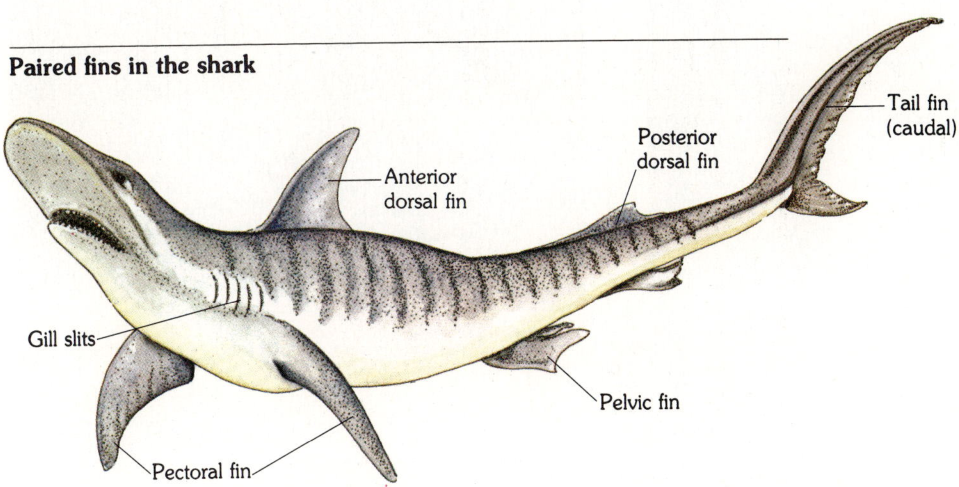

Sharks

Sharks have been scouring the oceans for millions of years. The body of the shark is protected by *placoid* (plak′oid) *scales,* which make the skin feel like sandpaper. Large placoid scales inside the mouth function as teeth.

Most sharks must keep moving to avoid sinking. Some species gulp air into their stomachs to stay afloat. When these species dive, they expel the air with a great roaring belch. Other species rely on oils in their livers to keep them from sinking. Muscle contractions, alternating from side to side, propel the shark through the water. In sharks, the brain is not essential for this movement. It is controlled largely by nerves in the spinal column. Internal ear canals give sharks a sense of balance.

Although sharks are reputed to eat almost anything, only a few attack humans and the largest shark is a filter-feeder. Sharks have a keen sense of smell and can pick up a scent over 400 meters away. Sharks are also sensitive to vibrations in the water. Small pits in the face and along the sides of the body are sensing organs. They sense changes in temperature and water pressure. They also pick up electric charges in the water.

Male and female sharks usually remain separated except during the mating season. The females of most species retain developing eggs within their oviducts and bear live young. Shark litters vary from 2 to 60. The number of shark young is small when you remember that one female lamprey can shed up to 65,000 eggs. The chances of survival of each young shark, however, are better than that of a young lamprey. Shark young are usually born fully formed, ready to fend for themselves.

Challenge!
Marine sharks have 100 times more of the waste product urea *in their blood than humans have. The urea retained in the blood makes the total osmotic concentration of the fluids inside a shark a little more than that of sea water. This prevents a movement of water from the cells of the fish out to the salt water. What would happen to the shark if more water flowed out of its cells than in?*

Leopard shark

Blacktip shark

Nurse shark

Skates and Rays

Skates and rays are similar to sharks, but have flattened bodies adapted to dwelling on the ocean bottom. Most species have smooth skins with patches of placoid scales. Flattened teeth enable most skates and rays to crush the shellfish upon which they feed.

The *pectoral fins,* extending from the chest area, are greatly enlarged in most species of skates and rays. The rippling motion of the extended pectoral fins carries the fish through the water.

Since skates and rays often bury their mouths in the ocean bottom to feed, they breathe through the *pharynx,* or throat region. The first gill slit is modified to form a *spiracle.* The water enters the spiracle via the pharynx. The spiracle closes and the water is forced out the gill slits.

Skates are harmless to humans, but some species of rays are dangerous. The manta, a filter-feeder, may measure up to seven meters across and can overturn a small boat. The scales of the sting ray form poisonous tail barbs. The sawfish has a long, blade-shaped snout armed with barbs. The torpedo ray can generate electrical shocks up to 220 volts—a stunning jolt!

Skate

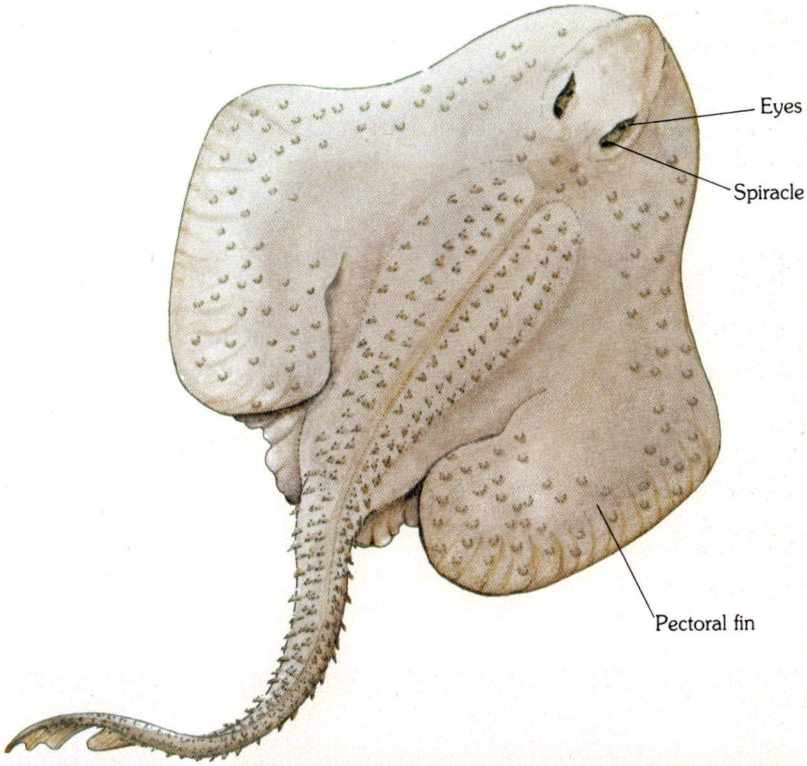

Manta ray

Sawfish

Moray eel

Class Osteichthyes

So far, you have studied two classes of living fish: Agnatha, cartilaginous fish without jaws, and Chondrichthyes, cartilaginous fish with jaws. The third class of living fish is **Osteichthyes** (os′tē ik′thē ēz′), bony fish with jaws. There are more species of bony fish than there are species of any other group of vertebrates in the world. About 20,000 species are known. Most of the fish you catch and eat—trout, catfish, salmon, perch, tuna, and cod—belong to the class Osteichthyes.

Fossil evidence suggests that bony fish developed in fresh waters, whereas the sharks ruled the oceans. Later the bony fish invaded the oceans as well. Like Chondrichthyes, Osteichthyes have biting jaws and paired fins. Bony fish have a bony skull that protects the brain and a vertebral column that protects the spinal cord. Bony fish have better eyesight, a more advanced brain, and a *swim bladder,* which enables them to adjust to different depths with minimal effort. In general, Osteichthyes share the following nine distinctive features:

1. The internal skeleton is mostly bone and includes a skull and vertebral column.
2. Blood is pumped through a two-chambered heart.
3. The eyes are large and have no lids.
4. The mouth is at the front of the head and is equipped with teeth and a hinged jaw.
5. The brain is larger than that of most cartilaginous fish.
6. The gills are supported by **gill arches** and covered by a plate, called the **operculum** (ō pėr′kyə ləm).
7. A swim bladder helps the fish change depths with ease.
8. Fertilization usually occurs outside the body of the female and numerous eggs are shed.
9. The body is covered by hard, thin scales.

The members of Osteichthyes are generally divided into two subclasses: the ray-finned fish and the fleshy-finned fish. The ray-finned fish are by far the most numerous. Although the fleshy-finned fish are small in number, they have a great deal of significance in evolutionary theory. Some scientists think the fleshy-finned fish exhibit traits similar to the precursors of land vertebrates.

Fish of the past

Ray-finned Fish

Nearly all living species of bony fish are ray-finned fish. They are an exceptionally varied group. They can be found at most temperatures and depths of fresh and salt water. They range in size from tiny minnows to sturgeons that weigh as much as 1,000 kilograms. Their shapes vary from streamlined tunas, platelike flounders, round porcupine fish, snakelike eels, and gliderlike flying fish to such unique forms as the sea horse. Every kind of food, ranging from microscopic organisms to other fish, is eaten by some species of ray-finned fish.

Sea horse

The Perch

The perch shows many features that typify Osteichthyes. If you were to look at the outside of a perch, you would see a streamlined, football-shaped body with distinct head, trunk, and tail regions. The mouth is at the front of the head. Directly above the mouth are two **olfactory** (ol fak′tər ē) **sacs,** or nostrils. Behind the nostrils, on either side of the head, are two large eyes without lids. The end of the head is marked by the rear edge of the bony flap called the *operculum.* There is an operculum on either side of the head. The opercula cover the gill slits. Directly behind each operculum is a small pectoral fin. The trunk and tail are covered with scales. Two *dorsal fins* stand erect on the back of the fish. A pair of *pelvic fins* and an *anal fin* are on the ventral side. Behind the anal fin is the tail, which ends with the *caudal fin.* The perch sweeps its tail from side to side to propel itself through the water. The fins help the perch keep its balance, turn, and stop. In addition, the perch has a row of tiny openings, called the *lateral line,* on either side of its body. The lateral line senses vibrations and changes in pressure. The ability of fish to swim together in schools depends on the sensitivity of the lateral line.

External structure of the perch

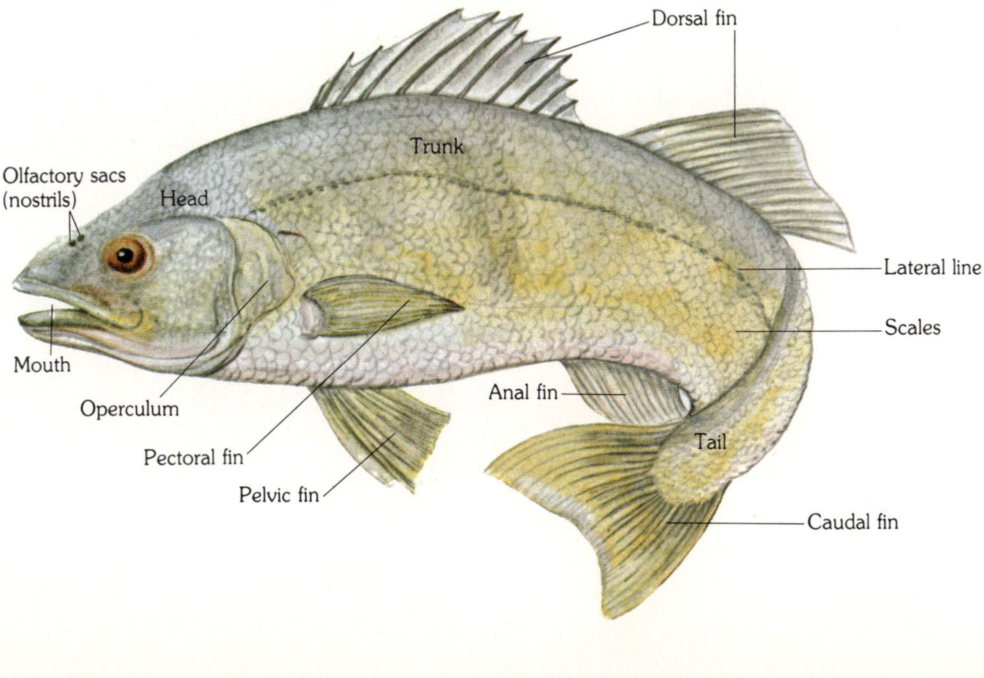

The Internal Structure of the Perch

The *brain* of the perch is protected by a bony skull. The skull is made up of bones similar to those found in the human skull. The perch cannot turn its head because the skull is closely fixed to its vertebral column. Vestiges, or remains, of the notochord can be found among the separate *vertebrae* that make up the vertebral column.

Blood is pumped through a two-chambered heart, which lies below the pharynx in the front of the body cavity. The blood is pale and not abundant. It contains both red and white blood cells. Food and oxygen are circulated by the blood. Wastes are removed from the blood by the kidneys and excreted as urine.

The perch breathes through its gills. Each gill is supported by a cartilaginous *gill arch* which is protected by the operculum. The perch breathes by taking water that contains oxygen into its mouth and expelling this water through its gill slits. Oxygen and carbon dioxide exchange occurs within the gills.

The *swim bladder* is found in the back part of the body cavity of the perch. The bladder is filled with oxygen, nitrogen, and carbon dioxide. To adjust to different water pressures, the perch takes in or gives off gases from its bladder.

The digestive pathway of the perch begins at the *mouth*. The *pharynx* is the throatlike area in back of the mouth. A short *esophagus* leads from the pharynx to the *stomach*. Digestive aids called *pyloric caeca* are located where the stomach meets the *intestine*. Digestion continues in the intestine with the help of the *liver* and *pancreas*. Absorption of nutrients also occurs here. Undigestible matter is passed through the ventral *anus*.

The reproductive organs in the perch consist of *ovaries* in the female and *testes* in the male. Fertilization is external. Most female perch lay eggs in long strands, which attach to plants along river bottoms. A single female may lay up to 2,000,000 eggs each season. *Milt*, a fluid containing sperm, is deposited by the male over the eggs.

Internal structure of the perch

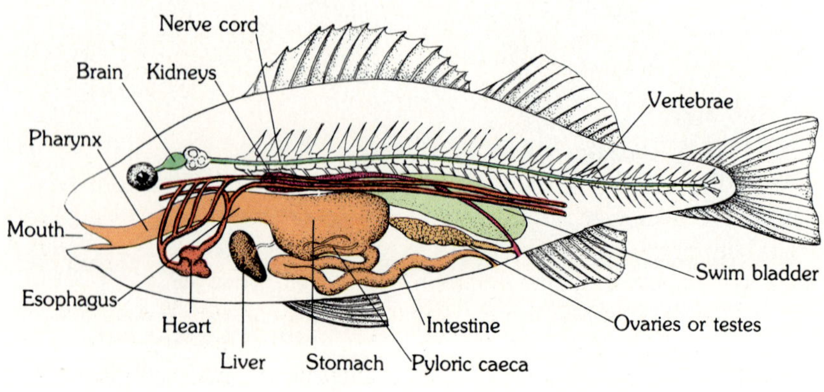

Other Ray-finned Fish

Ray-finned fish show many unusual adaptations to their environment. Flounders, for example, can change the color and design of their skin to match the ocean bottom. The archer fish captures insects by shooting them out of the air with small drops of water. The front rays of the pectoral fins of the sea robin are adapted for digging. When disturbed, the sea robin can quickly bury itself in the sand.

One of the most unusual adaptations to be seen in the entire animal kingdom occurs among angler fish. These fish live at depths of from 500 to 2,000 meters. It is extremely dark at such depths. Perhaps to ensure successful mating, the tiny male angler fish permanently attaches itself to the much larger female of the species. The bodies of the two fish fuse, with the male living as a parasite on the female.

Angler fish—male and female

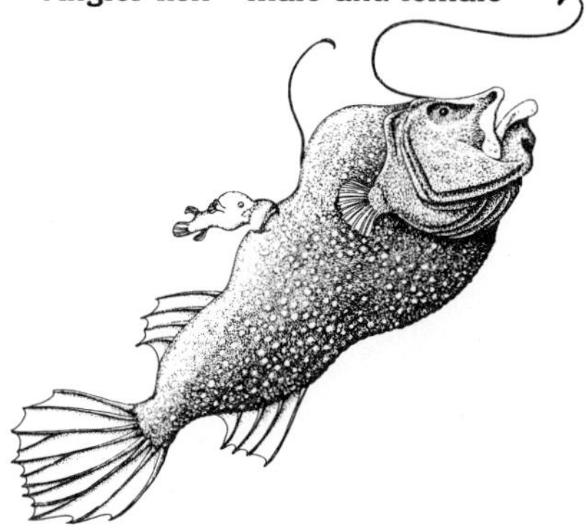

Camouflage in the flounder

Fleshy-finned Fish

Although you probably have seen many bony fish, it is unlikely that you have ever seen a fleshy-finned fish. Fleshy-finned fish have paired fins with fleshy lobes. They are quite rare. None occur in North America. Only a few zoos and aquariums have them on exhibit.

The fleshy-finned fish are divided into two groups: the *lungfish,* or *Dipnoi* (dip′noi), and the *lobe-finned fish,* or *Crossopterygii* (kro sop′tə-rij′ē ī). The fleshy-finned fish interest scientists because some species show characteristics that are like those of primitive land vertebrates.

Lungfish

As the name suggests, lungfish have lunglike air sacs that open from the throat. The areas of Australia, Africa, and South America where lungfish are found often experience dry seasons. During the dry periods, lungfish survive by gulping air or breathing it in through nostrils. Although the lungfish are the only living fish with lunglike structures, they are not thought to be directly related to primitive land vertebrates.

Lungfish

Lobe-finned fish (coelacanth)

Lobe-finned Fish

The lobe-finned fish are most like primitive land vertebrates. Fossils suggest that lobe-finned fish were once abundant in fresh waters. These fish had lunglike air sacs in addition to gills which could be inflated by taking in air through internal nostrils. The bones inside the fleshy lobes were arranged in a pattern like those inside the legs of land animals. The teeth showed similarities to teeth of land vertebrates.

Up until 1939, biologists thought that all lobe-finned fish were extinct. Then, off the coast of South Africa, the remains of a peculiar fish were brought ashore by a commercial fisherman. The fish was later identified as a **coelacanth** (sē′lə kanth), a species of lobe-finned fish thought to be extinct since the days of the dinosaurs. The search for living coelacanths was delayed by the outbreak of World War II. After the war, living coelacanths were found in deep waters off the shores of the Comoro Islands. Finding living coelacanths was as exciting to scientists then as the discovery of living dinosaurs would be today.

Review It

1. What five characteristics are shared by most members of Agnatha?
2. What are the distinctive features of Chondrichthyes and Osteichthyes?
3. Compare the lamprey, the shark, and the perch.
4. Why are the fleshy-finned fish of interest to many biologists?

19–3
Land Invaders: Amphibia and Reptilia

Fossil evidence suggests that the **Amphibia** (am fib′ē ə) were the first class of vertebrates to live at least part of their lives on land. On land, air must be breathed directly. Amphibians have developed special adaptations called lungs for this purpose.

The word *amphibian* means living in two ways. Most amphibians do live a double life: part of it in the water, part on land. Usually the gilled amphibian larvae live in the water. The larvae then *metamorphose,* or change, into a land-living adult form.

The approximately 2,500 living species of Amphibia are divided into two major groups: the tailed amphibians and the tailless amphibians. Most of the members of a third group are extinct, with the exception of one order, **Apoda** (ap′ə də). Apoda are small, wormlike amphibians with a short tail and no other appendages. Despite the many forms of Amphibia, most share the following nine characteristics during some part of their life cycle:

1. The notochord is not present beyond an early embryonic stage.
2. The skeleton is mostly bone.
3. Blood is pumped through a three-chambered heart.
4. Two nostrils connect with the mouth.
5. The eyes have movable lids.
6. Two pairs of limbs are used for walking and swimming.
7. Gills, lungs, and the skin are used for breathing.
8. The body is covered with moist, scaleless skin.
9. Fertilization may be inside or outside the body of the female.

Reptilia

Amphibia

Tailed Amphibians

Amphibians with tails include newts, salamanders, and mud puppies. Most species are under 20 centimeters long, although the giant salamander of China and Japan may reach a length of almost two meters. Tailed amphibians are found wherever the climate is relatively mild.

If you live in the eastern or southern part of the United States, you may have seen mud puppies in a nearby river. Mud puppies never leave the water for long. They remain larvalike with gills and a finned tail. They feed on small fish and invertebrates. In the spring, the female deposits her eggs in underwater nests and guards them until they hatch in the early or late summer.

Most true salamanders develop into lunged adults that live on land. A few species, however, lack both lungs and gills. They breathe through their skin and mouth. Depending on the species, eggs are deposited either in the water or in moist, protected nests on land. The larvae resemble the adult form and are carnivorous, feeding mainly on insects.

Mud puppy

Newt

Tailless Amphibians

Over 2,000 species of frogs and toads are included among the tailless amphibians. They range in size from a tiny species only one centimeter long to a West African species that is 30 centimeters in length. Most frogs live in or around water. Toads generally stay on the land, returning to water only to spawn.

The Frog

The most common laboratory animal is still the frog, even though the leopard frog is threatened because of overuse in classroom dissections. The frog is a popular animal for study because its organ systems show many similarities to human systems.

The Exterior of the Frog

If you were to hold a frog in your hands, you would certainly notice its soft, slippery skin. The skin is made slippery by the *mucus* secreted by glands in the skin. The color of the skin depends upon pigment granules in the cells. The change of color is under the control of a hormone. Changes in skin color allow the frog to blend with its environment and so escape the notice of predators. Frogs breathe primarily through the skin. Tiny blood vessels beneath the surface of the skin absorb oxygen.

Because of its slippery skin and strong hind legs, a held frog can easily escape an enemy. The muscles of the hind legs are adapted for jumping and swimming. A frog swims by kicking the hind legs together. The webbed foot provides a surface for pushing against the water. The weaker forelegs help the frog hold its head up and absorb the force of landing on firm ground.

The frog surveys its environment with its two large, movable eyes. The position of the eyes enables a floating frog to see things above the waterline. Each eye is protected above and below by two eyelids. A third, transparent eyelid, called the *nictitating* (nik'tə tāt'ing) *membrane,* keeps the eyes moist when they are exposed to air and protects them underwater. The disklike membrane behind each eye is an eardrum.

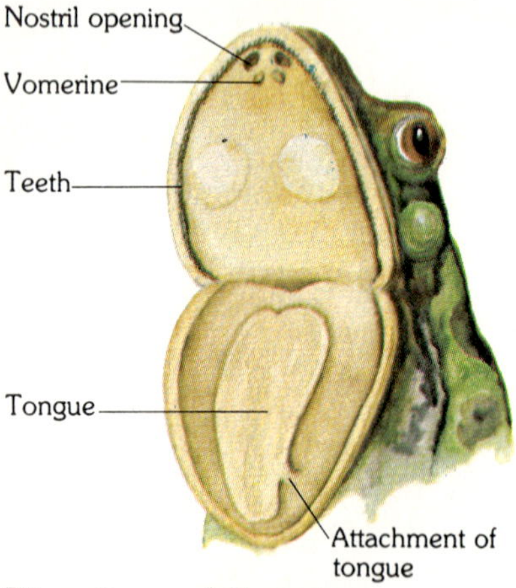

The Frog's Mouth

Inside the frog's large mouth is a sticky tongue attached at the front. The frog flips out its tongue, which catches food on its sticky surface. Food is held inside the mouth by one row of short, conelike teeth along the edge of the upper jaw and by the scalelike *vomerine* (vō'mə-rīn') *teeth* growing from the roof of the mouth. Food is forced down the throat by the downward pull of muscles on the eyes.

Digestive and Excretory Systems

Food passes down through the *esophagus* into the *stomach,* where digestive juices start breaking it down. The partially digested food then passes through a ring of muscles, called the **pyloric** (pī lôr'ik) **valve,** into the *small intestine.* The food is further digested in the small intestine with the help of the *liver* and the *pancreas.* The liver produces *bile,* which helps in the digestion of fats. The bile is stored in the *gall bladder,* which connects with the small intestine. The pancreas produces digestive enzymes. After digestion is completed, nutrients are absorbed by the blood through the walls of the small intestine.

Leftover wastes are pushed into the *large intestine* and out of the body through the **cloaca** (klōā'kə) and a small opening called the *anus.*

The *kidneys* filter out the wastes from the blood and form urine, which is stored in the urinary *bladder* and then eliminated through the cloaca. Carbon dioxide is eliminated through the skin, lungs, and mouth.

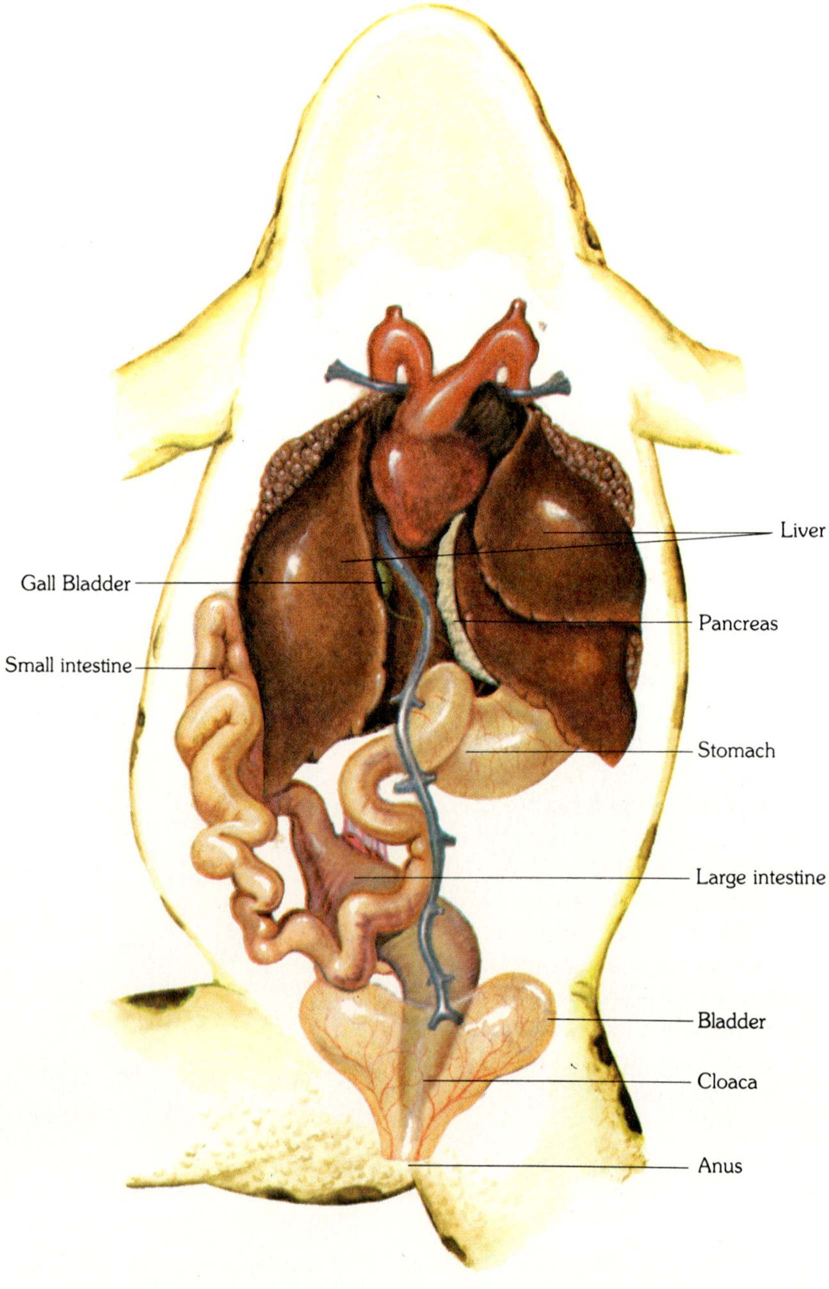

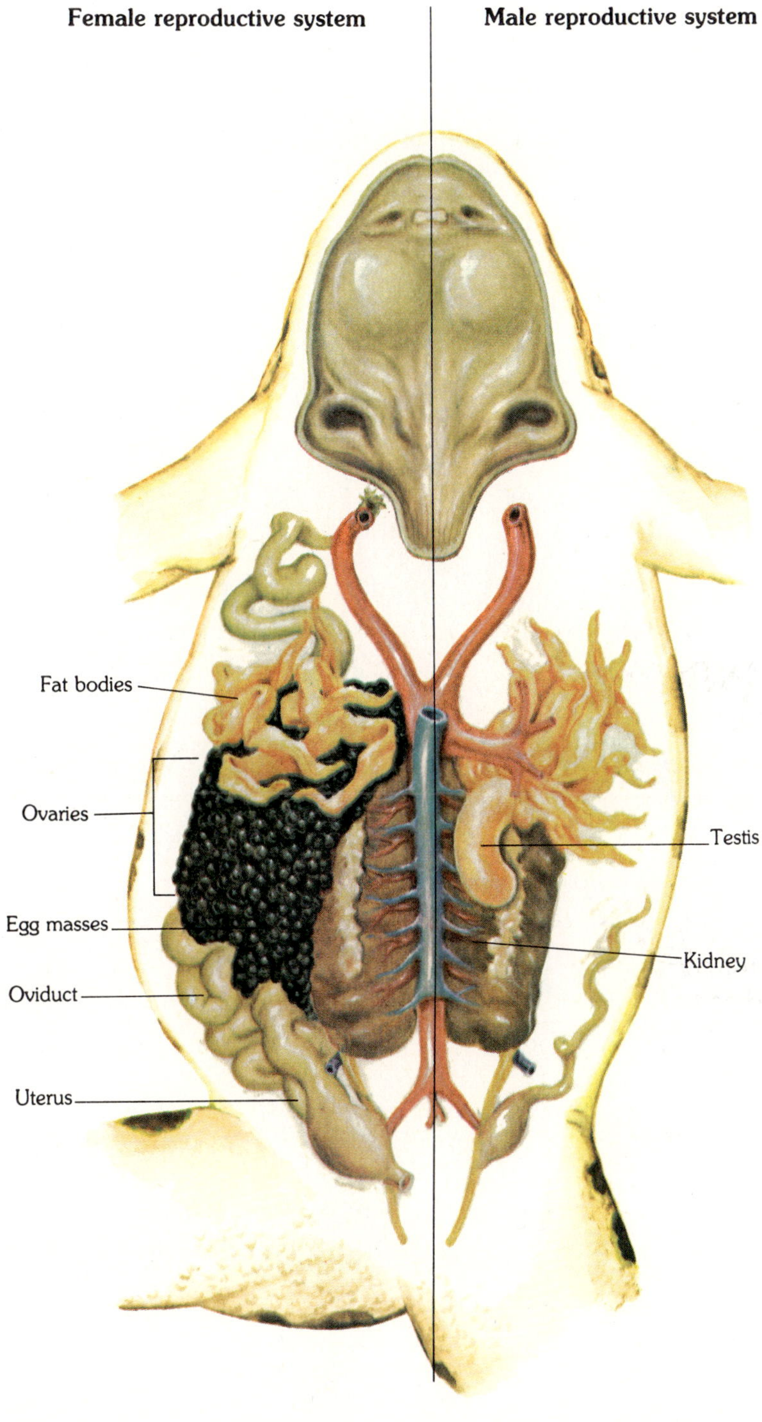

Reproductive System
Reproduction usually occurs sometime in the spring. Fertilization in the frog is external. The female frog produces eggs in the *ovaries*. The eggs soon fill much of the body cavity. Eggs are passed down coiled tubes called the **oviducts** (ō′və dukts), receiving a jellylike coating. The *uterus* temporarily stores the eggs until they are laid. Using the large thumb pads, the male frog clasps the abdomen of the female to help her release the eggs. As the eggs leave the cloaca of the female, the male fertilizes them with sperm produced in the *testes*.

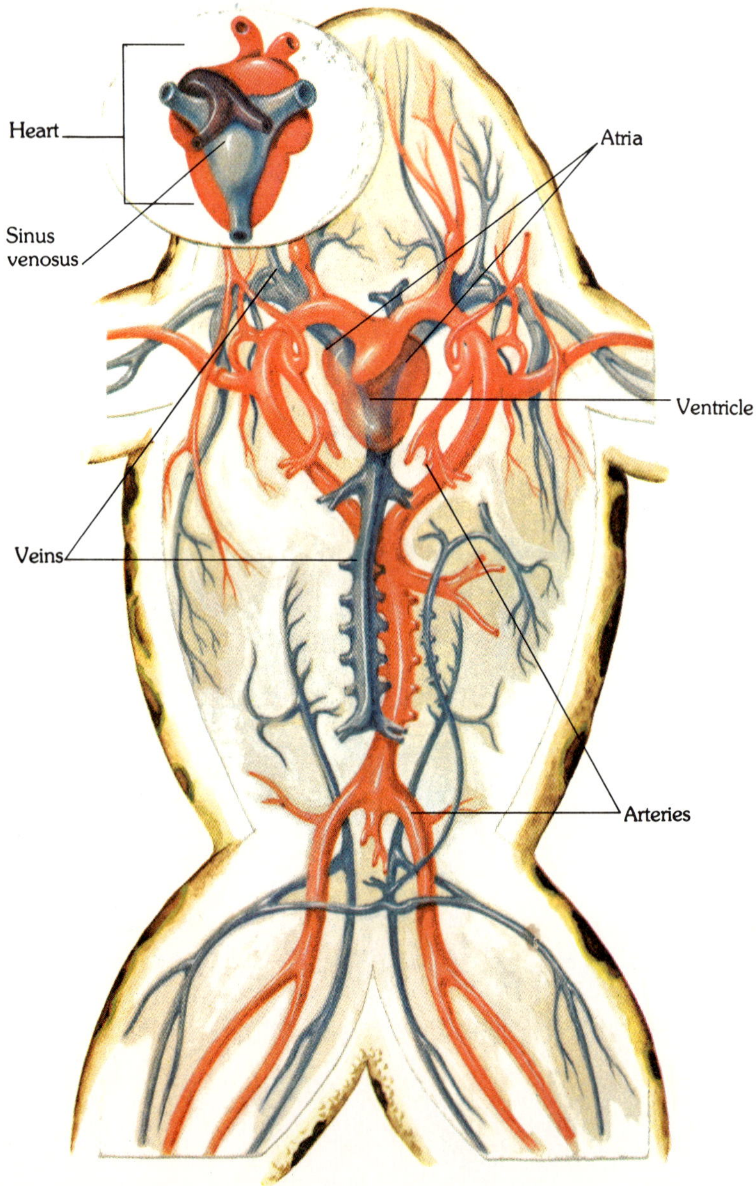

Circulatory System

The blood delivers nutrients and oxygen to all the cells in the frog's body. It also picks up wastes and carbon dioxide. Blood is pumped through the frog's body by a three-chambered *heart.* Arteries carry blood away from the heart. Veins carry blood toward the heart. One chamber of the heart, the right **atrium** (ā'trē əm), collects oxygen-poor blood from a large vein called the **sinus venosus** (vē-nō'səs). Oxygen-rich blood from the lungs collects in a separate chamber, called the left atrium. Both atria empty into a single **ventricle** (ven'trə kəl), which pumps the blood to other parts of the frog's body. Because oxygen-rich blood from the lungs mixes with oxygen-poor blood in the ventricle, the frog depends on its skin, not its lungs, for getting oxygen and eliminating carbon dioxide. This all-important exchange takes place in the tiny vessels, called capillaries, situated beneath the skin.

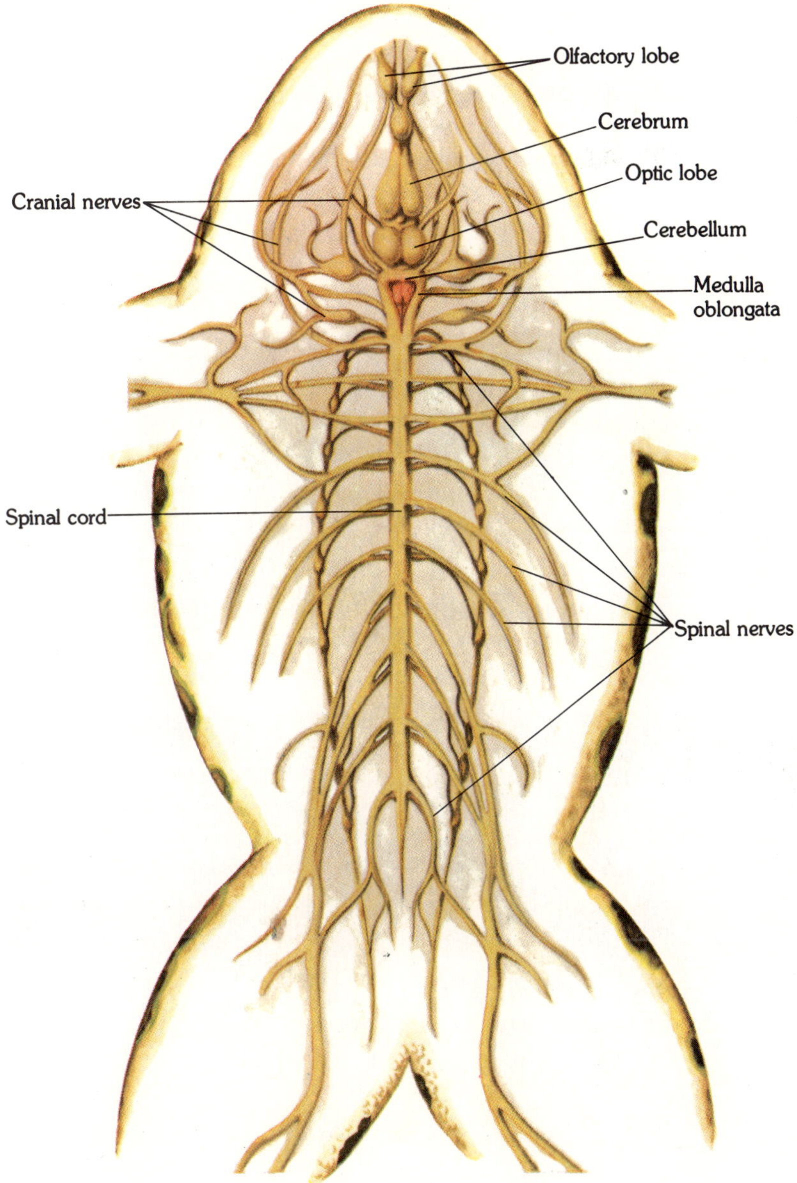

Nervous System

The dorsal nervous system of the frog consists of the *brain*, the *spinal cord* and the *nerves*. Ten pairs of *cranial nerves* connect the brain to parts of the head, such as the eyes, and to the abdomen. Ten more pairs of nerves connect the spinal cord to parts of the body other than the head.

Notice the lobes in the brain. The optic lobes control the frog's vision. The olfactory lobes control the frog's sense of smell. The cerebrum is the control center for sensory input and for voluntary muscle response. The cerebellum coordinates muscle actions and the medulla oblongata regulates the involuntary vital functions.

Development in the Frog

Frog eggs hatch into larvae, or *tadpoles*. The larvae have a fishlike shape with gills and a tail. Like fish, the larvae are equipped with only a two-chambered heart. The larvae feed on small plant life. After weeks or sometimes months, the larvae metamorphose into the adult form. The tail is absorbed while legs and forelimbs emerge. Air is breathed in through lungs and the heart becomes three-chambered.

Life cycle of the frog

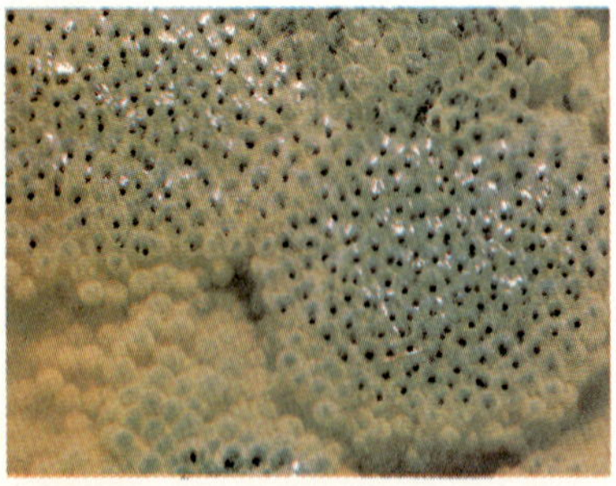

Egg mass

→

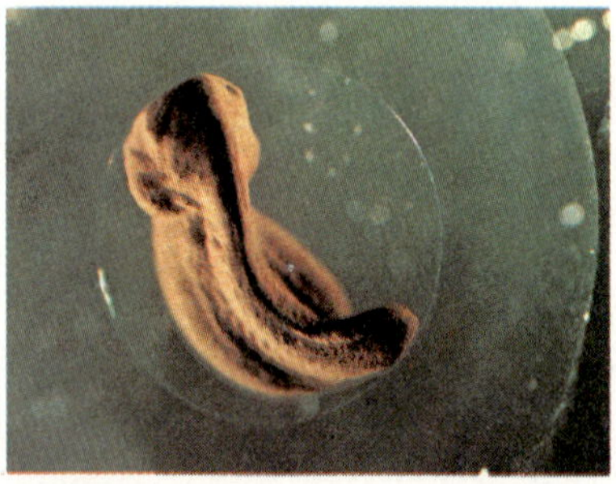

Embryo

→

Gill stage

↓

Tadpole

←

Development of the legs and lungs

←

Young frog

↓

Adult frog

Toads

Toads are much like frogs with a few minor exceptions. The skin of toads is often rougher than that of frogs. Toads usually lay their eggs in strings rather than masses.

Class Reptilia

Lizards, snakes, turtles, tortoises, crocodiles, and alligators are all *reptiles* (rep'təlz), members of the class **Reptilia** (rep til'ē ə). Unlike Amphibia, reptile species can complete their entire life cycle on land, although some reptiles spend most of their lives in water. A tough, leathery skin, a better developed circulatory system, and a type of egg that can survive on land enable reptiles to be true land vertebrates.

Most living reptiles belong to one of three groups. One group is made up of alligators, crocodiles, caimans, and gavials. A second group includes turtles, tortoises, and terrapins. Snakes and lizards make up the third group. A fourth group consists of only one living species: the New Zealand tuatara or *Sphenodon* (sfē'nə don). The 7,000 living species of reptiles generally share the following seven characteristics:

1. The body is covered with a dry, leathery skin.
2. The skeleton is made entirely of bone.
3. Blood is pumped through an incomplete, four-chambered heart.
4. Lungs are always present.
5. Fertilization occurs inside the body of the female.
6. The egg is covered by a waterproof shell and the embryo is protected by four membranes.
7. No metamorphosis occurs.

The reptilian egg is perhaps the single most significant advance over Amphibia. Certainly the reptiles could not have strayed far from water if their eggs were not able to survive on land. The successful adaptation of the reptile egg to land is demonstrated by species of snakes and lizards that live and lay their eggs in some of the driest regions on Earth. A porous, outer shell envelops the reptile egg and protects it. Inside the egg, the developing embryo is nourished and protected by four kinds of membranes. One membrane, called the *yolk sac,* contains the yolk upon which the developing embryo feeds. A second membrane, the **allantois** (ə lan'tō is), holds the wastes produced by the embryo. The allantois also provides for the exchange of oxygen and carbon dioxide. A third membrane, the **amniotic** (am'nē ot'ik) **membrane,** or **amnion** (am'nē ən), encloses the embryo in a fluid-filled bag, which acts like a tiny, artificial pond. The amniotic fluid keeps the embryo from drying out and protects it from injury. A fourth membrane, the **chorion** (kôr'ē on), encloses the other three membranes. Because of its special structure, the reptilian egg is often referred to as the **amniote egg** after the amniotic membrane.

Amniote egg

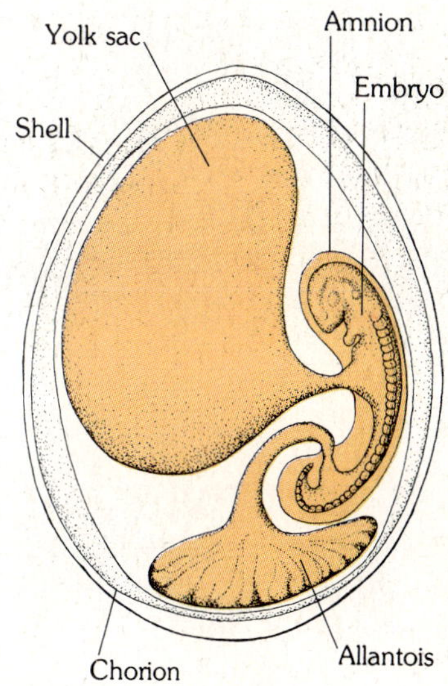

Early reptiles

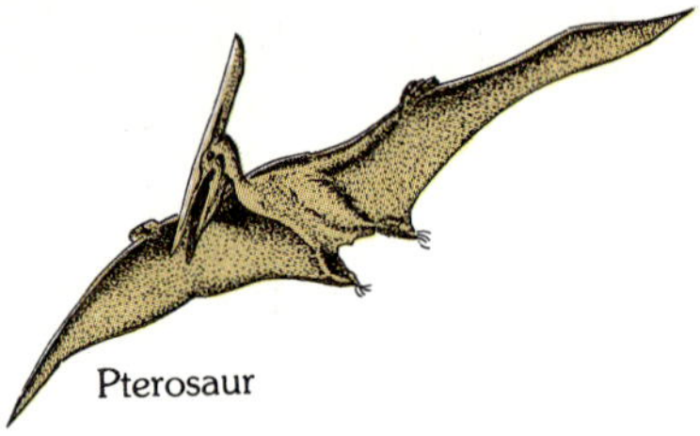

Pterosaur

Stegosaurus

Brontosaurus

Early Reptiles

The remains of many kinds of extinct reptiles suggest that class Reptilia was a much more successful group in the past than it is today. Altogether, ten major reptile groups have disappeared from the earth. Of the ten extinct groups, dinosaurs represent two. One dinosaur group had a solidly constructed skeleton and was more like living reptiles. *Brontosaurus* (bron'tə sôr'əs) is a representative of this group. The second dinosaur group was more birdlike in structure. *Stegosaurus* (steg'ə sôr'əs) is representative of this second type of dinosaur.

In addition to dinosaurs, there were other unusual types of reptiles that inhabited most of the areas that are today occupied by species of mammals. There were ocean-dwelling species called **ichthyosaurs** (ik'thē ə sôrz) and **plesiosaurs** (plē'sē ə sôrz). There were flying species called **pterosaurs** (ter'ə sôrz) and there were many land-living forms. These species of reptile are now extinct; the reasons for their extinction are still not completely understood.

Alligators

The alligator exhibits many traits that are typical of reptiles as a group. If you live in or have visited the South, you may have seen an alligator lying loglike, partially submerged in a river or swamp. Its eyes are placed high on either side of its head so it can see above water. A small ear opens beneath a flap of skin behind each eye. At the tip of the snout are two nostrils. Beneath the nostrils is a wide mouth edged with teeth. At the back of the mouth cavity are two folds. When pressed together the folds keep water from getting into the lungs of the alligator. The alligator's digestive system is similar to that of the frog.

The heart of the alligator is more like a human heart than an amphibian heart. It has four chambers: two atria and two ventricles. The additional ventricle keeps oxygen-rich blood from the lungs from mixing with oxygen-poor blood returning from other parts of the body.

Alligator

Turtles

There are about 200 species of turtles, which include tortoises (land dwellers), terrapins (freshwater dwellers) as well as the true sea turtles. Encased within a shell, many turtles can withdraw their head and feet into the safety of this armor. Although cumbersome, it is thought that the shell has provided such superior protection that it has enabled the turtle to survive.

Like all reptiles, turtles must lay their eggs on land. This action can make the turtle vulnerable. Several species are threatened with extinction due to severe predation by humans.

Turtle

Lizards

Lizards are an abundant group of living reptiles. Most species have four legs with five clawed toes on each foot. The eyes of most lizards are protected by movable eyelids and a nonmovable covering called a *spectacle.* Most lizards are carnivorous. They abound in many parts of the world.

Chameleons (kə mē'lē ənz) are familiar lizards found in southern United States. They possess hormones that control the ability to change their color. Effects of temperature and light may cause the external coloring of the chameleon to change. The color may also change when the lizard is suddenly aroused or excited. Like all lizards, the chameleon sheds its skin regularly.

Chameleon

Snakes

Snakes comprise another large group of living reptiles. They can be found in a variety of both land and water environments. Snakes, like all reptiles, are covered with scales. Usually the scales on the dorsal side are small and flexible. Ventral scales are more platelike and aid in locomotion. All snakes molt, shedding a layer of scales with each molt.

Chameleon on python

All snakes are carnivorous. A specialized feature, the modified jaw, allows a snake to engulf food many times its own body size. The jawbone can be distended to permit a snake to swallow its prey whole. There is no need to shred or break up the food into pieces. Double rows of teeth that slant back toward the throat *hold* rather than chew the prey.

The sense of hearing is not well developed in snakes. In fact, they lack external ear openings. But their sense of smell is acute. Paired nostrils and a forked tongue are suited to detect odors. The forked tongue gathers odor-bearing particles in the air and places them in pits located inside the mouth. These pits are lined with odor-sensitive nerve endings.

A snake's eyes are always open. There are no eyelids to cover them. Instead, a transparent scale protects each eye. When the snake molts, this scale is shed also.

Green snake laying eggs

Rattlesnake swallowing prey

Poisonous Snakes

Of the 3,000 species of snakes existing today, only about 300 species are poisonous. The poisons are introduced into the victim's body through needle-sharp fangs. The poison of snakes usually works in one of two ways. The poison of rattlesnakes, copperheads, and cottonmouths (water moccasins) affects the circulatory system, causing clotting and capillary destruction. The poison of cobras, seasnakes, mambas, and coral snakes affects the nervous system, paralyzing the victim.

Tuatara

Lizardlike in appearance, the **tuatara** (tü′ə tär′ə) is a member of a once widespread group of reptiles called the *Sphenodons.* At first classified as a lizard, it was later claimed to be the lone surviving species of this nearly extinct group. Isolated on some 20 small islands around New Zealand, this is the only natural area where they are found. Members of the population are thought to mature very slowly and may live over 100 years! Research is now being conducted to find out why this last representative of the Sphenodons has survived. Efforts are also being made to preserve the tuatara as a living relic of the past.

Tuatara

Review It

1. What are the nine traits that characterize class Amphibia?
2. Why can't amphibians live away from water their entire lives?
3. What characteristics of Reptilia enable some species to survive in desert habitats?
4. Describe the structure and function of the amniote egg.
5. Of what importance is the tuatara?

Issues in Biology

Were there warm-blooded dinosaurs?

All modern reptiles are cold-blooded. In general, reptiles are slow and sluggish unless warmed by environmental heat such as the sun. By contrast birds and mammals produce enough heat to maintain a high activity level. The table lists some traits that, in general, differ between warm- and cold-blooded animals.

For years scientists assumed that extinct reptiles such as dinosaurs had also been cold-blooded. But new lines of evidence suggest that some extinct reptiles were warm-blooded as are the birds and mammals of today. Knowledgeable vertebrate paleontologists differ in their interpretations of the most recent dinosaur data.

Those who argue that dinosaurs were *warm-blooded* point out that many of the smaller carnivorous dinosaurs had traits comparable to modern-day warm-blooded animals. 1) They were built for speed, with powerful hind legs for running in an upright position. 2) There is fossil evidence that carnivorous dinosaurs traveled in packs and attacked herbivores that were sometimes six times their own size. 3) Studies show that their bones contained numerous blood vessels as do bones of current warm-blooded animals. 4) Some supporters claim that a comparison between the number

of fossil prey animals with fossil predator animals found together indicates the predators ate more prey than they would need to eat to maintain the low body temperatures of a cold-blooded animal.

Scientists who favor the *cold-blooded* theory make these arguments. 1) A warm-blooded animal the size of the 72 metric ton *Brachiosaurus* would have to consume as much food as about 10 or 12 African bull elephants. 2) The predator/prey ratios for large carnivorous dinosaurs might be essentially the same as for warm-blooded animals and therefore predator/prey ratios in fossil deposits would not indicate warm-bloodedness. 3) A few present-day reptiles have bone structures with well-developed blood vessels whereas a number of small mammals and birds do not. Therefore, the fossil bone evidence does not necessarily mean warm-bloodedness.

Since dinosaurs cannot be collected and studied, the dispute may never be settled. Both sides have to depend on interpretations of the fossil record, which is incomplete.

Analysis

1. Some scientists now agree that both warm-blooded and cold-blooded dinosaurs existed. Can you find current evidence to support this position?
2. What evidence for warm-bloodedness do you think is the most convincing?

Trait	Cold Blooded	Warm Blooded
body temperature	heated by external source	heated by internal control
approximate food needs of a 90-kilogram animal	consume own weight every six months	consume own weight every month
method of food capture	lie and wait	chase and attack
structure of bone	few blood vessels	numerous blood vessels
posture	flat on ground	raised
movement	minimal	active

Chapter Summary

The most diversified, complex phylum on Earth today is the phylum Chordata. At some stage of their life cycle, all chordates have a dorsal hollow nerve tube, a notochord, and pharyngeal gill slits. Primitive chordates are represented by tunicates and lancelets. The largest group of chordates, Vertebrata, includes fish, amphibians, reptiles, birds, and mammals. All vertebrates are characterized by a cartilaginous or bony vertebral column.

The most primitive fish, represented by the lamprey and the hagfish, have notochords that persist in the adult form, cartilaginous skeletons, and jawless, round mouths. Sharks, skates, and rays also have cartilaginous skeletons, but their mouths are jawed. The most advanced of the fish, Osteichthyes, have bony skeletons, larger, more developed brains, and swim bladders. A lobe-finned bony fish is thought by some scientists to be similar to the ancestors of land vertebrates.

The first vertebrate group to invade the land were the amphibians. Amphibians, however, live only part of their life on land. They are spawned in the water and usually develop from larva to adult in the water. The amniote egg enabled reptile species to survive away from water. The dry, scaly skin and the four-chambered heart of reptiles also enhanced their success on land.

Questions to Think About

1. Fossil evidence suggests that amphibians were the first animals to invade the land and that reptiles developed from an amphibianlike ancestor. What kind of additional information would you seek in order to support or refute the theory that reptiles developed from an amphibianlike ancestor? Describe how an amphibianlike ancestor of the reptiles might have looked.
2. How is the form and structure of living fish, amphibians, and reptiles suited to their life in water or on land? Discuss at least one representative of each vertebrate class in your answer.
3. In this chapter, you have been introduced to one widely accepted way of classifying fish, amphibians, and reptiles. Think of at least one alternative way of grouping these animals. (Hint: Consider differences in habitats, diets, behaviors, body structures.)
4. In many parts of the United States, land and water development projects have been halted to protect an endangered species of fish, amphibian, or reptile. List the pros and cons of such actions to protect endangered species.

Extra Research

1. Dissect a preserved shark or frog. Label the different organs and explain your findings to your class.
2. Find or buy some tadpoles and keep them in an aquarium. Observe them as they mature into adults. Keep a log of how they change. Present your findings to your class.
3. Obtain a list of endangered species from the Environmental Protection Agency in Washington, D.C. Find out which species of fish, amphibians, and reptiles are endangered and what is being done to protect them. Find out why some people are against the protection of some species. Arrange for a student debate on this issue.
4. Use an aquarium or a terrarium to re-create the natural habitat of a particular species of fish, amphibian, or reptile.
5. Many museums and parks have displays of the remains of extinct reptiles such as the dinosaurs. Find out where in the United States reptilian fossils have been found and what is known about the life styles of extinct reptiles.
6. Identify the species of fish, amphibians, and reptiles that are common in your community. With the help of your teacher, plan a field trip to observe some of the species where they live.

Chapter Test

Part I

Copy the number of each test item. If the item is *true,* write *T* after the number; if the item is *false,* write *F.*

1. All chordates are characterized by a notochord, a dorsal nerve tube, pharyngeal gill slits, and a cartilaginous or bony vertebral column. (446)
2. The adult tunicate, or sea squirt, usually shows only one chordate feature—the presence of gill slits. The other chordate features can be seen in the larval form. (447)
3. The cephalochordates are the most familiar and widespread group of chordates. (447)
4. Vertebrata is generally divided into eight classes, one of which is extinct. (448)
5. All vertebrates have a bony vertebral column and jaws. (450)
6. All or part of the notochord persists in lampreys, hagfish, sharks, rays, and skates. (450)
7. Bony fish can adjust more readily to changes in water depth than sharks, rays, and skates. (454)
8. Some scientists believe that land vertebrates developed from a fishlike creature similar to a form of fleshy-finned fish. (457)
9. Most amphibians do not have a larval stage. (459)
10. Fossil evidence suggests that reptiles were more widespread in the past than they are today. (468)

Part II

Copy the number of each sentence below. After each number, write the word (or words) that completes the sentences correctly.

11. The phylum Chordata usually is divided into three subphyla: _____, _____, and _____. (446)
12. The sea squirt, or tunicate, is a member of the subphylum _____. (447)
13. The adult amphioxus, or _____, shows all three major chordate characteristics: _____, _____, and _____. (447)
14. Vertebrates usually share six basic characteristics: (1) _____, (2) _____, (3)_____, (4) _____, (5) _____, and (6) _____. (448)
15. The lamprey and the hagfish are the only known living representatives of the class _____, which have no _____. (450)
16. Sharks, rays, and skates show two major advances over the lamprey and the hagfish: _____ and _____. (451)
17. The presence of a _____ helps Osteichthyes change depths with ease. (454)
18. A group of fleshy-finned fish, called _____, show characteristics which are similar to those of primitive land vertebrates. (457)
19. The word *amphibian* means _____ because the larval form usually lives _____, whereas the adult form usually lives _____. (459)
20. The _____ egg, which enabled Reptilia to lay eggs on dry land, has four important membranes: _____, _____, _____, _____. (467)

Part III

Copy each numbered item from List A. After each item, write the letter and words from List B that best describe the item.

List A

21. shark (452)
22. endotherm (449)
23. ectotherm (449)
24. frog (464)
25. turtle (469)

List B

a. organism whose internal body temperature remains constant
b. cold-blooded
c. has a cartilaginous skeleton
d. organism that uses its moist skin for respiration
e. animal that has internal fertilization and external development of the young

Chapter 20

Birds and Mammals

Have you ever held a kitten in your hand? Have you ever held a frog? If so, you will remember they felt different to you. The kitten had soft fur whereas the frog had only skin. Also, the kitten felt warm, and the frog probably felt cool.

Birds and mammals are warm-blooded creatures. Because of this trait, they are often grouped together for purposes of study, as they are in this chapter.

Unlike most other animals, birds can fly. Humans throughout recorded time have envied this trait, so they studied the flight of birds.

Humans are interested in mammals for different reasons. Humans belong to the class Mammalia. The study of mammals gives humans insights into themselves.

In this chapter, you will learn more about special characteristics and adaptations of birds and mammals that place them among the most successful animals on earth.

Chapter Objectives

1. List the main characteristics of birds and explain how these characteristics help birds adapt to specific environments.
2. Demonstrate an understanding of the digestive, respiratory, circulatory, nervous, and skeletal systems of birds.
3. Describe the reproductive cycle in birds.
4. Give examples of the diversity of birds.
5. List the main characteristics of mammals and explain how these traits affect their environmental adaptability.
6. Explain differences between egg-laying mammals, marsupial mammals, and placental mammals and list one representative organism of each.
7. Describe the characteristics of warm-blooded animals.
8. Describe the mammalian brain. Explain how it differs from the bird brain and how the brain affects behavior.

Owls, such as this pearl-spotted owlet, have a wider range of binocular vision than any other bird. This allows depth of field, which is important in judging distance when an owl swoops down on its prey.

20–1
The Trait of Warm-bloodedness

Keeping warm

Cooling off

On a cold morning, a raccoon stalks its prey—a frog. It is likely to catch the frog, too, because the frog moves more slowly in cold weather than in warm weather. The raccoon has no such disadvantage—it moves at its usual pace, no matter what the weather.

Birds and mammals are warm-blooded, or *endothermic.* They are known as *endotherms.* Warm-blooded animals, such as raccoons, remain warm even when the temperature around them changes. The body temperature of frogs and other *ectothermic,* or cold-blooded, animals rises when the weather is hot and drops when the weather is cold. They are known as *ectotherms.*

In endotherms, as in ectotherms, body heat is produced when food and oxygen are combined in the cells. This *oxidation,* or burning, of food produces energy and heat. However, when an endotherm is exposed to cold, the rate of heat production goes up, and its body temperature remains constant.

Endotherms have good body insulation—feathers, fur, or fat. Their protective covering helps slow the loss or gain of body heat to or from the environment.

The part of the brain called the **hypothalamus** (hī′pō thal′ə məs) regulates the body temperature of endotherms. It triggers changes that increase or decrease the rate of heat production. The hypothalamus can also trigger increases or decreases in heat loss to the environment. Increases in muscle activity by shivering or exercise and in the rate of food oxidation increase the rate of heat production. When the body is at rest and less food is oxidized, heat production goes down.

Birds and mammals slow the loss of heat by fluffing their feathers or their hair for better insulation. The blood vessels near the skin also constrict. Heat loss may be increased by panting, by enlargement of blood vessels near the skin, and by sweating.

Challenge!
Make a list of the advantages that ectotherms and endotherms have for survival. Take into account as many environmental factors as possible.

	Endothermic	**Ectothermic**
Body temperature	constant	varies with environmental temperature
Insulation	good	poor
Heat production	high rate	low rate
Response to environmental temperature change	heat production varies	body temperature varies
Circulation	constant—affects body temperature	constant—does not affect body temperature
Food	large amounts needed	small amounts needed
Brain	hypothalamus regulates body temperature	no regulator of body temperature
Oxygen	large amounts needed	small amounts needed

Birds and Mammals

Because of their warm-bloodedness, birds and mammals are found in every environment on earth—polar regions, tropical rain forests, deserts, plains, high mountains, and oceans—wherever they can find the plants, insects, and other animals they need for food.

There are about 8,600 species of birds. Except for the larger ones, almost all of them can fly. They range in weight from the hummingbird, which weighs a few grams, to the ostrich, which weighs more than 130 kilograms. Birds eat various foods—seeds, insects, plant nectar, and fish, to name a few. Some birds, such as swifts and nighthawks, catch insects in midair; whereas others, such as pelicans and cormorants, go fishing underwater.

As the length of day changes, many species of birds migrate from one place to another in spring and fall. They may fly hundreds, even thousands of miles. Most species are beneficial to humans because they eat large quantities of harmful insects and weed seeds; but a few, such as blackbirds and starlings, damage property when they roost near human settlements and eat grain crops. Chickens, turkeys, and other bird species are raised for food by humans, who also eat the eggs of these birds.

There are about 5,000 species of mammals. Some mammals, such as bats, fly; others, such as moles and shrews, live underground. Whales and dolphins live in the ocean. Humans dominate the land.

Mammals range in weight from the 25-gram pigmy shrew to the more than 100,000-kilogram blue whale, the largest known animal. Some mammals, such as deer and cattle, eat only plants; lions and hyenas eat only animals. Squirrels, mice, and humans eat plants and animals.

The living patterns of mammals vary. Some live alone, others live as mated pairs, and still others live in herds and colonies. Most wild animals avoid contact with humans; but predators, such as coyotes and wolves, sometimes destroy domesticated sheep and cattle.

Most farmyard animals are mammals, and they are sources of food and clothing for humans. Humans use other mammals as well: dogs and cats as pets, mice and monkeys for medical research, deer and elk for hunting sport, and horses for racing and for some farm labor.

Birds and mammals show interesting similarities in some of their parts. Compare the skeletal forelimbs of the animals shown in the margin.

Did You Know?

One mammal has been particularly dangerous to birds and other mammals—humans. People have exterminated whole species of birds, for example. The dodo—a large, flightless bird first seen by sailors on islands in the Indian Ocean in 1507—slowly disappeared from Earth between 1681 and 1790.

Other bird victims of humans have been the great auk (1840s), the Carolina parakeet (1904), and the passenger pigeon (1914).

Among those birds now on the endangered list are the southern bald eagle, the masked bobwhite, the California condor, the Eskimo curlew, two species of peregrine falcon, the Aleutian Canada goose, the brown pelican, one species of prairie chicken, two species of warbler, and the ivory-billed woodpecker.

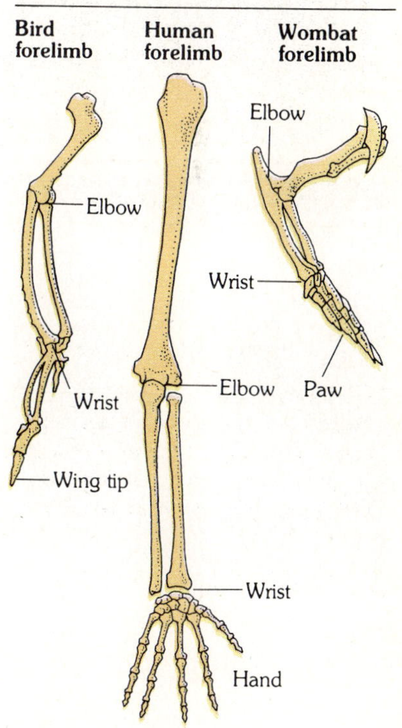

Comparison of forelimbs

Review It

1. Differentiate between endothermic and ectothermic animals.
2. Describe how the body of a mammal conserves heat when exposed to cold temperatures.
3. Describe how the body of a mammal responds to high temperatures.

20–2
Birds: Class Aves

Contour feather

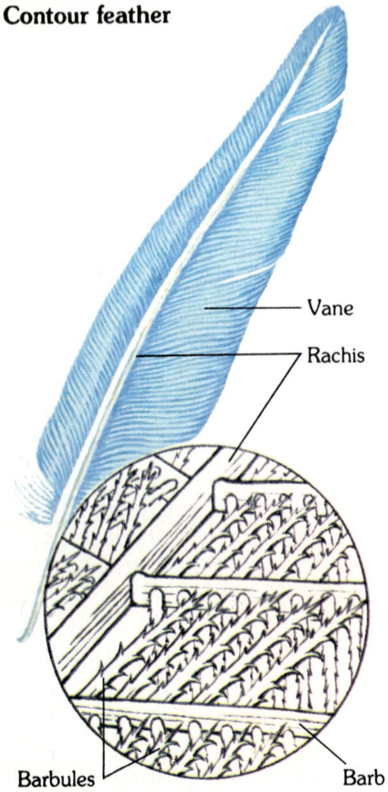

Preening

Class **Aves** (ā′vēz), the birds, has more species than any other class of vertebrates except the fish. All birds have one characteristic that no other animal has. This trait is *not* flight. Most birds can fly, but some cannot. All birds lay eggs, but so do snakes. All birds have forelimbs, or wings, and hindlimbs for walking, swimming, and perching. They have horny beaks. But the one thing they have that no other animal has is feathers.

Feathers

The bodies of endothermic animals are covered by skin. Hair grows on the skin of mammals, and feathers grow on the skin of birds. Each hair or feather grows from an opening in the skin called a **follicle** (fol′ə-kəl).

A feather has a central shaft called a **rachis** (rā′kis) that supports a row of barbs on each side. The barbs form the **vane** of the feather. Small branches, called **barbules,** which are arranged along the barbs, fill the area between the barbs. The barbules hook into projections on adjacent barbs so that the vane of a feather forms a thin, continuous sheet.

There are two main types of feathers: **contour feathers** and **down feathers.** Contour feathers are the largest feathers on a bird and include the quill or flight feathers on the wings and tail. Contour feathers form the shape of the bird.

Down feathers lie next to the skin and beneath the contour feathers. Down feathers have few barbs or barbules and no rachises. These feathers are almost weightless and keep the bird warm.

Feathers give a bird the ability to fly. Wing feathers lift the bird into the air, and tail feathers control the direction of flight. Feathers also protect against injury to the body.

Dead air trapped close to the body by the feathers insulates the bird in cold weather. The bird controls the rate of heat loss by changing the position of its feathers, thereby changing the amount of dead air trapped close to its body.

Feathers shed water. Waterproofing is best when the barbs and barbules are hooked together to form a tight web in the contour feathers. A bird repairs breaks in the arrangement of the barbs and barbules by shaking its feathers or by running its beak along the vanes. Some birds have glands at the base of the tail from which they take oil with their beaks and apply it to the feathers. Other birds have powderdown feathers that constantly disintegrate into a powder. These birds use the powder to clean their feathers. The rearrangement and oiling or powdering of feathers is called **preening.**

Birds lose their feathers, or **molt,** and grow new feathers at least once a year. Some birds molt twice a year, and in some species the color of the feathers changes to suit the season. In most cases, birds lose their flight feathers a few at a time so that they are still able to fly while molting. Ducks and some other birds, however, molt all at once during nesting season and are unable to fly during that time.

Plumage

The feathers that cover the body of a bird make up its **plumage** (plü′mij). Each species has a distinct pattern of color in its feathers and a distinct combination of feather types. Feathers in which the barbs and barbules do not interlock look like fine hair. The plumage of male birds is generally brighter than that of females. In many species, the male displays his brightly colored plumage during courtship of females during the mating season.

The color pattern in the plumage of some birds makes them difficult to see while they are on their nests. Even brightly colored birds living in the mottled light of forests blend into the patches of light and shadow and are protected from predators by their feathery coverings.

Brilliant plumage in some birds cannot be considered a benefit to them. The snowy egret is nearly extinct now because people have hunted it for its beautiful white feathers.

Pheasant

Sparrow hawk

Snake bird

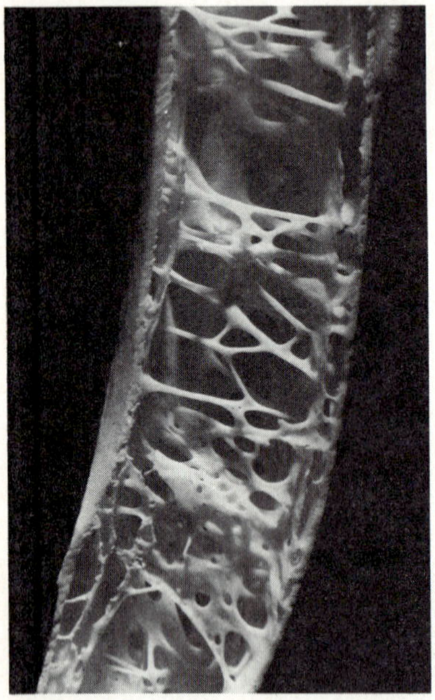

Internal structure of bird bone

Skeleton

The bodies of most birds are streamlined for flight. The skeleton of a flying bird is light and strong. Many of its bones are either hollow or air-filled between supporting trusses of bone. The head bone, breastbone, and pelvis are solid but thin. The vertebrae of the neck are many and flexible, whereas the vertebrae of the back are fused for rigidity. The flexible tail bones enable the bird to balance and to steer while in flight. A flying bird's weight is not in its bones; for example, a frigate bird weighs about 1,500 grams, but its skeleton weighs only about 110 grams.

A bird uses its beak for gathering food, building its nest, and preening its feathers. The beak is also used in feeding the young. The flexibility of the neck makes these tasks much easier.

In a flying bird, the **sternum,** or breastbone, has a large center blade called a *keel.* The *pectoral girdle* includes: the *coracoid* (kôr'ə-koid) *bones,* which support the shoulders; the *scapulas,* or shoulder blades, which lie on the back; and the *clavicles* (klav'ə kəlz), which make up the wishbone. The bones of the wings are similar to the arm bones in humans and end in a tip made up of fingerlike digits. At rest, the bird's wings are folded close to its body.

The hindlimbs attach to the body at the *pelvic girdle.* Most birds use their legs for walking or swimming and for perching. The hindlimbs most often end in scaly feet with four toes. There are three toes in front and one in back, which is turned the opposite way. This opposable back toe (called the first toe) enables birds to perch, to climb, and to grasp their prey.

Did You Know?
The ancient bird **Archaeopteryx** *(är'kē op'tər iks) must have been a poor flyer. It probably ran more than it flew. But the Archaeopteryx did have feathers, something the ancient flying reptiles did not have. The Archaeopteryx's bones were heavier and longer than the bones of present-day birds. Its skull and beak were heavier, too, and it had teeth. Its long tail and the large handlike bones at the ends of its wings could not have been much help in flying. Its flight muscles were small in proportion to its total body weight, and its brain was, at best, only one-fifth as large as the brain of present-day birds.*

Pigeon skeleton

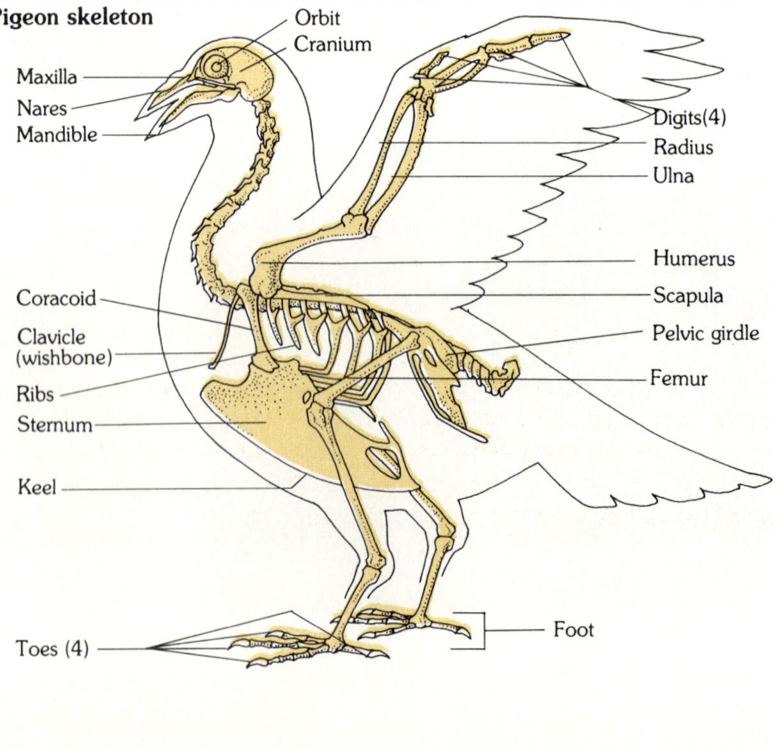

Beaks and Feet

Muscles

Birds have two exceptionally large flying muscles for each wing. These muscles lie on the sternum, and are attached to the keel. This is commonly called the *breast.* The larger outer muscle is attached by a tendon to the underside of the wing. The smaller muscle is attached by a long tendon to the front of the wing. In flight, these muscles alternately contract and relax. The small muscles raise the wings, and the large muscles lower them on the powerful downstroke.

Challenge!

Look at the beak and feet of each bird above. Determine where each bird would most likely live, and the type of diet each would have. Check your speculations with data from the library.

Bird brain

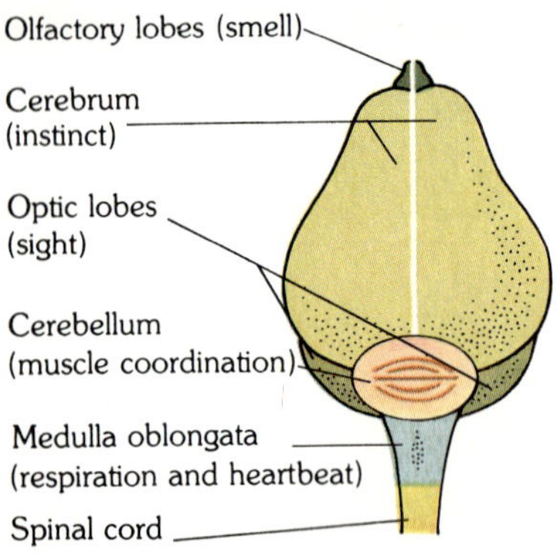

Nervous System

A bird's brain is quite large in proportion to its body weight. The **cerebrum** (sə rē'brəm) controls a bird's instinctive behavior. This includes food gathering, nesting, and migration. The **cerebellum** (ser'ə bel'əm) gives the bird excellent muscle coordination and enables it to fly through thick forests, take off and land in strong, gusty winds, and even drink from a lake while flying at almost full speed.

The best developed senses in birds are sight, hearing, and balance. Most birds have large eyes that are deep-set in the eye sockets. A bird's vision is far keener than that of a human. For example, a hawk can see a rabbit from a distance of two kilometers. Birds can see colors, too; most other animals cannot. A bird's eyes are protected by upper and lower eyelids and by a third lid called a **nictitating** (nik'tə tāt'ing) **membrane.** This transparent lid cleans and moistens the eye as it sweeps from the inner corner to the outer corner and back. The nictitating membrane also keeps wind and dust out of a bird's eye when the bird is flying.

Birds generally have a poor sense of smell. Two exceptions are the vultures and kiwis, which actually smell out their prey. The nostrils, or **nares** (ner'ēz), on a bird's beak serve largely as air passages.

Internal organs of bird

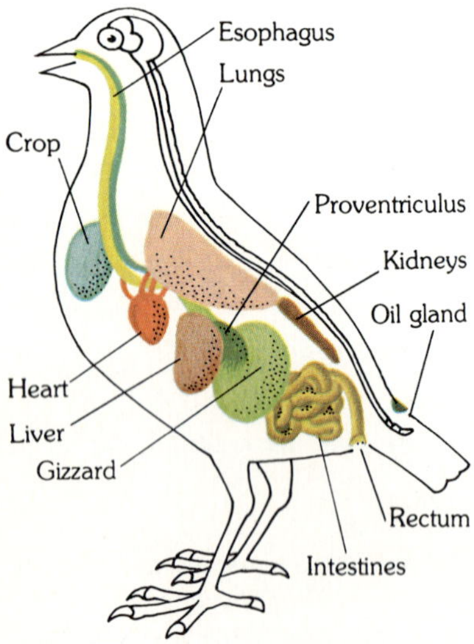

Digestion

Birds eat large amounts of food because they must maintain a constant body temperature and because flying uses enormous amounts of energy. For these reasons, a bird's digestion is rapid and efficient. A shrike digests a whole mouse in three hours, and a thrush excretes berry seeds only thirty minutes after eating the berries.

Birds have no teeth. Also they do not always have a steady food supply. Swallowed food is moistened and temporarily stored in the *crop* at the end of the *esophagus.* From the crop, the food moves to the first part of the stomach, the **proventriculus** (prō ven trik'yə ləs), where it is mixed with gastric juices. Then it passes to the second part of the stomach, the *gizzard,* which is a muscle lined with a tough, thick membrane. Coarse grit or pebbles swallowed by the bird lodge in the gizzard and help grind food as the strong muscle contracts. From the gizzard, the food moves into the *intestine,* where digestion continues. Here digested food products diffuse into the bloodstream.

Excretion

After usable food products have diffused from the intestine to the blood, unusable material moves to the *rectum* and then to the *cloaca.* The cloaca receives and excretes digestive wastes from the rectum as well as urinary pastelike wastes from the kidneys. *Sperm* from the *testes* in male birds also empty into the cloaca. During mating, sperm are transferred from male to female through contact of their cloacas.

Respiration

Birds, like other warm-blooded animals, have lungs in the chest cavity. Oxygen from the air is inhaled through the mouth and nares, down the *trachea* and *bronchial tubes* and into the *lungs*. In the lungs, oxygen is exchanged with carbon dioxide from the blood. Connected with the lungs are several *air sacs* that lie in different parts of the chest and abdomen. When a bird inhales, fresh air moves into the lungs and into the air sacs. When a bird exhales, carbon dioxide is carried out of the lungs, and some of the fresh air in the air sacs moves into the lungs. In this way, the lungs have a nearly continuous supply of fresh air. Birds need great quantities of air to help them metabolize their food and maintain a constant body temperature.

Some birds make sounds or series of sounds, called song. These birds have a sound-producing **syrinx** (sir′ingks)—an organ located on the trachea near the point where it divides into the two bronchial tubes. Bird songs fill the air during breeding season and probably attract mates for the male singer or warn off other birds of the same species and sex, or both. Certainly, sounds are used to establish territories among land birds. Differences in the songs of birds of the same species may help them recognize their mates or their own neighborhoods. Sounds may also keep pairs or flocks of birds together by helping them locate one another. Birds frequently sound alarm calls to warn of danger, and adult birds may be spurred on in their food gathering by the begging calls of their young.

Respiratory system of a bird

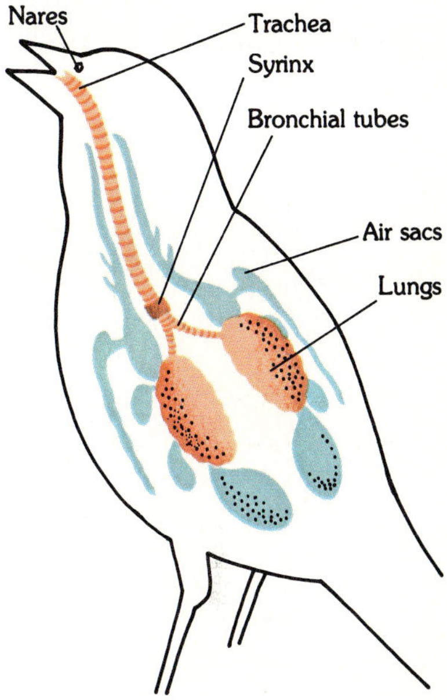

Circulation

At the center of the circulatory system in birds is a four-chambered heart composed of one **atrium** and one **ventricle** (ven′trə kəl) on the right side for *venous blood* and one atrium and one ventricle on the left side for *arterial blood*. Venous blood from the body enters the right atrium, is pumped to the right ventricle, and from there to the lungs. Oxygenated arterial blood returns to the left atrium, enters the left ventricle, and is pumped from there to the rest of the body.

The separation of oxygen-carrying arterial blood from carbon dioxide-carrying venous blood makes possible a very efficient metabolism which helps the bird maintain constant temperature. If the bird begins to overheat, heat loss is increased when the arteries containing blood from the heart begin to dilate near the skin. In cold weather, when a bird is losing heat to the environment, the arteries near the skin contract, and heat loss is slowed.

Four-chambered heart

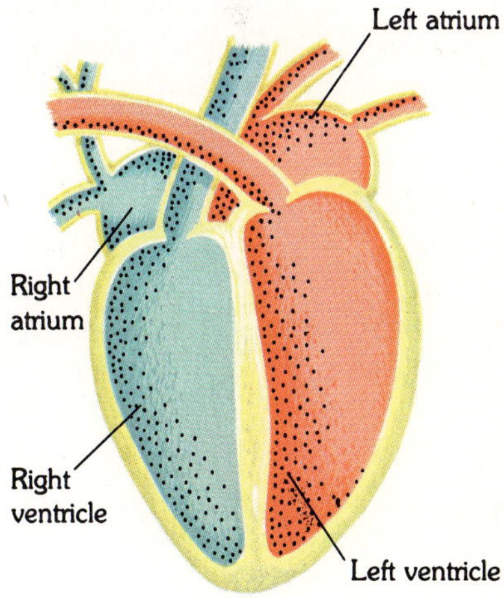

Chicken egg

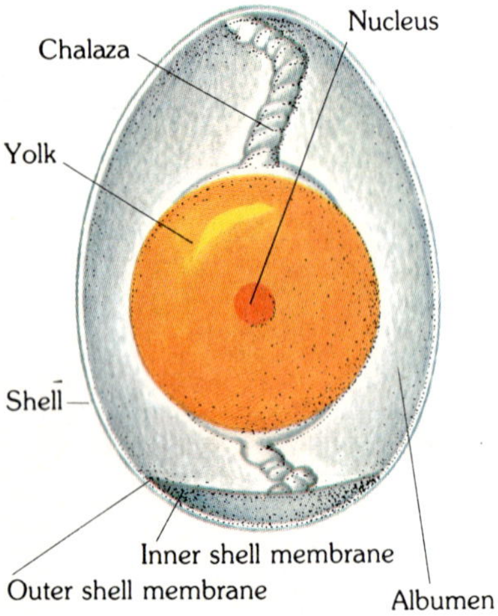

Reproduction

Reproduction in birds is similar to that in reptiles. In males, two oval testes are located near the kidneys. A pair of coiled ducts, the **vas deferens** (vas def′ər enz), carry sperm from the testes to each side of the cloaca. Males have no penis except in a few species, such as ducks and geese. During mating, the cloacas of the male and female birds are pressed together for an instant. Sperm travel from the male to the female cloaca, and then up the female oviduct.

Female birds have only one functioning ovary. Egg cells made up of a *nucleus,* a *yolk,* and *cytoplasm* are produced in the ovary. When an egg cell matures, it grows larger and breaks out of the ovary. Fine, hairlike cilia sweep it into the oviduct. Sperm traveling up the oviduct from the cloaca fertilize the egg cell. The fertilized egg begins to develop into an embryo and passes down the oviduct. Glands along the oviduct cover the yolk with the protein *albumen,* or egg white. Further down, a thin, tough inner shell membrane covers the albumen. Inside the membrane, the yolk is held in suspension by cordlike membranes called the *chalazas* (kə lā′zəz). The chalazas keep the yolk from settling to one side of the egg. Next, an outer shell membrane and a layer of calcium carbonate form the shell around the egg just before it leaves the oviduct and enters the cloaca.

Egg cells mature and pass down the oviduct in about twenty-five to forty hours, whether the egg cell is fertilized or not. Domestic hens lay eggs year-round, although they produce more eggs in the spring. Almost all wild birds lay eggs only in the spring, although some birds may hatch two clutches, or groups, of eggs in one season.

The embryonic chick develops from the egg cell, or nucleus, by using both the yolk and the albumen as food. The shell and shell membranes protect the growing embryo and admit enough air for it to breathe.

The eggshells of each species of bird are unique in color, size, and markings. Each species lays a characteristic number of eggs. Songbirds usually lay three to six eggs. Larger birds, such as quail, may lay as many as twenty eggs. Some hawks lay only one or two eggs.

The embryo in a bird egg usually does not generate enough heat to maintain a constant temperature. So parents must warm the eggs by sitting on them for a period—fourteen to fifty days, depending largely on the bird's size. Some newly hatched birds must be kept warm by parents for several weeks until the young birds' blood-warming mechanisms develop. Other young birds are able to keep warm without their parents, except in unusually cold weather.

Reproductive Behavior

In early morning in spring on the plains of Nebraska, a male sage grouse establishes his territory on a dancing ground. He raises the feathers on his neck, spreads his tail wide and raises it high. He spreads out his wings and drops them toward the ground. With head lowered, he stamps the ground and moves in circles several times. Then he inflates two air pouches on the sides of his throat and makes a hollow, booming sound. He is trying to attract a mate.

Each species of bird has its own way of behaving during courtship. Males usually arrive in a mating area before females arrive and set up a territory, which they then defend from other males. When the females arrive, the males may sing or call and display their bright plumage. This behavior is meant to attract the attention of the females. After a female enters the territory of the male of her choice, the male defends the territory and the female from the advances of other males.

After mating, each pair of birds usually stays together for the breeding season. Mated pairs of geese may stay together for life, which may be more than twenty years. In some species, mating groups may include a male and two or more females, or one female and two or more males.

The building of nests, incubation of eggs, and care of the young are tasks often shared by both parents. Some birds weave intricate nests from grasses and twigs. Some birds lay their eggs on bare ground. A female cowbird will lay an egg in the nest of another species of bird. After the eggs hatch, the foster parents feed all of the young birds, but the cowbird often grows so large so fast that it crowds out all of the other young birds. Instinct alone keeps the parents feeding the interloper.

While the eggs are being incubated, one parent stays on the nest nearly all of the time. Sometimes one parent does all of the incubating; sometimes the two take turns.

The young of such birds as ducks, chickens, quail, and ostriches are feathered, active, and ready to begin eating on their own when they emerge from the egg. Other young are hatched naked, weak, and helpless. For several weeks, the parents are busy stuffing food into the open mouths of the young birds, keeping them warm, and protecting them from predators.

Eggs in nest

Nesting

Care of young

Review It

1. How do birds shred their food without the help of teeth?
2. What are the functions of the cloaca in birds?
3. What purpose does the courtship ritual of birds serve?
4. Describe the functions of bird feathers.
5. How is the body of a bird adapted for flying?

Activity

Incubating Chicken Eggs

Materials
Incubator
Fertilized hen's eggs
Sand
Forceps
Yogurt carton
Stereomicroscope
Finger bowls
Prepared slides of chick embryos (if available)
Compound microscope

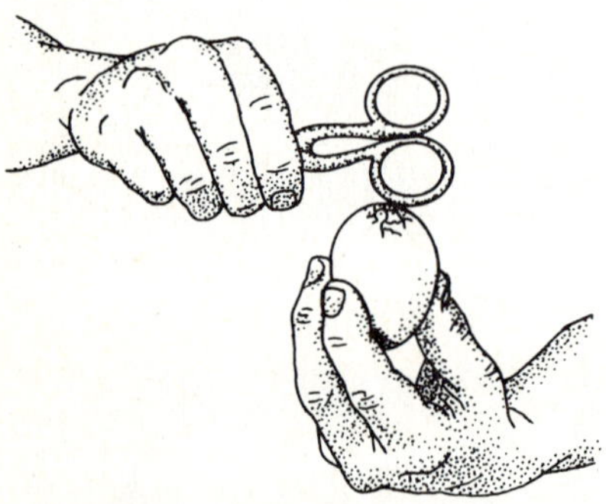

Crack open the egg

Procedure

1. Regulate the incubator to a temperature of 37°C. Place the bulb of the thermometer where the eggs will be. Add a pan of water to give a relative humidity of about 50 percent.
2. Obtain fresh eggs from a hatchery. Since some eggs may have incubated inside the hens' bodies longer than others, embryos will not necessarily be the same age.
3. Place eggs in the incubator. Turn each egg once a day to prevent the membranes from sticking together.
4. Fill the yogurt carton about three-fourths full of sand, and place it in the incubator.
5. You may wish to examine your first egg after twenty-four hours. Remove carton of warm sand and one egg. Bury the small, more pointed end of the egg in the sand. Carefully chip away the shell from the top of the rounded end of the egg and remove the shell pieces with forceps. The embryo should float to the surface. Examine the embryo with a stereomicroscope. Notice the heartbeat. In the warm sand the chick embryo should stay alive for several hours.
6. Continue from day to day to open and examine eggs. Chicks should hatch after about 21 days.
7. If prepared slides of chick embryos are available, study them with a compound microscope.

Analysis

1. How is the food transferred from the yolk to the embryo?
2. Which structures of the chick embryo develop first?
3. At what age during development does the animal show the definite characteristics of a bird?

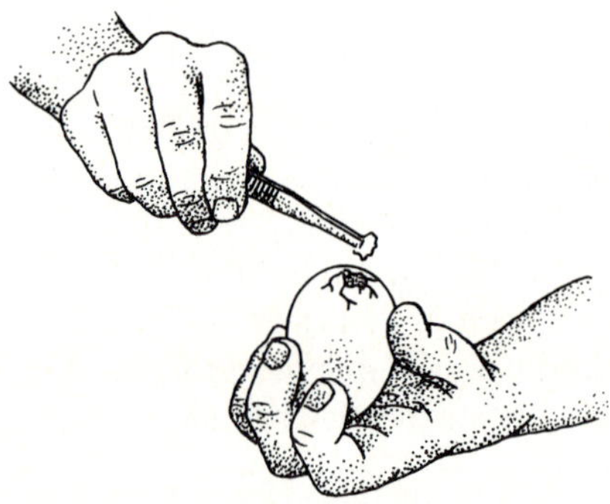

Remove pieces of shell

Challenge!
Windows made of cover glasses and wax or clear plastic can be made on the side of an incubating egg. In this way, you can observe the developmental stages of one chick.

Development of the Chick

In this sequence of pictures you can see the development of the chick embryo. The time is from 18 hours to hatching. Notice the development, change, and retention of structures.

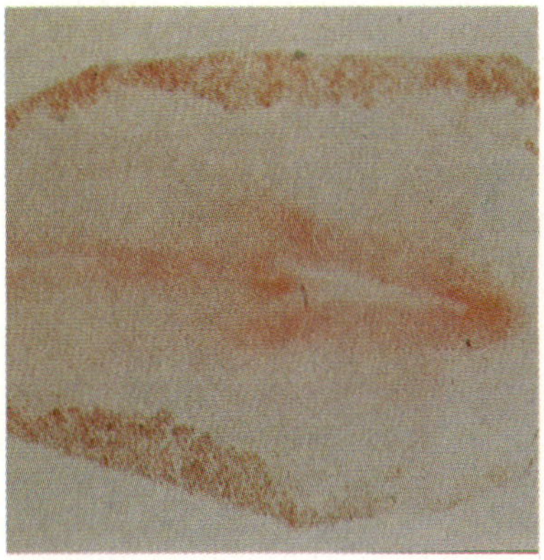

Courtesy Carolina Biological Supply Company

18 hours

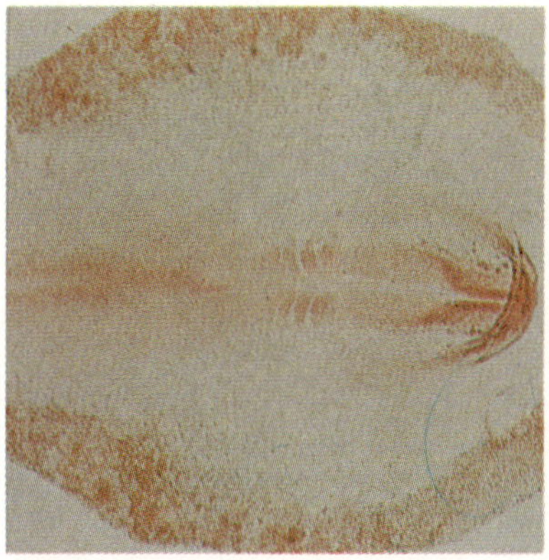

Courtesy Carolina Biological Supply Company

24 hours

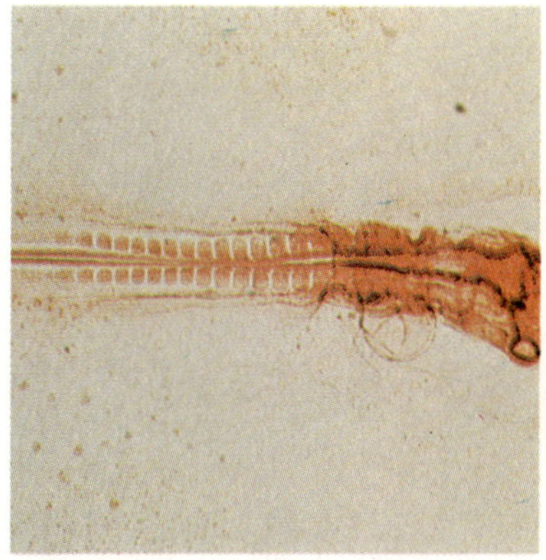

Courtesy Carolina Biological Supply Company

36 hours

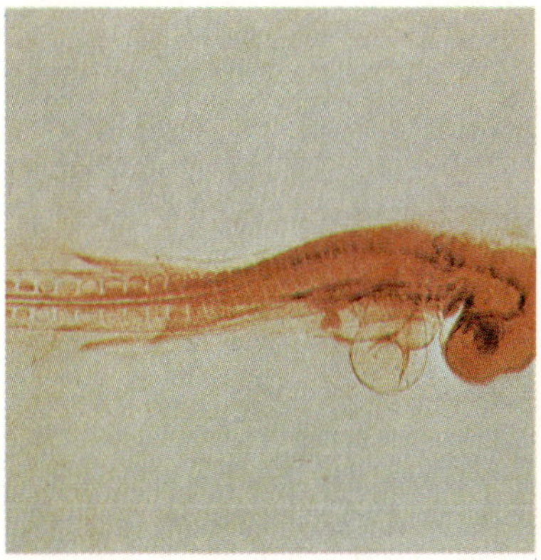

Courtesy Carolina Biological Supply Company

48 hours

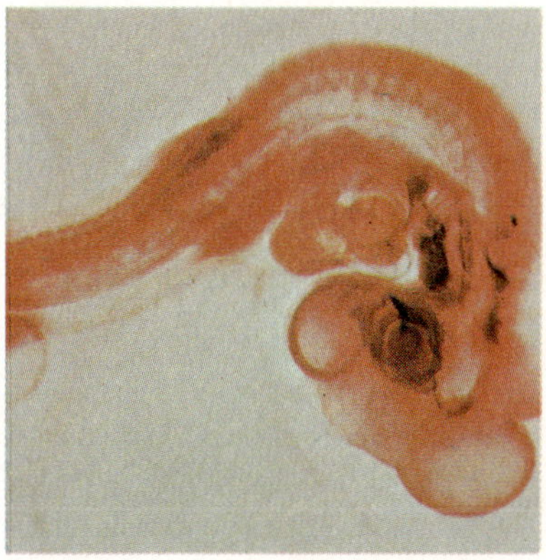

Courtesy Carolina Biological Supply Company

72 hours

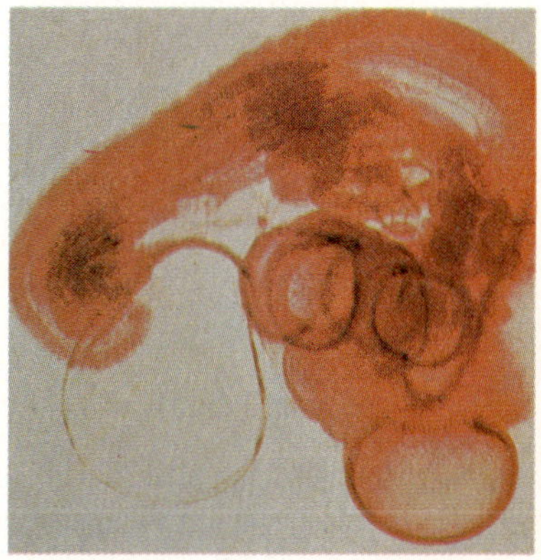

Courtesy Carolina Biological Supply Company

96 hours

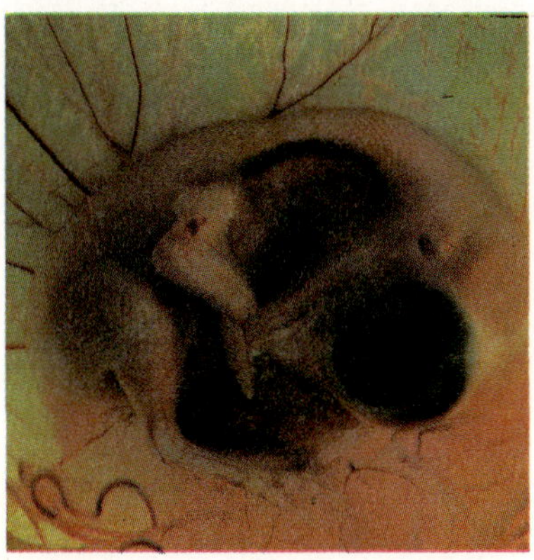

9 days

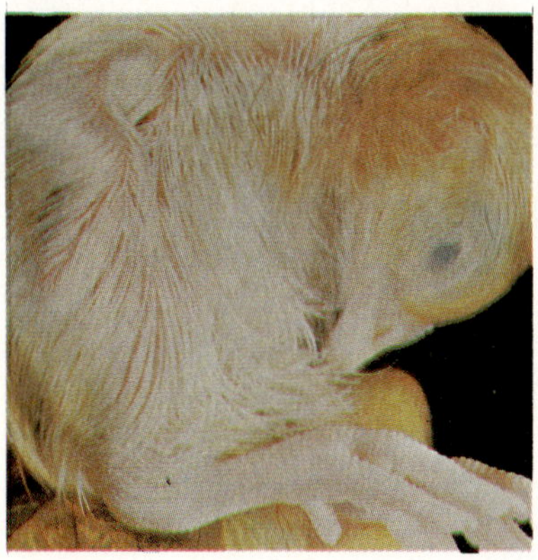

16 days

Hatching

20–3
Mammals: Class Mammalia

Binocular vision in humans

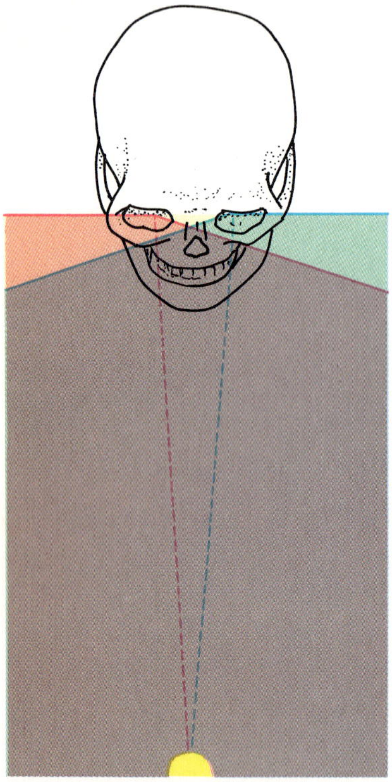

Members of the class Mammalia (ma mā'lēə) have one feature in common that gives them their class name. All species of mammals have mammary glands; and, in the females of the species, these glands secrete milk for the young. Most young mammals are neither able to look for food nor to digest an adult diet for a long time after birth. They live on mother's milk and are protected from predators by their parents until their brains and bodies are fully developed.

Most mammals are **placentals** (plə sen'tlz). The embryos of mammals are nourished through a temporary organ called the **placenta** (plə sen' tə) inside the body of the mother. Only two groups of mammals—the **monotremes** (mon'ə trēmz) and the **marsupials** (mär sü'pē əlz)—are not placentals. The monotremes reproduce by laying eggs, and marsupial young are born as partially developed embryos. However, both monotremes and marsupials are classified as mammals because they nurse their young.

Each species of mammal has a skeleton suited to the ways in which it feeds, moves about, and protects itself. Mammals that run well have long, strong leg bones. Mammals that climb trees usually have well-developed forelimbs. Mammals have strong skulls that protect their highly developed brains and flexible necks and backbones that make movement easy. Mammals also have strong jaws and teeth of different types and different arrangements suited to the food they eat.

Most mammals have good vision. Monkeys, apes, and humans have eyes that are located in the front of the head so that both eyes can see an object at the same time. This is called *binocular vision.* It enables these mammals to judge an object accurately without moving the head. Most mammals also have a good sense of hearing.

Most mammals have hair that conserves body heat, offers some protection from abrasive injury, and frequently sheds water. In whales, fat or blubber replaces hair as an insulator. The armadillo, a resident of warm climates, has no hair, but is covered instead with an armor of bony plates for maximum protection from predators. The porcupine is covered with modified hairs that form sharp quills.

Most mammals have a muscular wall called a **diaphragm** (dī'ə-fram) between the chest and abdominal cavities. This muscle increases the capacity and efficiency of the mammal's breathing apparatus.

Unlike some reptiles whose legs protrude from the sides of their bodies causing a waddling gait, land-going mammals tend to move on limbs swung directly under their bodies. Some mammals are exceedingly fast runners.

As a group, mammals are the most intelligent animals on earth. Mammals show the highest ability of all animals to learn in new situations and are least dependent on instinctive behavior. High intelligence, high activity, and efficient body systems enable mammals to live in different habitats.

Did You Know?
This is the Age of Mammals, the Cenozoic era, and it has been for about 70 million years. Before this age, reptiles dominated the earth. At that time, the continents emerged from the seas, oceans were formed, the mountains took shape, and the climate of the earth changed. Instead of the earth being warm and wet all over, some parts became cool or cold. Temperatures and humidity began to change from season to season. Ectothermic reptiles apparently were unsuited for these changes, but endothermic mammals flourished. Today, about 5,000 species of mammals live in all parts of the earth—on, above, and below ground, and in the seas.

Teeth

One distinguishing feature among mammals is their **dentition** (den-tish′ən)—the shape and arrangement of their teeth. Dentition tends to vary according to what the mammal eats. **Carnivores** (kär′nə vôrz) kill and eat other animals; **herbivores** (hėr′bə vôrz) eat plants; and **omnivores** (ôm′ni vôrz) eat both meat and plants. **Insectivores** (in sek′tə-vôrz) live on insects, and **frugivores** (frü jiv′ərəz) live on fruit. There are four basic types of teeth.

Incisors

The **incisors** (in sī′zərz), or front teeth, in most mammals are for plucking, cutting, and holding. Rats and rabbits have particularly prominent incisors. In gnawing mammals, such as *rodents,* the incisors grow continuously. In herbivores, they are adapted for cropping plants. Many carnivores have extra points on the incisors for holding their prey. Elephant tusks are extensions of the incisors.

Canines

The **canines** (kā′nīnz), or eye teeth, or dog teeth, in most mammals are often the largest in an animal's mouth, projecting beyond the level of the other teeth. They are used for tearing and are prominent in the carnivores. Often the canines interlock, limiting chewing action to an up-and-down motion. The tusks of wild boars and walruses are extensions of the canines.

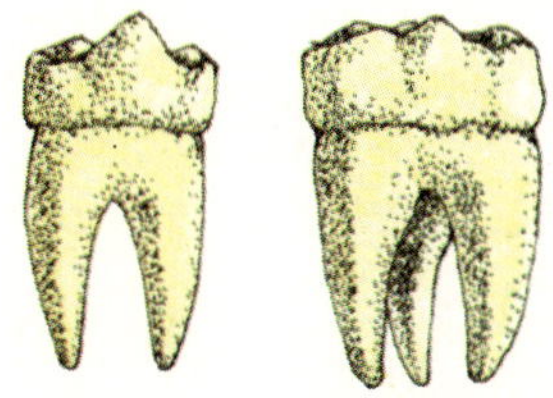

Premolars and Molars

The shape of **premolars,** or **bicuspids** (bī kus′pidz), tends to vary with their use. In herbivores, the surface is broad and cusped for grinding. In carnivores, premolars have a cutting edge for severing tendons and crushing bones. **Molars,** too, may be sharp in carnivores, but in herbivores and omnivores molars tend to be relatively flat, with cusped or ridged surfaces for grinding plant foods.

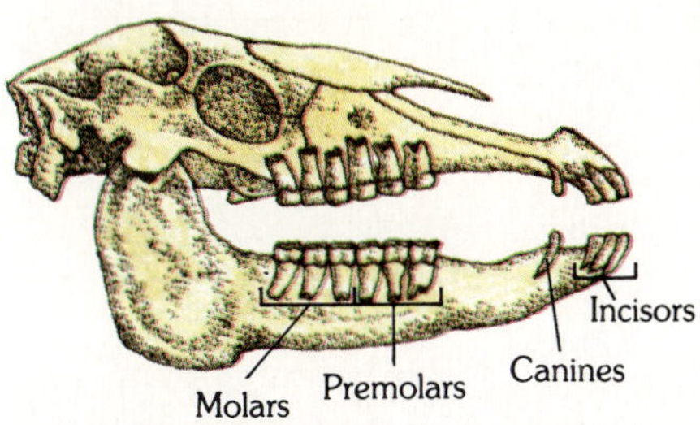

Herbivorous horse

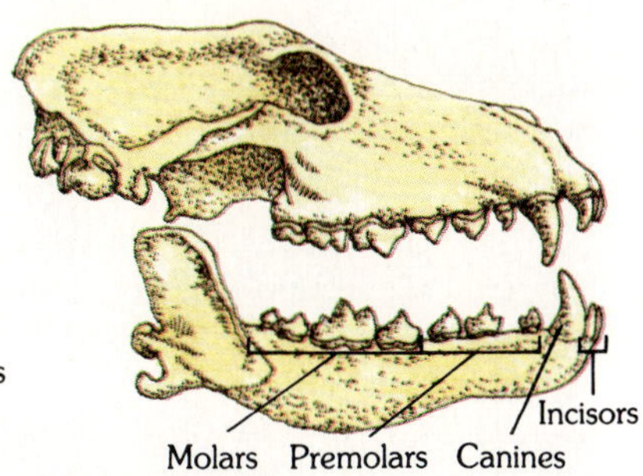

Carnivorous dog

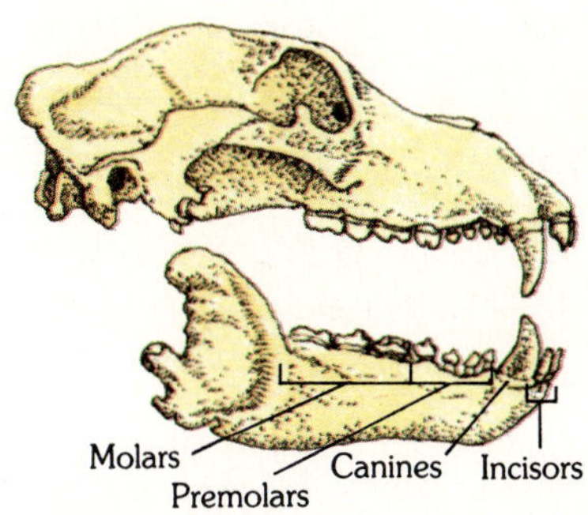

Omnivorous bear

Egg-Laying Mammals: The Monotremes

Of the three groups of mammals—monotremes, marsupials, and placentals—monotremes are the only egg-layers. Both of the monotreme species, the **duck-billed platypus** and the **echidna** (i kid′nə), or spiny anteater, are found in Australia. In many respects, monotremes resemble reptiles. Both reptiles and monotremes lay eggs. The name monotreme means "one hole," referring to their single cloacal opening. Like reptiles, as well as amphibians and birds, monotremes pass body wastes from the intestine and bladder, as well as sex cells, through the cloaca. Monotremes have no teeth, and they are not able to maintain as constant a body temperature as do other mammals. But they are warm-blooded, and they have mammary glands.

Adult spiny anteater

For two species that are biologically similar, the platypus and the spiny anteater are quite different in habits and appearance. The platypus has dense, soft fur, short legs, webbed feet, a flat, paddlelike tail, and a most unusual snout or bill of soft, moist skin over a bony frame. The platypus spends most of its life in water and feeds on shellfish, earthworms, and tadpoles.

The spiny anteater's body is covered with some hair and many sharp spines. When alarmed, it can form itself into a ball and dig quickly into the ground with its four clawed feet, leaving only a mound of spines as a target for its enemy. It lives on ants, worms, and termites, which it seeks out with its long snout and then laps up with its long, ridged, and sticky tongue.

Female platypuses incubate their eggs in nests on the ground. Female spiny anteaters carry their eggs in pouches on the abdomen. The mammary glands of monotremes have no nipples. The newly hatched young lick milk from the mother's skin near the openings of the mammary glands.

Adult duck-billed platypus

Pouched Mammals: The Marsupials

Unlike monotremes, marsupials do not lay eggs, and do not have a cloaca. Body wastes are excreted from two openings—one for the intestines and one for the kidneys. They give birth to living young that partially develop in a placenta in the uterus of the mother. Marsupials have mammary glands that secrete milk through nipples. They can maintain a constant body temperature.

Female marsupials have a pouch on the outside of the abdomen. The nippled mammary glands are located on the abdominal wall inside the pouch. Young marsupials are born without hair and are not fully developed. After birth, they crawl into the mother's pouch. There they stay for several weeks and feed on mother's milk. At the end of this feeding period, they begin to venture out for short periods. Kangaroos are marsupials. Young kangaroos remain in the pouch continuously for about two months after birth. A young kangaroo may return to the pouch for brief periods until it is about nine months old.

In North America, the only marsupial is the opossum. Opossum embryos stay in the uterus for only about twelve days after fertilization. Newborn opossums, about the size of honeybees, crawl to the pouch on their mother's abdomen where they stay attached to a nipple for about 100 days. After young opossums begin to leave the pouch, they stay with the mother and often ride around on her back.

Most marsupials are found in Australia and the surrounding islands. They live in a wide variety of habitats and take different forms. There is a marsupial mole that lives underground. Other marsupials look and act much as placental mice, squirrels, and wolves do. Next to the kangaroo and its relative the wallaby, the koala bear is probably the best known marsupial of Australia.

Marsupials have a full range of eating tastes. Kangaroos, wallabies, and koalas are herbivores. The opossum is an omnivore and a scavenger as well; it is willing to feed on decaying animal flesh if need be. The banded anteater is an insectivore, whereas the Tasmanian devil is a carnivore and a scavenger.

Kangaroo in pouch

Opossums in pouch

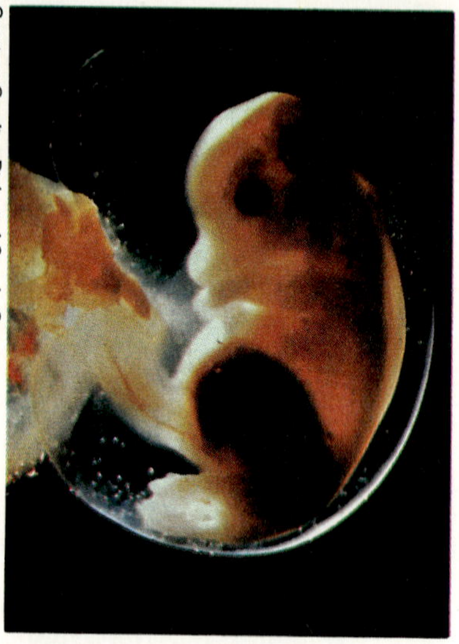

Embryo of pig

Placental Mammals

The largest group of mammals is the placentals, the group to which humans belong. Placentals live in every kind of environment—polar bears near the North Pole, dolphins in the oceans, beavers in rivers, moles underground, deer in woods, mice in fields, and camels in deserts. Among the placentals are such diverse members as bats and whales, shrews and elephants, armadillos and humans. Placentals move about by crawling, walking, running, leaping, gliding, climbing, swinging, swimming, and flying. Some are herbivores, some are carnivores, and some are omnivores. Placentals, which develop fully inside the mother, are more intelligent than other mammals. More than any other group of mammals, placentals are domesticated by humans and used by them for food, clothing, and as pets.

Embryonic Development

Placental mammals differ from all other mammals. The young are born fully developed from within a *placenta* that forms along with the growing embryo in the uterus of the mother.

After the egg cell has been fertilized by a sperm, it moves into the uterus and attaches to the wall there. As the embryo develops, an **umbilical cord** forms, which connects the embryo to the placenta around it. Through the umbilical cord run the veins and arteries of the embryo. These blood vessels branch into hundreds of smaller vessels, called **capillaries,** that lie inside hundreds of strands of tissue known as **villi.** Together with tissue from the uterine wall, these villi form the placenta.

The growing villi begin to rupture the mother's blood vessels in the placenta. This brings the capillaries in the villi close to, but never in direct contact with, the mother's blood. Oxygen and nutrients in the mother's blood diffuse through the villi and capillary walls to the embryo's blood system. Carbon dioxide and wastes return in the same way from the embryo to the mother. Until the embryo is born, the mother functions as the embryo's kidneys and lungs.

The Young Placental

When the young placental is born, the umbilical cord is severed, and the young begin to breathe oxygen through their lungs. They are nourished by milk from the mother's mammary glands. For each species of mammal, the milk from the natural mother is best for her own young, although milk from one species may sometimes be fed artificially to the young of another species.

Young placental mammals vary in their dependence on maternal care. Cats and mice are born helpless and naked, whereas guinea pigs and horses are born well developed and covered with hair. The newborn whale immediately begins to swim beside its mother.

After birth, the young placental is usually given more care than just nursing. It is often protected from predators as well as warmed by close contact with its mother's body. As the baby gets older, its own warming mechanisms take over, and it learns to eat food other than mother's milk. Young placentals are taught how to seek food and how to protect themselves. The skills needed for survival may have to be learned over many years. A young mountain lion may take more than two years to mature, and an Indian elephant may take more than eleven.

The length of time a developing embryo is carried in the uterus is known as the **gestation** (je stā'shən) **period.** For placental mammals, the gestation period ranges from about twenty days for a mouse to about twenty months for an elephant. Some placentals are born more than one at a time, while others are born singly. The table below shows the gestation periods and number of young born at one time for some common placentals.

	Average Gestation Period (in days)	Usual Number of Young at One Birth
House mouse	20	6
Dog	61	7
Cat	63	4
Lion	108	3
Pig	113	9
Bear	215	2
Cow	280	1
Human	280	1
Horse	336	1
Sperm whale	365	1
Giraffe	435	1
Indian elephant	645	1

Challenge!

In the book Born Free, *Joy Adamson describes how she raised a lion cub, Elsa, to maturity before returning her to the wild. Preparing Elsa for life among other lions involved long hours of training. Find out what skills Joy Adamson had to teach Elsa and how she went about it.*

Distribution of Marsupials

Except for two marsupials found in the Americas, almost all marsupials live in Australia, New Guinea, and other islands in that vicinity. At one time, marsupials lived in continental Europe, but after the separation of Australia from the larger continental land mass about 65 million years ago, marsupials disappeared in Europe. They never lived in Asia or Africa.

Marsupials

Placentals

Convergent Evolution

Many marsupials resemble placentals that are found elsewhere in the world. But scientists say that the ancestors of these animals differed in appearance from one another more than their descendants do today—that the similarities we see in marsupials and placentals are the result of what they call *convergent evolution.* As marsupials and placentals faced and adapted to similar environmental conditions in different parts of the world, they developed similar traits. Their evolutionary development brought about convergence from widely divergent beginnings. The striking similarities between the marsupials and placentals illustrated to the left are best explained as examples of convergent evolution.

Brain and Behavior

Mammalian brains and bird brains are similar. Compare the diagram of the mammalian brain on this page with the diagram of the bird brain. Both birds and mammals have large cerebrums. But, in mammals, the surface of the cerebrum is covered with a layer of "gray matter" called the **cerebral cortex.** The surface of the cortex is heavily folded. These folds increase the area of the gray matter. Size of brain is not nearly so important to intelligence as is the area of the cerebral cortex.

In birds, the cortex is smooth, and the cerebrum itself acts as a control center for many instinctive behaviors. These include flying, nest building, mating, and care of young. Birds do not learn their behaviors from other birds or by making independent adjustments to changes in environmental conditions. When it is time to build a nest, a bird instinctively begins building a nest that is just like the nests of other members of its species.

Mammals have more gray matter than birds. In the cerebral cortex, the mammal receives all messages from the senses and stores them as memories. Here, too, is where new sensory information comes to be compared with memories of past experiences and to be acted on.

Mammals exhibit many instinctive behaviors. For example, wolves do not have to learn how to catch their prey; squirrels automatically hoard food. Ways of hiding from predators and periods of sleep and wakefulness are not learned behaviors in many mammals. However, mammals learn from each other and from past experiences. The more intelligent mammals, such as apes and humans, learn almost all of their behaviors from parents, other members of their community, and their own independent actions taken in new situations. In addition, humans use language to share information with one another.

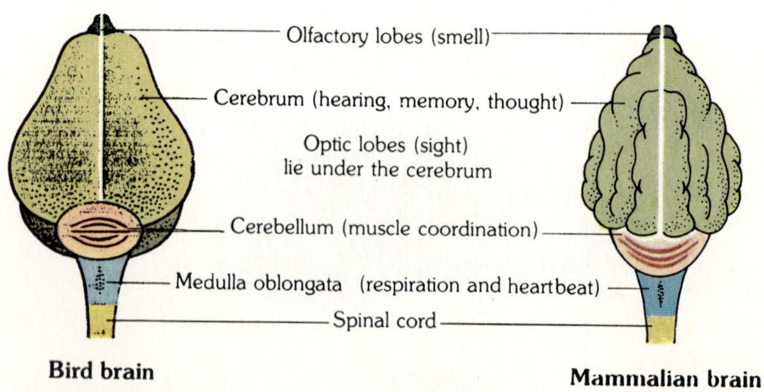

Review It

1. Describe some differences among the different mammal groups.
2. What is the major difference between behavior of birds and behavior of mammals?

Breakthroughs in Biology

Recent developments in ways to approach communicating with nonhuman primates have provided some insight into their capabilities in learning language.

Koko signs "Koko love baby."

One difference between humans and other primates is that the human vocal tract is developed for speech, whereas that of the other primates is not. A way had to be devised for communication that did not involve speaking.

Working in the late 1920s and 1930s, primatologists Ada and Robert Yerkes were convinced that gorillas had a great capacity to learn. But in language studies, the gorilla was largely ignored.

Chimpanzees, thought to be the most trainable nonhuman primates, were subjects of several studies in the 1960s. In one study Allen and Beatrice Gardner, working with Washoe, used American Sign Language, or Ameslan, the language of the deaf. In four years, Washoe was using 132 signs to communicate.

Then in 1972, when developmental psychologist Penny Patterson started graduate work at Stanford University, she took a chance on the gorilla and came up with a winner—Koko.

Using the approach to communication refined by the Gardners, Patterson began the painstaking task of teaching one-year old Koko that a sign she could make with her hands referred to an object, such as an orange; or even a feeling, such as "good." The technique involves taking the subject's hands and shaping the sign for a particular object that is in sight. At first, Koko didn't want to make signs. She didn't understand the purpose, and tried to bite Patterson. But when the understanding did come, she was a rapid learner. By age seven, she had used over 600 signs and had a vocabulary of 375 signs.

Koko can communicate about her feelings of happiness, sadness, anger; she can discuss the past; she lies to avoid trouble, such as a scolding for breaking something. (She'll blame someone else.) She asks for hugs, and makes up her own insults. When angry with someone, she forms the sign "bad toilet." She knows that one symbol can have two different meanings, and uses "nut" for both the food and for a person who is acting silly.

Koko also learned to use a computer synthesizer that produces a human voice to verbalize objects, actions, and feelings represented on keys by geometric symbols. With the synthesizer Koko can produce spoken English for words in her sign vocabulary.

Koko was officially "on loan" from the San Francisco Zoo. In 1977, the question came up about returning her. Patterson wants to continue the study through Koko's lifetime. And other people concerned with the project and Koko's welfare felt that a return to the zoo would be disastrous for her. She had been given language. Since language expands the possibilities for thinking, she was a different animal than when she left the zoo. The question was resolved when money was raised to buy Koko from the zoo.

Analysis

1. What kinds of changes do you think happen to an animal as it learns to use language?

2. Project Koko is an ongoing study. Try to find further information on Koko's progress.

Chapter Summary

Birds may be thought of as reptiles that became warm-blooded and learned how to fly. Archaeopteryx, the oldest fossil bird, certainly was reptilian in many ways. But in the descendants of Archaeopteryx, everything is adapted for flight—the bones, the feathers, the lungs, the heart, the brain. Birds are highly successful animals, as we can tell from the number of birds and bird species alive today all over the earth.

Mammals, which are also possible descendants of reptiles, show fewer ancestral reptile traits than birds do. Only the monotremes lay eggs and have cloacas. Warm-blooded like birds, mammals also have been exceptionally successful. Placental mammals bear live young, which have undergone prolonged development within the mother before birth. Mammals have body coverings of hair or fat that keep them warm and protect them. They feed their young on mother's milk. Above all, mammals have an advanced brain and nervous system. They can learn from experience and adapt to changes in their environment.

Questions to Think About

1. With rowboats and bicycles, humans can use the energy of their own bodies to travel on water and on land. Why has it been so difficult for humans to develop a machine or an apparatus that they can use to fly through the air?
2. Suppose all mammals except humans became extinct. In what ways would the loss of other mammals affect human life?
3. What are some flightless birds in the world today? Where are they found? Why have they been able to survive in those locations and not in others?
4. Suppose a mammal or a bird were born with a three-chambered heart or with an opening in the wall between the two ventricles in the four-chambered heart. This condition is normal in amphibians and reptiles. How do you think that condition would affect a higher vertebrate?

Extra Research

1. Find out what triggers bird migration. How do birds know in which direction to migrate?
2. Why are the monotremes sometimes considered the link between reptiles and mammals?
3. How have the more intelligent mammals, such as dolphins and chimpanzees, been valuable in experiments conducted by humans?

Chapter Test

Part I

Copy the number of each test item. If the item is *true*, write *T* after the number; if the item is *false*, write *F*.

1. Sweating decreases heat loss in mammals. (476)
2. Birds have two ovaries, just as mammals do. (484)
3. Down feathers provide insulation for a bird. (478)
4. Most birds have a poor sense of smell. (482)
5. Birds and mammals have lungs for respiration. (483)
6. All mammals nurse their young on milk. (488)
7. The gestation period is the time it takes for mammals to digest food. (493)
8. The behavior of birds is controlled more by instinct than is the behavior of mammals. (496)
9. The cerebrum of a bird has more folds than the cerebrum of a mammal. (496)
10. The diaphragm in mammals is an aid to digestion. (488)

Part II

Copy the number of each sentence below. After each number, write the word or words that complete the sentence correctly.

11. The part of the brain that regulates body temperatures of warm-blooded animals is called the ____. (476)
12. The body covering of most mammals consists of ____. (488)
13. The brain region that controls flying and muscle coordination in birds is the ____. (482)
14. In birds and mammals, excess heat is given off when blood vessels near the skin ____. (476)
15. The digestive organ that grinds up food in birds is called the ____. (482)
16. The organ that may contain wastes and sex cells in a bird is called the ____. (482)
17. The part of a bird's egg that keeps the yolk from settling to one side is the ____. (484)
18. Mammals that lay eggs belong to the group called ____. (490)
19. Animals that live on the flesh of other animals are ____. (489)
20. The group of mammals that provides the most care and protection for their young is the ____. (493)

Part III

Copy each numbered item from List A. After each item, write the letter and words from List B that best describe it.

List A

21. monotreme (490)
22. marsupial (491)
23. ectotherms (476)
24. oxidation (476)
25. placenta (492)

List B

a. Temporary organ that provides an exchange of materials between mother and embryo
b. Pouched mammal
c. Burning of food for energy and heat
d. Cold-blooded animals
e. Egg-laying mammal

Focus on Careers

Zoo Keeper

"I guess I've always been taking care of animals," Steve said. "At various times at home I've kept rabbits, a horse, cats, a goat, a sheep, and always dogs. I wanted to work with animals, but I couldn't afford to go to college right after high school. Now, as a zoo keeper, I'm working with animals and am also completing a special program for keepers here at the zoo.

"Right now I work in the primate house. Here, gorillas, chimpanzees, and orangutans are kept in tall, rounded towers that are meant to be more like their natural habitat than cages. Under the head primate keeper's supervision, I weigh, measure, and distribute food at specific times each day, recording everything as I go. I hose down the towers to keep them clean and examine the ropes and perching platforms the animals use. The animals are carefully observed for any signs of disease or injury. If there is a sick animal, we often move it to separate quarters where it is treated.

"All in all, the work is very routine, but it has high points too, for example, when babies are born. Sometimes the babies are moved to the zoo nursery where keepers feed and take care of them until they can be on their own. The nursery is a popular place at the zoo—for visitors and workers alike."

Zoo keepers, like Steve, are responsible for the care of all animals in a zoological park, or zoo. This includes feeding, cleaning, and providing any specialized attention the animals need. Since zoos are educational places, keepers often provide visitors with information about the animals. Many keepers specialize in the care of a specific kind of animal; he or she is then called bird keeper, cat keeper, reptile keeper, and so on.

Many zoos require keepers to have a year or more of college work in zoology, but many do not. Steve is working for a large zoo which offers formal training in animal keeping. While he completes his training, he gets valuable experience working with several different kinds of animals. In time, he can advance to head keeper, senior keeper, or maybe even to zoo director. Usually, however, the director of a zoo has an advanced degree in zoology and has had much experience working with animals.

The job market is expanding for people interested in working with animals. However, the number of applicants is also increasing. A person who has experience handling animals and has educational courses as well can compete effectively for these jobs.

Career Information

Fish Culturist: American Fisheries Society, 5410 Grosvenor Lane, Bethesda, MD 20014.

Livestock Production Technician: Treasure Valley Community College, 650 College Blvd., Ontario, OR 97914.

Veterinarian: American Veterinary Medical Association, 930 N. Meacham Road, Schaumburg, IL 60196.

Zoo Keeper: American Association of Zoological Parks and Aquariums, Oglebay Park, Wheeling, WV 26003.

Fish Culturist

Duties: Works in a fish hatchery raising fish, such as trout or salmon, for use in stocking streams and ponds or for commercial purposes. The culturist may collect, fertilize, and incubate spawn (fish eggs). He or she monitors the growth of the fish in large tanks and watches them for signs of disease or injury. When the fish reach the right size, the fish culturist helps deliver them to lakes and streams or to commercial fish packers. Some fish culturists investigate bodies of water to study where fish of different species breed to duplicate conditions in the hatchery.

Training: Qualifications vary. A two-year college program in fish technology is helpful because it includes courses in fish identification, aquatic environments, and fisheries equipment.

Opportunities: Outlook is good. Jobs are available in private and federal fish hatcheries, bait farms, and aquariums.

Livestock Production Technician

Duties: Assists farmers, ranchers, veterinarians, and others involved in raising livestock. Duties vary depending on the location and type of livestock handled. On a cattle ranch, for example, tasks may involve weighing, mixing, and storing feed; and assisting in breeding. The technician periodically examines all animals and cares for diseased or injured ones, administering treatment as directed by a veterinarian. The technician operates and repairs heavy equipment, such as dump trucks and manure-turning machines.

Training: Requires a two-year program in livestock technology. Previous experience working with animals is valuable.

Opportunities: Many jobs are open in ranching, farming, and other businesses, such as farm feed and equipment.

Veterinarian

Duties: Diagnoses and treats diseases and disorders of animals. Such a person is qualified to perform surgery and dental work and prescribe and administer drugs. He or she may carry out various laboratory tests to determine the presence of disease and likely source of infection. Veterinarians perform routine examinations on healthy animals and give advice on their care and breeding. Veterinarians in rural areas primarily treat farm animals.

Training: Requires a license and a Doctor of Veterinary Medicine degree. Veterinarians must attend a minimum of six years of college. Entrance to veterinary schools is competitive, and only those with high grades in science are chosen.

Opportunities: There is a continuing need for veterinarians, primarily in animal hospitals. Others carry out laboratory research in animal diseases; still others are inspectors in slaughterhouses.

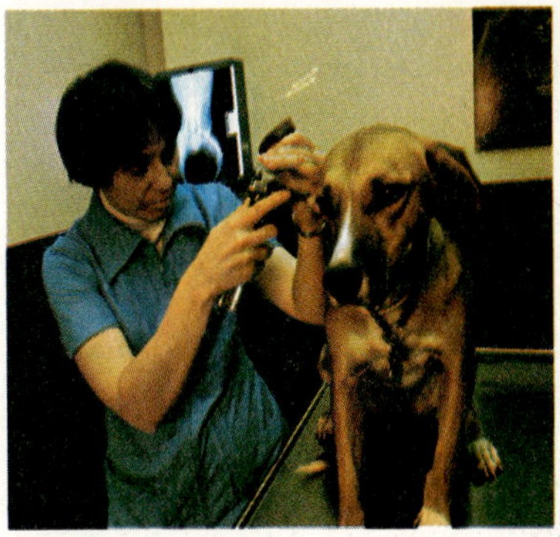

Bibliography: Unit Eight

Beebe, B. F. *American Desert Animals.* David McKay Co., Inc., 1966. Interesting facts about desert vertebrates.

Bellairs, Angus. *The Life of the Reptiles.* Universe Books, 1970. A two-volume reference on all aspects of the reptiles.

Boorer, Michael. *Mammals of the World.* Grosset and Dunlap, Inc., 1967. Survey of evolution, structure, and classification of mammals.

Breland, Osmond. *Animal Life and Lore.* Harper and Row, Inc., 1963. Fascinating and improbable facts about the ways of invertebrates, fish, mammals, and other vertebrates.

Burgess, Robert F. *The Sharks.* Doubleday and Co., Inc., 1970. Discusses history of man's fear of sharks, misconceptions about sharks, and current findings about these animals.

Burton, Jane. *The Wondrous World of Fishes.* Colour Library International Ltd., 1976. Brilliant photographs of fish along with descriptions of how colors of fish are produced and how these colors are used in the lives of fish.

Burton, Maurice. *The World of Reptiles and Amphibians.* Bounty Books, 1973. Fascinating photographs of amphibians and reptiles along with interesting facts about their lives.

Burton, Robert. *The Seashore and Its Wildlife.* G. P. Putnam's Sons, 1977. Examines shores around the world and describes the structure and behavior of both familiar and rare species.

Campbell, Andrew. *The Wonderful World of Life in the Sea.* Hamlyn Publishing Group, 1978. Beautifully illustrated book that describes the variety of life styles and body forms of animals found in the sea and on the shores.

Carrington, Richard. *The Mammals.* Time Inc., 1963. Pictures and information about mammalian evolution, body structure, defense mechanisms, and more.

Cochran, Doris. *Living Amphibians of the World.* Doubleday and Co., Inc., 1966. Picture essay volume on amphibians.

Cousteau, Jacques. *The Shark: Splendid Savage of the Sea.* Doubleday and Co., Inc., 1970. Fascinating account of adventures and discoveries made by Jacques Cousteau and his crew while studying sharks.

Cruickshank, Allan, and Cruickshank, Helen. *1001 Questions Answered About Birds.* Dodd, Mead and Co., 1958. Answers questions most frequently asked about birds: their structure, annual cycles, migration, and behavior.

Dozier, Thomas. *Dangerous Sea Creatures.* Time-Life Films, Inc., 1976. Beautifully illustrated volume on poisonous and venomous invertebrates and vertebrates.

Eimerl, Sarel, and DeVore, Irven. *The Primates.* Time Inc., 1965. Interesting information and pictures of monkeys and apes.

Flanagan, Geraldine Lux. *Window into an Egg.* William R. Scott, Inc., 1969. Photographs and account of day-by-day development of a chicken.

Minton, Sherman A. *Giant Reptiles.* Charles Scribner's Sons, 1973. Facts and myths about reptile evolution and behavior.

Peterson, Roger Tory. *The Birds.* Time Inc., 1963. Photographs accompany descriptions of evolution of birds, body structure, development, and migration.

Pritchard, Peter. *Living Turtles of the World.* T. F. H. Publications, Inc., 1967. Well-illustrated account of the history of turtles.

Reidman, Sarah, and Gustafson, Elton. *Home Is the Sea for Whales.* Rand McNally and Co., 1966. Traces the evolution of whales.

Sanderson, Ivan T. *Living Mammals of the World.* Doubleday and Co., Inc., 1965. Lavish illustrations and information about a variety of mammals.

Van Gelder, Richard G. *Biology of Mammals.* Charles Scribner's Sons, 1969. Describes how mammals reproduce; develop; seek shelter, food, and water; and how they protect themselves.

UNIT NINE

HUMAN STRUCTURE AND FUNCTION

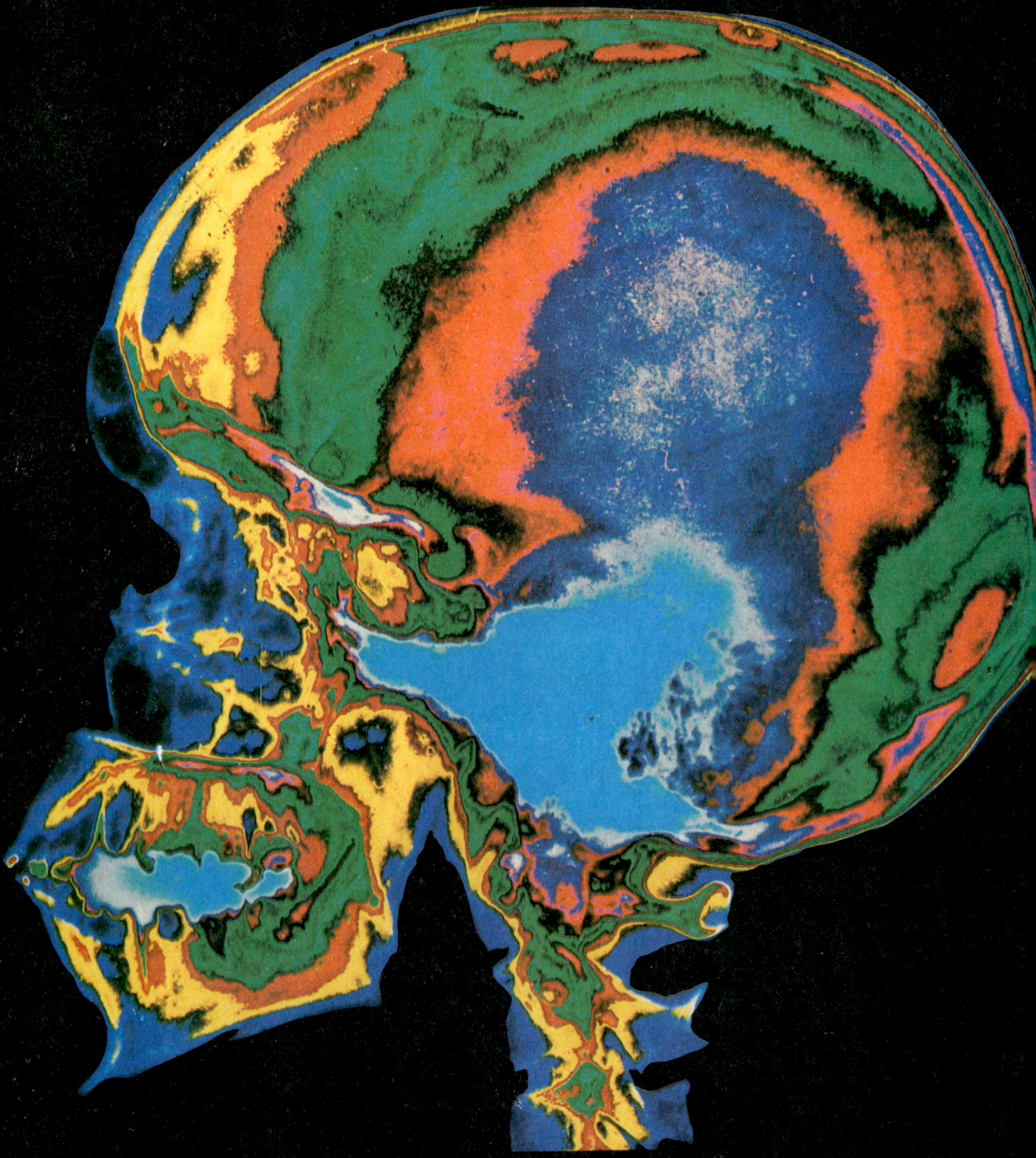

Chapter 21

Body Framework and Controls

Our bodies are complex structures which allow us form, movement, and the ability to respond to both our external and internal environments. The skeletal, muscular, nervous, and endocrine systems are responsible for the shape, movement, and control of our bodies.

The skeletal system is made up of 206 bones and some connective tissue. The bones provide shape, protection, and support for the muscles. Bones are joined in several ways to allow for a variety of movements.

About one-third of body weight is made up of muscles. It is the muscles that power the body's movements. Muscles bring the otherwise still skeletal system into lively action.

The nervous and endocrine systems are the control centers for all body functions. The endocrine system provides chemical control over some of the tissues within an organism as well as regulation of the metabolic processes within the cell. In addition, the nervous system controls various body functions by directing the action of the muscles. The nervous system is responsible for registering, and reacting to, any internal or external change. Learning, memory, emotion, and sensation are all possible because of our nervous system.

Objectives

1. List the functions of the skeletal system.
2. Describe the structure of bone.
3. Identify the various types of joints and describe the movement permitted by each type.
4. Compare the structures and functions of skeletal, smooth, and cardiac muscle.
5. Define the role of muscles in body movement.
6. Identify the structures and functions of the central, peripheral, and autonomic nervous systems.
7. Describe the transmission of a nerve impulse.
8. Describe the structure and function of each of the sense organs.
9. Describe the structure and function of the glands of the endocrine system.
10. Identify and describe various diseases associated with malfunctions of the endocrine system.

An X ray of a normal, healthy skull. For a medical examination, colors may be added to help define the area under study.

21–1
The Skeletal System

Our bodies combine flexibility and movement with strength and support. We can stand, run, walk, talk, bend, and write because of the ways our bones and muscles are connected and work together. The framework of the human body consists of 206 bones. Bones are described as long, short, flat, and irregular according to their shape. Each bone performs a specific function.

The skeleton has four major functions in the body.

1. The skeleton provides support for the soft tissues of the body. Without the skeleton, the body would not have a defined shape.
2. The skeleton protects internal organs such as the brain, heart, lungs, spleen, and spinal cord.
3. The skeletal system allows for the mobility of the body by providing places for the attachment of the skeletal muscles.
4. The bones of the skeleton serve as a storage area for minerals. The marrow inside the bones manufactures red blood cells, platelets, and some white blood cells.

Axial and Appendicular Skeletons

The skeletal system may be considered in two parts. The **axial** (ak'sē-əl) **skeleton** is made up of the skull, the vertebrae, the sternum, and ribs. The skull provides protection for the delicate brain tissues. The spinal cord is protected by the vertebrae. The sternum and ribs surround and protect the lungs, liver, heart, and spleen. The **appendicular** (ap ən dik'yə lər) **skeleton** consists of the bones of the arms and legs, and the *pectoral* (pek'tər əl) and *pelvic* (pel'vik) *girdles.* The girdles are the supports for the arms and legs. The pelvic girdle makes up the hips, whereas the pectoral girdle makes up the shoulders.

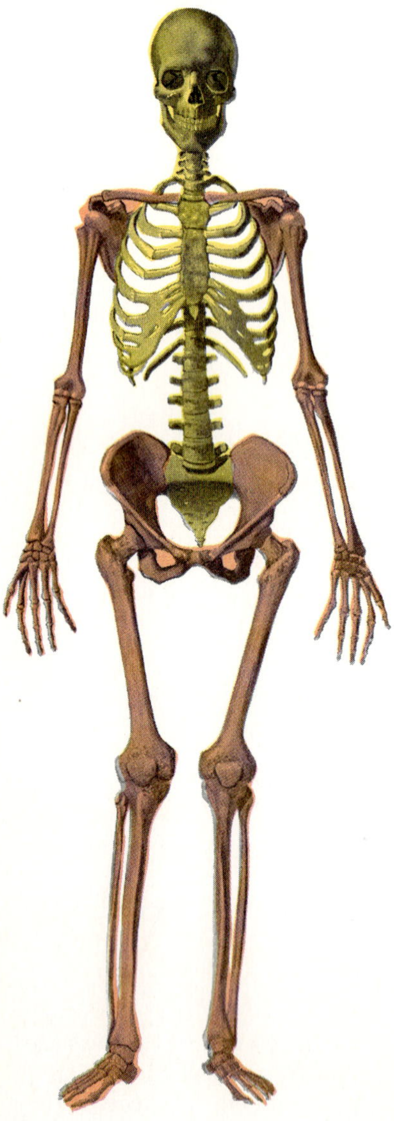

Axial skeleton
Appendicular skeleton

Human skeletal system

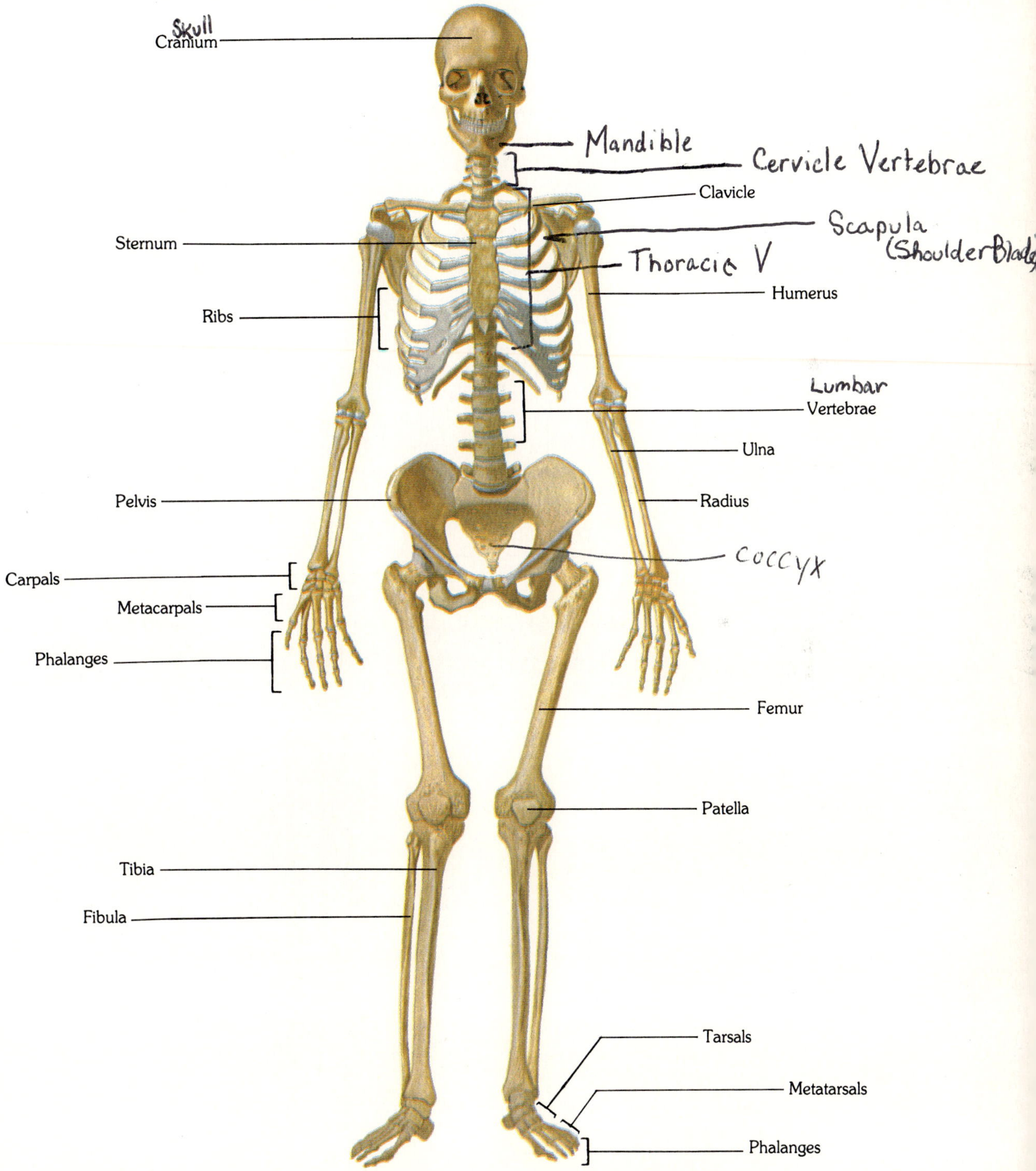

Challenge!
Broken bones or fractures occur commonly in both children and adults. Treatment for broken bones has been known as far back as the fifth century B.C. *Today doctors use X rays, steel pins, and plaster casts to set bones so that they will heal correctly. What were some of the early bonesetting methods?*

The Structure of Bone

The bones of a dead animal are as lifeless as a piece of concrete. The bones in a living animal are quite different. They are combinations of living cells and materials produced by these cells. If bones were made of hard materials all the way through, they would be strong, but very heavy. However, our bones are not solid.

Bones are covered by a tough, protective membrane, the **periosteum** (per'ē os'tē əm). Beneath the periosteum is a layer of compact bone made of calcium, phosphorus, salts, other minerals, and **collagen** (kol'ə jən), an organic protein. This compact bone provides strength to the bone and protection to the interior of the bone.

Channels called **haversian** (hə vėr'shən) **canals** run through the bony layer. Haversian canals contain blood vessels that are the only connection between the bone cells and other body systems. They bring food and oxygen to the bone cells and carry away metabolic waste products.

A bone such as the **femur** (fē'mər) of the leg is a tube with the outermost layer, the periosteum, enclosing the next innermost tube of compact hard bone which gives the bone strength. The interior of the bone contains a spongelike network of bone cells. This spongy layer gives additional strength to the bone without adding much weight. Some bones may also contain a soft tissue, the **marrow** (mar'ō). Bone marrow that contains mostly fat cells is known as *yellow marrow*. The yellow marrow stores fat but normally has no other function. However, when a great deal of blood has been lost, the yellow marrow can produce blood cells. Marrow that contains many red blood cells is called *red marrow*. Red marrow is found at the ends of long bones. Red blood cells, platelets, and some white blood cells are produced in the red marrow.

Bone structure

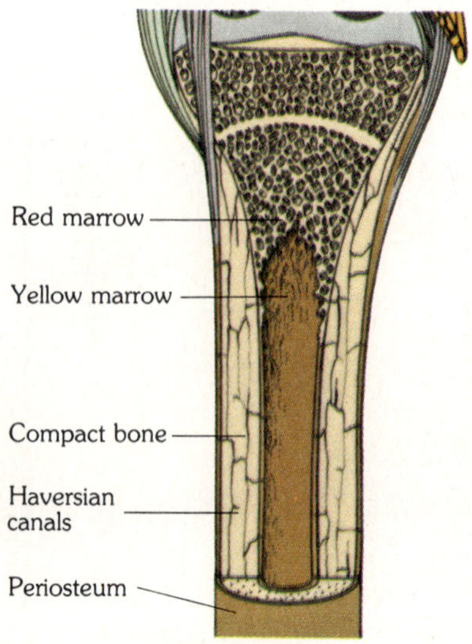

Joints

A **joint** is formed where two bones meet. Each type of joint allows certain movements or provides protection for internal organs. A thin yellow fluid, the *sinovial fluid,* is usually present in the joint. Sinovial fluid acts as a lubricant that helps the joint move more easily and reduces the wear on the bones of the joint. As a person ages, the amount of sinovial fluid decreases and the joints tend to stiffen.

There are several types of joints in the body. Each type works in a special way. Joints in the skull are locked together tightly. These *fixed joints* do not allow any movement. Fixed joints are very strong and provide maximum protection to the brain tissues. *Hinge joints,* found in the elbow and knee, allow movement in one direction only. They have the same type of movement as the hinge of a door. The *ball and socket joints* of the shoulder and hip permit movement in a circle. The movement of the skull on the vertebrae at the neck is the result of a *pivot joint.* The multiple movements permitted by the wrist and ankle are possible because of *gliding joints.*

Did You Know?
Sometimes a blow to the elbow sends a tingling sensation from the fingers up to the shoulder. When this happens we have struck our "funny bone." The sensation is really not funny, nor is it on a bone. The sensation is actually an electrical impulse caused by a blow to the ulnar nerve. This nerve runs the entire length of the arm. It is well protected everywhere except for one small spot at the elbow.

Types of joints

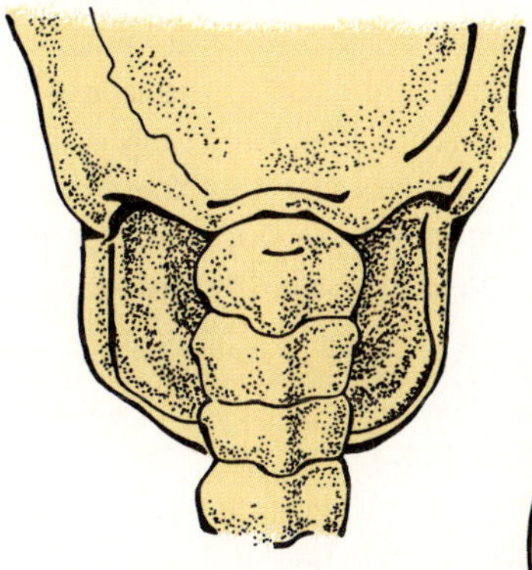

Pivot joint

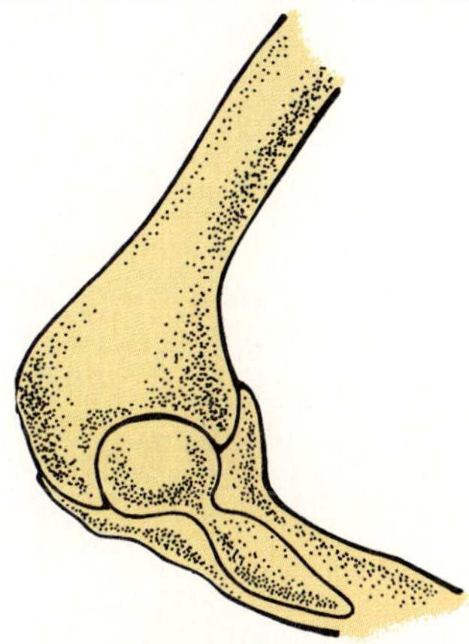

Hinge joint

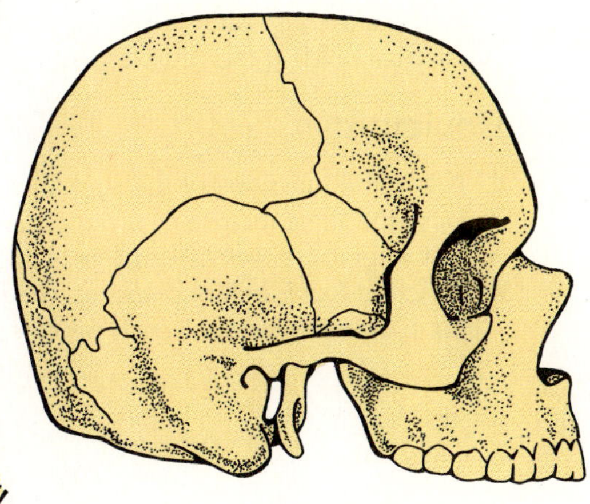

Fixed joint

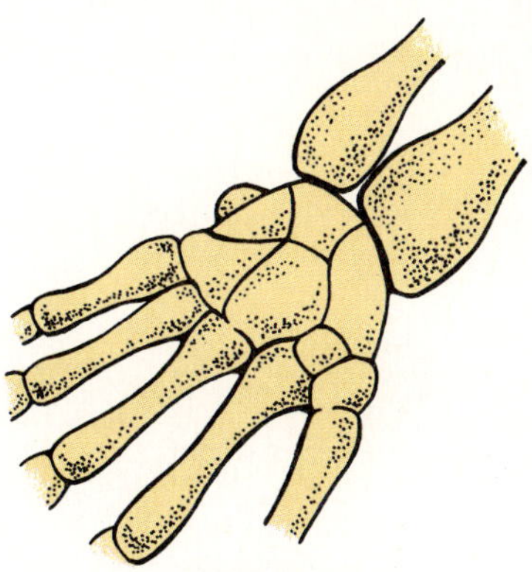

Gliding joint

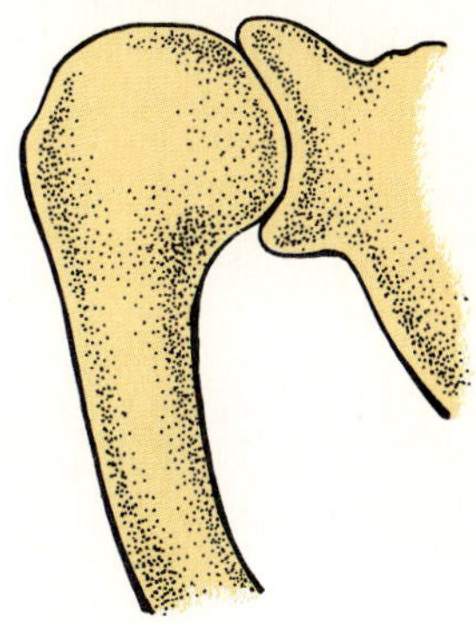

Ball and socket joint

Joint structure

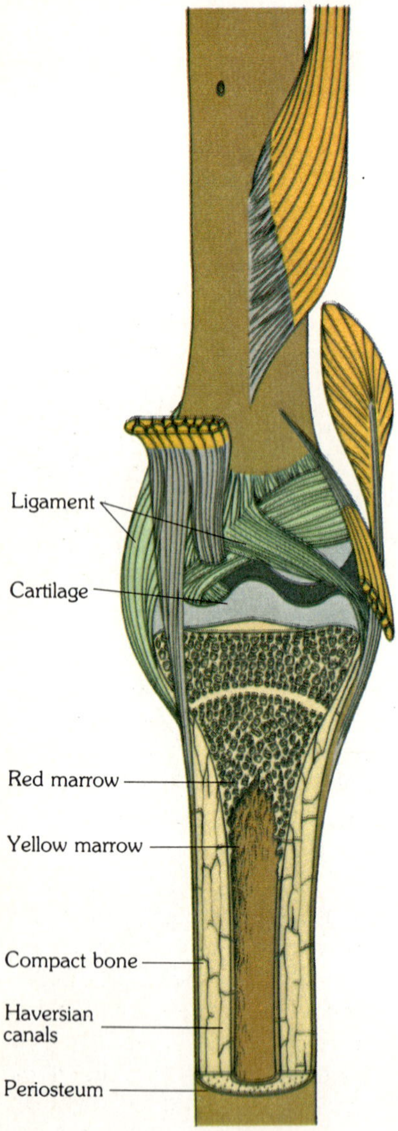

Cartilage

Infants and young children do not have bones like those of adults. Their bones are made mostly of **cartilage** (kär'tl ij), a firm elastic, fibrous material. As the individual grows and matures, the cartilage is gradually replaced by bone cells. These bone cells deposit crystals of calcium carbonate and calcium phosphate. This process, called **ossification** (os'ə fə kā'shən), greatly increases the strength of the bone.

Bones usually continue to grow through adolescence. During this time the bone ends still contain cartilage. Bones can grow without interfering with the way joints fit together. By the time a person reaches the mid-twenties, the cartilage in the bones has become ossified. Bone growth is then over.

Some cartilage remains flexible throughout a person's lifetime. This cartilage makes up the structure of the nose and outer ear, the larynx, and the trachea.

Cartilage forms fibrous pads at the ends of bones where one bone meets another. These cartilage pads reduce the friction between the bones and act as shock absorbers for the bones.

Connective Tissue

Individual bones are held together by strands of tough, connective tissues called **ligaments** (lig'ə məntz). Ligaments are made of flexible fibers, which are closely interwoven. Ligaments surround the joints, covering both the bones and the cartilage at the ends of the bones. Bones are held close together by the ligaments. The stretching of the ligaments allows the bones of a joint to move. Notice that the illustration on the left also shows muscle attachment to the bones. Connective tissues holding the muscles to the bones are called **tendons.**

Review It

1. Name the functions of the human skeleton.
2. Describe the structure of a bone.
3. List three purposes of cartilage.
4. Describe fixed, ball and socket, hinge, pivot, and gliding joints and the type of movement each permits.

21–2
The Muscular System

Whereas the skeletal system provides support and protection to the body, the force needed to move the skeleton, and thus to move the body, comes from the muscular system. Between one-third and one-half of a person's total body weight is made up of muscles. Hundreds of muscles are located in every part of your body. They range in size from the massive muscle in the thigh that helps you stand and walk to the tiny muscles in the middle ear. Your body has three types of muscles: skeletal, cardiac, and smooth.

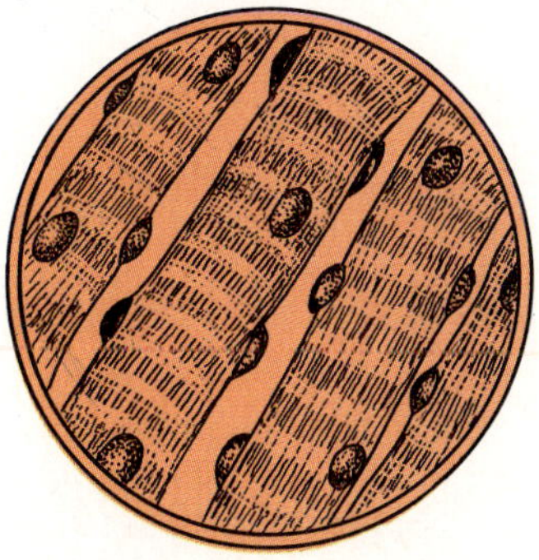

Skeletal Muscle
Skeletal (skel′ə təl) **muscles** cause the skeleton to move. Sometimes these muscles are called *voluntary muscles* because usually they are under conscious control. These muscles are composed of individual muscle fibers that are fastened together by connective tissue. Each muscle fiber consists of a large number of smaller threads. These threads, called **myofibrils** (mī ə fī′brəlz) are made up of thousands of protein filaments. The threads form overlapping, parallel patterns of light and dark markings. Skeletal muscle is sometimes called **striated** (strī′ā tid) muscle because of these characteristic bands. Sheets of connective tissue surround each muscle to eliminate friction as one muscle moves against another.

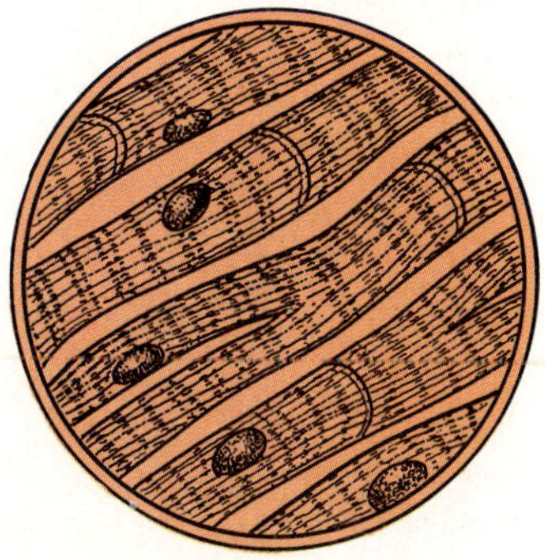

Cardiac Muscle
The heart is made up of **cardiac** (kar′dē ak) muscle. Cardiac muscle cannot be consciously controlled. It has the striated marking of skeletal muscle. All of the cells of cardiac muscle are held together by connective tissue. The rate of body activity determines the working rate of cardiac muscle.

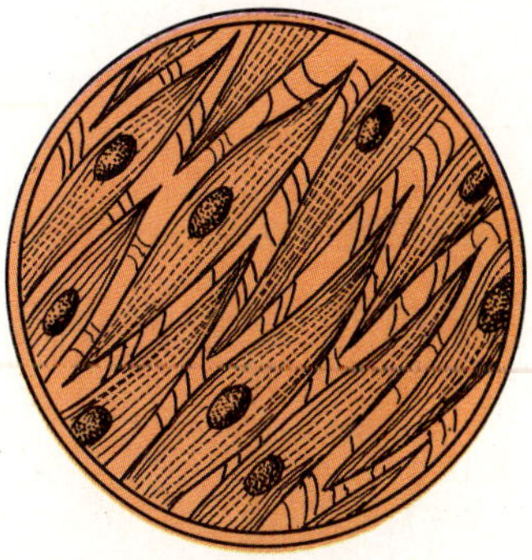

Smooth Muscle
Muscles of internal organs, such as the stomach, are **smooth muscles.** These muscles work without a person's conscious control and are sometimes called *involuntary* muscles. Smooth muscle fibers are long and spindle shaped. They are thick in the middle and thin at each end. These spindle-shaped fibers are joined together into sheets. The sheets are layered at right angles to each other. The movement of smooth muscle is slow compared to the rapid movements of skeletal muscle.

The functions of the smooth muscle include moving food through the digestive tract, controlling blood pressure, and maintaining and controlling other internal systems.

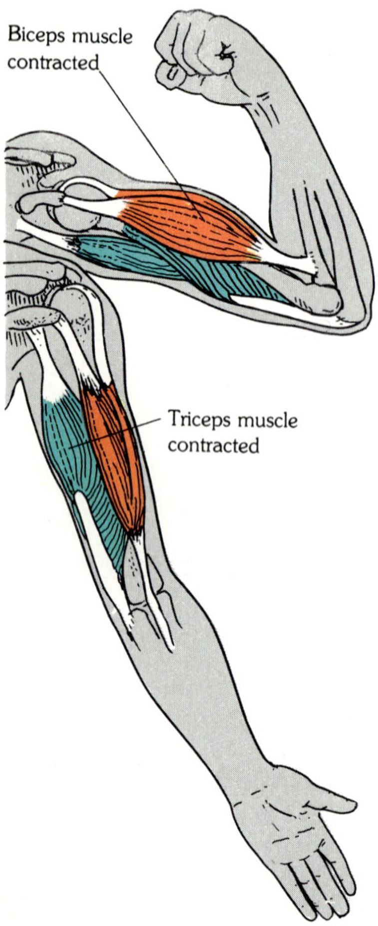

Muscle action

Challenge!
Muscular dystrophy, cerebral palsy, and Parkinson's disease are disorders that affect the nervous and muscular systems. Find out the characteristics of each disease. What do we know about the way the diseases develop? What can be done to treat each of the disorders?

Action of Muscles

All muscles—skeletal, smooth, and cardiac—move by contraction. The process of muscle contraction begins when a nerve stimulates a muscle. The muscle filaments slide past each other, shortening the muscle fiber. The *fiber,* composed of many *fibrils,* contracts. Every movement the body makes is a result of muscle contraction. Body movements appear to be caused by muscles pulling and pushing on a bone. In reality the bone moves in two directions because one muscle *pulls* it in one direction and another muscle then *pulls* it back to its original position. Movement is always the result of the contraction of a *pair* of muscles.

Muscle structure

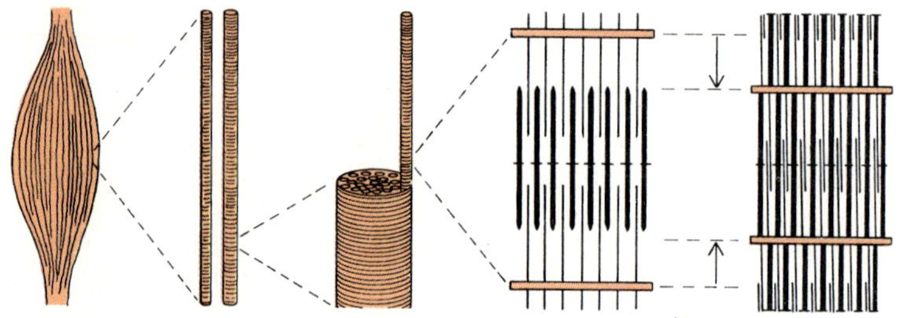

Muscle Contraction

Messages to skeletal muscles are delivered from the brain by way of the nervous system. A nerve cell stimulates several muscle fibers. The nerve cell and the muscle fibers it supplies are called a **motor unit.** When the nerve sends its message to the muscle, every muscle fiber that is stimulated by that nerve contracts over its entire length. Individual fibers never shorten part way. The action of an entire muscle, however, is not an all or nothing process. Muscles are composed of bundles of muscle fibers. Not all of the fibers in a muscle contract at the same time. More fibers contract when a muscle is doing heavy work than during light work.

All muscles, except the cardiac muscle, need a direct nervous impulse to begin contracting. When this nervous impulse is first received, nothing seems to happen. This moment of inaction is the *latent period* of muscle contraction. The latent period is followed by the beginning of the contraction. After the contraction has reached its strongest point, a period of relaxation begins. A muscle will not respond to another nervous impulse again until the muscle has finished relaxing.

Review It

1. What is the basic function of the muscular system?
2. Describe and compare the characteristics of the three types of muscle fibers.
3. Describe how muscles work in pairs to allow for movement.

Human musculature system

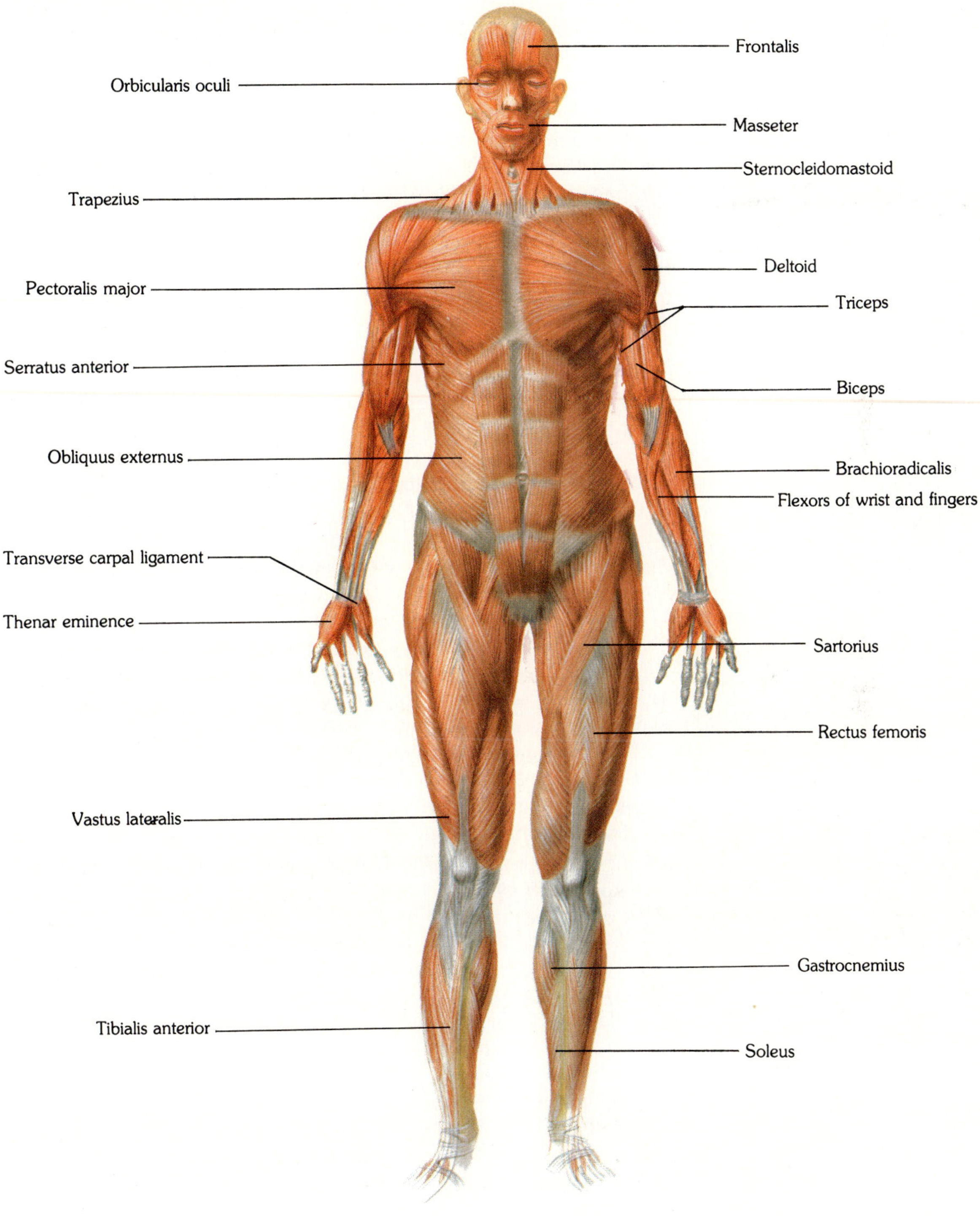

Human nervous system

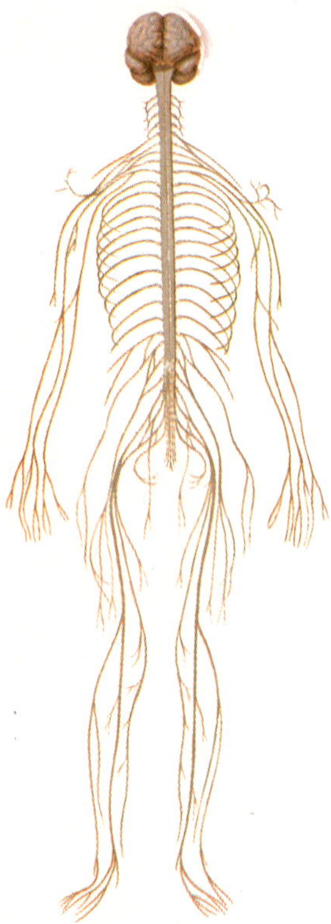

Structure of a neuron

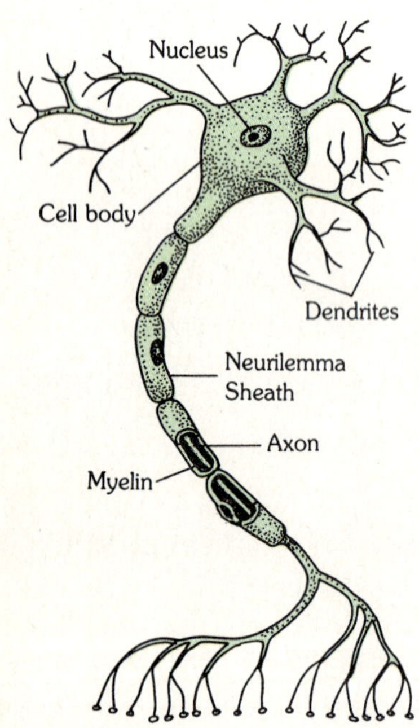

21–3
The Nervous System

There is more to body movements than muscles and bones. They provide us with the equipment we need to run, dodge, jump, laugh, and kick. However, a control system "tells" the parts of our bodies what to do and when to do it. That control system is the **nervous system.**

The human nervous system is composed of three parts. The first part is the *central nervous system,* which includes the brain and the spinal cord. Secondly, the *peripheral nervous system* is composed of all of the nerves leading from the spinal cord to all parts of the body. And thirdly, the *autonomic nervous system* is a system of regulator nerves that control vital body functions.

Structure of a Nerve Cell

The nerve cell, or **neuron** (nůr'on), is the basic working unit of the nervous system. The neuron is a single cell with several identifiable parts. The cell-body portion of the neuron contains the nucleus and most of the cytoplasm. Filaments called *nerve fibers* extend out from the cell body. These fibers include one long fiber, the **axon** (ak'son), and one or more shorter fibers, the **dendrites** (den'drītz). A single neuron may have up to 200 dendrites. The dendrites are responsible for carrying messages *into* the nerve cell. The axon carries messages *away from* the neuron. A neuron may be a fraction of a centimeter in length, or as long as a meter.

Fibers of a neuron may be covered by a layer of a fatty substance called **myelin** (mī'ə lən). The myelin may in turn be covered by a thin membrane, the **neurilemmal sheath** (nůr'ə lem'əl shēth). These two layers act as insulators. Fibers covered by myelin and a neurilemmal sheath transmit nerve impulses faster than fibers without coverings.

Neurons are classified according to their function. Those neurons that carry messages to the brain or spinal cord are **afferent** (af'ər ənt) or **sensory** neurons. Information or stimuli from either the internal or external environment is picked up by **receptors** (ri sep'tərz). Receptors may be neurons or specialized organs for detecting **stimuli.** A stimulus is anything that activates a response. Receptors detect odor, touch, taste, light, sound, as well as stimuli from internal organs.

Efferent (ef'ər ənt) or **motor** neurons relay messages from the brain or spinal cord. The messages they carry are transformed into action by the motor effectors, the muscles.

Neurons that are bridges between sensory and motor neurons are known as **associative** or **interneurons** (in'tər nůr'onz). The brain and spinal cord are mostly made up of interneurons.

When we speak of a *nerve,* we mean a bundle of nerve fibers bound together by a sheath of connective tissue. Nerves consist primarily of axons. The cell bodies of the neurons are also gathered together. If this group of cell bodies is outside the brain and spinal cord, it is called a *ganglion* (gang'glē ən). If the group of cell bodies is inside the brain or spinal cord, it is known as a *nerve center.*

The Nerve Impulse

Nerves send *impulses* in response to stimuli. A nerve impulse is a series of electrochemical changes in the neuron membrane. The impulse is relayed from one neuron to the next. This involves chemical reactions and the movement of sodium and potassium ions across the cell membrane. Changes taking place in one section of the neuron membrane result in changes in the adjoining parts of the neuron, a kind of chain reaction. An impulse that begins in the dendrites of the neuron passes through the cell body and then on to the axon. These changes occur rapidly: a nerve impulse can travel about 100 meters per second. There is a short time called a *refractory period* after a neuron has carried an impulse before it can transmit another impulse. The time that a neuron cannot transmit an impulse is normally about a thousandth of a second.

Every neuron needs a certain minimum amount of stimulus in order to begin to transmit an impulse. This minimum amount of stimulus is known as the neuron's **threshold** (thresh'ōld). If the neuron receives less stimulus than its threshold, no impulse is sent. If the stimulus is greater than the threshold, the neuron will respond by sending an impulse. If the stimulus continues, the neuron will continue to send impulses, one after another.

The axon and dendrites of one neuron are not in direct contact with the axon and dendrites of the next neuron. There is a small space, the **synapse** (si naps'), between the tips of the axon and the adjacent dendrite. Messages are able to cross the synapse because the tip of the axon secretes a chemical that receives and transfers the impulse across the synapse to the dendrites of the next neuron.

Nerve impulses can travel in only one direction across the synapse—from axon to dendrite. Only the end of the axon is able to produce the chemical that carries the message across the synapse. Only the ends of the dendrites are able to respond to the chemical and receive the message.

Different degrees of nerve responses are possible. We can tell the difference between lukewarm, warm, and hot because the sensory neurons produce more impulses per second as the temperature increases. We apply a light or heavy touch to an object depending upon the number of motor neurons involved in the response. The combined action of many motor units provides strength in muscular movement. The action of a selected number of motor neurons provides accuracy in the muscular movement.

Impulse travel in a neuron

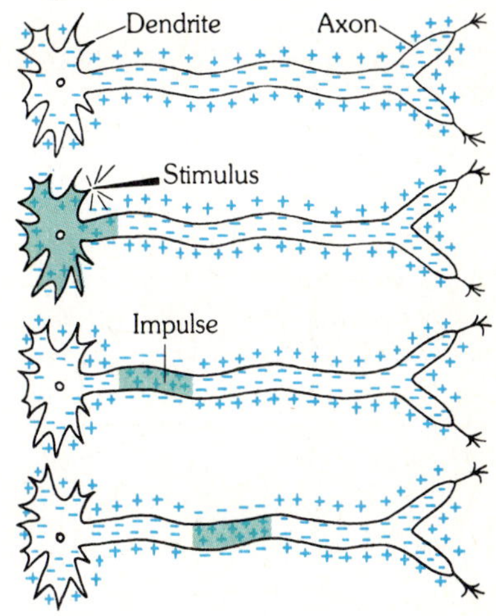

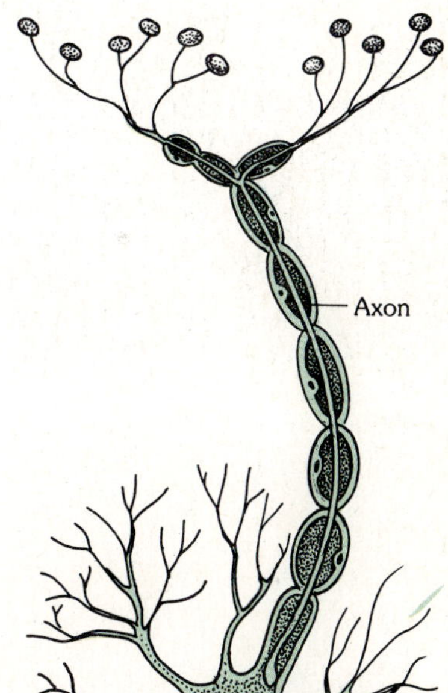

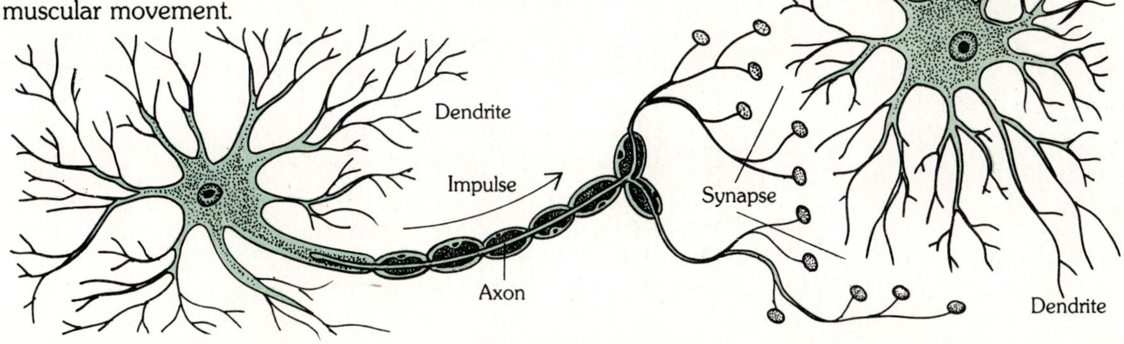

Did You Know?
Doctors may use the knee-jerk reflex to test the condition of the nervous system. While the knee is hanging limply, the tendon below the knee cap is tapped lightly. The blow causes the extensor muscle of the thigh to contract. A sensory neuron had sent a message to the spinal cord. A motor neuron was then automatically stimulated to counteract the stimulus to the tendon. The motor neuron will usually overcompensate. The extensor muscle contracts strongly, causing the foot to jerk forward with an involuntary kick.

Path of Nerve Impulses

Trains of nerve impulses travel along the sensory neurons to the brain, which analyzes the impulses. The brain then sends messages (nerve impulses) by way of motor neurons to parts of the body, which respond. Imagine you are enjoying a walk outside on a warm day. A sudden gust of cold air surrounds you. Receptors in your skin are stimulated to respond. An impulse is sent from the receptors to the sensory neurons. The impulse travels from one sensory neuron to another, to the spinal cord, and up the spinal cord to the brain. The neurons in the brain process the information of the impulse and send a message back to the skin by way of the motor neurons. The message passes back through the spinal cord and through motor neurons until it reaches the effectors in the skin. The action of these muscles may cause "goose bumps" to form, or the capillaries in the skin to constrict the flow of blood, as the body attempts to conserve its heat.

Reflex Reactions

In a reflex reaction, the nerve impulse travels only as far as the spinal cord. The impulse does not go to the brain. If you stepped on a sharp rock, the pressure receptors in your skin would send impulses through sensory neurons to the spinal cord. In the spinal cord, the sensory message is relayed directly to motor neurons, which activate an appropriate response. In response, the muscles in your leg and foot immediately pull back from the sharp stone. This short pathway of receptors, sensory neurons, motor neurons, and effectors is a **reflex arc.**

Other nerve impulses continue to travel along the regular pathway to the brain. When the messages reach the brain, pain is felt. An understanding of what has occurred is then clear to you. You reacted before you knew what had happened. The reflex reaction provides the body with the quickest possible response to a situation that may be harmful to the body.

Reflex arc

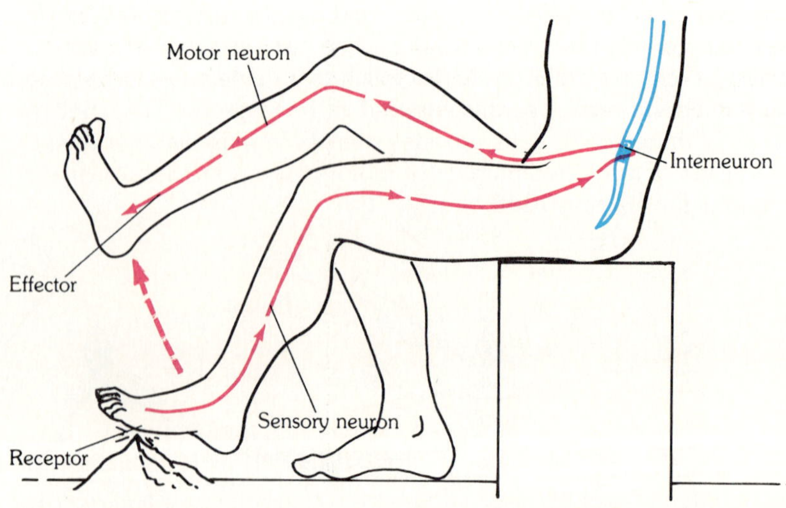

The Central Nervous System

The **central nervous system** is made up of the *brain* and the *spinal cord.* The brain is divided into several sections with each section responsible for particular functions. Each of these two crucial parts of the central nervous system is protected by bone. The brain is covered by the *skull,* and the spinal cord is encased by the *backbone.*

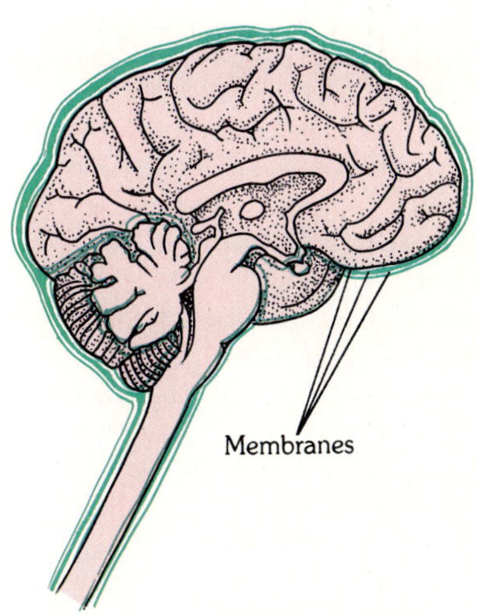

The Brain

Immediately inside the skull are three layers of membranes called the **meninges** (mə nin′jēz). The outermost membrane, the *dura mater* (dùr′ə ma′tər), is the toughest. The middle layer, the *arachnoid mater* (ə rak′noid ma′tər), is a meshlike network of cells. The space between the arachnoid mater and the innermost membrane, the *pia mater* (pī′ə ma′tər), is filled with a fluid. This fluid, the **cerebrospinal** (sə rē′brō-spī′nl) **fluid,** acts as a shock absorber in both the skull and the spine. The pia mater provides both protection and nutrition to the brain. Masses of blood vessels in the pia mater carry food and oxygen to, and remove wastes from the brain cells. The brain contains about ten billion interneurons.

The Cerebrum

The **cerebrum** (sə rē′brəm) makes up about 80 percent of the total brain mass. The cerebrum is divided into two parts: the right hemisphere and the left hemisphere. The right hemisphere controls the left side of the body; the left hemisphere controls the right side. The two hemispheres are connected by the *corpus callosum* (kôr′pəs kal-ō′səm), which sends messages from one side of the brain to the other. Messages from sensory neurons are processed in the cerebrum. Much of the processing actually takes place in a thin layer of tissue that covers the cerebrum. This tissue is the **cerebral cortex** (kôr′teks). Only about one and a half to three millimeters thick, the cerebral cortex is deeply folded to greatly increase the area of the cerebrum. Within the cerebrum are the billions of neurons needed for advanced intelligence. Most of the major mental processes—thinking, memorizing, and decision making—take place in the cortex. Specific sections of the cortex control specific functions. Motor areas control voluntary movement. Sensory areas interpret various senses.

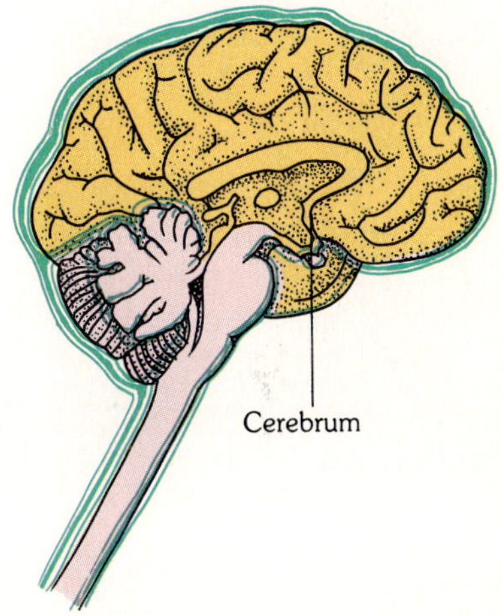

Did You Know?

You have heard about using your "gray matter" to solve a problem. "Gray matter" is the grayish tissue of the cerebral cortex. In contrast, the tissue found in the rest of the cerebrum—mainly nerve fibers and the fatty insulation surrounding them—is lighter in color. It is called "white matter."

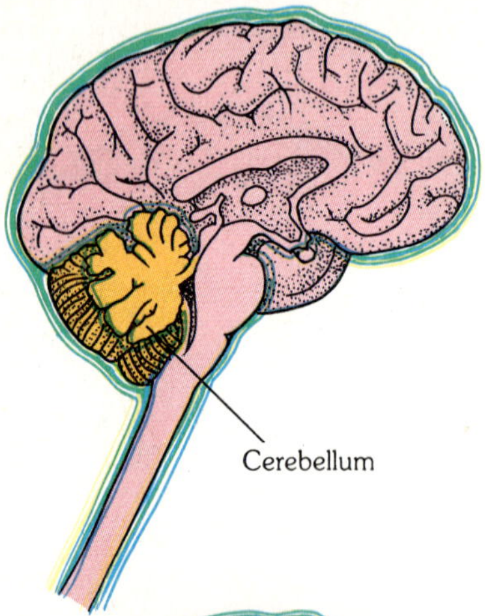

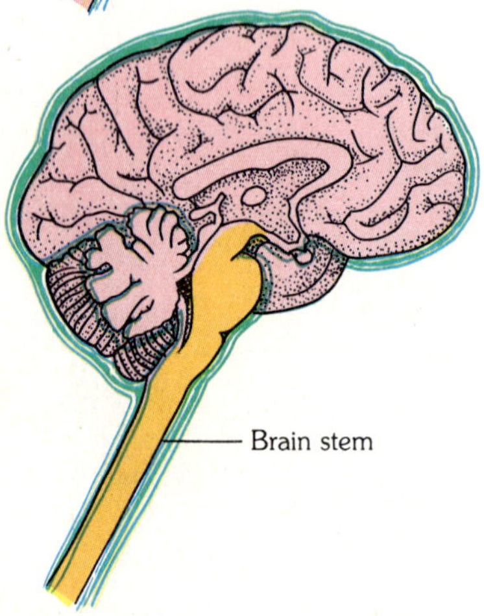

The Cerebellum

The **cerebellum** (ser′ə bel′əm) is located in back of the cerebrum. Its structure is similar to that of the cerebrum, but it is smaller in size. The cerebellum also has many infoldings that are shallow and quite regular.

The function of the cerebellum is not to originate responsive motor actions, but to *coordinate* them. A motor response developed in the cerebrum is first directed to the cerebellum before being sent to the body. Muscle movements are then coordinated into a proper sequence to ensure a smooth delivery of motor actions.

The cerebellum also helps to maintain balance in the body. Input from the senses tells the cerebellum about the body's position in relation to its surroundings.

The Brain Stem

An enlarged area at the base of the brain is the **brain stem.** The lowest segment of the brain stem is called the **medulla oblongata** (mi-dul′ə ob′long gä tə). This section is responsible for the control of several vital functions. Automatic activities such as the heart rate, blood pressure, and respiration are controlled by the medulla oblongata.

For example, if the CO_2 concentration in the blood increases, the medulla oblongata monitors this increase and directs actions to balance the change. Messages are sent to the lungs and heart to work harder. This means breathing and circulation rates increase to rid the body of excess CO_2.

The **thalamus** (thal′ə məs), located near the base of the brain, transmits sensory impulses to the cerebral cortex. Below the thalamus is the **hypothalamus** (hī′pō thal′ə məs). It controls automatic body responses such as hunger and thirst. It also analyzes messages that call for changes in digestion, body temperature, sexual behavior, and other functions.

Emotions and certain subconscious drives also originate in the human brain. The *limbic system* consists of a group of structures that are related to anger, pleasure, depression, and pain. Most of the structures of this system are in the lower part of the brain, although the hypothalamus and parts of the cerebrum are also involved.

The Spinal Cord

The tubelike structure extending down from the medulla is the *spinal cord.* It is surrounded by the meninges and is protected by the vertebrae. Its main functions are the transmission of impulses to and from the brain and to act as a reflex center. The spinal cord is composed of white matter except for an "H" of gray matter in the center. A small canal runs the length of the spinal cord. This tube is filled with cerebrospinal fluid.

The Peripheral Nervous System

The nerves that connect the central nervous system with the various parts of the body make up the **peripheral nervous system.** The twelve nerves that connect with the brain are *cranial* (krā′nē əl) *nerves.* Thirty-one pairs of spinal nerves connect with the spinal cord.

Most of the cranial nerves connect the brain with the glands, muscles, and sense organs of the head. Some cranial nerves, the *olfactory* (ol fak′tər ē) nerve to the nose, the *optic* (op′tik) nerve to the eyes, and the *auditory* (ô′də tôr′ē) nerve to the ears, contain only sensory neurons. Other cranial nerves are composed almost totally of motor neurons. Motor nerves control the lenses, pupils, and muscles of the eyes and some shoulder muscles. The remaining cranial nerves connect the brain with the heart, face, and intestinal tract. They contain both sensory and motor neurons.

Each of the thirty-one pairs of spinal nerves responds to the receptors and effectors of one part of the body. The spinal nerves have both sensory and motor neurons. Each spinal nerve has two roots that meet just beyond the spinal cord. The dorsal root primarily contains sensory neurons, bringing information to the central nervous system. The ventral root contains mostly motor neurons, which carry nerve impulses away from the central nervous system.

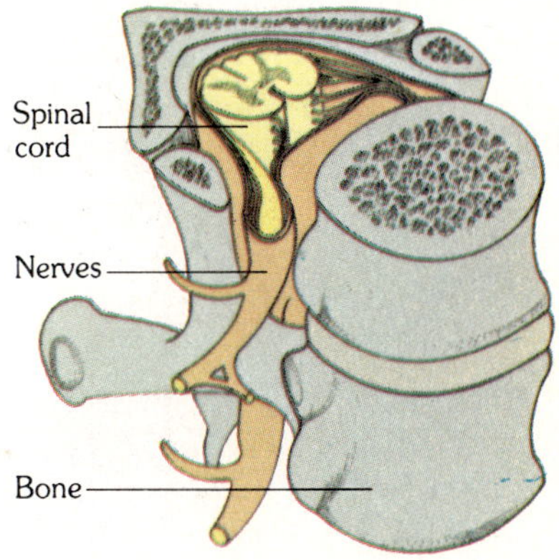

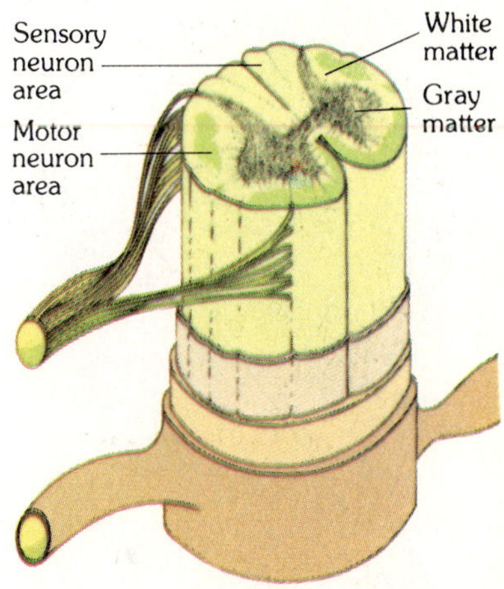

Cross section of the spinal cord

Functional areas of the brain

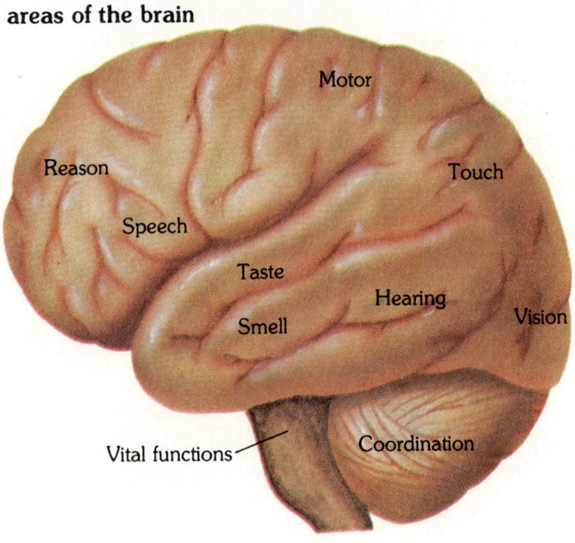

The Autonomic Nervous System

Many essential body functions go on without any thought or effort on our part. Food passes through the digestive system, wastes are removed in the kidneys, and blood travels through the heart. These and other involuntary functions are controlled by the **autonomic** (ô'tə-nom'ik) **nervous system.** The autonomic nervous system consists of two parts, the **sympathetic** (sim'pə thet'ik) and the **parasympathetic** (par'ə sim'pə thet'ik) nervous systems. Both systems are connected with the heart, intestines, bladder, and other organs.

The sympathetic system consists of two rows of nerve cords, lying on either side of the spinal column. The major nerve of the parasympathetic system is the **vagus** (vā'gəs) nerve, running from the medulla oblongata to the chest and abdomen.

When most sympathetic nerves are stimulated, they release **noradrenalin** (nôr'ə dren'ə lən). This chemical causes certain muscle cells to relax and others to contract. Stimulation of a parasympathetic nerve causes the release of **acetylcholine** (ə sē'tl kō'lēn'). The effect of this chemical is the opposite of that of noradrenalin. A muscle that contracts with noradrenalin relaxes with acetylcholine. A muscle that relaxes with noradrenalin will contract with acetylcholine.

This means that, for every effect produced by a sympathetic nerve, the opposite effect can be produced by a parasympathetic nerve. For example, if the blood pressure is too high, a message to the brain (medulla) stimulates a message to a parasympathetic nerve. The response will slow the heart rate and decrease the blood pressure. Conversely, if the blood pressure lowers too much, a message is sent to a sympathetic nerve to stimulate the heart to work faster and increase the pressure.

Usually both systems operate at the same time. One system may predominate at a given time. However, the other system does not shut down completely, except in the case of an emergency. It is the sympathetic and parasympathetic nervous systems that are directly involved in maintaining the body's internal environment at the best possible level for cells to function.

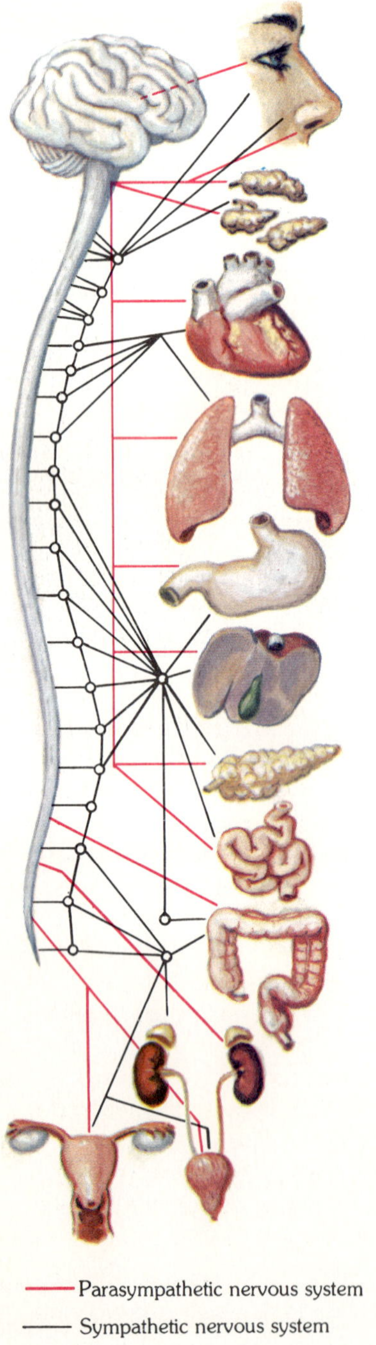

Autonomic nervous system

Review It

1. Diagram the path by which a nerve impulse is transmitted from receptors to effectors. Identify each part.
2. Describe a reflex arc and explain what advantages it offers to the body.
3. Identify the regions of the brain and explain the function of each.
4. Explain how the sympathetic and parasympathetic nervous systems work together to control smooth muscle movement.

Breakthroughs in Biology

The Right and Left Hemispheres of the Brain

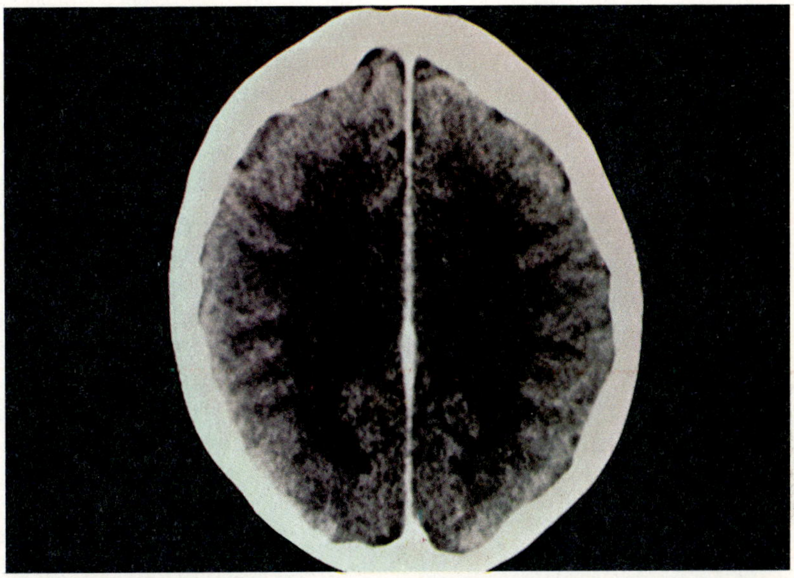

Cross-section view of left and right brain hemispheres.

Scientists have known for over a hundred years that the two halves of the brain are not identical, but they did not have much information about the differences. Recent research has shown that differences between the two hemispheres are concerned with the most basic mental processes. In general, the left hemisphere is responsible for logical, orderly thinking, while artistic and creative functions originate in the right hemisphere. Brain lateralization is not exactly the same for everyone, but the information is helping scientists to better understand various brain disorders and behavior differences.

An electroencephalograph is an instrument used to measure the electrical activity of the brain. Recordings can be taken from both sides of the brain while various activities are being performed, and can show which hemisphere is more active during a task. When recorded subjects were drawing a picture or humming a tune, the right brain was more active. Scientists have found that remembering and recognizing visual images, colors, and spatial forms, and responding to and producing musical sounds are the major functions of the right hemisphere.

When subjects were writing, speaking, or doing tasks that required fine muscle coordination, the left hemisphere recorded more activity. The left half of the brain regulates such functions as the ability to recognize and organize words, to understand numerical relationships, to process information in a logical sequence, to pay attention, and to express thoughts in speech and writing. "Intelligence" tests usually measure primarily left-hemisphere activities, although newer tests have been devised to measure right-brain skills as well.

Although the hemispheres are specialized, they work together. When you see someone you know, for example, recognition of the face occurs in the right hemisphere and remembering the name, in the left. When the two halves do not coordinate properly, problems result. Some researchers now think that *dyslexia,* an impaired reading ability in otherwise normal individuals, may develop because the left brain has taken over certain functions usually regulated in the right half, such as spatial recognition. This would "overload" the left hemisphere, they explain.

Analysis

1. The study of brain lateralization is an active field of research. See what you can learn about recent discoveries.

2. Do you think that one of your brain hemispheres may be more dominant than the other? Why?

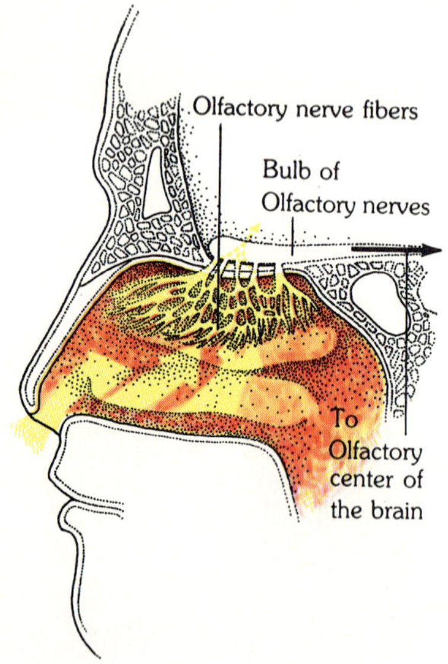

Olfactory cells

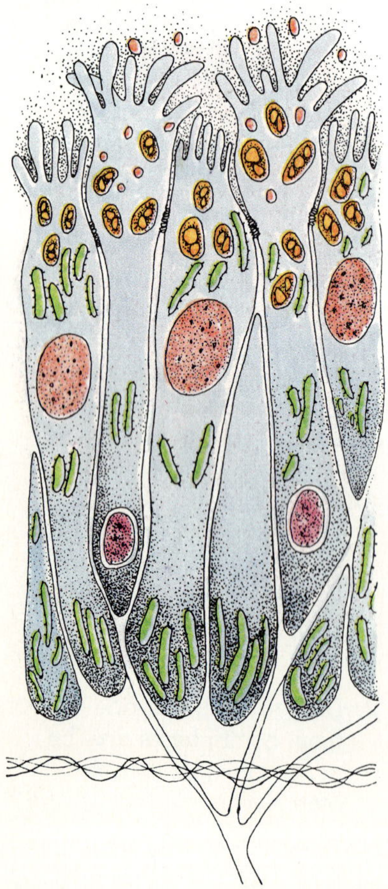

21–4
Sense Receptors

Humans have some of the most highly specialized and developed sensory receptors in the animal kingdom. Some of these receive conscious sensations and are located in the skin (for touch, pressure, cold, warmth, and pain), eyes (for sight), ears (for sound and balance), mouth (for taste), and nose (for smell). Other receptors receive sensations below the conscious level and are located in muscles, blood vessels, the brain, the lungs, and other tissues.

Receptors have two important functions. First, they have the ability to trigger nerve impulses. Secondly, they contribute to the way in which we understand the world. Everything we know about the world comes to us through our senses. Everything we know, then, depends to some extent on the way our senses work.

A receptor may be a specialized cell or simply a bare nerve ending. Each receptor is particularly sensitive to one kind of stimulus. It responds to this stimulus much more readily than other receptors will. Some receptors respond to more than one stimulus, but their threshold is very high. For example, receptors in the eye that respond to light will also respond to pressure, but not as readily as do the pressure receptors in the skin.

Receptors in various parts of the body relay information about joint position and movement. Other receptors transmit information that does not reach the conscious level. This includes information about muscle tension, muscle length, blood pressure, temperature of the blood in the head, the chemical balance of the blood and cerebrospinal fluid, among other functions.

Smell Receptors

You have noticed how dogs, rabbits, and other animals find out about their environment by sniffing. In many lower animals, smell is the most important of the senses. For humans, the sense of smell has become less useful than other senses.

Our system for detecting odors is built into our nose. Odors are always gases. When an odor/gas enters the nose, it dissolves in the mucus lining of the nasal cavity. The gas/mucus solution stimulates olfactory cells. Impulses are carried in the olfactory nerve to the olfactory part of the cerebral cortex. There, the impulses are "read" and then acted upon.

Taste Receptors

Our sense of taste is controlled by receptors called taste buds on the tip, sides, and back of the upper surface of the tongue. A few taste buds are also found on the epiglottis and the soft palate. Babies and young children have more taste buds than do older people.

Each taste bud consists of a cluster of *chemoreceptor* (kem'ō ri-sep'tər) *cells* that contain a number of hairs. A network of nerve fibers carries impulses from the taste buds to two of the cranial nerves. These nerves are the pathways that transport taste sensations to the taste centers in the thalamus and cortex of the brain.

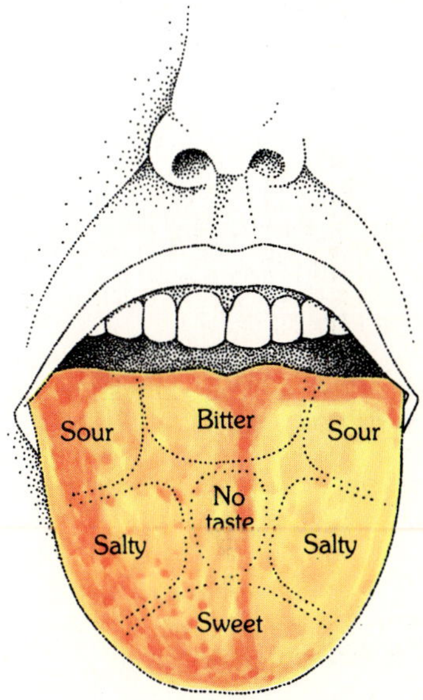

Location of Taste Buds

If your mouth were completely dry, you would have no sense of taste. You can detect a taste only after a substance has dissolved in the saliva. In the taste buds, each food is recognized as one or a combination of the four basic tastes. Receptors for each of the four primary tastes—sweet, sour, salty, and bitter—are located in distinct regions of the tongue. Sweetness is recognized by the taste buds on the front of the tongue. Receptors for sour are located on the sides. Those receptors for bitter are found at the back of the tongue. Taste buds at two places along the sides, at the tip and also at the back of the tongue produce the salty sensation.

It is difficult to separate the taste of food alone from the usual taste-odor combination. Our sense of taste is closely influenced by our sense of smell. Most foods have odors. When we put food into our mouth, the gases they contain pass into our nose. There they stimulate the olfactory receptors. The combination of taste and smell is known as *flavor.*

Pain and Touch

Receptor cells in our skin respond to touch, pressure, pain, heat, and cold. For each stimulus, there is a specific type of receptor. Certain cells, for example, react to gentle touch. Whenever our body contacts something, these receptors transmit the "touch" message to the sensory cortex. Certain stronger pressures are detected by other receptors buried deeper in the skin. They respond only to pressures greater than a gentle touch.

Messages for pain originate in free nerve endings. Pain is usually a danger sign. The body reacts as quickly as possible to pain. Impulses of pain travel from the skin to the spinal cord. Directions for getting rid of the pain are then sent directly from the spinal cord to the appropriate muscles. The brain finds out about the pain only after the spinal cord has reacted. Short circuiting the brain through this kind of reflex arc results in the fastest possible response.

Receptors for touch often occur close to receptors for other senses. Warm milk has a different taste from cold milk partly because different receptors are stimulated by the "touch" of the two liquids.

Sensory receptors

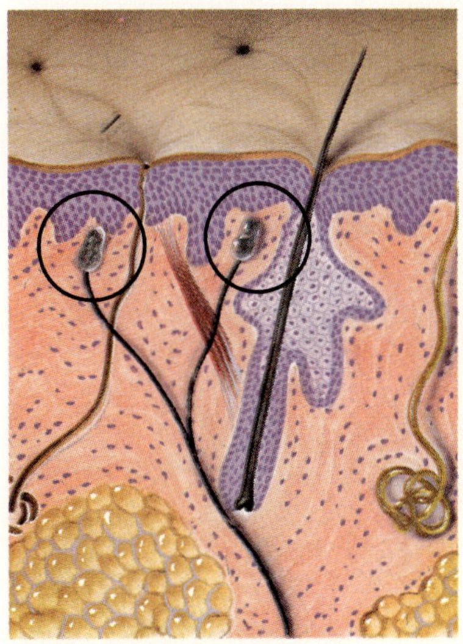

Sight Receptors

Humans probably obtain more information about the outside world through their sight receptors—the eyes—than in any other way. The wall of the eyeball is composed of three layers. The outside of the eye, the *sclera* (sklir′ə), is made of a tough white layer of connective tissue. At the front of the eye, the sclera becomes thinner and transparent. This section is called the **cornea** (kôr′nē ə). Light enters the eye through the cornea. The middle layer is the *choroid,* which contains a pigment for absorbing light. The choroid also is a network of blood vessels that supply the eye with oxygen and nutrients.

Behind the cornea, the choroid is modified into the **iris** (ī′ris). The iris controls the amount of light that enters the eye by changing the size of the **pupil,** the opening in the iris. The pupil is adjusted automatically to changes in light. Pigments in the iris are responsible for eye color.

Directly behind the iris is an elastic and transparent lens. The lens focuses the light that comes into the eye through the pupil. Small muscles attached to the lens change the shape of the lens to bring the light into focus on the third layer of the eye, the light sensitive **retina** (ret′n ə).

The chamber between the cornea and the lens is filled with a clear salty fluid known as the *aqueous humor.* The eyeball behind the lens is filled with a jellylike material, the *vitreous humor.* The aqueous and vitreous humors help to maintain the shape of the eyeball.

The innermost layer of the eye, the *retina,* is the only part of the eye that is sensitive to light. The retina contains two specialized types of receptor cells, the rods and cones. The names of these cells refer to their shape when viewed under a microscope. The rods and cones are connected to the brain by sensory neurons that form the **optic** (op′tik) **nerve.** The point at which the optic nerve enters the eye is called the **blind spot.** It contains no rods or cones.

Human eye structure

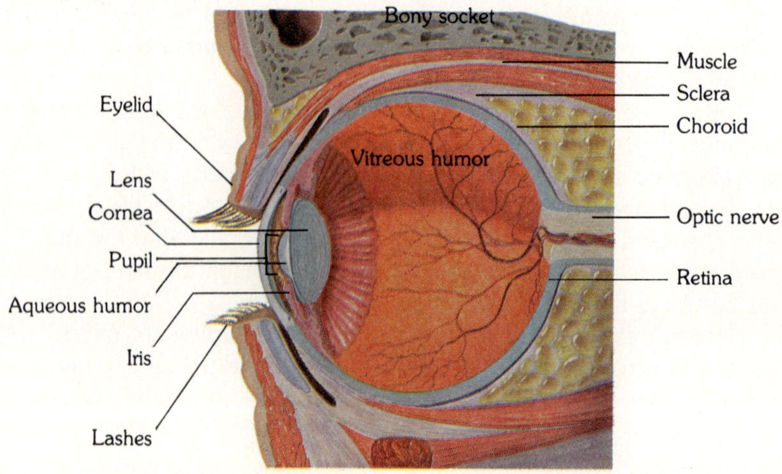

Photoreceptor Cells

Rods and cones are **photoreceptor** (fō'tə ri sep'tər) **cells**—they are stimulated by light. The rods are more sensitive to light than are the cones. The cone cells are sensitive to colors. Cone cells function in bright light and allow us to perceive details. When we are trying to see at night or in a darkened room, we depend upon the rod cells to transmit nerve impulses to the brain that produce a visual image. When more light is available, we rely on the cone cells.

Rods and cones are not distributed evenly throughout the retina. The part of the retina that is directly behind the center of the cornea and the lens is called the **fovea** (fō'vē ə). The fovea has only cone cells and is the area of best vision. The surrounding area of the retina is composed of both rods and cones.

Rods and cones are photoreceptor cells because they are stimulated by light. This stimulation is a chemical one in both rods and cones, although more is known about the response of rods than of cones. Rods contain **visual purple,** a substance composed of a protein and a chemical compound, **retinene** (ret'n ēn'). As light strikes the visual purple, this compound breaks down into protein and retinene. This causes a nerve impulse to be sent through the optic nerve to the visual cortex of the brain for analysis and action. In the retina, the visual purple is quickly restored.

Color vision is not fully understood. Most explanations of color vision assume there are three types of cone cells that respond to red, green, and blue, the primary colors of light. All colors, including white, can be made by various combinations of these three primary colors. Different color images may be produced by various stimuli to these three types of cones.

The eye is well protected. The eyeball is located in a bony socket of the skull. This socket is lined with a shock absorbing cushion of fatty tissue. The front of the cornea is protected by the **conjunctiva** (kon'jungk tī'və). A membrane similar to the conjunctiva covers the inside of the eyelid. Eyelids also contain small glands that secrete an oil to lubricate the eyeball and the eyelid. The **lacrimal** (lak'rə məl) **glands** secrete tears that help to moisten and clean the outer surface of the eyeball.

Rod and cone cells

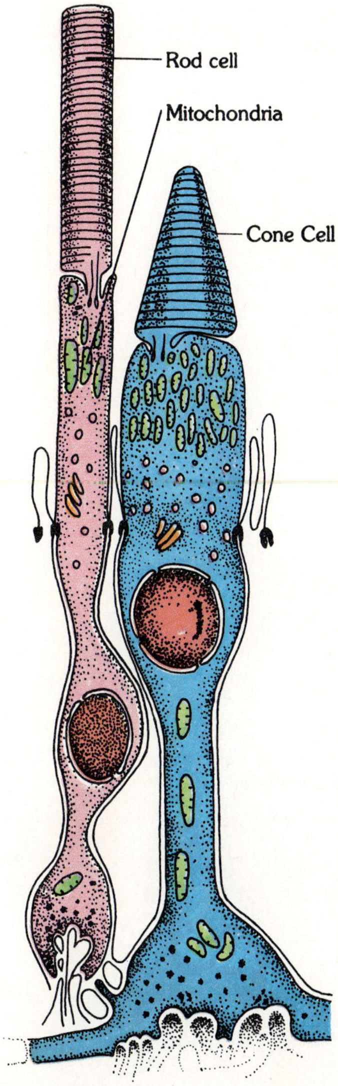

Hearing and Balance

The human ear has three parts: the outer ear, the middle ear, and the inner ear. The outer ear is the **pinna** (pin′ə), a skin-covered flap of cartilage. The pinna helps direct sound into the ear and provides some protection for the ear.

At the base of the pinna is a small opening, the **auditory** (ô′də-tôr′ē) **canal.** This canal is lined with hairs and glands that secrete wax. When sound waves enter the funnel-shaped auditory canal, they strike a membrane, the **eardrum,** and start it vibrating. The frequency at which the eardrum vibrates is the same as the frequency of the incoming sound.

The eardrum separates the outer ear from the middle ear. The middle ear contains three tiny bones, the **malleus** (mal′ē əs), the **incus** (ing′kəs), and the **stapes** (stā pēz′). The **eustachian** (yü stā′kē-ən) tube connects the middle ear and the throat. The eustachian tube is a device to equalize the air pressure on both sides of the eardrum. Vibrations of the eardrum vibrate the malleus, which is attached to it. The malleus in turn vibrates the incus, which then vibrates the stapes.

The vibrations of the stapes push against a membrane, the oval window, at the beginning of the inner ear. The oval window transmits vibrations to a fluid inside the **cochlea** (kok′li ə), a circular tunnel in the inner ear. The cochlea contains the **organ of Corti** (kôr′tē). This complicated structure receives vibrations from the fluid in the cochlea. The organ of Corti transmits nerve impulses to the brain by way of the auditory nerve.

Human ear structure

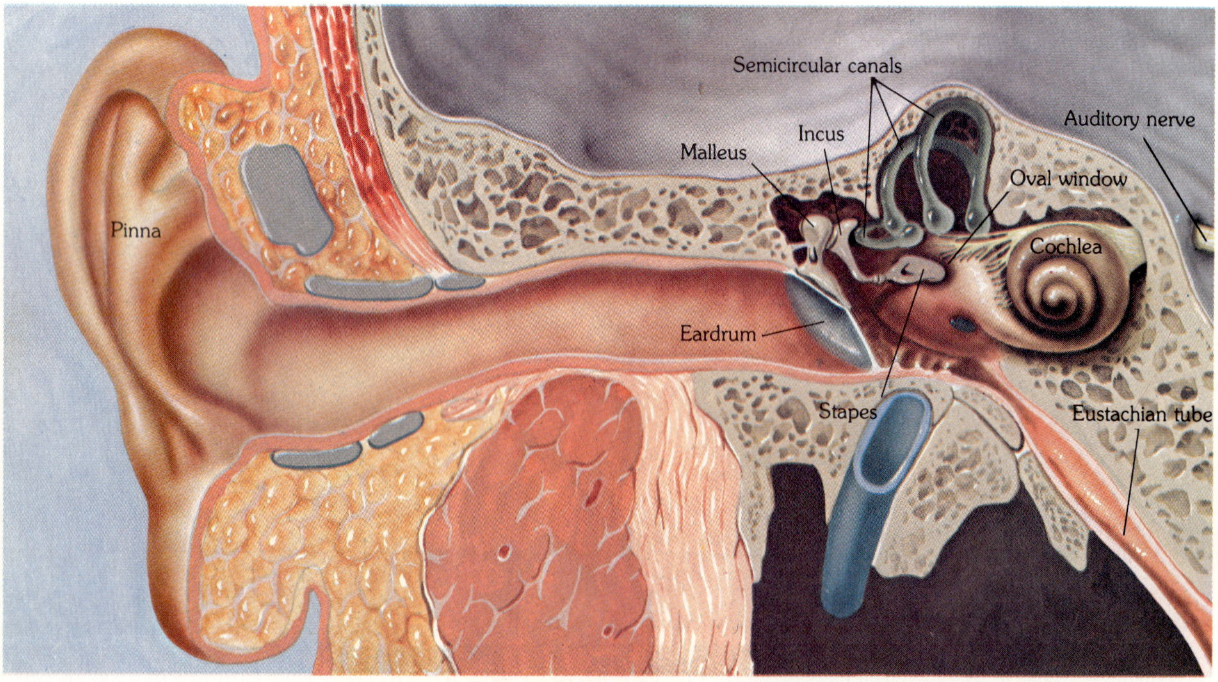

Organs of Balance

The sense of balance is dependent upon sight and stimuli from skeletal muscles, the soles of the feet, and the parts of the inner ear that are not involved in hearing. These parts consist of three **semicircular** (sem′i sėr′kyə lər) **canals** and two small chambers, the **utricle** (ū′trə-kəl) and the **saccule** (sak′ūl). The semicircular canals respond to changes in the rate and direction of head movements. The utricle and saccule note the position of the head.

The semicircular canals contain fluid. The **ampulla** (am pul′ə), a swelling at the base of each canal, contains tufts of hair cells. When the head is moved, the fluid inside the semicircular canals begins to move. The hair cells in the ampulla are stimulated by the movements of the fluid to send impulses to the brain. These impulses are interpreted in the brain as movements of the head.

The utricle and saccule respond to the position of the head in space. These chambers are lined with sensory hairs and are filled with a fluid. Each chamber also contains tiny grains of calcium carbonate, the **otoliths** (ō′tə lithz). When the head is tipped, the otoliths move. When the otoliths move, they stimulate the hair cells. Nerve impulses are sent to the brain, where they are analyzed as changes in the position of the body.

Organs of balance

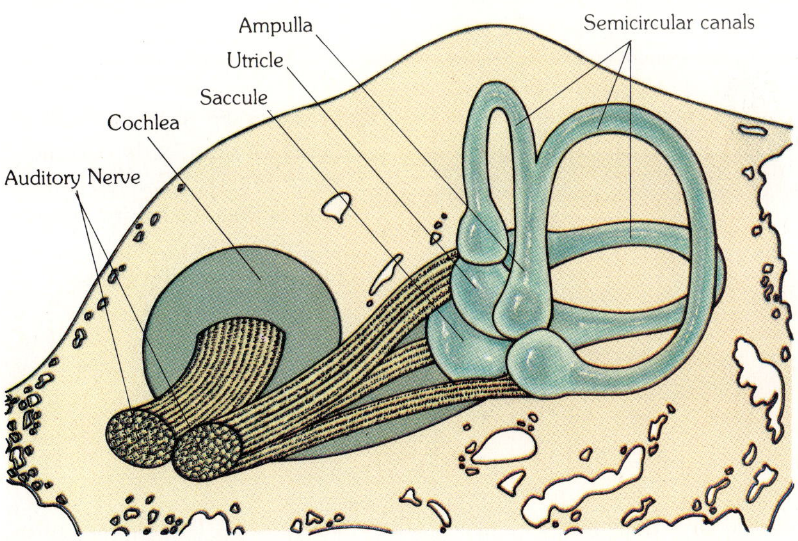

Review It

1. Explain how receptor cells in the nose and mouth detect odors and tastes.
2. List some of the ways the skin detects various types of touch.
3. What are the major parts of the ear? How are they involved in the process of hearing?
4. Make a sketch of the eye. Indicate the part played by each area in the process of seeing.

Activity

Touch Reception

Purpose
To measure the sense of touch and spatial discrimination at different locations on the skin.

Materials
5 pieces of cardboard, 1.5 cm by 6 cm
9 straight pins
Ruler

Procedures

1. Arrange pins in cardboard as shown in diagram.
2. Students should work in pairs. Blindfold a laboratory partner, select one of the pieces of cardboard, and lightly touch the surface of his or her fingertip. (Caution: Do not press hard enough to break the skin.)
3. Record in the chart whether the partner feels one or two pins.
4. Repeat the procedure for each of the pieces of cardboard with pins on the different locations listed in the chart.

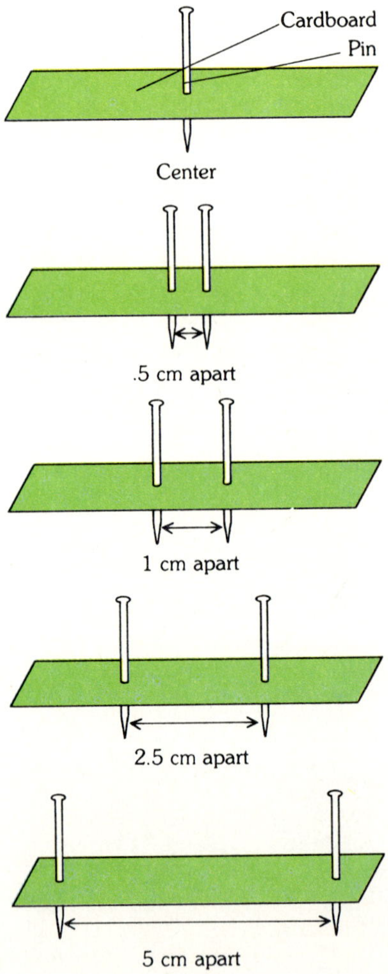

	Number of pins	1	2	2	2	2
	Spacing in cm		.5	1	2.5	5
Number of pins felt	Fingertips					
	Cheek					
	Lips					
	Forearm					

Analysis

1. List the areas of skin in the order of decreasing sensitivity.
2. What would happen if all the areas on your body had the same sensitivity? What advantage is there in having varying degrees of sensitivity over different parts of the body?

21–5
The Endocrine System

In addition to the nervous system, the **endocrine** (en′dō krən) **system** is a known regulator of body activities. A group of tissues named the *endocrine glands* secrete chemical substances called *hormones,* which influence the actions of other body tissues. Hormones are released directly into the bloodstream and are transported via the circulatory system. Because endocrine glands have no *ducts,* or tubes, they are often referred to as the *ductless* glands.

Human endocrine system

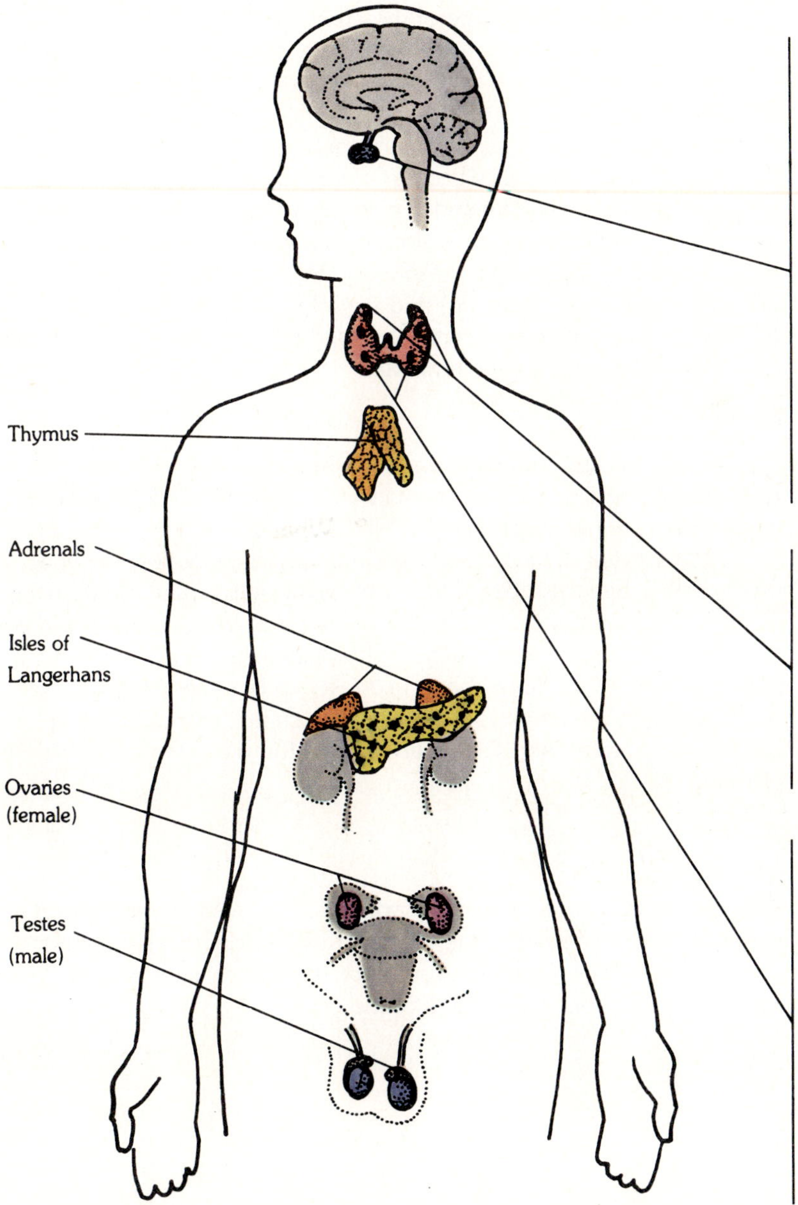

The Pituitary
The **pituitary** (pə tü′ə ter′ē) gland is well protected at the base of the brain. It produces numerous hormones such as **somatotropin** (sō mət′ə trō pin), a growth regulator. **FSH,** which controls the production of egg and sperm, and **ACTH,** which stimulates the *adrenal glands,* are two of the many other hormones produced. Because the pituitary regulates the actions of many of the other glands, it is sometimes called the *master gland.*

The Thyroid
The **thyroid** (thī′roid) gland, located at the base of the neck, secretes the hormone **thyroxin** (thī rok′sən). This hormone controls the metabolic rate of the body. The element *iodine* is an essential component of thyroxin.

The Parathyroids
Four small glands named the **parathyroids** (par ə thī′roidz) are embedded within the thyroid gland. The concentration of calcium in the blood is controlled by **parathormone** (par ə-thôr′mōn), which is secreted by the parathyroids. Bone and teeth development, blood clotting, and nerve functioning are all influenced by calcium concentrations.

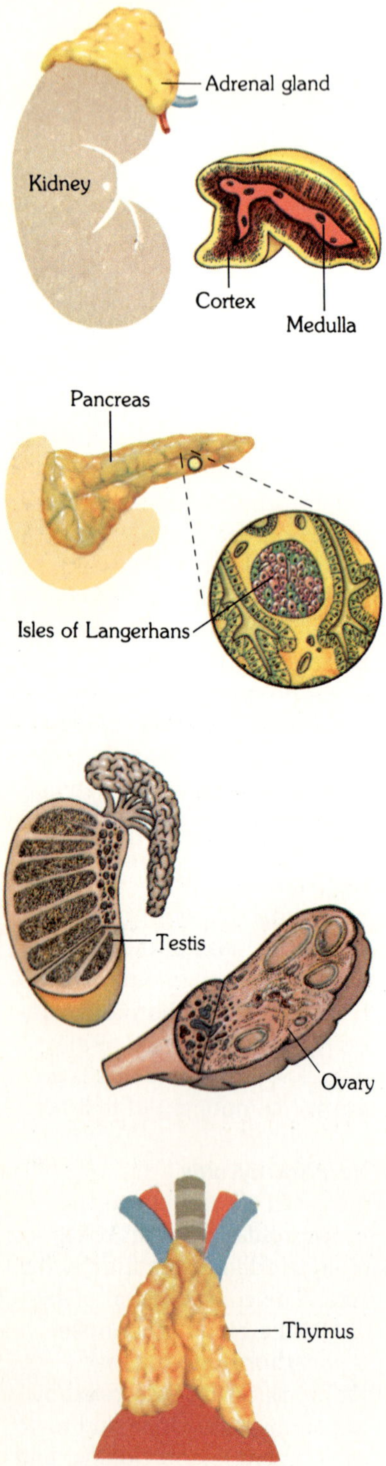

Several endocrine glands

The Adrenals

Composed of two separate glands, the **adrenals** (ə drē′nlz) are located on top of the kidneys. The outer section, called the **cortex,** is an essential gland, producing hormones that regulate the balance of water and salts in the blood. Blood pressure, white blood cell production, and the metabolism of fats, proteins, and carbohydrates are also controlled by cortex hormones. The inner gland, or **medulla** (mi dul′ə), produces **adrenaline** (ə dren′l ən). This hormone stimulates a rapid increase in body activity. Sometimes called the "fight or flight" hormone, adrenaline prepares the body in times of emergency for immediate action. Increases in heart rate, circulation, and amount of sugar available for energy are some of the changes caused by adrenaline production.

The Pancreas

The **pancreas** (pan′krē əs), normally thought of as an organ of digestion, contains specialized cells called the *Isles of Langerhans,* which function as an endocrine gland. They produce the hormones *insulin* and *glucagon.* **Insulin** (in′sə lən) controls the sugar level (available for cellular use as energy) by regulating the liver. Excess sugar is converted into *glycogen* (a starch) and stored in the liver. **Glucagon** (glu′kə gon) balances the action of insulin by converting glycogen into sugar for use as energy when needed.

The Sex Glands

The *sex glands* are endocrine glands. Hormones that influence the traits of maleness or femaleness, as well as gamete production are produced in the sex glands.

Testosterone (te stos′tə rōn′) is one of several male hormones produced in the *testes.* Secondary sex characteristics such as voice change; body hair, including beard growth; and increased musculature are direct results of the production of testosterone.

The *ovaries* of the female produce a group of hormones, among them **estrogen** (es′trə jən). Secondary sex characteristics including breast enlargement, and the broadening of the hips, are influenced by estrogen.

The Thymus

Although unknown as to its function for many years, the **thymus** is located below the thyroid in the central chest cavity. It secretes a hormone that enables the body to produce *antibodies,* substances used in defense against foreign invaders of the body.

The Action of the Endocrine System

A normally functioning endocrine system most often works on a *feedback system.* As a hormone triggers one specific action, the result of that action effectively shuts off the production of the original hormone. For example, the pituitary may produce a large amount of ACTH. This stimulates the cortex of the adrenal glands to release its hormones. As these hormones increase in blood concentration, further production of the pituitary hormone ACTH is inhibited. This process is known as *negative biofeedback,* and is a crucial part of balancing the control mechanisms.

It is also important to note that the nervous system and the endocrine system influence each other. For example, the hormone called adrenaline may be produced as a result of nerve stimulation, and its effects, such as increased heart rate, may be directed by nerve action.

Malfunctions of the Endocrine System

Occasionally, glands of the endocrine system are defective. They may secrete abnormal amounts of hormones; too much, too little, or none at all. The following chart lists some of the diseases associated with several glands of the endocrine system.

Gland	Hormone	Defect	Resulting symptoms or disease
Pituitary	Somatotropin	too much	Giantism—super-secretion during growth years results in a taller-than-average person
	Somatotropin	too much	Acromegaly—super-secretion in adult years leads to thickening of facial bones and fingers
	Somatotropin	too little	Midget—too little secretion results in stunted bone growth but normal mental development
Thyroid	Thyroxin	too much	Hyperthyroidism—increase in metabolic rate, heart beat, circulation, and blood pressure. Often, person exhibits irritability and nervousness.
	Thyroxin	too little	Hypothyroidism—very low metabolic rate and sluggish activity. The person may become overweight
	Thyroxin	too little	Cretinism—too little thyroxin during childhood years causes a dwarfing effect plus mental retardation
	Thyroxin	lack of iodine	Goiter—abnormal growth of thyroid tissue due to a lack of iodine
Adrenals	Cortex compounds	damage or destruction of gland	Addison's Disease—person affected becomes weak, nauseous, has lowered circulation, and bronzing of the skin
Pancreas	Insulin	too little	Sugar Diabetes—sugar in the blood is not utilized by the cells for energy, so excess amounts of blood sugar are excreted in the urine. Frequent thirst and urination are common symptoms.

Review It

1. What is the significance of the endocrine system?
2. Why is a feedback system important in the endocrine system?
3. Name and give the function of the hormones secreted by the following glands: pituitary, adrenals, thyroid, pancreas.

Chapter Summary

The skeletal system, our body's framework, helps protect our internal organs and produces certain needed materials such as red blood cells. Bones provide strength and rigidity as well as lightness and mobility. A variety of different kinds of joints make it possible for bones to move in different directions.

Three types of muscles are found in the body. Bones move when skeletal muscles contract. Smooth muscle comprises the digestive tract, the kidney, and other internal organs. The heart is composed of cardiac muscle.

The nervous system is divided into three systems. The central nervous system is composed of the brain and the spinal cord. The peripheral nervous system is composed of the cranial and spinal nerves. And the autonomic nervous system consists of the parasympathetic and sympathetic nerves. Nerves are made up of individual nerve cells called neurons. Chemical changes inside and outside the neuron membrane create nerve impulses, or messages.

The brain has three major regions of importance. The cerebrum is the center of advanced thinking. The cerebellum coordinates movements and provides for balance. The brain stem, which includes the medulla oblongata, controls involuntary vital reflexes.

Signals from our external and internal environments reach the brain by way of our senses. Sensory receptors are located in the skin, eyes, ears, nose, and mouth. These receptors send impulses that are analyzed and acted upon by the brain. Internal receptors send information about blood pressure, muscle tension, and other conditions of the body that the brain must know to regulate body function.

The endocrine system exerts chemical control over our bodies. Substances called hormones are produced in ductless glands, transported by the bloodstream, and affect the activities of tissues elsewhere in the body.

Questions to Think About

1. Animals that sleep during the day and forage at night are called **nocturnal** (nok tėr′nl). Owls and rats are examples of nocturnal animals. This type of animal has only rods in the retina of its eyes. What advantage would this offer to the animals? How do the life styles of such animals reflect the structure of their eyes?
2. We have been able to identify five areas of sensory perception: sight, hearing, touch, smell, and taste. Some people believe that other types of sensory perception may exist but have not as yet been recognized and understood. What is meant by extrasensory perception? What examples of extrasensory perception have been explored? What research on ESP is presently taking place?
3. What might happen if the production of estrogens or testosterones never occurred in an individual? Would you consider this a fountain of youth? Why or why not?

Extra Research

1. When you were a small child, your parents may have told you to eat your carrots so that you would be able to see well at night. Carrots are rich in Vitamin A. What role does Vitamin A play in vision?
2. Most adults can hear sound of between 40 and 20,000 vibrations per second. Some animals are able to hear vibrations at much higher frequencies. What frequencies of sound can be heard by animals such as bats, owls, and dogs? Why is a dog whistle audible to dogs but not to humans?
3. A stroke, or cardiovascular accident, may result in the loss of speech, paralysis, loss of memory, or death. What are strokes? How are they caused? What effects does a stroke have on the central nervous system? What treatment is available for stroke victims?
4. Sugar diabetes is one of the most commonly occurring diseases of the endocrine system. Find out some statistics about this disease such as: Whom does it affect? How is it detected? What treatments are available? What additional problems or side effects may diabetics encounter?

Chapter Test

Part I:
Copy the number of each test item. If the item is true, write *T* after the number; if the item is false, write *F*.

1. The action of joints causes bone movement. (509)
2. Of the three types of muscle, only skeletal muscle is under conscious control. (511)
3. The amount a muscle contracts depends upon the strength of the nerve impulse reaching it. (512)
4. Nerve impulses pass from the axon, through the cell body, and out of the dendrites of a neuron. (515)
5. Messages from the sense organs are processed in the cerebellum. (517)
6. The operation of smooth muscle is under the control of sympathetic and parasympathetic nerves. (520)
7. The ear structures used in hearing also help to maintain balance. (527)
8. Negative biofeedback refers to an emergency shut-off system in the endocrine glands. (531)
9. Thyroxin is absolutely essential in maintaining a proper metabolic rate. (529)

Part II:
Copy the number of each sentence below. After each number, write the word (or words) that completes the sentence correctly.

10. The skeleton in a newborn baby is primarily made of ____. (510)
11. Padding and lubrication in joints is provided by ____ and ____. (509-10)
12. The gap which separates nerve fibers is called a (an) ____. (515)
13. The strength of a muscle action depends upon ____. (512)
14. The two parts of the central nervous system are ____ and ____. (517)
15. Internal organs are made of ____ muscle and are under the control of the ____ nervous system. (511)
16. The three types of neurons are ____, ____, and ____. (514)
17. The adrenal glands are actually two separate glands, the ____ and the ____. (530)
18. Messages about body movement are coordinated in the brain by the ____. (518)
19. Fine focusing in the eye is provided by the ____. (524)

Part III:
Copy each numbered item from List A. After each item, write the letter and words from List B that are most closely associated with it.

List A

20. skeletal system (506)
21. muscular system (511)
22. central nervous system (517)
23. endocrine system (529)
24. autonomic nervous system (520)
25. reflex arc (516)

List B

a. control by hormonal secretions
b. pathway of fastest response to stimuli
c. brain and spinal cord
d. sympathetic and parasympathetic networks
e. skeletal, cardiac, and smooth muscle
f. protection, support, and metabolic functions

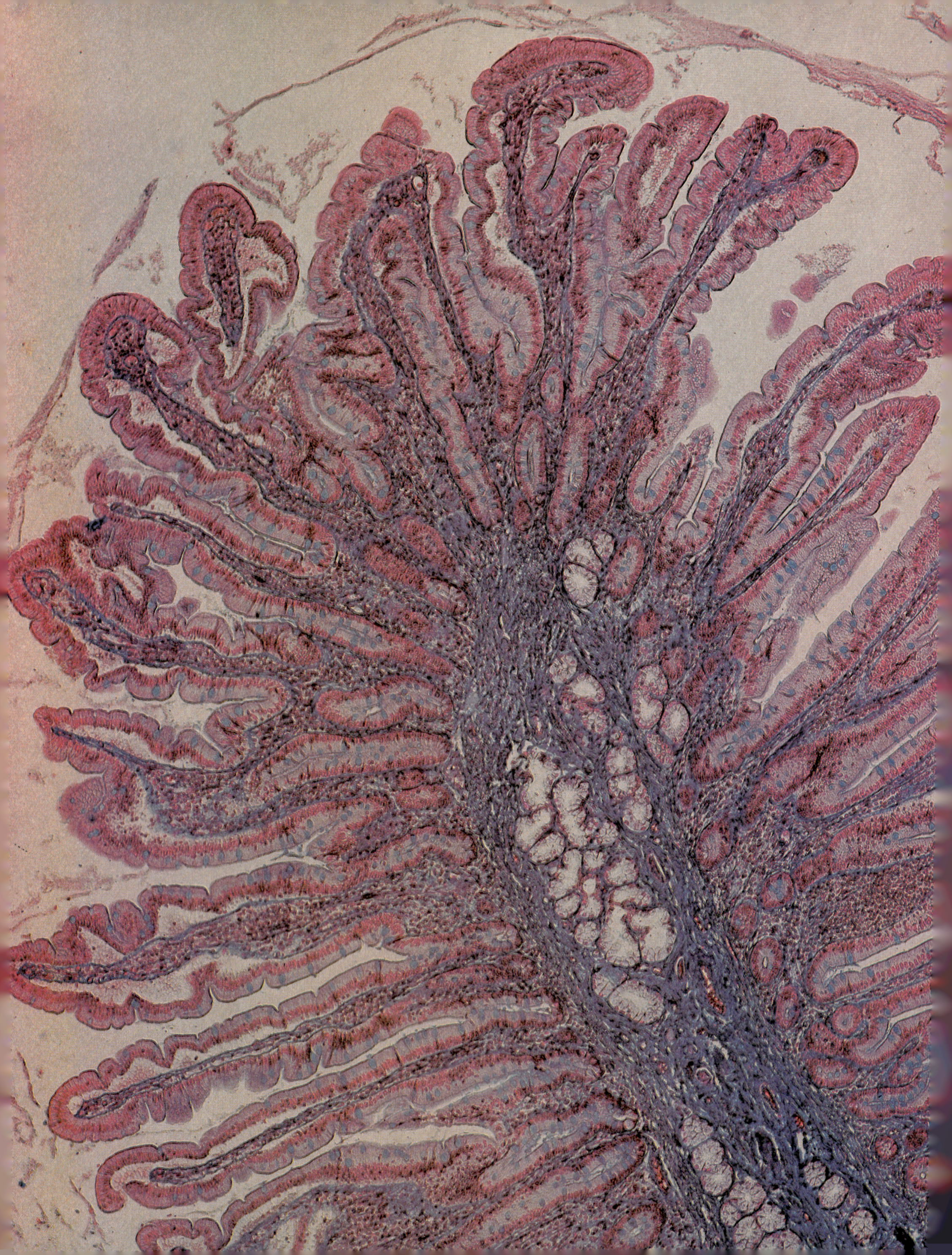

Chapter 22

Nutrition and Digestion

Food you eat is converted into energy for all body functions and is used to form new cells for body tissues. Food is the raw material the body needs to grow and stay healthy. However, eating only one kind of food does not provide everything the body needs. You need to eat foods that provide all essential nutrients: carbohydrates, fats, proteins, vitamins, minerals, and water. An absence or inadequate intake of any one nutrient in the diet can lead to several conditions ranging from simple fatigue to death.

The food you eat must be digested before it can be used by the cells. During digestion, food is broken down by both mechanical and chemical actions. After food is digested, it is absorbed into the blood and lymph and carried throughout the body.

Eating foods that supply the right amounts of the essential nutrients is the key to good nutrition. Do you know what foods your body requires? Do you know how foods are broken down so your body can use them? This chapter will help you find out.

Chapter Objectives

Here are some things you should be able to do after reading this chapter:

1. Define nutrition and malnutrition.
2. List the functions of the six essential nutrients.
3. Explain how a balanced diet can be obtained.
4. Explain why digestion is necessary.
5. Describe what happens to food in the mouth, the stomach, the small intestine, and the large intestine
6. Explain how digested food is absorbed.

The lining of the small intestine looks like this when magnified 100 times under a microscope. The fingerlike projections are called villi, sites where digested food enters the blood.

22-1
Nutrition and Malnutrition

Challenge!
For many Americans, the major nutritional problem is not lack of food, but too much food of the wrong kind. What are some health problems associated with being overweight? When is a person obese? Why is obesity a problem for so many people?

Why do you eat? The easiest answer to that question is that you are hungry. Being hungry is the body's way of saying that it needs more fuel to keep going. That fuel is provided by the food you eat. Scientists now know that the human body needs six different substances to function properly. These substances are carbohydrates, fats, proteins, vitamins, minerals, and water. They are known as **nutrients** (nü'trē-ənts). Nutrients in food function in several ways. In the body, they (1) provide energy, (2) provide materials for growth and maintenance, and (3) provide materials for regulating body processes. A nutritious food contains substances that carry out one or more of these functions.

Nutrition is the study of the processes by which food is taken in and used for growth, energy, and all body functions. Today we know a great deal about what nutrients the body needs and what foods provide certain nutrients. This makes it possible to plan a diet that will help the body grow and develop properly.

When an essential nutrient is missing from the diet for a long time, normal growth may be affected or illness may result. **Malnutrition** refers to any condition in which the body does not obtain one or more essential nutrients in the right amount.

Malnutrition, even for a short time, can affect many areas of your life. For example, it might slow down your skeletal or muscular development. It could also cause you to feel tired and listless. And it might affect how well you do in school and how well you get along with others. Have you ever felt tired and cranky on those days you didn't eat breakfast? This could have happened because your body had been deprived of energy-rich nutrients. However, a good diet can help you to look, feel, and behave the way you want to.

A Balanced Diet

For good nutrition, you need to eat foods that provide enough of the essential nutrients. However, there is no one correct diet. Most foods provide at least small amounts of each kind of nutrient. Also, there are many concentrated sources for each nutrient. For example, large amounts of protein are found in meats, fish, dairy products, whole grains, certain vegetables, nuts, and eggs. Eating a variety of foods every day can usually provide a balanced diet. Study the photos on the facing page. Use them to help you choose those foods that will provide you with essential nutrients each day.

Meat Group: Two or more servings of beef, veal, pork, lamb, poultry, fish, eggs. As alternates—dry beans, dry peas, nuts, peanut butter.

Bread-Cereal Group: Four or more servings whole grain, enriched, or restored

Vegetable-Fruit Group: Four or more servings, including a citrus fruit or other fruit or vegetable for vitamin C. A dark-green or deep-yellow vegetable for vitamin A—at least every other day. Other vegetables and fruits.

Milk Group: Children under 9: 2 to 3 cups
Children 9-12: 3 or more cups
Teen-agers: 4 cups
Adults: 2 cups

Carbohydrates

Structure of glucose

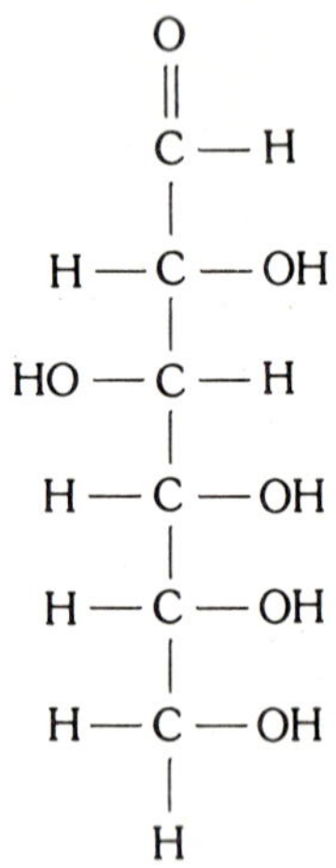

$C_6H_{12}O_6$

Each of the major nutrients has one or more specific functions to perform in the body. **Carbohydrates** are the chief source of energy for all body functions. They also assist in the digestion and absorption of other foods. For example, carbohydrates help regulate the breakdown and use of proteins and fats in the body.

The three main kinds of carbohydrates present in foods are *sugars, starches,* and *cellulose.* All three are made up of the elements carbon, hydrogen, and oxygen. You can read about each kind of carbohydrate below.

Monosaccharides (mon′ə sak′ə rīds′) are simple sugars. They are the most easily utilized by the body. The structural formula of one unit of *glucose,* a simple sugar, is shown here. Two other simple sugars are *fructose* and *galactose.* Simple sugars are found in fruit juices, honey, and some vegetables.

Disaccharides (dī sak′ə rīds′) are double sugars. They require digestive action to be broken down into simple sugars. Some examples of double sugars are *sucrose* (table sugar), *lactose* (milk sugar), and *maltose.* Double sugars are often added to packaged foods such as bread, soup, canned vegetables and fruit.

Structure of maltose molecule

$$C_6H_{12}O_6 + C_6H_{12}O_6 \rightarrow C_{12}H_{22}O_{11} + H_2O$$

Polysaccharides (pol′ē sak′ə rīds′) are complex starches and cellulose. They consist of chains of glucose units. Starch is found in grains, products made from grains, beans, peas, and potatoes. Starches require a period of digestive action before they can be used by the body.

Cellulose is found in the cell walls of all plant tissues. It is in whole grains and in the skins and seeds of fruits and vegetables. Cellulose is not digested by the human body and so is not a source of energy. However, cellulose provides bulk which stimulates the muscles that move food through the digestive system. It is important to have at least some cellulose in the diet.

Section of starch (or cellulose) molecule

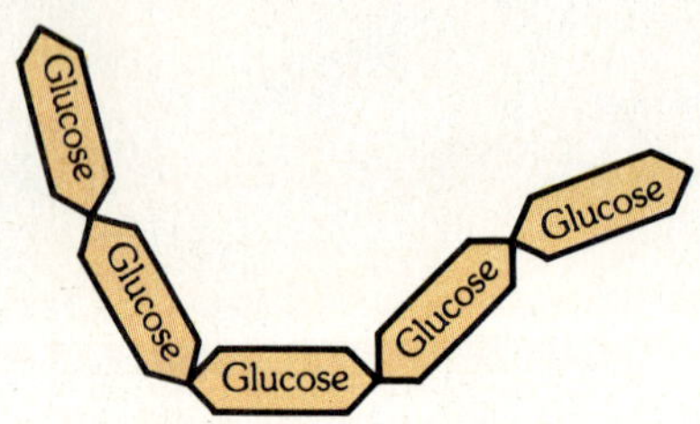

Role of Carbohydrates

In body cells, disaccharides and the digestible parts of polysaccharides are broken down into glucose units. Then each glucose unit is taken apart in a series of complex reactions. The energy produced in these reactions is stored in the form of ATP molecules. For each digestible glucose unit, or molecule, 38 molecules of ATP are produced. As you know, ATP supplies energy for almost all of the biochemical reactions in the body.

Some of the energy from glucose is used as fuel by tissues of the brain, nervous system, and muscles. A small portion is converted to *glycogen* and stored in the liver and muscles. The excess is changed to fat and stored throughout the body as a reserve source of energy.

Carbohydrates provide the body with almost "instant" energy. They cause a sudden rise in blood sugar level. However, the blood sugar level drops again rapidly if other kinds of food, such as meat, nuts, fresh fruit, and vegetables are not also eaten. A drop in blood sugar level may cause fatigue, dizziness, nervousness, and headache. Too many carbohydrate foods can also crowd out other essential foods from the diet and cause malnutrition. And continued intake of excess carbohydrates can result in overweight. You may know that being overweight is strongly associated with diseases of the heart and blood vessels and with diabetes.

Approximate Percent of Carbohydrates in Typical Foods

99%	sugar, honey, syrup
85%	chocolate candy
75%	cereals
70%	cookies, crackers, jams, dried fruits
50%	breads, rolls
30%	rice, spaghetti
20%	beans, peas, sweet potatoes
15%	fresh fruits
5%	leafy vegetables

No more than 66 percent of your total caloric intake for one day should come from carbohydrates.

Calories

The energy that food provides is expressed in units of heat, or **kilocalories.** These units are also called **Calories,** spelled with a capital *C.* One Calorie is the amount of heat needed to raise the temperature of 1 kilogram of water 1° C. Total calorie needs vary from person to person. For example, a growing child and a hard-working laborer need more high-energy foods than an older or inactive person. You can see some typical carbohydrate foods in the table to the right.

Foods high in carbohydrates

Fats

Like carbohydrates, fats are composed of carbon, hydrogen, and oxygen, but in different amounts. A fat molecule consists of two parts: glycerol and fatty acids.

Structure of fat molecule

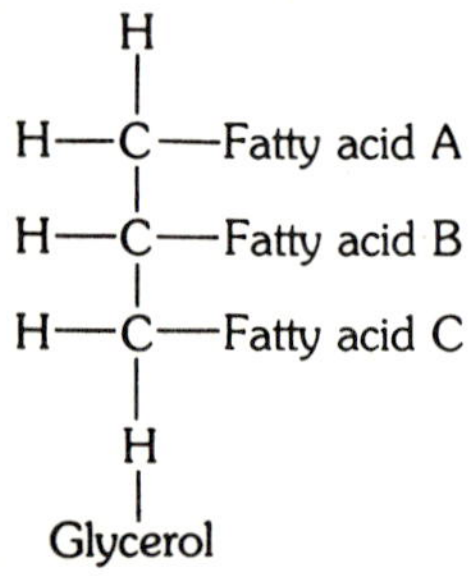

Fats provide the body with more than twice as many calories per gram as do carbohydrates. As a result, fats provide, among other things, a concentrated source of energy. Fats also enable the body to use vitamins A, D, E, and K. Fats are also a source of **fatty acids** which are essential for normal growth and healthy nerves, blood, arteries, and help to keep the skin from drying out.

Fatty acids give fats their flavor, texture, and melting points. They may be either **saturated** or **unsaturated.** Saturated fatty acids are usually solid at room temperature and come mainly from animals. Some examples of saturated fats are butter, lard, and tallow. Unsaturated fatty acids are usually liquid at room temperature and come from plants. Corn, olive, soybean, peanut, and most other vegetable oils are unsaturated.

Many physicians recommend that we include more unsaturated fats and fewer saturated fats in our diets. Saturated fats seem to increase the level of **cholesterol** (kə les'tə rol') in the blood. Cholesterol is a fatlike substance normally found in the body. High blood cholesterol levels may be associated with heart and blood vessel problems. But it has not yet been shown that lowering the level of cholesterol will lower the risk of such problems.

Approximate Percent of Fats in Typical Foods

100%	oils, shortenings
80%	butter, margarine, mayonnaise
60%	peanuts, walnuts, almonds, and other nuts
50%	peanut butter, unsweetened chocolate, bacon, doughnuts
30%	cheeses, roast beef
10%	ice cream, cakes, pies
5%	milk, rolls, bread
$<1\%$	most fruits and vegetables

Fats should make up about 25% of your daily caloric intake.

Role of Fats

During digestion, fat is broken down into its main parts. Fats that are not oxidized to produce energy are stored in fatty tissues in various parts of the body. (Excess carbohydrates and proteins are also converted to body fat and stored.) Fat deposits are used to provide fuel for future energy needs, conserve loss of body heat, and protect vital organs against injury. What sources of fats do you eat regularly? Are the fats you eat mostly saturated or unsaturated?

Proteins

Every living cell contains **protein** (prō'tēn'). In fact, protein provides building blocks of the human body. It is necessary for building and maintaining all body tissues. It is the main source of raw materials for the body's muscles, blood, skin, hair, nails, heart, and brain. In addition, protein is a source of energy when carbohydrates and fats are not available. *Enzymes,* important to chemical reactions in the body, are also proteins. And processes which control such body functions as growth, sexual development, and metabolism make use of chemical groups that proteins provide.

Proteins are made up of long chains of **amino** (ə mē'nō) **acids.** Like carbohydrates and fats, proteins contain molecules of carbon, hydrogen, and oxygen. In addition, proteins always contain nitrogen. And they may contain other elements, such as sulfur. The structural formula for a typical amino acid is shown in the margin. A section of a protein, indicating how different amino acids are joined together, is also shown.

Section of protein molecule

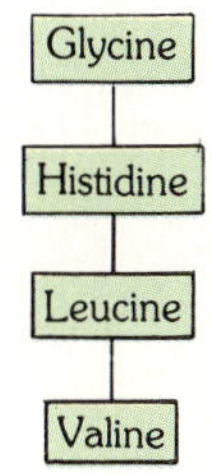

Structure of an amino acid

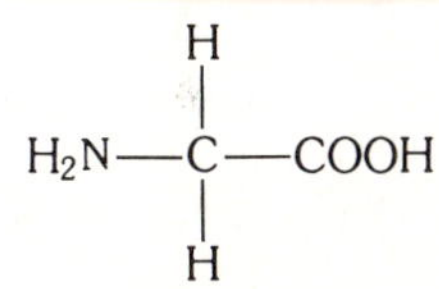

Glycine

Role of Proteins

Amino acids are necessary to produce body protein. During digestion, proteins you have eaten are broken down into amino acids. In this form, they can be transported to body cells, where they are used. Approximately 20 different amino acids are required to make all of the proteins needed by the body. All but eight of these can be put together in the body. The other eight must be supplied by foods you eat. These are called *essential amino acids,* meaning that it is essential they be in the diet.

Some of the foods you eat may contain **complete proteins.** This means that they contain all of the eight essential amino acids. Most meats and dairy products contain complete proteins. Sunflower seeds also contain complete proteins. **Incomplete proteins** lack one or more essential amino acids or are very low in any one of them. Most vegetables and fruits contain incomplete proteins.

What sources of complete proteins do you eat regularly? How much protein should you have daily? You can find the answers to these questions by studying the picture and table below.

Protein Requirements of Different Age Groups

Sex	Age	Height (Centimeters)	Protein Requirements (Grams per day)
Males	11-14	158	44
	15-18	172	54
Females	11-14	155	44
	15-18	162	48

Vitamins and Minerals

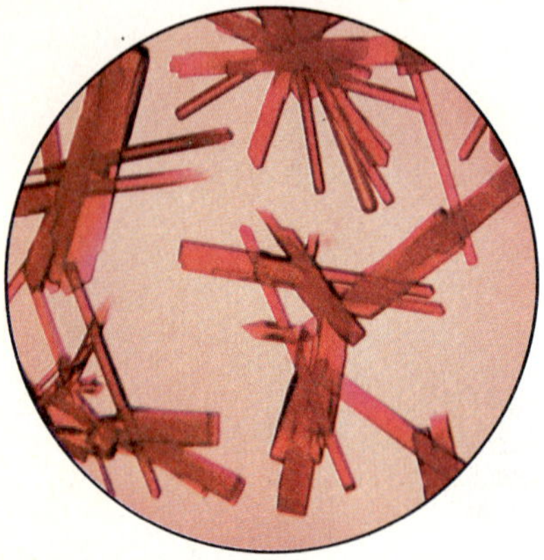

Vitamin B_{12} crystals

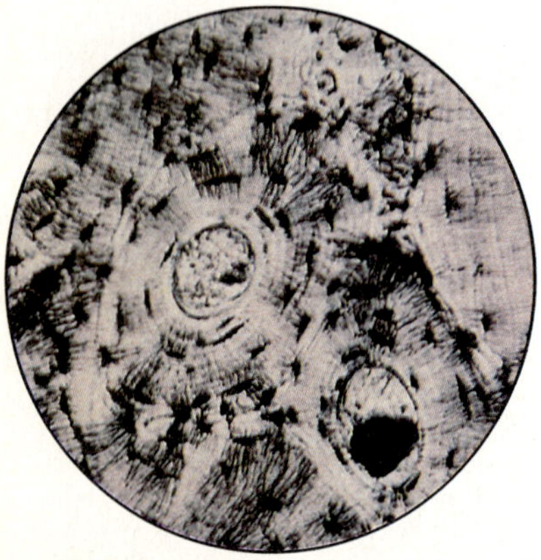

Calcium in bone

Very small amounts of about twenty substances called vitamins and minerals are needed by the body. **Vitamins** are organic compounds that are necessary for normal growth, maintenance, and reproduction. The functions of vitamins in the body are still not fully understood. The body does not metabolize them and use their products as it does carbohydrates, fats, and proteins. That is, vitamins are not a source of energy. Instead, vitamins serve several regulatory functions. Some vitamins work *with* enzymes to speed up certain chemical reactions. Some act *in place of* enzymes to speed up chemical reactions.

Many **minerals,** also referred to as salts, are vital components of the body. Some minerals are important parts of hormones. Others are parts of cells, bones, teeth, and blood. Iron, for example, is part of the hemoglobin molecule in red blood cells. Many minerals help regulate body functions, such as growth and metabolism. Most minerals are easily obtained in the diet. Those that may be deficient and must be consciously supplied include calcium, iron, phosphorus, and iodine. Study the table on the next page to find out what foods provide important vitamins and minerals.

Water

Although not considered a nutrient in the same sense as carbohydrates, fats, proteins, vitamins, and minerals, *water* is essential to life. Water transports nutrients and wastes throughout the body. It takes part in many chemical reactions in the body. It is the solvent in which all biochemical reactions take place. Water is lost daily from the body through respiration, perspiration, and elimination in urine and feces. You replace it by drinking water and other liquids and by eating foods, such as fruits, vegetables, and meats, that contain water.

Review It

1. What is nutrition? malnutrition?
2. What is the function of each of the six groups of nutrients?
3. What is the difference between complete and incomplete proteins?
4. What is the relation between diet and disease? Give examples.

Vitamins	Functions in the body	Food sources
A	• Helps develop and maintain healthy skin, hair, and mucous membranes • Helps prevent eye disorders • Aids in resisting infections	Whole milk; liver; green and yellow vegetables; egg yolk; butter; margarine
B_1 Thiamin	• Helps release energy from foods • Promotes appetite • Keeps digestive tract and nervous system healthy	Meats, especially pork; liver; enriched and whole-grain breads and cereals; legumes; nuts; vegetables; milk
B_2 Riboflavin	• Helps maintain skin, eyes, and nerves • Helps the body use carbohydrates, fats, and proteins to release energy	Liver; milk; green, leafy vegetables; enriched and-whole grain breads and cereals; cheese; meats; eggs; legumes
B_6 Pyridoxine	• Is involved in protein metabolism • Aids in making red blood cells • Helps the nerves function	Meats; poultry; fish; whole-grain breads and cereals; potatoes
B_{12}	• Is needed to produce red blood cells • Is needed for normal growth of all cells • Helps nervous system function	Meats; fish; eggs; milk and milk products
Folic acids	• Is needed to produce RNA, DNA, and normal red blood cells • Helps maintain digestive tract	Liver; green, leafy vegetables, legumes; enriched whole-grain breads and cereals; nuts
Pantothenic acid	• Aids in metabolism of carbohydrates, fats and proteins	Milk; eggs; enriched and whole-grain breads and cereals; liver; vegetables; fruits
Niacin	• Aids in producing energy from carbohydrates • Helps maintain all body tissues	Enriched and whole-grain breads and cereals; fish; liver; meat; poultry; legumes
Biotin	• Aids in metabolism of carbohydrates, fats, and proteins	Liver; kidneys; egg yolks; legumes; nuts
C	• Helps form and maintain the "cementing" material between cells • Is essential for healthy teeth and gums • Aids in resisting infections	Citrus fruits; berries; melons; tomatoes; potatoes; green peppers; green, leafy vegetables
D	• Helps build strong bones and teeth	Milk; fish-liver oil; liver
E	• Prevents damage to cell membranes • Protects vitamin A	Vegetable oils; margarine; vegetables; liver; enriched and whole-grain breads and cereals
K	• Promotes normal blood clotting	Green, leafy vegetables; peas; grains
Minerals		
Calcium	• Builds and maintains teeth and bones • Helps blood clot • Helps nerves and muscles function	Milk; cheese; ice cream; leafy vegetables
Phosphorus	• Helps build and maintain teeth and bones • Helps nerves and muscles function	Milk; cheese; whole grains; nuts; liver; fish; eggs; poultry
Magnesium	• Activates many enzymes reactions • Helps make proteins • Helps body use calcium and phosphorous	Red meats; potatoes; nuts; corn; green, leafy vegetables; cereals
Iron	• Combines with protein to make hemoglobin • Aids in growth of muscles, glands, and nerves	Liver; red meat; egg yolk; dried fruits; enriched cereals; green, leafy vegetables; legumes; nuts
Iodine	• Helps regulate the rate at which foods are burned	Seafood; iodized salt
Copper	• Aids in several reactions in the body, such as the making of hemoglobin	Nuts; raisins; liver; kidneys; legumes; shellfish
Zinc	• Necessary for normal taste and smell • Aids in general growth of all tissues • Aids in healing wounds	Meat; liver; eggs; seafood; milk; whole grains

Issues in Biology

Food additives: a danger, a safeguard, or both?

It has been proven that certain chemicals added to our food can cause cancer in laboratory animals. If chemical additives represent a danger to health, why are food processors allowed to use them? Why are chemicals added to food? How could such a procedure become established?

The history of nitrites as additives provides an example. At first they were added to meat, fish, and poultry because their interaction with bacteria gave the food an attractive pink or red color. Later it was found that nitrites act as a preservative, slowing down the growth of the spores of a bacterium that is responsible for a food poisoning in humans that is often fatal. This is important when refrigeration is unavailable. It is recognized that with the addition of nitrites, the risk of botulism poisoning from fish and poultry in the United States dropped to almost zero.

Both nitrates and nitrites are common food additives. But the danger to animals and man of excessive use of these chemicals in the diet has been known for a long time. N. S. Mayo reported in 1895 on the deaths of cattle that had eaten cornstalks containing large amounts of nitrates. Later scientists confirmed that nitrates are changed in the mouth or in the digestive tract into nitrites, many of which combine with by-products of amino acids called amines to form nitrosamines. Nitrosamines have caused cancer in laboratory animals. A 1978 study provided evidence that nitrites themselves can cause cancer in rats.

Saccharin is another additive that remains in our food supply in spite of evidence that it causes cancer in laboratory animals. But is saccharin really dangerous to humans? Some scientists don't believe it is. Large doses given to laboratory animals over two generations will produce bladder tumors. A recent study indicated a direct relationship between saccharin consumption and bladder cancer in males.

There is disagreement regarding the safety of using saccharin. The presence of saccharin in foods instead of sugar makes the foods available to dieters and diabetics. How can we balance the benefits against the risks?

In the United States, foods are often shipped long distances and then stored. Additives are used to protect many of these foods from their own enzymes as well as bacteria and fungi. Would you be willing to limit your diet to foods which can be grown in your section of the country to eliminate some additives from your diet?

Other additives cause baked goods to rise or give them a finer texture; prevent foods like shredded coconut from becoming wet and sticky; provide flavor or color; increase the vitamin and mineral content of food. Are these alterations necessary?

Analysis

1. What standards or criteria would you set for the use of food additives?
2. Look at labels and find what additives are in some of your favorite foods.

22–2
Digestion

During digestion, food is changed into a form the body can use. Food not yet digested is useless for several reasons. First, pieces of foods are too large and too complex to pass directly into the bloodstream. Molecules of starch, for example, are larger than the tiny openings in cell membranes. They have to be broken down into smaller units before they can enter the cells. Second, some foods are not soluble in water. For example, fats must be changed into a water-soluble form before they can be transported to body cells and used. Third, the body cannot use most nutrients as they occur in food. Some carbohydrates are consumed in the form of *sucrose,* a double sugar. However, the body can only use it as *glucose,* a simple sugar. Digestion converts all carbohydrates you eat into usable simple sugars. Fats and proteins are also broken down during digestion. Only vitamins, minerals, and water can be used by the body without being changed chemically.

The digestive system consists of the mouth, esophagus, stomach, small intestine, and large intestine. As food passes through the digestive system, its form and composition are changed. Along the way, many secretions and enzymes help to break down foods chemically. Mechanical actions, such as chewing, are also part of digestion. You can see the major organs of the digestive system in the X ray below. You will examine the structure and function of each organ on the pages that follow.

X-ray of human digestive system

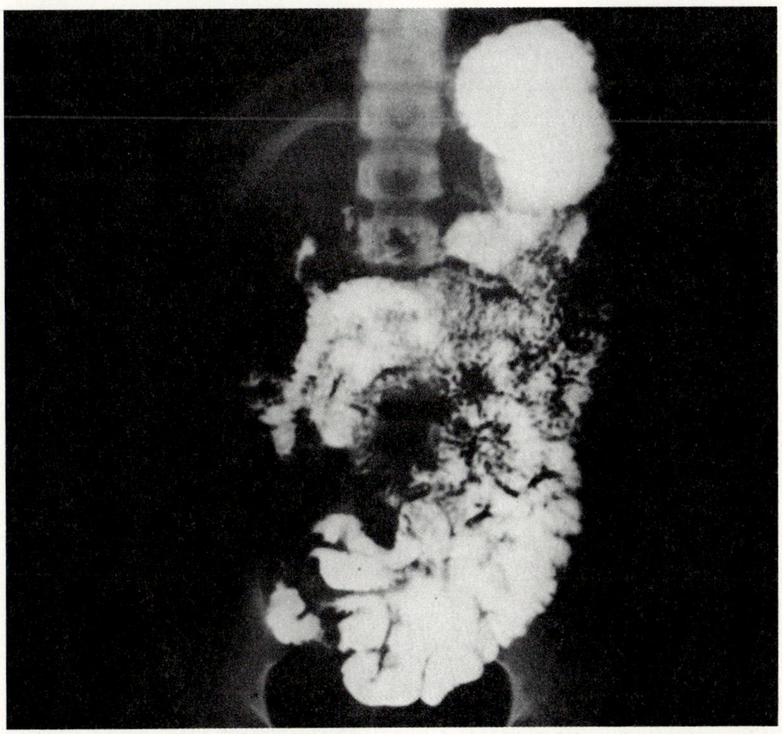

Diagram of human digestive system

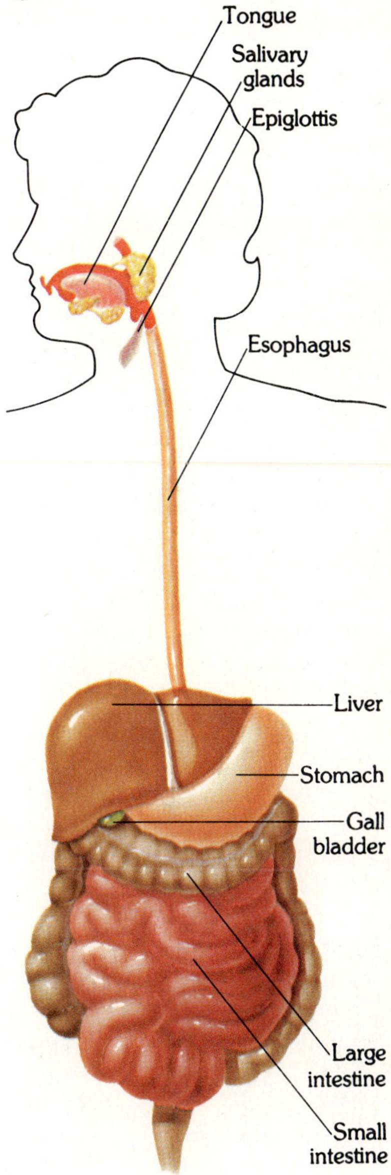

Organs of Digestion

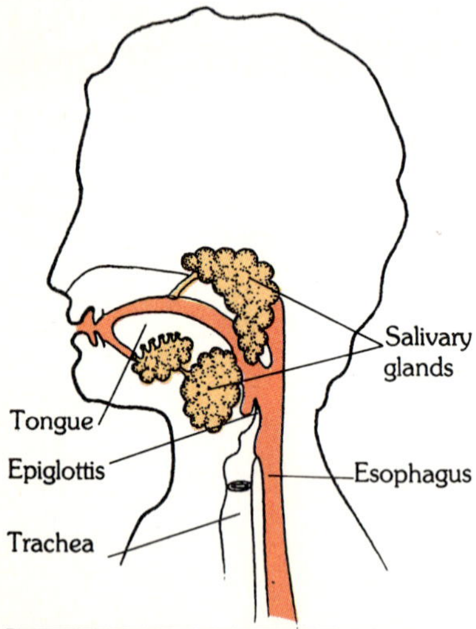

Digestive organs of the mouth

Digestion begins as soon as you put food into your mouth. Your teeth chop and grind food into pieces small enough to swallow. As you chew, food is moistened by **saliva** (sə lī′və) released from **salivary** (sal′ə ver′ē) **glands.** These glands begin to function at the sight, smell, or even the thought of food. Think of a nice hot turkey dinner and see if your salivary glands respond. Does your mouth "water"? That is saliva being released. Saliva is mostly water mixed with mucus and **amylase** (am′ə lās), an enzyme. This enzyme begins digestion of starches in the mouth.

Your tongue moves chewed food around against the roof of your mouth until it forms a small, wet ball called a **bolus** (bō′ləs). Moving the bolus toward the throat in preparation for swallowing is a voluntary act. You can start to swallow any time you want. Once you start, however, you cannot stop. As food passes from the mouth, a small flap of cartilage moves downward. This flap of tissue is called the **epiglottis** (ep′ə glot′is). Its function is to close off the *trachea,* or windpipe, and prevent food from getting into the lungs.

Esophagus

From the mouth, food passes into the *esophagus.* This is a muscular tube about 25 centimeters long that carries food from the mouth to the stomach. Two layers of smooth muscle line the esophagus. One layer is horizontal and encircles the esophagus. These muscles contract inwardly. The second layer is vertical and runs down the inside of the esophagus. These muscles contract with a wavelike motion. When the two sets of muscles contract together, food is forced down the esophagus and into the stomach. The muscular contractions are called **peristalsis** (per′ə stal′sis).

Peristalsis

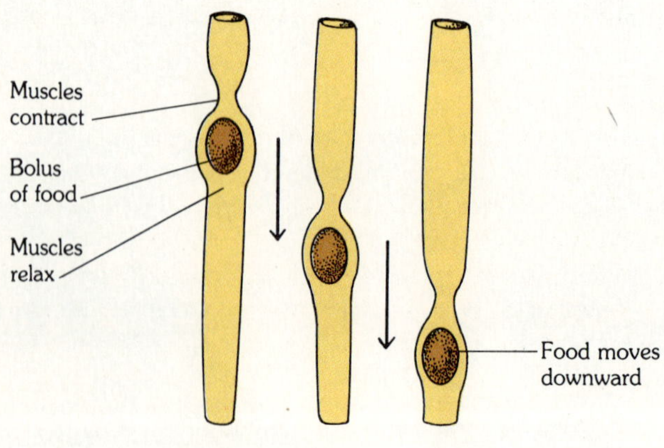

Stomach

Food passes from the esophagus into the stomach. The stomach is a collapsible sac that can hold a little over a liter of fluid. The stomach is lined with a very thick membrane that contains many **gastric** (gas'trik) **glands.** These glands secrete **gastric juice,** which contains hydrochloric acid, digestive enzymes, and mucus. Hydrochloric acid kills bacteria in the stomach. It also activates **pepsin,** a digestive enzyme. Pepsin begins the digestion of proteins. Mucus protects the stomach lining from abrasion and attack by acid.

The stomach lining contains the horizontal and vertical muscles found in the esophagus. In addition, there is another layer that lies at an angle to these muscles. When the three sets of muscles contract together, the food is moved around and so is mixed with gastric juice. After about three hours in the stomach, food becomes a pastelike mixture known as **chyme** (kīm).

Chyme is gradually pushed out of the stomach by peristalsis and into the **small intestine.** Water and some glucose may be absorbed directly into the body from the stomach. Most foods, however, cannot be used by the body until they undergo further digestion in the small intestine.

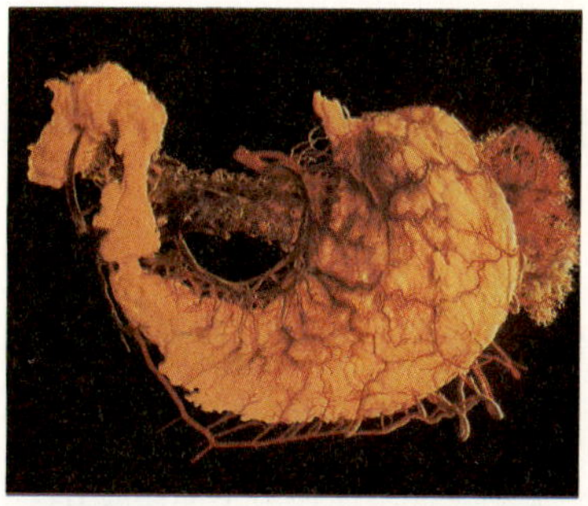

Stomach and blood vessels of the stomach

Muscular contractions of stomach during digestion

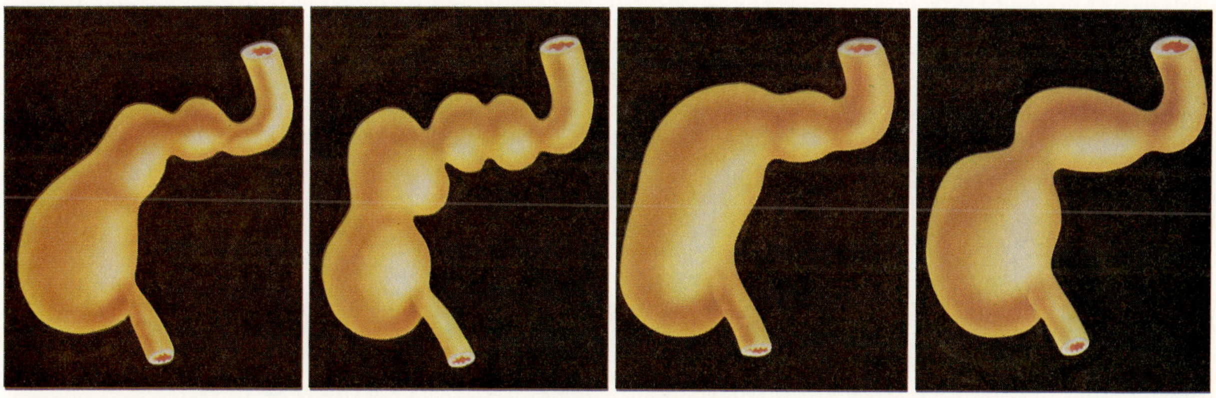

Small Intestine

Chyme is pushed by peristalsis into the small intestine. It is now ready for the final stages of digestion. The small intestine is a muscular tube 7 to 8 meters long. Digestion is essentially completed in the first section of the small intestine called the **duodenum** (dü'ō dē'nəm). Here, secretions from the **pancreas** (pan'krē əs), **liver,** and **intestinal glands** complete the digestion of food. Proteins are broken down into amino acids. Starches and complex sugars become simple sugars. Fats are changed to glycerol and fatty acids.

Liver and gall bladder

Pancreas and Liver

The pancreas and the liver are of special importance in digestion. The pancreas is made up almost entirely of digestive glands. It secretes pancreatic juice directly into the duodenum. Pancreatic juice, which contains several enzymes, functions to break down fats and continues to break down carbohydrates and proteins. The pancreas also releases a fluid that neutralizes the acidity of the chyme in the duodenum.

A primary function of the liver is to produce **bile.** Bile is a yellowish, bitter mixture of various compounds in water. Bile is released either directly into the duodenum or into the **gall bladder** for storage.

Bile *emulsifies,* or breaks up, fats. Since fats are not soluble in water, it is nearly impossible for enzymes to digest them. Bile breaks fats into tiny droplets. In this form, fats can be much more easily digested. Bile also contains a small amount of sodium bicarbonate. This helps to neutralize the acids in gastric juice. In addition, bile aids in the absorption of vitamins A, D, E, and K, which are soluble in fats.

Up until this point, most foods you have eaten have been useless to the body. Even though digestion begins in the mouth, food is not ready to be absorbed and carried to the cells until it reaches the lower portion of the small intestine. You can review in the table below the chemical actions that take place to digest food.

Summary of Digestive Process

Organ	Source	Secretion	Action
Mouth	Salivary glands	Salivary amylase	Begins starch digestion
Stomach	Gastric glands	Pepsin	Begins protein digestion
	Gastric glands	Lipase	Begins fat digestion
Small intestine	Liver	Bile	Emulsifies fats so they can be digested
	Pancreas	Trypsin	Continues protein digestion
		Amylase	Continues starch digestion
		Lipase	Completes fat digestion
	Intestinal glands	Peptidases	Complete protein digestion
		Lactase, maltase, sucrase	Complete starch digestion

Absorption

Digested food is pushed by peristalsis to the lower portion of the small intestine. From here it enters the bloodstream and lymph system and is carried to all body cells. This process is called **absorption.**

The job of absorbing all the food one consumes is staggering. Think about everything you eat and drink in one day. About four to six hours after you finish one meal the stomach is empty. It collapses like a balloon with a slow leak. Then the muscles of the stomach lining begin to contract. You feel hungry again. Meanwhile, the food you ate hours ago has been pushed along into the small intestine where it undergoes further digestion. About 90 percent of the food will then be absorbed by structures in the small intestine.

The lining of the small intestine is covered with millions of tiny fingerlike projections called **villi** (vil'ī). The effect is one of a soft, thick carpet. Villi contain blood and lymph vessels. Most of the digested food enters the blood vessels of the villi. The larger fat particles may pass directly into the lymph system through the villi. The many projections greatly increase the surface area of the intestine. This allows absorption to occur easily. Once absorbed, nutrients are carried to cells where they will be used for the production of energy and body tissues. The structure of villi is shown in the margin.

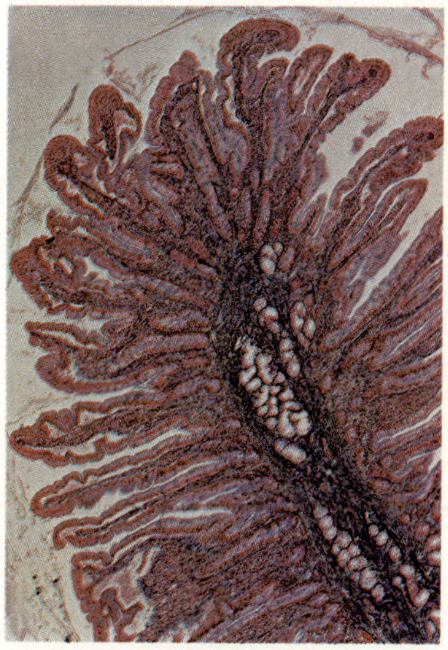

Villi lining the small intestine

Large Intestine

Material that is not absorbed leaves the small intestine and enters the **large intestine,** or **colon** (kō'lən). This material consists of water and food and digestive wastes, such as cellulose and inorganic salts, which the body can neither digest nor use. In the large intestine, most of the water is reabsorbed. This means it is taken into the body through the wall of the intestine. Eventually, the water is eliminated from the body as *urine.* Solid wastes that remain after the water has been removed are called **feces** (fē'sēz). Periodically, feces are moved by peristalsis into the **rectum** and eliminated through the **anus.**

The whole process of digestion is diagramed for you on the next page.

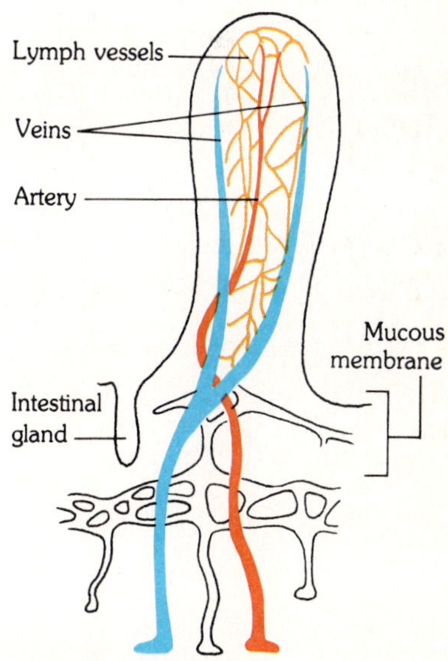

Diagram of a villus

Review It

1. What mechanical actions take place during digestion?
2. What chemical actions take place during digestion?
3. Where are carbohydrates, fats, and proteins digested?
4. How do villi aid in absorption of digested food?

Summary of Human Digestion

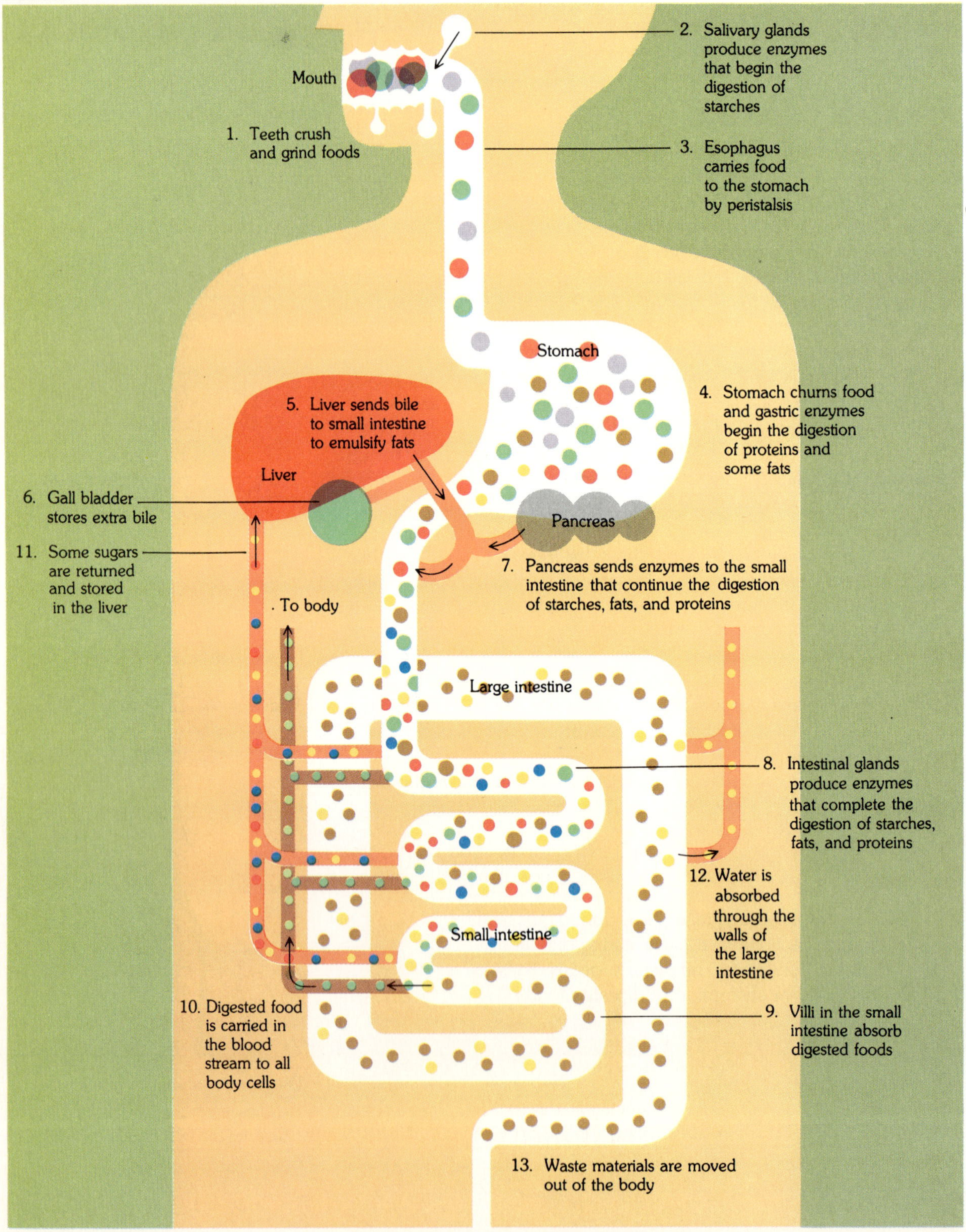

Activity

Studying Digestive Enzymes

Materials
Artificial gastric fluid
Artificial pancreatic fluid
Hard-boiled egg white
Cracker
Butter or lard
6 test tubes
Marking pencil
Test-tube rack

A. Procedure
Gastric fluid contains hydrochloric acid and the enzyme pepsin. You are going to test the effects of artificial gastric fluid on different foods.

1. Number three test tubes and put them in a rack. Put some pieces of cracker (carbohydrate) in test tube 1, a small piece of egg white (protein) in test tube 2, and a small piece of butter or lard in test tube 3.
2. Fill each test tube 1/3 full of artificial gastric fluid.
3. Allow the test tubes to stand overnight in a warm place. Then check to see how much digestion has occurred by observing the cloudiness of each solution.
4. Make a table like the one on this page. Mark a (+) in the table if any digestion has occurred. Mark a (−) if no digestion has occurred.

Analysis
1. Which substance(s) were digested by the gastric fluid?
2. What is the major type of digestion that occurs in the stomach? Give evidence to support your answer.

B. Procedure
Pancreatic fluid is secreted from the pancreas into the small intestine. This fluid contains several enzymes necessary to digestion. You are going to test and compare the effects of pancreatic fluid on those foods you tested in Part A.

1. Number three more test tubes and put them in a rack. Put a few pieces of cracker in tube 4, a piece of egg white in tube 5, and some butter or lard in tube 6.
2. Fill each tube 1/3 full of artificial pancreatic fluid.
3. Allow the tubes to stand in a warm place overnight. Check the amount of digestion that has occurred by noting the extent to which each food has been dissolved.
4. Mark down your observations in a table such as the one on this page.

Analysis
1. Which substances are digested by pancreatic fluid?
2. What kind(s) of foods are digested in the small intestine? Give evidence to support your answer.
3. Compare your results in Part A with those in Part B. What conclusions can you draw about digestion in the stomach and the small intestine?

Food Tested	Gastric Fluid	Pancreatic Fluid
Cracker (carbohydrate)		
Egg white (protein)		
Lard or butter (fat)		

Chapter Summary

The food you eat has three major purposes: (1) it provides the energy you need for all body functions, (2) it contains the basic molecules from which new body cells are made, and (3) it provides materials that help to regulate body functions. The study of how people take in and use food is known as nutrition.

The body needs six different kinds of substances, called nutrients, to maintain good nutrition. These are carbohydrates, fats, proteins, vitamins, minerals, and water. Each kind of nutrient is important. Failure to include enough of any one of these nutrients generally causes ill health. A severe lack of one or more can lead to malnutrition.

Food is changed into forms body cells can use through the process of digestion. Digestion involves both mechanical and chemical changes. It begins in the mouth and continues in the stomach. However, most digestion takes place in the small intestine.

As a result of digestion, each kind of nutrient is changed into simpler units by enzymes or other digestive juices. Complex carbohydrates become simpler sugars. Fats become glycerol and fatty acids. Proteins become amino acids. These nutrients are absorbed in the intestine and transported to body cells. Waste products of digestion are moved through the large intestine and out of the body.

Questions to Think About

1. Suppose you were asked to suggest a diet that would provide four people with good nutrition at the lowest possible cost for one week. What kinds of foods would you suggest?
2. Since the mouth is the beginning of both the digestive and respiratory systems, how are their functions kept separate? What happens when the "safety device" that separates the two functions doesn't work properly?
3. A person with stomach ulcers may have to have part of the stomach removed. What is an ulcer? How does the patient survive without a complete stomach?
4. Should the pancreas and the liver be thought of as parts of the digestive system? Why or why not?

Extra Research

1. Design a study that will help you find out whether the students in your class are eating nutritious meals. You may want to ask your health teacher or school nurse for help in getting the necessary materials.
2. Diets vary in different parts of the United States and in different parts of the world. Select three groups of people and find out what is a typical diet for each group. Some choices are: an Amish community, a vegetarian commune, or a community in Africa or Oceania. Do the diets provide all essential nutrients?
3. One of the standard tests for water pollution is to measure the amount of the intestinal bacteria, *E. coli,* present in water. What does this test show about the purity of water?
4. Milk is sometimes said to be a perfect food. However, some people cannot digest milk. Find out why this is so.
5. In recent years there has been a lot written about the production of concentrated food tablets. The intention is to produce a whole meal that can be held in the palm of your hand. Why would there be any need for this? What would be the advantages and disadvantages of producing food in this form?

Chapter Test

Part I

Copy the number of each test item. If the item is true, write *T* after the number; if the item is false, write *F*.

1. Many different kinds of foods can provide a balanced diet. (536)
2. The only form in which cells can make use of double sugars and starches is in the form of glucose. (538)
3. Animal fats are more unsaturated than fats and oils from plants. (540)
4. The most important function of proteins in the human body is to provide raw materials for body tissues. (541)
5. Vitamins are an important energy source. (542)
6. Fats are the only foods that are partly digested in the mouth. (546)
7. Most digestion takes place in the stomach. (547)
8. Peristalsis refers to all the chemical changes that take place during digestion. (546)

Part II

Copy the number of each sentence below. After each number, write the word or words that complete the sentence correctly.

9. _____, _____, and _____ are essential substances the body requires, even though none of them provides energy or body-building materials. (542)
10. Fats are broken down during digestion into _____ and _____. (540)
11. The body can use protein in foods only in the form of _____ _____. (541)
12. The muscular contractions that pass food through the esophagus and into the stomach is called _____. (546)
13. The major function of the liver in digestion is to produce _____, a substance that emulsifies fats. (548)
14. The _____ produces juices that help break down fats and continue the digestion of carbohydrates and proteins. (548)
15. The small intestine is lined with _____, which aid in absorption. (549)

Part III

Copy each item from List A. After each item, write the letter and words from List B that best describe it.

List A

16. elimination (549)
17. gall bladder (548)
18. nutrition (536)
19. malnutrition (536)
20. absorption (549)
21. polysaccharides (538)
22. Calorie (539)
23. amino acids (541)
24. salivary amylase (546)
25. pepsin (547)

List B

a. condition in which a person does not receive one or more essential nutrients
b. digestive enzyme that begins the digestion of proteins
c. transfer of digested food to bloodstream and lymph system
d. study of how people consume and use food
e. product of protein digestion
f. complex starches and cellulose
g. removal of wastes from the body
h. storage area for bile
i. unit of measure for determining energy value of food
j. digestive enzyme that begins the digestion of starches

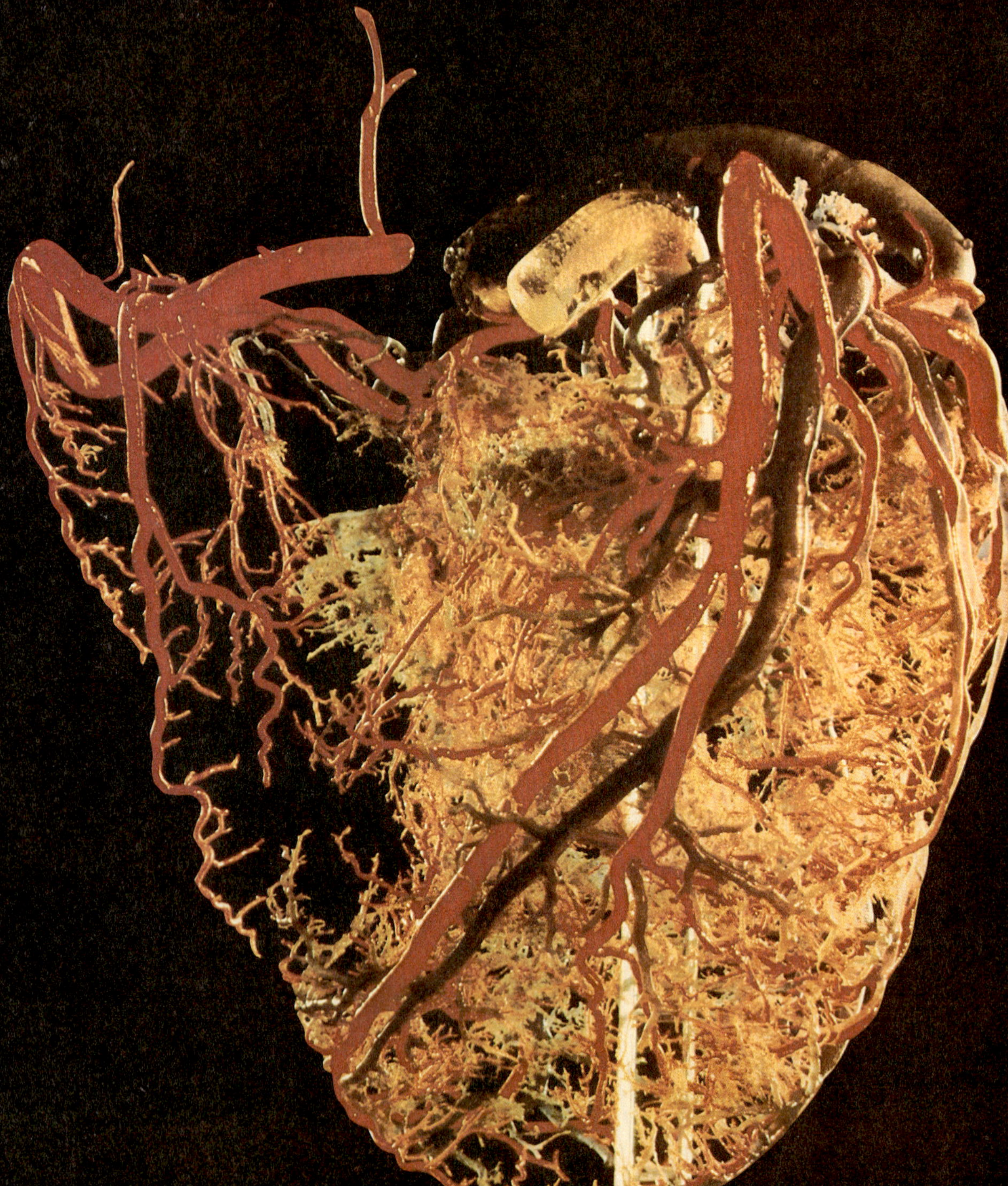

Chapter 23

Transport and Body Defense

Digested food, oxygen, metabolic wastes, and other materials essential to cell life are transported within the body by the *circulatory* and *lymphatic systems.* These systems work separately and also together to perform the vital functions of transport and defense in the human body. The *circulatory system* consists of the heart, arteries, veins, and the blood. The lymphatic system is composed of a network of tubes and a liquid which carries some digested materials and special disease-fighting cells.

The human body has several complex ways of defending itself against disease. Protective structures and special cells guard the body against certain diseases. Exposure to some diseases causes an immune response in the body that may provide protection against those diseases for many years. Artificial methods of producing immunity have also been discovered.

Chapter Objectives:

1. Identify and define the functions of the blood.
2. Describe the flow of blood through the body.
3. Identify and define the composition of blood and lymph.
4. Describe the functioning of the heart.
5. Identify and define the methods the body uses to protect itself against infection and disease.
6. Describe the development of immunity to disease.
7. Describe the function of vaccination in the development of immunity.

This is a model of the heart, showing the blood vessels.

23–1
Blood: The Liquid Tissue of Life

Millions of years ago all organisms lived in the sea. The food and oxygen they needed were dissolved in the seawater that surrounded them. Through the process of diffusion and active transport, food and oxygen entered the organism directly through the cell membrane. Wastes left the organism in the same manner.

Human cells take in food and oxygen and eliminate wastes in a similar way. However, in the body, blood is the fluid that carries these materials. The functions of blood in the body are complex and varied.

Functions of Blood

In the human body, blood functions in the following ways:

1. To carry partially digested food material from the small intestine and oxygen from the lungs to the cells.
2. To transport the waste products of metabolism from the cells to the kidneys, lungs, and skin for elimination from the body.
3. To carry hormones from the endocrine glands to the cells they control.
4. To control the amount of water and the level of acidity inside the cells.
5. To provide antibodies and white cells that protect the body against infection.
6. To provide substances that will clot blood if blood vessels are damaged.
7. To carry enzymes throughout the body.
8. To distribute heat produced during cell metabolism.

Did You Know?
The rapid spinning of a centrifuge can separate blood particles from the liquid plasma. After centrifuging, the heavier particles will settle to the bottom of the test tube. Plasma will remain on top.

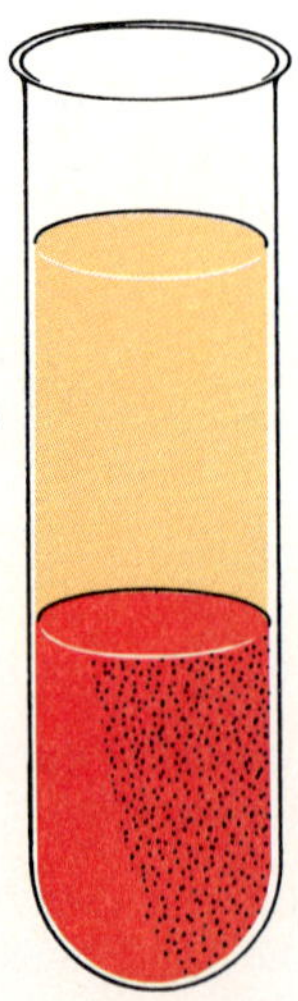

The Composition of Blood

Whole blood contains cells suspended in a liquid. These cells—red cells, white cells, and platelets—make up 45 percent of the blood. The remaining 55 percent of blood is the liquid portion called the **plasma.**

When plasma is separated from the cells, it is a straw-colored fluid. Plasma is between 91 and 92 percent water. Proteins, such as albumin, globulin, and fibrin, make up between 7 and 8 percent of the plasma. Very small amounts of fats, sugars, and mineral salts are also contained in the plasma.

Blood Cells

The cell portion of blood contains a mixture of red blood cells called **red corpuscles** (kôr'pus'əlz) or **erythrocytes** (i rith'rō sīts), white blood cells called **white corpuscles** or **leukocytes** (lü'kə sīts), and particles called **platelets** (plāt'lits) or **thrombocytes** (throm'bə sīts).

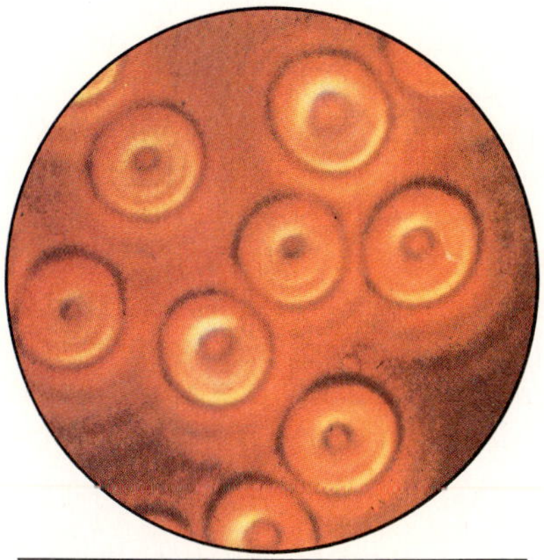

Red Blood Cells

Red blood cells look like red disks or saucers with pale centers. Red blood cells are in fact not cells, but corpuscles because they do not contain nuclei that are necessary for cell division. About one-third of a red blood corpuscle is made up of **hemoglobin** (hē'mə glō' bən), an iron-rich protein, that combines with oxygen as the blood passes through the lungs.

Produced in the red marrow of bones, red blood cells live from 100 to 120 days. Old red blood cells are removed by the liver and the spleen.

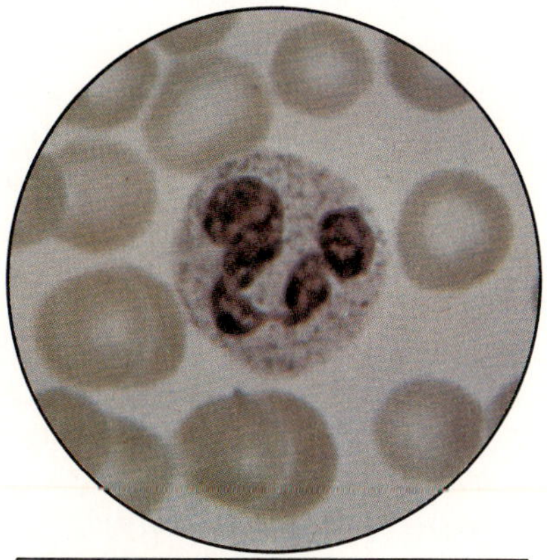

White Blood Cells

White blood cells have well-formed centers or nuclei and are really colorless. They may be seen, however, if they are stained. White blood cells are larger than red blood cells. There is approximately one white cell to every 600 red cells.

White cells are important agents in defending the body against disease. The ability of white cells to move and surround solid particles enables them to attack and destroy bacteria by the process of phagocytosis.

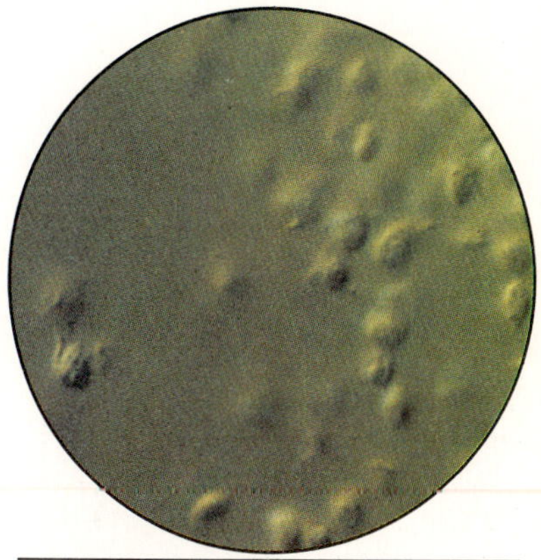

Platelets

Blood platelets are small, granular fragments of cells also thought to be formed in the red marrow of bones. Floating in the bloodstream with numbers far less than the red blood cells, platelets perform a vital function in tissue repair by aiding in the formation of blood clots.

The Formation of a Blood Clot

When blood vessels are injured, a sequence of reactions, both physical and chemical occurs rapidly to prevent major losses of blood.

Blood platelets contain a substance called **thromboplastin** (throm'bō plas'tən) which is released as the platelets rupture against the jagged edge of the broken vessel. Thromboplastin, with the help of calcium, reacts with a blood protein called **prothrombin** to form **thrombin.** Thrombin, in turn, reacts with another blood protein called **fibrinogen** (fī brin'ə jen) to form **fibrin.** Fibrin forms long, thin strands of material that form a net over the wound trapping red blood cells. A *clot* or scab is quickly formed and prevents further blood loss.

Blood clotting process

Thromboplastin (platelet) + **Prothrombin** (plasma protein)

↓

Thrombin

\+

Fibrinogen (plasma protein)

↓

Fibrin

Bone marrow

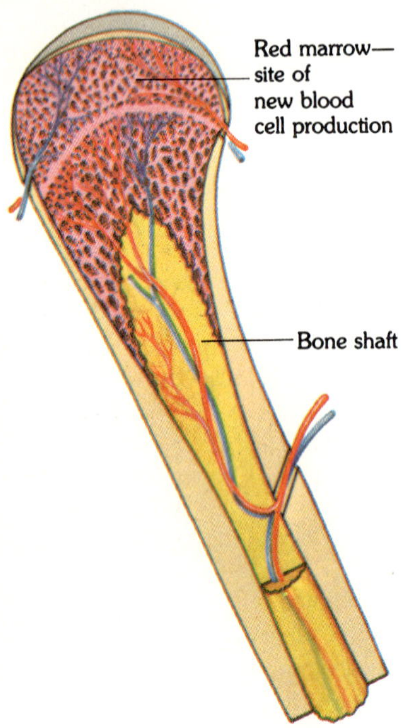

Blood Formation

Most blood cells are formed in the soft center portion or **marrow** of the long bones. Special **stem cells** in the marrow are the parent cells of both red and white cells. Stem cells change gradually into either red or white cells which then enter the circulatory system. Early in their development red blood cells lose their nuclei. The hemoglobin protein is added to red cells as they reach maturity. The supply of red cells in the blood increases during periods of strenuous muscular exercise, emotional stress, at high altitudes or high temperatures.

White cells are produced in the lymphoid tissues of the body and from parent stem cells in bone marrow. At times when the body is in great need of white cells for fighting disease and infection, white cells are produced rapidly in large numbers.

Platelets are produced in the bone marrow also. Unlike red and white cells, they do not develop from parent stem cells. They are formed from tiny pieces of **cytoplasm** (sī′tə plaz′əm) that break off from stem cells. Platelets have a short life-span of about 10 days.

Blood Types and Factors

In Chapter 7 you learned that blood may be identified chemically as belonging to four main groups—A, B, AB, and O. Blood types are determined by multiple alleles and remain constant throughout life. Red blood cells also contain a second pair of chemical substances known as M and N. These are inherited separately from the A and B pair. Blood may be identified by using the A, B and M, N substances.

The presence or absence of a blood protein, the Rh factor, is another way of identifying blood. The Rh stands for **rhesus** (rē′səs) after the rhesus monkey in which it was first identified. People who have this blood protein are Rh positive (Rh+), and those people who do not have it are Rh negative (Rh−). Rh− blood will form protective antibodies against Rh+ blood cells. For example, the blood of an Rh+ baby whose mother is Rh− is in danger of severe clotting because of the antibodies produced by the mother against the Rh+ blood cells of the baby. A potential problem exists only if the mother is Rh− and the baby is Rh+.

Did You Know?
About 85 percent of the white population is Rh+. The percentages of Rh+ and Rh− are not the same for all of the ethnic groups. For example, the Rh− factor is almost never found in Oriental peoples.

Review It

1. List the eight functions of blood in the human body.
2. Describe and compare red blood cells, white blood cells, and platelets.
3. What is the composition of plasma?
4. How are red blood cells, white blood cells, and platelets formed in the body?

Challenge!
Transfusions of whole blood, plasma, and plasma fractions are extremely important in saving lives. "Blood banks" have been established to store blood until it is needed. Find out about how blood is donated, preserved, and used. What fractions of blood have been isolated? How are these fractions specifically used?

23–2
The Circulatory System

Blood is transported through the body by a pump, the heart, and a connecting pathway of tubelike blood vessels. Blood vessels that transport blood away from the heart are **arteries** (är′tər ēz); blood vessels that transport blood toward the heart are **veins** (vānz). **Capillaries** (kap′o ler′ēz) serve as links or connectors between the arteries and veins.

Blood flow

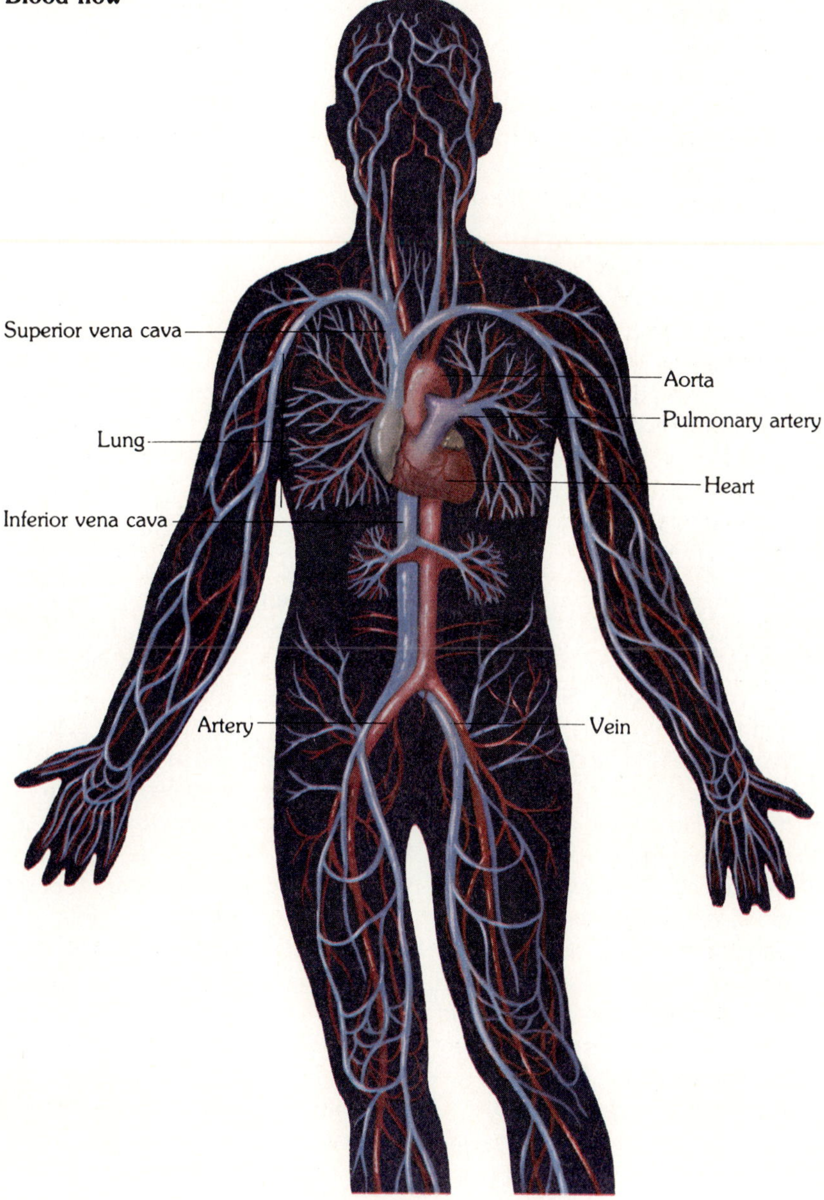

Pathway of Blood

From the heart, blood is pumped into the **aorta** (ā ôr′tə), the largest blood vessel in the body. The aorta distributes the blood throughout the body. The aorta branches into arteries which in turn branch into the smallest arteries, the **arterioles** (är tir′ē ōlz). From the arterioles the blood flows into the smallest vessels, the capillaries. The capillaries carry the blood to the individual cells of the body where oxygen and food are delivered and metabolic waste products are collected for disposal. Capillaries then connect with **venules** (ven′yülz) which run into veins. The veins in turn flow into the venae cavae that return the blood to the heart. The entire cycle of the blood through the body has taken about 20 seconds.

Capillary bed

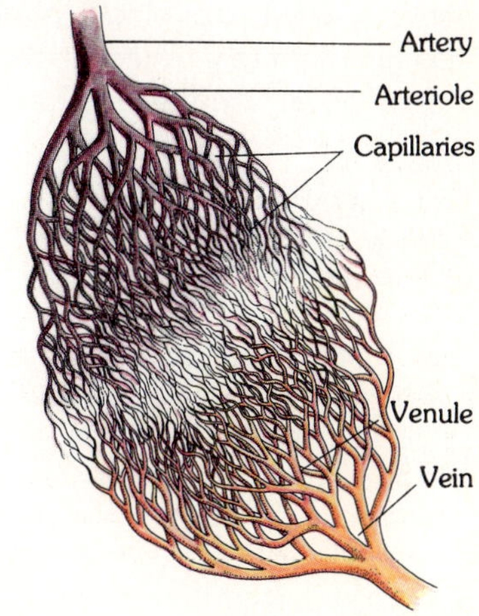

Blood Vessels

Each type of blood vessel: artery, vein and capillary, has a particular structure well suited to its specific function. The illustrations below show the structure of each.

Structure of an artery

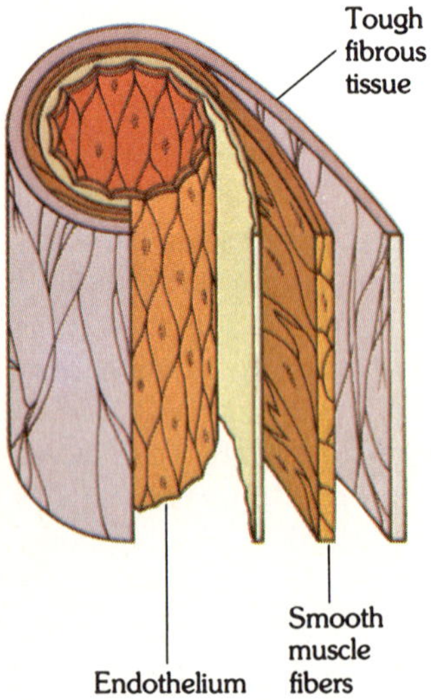

Structure of a vein

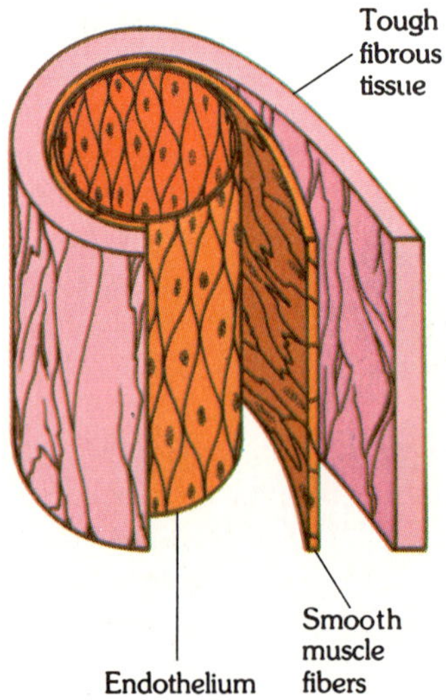

Structure of a capillary

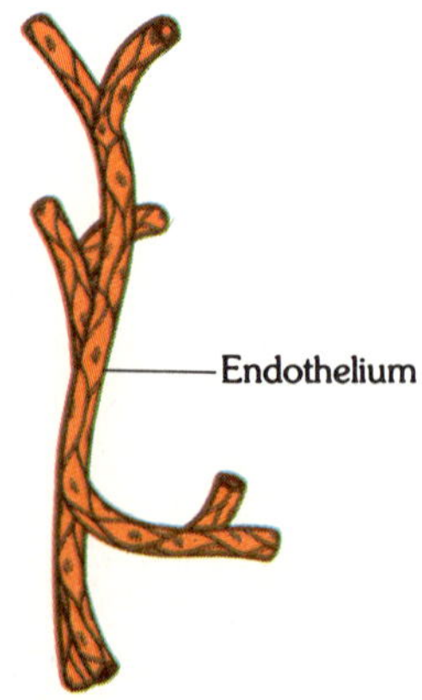

Arteries

The walls of arteries are made of three layers of tissue. The outer layer is made up of a tough, fibrous tissue that is not very elastic. This outer layer limits how much the artery can stretch. The middle layer of the artery consists of elastic smooth muscle fibers. This layer of muscle expands as a surge of blood comes through the artery. The smooth muscle fibers also prevent the artery from collapsing. The innermost layer is a thin, smooth sheet of cells, the **endothelium** (en′dō thē′lē əm).

Veins

The walls of veins are made up of the same three tissue layers as the arteries. However, since the veins do not have to withstand the same surges of pressure as the arteries, vein walls are much thinner. The middle layer of a vein has less smooth muscle than in an artery and a layer of endothelium lines all veins. The outer layer of tough, fibrous tissue is well developed in veins. As blood returns to the heart through the veins it is no longer under great pressure from the contractions of the heart. Blood flow in the veins is assisted by the contractions of muscles around the veins. One-way valves in the larger veins prevent the blood from backing up as it moves through the veins.

Capillaries

The walls of the capillaries are made up of a single layer of cells, the endothelium. This is the same layer of endothelium that lines the heart, arteries, and veins. Capillaries are very small—actually only one cell thick. Every cell in the body is next to or very close to a capillary. This eases the exchange of materials between the cell and the blood in the capillary. Food, oxygen, hormones, enzymes, and other substances pass through the thin capillary walls into the tissue spaces. White blood cells pass through the cells of the capillary walls. Waste products from cell metabolism leave the tissues and enter the blood in the capillaries.

Regulation of Blood Flow into the Capillaries

There is not enough blood to fill all of the body's capillaries at the same time, so the amount of blood in the capillaries must be controlled. Tiny "gate-keeper" muscles in the walls of the arterioles that open into the capillaries regulate the flow of blood. During strenuous exercise, for example, a great deal of blood is needed by the capillaries of the muscular system. The arterioles of the muscles open to allow blood to flow into the capillaries. At the same time very little blood is let into the capillaries of the digestive system.

Blood Pressure

Blood in the arteries is under pressure from the contractions of the ventricles of the heart. The pressure of the blood at the time the ventricles contract is referred to as **systolic** (si stol′ik) **pressure.** During the resting stage of the heart's rhythm the pressure drops. This lower pressure is the **diastolic** (dī′ə stol′ik) **pressure.** Information concerning the health of the heart and circulatory system can be gained by measuring the difference between the systolic pressure and the diastolic pressure.

To measure blood pressure, a device called a **sphygmomanometer** (sfig′mō mə nom′ə tər) is used. An airtight rubber cuff is wrapped around the arm. The cuff is connected to a glass tube of mercury or pressure meter. Air is pumped into the cuff until it stops the flow of blood into the artery. As the air is gradually let out of the cuff, just enough blood spurts into the artery to make a sound that can be heard through a stethoscope placed on the arm. The pressure reading when the first sound is heard is recorded as the systolic pressure. More air is gradually let out of the cuff. Finally, as the pressure in the rubber cuff is almost gone, the sounds will stop. The point just before the sound stops is recorded as the diastolic pressure.

A reading of 120/70 means that the person's systolic pressure is 120 millimeters of mercury and the diastolic pressure is 70 millimeters of mercury. A typical reading for a healthy adult is 120/70.

Hypertension is the term used to indicate that a person's blood pressure is higher than what would normally be expected for the person's age, height, and weight. Hypertension may result from the effects of age, disease, weight gain, and diet.

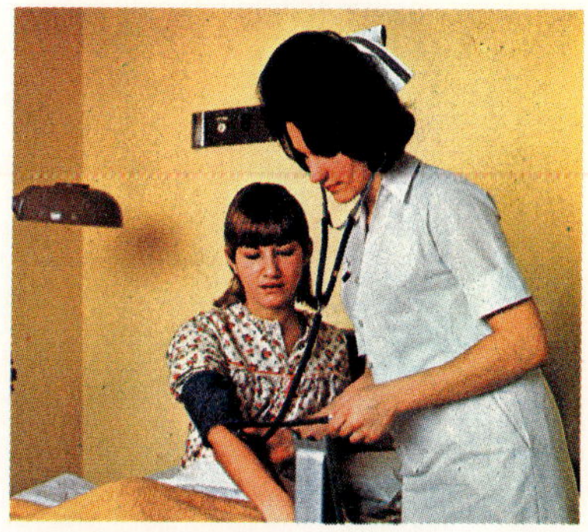

Measuring blood pressure

Did You Know?
William Harvey, a 17th-century English physician, studied the heart and circulatory system of many animals. Through his observations he concluded that the heart was a pump that recirculated the blood through the body. He was unable to prove his theory because he could not see the capillaries, the links between arteries and veins. Harvey's ideas were not generally accepted in his own lifetime because they contradicted the medical beliefs of the time.

An Italian physician, Marcello Malpighi, using a new instrument, the microscope, proved Harvey's theory of circulation when he identified capillaries.

Review It

1. Trace the flow of blood from the heart, through the body, and back to the heart.
2. Describe the structure of the walls of arteries and veins.
3. What is the role of the capillaries in circulation?
4. What do the figures of a blood pressure reading of 118/92 represent?

23–3
The Heart

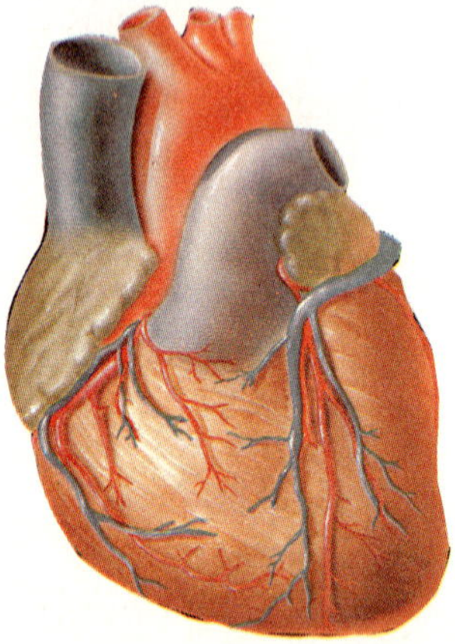

The human heart is a fist-sized organ located along the mid-center line of the body. The heart is composed of cardiac muscle that is held together by thick connective tissue. The **pericardium** (per′ə kär′dē əm), a tough protective sac, surrounds the heart.

The heart has four chambers: the left and right **atria** (ā′trē ə) and the left and right **ventricles** (ven′trə kəlz). A thick wall of muscle, the **septum** (sep′təm), separates the right and left sides of the heart. A thin smooth sheet of endothelium lines the heart. The atria, as the upper chambers of the heart, have thin muscle walls. Their function is to act as receiving chambers for blood returning to the heart. The atria move the blood into the ventricles. The lower chambers of the heart, the ventricles, have thick muscular walls. Their function is to pump the blood away from the heart to the lungs and the body.

Structure of the heart

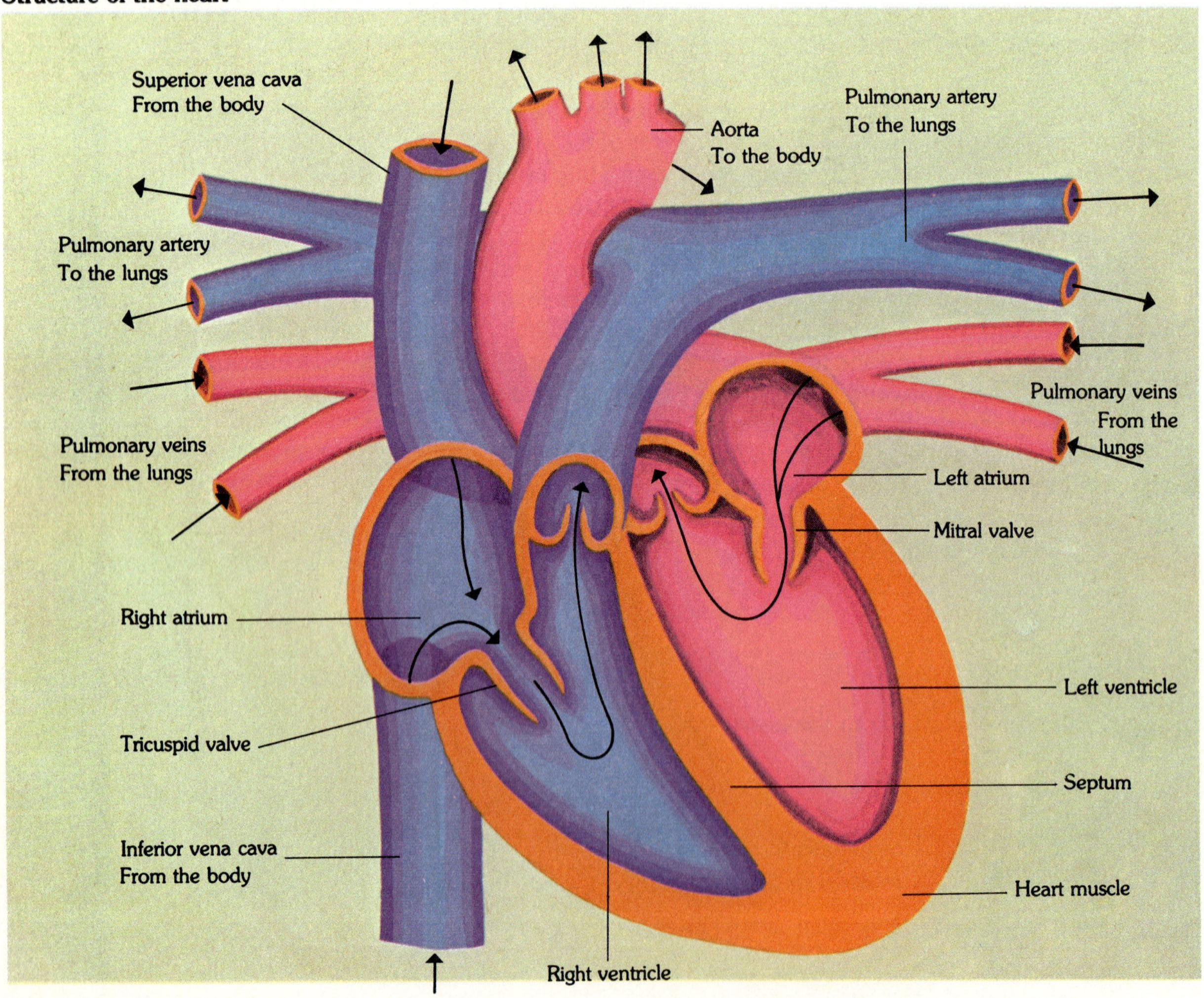

Blood Flow Through the Heart

Blood that has been circulated through the body has lost much of the oxygen that it carried. This blood is returned to the heart by the **venae cavae** (vē'nə kā'və), large veins that enter the right atrium. The right atrium contracts, forcing the blood into the right ventricle. As the right ventricle contracts, the blood is pumped through the **pulmonary** (pul'-mə ner'ē) **artery** to the lungs.

As the blood flows into the lungs it circulates through beds of capillaries that line the microscopic air sacs. In these sacs the red blood cells exchange the carbon dioxide from the body's cells for oxygen. This freshly *oxygenated* blood is returned to the left atrium by the pulmonary vein. This pathway of the blood from the heart to the lungs and back to the heart is the **pulmonary system** of circulation.

From the left atrium the blood is passed into the left ventricle. The left ventricle pumps the blood into the aorta from which it is distributed throughout the body. The pathway of the blood from the heart to the various systems of the body and back to the heart is the **systemic** (si stem'ik) **system** of circulation.

Pulmonary circulation

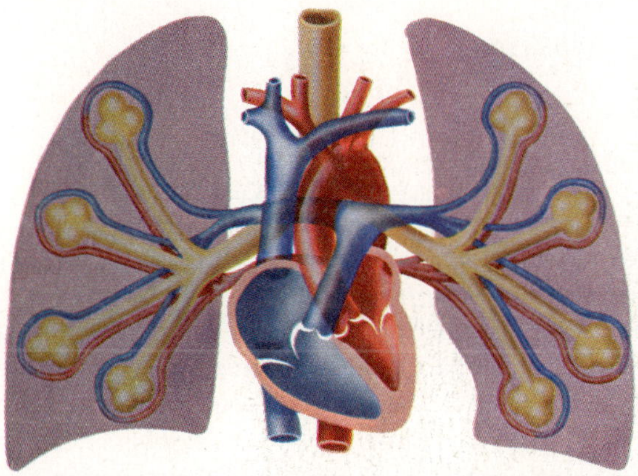

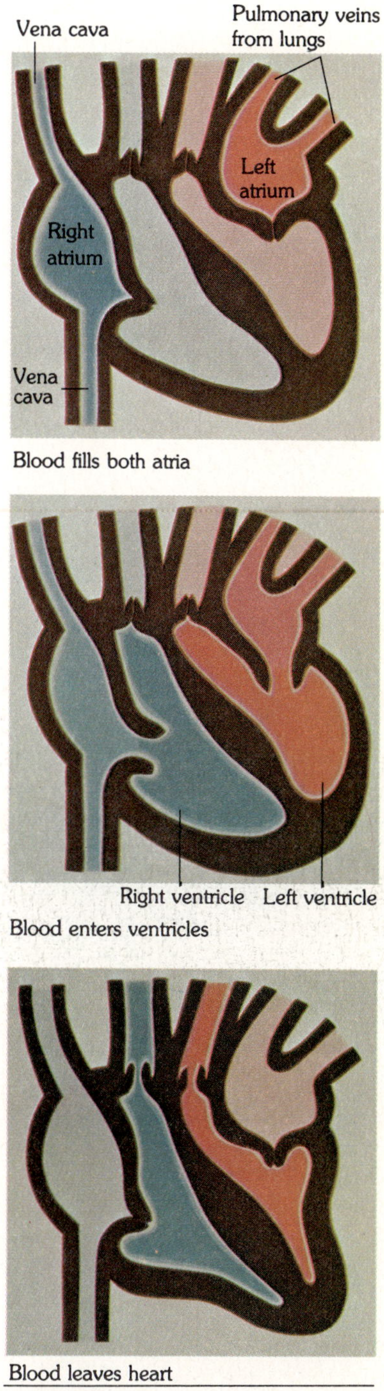

Pumping cycle of the heart

Valves in the Heart

The blood always flows through the heart in one direction. This flow is controlled by valves between the atria and ventricles. When the ventricles begin to contract, these valves close. The closed valves prevent the blood from flowing back into the atria. There are also valves at the base of the aorta and the pulmonary artery. As the ventricles relax after contracting, these valves close to prevent blood in the aorta and pulmonary artery from flowing backward into the ventricles.

The familiar "lub-dup" sound of a heartbeat is made by the actions of these valves. The "lub" is caused by the shutting of the valves between the atria and the ventricles. At almost the same time the valves at the base of the aorta and pulmonary artery close, making the "dup" sound. There is a brief pause between the "dup" sound and the next "lub." During this pause the heart is at rest.

Location of the pacemaker

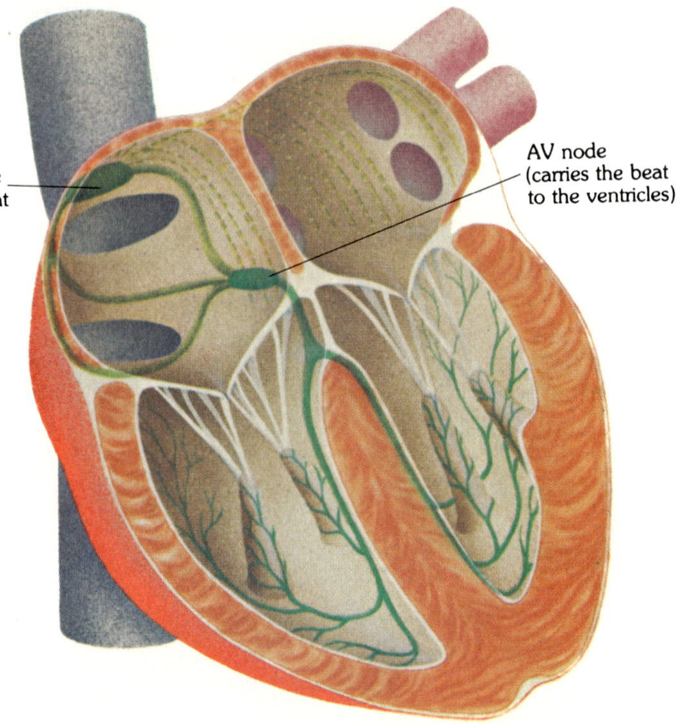

The Pacemaker

The heartbeat is controlled by a knot of tissue in the right atrium known as the **sinoatrial** (sī′nō ā′trē əl) **node** or *SA node.* The SA node is called the pacemaker because it produces an impulse that starts the heart's beat. This impulse spreads quickly over the entire atrial muscle. The impulse reaches a second node, the **atrioventricular** (ā′trē ō ven trik′yə lər) **node,** or *AV node,* which lies between the atria just above the ventricles. The AV node then passes the impulse to the ventricles which contract together.

Many people have pacemakers that do not work properly because of disease, birth defects, or the effects of aging. Artificial pacemakers have been developed that can regulate the heartbeat when a person's own pacemaker is unable to do so. These tiny devices are implanted in the chest to control the actions of the heart.

Review It

1. How do the atria and ventricles differ in structure and function?
2. What is the pulmonary system of circulation?
3. What is the systemic system of circulation?
4. What role do the valves in the heart play in circulating blood?

Activity

Observing Capillaries

Materials
Small goldfish or platy in an aquarium
Half of a Petri dish
Half microscope slides
Absorbent cotton pads
Compound microscope

Procedures

1. Remove the fish from the aquarium and wrap it loosely in wet absorbent cotton pads. Be sure to keep the gills wet.
2. Place the fish in a Petri dish. Its tail should be on a half slide.
3. Place another half slide over the tail.
4. Place the Petri dish on the stage of the microscope and focus on the tail.
5. Move the dish around until you find a part of the tail in which you can see the blood vessels clearly.
6. Look for a point where a small artery divides into capillaries. Notice the blood cells as they pass through the capillaries. **Remember to keep the cotton wet.**

Analysis

1. Does the blood move more rapidly in arteries or capillaries?
2. What is the advantage of red blood cells passing through the capillaries one at a time?
3. Why is it an advantage for the capillaries to have such thin walls?

23–4
Systems of Defense

The human body is equipped with defense mechanisms that protect it from organisms that might enter the body and cause disease or even death.

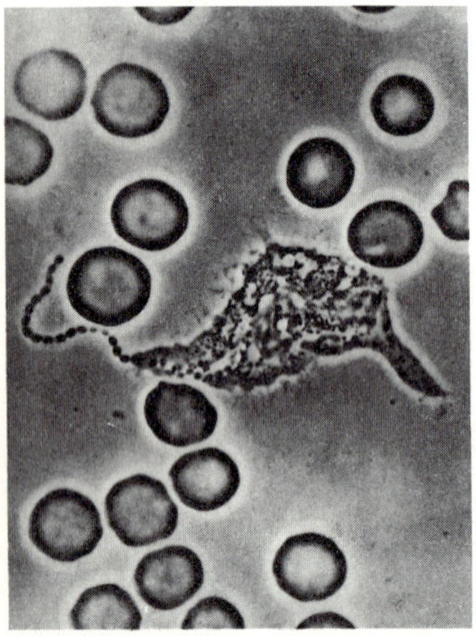

White blood cell attacking microorganisms

Protecting Against Infection

The skin is part of the body's primary defense system. Although the skin is only a half-centimeter thick, it is remarkably effective at keeping harmful microorganisms from entering the body. A cut or wound might provide a way of entry into the body. However, blood clots close these openings very quickly.

Skin does not provide protection for the eyes, ears, nose, mouth, or digestive, excretory, or genital passages. These areas are lined with a modified skin, the *mucous membrane,* or **mucosa** (myü kō′sə). The mucosa usually contains glands that secrete **mucus** (myü′kəs), a thick, slippery substance. Bacteria that reach the mucosa are often trapped in the mucus and then destroyed or eliminated.

Harmful organisms may also be met by special defense mechanisms. **Lachrymal** (lak′rə məl) **glands** in the eyes secrete tears that wash away dust and bacteria. Cilia, tiny hairlike structures in the respiratory system, trap bacteria and pass them back up the throat in wavelike motions. Most of these bacteria are then swallowed or expelled by coughing. Hairs in the nose trap dust, particles, and bacteria which are then expelled by sneezing. The flow of urine helps to prevent microorganisms from entering the urinary tract.

Most bacteria that enter the digestive system are destroyed by hydrochloric acid produced in digestion. The enzymes that digest foods may also digest bacteria. Any bacteria that survive the effects of acids and enzymes pass into the bloodstream and are carried to the liver. In the liver, special **phagocytic** (fag′ə sit′ik) cells ingest bacteria, viruses, and other particles.

The body's most reliable defense against bacteria, the white blood cells, are especially designed to kill bacteria everywhere in the body. Unwelcome bacteria may be carried to places in the body where white blood cells are especially plentiful such as the lymph nodes, **adenoids** (ad′n oidz), or **tonsils** (ton′səlz). White blood cells from throughout the body may swarm to the site of an infection to attack bacteria. When there are not enough existing white cells to hold back the progress of an infection, the body meets the emergency by manufacturing more white blood cells and releasing them into the blood. Even very young and immature white cells can be released as a last resort to help fight the disease.

The Lymphatic System

Some of the liquid portion of blood is continually passing through capillary walls to enter the spaces between body cells. Once outside the capillaries this fluid is called **lymph** (limf). The lymph contains dissolved oxygen, digested food particles, plasma proteins, and many white blood cells. Lymph is the clear, colorless liquid found in a blister.

Lymph bathes the cells, helping to bring dissolved oxygen and food to the cells that are not directly in contact with capillaries. It also helps to remove wastes from the cells. Lymph is returned to the circulatory system through irregular channels known as lymph vessels. The lymph vessels eventually connect with the large veins near the heart. The flow of lymph through these vessels is similar to the flow of blood through the veins. It depends on muscle contractions, breathing movements, and a system of valves to pass the lymph along and to prevent a backward flow.

Lymph vessels pass through clumps of specialized tissues called **lymph nodes** (limf nōdz). These lymph nodes are concentrated in the neck, armpit, elbow, and groin. Tonsils and adenoids are examples of lymph nodes. Many white blood cells are produced in the lymph nodes. Lymph is purified as it flows through the lymph nodes. Harmful bacteria in the lymph are destroyed by the white blood cells in the nodes before the lymph is returned to the bloodstream.

The lymph nodes may become swollen and tender when an infection is present in the body. This is caused by the increase in production of the white blood cells which are needed to fight the infection. When a person has a sore throat, the lymph nodes in the neck may become swollen. A major infection may cause all of the lymph nodes in the body to enlarge.

The **lymphatic** (lim fat′ik) **system,** consisting of the lymph, lymph nodes, and vessels, is an auxiliary system that provides for the return of fluid and proteins from tissue spaces to the circulatory system. There are no lymphatic arteries.

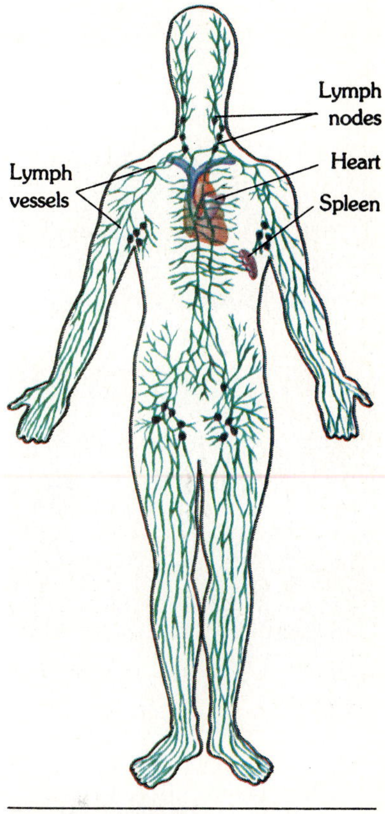

Lymphatic system

Blood and lymph in tissue

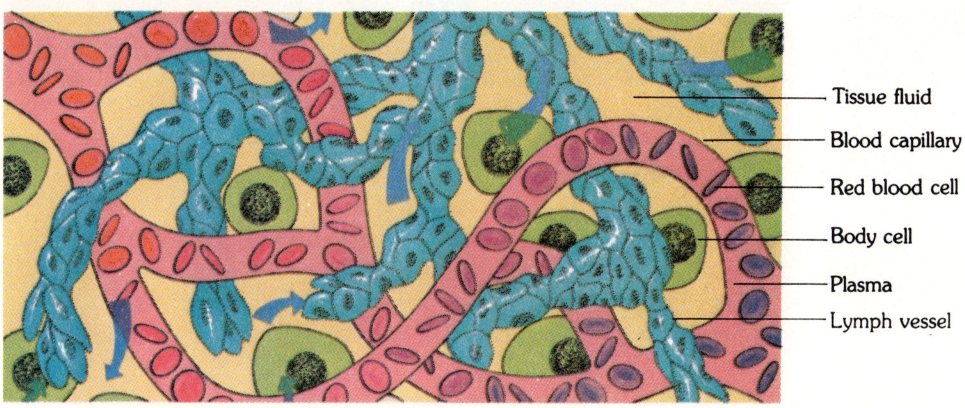

Drug Therapy

The body's own defense system against infection may not be enough to prevent disease. Long ago people learned that certain substances were effective against some diseases and could be used to help the body's natural defenses. Roots, leaves, bark, flowers, seeds, and other natural materials were originally used as **drugs** to treat disease.

Some microorganisms are natural enemies of disease-causing organisms. Substances produced by microorganisms that kill disease-causing organisms are **antibiotics** (an′ti bī ot′iks). Certain chemicals that kill disease-causing organisms have also been found. These chemicals can be produced in the laboratory. Thousands of chemical compounds have been identified for treating diseases.

Dr. Alexander Fleming discovered that substances released by the blue-green mold *Penicillium* stopped the growth of the *Staphylococcus* bacteria. Today dozens of similar antibiotics have been identified and are used to treat infection and disease.

The first chemical compound used to treat disease was Salversan, an organic compound containing arsenic. It was developed by Paul Erlich in 1910. Salversan has been especially effective against the microbe that causes the venereal disease syphilis.

Antibiotics and chemicals work by killing the disease-causing organism directly or by preventing the growth of bacteria so that no new bacterial cells are available to replace those that die naturally.

Untreated staphylococcus aureus culture

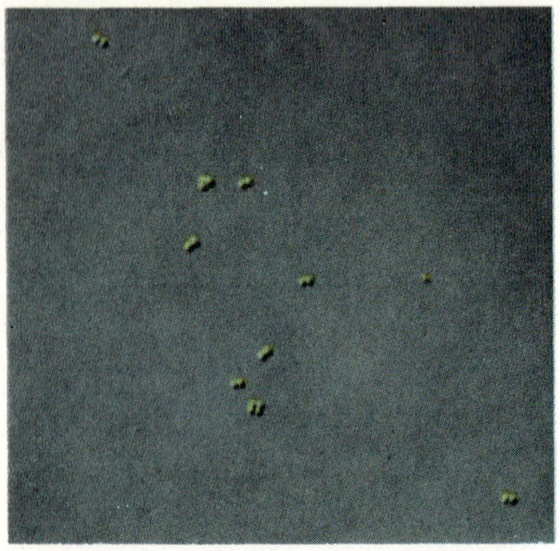

0 hours

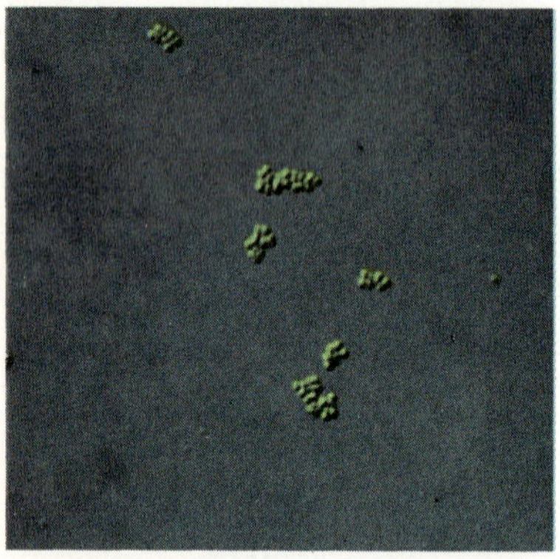

1 hour

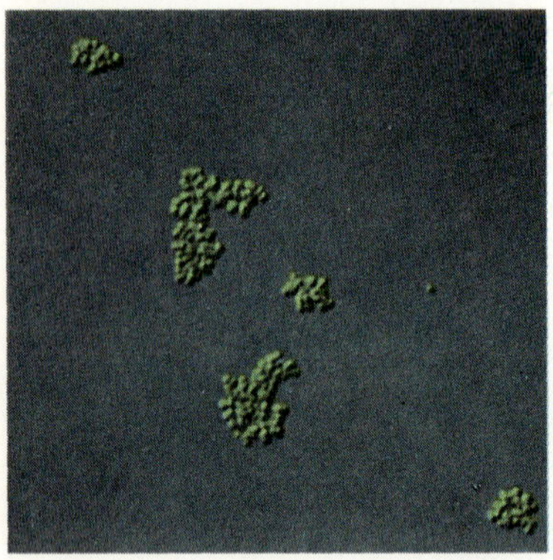

2 hours

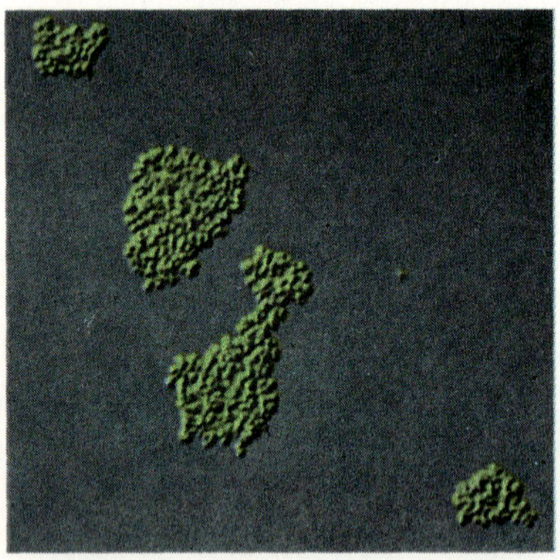

3 hours

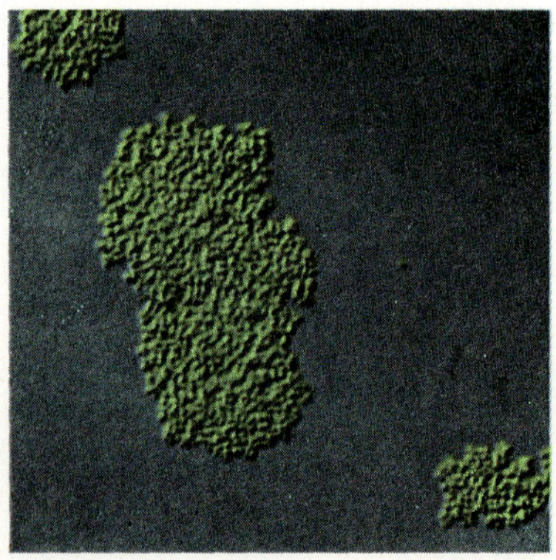

4 hours

Review It

1. What role do the mucous membranes play in defending the body?
2. How are bacteria destroyed in the digestive system?
3. Describe lymph and how it functions to protect the body.
4. Why are white blood cells the body's most reliable defense against harmful, invading bacteria?

23–5
Immune Responses

Certain organisms that enter the body automatically cause the body to produce chemicals to defend itself.

Antigens and Antibodies

Any substance that causes this type of response is known as an **antigen** (an′tə jən). Antigens may be bacteria, viruses, or **allergens** (al′ər-jənz), which are anything that causes an allergic reaction. Antigens cause the body to protect itself by producing substances called **antibodies.**

Antibodies are manufactured in the tissues of the lymphatic system, bone marrow, the spleen, and in the intestines. When an antigen enters the body, formation of antibodies begins throughout the lymphatic system. The antibodies rush to the site of infection and attack the invading antigens in various ways. Some antibodies may cause antigens to fall apart by destroying the cell membrane of the antigen. Other antibodies may soften the cell membrane, which allows protective blood proteins inside the antigen. Antigens may be prevented from doing damage by antibodies which attach themselves to the active area of the antigen. Groups of antibodies may clump together with antigens. This creates a particle that can be destroyed by white blood cells.

Each type of antigen causes the body to make antibodies specifically designed to combat that antigen. Antibodies produced to destroy measles will only destroy measles and are not effective against rabies, smallpox, typhoid, mumps, or any other disease antigen.

Edward Jenner

Challenge!
Edward Jenner (1749–1823) is credited with discovering a vaccine that could protect an individual against the serious disease smallpox. Smallpox may have been eliminated at last through a worldwide vaccination program sponsored by the World Health Organization, a branch of the United Nations.

Find out how Jenner first thought about a vaccine for smallpox. When and how was he first able to test its effectiveness?

The Production of Antibodies

The ability of the body to produce antibodies is a result of the ability of cells to recognize foreign substances. The body reacts to any material that is not genetically identical to the body cells. This ability to recognize and react against any foreign substance is helpful in protecting the body from disease.

When an antigen enters the body, almost a week passes before any antibodies are produced and appear in the blood. In the next week the number of antibodies slowly rises to a peak and then begins to decrease. Some antibodies remain in the blood after the original response. If the same antigen returns to the body, antibodies are produced very rapidly. Enough antibodies are usually produced to prevent any of the disease symptoms from appearing. More antibodies will remain in the blood after the second exposure of the body to the antigen.

Jonas Salk

Natural and Artificial Immunity

A person who has had a disease such as mumps, measles, or smallpox is unlikely to have that disease again. Antibodies that were produced during the first attack of the disease remain in the body for a long time. These antibodies may remain for a person's entire lifetime. If the same disease antigen enters the body at a later time, the antibodies needed to fight them are already present, and the person is protected from that disease. This protection is called **immunity** (i myü'nə tē). Immunity may occur naturally as a result of a disease or it may be stimulated artificially by a **vaccine** (vak'sēn'). A vaccine is a sample of an antigen that is too small to cause the disease but large enough to cause the body to produce antibodies. Vaccines have been developed for many diseases: polio, mumps, measles, tetanus, diphtheria, typhoid, and cholera. Young children are routinely **vaccinated** (vak'sə-nāt ed) against many diseases.

Vaccines may contain a very weak dose of a living antigen, or they may contain a dead, **detoxified** (dē tok'sə fīd) antigen. In this case the presence of the structure of the dead antigen is enough to cause antibody production. The Sabin polio vaccine is an example of a live antigen vaccine. The Salk polio vaccine is a detoxified antigen vaccine.

All of us have a natural immunity to certain diseases that affect other kinds of organisms. For example, human beings do not catch distemper from dogs or cats. People may also become immune to a disease or diseases during their lifetime. This type of immunity is *acquired immunity*. There are two types of acquired immunity: active and passive. Active acquired immunity results when a person has had a disease and becomes well. The body has produced antibodies that remain in the blood to prevent further infection by that particular disease antigen.

Passive acquired immunity is produced through the injection of antibodies formed in the blood of other human beings or animals. This type of immunity lasts for only a few weeks or months. It is used when a person has been exposed to a disease and there is not enough time to actively acquire immunity through vaccination. It is also used when the disease is so severe that the person is unlikely to survive the disease unless antibodies are given. Passive immunity provides immediate protection from the disease antigen.

Review It

1. Define antigens.
2. How do antibodies defend the body from antigens?
3. How does a person acquire natural immunity to a disease?
4. How does a vaccination prevent a person from getting the disease?

Pioneers in Biology

Charles Richard Drew: Pioneer in blood plasma research.

Imagine this scene: A patient is brought to a hospital—in shock and near death from blood loss. Some friends and relatives offer their blood for transfusion. Quickly, the doctors check and cross-check blood types for a donor that will match. They haven't much time. The patient's life is slipping away. Finally, an intern walks in who is known to have the blood type needed. The intern is rushed into the operating room and transfusion begins. They were lucky.

That's how it was in the 1930s when Charles Drew was studying and later practicing medicine. He was consumed with the thought that there *must* be a way to store blood for these emergencies.

Charles Drew was born on June 3, 1904. As he grew up, he was recognized as an able leader. Among his honors while at Amherst College were the National A.A.U. high-hurdle championship and the Mossman Trophy for the athlete bringing the greatest honor to the school.

It might have been due to the death of his sister during the 1918 influenza epidemic, or the weeks he spent in a hospital recovering from a football injury, but sometime during his senior year in college, Charles made up his mind to become a doctor. He would have to work for two

years before entering medical school at McGill University in Canada. There, his interest in blood research began.

After graduating from McGill in 1935, Charles taught pathology at Howard Medical School in Washington, D.C., and trained in surgery at Freedman's Hospital. In 1938, with a two-year fellowship from the Rockefeller Foundation for Columbia-Presbyterian Medical Center in New York, he was able to work in his own field of interest—blood research.

Through meticulous study of worldwide blood research, and careful testing of hypotheses, Dr. Drew formed an idea about blood plasma. These were the facts: 1) When whole blood is stored, the red cells break down, releasing toxic potassium. 2) Plasma contains everything that whole blood does—*except* the red cells. 3) Plasma does not deteriorate if it is prepared properly. His hypothesis: Plasma can be used for transfusion. It *can* be stored!

Dr. Drew tried out his idea and found it workable. It was none too soon. World War II was starting in Europe, and plasma was desperately needed on the battlefields. But there were problems. Contamination of blood supplies, inadequate collection, and poor organization plagued the efforts. Dr. Drew was eventually called to direct the plasma project for Great Britain. With his scientific background and ability to inspire cooperation from fellow workers, the project became a life-saving operation. Americans sent 17,000 pints of plasma overseas before the British were able, in 1941, to rely on their own blood banks. Dr. Drew was the natural choice to set up the National Blood Bank program with The American Red Cross, which he did with his usual competence.

Charles Drew had always been involved in paving the way for other black people in medicine. His standards were high. After resigning his position with the Red Cross, he returned to teaching and worked tirelessly with the surgeons he trained at Howard University Medical School. He wanted them to prove to themselves that they could excel—and they did.

In 1950, at age 45, Charles Drew died in an automobile accident. He left a legacy not only as a pioneer in blood research, but as a pioneer in the development of humanity.

Chapter Summary

The circulatory system transports blood throughout the body. The movement of blood allows for the transport of food, oxygen, wastes, heat, hormones, enzymes, and other materials used and produced by the body. The blood is composed of red and white blood cells and platelets in a thin, yellowish solution called plasma.

Blood leaves the heart by way of the aorta and travels through progressively smaller arteries and arterioles. Blood next passes through the capillaries where it is utilized by the cells. It then enters the venules and smaller veins, finally collecting in larger veins and returning to the heart. The heart itself is a double pump which collects deoxygenated blood from the body, sends it through the lungs by way of the pulmonary circulatory system, and pumps oxygenated blood back to the body by way of the systemic circulation.

The body protects itself against disease-causing organisms in many different ways. The skin is an effective barrier. Body structures provide defense against microorganisms that do manage to enter the body. These structures include hair, cilia, mucous membranes, and chemicals. White blood cells are specially adapted to capture and destroy harmful organisms.

Antigens cause the body to produce chemicals automatically to defend itself. Antibodies fight antigens at the time of initial infection and may also remain in the bloodstream for many years to provide long-term immunity. Immunity can also be provided by vaccination.

When the body's natural defense mechanisms fail and a disease develops, it can often be treated with drugs. Drugs may be naturally occurring products of plants and animals, antibiotic products of microorganisms, or chemical compounds that destroy bacteria, viruses, and other disease-causing agents.

Extra Research

1. The term anemia (ə nē′mē ə) refers to a group of diseases that are caused by the failure of the red blood cells to carry enough oxygen to other cells. Find out more about each of the following types of anemia: iron-deficiency, Thallasemia, pernicious, and sickle cell anemia.
2. Families in which the father and mother have different Rh blood types may face problems in having children. What does the term Rh factor mean? What is the problem that concerns such families? What recent discovery has lessened the problems faced by parents with differing Rh blood types?
3. Cardiovascular disease is a leading cause of death in the U.S. today. What kinds of medical problems are included in this category? What can be done to prevent cardiovascular disease?
4. Allergies are a familiar problem to many people. Find out how allergic reactions occur. How do antihistamines help in treating an allergic reaction? What other methods are used to treat an allergic reaction?

Questions to Think About

1. A person goes into shock when the tiny muscles in the arterioles that control circulation in the capillaries fail. What effects would this have on the body? What treatment is recommended for shock?
2. Coughing, sneezing, and crying are all involuntary reactions to the presence of some foreign substances in body openings. Explain what happens in each of these cases and how each is a defense against infection.
3. Antibodies are produced to fight against any foreign invader in the body. This poses special problems in the body after an organ transplant. The most successful organ transplants are those that take place between identical twins. Using your understanding of the immune reaction system of the body why do you think this is true?
4. Some people believe that the human circulatory system reflects our evolution from aquatic organisms. What reasons would there be for believing such a statement?

Chapter Test

Part I

Copy the number of each test item. If the item is *true*, write *T* after the number; if the item is *false*, write *F*.

1. Blood plasma consists primarily of red and white blood cells and platelets. (556) F
2. Blood normally contains about 600 times as many red blood cells as white blood cells. (557) T
3. Blood is bright red in arteries and blue in veins. (563) F
4. The largest vein in the body is the aorta. (559) F
5. The heart consists of four chambers: the right and left atria and the right and left ventricles. (562) T
6. In a blood pressure reading of 100/80, the two numbers give the pressure of blood in the arteries when it is under pressure and when it is not. (561) F
7. Lymph flows to the heart in a system very much like the systemic circulatory system. (567) F
8. Antibodies produced in response to smallpox antigens are only partially effective against the antigens that cause measles. (569) T
9. A vaccine against measles could be made from weakened antigens or from "dead" measles antigens. (570)

Part II

Copy the number of each sentence below. After each number, write the word or words that complete the sentence correctly.

10. Blood delivers food, oxygen, and _____ to cells. (556) hormones
11. Blood consists of _____ percent plasma and _____ percent solids. (556) 55
12. Red and white blood cells and platelets are manufactured in the _____. (558) Marrow
13. As red blood cells pass through the lungs, oxygen becomes chemically attached to a compound called _____. (557) hemoglobin
14. The major structures that veins have but arteries do not need are _____. (560) valves
15. The "lub-dup" sounds during a heart beat are caused by _____. (563) valves
16. The particles in blood that initiate the process of clotting are _____. (557) platelets
17. Microorganisms that enter the throat may be trapped by the _____. (566) cillia
18. The chemical in the digestive system that is most effective in destroying disease-causing microorganisms is _____. (566) hcl
19. Antibiotics are substances that are produced by _____. (568)

Part III

Copy each numbered item from List A. After each item, write the letter and words from List B that best describe it.

List A

20. lymphatic system (567) D
21. pulmonary system (563) B
22. immune system (570) C
23. circulatory system (559) F
24. systemic system (563) E
25. primary defense system (566) A

List B

a. provides antibodies for defense against disease
b. carries blood between heart and lungs
c. consists of skin, mucous membranes, and specialized structures
d. consists of nodes, tissue fluids, and white blood cells
e. carries blood to parts of the body and back to the heart
f. provides transport for blood, food, metabolic wastes, and other materials

White Blood Cells

General { Phagocytes, Macrophage, Granulocytes }

Lymphocytes { "B" - Bone, "T" - Thymus } Custom Specific

Phagocytes produce Interluchen I which raises body temperature and stimulates Appert cells
"T" cells produces Interluchen 2 which produce more "T" cells.
"B" cells tagg virus or antigen
Toxic "T" cell punctures

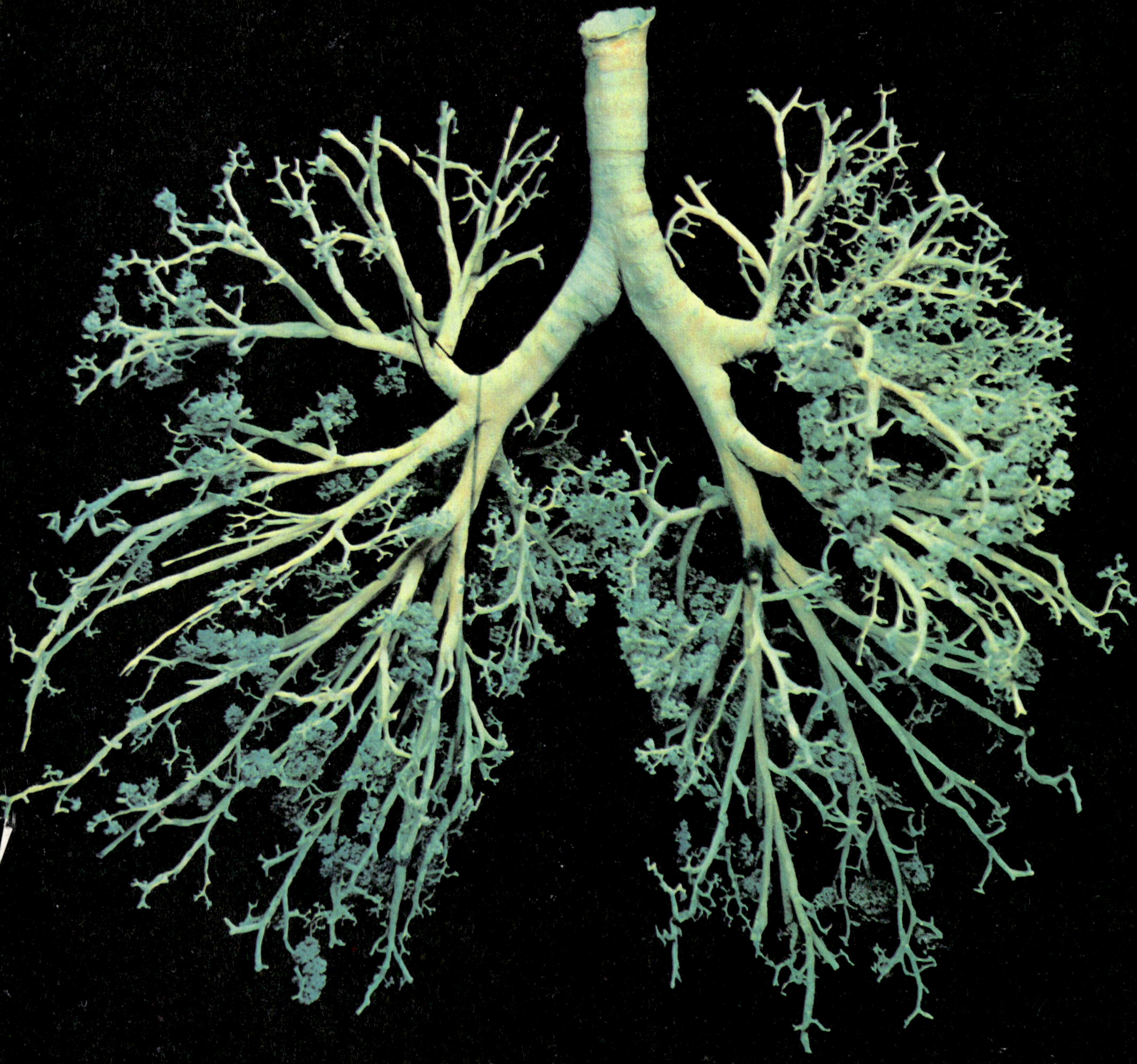

Chapter 24

Respiration and Excretion

Cells in your body need oxygen to release energy. When you inhale, oxygen enters your lungs. It is delivered to the cells by the bloodstream. Carbon dioxide that has been produced by the cells as a waste product is returned to the lungs by the bloodstream. It leaves the body when you exhale.

Cells also use food to release energy for all body functions. The chemical reactions involved in all of the body's functions are called *metabolism.* Metabolism produces wastes that must be removed from the body. These wastes are excreted by the lungs, kidneys, skin, and liver. The kidneys are important organs of excretion because they filter wastes from blood and maintain the proper level of water, salts, proteins, and other materials in the blood.

In this chapter, you will study the processes of respiration and excretion. These processes are vital to life and take place continuously.

Chapter Objectives

1. Describe the functions of the respiratory system.
2. Identify the organs of the respiratory system.
3. Describe the process of breathing.
4. Describe two kinds of cellular respiration.
5. Describe the functions of the kidneys.
6. Identify the structures of the kidney and their role in excretion.
7. Describe the role of the liver and of the skin in excretion.
8. Describe homeostasis in the processes of respiration and excretion.

This model of the lungs shows the bronchial tubes in detail.

24–1 Respiration

Respiration includes all of the processes involved in using oxygen to release energy within a cell. A study of respiration includes the inhaling and exhaling of air, the exchange of gases between cells and the environment surrounding them, and the combining of glucose and oxygen to produce energy with carbon dioxide and water as waste products.

Oxygen is available in the air around us. The **respiratory** (res′pər-ə tôr′ē) **system** is responsible for getting oxygen from the air to the cells and for removing carbon dioxide from the blood and eliminating it from the body.

Organs of Respiration

Air enters the body through the **nostrils** (nos′trəlz) or the mouth. Tiny hairs at the entrance of the nostrils prevent large unwanted particles from entering. Air passes into the **nasal** (nā′zəl) **cavity,** where it is filtered. The nasal cavity is covered with a mucous membrane and cilia. Mucus traps unwanted materials such as dust and bacteria. The cilia, in wavelike motions, push trapped particles toward the throat where they are swallowed or coughed out. A rich flow of blood through the mucous membrane also warms and moistens the incoming air. The cleaned, warmed, and moistened air is now ready to be sent to the delicate tissues of the lungs.

The **pharynx** (far′ingks) at the back of the nasal passage divides into two passages: one to the stomach and the other to the lungs. The **epiglottis** (ep′ə glot′is), a flaplike valve, protects the passage to the lungs when food is being swallowed. Air is next passed through the **larynx** (lar′ingks), a stiff box where the vocal chords are located. When leaving the lungs, passing air through the tightened vocal chords causes them to vibrate and produce the sounds of speech. Incoming air is not involved in speech.

Did You Know?
The suffix -itis *means inflammation of an organ. When you see the words* bronchitis, appendicitis, laryngitis, or tonsillitis *you know what part of the body is affected. Respiratory tissues are particularly vulnerable to infection.*

Air enters the lungs by passing through the *trachea,* or windpipe. The trachea has rings made of cartilage that give it shape and support. The lower end of the trachea divides into two tubes, the **bronchi** (brong′kī), which enter the lungs. Both the trachea and the bronchi are elastic and expand when air is inhaled. These structures contain cilia and a mucous membrane to further prevent foreign particles from entering airways.

One bronchus enters each saclike **lung,** where it subdivides into smaller and smaller vessels called **bronchioles** (brong′kē ōlz). The smallest bronchioles lead to millions of tiny air sacs, the **alveoli** (al vē′-ə lī). Each lung has approximately one hundred million of these complex air sacs. The alveoli are clustered together in bunches, giving the lungs the appearance of a sponge.

Respiratory system

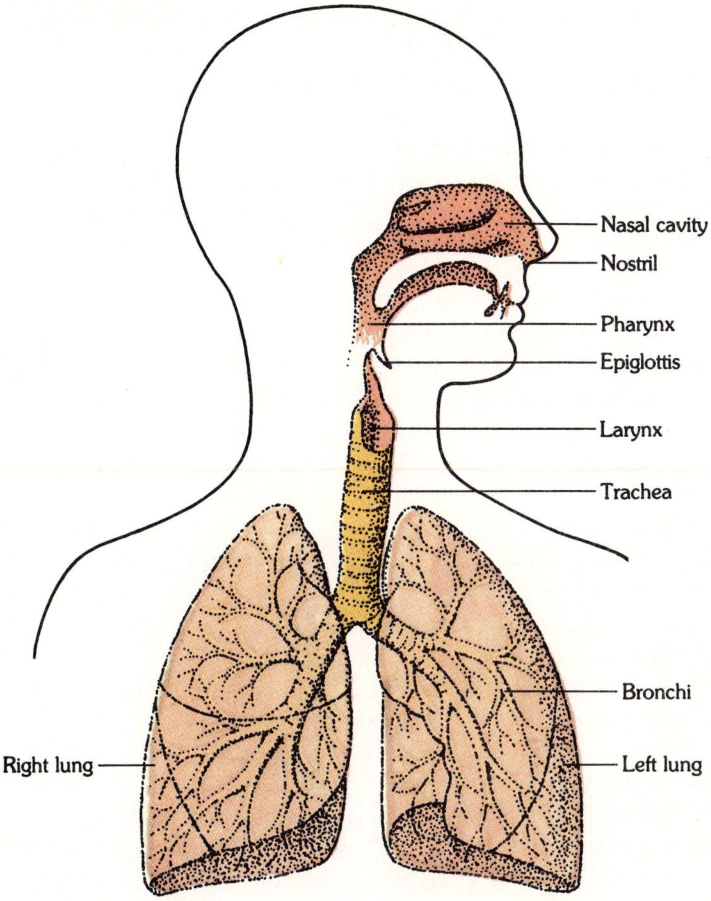

Alveoli and bronchioles

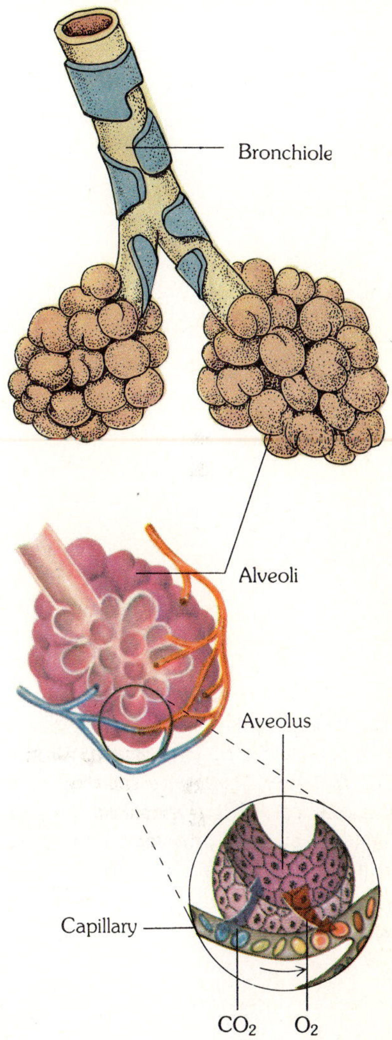

Exchange of Gases in the Alveoli

Each alveolus is lined with a single layer of flat cells that are richly supplied with capillaries. Through these capillaries, gases pass into and out of the bloodstream. Oxygen and carbon dioxide easily cross the moist, living cell membranes. Large amounts of oxygen are carried to the body cells in combination with hemoglobin in red blood cells. Similarly, large amounts of carbon dioxide are carried from the body cells to the lungs. Blood flowing through capillaries near the alveoli picks up oxygen and carries it to body cells. Carbon dioxide diffuses from the capillaries into the air sacs. It is then removed from the body by exhaling.

Mechanics of Breathing

The lungs themselves cannot pump air into or out of the body. Lung tissue is soft, not muscular. Most of the work of breathing is done by the muscles of the chest wall and the **diaphragm** (dī'ə fram), a sheet of muscle that stretches across the lower chest cavity at the base of the lungs.

When the diaphragm is relaxed, it is dome-shaped. When it contracts, the diaphragm flattens. As you inhale, the diaphragm contracts and pulls downward. At the same time, the **intercostal** (in'tər kos'tl) **muscles** between the ribs pull the ribs outward and upward. These movements enlarge the chest cavity. As the volume of the chest cavity increases, the air pressure inside the lungs decreases slightly. The outside air pressure is now greater than the pressure inside the lungs. Air moves into the lungs until the pressure is equalized.

Inhalation

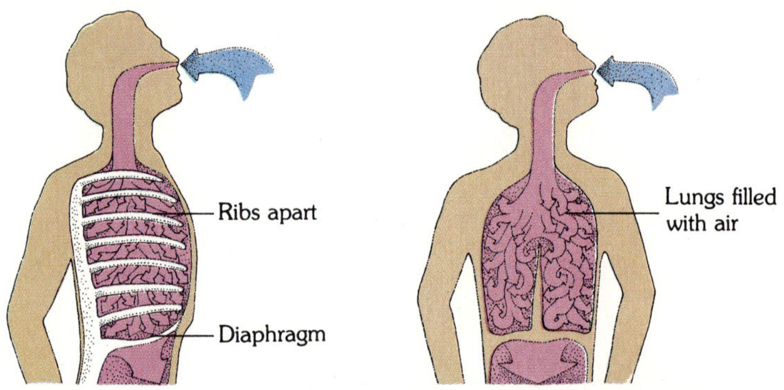

The process is reversed when you exhale. The intercostal muscles relax. This lowers the chest wall and moves it inward. The diaphragm becomes dome-shaped as it relaxes. These movements increase the air pressure inside the lungs. The lungs deflate, and air is forced out.

Exhalation

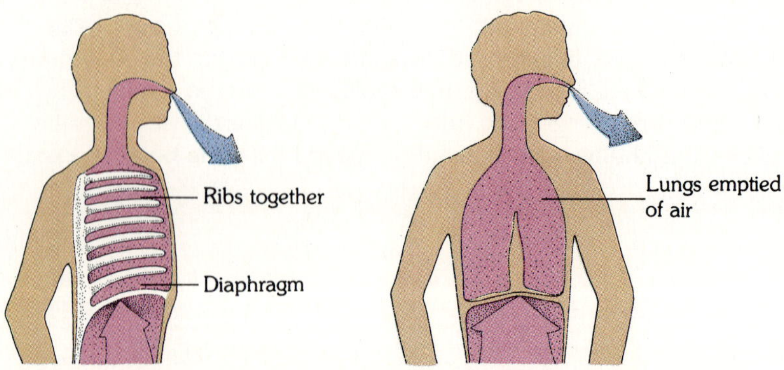

Transfer of Gases

In the blood, hemoglobin molecules combine with oxygen to form the loose compound *oxyhemoglobin.* This compound is formed in the lungs where there is a high concentration of oxygen. In areas of low oxygen concentration, such as cell tissues, oxyhemoglobin releases oxygen. By diffusion, oxygen leaves the blood and enters the cells for use in metabolism. Carbon dioxide will temporarily combine with hemoglobin in areas of low oxygen concentration. In the lungs, the carbon dioxide is exchanged for oxygen.

Each hemoglobin molecule can combine with four oxygen molecules. Only one or two of these molecules may be diffused into the body cells during each cycle of the blood through the systemic circulatory system. Oxygen that was not diffused into the cell tissues remains attached to the hemoglobin molecules as oxyhemoglobin. It may be used during another cycle.

Did You Know?

Did you know that carbon monoxide poisoning is one of the most common causes of respiratory failure? Carbon monoxide combines with hemoglobin in blood over 200 times faster than does oxygen. This means that when there are high levels of carbon monoxide, cells are deprived of the oxygen they need. Two sources of this deadly gas are automobile exhaust and cigarette smoke.

Gas Transport and Exchange

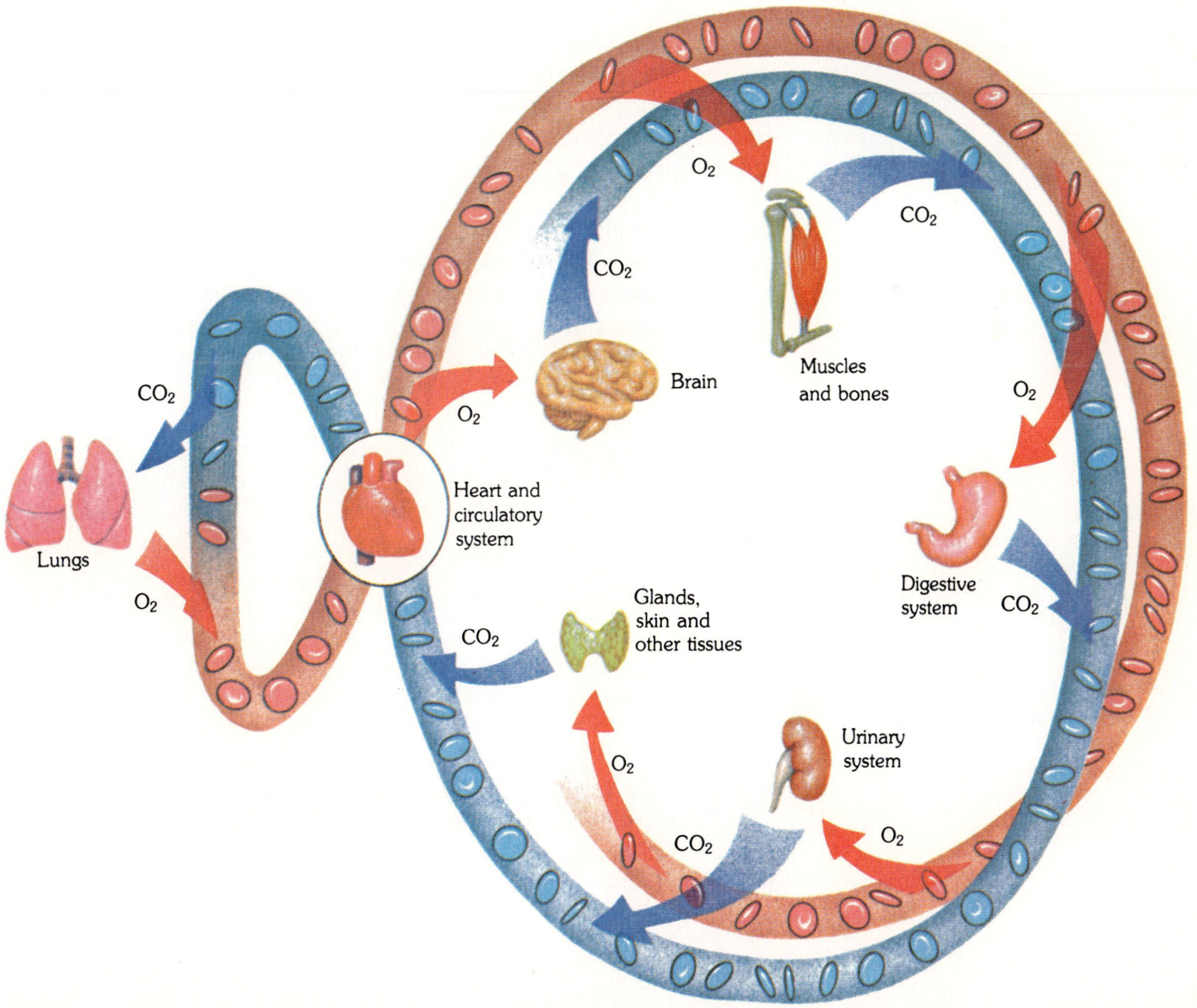

Cellular Respiration

The energy your body needs to carry on its many functions is produced in the cells. Oxygen and glucose, a product of digestion, are necessary to produce body energy. The process in which cells use food molecules and oxygen to release energy is called **cellular respiration.**

In the mitochondria of cells, glucose is oxidized to release energy. This oxidation process produces ATP, the complex organic compound that stores energy, and water and carbon dioxide.

$$C_6H_{12}O_6 + 6O_2 \rightarrow 6CO_2 + 6H_2O + \text{energy}$$

ATP is then released to other parts of the cell where it is used as fuel for other cell reactions or to provide body heat.

When you exercise vigorously, the muscle cells may not receive enough oxygen to react with glucose to produce all of the energy needed to keep up the pace of activity. In this situation your muscle cells can produce energy without oxygen. When this occurs, glucose stored in the cells is changed to **lactic** (lak'tik) **acid,** and small amounts of energy in the form of ATP are released. Lactic acid accumulates in the muscles and causes fatigue and soreness. As more oxygen becomes available, lactic acid is either broken down or reassembled into glucose. The feeling of soreness gradually goes away. This method of releasing energy is not as efficient as combining oxygen and glucose.

Homeostasis in Respiration

Breathing rate is influenced by changes in body activity, which in turn cause changes in the level of oxygen, carbon dioxide, and hydrogen ions in the blood. Receptors in major arteries and in the central nervous system send information about changes in either the concentration of carbon dioxide or in the acidity of the blood to the respiratory center of the brain. This center, which is located in the medulla, stimulates an increase or decrease in the rate of breathing.

Homeostasis (hō'mē ə stā'sis) is the regulation of body processes that keeps the internal conditions of the body the same, or nearly the same, when there are external changes that affect the body. Homeostasis allows the body cells to function well under widely varying conditions. In respiration, homeostasis means adjusting breathing according to the needs of the body for oxygen and removal of carbon dioxide. Breathing is regular and rhythmic. Messages from the brain activate the muscles that do the work of breathing.

During periods of strenuous exercise, the body needs an increased amount of oxygen. Carbon dioxide is also produced in greater quantities. The breathing rate increases as the body attempts to provide the cells with more oxygen and to remove more carbon dioxide. When the strenuous activity is stopped, the breathing rate slows and eventually returns to normal.

Challenge!
The Surgeon General of the United States has found cigarette smoke to be harmful to the health of the smoker. What respiratory diseases do cigarette smokers suffer from? Investigate the following diseases to learn how delicate tissues can be harmed by cigarette smoke: chronic bronchitis, lung cancer, emphysema.

Metabolism in Individual Cells

The chemical reactions of metabolism take place in the individual cells. Two types of metabolism occur. The release of energy by chemically breaking down nutrients is called *respiration.* The complex processes of respiration occur in the mitochondria of the cells. In this reaction, nutrient materials are combined with oxygen to release water, carbon dioxide, and energy in the form of ATP. The metabolic reactions in respiration are called **catabolic** (kat′ə bol′ik) **reactions.** The energy released by respiration is used to carry on cell functions, to warm the body, and to power muscles.

Chemical reactions of metabolism that form new substances are **anabolic** (an′ə bol′ik) **reactions.** The making of new components for growth, maintenance, and repair of body tissues is the **synthesis** (sin′thə sis) **process** of metabolism. Synthesis of proteins takes place in the ribosomes of cells. Energy that has been made available by respiration is used to build new and complex proteins or organic compounds. These newly formed substances provide energy or materials for growth and repair of the body. They may be used by the body immediately or stored for later use. For example, glucose that is not used immediately by the body is sent to the liver where it is formed into **glycogen** and stored. When the body needs glucose, the liver breaks down the glycogen into glucose. The liver then releases the glucose into the bloodstream a little at a time. In a similar way, the liver picks up amino acids that are produced when proteins are digested. In the liver these amino acids are converted to simple plasma proteins which are returned to the blood.

Metabolism is affected by vigorous activity, anxiety, fear, or a rise in body temperature. It can also be changed to meet a specific body need, such as repairing a broken bone. Changes in metabolism are controlled by *hormones.* Hormones alter cell activity in various ways. You may recall that *thyroxin* is a hormone produced and released by the *thyroid gland.* Thyroxin affects cell mitochondria and triggers an increase in oxygen consumption. An increased amount of energy is released, and the rate of metabolism goes up. A drop in the level of thyroxin results in a lower rate of oxygen consumption and consequently less energy is released.

Metabolic processes

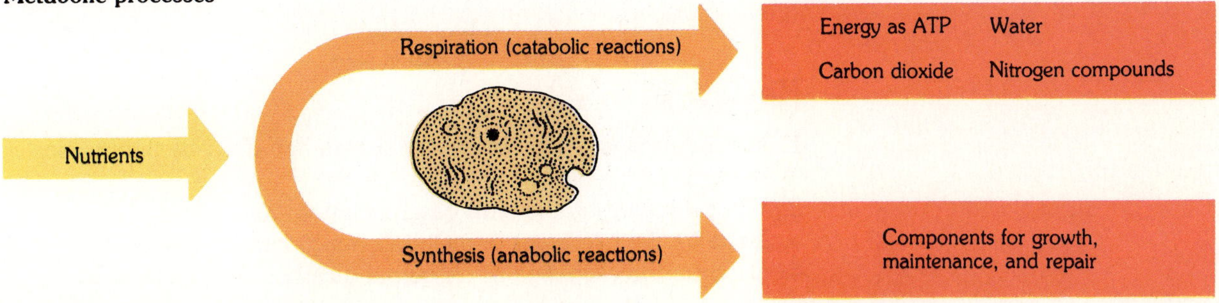

The Use of Energy

Energy is used by the body to move, to digest food, and to carry on basic life processes. Those life processes include circulating blood, pumping air through the lungs, and maintaining a constant body temperature. In addition, they also include activities of the excretory, glandular, and nervous systems.

The rate at which the body uses energy is called the **metabolic rate.** A person's normal metabolic rate varies from day to day, hour to hour, and even minute to minute depending on various factors. Those factors include exercise, activity of the glands and digestive system, illness, strong emotions, and use of stimulant and depressant drugs. For, example, a person may use 65 kilocalories an hour while sleeping, 100 kilocalories an hour while watching television or reading, 200 kilocalories an hour while walking, and 1,100 kilocalories an hour while climbing stairs. The metabolic rate is also increased by a fever, by the use of caffeine, and by eating a meal. It may be decreased by the use of a depressant or by inactivity of certain glands.

Measuring the Use of Energy

The metabolic rate may be found by measuring the use of oxygen. Another less common method is by measuring the amount of heat produced by the body. Measuring metabolic rate for the purpose of comparison is generally done after a restful night of sleep and at least 12 hours after the last meal. A metabolic rate measured under those conditions is called the **basal metabolic rate.** Basal metabolic rate varies with age, sex, and body size. In general, the basal metabolic rate declines as a person grows olders. The normal basal metabolic rate in men is about 10 percent higher than it is in women.

Basal metabolic rates in humans

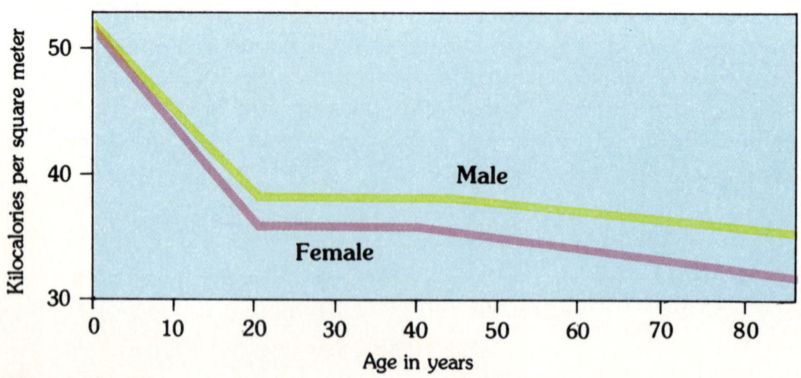

Review It

1. List the organs of the respiratory system and describe the function of each.
2. What is meant by the term basal metabolic rate?
3. Describe the two types of metabolic reactions in the body cells.
4. Describe the exchange of oxygen and carbon dioxide in cells.

Activity

Respiration

Purpose
To determine the effects of different kinds of activities on breathing rate and pulse rate.

Materials
3 small beakers with Bromthymol blue (BTB) solution
3 straws
Pencil and paper

Procedures

1. In Trial 1, establish your normal breathing rate and pulse rate at rest. Record the data in a chart like the one shown.
2. In Trial 2, do some type of mild exercise such as walking for one minute before taking each reading. Record your data.
3. In Trial 3, do some type of strenuous exercise such as running in place for one minute before taking each reading. Record your data.
4. In Trial 4, take in ten deep breaths before taking each reading. Record your data.
5. In Trial 5, repeat the strenuous exercise for one minute. Then immediately bubble your breath through a straw into the BTB solution in the first beaker. Wait one minute and bubble your breath for ten seconds into the second beaker. Wait one more minute and bubble your breath for ten seconds into the third beaker. Note the resulting color of the solution, as well as its intensity. Record your results.

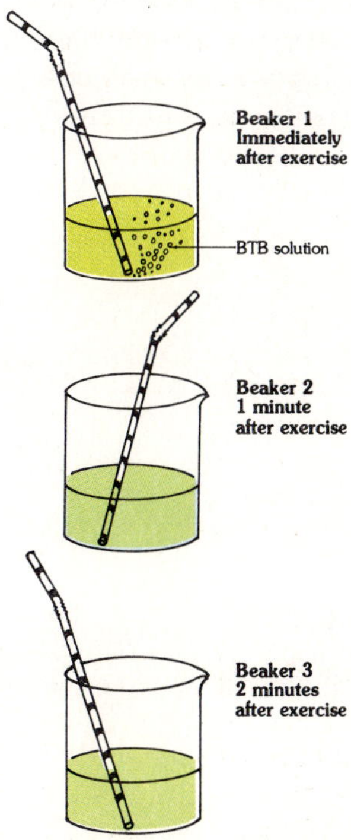

Trial	Breaths/min	Pulse/min
1		
2		
3		
4		

Trial	Color Change		
	Beaker 1	Beaker 2	Beaker 3
5			

Analysis

1. Which trial acts as the control for the experiment?
2. Why does the breathing rate change with each trial?
3. How does the pulse rate change with each trial?
4. The color change noted in the BTB solution indicates the presence of carbon dioxide. Explain the differences in the color of the solutions in each of the beakers.

24–2 Excretion

Within the cells the chemical reactions of metabolism build the cell structure, transfer energy necessary for cell life and activity, and produce substances used throughout the body. Some of the chemical products of these reactions are wastes and can be potentially harmful to the body. Disposal of wastes from the body is called **excretion** (ek-skrē′shən). These wastes include carbon dioxide, water, and compounds containing nitrogen, such as ammonia, urea, and uric acid. Some of the energy that is transferred in cell reactions is changed to heat. This energy cannot be used by the body for other purposes.

The Role of the Lungs, Liver, and the Skin in Excretion

Carbon dioxide and some water and heat are removed from the body during exhalation. A small number of substances are removed from the blood by the liver. These substances are passed from the liver as part of the bile. They enter the intestines and are passed out of the body with the feces. The liver also forms urea, a substance produced when amino acids are metabolized. It is a chemical combination of ammonia and carbon dioxide. Urea is eliminated through the kidneys and sweat glands of the skin.

The skin serves to eliminate wastes and regulate body temperature. Approximately two million sweat glands are located in the skin. These tiny coiled *tubules* are found in the **dermis** (dėr′mis), a layer of skin just below the surface. These glands continually secrete **sweat.** Sweat contains primarily water along with salts, urea, and other wastes. These materials pass from the blood vessels in the skin into the sweat glands. They are deposited on the surface of the skin through *pores,* or small openings in the skin. The water evaporates as perspiration. The solids remain on the surface of the skin or on clothing. Perspiration itself has no odor. However, bacteria living on the skin decompose urea to form new compounds that do have odors.

Section of skin

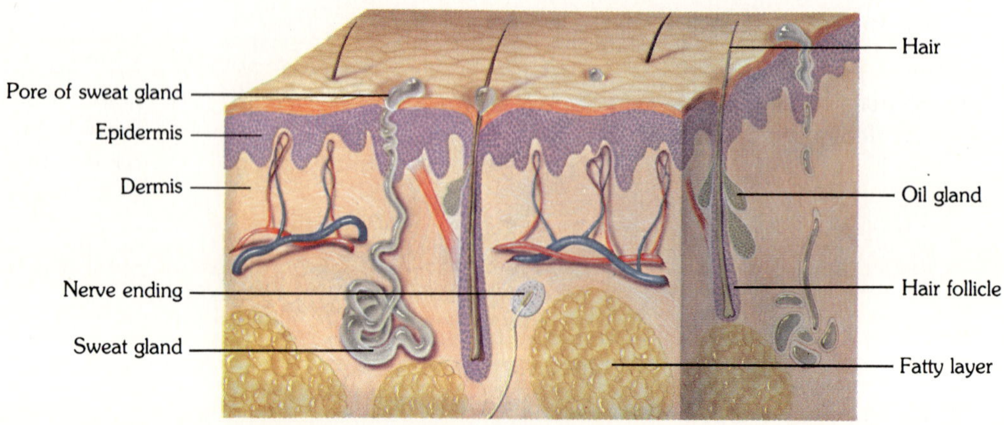

Skin and the Regulation of Body Temperature

Some body heat is lost through the lungs and during excretion. However, over 80 percent of body heat is lost through the skin. The body may become overheated from strenuous exercise or being in a warm environment. As a result, blood vessels in the skin dilate (or enlarge) allowing more blood to flow through them. This produces two effects. First, more heat is lost directly through the skin, and the sweat glands become more active. Second, as the sweat evaporates, the increased amount of perspiration cools the body.

If the body becomes cold, the blood vessels in the skin contract, or become smaller. Sweat production decreases, and body heat is conserved. The rate of heat production in the body is increased by shivering, chattering of the teeth, and tightening of the body muscles.

The Kidneys

The **kidneys** are the major organs of excretion in the body. They are located just above the pelvis on either side of the spine. Pads of fat serve as protection for the kidneys.

The kidneys function to filter blood and to remove wastes. They also regulate the volume and composition of blood. The kidneys also maintain a proper balance of water and salts in the body. The kidneys work continuously, processing all of the blood in the body every 30 minutes.

Did You Know?
The type of picture above is called a thermograph. It shows the heat patterns on the surface of this child. The temperature is indicated by color. The red areas are warmest, and the blue areas are coolest. Thermographs are used in a variety of ways, but one of the most important ways is to detect tumors. Tumors are warmer than surrounding tissue and can sometimes be detected by using this process.

Urinary system

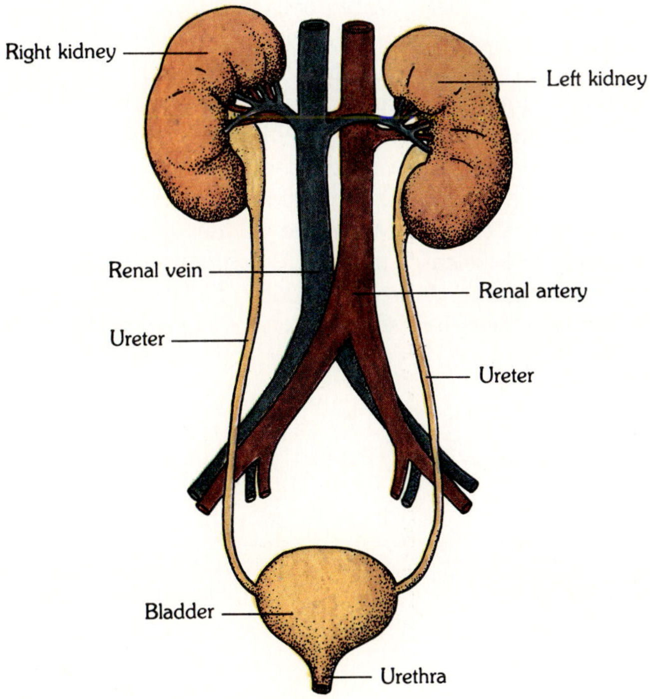

Challenge!
Find out what a "kidney machine" is. Trace the operation of this machine through one cycle. Explain how it functions and how it is used to benefit humans.

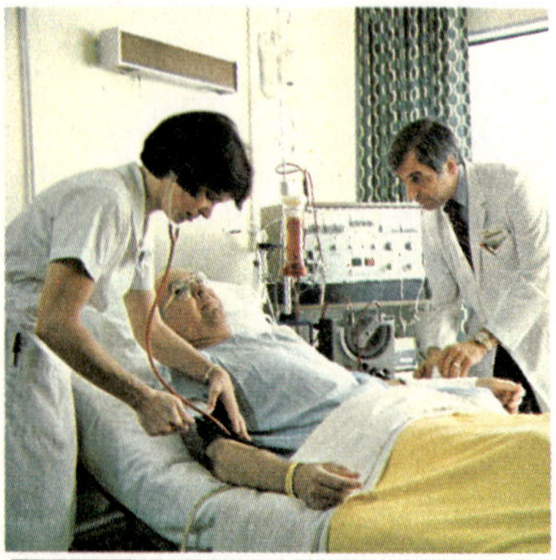

Artificial kidney machine

Filtering the Blood

Blood enters each kidney through the **renal** (rē'nl) **artery,** which branches directly from the aorta. The blood passes through smaller and smaller blood vessels which eventually lead to tiny tubelike structures called **nephrons** (nef'ronz). Each kidney is composed of about one million nephrons. The important function of **filtration** is performed in the kidneys by the nephrons.

The nephron is a long, coiled tubule. One end of the nephron is expanded to form a cup-shaped structure called **Bowman's capsule.** The Bowman's capsule almost completely surrounds a network of capillaries, the **glomerulus** (glə mer'ə ləs). Other capillaries surround the tubule along its entire length. From Bowman's capsule, the tubule extends and eventually opens into a duct, the **ureter** (yů rē'tər), that collects urine.

Blood is filtered through the capillary membranes of the glomerulus in the nephrons. These membranes act as selective filters. Blood cells and plasma proteins of the blood do not pass through the membranes but continue through the capillaries. The fluid that has been filtered from the blood is very much like blood plasma except it does not contain plasma proteins.

As this fluid moves along the tubule of the nephron, glucose, amino acids, sodium, water, and other substances that are used by the body are returned to the bloodstream by the cells of the nephrons. This process is called **reabsorption** (rē'ab sôrp'shən). Active transport, using energy, rather than simple diffusion, returns the useful materials to the blood. Reabsorption continues until all of the glucose and other helpful substances have been passed back into the blood.

Two-way traffic occurs across the walls of the nephron. This permits the nephron to sort out and return important substances and to discard useless, harmful, or excess material from the blood. The undesirable substances are wastes, consisting of urea, dissolved salts, and excess water. The solution of wastes called *urine* moves through the *collecting tubule* and passes out of the kidney through the *ureter.* Urine is stored in the **bladder.** When the bladder is full, a tiny **sphincter** (sfingk'tər) muscle releases the urine into the **urethra** (yů rē'thrə), a tube that passes the urine out of the body. The purified blood leaves the kidney through the **renal vein** and returns to general circulation.

Cross section of left kidney

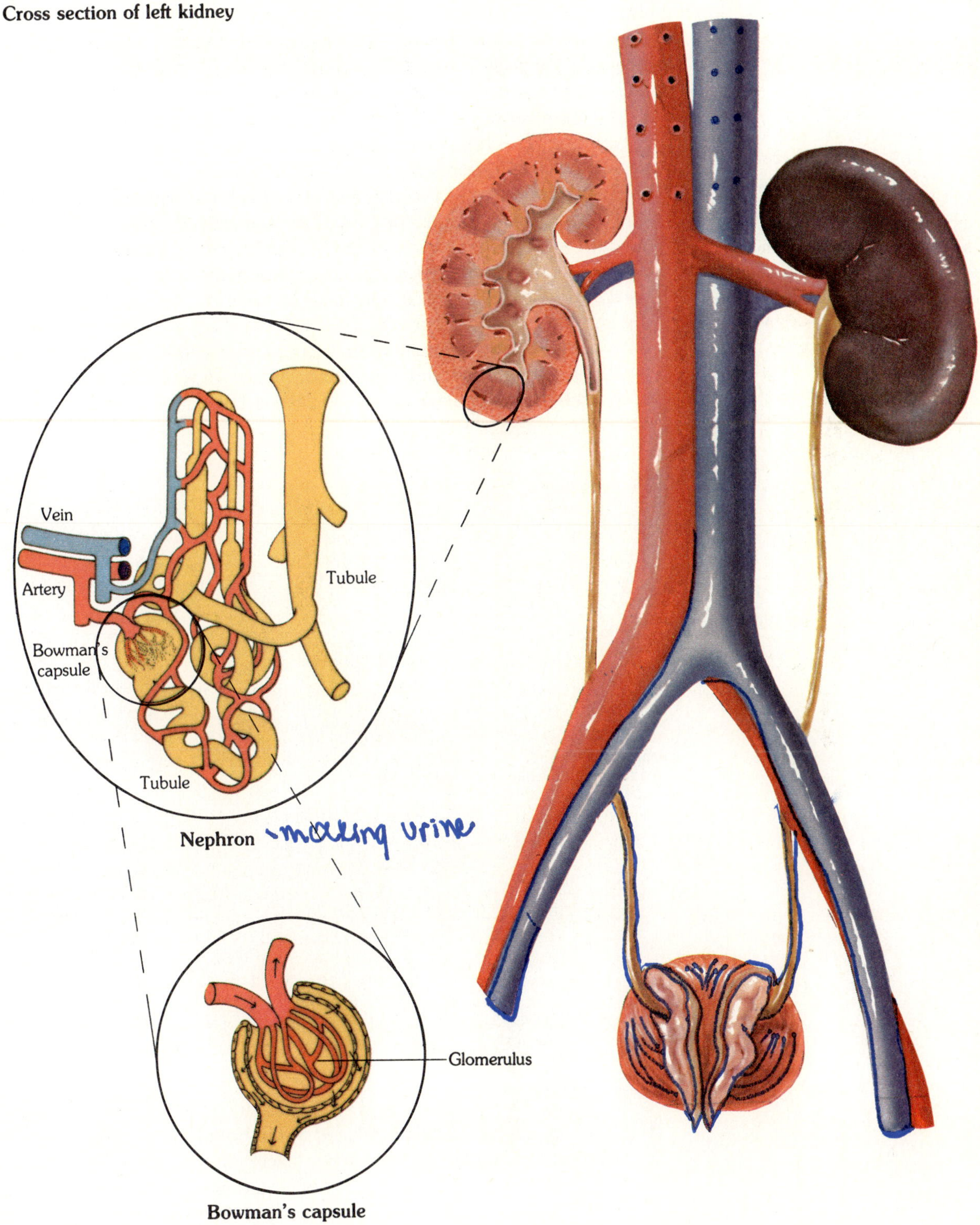

Did You Know?
The liver breaks down old red blood cells, recycling some of the materials, and providing for the removal of others. As hemoglobin breaks down, the pigments of the hemoglobin, which give it color, are passed into the intestine in bile. Bacteria in the intestine make changes in the pigments. Some of the pigments stay in the intestines and are eliminated with the feces. Other pigments are sent to the kidneys. These pigments are excreted from the body in the urine. The pigments give urine its yellow color and feces its brown color.

Homeostasis and Excretion

The actions of the kidneys keep the composition of the blood constant and maintain the proper balance of salts and water in the body. The needs of the body vary, so kidney activity must be regulated.

When the body's salt and water balance needs adjustment, the pituitary gland receives a message from a portion of the brain to secrete a hormone. This hormone stimulates the kidney to eliminate the excess salt or water, depending on the body's needs. When the body's salt and water balance has returned to normal, the brain signal stops, and the kidneys return to their usual level of activity.

If, for example, you drink large quantities of liquids, the amount of urine increases as the kidneys attempt to maintain a constant water level in the blood. Excess salt in the diet may result in a decrease in the amount of urine produced. By absorbing less water, the kidneys allow more water to remain in the blood, and the level of salt concentration is reduced. A common cause of death in humans is the failure of the kidneys to regulate the salt and water balance.

Homeostasis

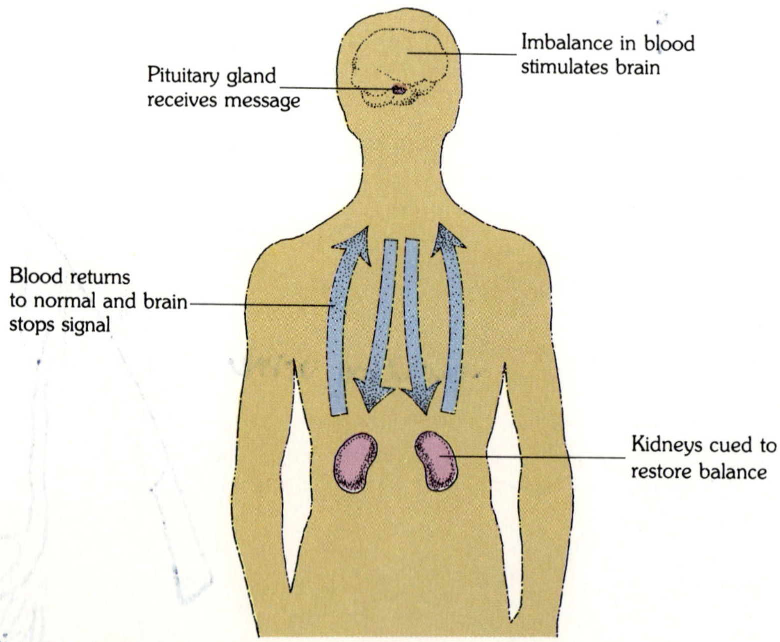

Review It

1. How are waste materials produced in the body?
2. Describe the functions of the skin in excretion.
3. How are helpful substances returned to the blood in the nephrons?
4. List the functions of the kidneys in body maintenance.

Pioneers in Biology

Annie Dodge Wauneka:
Pioneer in teaching health-care methods.

In the 1920s and 30s, when Annie Wauneka was growing up on the vast Navajo reservation, tuberculosis killed more of her people than any other disease. In 1951 Wauneka was elected to the Navajo tribal council. As chairperson of the Health Committee, she tackled the problem of tuberculosis and other diseases. Not only did she bring down the death rate of the dreaded lung disease, she served as a model of the most effective way to be a bridge between two cultures.

Annie Wauneka was born April 10, 1910, in her mother's hogan.

When she was 8, Annie was sent to the government school at Fort Defiance. During the following winter of 1917–18, the terrible influenza epidemic struck, killing thousands of people throughout the country. She developed a mild case of the flu, but recovered to help care for the sick children at her school. That epidemic, and another disease that hit the school a few years later, left an impression upon her of the need for better health care. She completed the eleventh grade at the Albuquerque Indian School. Then, at age 18, Annie returned home.

The Navajo nation includes about 130 thousand people. As chairman of the 74-member tribal council, her father, Henry Chee Dodge was a respected man. He took Annie to meetings of the Navajo people and members of the Bureau of Indian Affairs. Annie learned much about leadership from her father. She learned how to speak carefully. Henry Chee Dodge always tried to leave a clear impression the first time he spoke.

In 1951, when Mrs. Wauneka became the first woman member of the tribal council, she used her abilities as a leader to help improve the health care for the Navajo people. First she went to the government doctors on the reservation and learned as much as she could about tuberculosis (TB). The doctors cooperated with her, and for three months she studied in their laboratories, using microscopes, reading X rays, and watching patients. Next she talked with the tribal medicine men, explaining that TB was an "outside" disease that might be too powerful for their medicine. She suggested that if the medicine men and doctors shared their knowledge and skills, much could be gained.

The Navajo people had not liked going to hospitals because of the impersonal atmosphere and strange food. The ways of the medicine men included the support of friends and family to help a person get well. So Annie made hospital visits, bringing news from the patients' families. She encouraged them to stay until their health improved.

With Annie Wauneka's encouragement, 20,000 of her people reported for X rays. Of those, 2,000 were hospitalized for treatment. In 1958, the Arizona Public Health Association recognized her work with the Outstanding Worker in Public Health award. In 1963, Mrs. Wauneka was presented the Presidential Medal of Freedom for her campaign to bring modern health care to the Navajo nation.

Because of her success in retaining the best from both cultures, Annie is consulted on health-care matters concerning the Eskimos and other groups of people in the United States. She was awarded an honorary doctorate from the University of Albuquerque.

Chapter Summary

Human cells need oxygen to release energy. Every time you inhale, oxygen passes into the nasal cavity and proceeds down the pharynx and trachea through the bronchi and into the lungs. In the alveoli of the lungs oxygen passes into the blood. As blood circulates, it carries carbon dioxide from the cells to the lungs, where it leaves the body by exhalation.

Oxygen is carried in the blood as a chemical compound called oxyhemoglobin. This compound is formed as blood passes through the lungs. Oxyhemoglobin breaks down to release oxygen as it reaches tissue capillaries.

Cells combine glucose and oxygen to form energy by means of cellular respiration. Energy is released in the form of ATP. Water and carbon dioxide are also produced.

Metabolism is the sum of all of the chemical reactions that occur in body growth, maintenance, and tissue repair. Cell respiration (catabolic reactions) releases energy. Synthesis (anabolic reactions) uses energy to produce substances used or stored by the body for growth and repair. Metabolism is controlled by hormones and may vary during a person's lifetime. The rate at which metabolism occurs in a resting person is one indication of general health.

Waste products that result from the processes of metabolism are removed from the body in several ways. The lungs, kidneys, liver, and sweat glands of the skin all take part in the excretion of wastes. The lungs remove carbon dioxide and water. Urea and other waste materials are processed by the liver. The skin functions to remove excess water, salts and urea while helping to regulate body temperature. The kidneys filter blood to remove wastes, regulate the volume and composition of blood, and maintain the balance of water and salts in the blood.

Homeostasis is the process by which the body attempts to maintain its internal conditions at a constant level. In respiration, homeostasis increases or decreases breathing according to the body's needs for oxygen and the removal of carbon dioxide. Homeostasis in excretion allows for changes in the amount and types of substances that are reabsorbed into the blood after filtering.

Questions to Think About

1. How is the transfer of oxygen in the blood different from the transfer of carbon dioxide? In what ways are the two processes alike?
2. On a hot summer day, a person who does strenuous work may need to take salt tablets and increase the intake of fluids. Why would these actions be necessary?
3. Sometimes a child throws a temper tantrum, holding the breath until he or she faints. At this point, the child starts breathing again, even though the child is still unconscious. Why?
4. What differences in breathing rate and fatigue would a marathon runner encounter between a race in LaPaz, Bolivia, at 4000 meters above sea level and a race in Los Angeles, California, at sea level?
5. Artificial heart-lung machines have been developed to take over the job of these organs during some types of surgery. How do these machines function? What breakthroughs in surgical procedures have these machines permitted?

Extra Research

1. Normal breathing is interrupted from time to time by certain abrupt changes in the respiratory process such as yawning, sighing, hiccupping, coughing, sneezing, laughing, and crying. What effects does each of these have on the rhythm of breathing?
2. One of the major problems encountered by deep-sea divers is "the bends." What is this condition? What can be done to prevent and treat it?
3. A urinalysis (yůr'ə nal'ə sis), a test of urine, is a common medical procedure. What information could be gained by studying a urine sample? How could these results be used to determine the health of the person being examined?
4. Some of the earliest research on respiration was done during the 17th and 18th centuries. Learn about the work of one or more of the following people: Boyle, Scheele, Priestley, and Lavoisier.

Chapter Test

Part I

Copy the number of each test item. If the item is *true*, write *T* after the number; if the item is *false*, write *F*.

1. When you inhale and exhale, oxygen enters the lungs and carbon dioxide leaves the lungs. (575)
2. Cilia prevent food from going down the windpipe. (576)
3. During inhalation, the diaphragm pulls downward on the lungs, and the ribs move down and inward. (578)
4. In the alveoli, hemoglobin found in the red blood cell, combines with oxygen to form oxyhemoglobin. (579)
5. Cellular respiration releases energy. (581)
6. Synthesis metabolism produces materials for growth and repair of body structures. (581)
7. Metabolism refers to the process by which materials are broken down into simpler forms. (581)
8. Your metabolic rate differs from time to time throughout your life and also from time to time during a single day. (582)
9. Some waste products of cell functions are harmful to the body. (584)
10. One way in which body temperature is controlled is through the rate of perspiration. (585)

Part II

Copy the number of each sentence below. After each number, write the word (or words) that completes the sentence correctly.

11. Oxygen passes from the lungs into the blood through tiny air sacs called ____. (576)
12. Oxygen travels in the blood as ____. (579)
13. ____ is the effort the body makes to control its internal environment by regulating the intake and loss of substances. (580)
14. The final product of kidney excretion is ____. (586)
15. The ____ function to maintain a normal balance of water and salts in the body. (588)

Part III

Copy each item from List A. After each item, write the letter and words from List B that best describe it.

16. nephrons (586)
17. bladder (586)
18. anabolic (581)
19. sphincter (586)
20. diaphragm (578)
21. sweat glands (584)
22. urea (584)
23. ATP (581)
24. homeostasis (580)
25. catabolic (581)

List B

a. controls the release of urine from the bladder
b. complex organic compound that stores energy
c. provides for the body to function well under widely varying conditions
d. reactions that require energy, such as changing amino acids to plasma proteins
e. reactions that release energy, such as digestion
f. stores urine until it passes from the body
g. waste product produced by the liver
h. organs that secrete wastes out of the body through the skin
i. organs within the kidneys that filter blood
j. flat sheet of muscles across the lower chest cavity

Chapter 25

Reproduction, Growth, and Development

Most organisms can stay alive without ever reproducing. But if all of the individuals of a population failed to reproduce, the population would become extinct. Species survive through the process of reproduction. If humans were to stop producing babies, our species would disappear.

Human sexual reproduction is a complex biological process. Reproduction and sexuality are further complicated by personal and social factors.

How do reproductive hormones affect your body? How does a fertilized egg develop and grow in the human female reproductive system? What is involved in the process of birth? What is the aging process? These and other questions about human reproduction are discussed in this chapter.

Chapter Objectives

1. To describe the importance of reproduction to the survival of a species.
2. Describe the reproductive systems and their functions in the male and female.
3. List the major reproductive hormones, their origins, and their effects.
4. Describe the various events of the menstrual cycle.
5. Describe the events of human development from fertilization to birth.
6. Describe human growth and development from birth to adulthood.
7. Define aging.

The various stages of life are well represented by the four generations of the Greene family—The Reverend Thomas W. Greene, Jr., born in 1881; Thomas W. Greene III, born 1908; Thomas W. Greene IV, born 1942; and Thomas W. Greene V, born 1966.

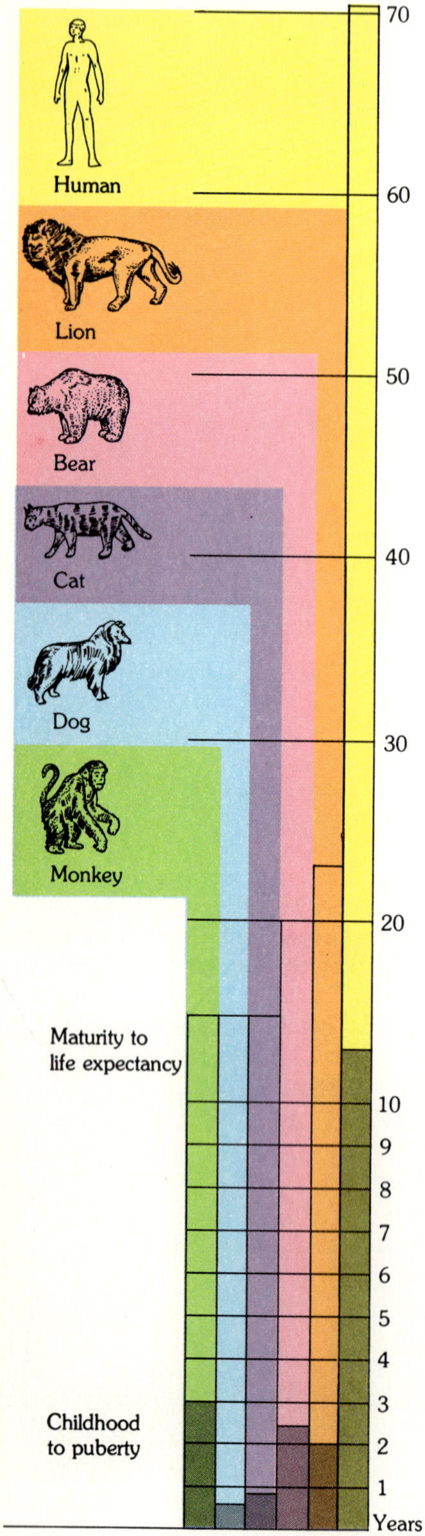

25–1
Survival and Reproduction

Some plants live thousands of years. Some organisms live only a few hours. The longest time an individual of a species has been known to live is spoken of as the *life-span* of that species.

The recorded 114 year life-span of a human is the longest for any mammal. Among all of the animals with backbones, only the tortoise lives longer than humans. Death comes to organisms at all ages. Not every human can hope to live to be 114 years old. Organisms, including human beings, usually die long before reaching their maximum possible life-span because of accidents, disease, and other causes.

Life expectancy is the age to which half of all individuals born at one time can expect to survive. Thus, a child born in the United States in 1974 had a fifty percent chance of living 72 years. That child's life expectancy is 72 years. A child born in Ethiopia in 1971 had a life expectancy of less than 39 years. The difference in life expectancies is due to the different conditions for life in the two countries.

Many things affect life expectancy. Two major factors are *diet* and *health care,* especially during infancy. As diet and health care improve, the life expectancy of the population usually increases. Some organisms will live longer than their life expectancy. Eventually, however, every living thing dies. Without a new generation, life would not continue. Reproduction helps ensure the survival of all species, including humans.

Human Reproduction

At first glance, human reproduction appears to be similar to sexual reproduction in other members of the animal kingdom, especially other mammals. Organs that produce human sex cells function much like the sex organs of other mammals. The hormones that regulate human reproduction are chemically similar to the hormones that regulate reproduction in other mammals.

Yet, human reproduction and sexuality differ in several ways from reproduction and sexuality of other mammals. In other mammals, breeding is seasonal and is regulated by hormones. Humans do not have special breeding seasons. Biologically, humans can be sexually active from youth to old age.

For humans, sexuality is far more than a means of having children. Human sexuality is a kind of glue that helps to hold our society together. Our sexuality strongly influences what we think about ourselves and how we behave with other people—male and female. Attitudes and behaviors about sexuality and reproduction are shaped in part by laws and customs, which vary among societies. From time to time, these laws and customs change.

In humans, the ability to reproduce begins with **puberty** (pyü'bər tē). At puberty, changes occur in the body that make it possible for the individual to become physically able to reproduce. These changes include the development of sex cells and the development of certain body structures associated with reproduction.

Male Reproductive System

An abdominal outpocketing of the body wall is the location of the *testes*. This wrinkled sac of skin and muscle is called the **scrotum** (skrō'təm). Since sperm production does not occur at normal body temperature, the temperature of the scrotum is lowered and controlled by the blood flow through it. At puberty, the hypothalamus stimulates the pituitary gland to produce hormones that activate the testes. The development of secondary sex characteristics, such as deepening of the voice, growth of body hair, bones, and muscles, occurs at this time. Male hormones are also partly responsible for the sexual drive and behavior of males.

The testes are largely made up of **seminiferous** (sem'ə nif'ər əs) **tubules.** At puberty, these tiny tubules become hollow and produce sperm in great numbers. Cells in the ducts of the testes produce a fluid called the **seminal** (sem'ə nəl) **fluid.** Seminal fluid provides food and a liquid environment in which the sperm can easily move. The sperm are active in an alkaline solution. The normal acidity of the testes is neutralized by the alkaline seminal fluid. **Semen** (sē'mən) is the name given to the mixture of the sperm and seminal fluid. Hundreds of millions of sperm are stored in the area called the **epididymis** (ep'ə did'ə mus). A duct, the **vas deferens** (vas def'ər enz) connects each testis to the urethra. Additional fluid is added by **seminal vesicles** (ves'ə kəlz), **Cowper's gland,** and the **prostate gland** as sperm pass into the urethra. When the male becomes sexually excited, the **penis** (pē'nis), the reproductive organ by which sperm are transferred to the female, becomes enlarged. Muscles hold blood in spongy tissue of the penis, causing it to become hard and erect. If a series of contractions called an *ejaculation* (i jak'yə lā'shən) occur, the semen is released from the external opening of the urethra. If the sperm are not released, they are absorbed by the male's body.

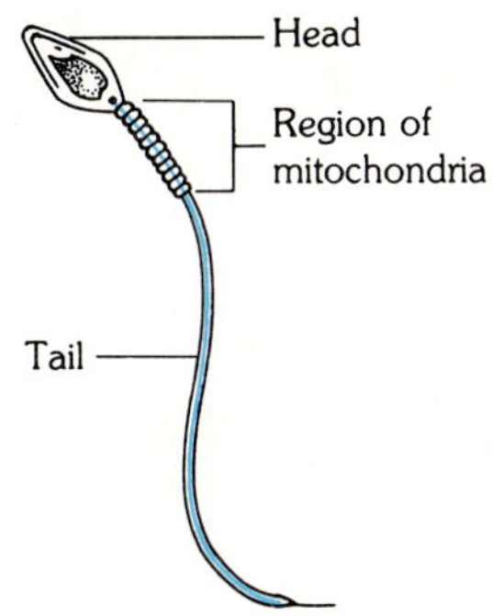

Structure of a sperm

Did You Know?

The temperature of the testes is regulated by arteries that supply the testes with blood and veins that carry blood away from them. These arteries and veins pass very close to each other. Heat from the arteries flows into the cooler veins. In this way the temperature of blood entering the testes may be several degrees lower than body temperature. The skin of the scrotum also contains numerous sweat glands that help in the cooling process.

Male reproductive system

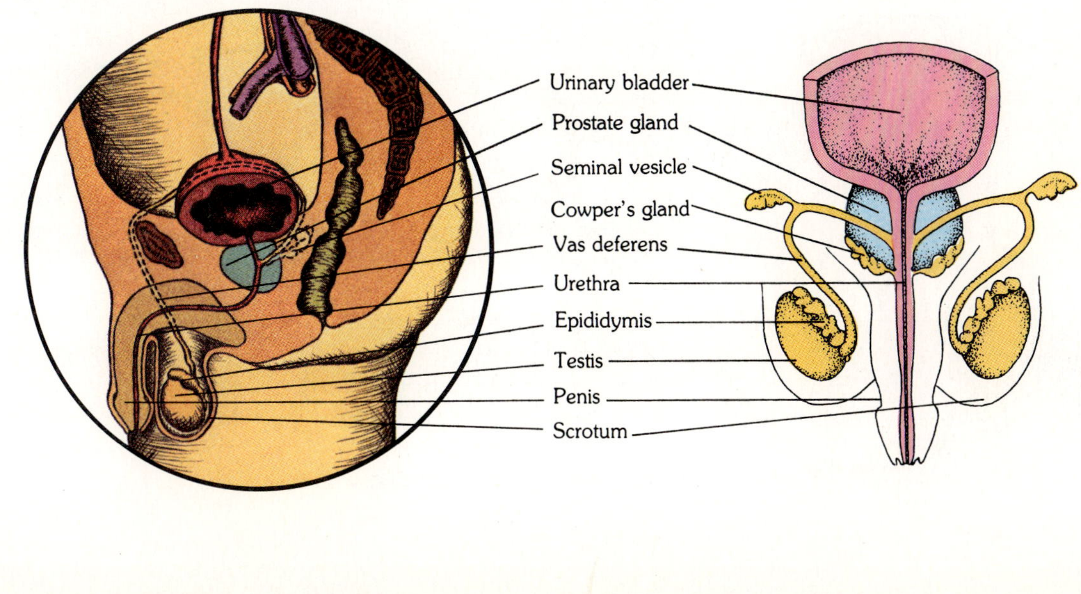

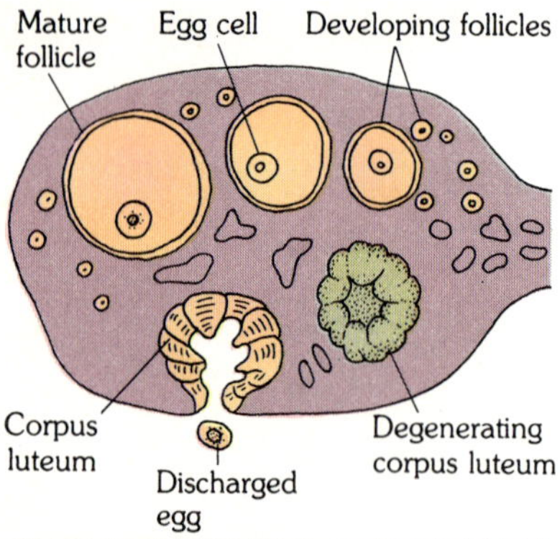

Egg development in the ovary

Female Reproductive System

The female reproductive cells are produced by a pair of almond-shaped organs, the *ovaries,* which are located within the abdominal cavity. At puberty, the ovaries are stimulated by the pituitary to become functional. They produce hormones that result in the development of the secondary sex characteristics of females such as the growth of milk-producing (*mammary*) glands in the breasts, the depositing of fat beneath the surface of body skin, the widening of the pelvic girdle, and the development of body hair and oil and sweat glands in the skin. Puberty in females usually begins several years earlier than it does in males.

After the beginning of puberty, about once each month an egg in one of the ovaries begins to ripen. At birth, the ovaries contain an estimated 400,000 immature eggs. However, only about 400 of these eggs will ever mature.

The egg develops inside a **follicle** (fol′ə kəl), or cavity. The follicle produces the hormone *estrogen.* When the egg is mature, it breaks out of the ovary into the abdominal cavity. This is called **ovulation** (ō′vyə lā′ shən). Slightly to the side and a little above each ovary is an **oviduct** (ō′və dukt), or **Fallopian** (fə lō′pē ən) **tube.** When the egg is released from the ovary, it is drawn into the opening of the oviduct by cilialike structures. The follicle in which the egg developed now becomes the **corpus luteum** (kôr′pəs lü′tē əm), or yellow body. The corpus luteum produces the hormone **progesterone** (prō jes′tə rōn′).

Progesterone stimulates the **uterus** (yü′tər əs) to prepare for the reception of a fertilized egg. The uterus is a muscular, pear-shaped organ that is about the size of a fist. The oviducts are attached to the upper end of the uterus. The lower end of the uterus is called the **cervix** (sėr′viks). The uterus is attached to the **vagina** (və jī′nə), or birth canal, which leads to the outside of the body.

Female reproductive system

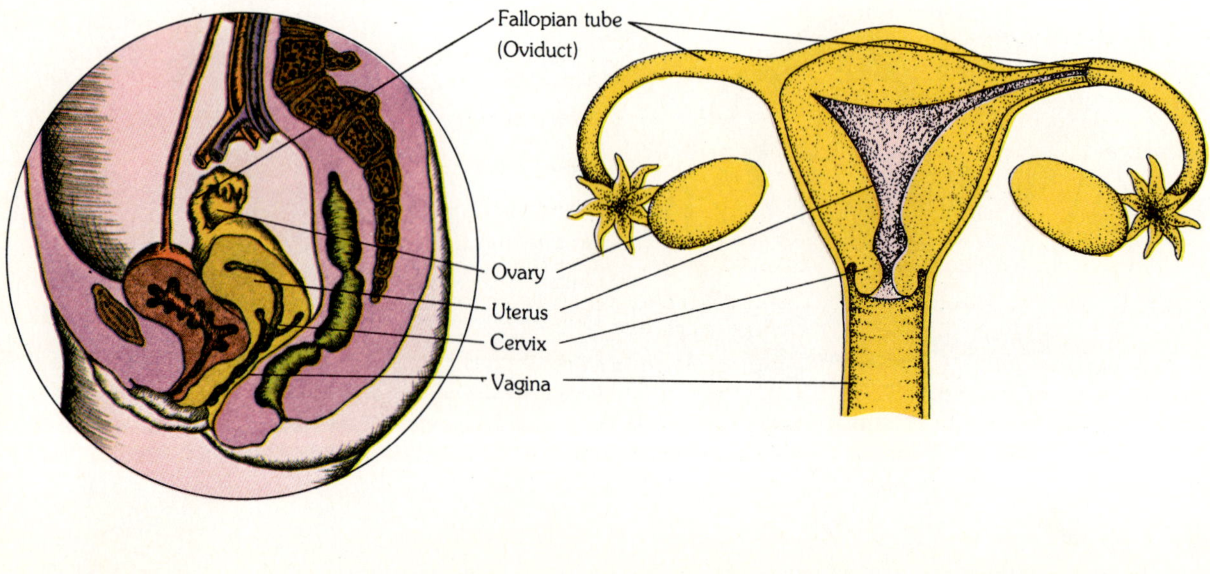

Menstruation

The egg will live for only about 24 hours after it is released from the ovary. If the egg is fertilized by a sperm, the egg, which would then be called a *zygote,* would attach itself to the lining of the uterus about the seventh day after fertilization. If the egg is not fertilized, it will simply pass out of the body.

Progesterone stimulates the uterus to produce a spongy tissue that is rich in blood. This tissue lines the inside of the uterus. The purpose of this lining is to provide nourishment for a fertilized egg. If the egg passes out of the body without being fertilized, there is no longer any need for the uterine lining of blood-rich tissue. The tissue breaks down, and the disintegrating blood vessels then pass out of the uterus through the cervix, into the vagina, and on out of the body. This flow of blood and tissues out of the uterus is called **menstruation** (men′strü ā′shən). After menstruation has occurred, the uterine lining begins to build up again. The repeated building up and shedding of the uterine lining and release of an egg is called the **menstrual** (men′strü əl) **cycle.**

Hormonal Control of Reproduction

Reproductive processes are controlled by a variety of hormones. They are produced in certain glands and travel in the blood to the target tissues. At puberty, the hypothalamus produces a chemical substance that travels to the pituitary gland through a network of blood vessels. This substance stimulates the pituitary gland to secrete two hormones—**follicle stimulating hormone** (FSH) and **luteinizing** (lü′tē ə nīz ing) **hormone** (LH).

In males, FSH stimulates the seminiferous tubules of the testes to manufacture sperm cells. The amount of FSH is kept at a constant level in males. The constant level of FSH allows sperm to be produced at a continuous rate. The target tissues of LH in males are comprised of cells that lie between the seminiferous tubules of the testes. These cells produce *testosterone,* the male sex hormone.

Reproductive Hormones

Hormones	Main Source	Main Effect
1. Estrogen	Ovary (follicle)	Thickens uterine lining Female sex characteristics
2. FSH	Pituitary	Stimulates gamete production (egg and sperm) Stimulates estrogen production
3. LH	Pituitary	Stimulates production of testosterone and progesterone
4. Progesterone	Corpus luteum	Prepares uterine lining for implantation of fertilized egg
5. Testosterone	Testes	Sperm production Male sex characteristics

Female Hormone Cycle

In females, FSH stimulates the growth of follicles in the ovaries. As an egg develops within the follicle, the cells that surround it produce the sex hormone *estrogen*. When the egg has matured, there is a large surge of LH and a slight surge of FSH, and ovulation occurs.

After ovulation, the follicle becomes the yellow corpus luteum. The cells of the corpus luteum, under the influence of LH, produce another hormone, *progesterone*. This hormone keeps the lining of the uterus from passing out of the body. If the egg is not met by a sperm and fertilized, the corpus luteum breaks down and no longer produces progesterone. The spongy lining of the uterus is then sent out of the body. Note the relationships between the menstrual cycle and the production of hormones in the following chart.

Hormonal control of the menstrual cycle

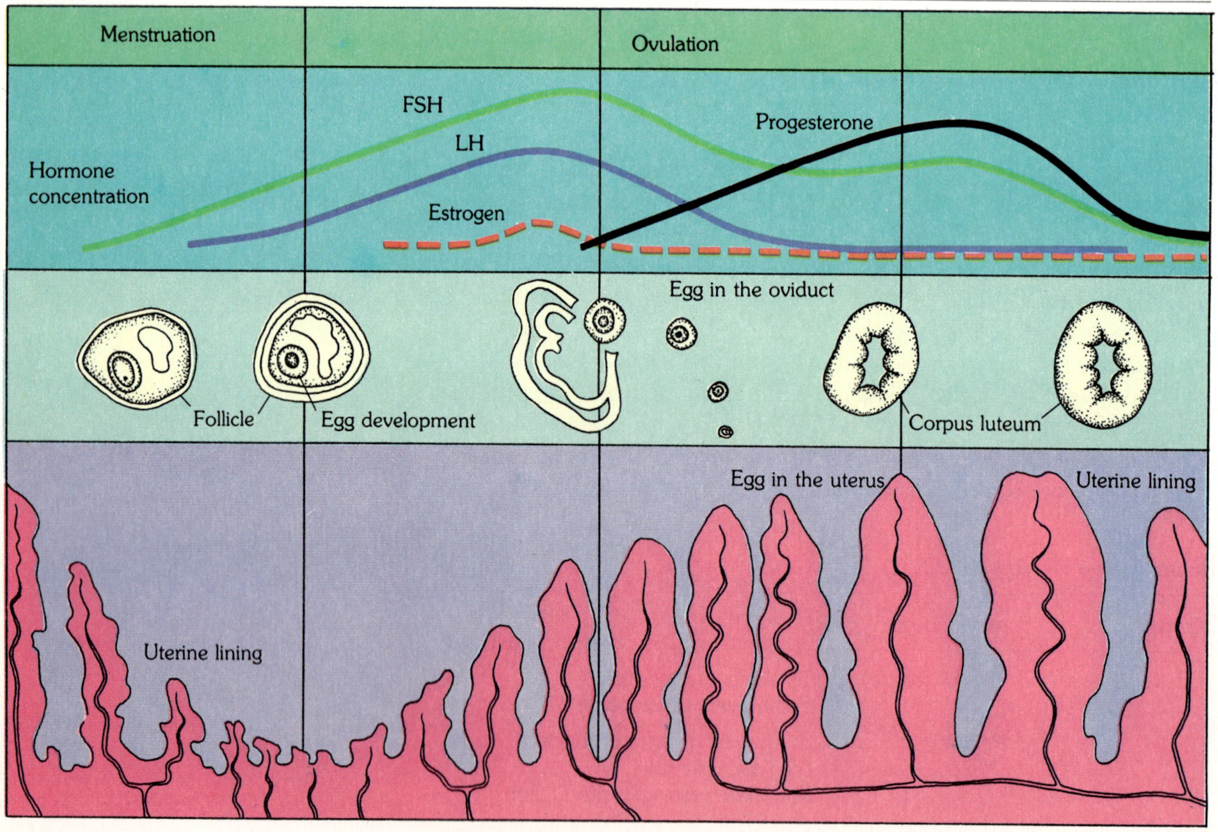

Hormone Feedback System

You may have wondered how the body regulates a gland to produce just the right amount of a hormone at just the right time. How, for example, do the testes secrete a constant level of testosterone? What prevents them from producing too much or too little?

The production of testosterone is controlled by a *negative feedback system* as discussed in Chapter 21. The hypothalamus stimulates the pituitary gland to release luteinizing hormone (LH). LH stimulates the production of testosterone in the testes. As the testosterone builds up in the blood, it acts on the pituitary. The pituitary then acts to slow the production of LH. In effect, the production of testosterone is "turned off" when the level of testosterone in the blood becomes too high. This turning off of the production of hormones is what gives this system of regulation its name. A negative feedback system shuts something off. A positive feedback system turns something on.

In males, this on-off feedback system produces a nearly constant level of hormones. However, in females the hormones are produced in cycles. The menstrual cycle is a very complex feedback system. Estrogen and progesterone build up slowly in the blood. As a result, they turn themselves off and on slowly. When the production of estrogen from the follicle reaches its peak, the follicle stimulating hormone (FSH) is turned off. Production of the luteinizing hormone begins. The LH then stimulates the formation of the corpus luteum. The corpus luteum in turn produces progesterone and some estrogen. As the level of progesterone in the blood rises, it turns itself off. This entire feedback system repeats the cycle about every 28 days.

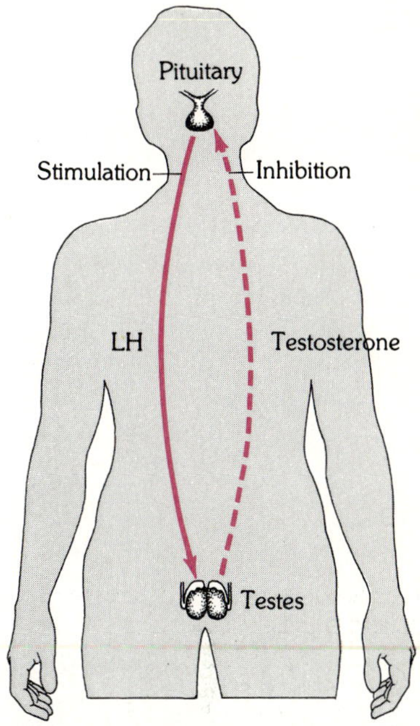

Negative feedback control of testosterone production

Review It

1. List and describe the function or functions of each of the structures of the male reproductive system.
2. List and describe the function or functions of each of the structures of the female reproductive system.
3. What are the functions of the follicle stimulating hormone and the luteinizing hormone in the male and the female?
4. What is the importance of reproduction to the survival of the species?

25–2
Human Development

Fertilization occurs when a sperm cell meets and joins an egg cell. This usually occurs in the upper part of the oviduct. Released into the vagina, several hundred million sperm cells begin to travel through the vagina, the uterus, and into the oviducts. Only a few sperm cells will reach the area near the egg, and only one sperm will fertilize the egg cell. Sperm are believed to be able to live within the female reproductive ducts for several days. However, the period of time that fertilization can take place is relatively short, because the egg cell degenerates after about 24 hours following its release from the ovary.

If fertilization does occur, **pregnancy** (preg′nən sē) begins. Pregnancy is the **gestation** (je stā′shən) period, or the time from the fertilization of the egg to the birth of the baby.

Sperm cells contain 23 chromosomes, or half the number of chromosomes in a human body cell. The egg also contains 23 chromosomes. The head of a sperm cell is covered with a cap containing digestive enzymes. As sperm approach an egg, they are stimulated to release this digestive enzyme. It dissolves pathways through the small cells that surround the egg. When a sperm and the egg meet, their membranes fuse. The nucleus of the sperm enters the cytoplasm of the egg. Changes in the membranes surrounding the egg keep other sperm from entering. The fertilized egg is now called a *zygote.* The zygote has the normal human set of 46 chromosomes. It has received 23 chromosomes from the father's sperm cell, including an X or a Y chromosome that determines the sex of the zygote, and 23 chromosomes from the mother's egg cell.

Implantation of egg after fertilization

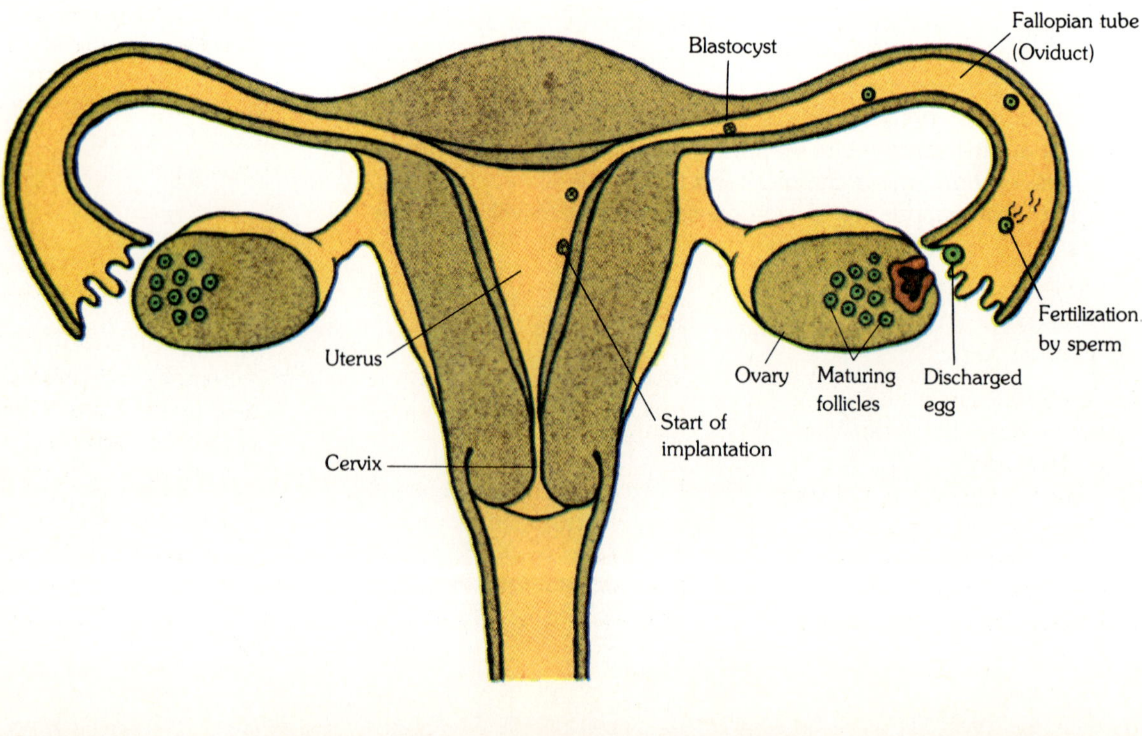

Activity

Sex Determination

Purpose
To examine the probability of obtaining a ratio of 50:50 of male and female children.

Materials
2 pennies
Masking tape

Procedure A
1. Put a small piece of masking tape on both sides of each coin. Mark an X on both sides of one coin. On the other coin mark X on one side and Y on the other side.
2. On a separate sheet of paper draw a chart to record the results of your coin tosses.

Possible Combinations	Number of times the combinations appear
XX	
XY	

3. Predict how many XX combinations will occur in 50 tosses of both coins.
   Predict how many XY combinations will occur in 50 tosses of both coins.
4. Flip both coins 50 times. Record the results of each flip on the chart.
5. Obtain the results of the coin tosses from 9 of your classmates. Record these results on charts similar to the one below. Total the results.

Results of Coin Tosses	XX Combinations	XY Combinations
1. (your own results)		
2.		
3.		
3.		
4.		
5.		
6.		
7.		
8.		
9.		
10.		

Analysis
1. How many of your tosses resulted in XX, in XY? Does this agree with the results you expected?
2. What are the totals of the XX and XY combinations that were obtained from adding your results and those of your classmates?

   Are these results close to the expected ratio of 50:50 XX and XY?

   If the chance of having an XX combination (a girl) or an XY combination (a boy) is 50:50, in a family of four children, one would expect to find two boys and two girls. However, not all families of four children consist of two boys and two girls.

Procedure B
1. On a separate sheet of paper copy the following chart.

Possible Combinations	Number of times combination appears
4 boys-0 girls	
3 boys-1 girl	
2 boys-2 girls	
1 boy-3 girls	
0 boys-4 girls	

2. Toss the coins a total of four times. Make a mark on the chart showing the combination of children represented by the tosses. Repeat the coin tosses in groups of four 24 times for a total of 100.

Analysis
1. Which combination of children occurred most often? Is this what you had expected?
2. Of the 100 total children indicated by the tosses of the coins, how many were boys? How many were girls?
3. Explain why the actual number of boy and girl children born into a family may be different from the expected 50:50 ratio.

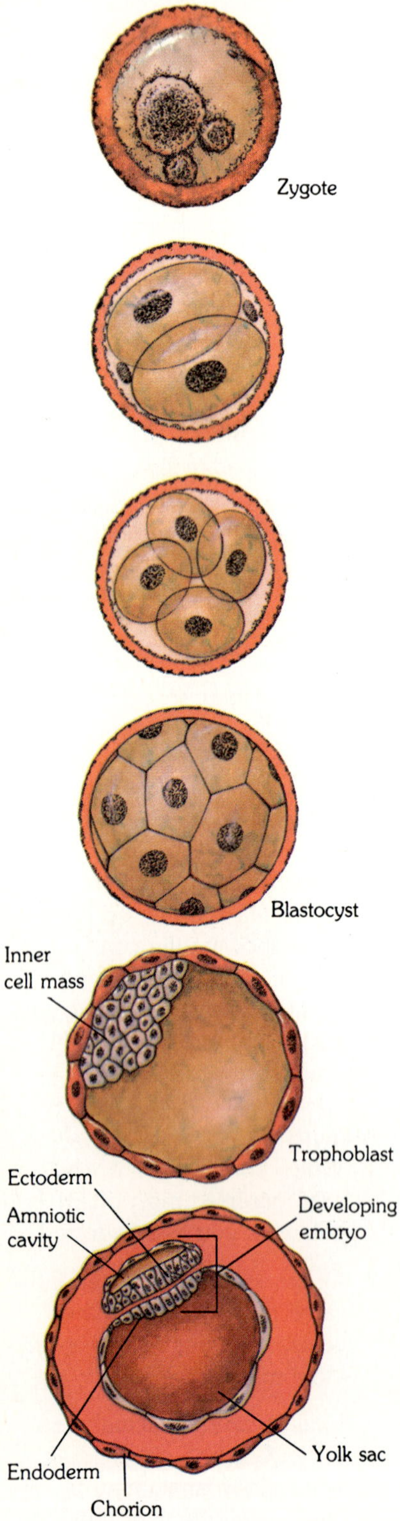

Early Development

About thirty hours after fertilization has taken place, the zygote divides into two cells. During the three days that it takes the zygote to move through the oviduct, the cells continue to divide by mitosis. As each division is made, the cells become smaller. These cells are held together within a membrane. On about the third or fourth day, the zygote is made up of a ball of 16 cells. However, it is still no larger than the egg was at the time it was released from the ovary.

The zygote continues to divide after it has entered the uterus. About the fourth or fifth day it takes on the form of a hollow ball known as a **blastocyst** (blas'tə cist). At this stage, it obtains its food from fluids secreted by tissues of the uterus. The blastocyst contains two distinct kinds of cells. Each type of cell will follow a separate course of development. The outer, hollow ball of cells of the blastocyst will attach itself to the lining of the uterus and will help it to become buried in the uterine tissue. This outer ball of cells will eventually develop into a membrane called the **chorion** (kôr'ē on). The inner cells of the blastocyst will develop into the **embryo** (em'brē ō) and additional embryonic tissues.

On the seventh day after fertilization, the **trophoblast** (trō'fō blast) as it is now called, begins to bury itself in the inner wall of the uterus. This process is called *implantation* (im' plan tā'shən). Soon after implantation, the cells of the embryo produce an extremely important hormone. This hormone from the embryo signals the corpus luteum to continue to secrete progesterone and estrogen. If estrogen and progesterone production were lowered, the uterine lining—including the implanted embryo—would be shed as it is at the end of the menstrual cycle.

As the inner clump of cells of the trophoblast continues to develop, several embryonic membranes begin to form. The amnion lines the amnionic cavity and the yolk sac develops even though no "yolk" is available for nourishment. Three primary germ layers, the **ectoderm** (ek'tə dėrm'), **mesoderm** (mes'ə dėrm'), and the **endoderm** (en'dō dėrm'), will eventually differentiate to form all of the tissues, organs, and systems of the body. The ectoderm will help to form the brain, spinal cord, lining of the mouth, hair, nails, lenses of the eyes, and the epidermal layer of the skin. The mesoderm differentiates into the skeleton, the muscles, circulatory system, reproductive organs, kidneys, and the dermal layer of the skin. The endoderm will form the lining of the digestive tract, the bladder, urethra, liver, pancreas, and lungs. About two weeks after fertilization, all of the major parts of the body have started to form. By definition, the developing organism is an embryo until seven or eight weeks after fertilization. It is thereafter known as a **fetus** (fē'təs).

The Embryonic Membranes

Humans develop from an *amniotic egg* as do other mammals, reptiles, and birds. You may recall that embryos developing from such eggs produce four kinds of membranes. The *amnion* lining the amniotic cavity, fills with a liquid called the *amniotic fluid.* The amniotic sac and fluid provide a protective, shock-absorbing cushion for the developing young.

Another sheet of membrane surrounds the entire embryo. This membrane, as previously mentioned, is known as the chorion. It develops fingerlike projections called *villi,* that penetrate the tissues of the uterus. Tissues of the mother's uterus and the chorion grow together to form a disk-shaped organ known as the **placenta** (plə sen'tə). Through the placenta, food and wastes are exchanged between the blood of the embryo and the mother by diffusion. A structure known as the **umbilical** (um bil'ə kəl) **cord** connects the embryo with the chorion.

A third membrane, the yolk sac, forms from the gut of the embryo. The yolk sac has no known feeding function in humans as it does in reptiles and birds. The fourth membrane, the *allantois,* is also a sac that grows out of the gut of the embryo. The tissues of the allantois become the blood vessels that connect the embryo with the chorion in the umbilical cord and placenta.

Human embryo in embryonic membranes

Cross section of placenta and uterine wall

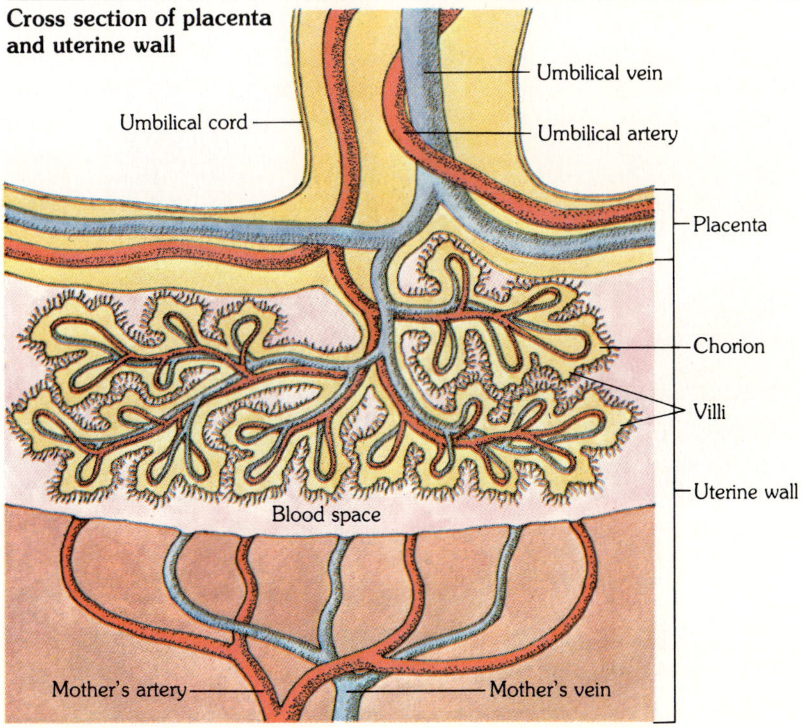

Did You Know?

Sex hormones are also produced by organs other than the testes and the ovaries. The organ that attaches the embryo to the wall of the uterus of its mother produces estrogen and progesterone. The adrenal glands on the kidneys produce testosterone in both males and females. The bearded ladies seen in circuses often had tumors of the adrenal glands. These tumors caused the adrenal glands to produce large amounts of testosterone, which was responsible for the development of secondary sex characteristics usually found only in males.

Challenge!
Much new information is now available about the effects of alcohol, smoking, drugs, and diet on an unborn child. Investigate the studies that have been done in these areas. Why is prenatal (before birth) care important to both the mother and her child?

The First Trimester of Pregnancy

Each of the three-month periods before birth is called a *trimester* (trī-mes'tər). During the first trimester, the embryo changes from a single cell to a miniature mammalian form firmly attached to tissues of the uterus. During the first several weeks of development, the embryo forms from a disk of cells. A groove forms down the middle of the disk. Folds develop over the groove, forming the beginnings of the spinal cord and brain. The growth of the brain is rapid. Soon the anterior portion of the embryo bulges and bends toward the tail. As the sides of the disk grow down and around the body, the embryo takes on the shape of a cylinder.

By the end of the first month, the embryo is about 5 mm long. It possesses such vertebrate characteristics as a tail, gills, notocord, and segments. A very simple heart has begun beating. Arm and leg buds are present.

Development of the fetus

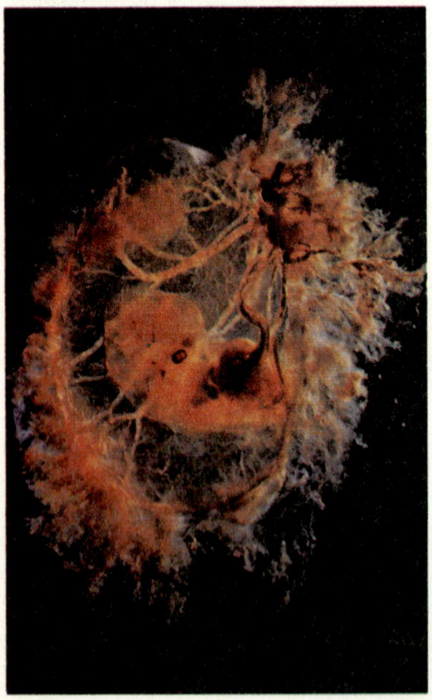

5 weeks

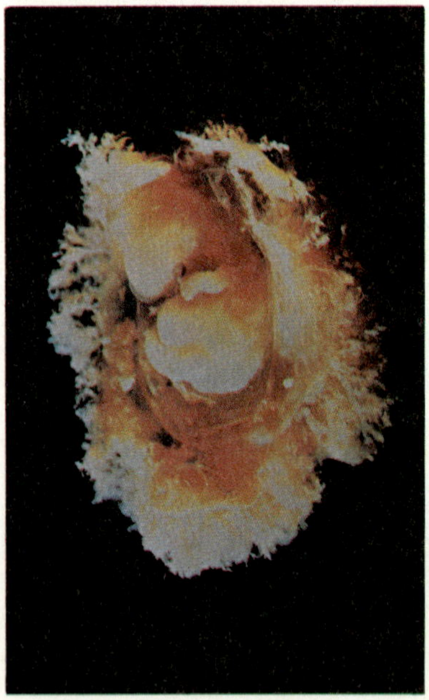

6 weeks

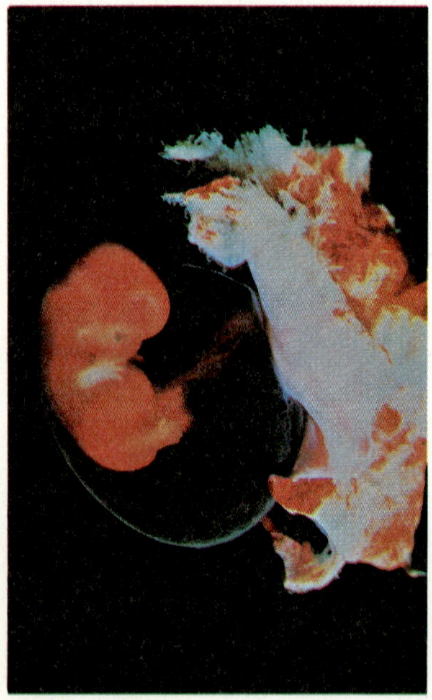

7 weeks

Actual size of embryo

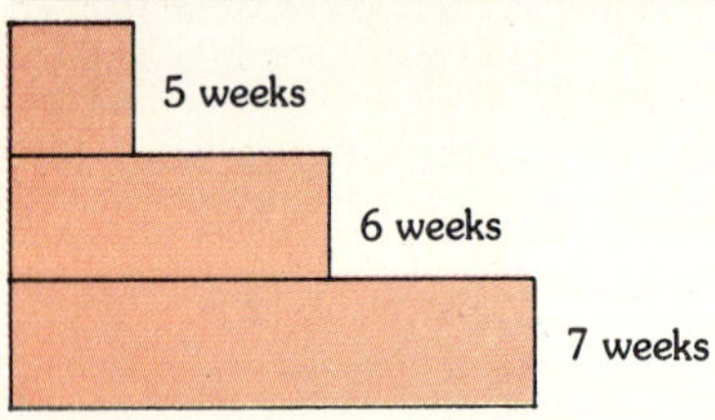

Major developments

5 weeks: Eye, extensive vascular system with looped heart

6 weeks: Fingers apparent, bulbous brain, tail

7 weeks: Rapid growth of liver, the main blood forming organ at this time

The Second and Third Months

At about two months, the embryo is 25 mm long. The face is unmistakably human, with clearly defined eyes, ears, and nose. The embryo is now considered to be a fetus. Fingers, toes, elbows, and knees have formed. The tail has begun to shorten. The sex organs have formed; however, it is difficult to tell if the fetus is male or female.

By the end of the third month, the fetus is about nine centimeters long and weighs up to 30 grams. Female fetuses have developed a clearly defined uterus. Buds for the baby teeth are present. Fingernails have begun to form. The fetus can frown, squint, and make other facial movements. It can also move its arms and legs, but the movements are too weak to be felt by the mother.

Special care must be taken to protect the fetus during the first trimester. It is now well known that seemingly harmless substances can cause birth defects or cause cancer many years after birth. The developing embryo is sensitive to drugs, medicines, diseases, and X rays that may pass to it through the mother's body.

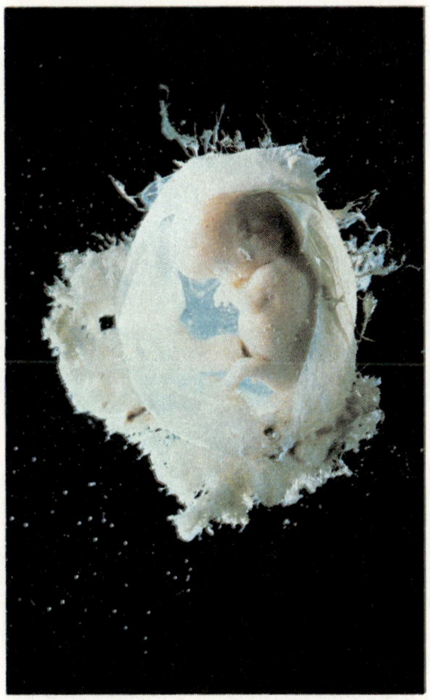

9 weeks

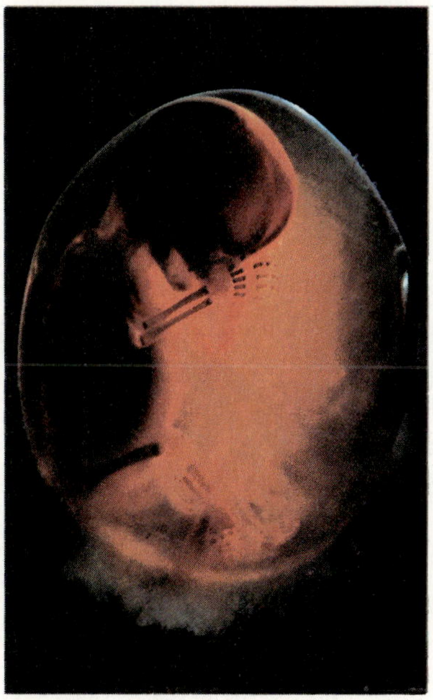

10 weeks

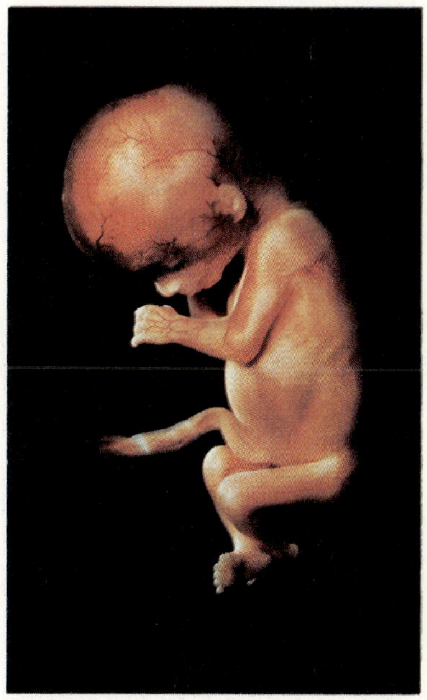

14 weeks

Actual size of embryo

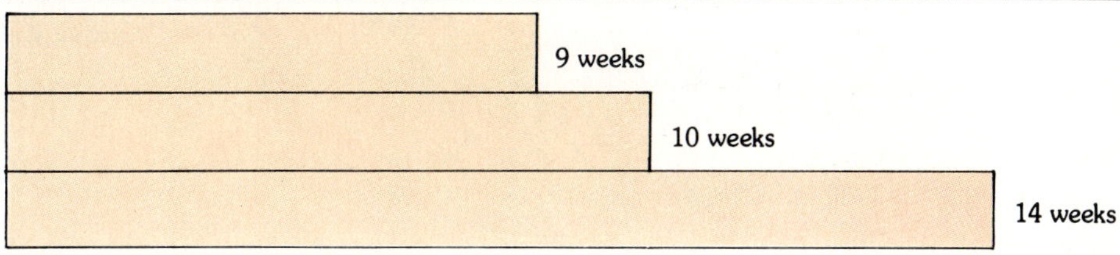

Major developments

9 weeks: Eye lids and outer ear forming, movement, may suck thumb

10 weeks: Bones developing, no joints yet

14 weeks: Fingerprints, fingernails present, vigorous movement

The Second Trimester

From the beginning of the fourth month to the end of the sixth month of development, the fetus increases from 9 to 36 centimeters in length and from 30 to 680 grams in weight. During this time the skeleton begins to form. Tiny, fine hairs appear on the skin. The skin is red and wrinkled. The mother can occasionally feel faint movements of the fetus. The heart beat becomes strong enough to be heard with the aid of a stethoscope placed on the mother's abdomen. A six-month-old fetus may be able to survive outside the body of the mother. To do so, however, it needs special medical care.

The Third Trimester

During the last three months before birth, the fetus grows considerably in size and weight. Fat is deposited beneath the skin, causing the wrinkles to disappear and the body parts to become rounded. Most of the fine hair on the body disappears. Nerve tissues, especially the cerebral cortex of the brain, form and develop during the last several months of pregnancy. Brain waves from the fetus are detectable at this time.

Recent studies suggest that mothers should be particularly careful to eat enough protein during this period. Research also indicates that, if the mother drinks alcohol during this period, there may be damage to tissues of the fetus. Pregnant women should avoid drinking alcohol. They should also avoid smoking. Mothers who smoke face a much greater risk that their babies will die before birth or shortly thereafter. Also, the babies of smoking mothers are more likely to weigh less than normal at birth and therefore will have a poorer start in life.

Third term fetus

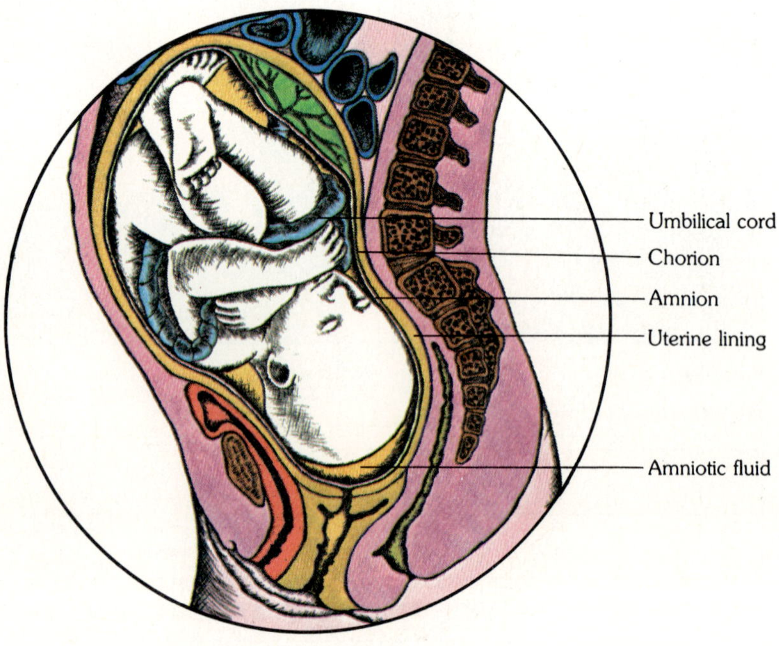

Birth

The human gestation period is normally thirty-eight weeks from fertilization to delivery. However, live births can occur as early as 28 weeks or as late as 45 weeks. During pregnancy the uterus usually increases to about 24 times its normal size. By the end of the pregnancy, the fetus may weigh 3000 grams, or about one-twentieth (1/20) the weight of the mother.

The process of childbirth, or labor, is started and controlled by a variety of hormonal events. Labor is divided into three stages: dilation, expulsion, and the placental stage.

During the first stage of childbirth, the cervix dilates, or becomes larger. This is the longest stage of childbirth. The opening of the cervix is initially quite small. It must enlarge from a diameter of one to two centimeters to 10 to 11 centimeters before the fetus can leave the uterus. Dilation begins when the muscles of the uterus begin to contract. These muscle actions of the uterus gradually force the fetus and the amniotic sac downward. At first the contractions are mild and occur about 20 minutes apart. Later, the contractions become stronger and may occur as close as a minute apart. Often, toward the end of this stage, the amniotic sac breaks.

Dilation stage

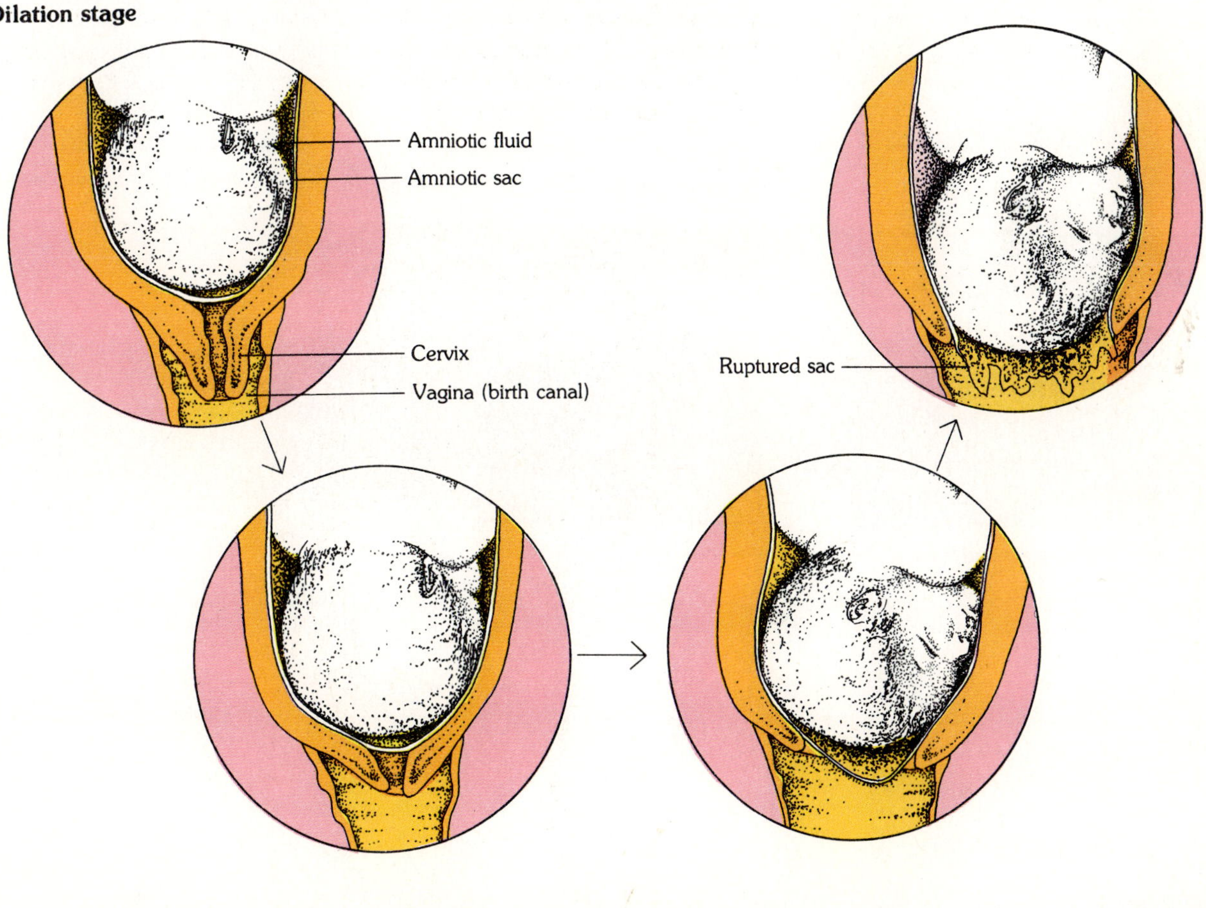

The Final Stages of Birth

The expulsion stage begins when the cervix is completely dilated and ends with the birth of the baby. This stage is characterized by strong and frequent contractions of the uterus. The mother can help by contracting the muscles of her abdominal wall. The expulsion of the fetus from the uterus to the outside of the mother's body usually takes from half an hour to two hours.

In the last stage of birth, the placental stage, contractions of the uterus dislodge the placenta from the wall of the uterus. Continued contractions expel blood, other tissues, fluids, and the placenta through the vagina. The placenta is squeezed as it passes out of the body. This action forces fetal blood from the placental blood vessels into the body of the newborn baby. After the umbilical cord stops pulsating, the cord is tied and then cut. The placental stage of birth lasts from ten to fifteen minutes.

Stages of birth

1. First stage of labor begins

2. Cervix dilates

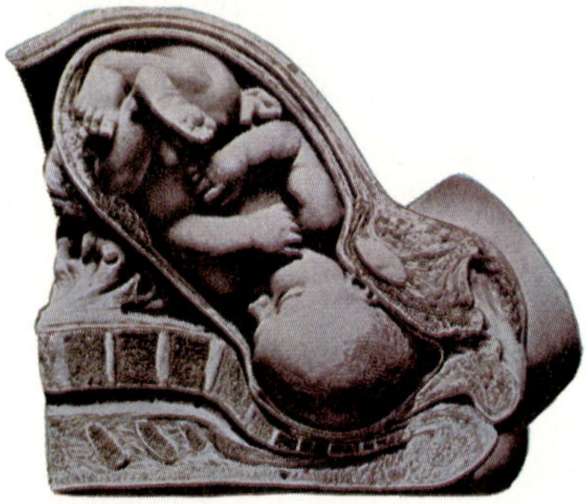

3. Second stage begins: cervix is fully dilated

4. Baby's head becomes visible

5. Second stage continues

6. Birth of baby

Growth and Development After Birth

During its nine-month gestation period, a human grows from a zygote 0.1 centimeter in diameter to an average length at birth of 50 centimeters. From birth to adulthood, at about age 20, the human being grows to an adult height that is about three and a half times what it was at birth.

All parts of the body do not grow and develop at the same rate. When a baby is born, about 25 percent of the length of the baby is its head. The head of an adult is about one seventh of the total body length. Humans do not grow at a uniform rate. Between birth and the end of the third year, body growth is rapid. Growth then levels off until puberty, when sudden spurts of growth occur. The average maximum spurt of growth for girls is about nine centimeters a year at age 12. The average maximum spurt of growth for boys is about 10 centimeters a year at the age of 14. After puberty, the rate of growth drops rapidly. By age 18, it is nearly zero. In North America today, adult females average 170 centimeters in height. Adult males average 179 centimeters.

Different parts of the body reach adult size at different ages. The brain and head reach nearly their adult size by about age nine. The reproductive system grows slowly until puberty, when it begins to grow rapidly.

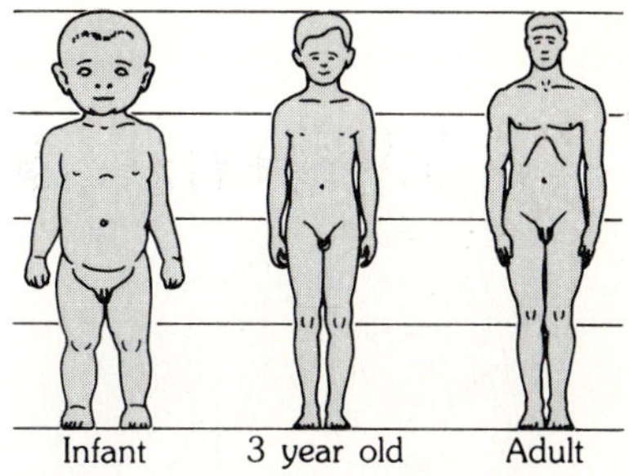

Change in body proportions during growth

Development

Aging

We usually think of aging as certain obvious biological changes that occur in our grandparents. Aging is often thought of as the graying of hair, wrinkling of the skin, weakening of muscles, and a slowing of the reflexes. Actually, some parts of the human body begin to age before birth. For example, the ovaries of a human female fetus contain several million eggs during the sixth month of development. By the time the fetus is born, 90 percent of these eggs have been reabsorbed.

Humans peak at different times in different activities such as long-distance running, childbearing, and intelligence. The physical conditions of the body that are best for intense physical activity such as speed swimming often peak during the teenage years. Olympic swimmers in their twenties are considered old men and women in this sport.

Aging is a decrease in the capacity of an organism to carry out life processes. As organisms become older, they do not see, hear, smell, walk, breathe, and digest food as well as they did when they were younger. For every species there is a maximum period of life. Aging is a genetic trait of all species. Different aging patterns vary with different species.

Chris Evert in her 20's
Women's tennis champion

Achievements continue through life

Dr. Martin Luther King in his 30's
Nobel Prize recipient

Dame Margot Fonteyn in her 50's
and Rudolf Nureyev in his 30's
Famed ballet performers

Arthur Fiedler in his 80's
Renowned conductor

Challenge!
With the life expectancy of human beings increasing on a regular basis, the population has a larger percentage of older people. What facilities are available for older people in your community? What problems are faced by people who are 60 years of age and older? What special benefits might older persons offer to a community?

Review It

1. Trace the development of the fetus in the first, second, and third trimesters of pregnancy.
2. How is the sex of the fetus determined?
3. Describe the events of birth as they occur in the dilation, expulsion, and placental stages of birth.
4. What precautions should a pregnant woman take to protect the health of her unborn child?

Breakthroughs in Biology

Recent research into fetal alcohol syndrome gives prospective parents a better understanding of the dangers to a fetus from alcohol.

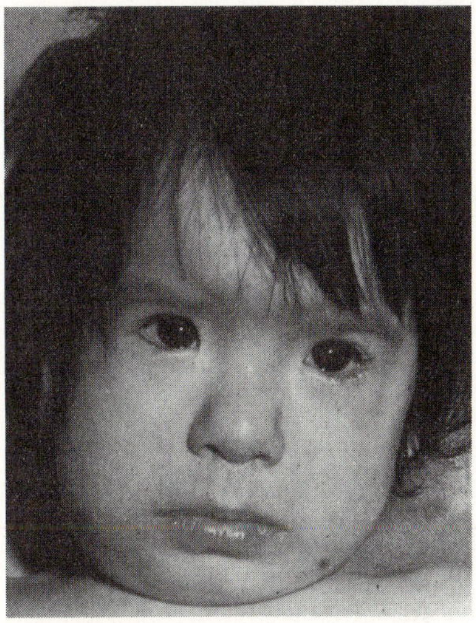

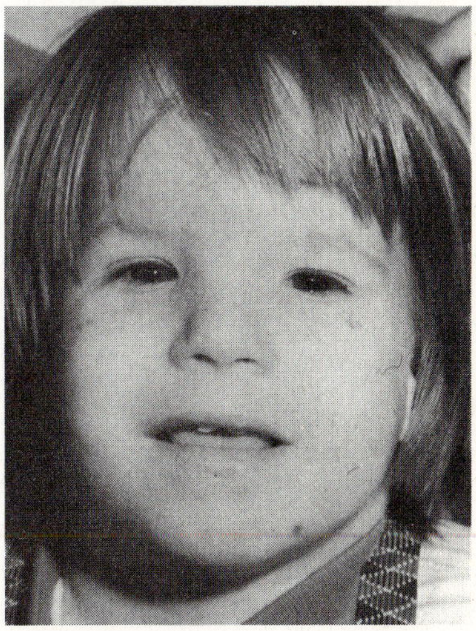

People used to think that the placenta protected an unborn child from the effects of drugs taken by its mother. It is now known that most drugs pass from the mother's bloodstream, through the placenta, and on to the fetus. The unborn are very vulnerable to the effects of drugs. And whether or not they will be damaged depends upon the drug, the dose, and the month of pregnancy.

Many people are aware that certain medications and illegal drugs are *teratogens,* agents that cause birth defects. But more babies are harmed by a substance that is widely used and isn't often thought of as a drug—alcohol.

In an adult, alcohol affects the brain and various other organs. The severity of the effect will depend on the body's ability to metabolize the alcohol, a factor that varies for different people. Some scientists believe the damage to the fetus is due to its inability to metabolize the alcohol as an adult would.

Babies born to mothers who are heavy drinkers are often severely damaged or stillborn. There is a higher than average risk of premature birth; mental retardation; physical deformities such as small heads, narrow eyeslits, heart defects, and joint malformations. Trapped for nine months, the fetus can become a victim.

Dr. David Smith of the University of Washington, where fetal alcohol syndrome was first identified in the U.S., estimated that about one to two of every 1,000 infants are born with some type of defect caused by the mother's drinking during pregnancy. In a study of 1,500 women it was shown that even moderate drinking can have mild effects on unborn babies. It is difficult to know exactly what is a dangerous amount of alcohol for an expectant mother to drink, but one to two ounces a day is considered risky during any stage of pregnancy.

Analysis

1. Why would a fetus not be able to metabolize alcohol as an adult would?
2. Write to an agency such as the Federal Drug Administration for a list of other teratogens.

Chapter Summary

The longest time any type of plant or animal has been known to live is called the life-span of that species. Many organisms die long before reaching their full life-span. Life expectancy is the age to which 50 percent of a population can expect to live.

The growth and development of a human being starts at the moment the egg and sperm cell unite. Growth generally stops by adulthood, but development continues throughout life. Different organs and systems of the body develop and grow at different rates. As individuals become older, various tissues of the body lose the ability to function as well as they formerly did. This phenomenon of aging actually begins in some tissues before we are born.

At puberty, humans undergo specific physical changes that prepare their bodies for reproduction. These changes are brought about by the production of chemical substances called hormones. The production and transport of sex cells are controlled by hormones. Reproductive hormones are made in, and released from, the pituitary gland and other glands, including the testes and ovaries. Sex hormones are responsible for many of the physical differences between males and females. The regulation of the amounts of hormones in the blood during the reproductive cycle is managed by feedback systems. At puberty, males begin to produce sperm continuously. At puberty, the ovaries of females usually release one egg each month as part of the menstrual cycle.

Fertilization of an egg occurs in an oviduct. The resulting zygote divides by mitosis while it moves toward the uterus. About seven days after fertilization, the blastocyst attaches itself to the lining of the uterus. With protection and support from the membranes of the egg and the development of the placenta, the fetus is nourished through the gestation period of approximately 40 weeks. Through the process of birth, the fetus leaves its mother's body and begins life as an independent being.

Questions to Think About

1. How is it possible for a mother to provide food, oxygen, and excretion for a fetus if there is no direct connection between the circulatory system of the mother and the fetus?
2. Having fraternal, or two-egg twins, appears to be an inherited genetic trait. Would mothers of fraternal twins produce more or less FSH than average? Explain.
3. The age at which puberty begins has been decreasing during the last fifty years. What factors may account for this change?
4. Based on what you know about the production of sperm and eggs, why might a man or woman be unable to produce children?
5. What are some of the common symptoms of advancing age?
6. Children usually go through a period of "latency" prior to puberty. What is meant by this term? What does it mean in terms of a child's development?
7. In recent years there has been a marked increase in the number of multiple births. What medical discoveries have made multiple births more common?

Extra Research

1. Human chorionic gonadotropin (HCG) is a hormone secreted by an embryo just after implantation. How does this hormone affect the mother? How is it useful in pregnancy tests?
2. What is an ectopic pregnancy? Is it possible for a fetus to develop and survive under these conditions? What is the usual outcome of such a pregnancy?
3. Recent research has linked smoking with premature wrinkling of the face. Investigate these studies. What may be the causative agent in cigarettes that brings about this early wrinkling?
4. What are the benefits of breast-feeding to both the newborn infant and the mother?

Chapter Test

Part I

Copy the number of each test item. If the item is true, write *T* after the number; if the item is false, write *F*.

1. Reproduction ensures the survival of the individual. (593)
2. Testes in the scrotum are kept at a temperature lower than body temperature. (595)
3. Sperm are stored in the epididymis between the testes and the urethra. (595)
4. The corpus luteum forms in the ovary from cells of the follicle. (596)
5. At the time of ovulation there is a surge of estrogen and progesterone. (598)
6. Once a sperm cell enters an egg, additional sperm cells are prevented from entering the egg. (600)
7. A human zygote contains 23 chromosomes. (600)
8. The human blastocyst obtains its food from the yolk before it embeds itself in the uterus. (602)
9. The amnion forms a sac that fills with fluid and surrounds the embryo. (603)
10. Most of the development of an embryo occurs during the second trimester of pregnancy. (606)

Part II

Copy the number of each sentence below. After each number, write the word (or words) that completes the sentence correctly.

11. The longest time any one kind of animal has been known to live is called the ____ of that organism. (594)
12. A fertilized egg is called a ____. (600)
13. Chemicals that help to determine human sex traits are sex ____. (595)
14. The repeated shedding and rebuilding of the human female's uterine lining is called the ____. (597)
15. During the first stage of labor, the ____ dilates from one centimeter to 10 centimeters. (607)
16. The ____ is the disk-shaped organ that is expelled through the vagina after birth. (608)
17. The hormone ____ is secreted by the corpus luteum and prepares the uterus for implantation. (602)
18. After fertilization, the embryo spends about ____ days in an oviduct, and then about ____ more days in the uterus before implantation. (602)
19. A negative feedback system is one in which a high concentration of a hormone causes the hormone production to ____. (599)
20. The part of the brain that stimulates the pituitary gland's production of reproductive hormones is the ____. (595)

Part III

Copy each numbered item from List A. Write the letter of the phrase in List B next to the item from List A that it best matches.

List A

21. testosterone (597)
22. estrogen (597)
23. ovulation (596)
24. umbilical cord (603)
25. amniotic fluid (603)

List B

a. protective material around fetus
b. tubelike structure linking the fetus and the uterus
c. male sex hormone
d. female sex hormone
e. release of an egg from the follicle of an ovary

Focus on Careers

Medical Assistant
Can you remain calm when you've got a dozen things to get done *right now?* Are you organized? Can you listen carefully and follow directions *exactly?* If you can answer yes to all these questions, you might think about a career as a medical assistant.

Medical assistants help doctors in their offices. The work they do includes measuring blood pressure, height, weight, and temperature. An assistant may be trained to take X-ray pictures, use an electrocardiograph machine to record heartbeat patterns, and apply bandages. He or she may take blood samples and do routine laboratory tests. In addition, the medical assistant keeps records of patients, accepts payment for services, schedules appointments, answers the telephone, and greets patients. The exact duties vary with the type of office. However, a survey completed by the American Association of Medical Assistants listed over 400 separate tasks medical assistants perform on the job!

Most medical assistants are trained on the job. However, there is a one- or two-year program in medical assisting available in certain vocational schools and junior colleges. All medical assistants need a high-school diploma, and high-school courses in science, mathematics, and office practices are helpful.

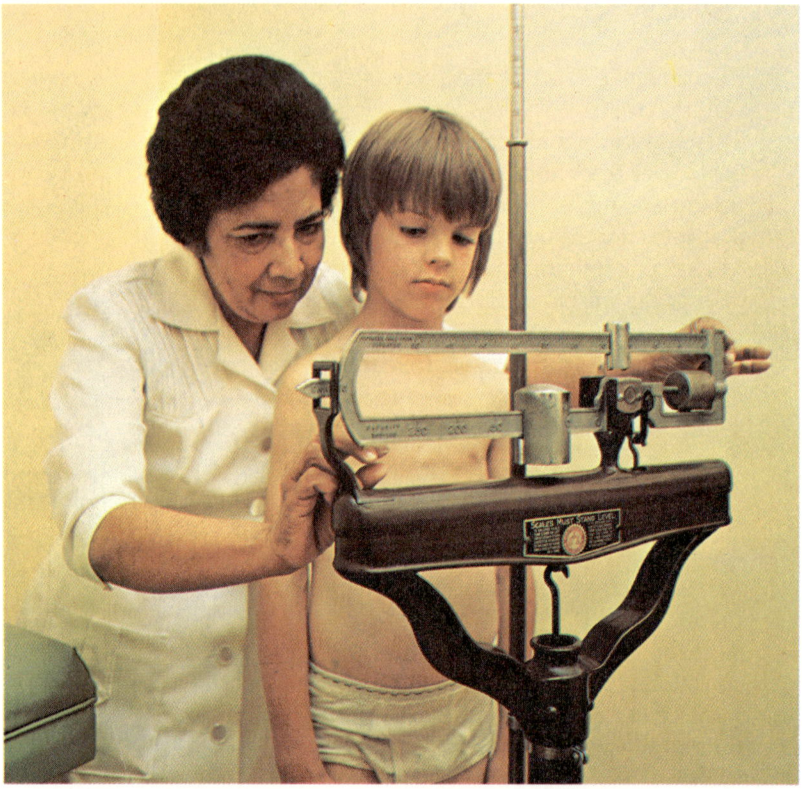

After several years of experience, a medical assistant can become certified or registered by passing an exam. Certification or registration is not required for employment. However, doctors usually regard medical assistants who have these credentials as being better qualified. There is a need for medical assistants in most parts of the country.

Career Information
Food Inspector: Interagency Board of U.S. Civil Service Examiners for Washington, D.C., 1900 E Street NW, Washington, DC 20415.

Medical Assistant: American Association of Medical Assistants, 1 E. Wacker Drive, Suite 1510, Chicago, IL 60601.

Medical Writer: American Medical Writers' Association, 5272 River Road, Suite 290, Bethesda, MD 20014.

Nurse: American Nurses Association, 2420 Pershing Road, Kansas City, MO 64108.

Food Inspector
Duties: Visits businesses that produce, handle, store, or sell food to make sure these products are being handled in a safe, sanitary manner. The inspector looks for anything that might harm someone's health: untruthful labeling, decomposition of food, chemical or bacteriological contamination. He or she may use such equipment as portable scales, cameras, and chemical testing kits. Written reports are prepared and sometimes are used as evidence in court.

Training: Usually a bachelor's degree or three years' work experience is required. Courses in biology and chemistry that include laboratory work are helpful. Learning rules, regulations, and inspection procedures is part of on-the-job training.

Opportunities: The need for food inspectors is expected to increase faster than the average for all occupations. Most are employed by the federal government. A passing score on the Professional and Administrative Career Examination (PACE) is required in order to work for the federal government.

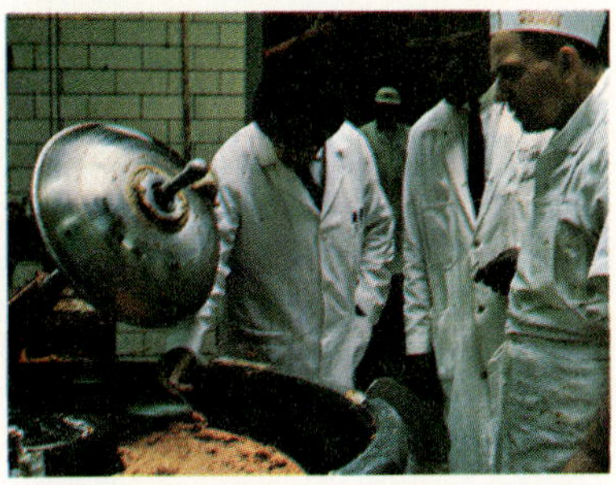

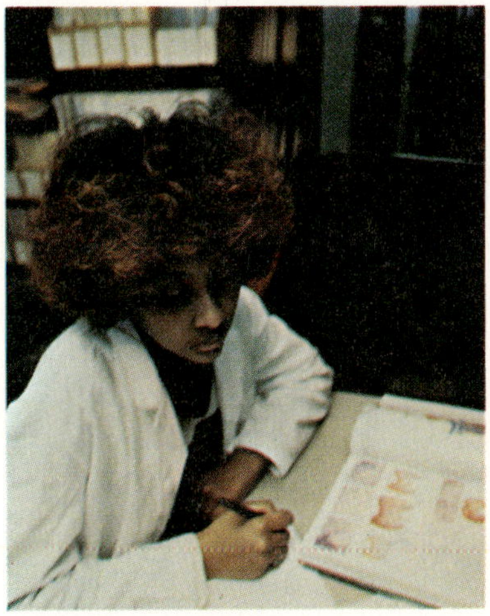

Medical Writer
Duties: Researches, writes, and edits materials related to health and medicine. For example, a medical writer might update a set of pamphlets on heart diseases. A writer must know how to fully research a topic, organize useful information, and present it in a clear, concise way. The writer often conducts interviews as a way of gathering information. Some writers use information gathered by someone else to prepare their finished products.

Training: No rigid requirements. Generally a college degree in science, plus courses in English, journalism, or technical communications is needed.

Opportunities: Look for jobs in research laboratories, textbook publishers, and government agencies. Experienced writers may work on a free-lance basis.

Registered Nurse (RN)
Duties: Cares for sick people and generally promotes good health. An RN has technical skills which include giving doctor-prescribed drugs and treatments, monitoring the progress of a patient in a hospital, and giving intravenous feedings and blood transfusions. Many RNs specialize, for example, in pediatrics or surgery. In hospitals, they supervise licensed practical nurses (LPNs), nurse's aides, and orderlies.

Training: RNs train in three ways: a four-year bachelor's degree, a two-year associate degree, or a two- to three-year hospital diploma program. Courses include biology, chemistry, anatomy, and supervised practice with patients. RNs with bachelor's degrees have the greatest opportunities for advancement.

Opportunities: Increased medical services, interest in rehabilitation and preventive medicine will increase the need for RNs.

Bibliography: Unit Nine

Asimov, Isaac. *The Human Body: Its Structure and Operation.* Houghton Mifflin Co., 1963. Good discussion of most body systems.

Asimov, Isaac. *The Human Brain.* Houghton Mifflin Co., 1963. Good discussion of endocrine glands, the brain, and sense organs.

Brooks, Stewart. *The Sea Inside Us: Water in the Life Processes.* Meredith Press, 1968. Describes the composition and functions of body fluids and explains the diseases that an imbalance of fluid may cause.

Clark, Randolph Lee, and Cumley, Russell W., Editors. *The Book of Health,* 3rd ed. Comprehensive volume containing accurate information on diseases and human physiology.

Flanagan, Geraldine Lux. *The First Nine Months of Life.* Simon and Schuster, 1962. Combines a clear narration with authentic illustrations of human development.

Froman, Robert. *The Many Human Senses.* Little, Brown and Co., 1966. Explains in understandable terms how human senses function. Also probes ESP.

Glasser, Ronald. *The Body Is the Hero.* Random House, 1976. Describes the history and development of knowledge about the immune system.

Hyde, Margaret O. *Your Brain: Master Computer.* McGraw-Hill, 1964. Describes in simplified terms the regions of the brain and their functions including memory and sleep.

Kahn, Fritz. *The Human Body.* Random House, 1965. Picture essay volume describing organ systems and their functions.

Kelly, Patricia. *The Mighty Human Cell.* The John Day Co., 1967. Discusses in simple terms the different types of tissues and their functions.

McNaught, Ann, and Callander, Robin. *Illustrated Physiology.* E. and S. Livingstone Ltd., 1967. Good collection of drawings that show and explain in simple terms the physiology of the human body.

Morrison, Thomas, and others. *Human Physiology.* Holt, Rinehart and Winston, Inc., 1967. Basic textbook of structure and function of the human body.

Nourse, Alan E. *The Body.* Time Inc., 1971. Picture essay volume describing human anatomy and physiology.

Pines, Maya. *The Brain Changers.* Harcourt Brace Jovanovich, Inc., 1973. Describes the exciting research in the attempt to solve the mysteries of the brain and discusses how scientists might use their power to change the brain.

Ratcliff, J. D. *Your Body and How It Works.* Reader's Digest Press and Delacorte Press, 1975. Excellent, interesting collection of publications based on the popular *Reader's Digest* series, "I Am Joe's . . . I Am Jane's . . ."

Reuben, David. *Everything You Always Wanted To Know About Nutrition.* Simon and Schuster, 1978. Describes the role of vitamins, fats, carbohydrates, and proteins in the human diet. Also discusses the relationship between food additives and chemicals to human ailments.

Rorvik, David. *Brave New Baby.* Doubleday and Co., 1971. Explores the fascinating, sometimes frightening possibilities and problems of genetic engineering, creating test-tube babies, and creating a mechanical body for a human brain.

Seeman, Bernard. *The River of Life.* W. W. Norton and Co., Inc., 1961. Traces the development of man's knowledge of blood from primitive religious rites to modern research.

Seeman, Bernard. *Your Sight: Folklore, Fact, and Common Sense.* Little, Brown and Co., 1968. Discusses various vision defects or mishaps that threaten vision, the superstitions and occult powers attached to the eye, and proper care of the eyes.

Snively, William D. *Sea of Life.* David McKay Co., Inc., 1969. Interesting discussion of the role of the body's fluids and the substances contained in them.

The Human Machine: Mechanisms. Doubleday and Co., Inc., 1965. Picture essay encyclopedia volume describing human body mechanisms.

Woodburn, John H. *Know Your Skin.* G. P. Putnam's Sons, 1967. Describes the structure of skin and probes the causes and cures of the most common skin disorders.

UNIT TEN

ECOLOGY

Chapter 26
Populations and Communities 619

Chapter 27
Ecosystems and Biomes 643

Chapter 26

Populations and Communities

Ecology is a study of the interactions among different kinds of organisms and between organisms and their physical environment. In this chapter you will study the behavior patterns within populations, the ways organisms are affected by their environment, and the food and energy flow in a community.

These Kodiak bears in Alaska are eating salmon, while the scavenging gulls wait patiently for leftovers. This relationship is part of an ecological community.

Chapter Objectives

1. Define and give an example of a population, a community, an ecosystem, a biome, and define biosphere.
2. Describe the dynamics of population growth.
3. Give examples of special feeding relationships within a community.
4. Describe the flow of energy through a community.

26-1

Human influence on environment

Overview of Ecology

In recent years the term **ecology** (ē kol′ə jē) has become a part of everyday conversation. To some people it represents environmental problems: oil spills, air and water pollution, and the misuse of natural resources. To others it represents undisturbed wilderness with blue skies, clear streams, and blooming wildflowers.

The field of ecology is broader than these views represent. The term *ecology* comes from the Greek word for house, *oikos*. Our planet Earth is the house in which all organisms reside. It is the house that contains all of the materials necessary to sustain life.

Ecology is not only concern over air pollution, oil spills, and pesticides. Rather, it is the study of the related interactions of living organisms with one another and with nonliving aspects of their surroundings. The environment is the living and nonliving components surrounding an organism. Ecology includes the influence this environment has on the living things and how living things can, in turn, influence their environment. It is also the study of the flow of energy from the sun through the plants and animals on Earth.

You have already recognized that topics like air pollution and oil spills have something to do with ecology. This is because such problems affect the environment and the living things in it. When a problem affects the environment, it also affects the organisms that live there because of their interrelationship.

Understanding Ecology

An understanding of ecology is important to us in two major ways. First, ecology provides the knowledge people must have to deal effectively with environmental considerations. Second, as living organisms, human beings affect and are affected by the environment.

Whenever human beings make changes in the environment, they influence the relationships of all the living organisms within that environment. Changes can also affect the physical setting of the environment, such as air and water quality and the shape of the land. In the past most of the changes people have made in the environment were done without understanding that altering one aspect of the environment changes the entire system. This is true of strip-mining for coal, plowing up prairie sod for agriculture, or paving a meadow for a parking lot.

Overview of Ecosystem Structure

Much of our knowledge of ecology comes from the study of ecosystems. To understand ecosystems, we first need to know the levels within an ecosystem.

The organization of an ecosystem starts with the **organism.** The organism may be simple and one-celled or complex and multicellular. A group of organisms of the same species living in a particular place at the same time is a **population.** For example, a population may be a single flock of animals or all the members of a plant species growing on a hillside or on a mountain range, or a mass of slime mold. It is seldom possible to study all the members of a species so the ecologist must decide on the size of the sample group to be studied. It may be the prong-horned antelope of a certain national park or the rats of a particular neighborhood.

Challenge!
Consult your local newspaper for current environmental problems. You may see concerns such as air and water pollution, soil erosion, pesticides, energy shortages, waste disposal, sewage and water treatment, and land use. What factors are important to understanding the problems? What groups of people are involved in creating and in solving the problems? Are any particular groups well informed about these factors? Do the different groups agree on which factors are important? What stand would you take if you were to become involved?

Snow goose—organism

Flock of snow geese—population

Community

Ecosystem

Groups of Living Things

All populations of organisms that live together in a specific area and depend upon one another for food and shelter make up a **community.** Different types of environments support different populations. A desert community consists of populations of plants and animals different from those in a tropical rain forest. Your backyard, Lake Erie, and a rotting log in the woods all harbor communities.

Populations also interact with the nonliving components of the environment, such as water, heat, light, soil, air, and the landforms of the area. The interactions of a community with the nonliving parts of its environment make up an **ecosystem** (ē′kō sis′təm). Central to the study of ecology is the ecosystem and the interactions among the living and nonliving elements in the environment.

Around the Earth similar ecosystems can be found where the environmental conditions are similar. Ecosystems can be grouped according to similarities in their most common plants. These groups of major ecosystems are called **biomes** (bī′ōmz). Deserts, grasslands, and rain forests are examples of biomes.

When all of the biomes of the earth are considered together, they make up the nonliving surface of the earth and the living organisms that inhabit it. These biomes, together with soil, minerals, water, and atmosphere, are called the **biosphere** (bī′ə sfir).

Biome

Biosphere

Review It

1. Define ecology.
2. How is an understanding of ecology helpful?
3. What is the difference between a population and a community? Give an example of each.
4. List the following terms so that the term that includes the fewest parts is first and the term that includes the greatest number is last:

biome	population	community
ecosystem	organism	biosphere

26-2
Characteristics of Populations

Populations are groups of individuals of the same species living in a particular place at the same time. Populations may differ in their number, patterns of living together, behavior, and organization. All of these factors are important when considering the interactions of a population with its environment and other populations.

Density

The number of individuals of a species living in a particular area is the **density** (den'sə tē) of that population. Density is dependent upon many factors, such as danger of predators and the availability of food, water, light, heat, and nesting places. The density of a population increases when conditions are favorable to the population and decreases when they are not favorable.

You might find 1,000 or more grass plants in a square meter of well-maintained lawn. In a semi desert where water is limited, less than 10 grass plants might be found per square meter. The study of population density and factors that limit it are important in ecology.

Challenge!
The dandelion is a common plant in most regions. Investigate the density and distribution of dandelions in a vacant lot next to a building, and in a well-tended lawn. How do the density and distribution vary in each of the locations? What factors do you think account for these differences?

Distribution

The arrangement of the individuals of a population within a particular space is called the **distribution** (dis trə byü'shən). While population density tells us how many individuals exist in an area, distribution tells us how those individuals are arranged in that area. The patterns of distribution are *random, uniform,* and *clumped.*

Random distribution in flowers

In random distribution organisms occur in no particular order. The appearance of the organism is random throughout the area without an overall pattern.

Uniform distribution in a prairie dog town

Organisms that are distributed in a uniform manner are spread evenly over an area. No one place has more or less of the organisms than any other place.

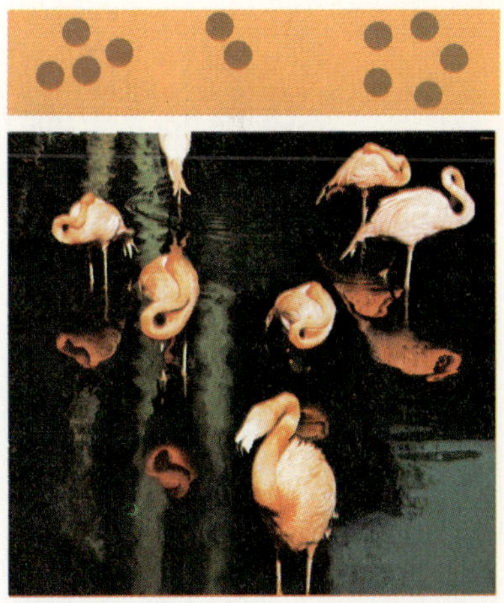

Clumped distribution in flamingos

Clumped organisms are concentrated at points throughout an area. Clusters may offer the population protection from enemies, wind, and cold.

Behavior of Populations

An important characteristic of animal populations is their behavior. Animal behavior consists of all of the movements that the animals make: movements of motion, feeding, mating, nesting, and defense. Each animal population has its own particular set of behavior patterns.

Communication

The survival and well-being of an animal population is largely dependent upon the communication among the members of the population. Through specific methods of communication, members of a population indicate dominance, submission, the source of foods, the presence of danger, hunger, the urge to mate, and the limits of territory. Even simple organisms have developed methods of communication.

Aggression

Visual Displays

Animals may use visual displays to communicate. The visual display may be the way the animal stands, moves, or shows body coloring. Notice in the picture of the cobra how aggressive behavior is shown when the snake arches its back, bares its fangs, and displays the neck hood. Some species of animals have special adaptations for communicating their message. The porcupine fish turns into a ball of quills to tell an enemy to beware. Male birds are often brightly colored to attract females of their species and to draw the attention of enemies away from the female and young. The wood duck shows this type of coloration.

Coloration

Defense

Communicating by Sound

Animals may also communicate by making sounds and sometimes combine their visual displays with sounds. Each call or cry carries its own special message. Even animals such as rabbits and giraffes, usually thought of as mute, may utter sounds when they are intensely afraid.

Eagle calling

Use of Chemicals

Chemicals, including *pheromones*, as mentioned in Ch. 18, may be secreted by some animals. These substances communicate territorial rights, fear, or a desire to mate. Chemicals can be produced and secreted by specialized glands or tissues and may be included in the urine or saliva. The male lion marks his territory by spraying urine at points along the outer edge of his territory. The skunk is a well-known example of an animal that secretes chemicals to communicate.

Skunks

Physical Contact

To convey concern, hunger, nurturing, play, or aggression, animals may use physical contact. A young bird pecks at the parents' bill to communicate hunger. Mother animals lick their young to communicate caring as well as to groom the offspring. A pet cat will rub up against a person to communicate pleasure. Giraffes use their long necks in struggles for dominance by slowly and forcibly knocking heads and necks. Animals seldom fight other members of their species to the death. Rather, they fight until one animal admits defeat by communicating signs of submission or by leaving the area.

Mountain lion and her cubs

Challenge!

Three early investigators of animal behavior were Charles Darwin (1809–1882), J. Henri Fabré (1823–1915), and Ivan Pavlov (1849–1936). Investigate the contributions each made in the study of animal behavior.

Musk oxen in defensive circle

Social Behavior and Organization

Another characteristic of populations is their social behavior. This is the interactions of animals among their own kind. These interactions depend upon communication between the individuals, and all contribute to the success of the population. Organized social interactions offer the advantages of cooperation in reproduction, caring for the young, obtaining food, and in defense. A herd of musk oxen form a circle for protection, as shown at the left.

Social organization may be very simple or very complex. The simplest social behavior is herding, schooling, or flocking. You can see in the photograph of the school of fish that there is "safety in numbers." Individuals are less likely to be captured by a predator while in a school than if alone.

Social organization also aids in reproduction and rearing. For the seal, meeting at the breeding area makes contact between male and female easier. The baboons, as shown below, live together for such advantages as grooming each other and caring for the young.

Seals

Schooling

Baboons

Review It

1. How do population density and population distribution differ?
2. List the three types of population distribution and describe each one.
3. List and give an example of each type of communication within animal populations.
4. List the ways that social organization among animal populations helps ensure the survival of the populations. Give examples of how each way works to the benefit of the population.

26-3
Population Dynamics: How Populations Change

Populations can change in many ways. They can get larger or smaller. The areas they occupy can increase or decrease. The ratio of male to females can vary. Changes in a population occur in response to some change in the factors that influence the population.

Biotic Potential and Environmental Resistance

Every kind of organism produces more offspring than are necessary for that species to survive. The maximum reproduction rate of a population, under ideal conditions, is called **biotic potential** (bī ot′ic pə-ten′shəl). Biotic potential assumes that all of the young will live long enough to reproduce. To reach its biotic potential a population must have all of the food or raw materials, water, and space it needs to survive. It also needs an ideal climate and the absence of competitors and disease. The age at which an individual is able to reproduce and the number of young born at any one time influence the biotic potential.

Dramatic examples of biotic potential can be calculated. If a pair of houseflies produced one generation every two weeks, they could have 391,000,000,000 descendants at the end of one year. Some bacteria divide about every 20 minutes. At this rate, 72 generations could be produced in 24 hours. The number of individuals produced would be enough to cover the entire surface of the earth to a depth of over 20 centimeters.

Populations do not reach their biotic potential. Many factors influence the growth rate of a population. **Environmental resistance** is the name given to the collection of factors that reduce the growth rate of a population. The advantage of a high biotic potential, such as that of the housefly, is to counteract the effects of environmental resistance.

Biotic factor: presence of predators

Abiotic factor: freezing temperatures

Biotic factor: Availability of food

Abiotic factor: slope of terrain

Biotic and Abiotic Factors That Influence Population Growth

Factors that influence population growth are either biotic or abiotic. **Biotic** factors are living organisms or their characteristics. Examples of biotic factors that influence a population are the presence of predators, or hunters, the food supply, the density of the population itself, the biotic potential of the species, and disease. **Abiotic** factors are nonliving physical elements in the environment. Abiotic factors include temperature, humidity, water, light, soil, slope, air, and wind. The pictures on the right show some biotic and abiotic factors.

In the case of the housefly, one reason it does not reach its biotic potential is that its predators are numerous. Not only is it an important food supply for other insects and birds, but it is so annoying to humans that the population is reduced by fly swatters and other means.

Density Dependent and Density Independent Factors

Another way to think about the factors that influence population growth is to group them as either density dependent or density independent. **Density-dependent** factors influence a population differently if the population is crowded than if it is not crowded. Biotic factors are usually dependent on the density of the population. The effects of predators, disease, and competition vary with the concentration of the population. For example, a disease spreads more rapidly through a crowded population than through a sparse one. The loss of a food source affects a crowded population severely, because each individual has less area available in which to seek another food supply.

The series of photos at the left shows two predators, a coyote and a badger, competing for food; in this case, a rattlesnake.

Abiotic factors, the factors in the environment that are not directly associated with living things, influence population growth by changes in temperature, humidity, the availability of air, water, and light, soil conditions, wind rate, and the slope of the environment. These factors influence all populations, regardless of their density, and are said to be **density-independent.** A forest fire affects the food supply and shelter of all organisms in the area whether they live together closely or are scattered apart.

Coyote and badger compete for a rattlesnake

40 years after a forest fire

Carrying Capacity

The number of individuals of a population a particular area can support in terms of space, food, and shelter is the **carrying capacity** of the area. The interaction of biotic potential and environmental resistance tends to hold most populations at a fairly stable level that matches the carrying capacity of the area.

The natural interplay of biotic potential and environmental resistance is sometimes disturbed by humans. People make changes in the environment or attempt to control the environment in order to increase the environmental resistance against organisms they do not want. These changes alter the carrying capacity of the area. When the land area and the food supply it contains are reduced, the remaining animal populations become more concentrated. The populations must get smaller if all members are to have enough food to survive. This is happening with the trumpeter swan. Their numbers are decreasing as available food supply and living area are eliminated through the destruction of marshlands by humans.

Challenge!
See what you can learn about human population growth in various countries. The Population Reference Bureau each year publishes the "World Population Data Sheet." It gives important human population information for more than 150 countries of the world. Notice the "Rate of Population Growth" and "Number of Years to Double Population." Compare Thailand and Mexico, for example, with Australia and West Germany. Can you give some reasons for the differences?

Trumpeter swan

Review It

1. Define biotic potential and explain why it is useful to living organisms.
2. List the abiotic and biotic factors of environmental resistance.
3. Define density-dependent factor and density-independent factor. Give an example of each.
4. Under normal conditions what is the result of the interaction of biotic potential and environmental resistance?

26-4
Community Structure and Function

Desert community

Coastal community

A given space is seldom occupied by a single population. Rather, numerous species of plants and animals occupy an area. A community is made up of a number of populations that live together in the same area such as a pond, a forest, or a meadow. These populations depend on one another for food and shelter. The dynamic relationships in the community change with and adapt to changes within the populations themselves.

Species Diversity and Dominance

Communities differ in **species diversity.** This means they differ in the number and kinds of populations that exist in the community. A desert community has different species than a coastal rain forest.

Usually a community has a few populations that are more important in determining the nature of the community than the others. These **dominant species** strongly influence the food supply and the environment of the other species. In a forest, the large trees are usually dominant species. A common dominant tree species in the eastern United States is the maple. In Pacific coast forests, dominant tree species are different. In northern California and southern Oregon, for example, coastal redwoods are often dominant. Large trees dominate forests because their canopy of leaves over the other populations determines how much light is available to other plants. Also, their leaves or needles make up much of the litter on the ground and therefore influence the make-up of the soil. The fallen leaves provide food for soil organisms that cause the leaves to decay. Their trunks and branches are the places where insects, birds, lichens, mosses, and fungi live. Some of them use the living tree for food.

Maple forest

Redwood forest

Habitats and Niches

Every population in a community has its own special place to live within the community. This living place is known as its **habitat** (hab'ə-tat). Some populations can live in a variety of habitats, whereas others can survive within only very limited conditions. For example, rats can live in cities or rural areas. Purple marsh crabs live only in saltwater marshes. The habitat supplies all of the biotic and abiotic elements necessary for the survival of the population. The habitat can be considered to be the population's address within the community.

Many plants and animals share the same general habitat. The life style of an individual population in the habitat is known as its **niche** (nich). Crickets and grasshoppers are closely related insects that live in the same habitat yet each occupies a different ecological niche. The grasshopper is very active during daylight. It can usually be found on a plant, its head pointing upward as it feeds on the plant parts. Although the cricket lives in the same field, it is quite different. During the day, the cricket hides under leaves or plant debris and is usually inactive. However, the cricket is active at night when it feeds, moves about, and makes sounds. The cricket and the grasshopper do not interfere with each other's activities in the same habitat. In general, the niche of an organism is the total position and function of an individual in its environment.

Habitat of the purple marsh crab

Cricket

Grasshopper

Special Relationships Among Organisms

Living things continually interact with other living things. Interaction may be between individuals of the same species or between individuals of different species. Types of interactions differ among populations in a community.

Competition

The amount of food, light, space, minerals, and water that are available in a particular habitat is limited. As a result, organisms are in *competition* with one another for one or more of these factors. Populations with the most similar niches will have the most intense competition. Populations may compete directly, leading to the extinction or adaptation of one of them. Many animals establish territories within which they live and which they will defend against others of their species who try to intrude. By staying in their own territories, competition and combat are lessened. In one area, five species of warblers live together in the same tree. You can see in the illustration on the left that each warbler has a specific niche, so they do not compete directly for food or space.

Predation

The relationship between *predators,* or hunters and their kill, or *prey,* is called *predation.* Predators have developed characteristics that make it easier to capture their prey. Exceptionally keen eyesight has made it possible for some birds to fly high over the oceans, spot a tasty fish, and dive for their dinner. Prey have also adapted to avoid capture. Some, such as the chameleon, use protective coloration to blend with their environment.

Predation is often density dependent. When the population of prey becomes very dense, the individuals are often weaker due to competition within the population for common resources. The prey is then more easily captured by predators. Predation is a means of reducing an overly large population and controlling its size. It should be remembered that predation is one of the biotic factors of environmental resistance.

Symbiotic Relationships

Symbiosis (sim'bē ō'sis) actually means living together in close association. Forms of symbiosis differ among populations in a community.

Oxpecker bird on rhinoceros

Wasp cocoons on caterpillar

Remora fish and shark

Mutualism

When two species live together in a close association that is helpful to both species, the relationship is known as **mutualism** (myü'chü ə liz'əm). The oxpecker bird and the rhinoceros exhibit this relationship. The oxpecker receives protection and obtains food from the ticks and other pests infesting the rhino's skin. The rhino receives cleaning and early warning of approaching dangers. Algae and fungi live together in a mutualistic relationship in lichens. The algae are able to produce food by photosynthesis. The fungi obtain water and minerals and provide attachment for the lichen.

Parasitism

Parasitism (par'ə sī'tiz'əm) is a relationship in which one organism, the **parasite** (par'ə sīt), spends much or all of its life living in or on another organism, the **host.** The parasite is dependent upon the host for food. The parasite benefits from the relationship, and the host is always harmed. Parasites may bring about the death of their host, but most often only weaken their host. Some wasps inject their eggs under the skin of caterpillars. The wasp larvae then chew their way through the skin of the caterpillar and spin their cocoons, as shown in the picture. Parasitism is a very common symbiotic relationship.

Commensalism

Some organisms live together so that one organism benefits by the relationship while the other organism is neither helped nor harmed. This type of relationship is known as **commensalism** (kə men'sə liz'əm). An example of commensalism is the relationship between the shark and the small remora fish. The remoras may attach themselves to the shark as its swims through the water. When the shark finds food, the remoras eat some of the food not consumed by the shark. The shark is not harmed by the remora, whereas the remora is definitely helped.

Review It

1. Define dominant species and give an example.
2. Define and give an example of each of the following interactions: parasitism, predation, commensalism, competition, and mutualism.
3. Explain the difference between habitat and niche.

Challenge!

Research the anatomy of the tapeworm. It is an intestinal parasite of vertebrates, including the human. What are some of the interesting adaptations of the tapeworm to its unusual environment? How does this parasite influence its host?

26-5
Food Chains and Energy Flow

Challenge!
Do some research in your community on food chains. Make a chart showing a commercially important food chain from your own community that includes a producer population, a primary consumer population and at least one secondary consumer population. Possible resource people include your County Cooperative Extension Service, U.S. Department of Agriculture in your area, State Department of Agriculture, State Department of Ecology, State Game Department, and fisheries and wildlife services in your area.

Energy enters the world of living things through the process of photosynthesis carried on by green plants. Energy is transformed from solar radiation to the chemical energy of food molecules. In this chapter, we have been studying some special relationships among populations. An important part of these special relationships is food-getting.

Food-getting interactions are made up of several feeding, or **trophic** (trof′ik), levels. The first trophic level is made up of populations of green plants and other autotrophs which use energy from sunlight to make food through the process of photosynthesis. These organisms, which are mostly green plants, are called **primary producers.** All other populations in a community are dependent upon green plants and are referred to as **consumers.** Consumers are heterotrophs and do not produce food themselves. They depend directly or indirectly on green plants for food. Populations that eat only plants are **primary consumers,** or *herbivores.* Animal populations that feed upon other consumers are **secondary consumers.** Secondary consumers that only eat other animals are *carnivores.* Within a food chain, primary and secondary carnivores can be found. **Primary carnivores** eat animals and are in turn eaten by other animals, **secondary carnivores.** *Omnivores* are consumers that eat both plants and animals. Humans, for example, are omnivores. Below are examples from each of the trophic levels.

Primary producer

Herbivore (primary consumer)

Primary carnivore (secondary consumer)

Secondary carnivore

Omnivore

Food Chains

The energy in a community is passed from one organism to another in feeding patterns. These patterns of energy transference together form a **food chain.**

The links of one food chain are a flower, a butterfly, a dragonfly, a frog, a snake and a hawk. The food chain begins with the flower, containing food produced by photosynthesis. The flower nectar is eaten by the butterfly, the butterfly is eaten by the dragonfly, and so on through each successive link of the food chain. The chain appears to end with the hawk, since hawks are not ordinarily used for food by other organisms. When a hawk dies, if it is not consumed by another carnivore, such as a vulture, it decays. Its cells and tissues are digested and used as food by bacteria and other organisms of decay, that is, the decomposers. In this way, the organic materials of the food chain are returned to the soil and recycled through another chain.

Communities are not made of unrelated food chains. A consumer often has several major sources of food. A predator such as a mountain lion may feed on several different species. Similarly, a population of rabbits may be the prey of several different predators. The collection of connecting feeding patterns in a community tends to look more like a web than a chain and is in fact called a **food web.**

Hawk

Flower

Sun

Snake

Butterfly

Dragonfly

Bullfrog

Food web

Food Webs and Energy Transformation

Food webs represent the feeding relationships in a community. These webs also represent the flow of energy through the community. Through photosynthesis producers transform the energy of sunlight to chemical energy of foods. These foods are the carbohydrates, proteins, and fats that make up cells, tissues, and organs of green plants. The food molecules are packages of energy. When a rabbit feeds on grass, it does three main things with the energy from food. First, through respiration, energy is converted into energy of motion as the animal moves about. Second, through respiration, some of the energy is converted to heat. The heat maintains the body temperature of the rabbit and also is continuously lost from the animal to the environment. Third, some of the energy from food is built into the structure of the rabbit's body as muscles, fat, and so on. The energy may be used for growth of tissue or body maintenance.

Energy Transfer in Food Webs

Studies of food webs have shown that only about one tenth (10%) of the food energy in green plants will actually show up as food energy at the primary consumer level. Much of the other nine tenths (90%) of the energy goes to heat and movement. Similarly, only around one tenth of the food energy that makes up the body tissues at the primary consumer level (the herbivore) will occur as cells and tissues at the secondary consumer (the primary carnivore) level. The actual fractions or percentages of energy from food at one level converted to energy from food at the next level may be more than 10% or less than 10%. This general pattern forms an **energy pyramid,** because, as you can see, each successive trophic level has less energy available to it.

In studies of energy flow, it is often helpful to keep track of the number of individuals in each of the trophic levels. For example, many aphids (primary consumers) may feed on a single rose (producer). On the other hand, a single cow (primary consumer) may eat many corn plants (producers). In both cases, consumers are feeding on producers, but the numerical relationships in the two cases are different. By understanding the differences in these relationships, you can get a clearer idea of what biotic interaction is all about.

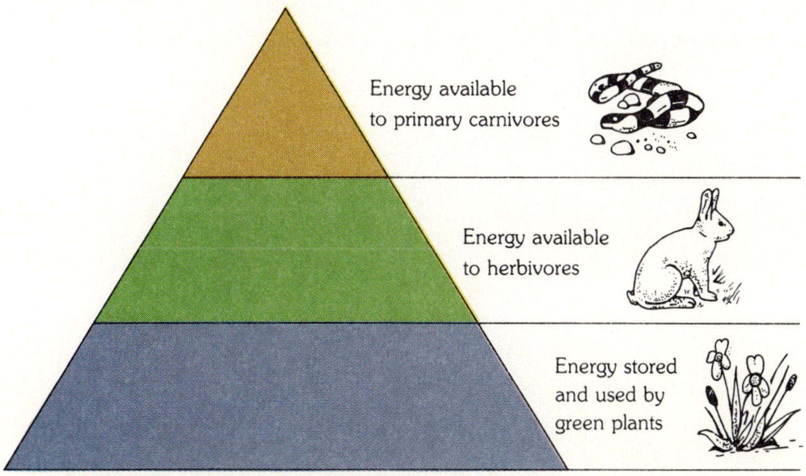

Review It

1. Define the following terms: food chain, food web, trophic level, producer, consumer, herbivore, carnivore, secondary carnivore, biome.
2. In what form is energy used in photosynthesis? How is the energy stored by photosynthesis?
3. In what form is energy used in respiration by plants and animals? What happens to this energy as a result of respiration?
4. Describe the flow of energy from sunlight through the secondary carnivores in a community.

Activity

Surveying a Biotic Community

In this survey, you will have the opportunity to observe the relationships of organisms in a particular community. You will be able to consider the interaction of the organisms with one another and with the abiotic factors of the area.

Materials
String, about four meters long
Four wooden stakes
Hammer
Large jars
Plastic bags
Elastic bands
Trowel
Notebook
Microscope

Procedure
1. Choose the area you wish to survey. It may be a field, forest ground, garden, or some other nearby area. Measure off one square meter by putting the wooden stakes into the ground at the corners one meter apart. Tie the string around the stakes to fence off the square. Remember that the community you will be observing is *below* the ground as well as above it.
2. Note the physical (abiotic) factors of the environment, such as light, water, slope.
3. Observe the visible members of the community and estimate their number. Next, collect samples of these organisms in the jars and plastic bags. Finally, collect samples of the soil that will be used in observing microscopic organisms.
4. In your laboratory, view the soil samples under the microscope. Identify as many of these and the other organisms you have collected as you can.

Analysis
1. What is the dominant species of your observed community? Why do you think this is so?
2. Classify the organisms as producers, primary consumers, carnivores, and so on. Draw an energy pyramid and indicate the trophic level of each organism.
3. Describe the characteristics of the populations in the community. Include density and distribution.

Issues in Biology

North America's wetlands: to change or to preserve?

The edges of the world's oceans and sea inlets are largely bordered by a narrow green strand of wetland vegetation. Over centuries of development by humans, many of these coastal areas have been severely altered or destroyed. Recently, a better understanding of the ecology of the wetlands raises questions about their future use.

Historically, human settlements have thrived where rivers meet the seas. Boston, New York City, Philadelphia, Norfolk, Liverpool, San Francisco, Venice, and Amsterdam are examples of such cities. Development often results in widespread land filling and the disappearance of marshes and swamps. In some swampy areas land may be low in price, and a breeding ground for mosquitoes. Industry buys and drains the land for factory sites. Because the soils are fertile, other large areas have been drained and converted into productive farms. In Asia, drained marshes are used for salt production or the harvesting of mangrove timber for charcoal. Petrochemical complexes, power stations, and huge dikes are being planned along the coast of Northern Europe. These activities, some critical to human welfare, affect the wetlands in which they occur.

Nitrogen plant on a swamp

Swampland in Georgia

What are the consequences of widespread changes in wetland communities? Marshes are nurseries for many kinds of commercial fish. Without the marsh, these species will not reproduce. The U.S. Department of the Interior estimates that remaining coast marshes produce almost a million ducks each year. A major part of the commercial catch of clams, scallops, smelt, and other food species come from our salt marshes. Wetlands release rich organic matter to coastal waters, thus maintaining the aqua culture of shellfish such as oysters. In addition to being a food resource, coastal marshes scrub contaminants from tidal waters and retain them in their muddy bottoms, thus contributing to water purification. Biologists warn that tidal wetlands are a rich but diminishing resource.

Analysis

1. Is the need to develop wetlands for farms or industry as great as the benefits now derived from them?
2. How would you decide which wetland areas to change and which ones to preserve?

Chapter Summary

Ecology is the study of related interactions of living organisms with one another and with the nonliving aspects of their environment. A thorough understanding of ecological interactions is necessary if environmental problems are to be solved. Ecological systems include organisms, populations, communities, ecosystems, biomes, and the biosphere.

Populations are groups of individuals of the same species living in a particular place at the same time. Population density is the number of individuals living in a unit of space. Their arrangement in this space is the population distribution. Each population has a set of behavior patterns characteristic of that population.

Population dynamics is concerned with changes in the number of individuals in a population. Such changes are the combined result of biotic potential and environmental resistance. The biotic potential of a population is its maximum reproduction rate under ideal conditions, assuming all of the young survive to reproduction age. Environmental resistance is the total of all factors—both biotic (living) and abiotic (nonliving)—that influence population growth.

Communities are made up of a number of populations. Species diversity and dominant species within the community help to determine the characteristics of the community. Special relationships between populations in communities include competition, predation, parasitism, mutualism, and commensalism.

A major part of the functions of a community are the food-getting interactions of the populations. Usually a population feeds upon another population and is in turn fed upon by another population. Trophic levels in a community are producers, primary consumers, and secondary consumers. A diagram of all of the food-getting interactions of a community shows a highly complex food web.

Living organisms require energy to survive, and they get this energy from food. The energy enters the community as sunlight and is converted to the chemical energy of foods through photosynthesis. At each trophic level, energy is transformed into heat and energy for movement or is stored as food. The general pattern of energy transfer is called an energy pyramid.

Extra Research

1. Many methods have been devised for studying populations. Investigate and describe how each of the following methods works. What information does each method give about the population or populations being studied?
   a. Quadrant method of vegetation analysis
   b. Use of plankton net
   c. Use of insect netting and surveying
   d. Use of berlese funnel
   e. Use of increment borer
2. When the last individual of a species dies, that species is said to be extinct. What factors cause a species to become extinct? What are some endangered or recently extinct species?
3. DDT is a pesticide that was once widely used. Investigate DDT and try to answer the following questions about it:
   a. What is DDT and how does it function?
   b. How has it been used to benefit people?
   c. What long-term effects have been attributed to DDT?
   d. Why has the use of DDT now been abandoned in the U.S.?
4. Find examples of "alien" species that have been introduced into favorable environmental settings. What effect has introducing these species had on the native or original populations? What efforts have been made to control these alien species?

Questions to Think About

1. Explain why energy pyramids must be smaller at the top than at the bottom.
2. In this chapter, you have studied communication within populations by chemicals, visible displays, sound, and physical contact. On a separate sheet of paper make three columns as shown below. In the first column, labeled Behavior, briefly describe what is actually done. In the next column, Communication, state what message is communicated. In the last column, Function, state how this helps the population. Include at least ten animal behaviors.

Behavior	Communication	Function

3. Do you think it is important to protect endangered animal species? Why?

Chapter Test

Part I

Copy the number of each test item. If the item is *true*, write *T*, after the number; if the item is *false*, write *F*.

1. Communities are groups of individuals of the same species occupying a particular place at a particular time. (622)
2. Population density is the number of individuals of the same species occurring in a certain unit of space. (623)
3. Populations can reach their biotic potential under any conditions. (627)
4. Some animals communicate to other members of a population through visual displays, such as spreading feathers. (624)
5. The carrying capacity of an area does not take into consideration the environmental resistance factors. (629)
6. Parasitism is a special kind of predation in which the parasite is harmed by the host. (633)
7. In photosynthesis, the energy of solar radiation is converted to the chemical energy of food molecules. (636)
8. Primary consumers are herbivores. (634)
9. There are always more producers than primary consumers. (637)
10. In actual living communities, energy is transferred from organism to organism through the food web. (636)

Part II

Copy the number of each test item. After each number, write the word (or words) that correctly completes the sentence.

11. The style or way of living of an organism in its surroundings makes up its ____. (631)
12. An organism that captures and eats its prey is called a ____. (632)
13. As energy in a community moves upward in the energy pyramid, it may be stored as food or converted to ____ or ____ energy. (637)
14. In photosynthesis, the energy for food production comes from ____. (636)
15. A ____ is a group of organisms of the same species which live in a particular place. (621)
16. The organisms that feed on dead organisms, causing them to decay, are called ____. (634)
17. ____ are usually the most plentiful organisms in a community. (637)
18. ____ factors of a community are elements such as water, air, and slope. (627)

Part III

Copy each numbered item from List A. After each item, write the letter and words from List B that best describe it.

List A

19. omnivore (634)
20. mutualism (633)
21. carrying capacity (629)
22. niche (631)
23. habitat (631)
24. carnivore (634)
25. commensalism (633)

List B

a. a population's function or role in an environment
b. a consumer that eats both plant and animal material
c. a relationship between individuals that is beneficial to both populations
d. the maximum number of individuals of a population a particular area can support in terms of space, food, and shelter
e. a relationship between two individuals that is beneficial to one and neither helpful nor harmful to the other
f. the address or living area of a population in a community
g. consumer that eats animals only

Chapter 27

Ecosystems and Biomes

In Chapter 26, you studied populations of organisms and how they interact with each other in biological communities. In this chapter, you will extend your study to include the *abiotic,* or nonliving, environment. You will recall that a biological community interacting with the nonliving environment is called an ecosystem. Groups of ecosystems that have similar types of vegetation are called *biomes.* Some examples of biomes are: desert, grassland, temperate-deciduous forest, and tundra. All the biomes of the earth taken together make up the *biosphere.*

This chapter examines the movement, or cycling, of chemicals in ecosystems. You will also learn the ways that ecosystems develop and the characteristics of some major biomes.

Chapter Objectives

1. Describe how the water cycle functions in the biosphere.
2. Describe the oxygen cycle, the carbon cycle, and the nitrogen cycle.
3. Describe the events that take place during succession in an ecosystem.
4. Compare the different kinds of biomes.
5. Give examples of the human impact on ecosystems and the biosphere.

In this view of Mt. Shuksan, Washington, there are several ecosystems and biomes, in which many biotic and abiotic factors are at work.

27–1
Characteristics of Ecosystems

The living parts and the nonliving parts of ecosystems are very closely interconnected. In Chapter 26 you studied how energy in the form of food flowed from one organism to another organism within a community. This was done through feeding relationships such as grazing, predation, and parasitism. The community is only one part of the ecosystem. Another important part of the ecosystem is the abiotic portion. The **abiotic** (ā'bī ot'ik) part of the ecosystem includes chemical elements such as oxygen, carbon, and nitrogen. It also includes physical elements such as wind, fire, and climate. Together, the biotic and abiotic parts determine the characteristics of the ecosystem.

Energy moves through the food chain in one direction only. The movement of matter, however, is **cyclic** (sī'klik), or circular. Energy flows through the ecosystem and is finally lost to space in the form of heat. Matter may be changed from one form to another, but it always remains in the ecosystem. Let us consider the cyclic movement of some important chemicals in the ecosystem.

The Hydrologic Cycle

Without water, life as we know it would not exist. The unique chemical properties of water make it the supporter and carrier of life, and all biological processes depend upon it.

The table shows how the earth's water is distributed. A given molecule of water does not necessarily stay in the same place forever. It may evaporate from a lake into the atmosphere or precipitate from the atmosphere as rain or snow. It may be taken up by plant roots and become part of plant tissues. It may be frozen into a glacier or run off the land to join the ocean. This movement of water throughout the biosphere is called the **hydrologic** (hī'drə loj'ik) **cycle.**

Location	Percentage of Water
Oceans	97.2
Icecaps and glaciers	2.15
Atmosphere	0.01
Freshwater lakes	0.009
Saline lakes and inland seas	0.008
Average in stream channels	0.0001
Soil moisture	0.005
Ground water within depth of half a mile	0.31
Ground water—deep lying	0.31

Water Balance

What actually happens to water in the cycle is quite complex. For example, some water does not evaporate directly from the soil but is absorbed by plant roots and eventually is lost from plants, especially through the stomates of the leaves. Also, some water does not run off directly from land to the sea, but instead it is soaked deep into the ground and becomes part of the ground water.

The cycle diagram shows that the total evaporation is balanced by **precipitation** (pri sip′ə tā′shən), or rainfall. We must remember, however, that rainfall is not evenly distributed over the surface of the earth. The air circulation patterns and landforms determine the amount of precipitation each area receives.

Water cycle

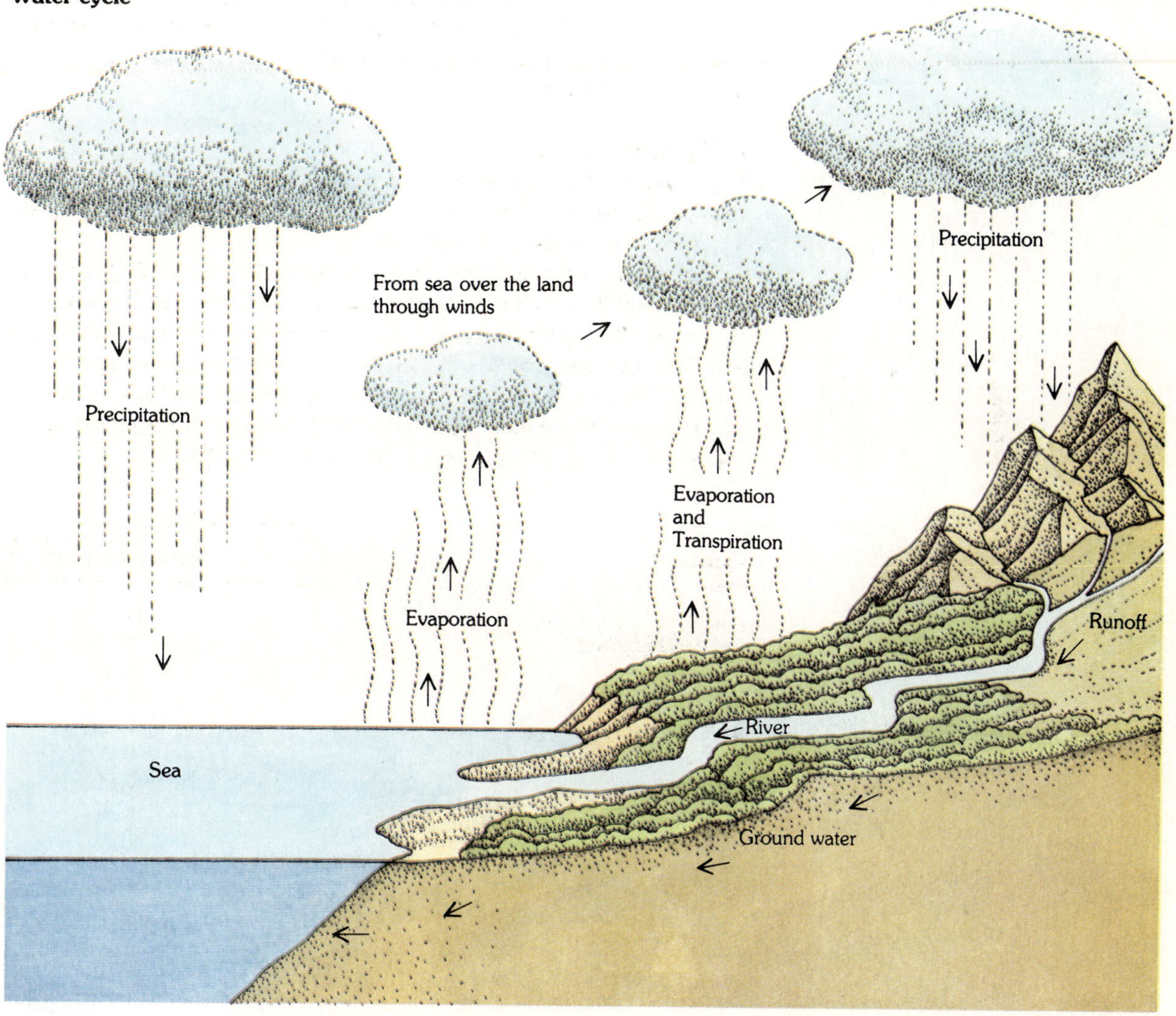

Rainfall Distribution

Mountain ranges affect rainfall distribution by what is known as the rain shadow effect. For example, in the northwestern part of the United States, the climate west of the Cascade Mountains in Oregon and Washington is quite wet, averaging around 100 cm (40 inches) of precipitation per year in Portland, Oregon, and Seattle, Washington. During much of the year, moist air is blown onto land from the Pacific Ocean. As it meets the mountains, the air is forced to rise, cooling as it rises. Because the cooler air cannot hold as much moisture as the warmer air, some of the moisture is lost as the air rises to pass over the mountains. This moisture falls as rain or snow on the western slopes of the Cascades. After it crosses the mountain range and goes down the eastern slopes of the mountains, the air is relatively dry. The result is that eastern Washington and eastern Oregon have a much drier climate than the western portion of these states. As you might expect, the types of plants and animals living on the two sides of the Cascade Mountains differ considerably.

Western Oregon

Eastern Oregon

Rain-shadow effect

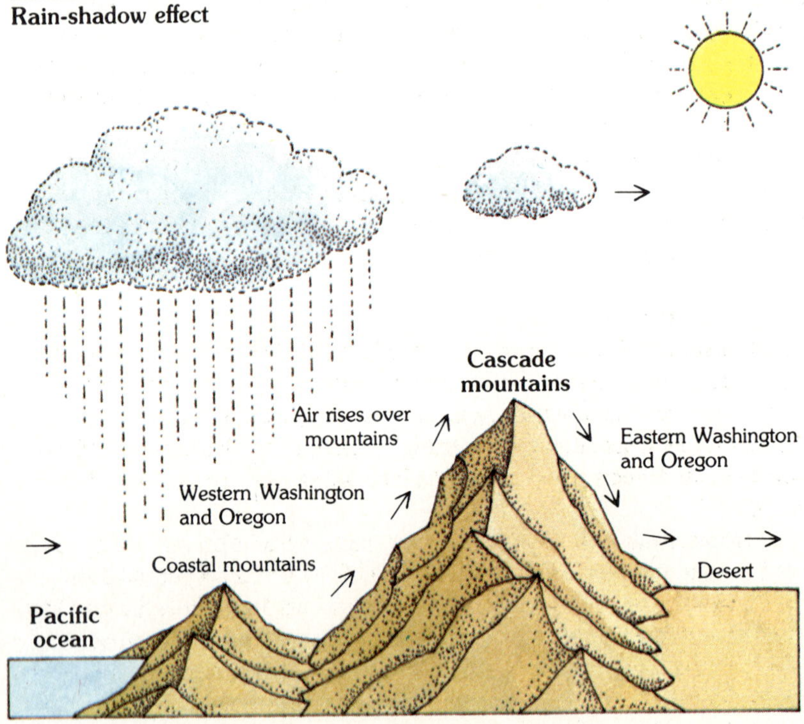

Water and Organisms

A major part of the weight of most organisms is water. Water accounts for about 75% of the weight of most living tissues. In plants as well as animals, water is important because it is the medium by which other materials, such as nitrogen and phosphorus, are transported internally.

Water in Food Production

Remember that water is one of the substances used in photosynthesis and a waste product of respiration. The hydrogen (H) of water molecules (H_2O) is built into food molecules during photosynthesis. Food molecules, in turn, are modified and rebuilt into structural parts of living tissues some of which will be consumed by other organisms. When food is used in respiration for energy, water is released as a waste product. The general chemical equations for these two processes are:

PHOTOSYNTHESIS

$$6\ CO_2 + 12\ H_2O \xrightarrow[\text{chlorophyll a}]{\text{sunlight}} C_6H_{12}O_6\ \text{(food)} + 6\ O_2 + 6\ H_2O$$

RESPIRATION

$$C_6H_{12}O_6 + 6\ O_2 \longrightarrow 6\ CO_2 + 6H_2O + \text{energy released}$$

By far the greatest amount of water entering green plants is not used for making food through photosynthesis. For example, in a tree water enters the roots, then moves up the tree through specialized conducting tissues, and finally is lost to the atmosphere mainly from the stomates of leaves. This process of *transpiration* serves to keep the plant cool and also to bring a continuous supply of essential nutrients from the soil into the leaves.

These are just a few examples of how water is important to life. Later in this chapter when you study biomes, you will notice that the availability of water plays an important part in determining what type of biome will occur in a certain area.

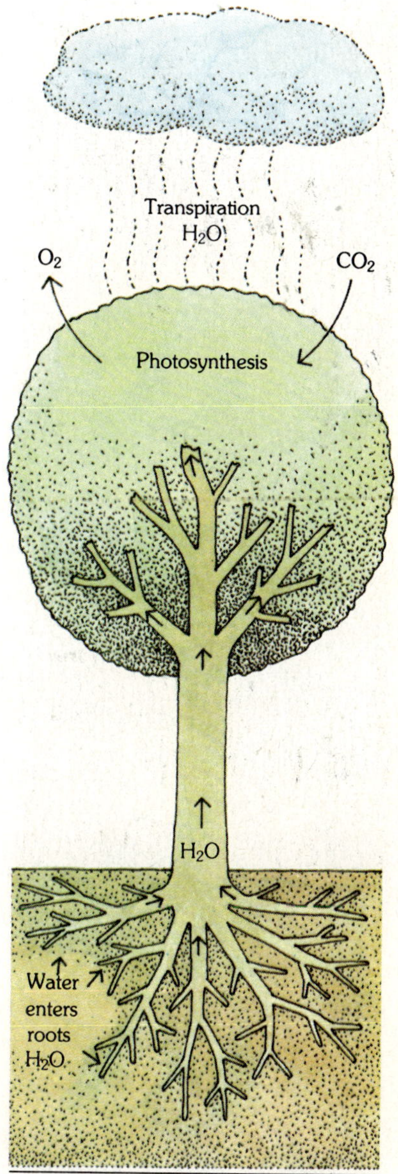

Path of water through a tree

Water and Humans

For thousands of years, humans have located their settlements near water. They have depended upon water for food, transportation, waste removal, washing, energy, and as part of industrial processes. Inevitably, as water has had a strong impact upon civilization, civilization has had effects upon the hydrologic cycle in various parts of the world.

Tropical rain forests serve as vast water reservoirs. As forests are cut to make way for agriculture, city building, or mining operations, more water runs off the land to the oceans and less rain is returned to the atmosphere through transpiration. Every year people destroy enough rain forests to equal an area larger than the state of Ohio. If this rapid rate continues, it is possible that one day these areas of lush vegetation will become barren.

Changes in the Water Supply

Some scientists think the rising amount of dust in the atmosphere resulting from the overuse of already dry lands can actually reduce the amount of rainfall in those areas. Other scientists think the increased atmospheric dust may even cause global climatic shifts. In recent years, the monsoon rains of central Africa and southern Asia have fallen farther south than normal. For India, this means that rains that would normally fall on the land are falling into the ocean. The result can be drought and starvation.

The once fertile lands of the Middle East became deserts as the grasslands were overgrazed by sheep and goats and the forests were cut for lumber. Experiments in Israel have shown that when grass and forests are replanted and protected from destruction, rainfall in that area increases.

As human activities are changing the rainfall patterns, they are also changing the quality of water in lakes, streams, and the oceans. **Pollutants** (pə lüt′nts), materials humans add to water in quantities greater than would occur there naturally, make the runoff and in some cases ground water unfit for drinking, or for industrial uses, or even for recreation. For many years, humans have used rivers and lakes as their sewers, assuming that the water would clean itself. Unfortunately, so much waste has been dumped into the water in some areas that few organisms can live in it.

Heavy rainfall on Nebraska plains

Drought effects on pond in Minnesota

Water pollution in California

Pollution

New York City has hauled its garbage out to sea for many years. Recently, an area of many square miles in the Atlantic Ocean off northern New Jersey was found to be very deficient in oxygen. Marine scientists believe that the reason for this is that too much garbage had been dumped into the ocean for too many years. Another severely polluted body of water is the Cuyahoga River, which flows through Ohio into Lake Erie. So much oil has been dumped into its water that several times the river actually caught fire! Of course these are extreme cases, but many lakes and rivers around the world are sufficiently polluted to cause illness to anyone who drinks the water without first purifying it. Poisonous compounds dumped into rivers may become part of the bottom sediments or enter the river food chain, causing trouble many years after they are dumped.

The water stored in underground aquifers poses a further problem for humans. The aquifer water, some of it stored under pressure, accumulated by slow seepage of rainwater into the ground over thousands of years. In areas of low rainfall, cities have been built around wells tapping these aquifers. Water is now being pumped out of the aquifers at rates far greater than it can be returned to them. The ground above some aquifers has already sunk several feet because of the reduced water level. The thriving cities that depend upon aquifer water may become ghost towns when the wells run dry.

The Oxygen Cycle

There are many chemical substances in addition to water that are ecologically important. Oxygen is one of these substances. It makes up nearly 21% of the earth's atmosphere. Oxygen occurs in the atmosphere in a variety of forms: as molecular oxygen, as ozone (ō'zōn), and in chemical combination in carbon dioxide and water vapor. Oxygen is also an important element in all living tissues. It makes up roughly one-fourth of their atoms, mostly contained in water molecules.

The equations for photosynthesis and respiration show how oxygen cycles between living tissues and the atmosphere. But the equations do not show how the chemicals interact during these processes. Although water enters into photosynthesis and is released by respiration, the molecules of water are not the same. The oxygen molecule produced by photosynthesis comes from the splitting of water. The oxygen molecule used in respiration is combined with hydrogen to form water.

The cycling of oxygen back and forth between living tissues and the atmosphere is not the only way in which oxygen moves. Since oxygen atoms are part of water molecules, oxygen movement is strongly influenced by the hydrologic cycle. More than 90% of the oxygen in the crust of the earth is tied up in calcium carbonate. It is not available to living organisms except when it is released in small quantities by volcanic activity. During sediment deposition and the weathering of rocks, oxygen is removed from the atmosphere and stored in the form of salts and oxides.

Ozone (O_3) is formed at high altitudes from molecular oxygen and atomic oxygen. The ozone layer in the upper atmosphere is very important in that it reduces the amount of ultraviolet radiation from sunlight entering the atmosphere. Ultraviolet radiation is harmful to life. There has been concern among scientists that the fluorocarbons used as propellants in spray cans and exhaust products from automobiles, jets, and rockets may be destroying this protective ozone layer. The highly reactive ozone may combine with the pollutants and therefore not be available to trap the ultraviolet rays. This would allow the dangerous ultraviolet rays to enter the lower atmosphere and pose a threat to life.

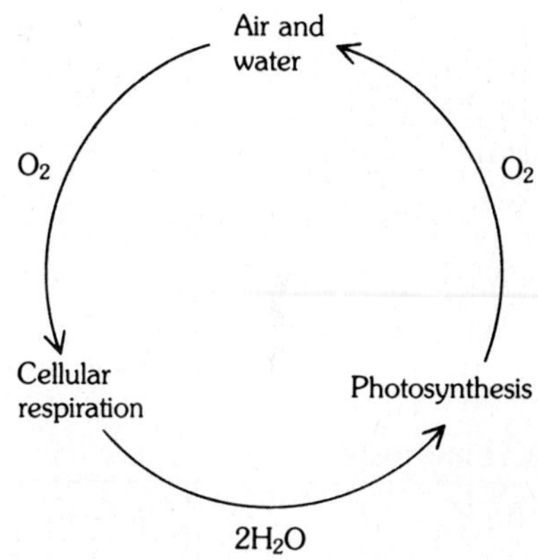

Oxygen cycle

The Carbon Cycle

The events of the carbon cycle are very closely related to the events of the oxygen cycle. In fact, many of the processes are the same. The diagram below illustrates how carbon is continually recycled through the environment.

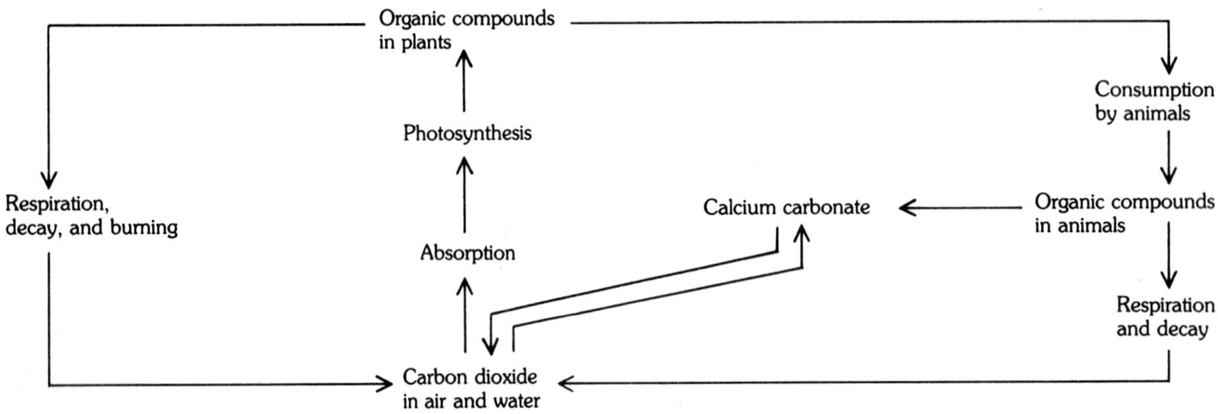

Carbon cycle

Again, photosynthesis and respiration are the most important processes in the cycle. During photosynthesis, carbon is taken from carbon dioxide. It becomes part of carbohydrate molecules—food for plants and all the animals, which depend upon them. The process of respiration returns carbon dioxide to the atmosphere.

Some organisms use carbon to build skeletal materials such as the calcium carbonate (limestone) shells of **coralline** (kôr′ə līn) **algae**, clams, and mussels. When the bodies of organisms do not completely decay after death or when the carbon is in a relatively stable form, such as carbonate, molecules containing carbon may be stored for long periods of time as peat, coal, and carbonate rocks. The burning of peat, wood, coal, and oil by humans is an important way in which carbon dioxide is returned to the atmosphere and water.

The Carbon Cycle and Humans

During the past 100 years or so, the industrialized nations of the world have burned enormous amounts of fossil fuels such as coal, gas, and oil. When they are burned, oxygen is used and carbon dioxide is produced. Since the beginning of this century, the average atmospheric concentration of carbon dioxide has risen. This means that carbon dioxide is being released into the air faster than it is being removed. Many scientists think this change has been caused by the massive burning of fossil fuels. If this is correct, this widespread human activity may not only continue to raise the carbon dioxide levels in the atmosphere but may also exhaust the supply of concentrated energy in fossil fuels.

The Nitrogen Cycle

Molecular nitrogen (N_2) makes up about 78% of the earth's atmosphere. It is the fourth major element found in living tissues after carbon, oxygen, and hydrogen. It is an essential part of amino acids.

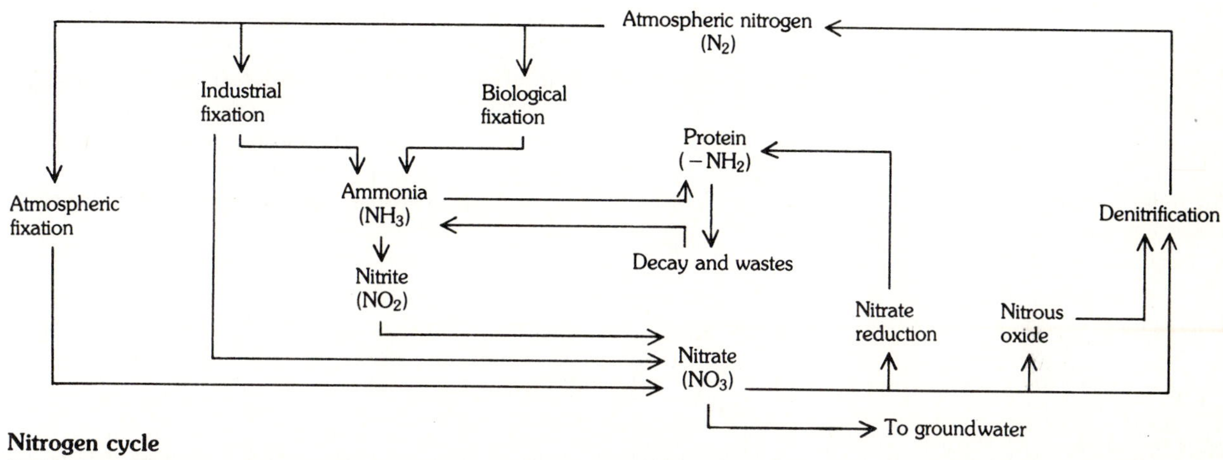

Nitrogen cycle

The complex nitrogen cycle differs from the oxygen and carbon cycles in several ways. First, atmospheric nitrogen cannot be used directly by most green plants. It must be changed into a usable form—either ammonia or nitrate—before green plants use it. Second, once nitrogen has been incorporated into nucleic acids and proteins, it can only be returned to inorganic forms through several steps, some of which require specialized types of bacteria. Third, most of these reactions occur in the soil, where availability of nitrogen is influenced by the solubility of inorganic nitrogen compounds.

The process by which atmospheric nitrogen is changed to ammonia is called **nitrogen fixation.** Only certain species of bacteria are capable of nitrogen fixation. For example, *Rhizobium,* which lives in symbiosis with legumes such as beans, is a bacterium of plant root **nodules** (noj'ülz). Some blue-green bacteria are important for nitrogen fixation in aquatic systems.

Molecular nitrogen may also be converted to usable forms by nonbiological processes. Atmospheric nitrogen can be converted to nitrate (NO_3) through energy provided by lightning. It may also be fixed industrially to produce the chemical fertilizer ammonium nitrate.

Remember, biological fixation produces ammonia. Notice that ammonia also comes from the decay of dead organisms and wastes. This process is called **ammonification** (ə mon'ə fə kā'shən).

The conversion of ammonia to nitrate is called **nitrification** (nī'-trə fə kā'shən). This process occurs during decomposition. For example, some soil bacteria convert the nitrogen of the proteins in the manure to ammonia (ammonification). Other soil bacteria may then convert the ammonia, first to nitrite (NO_2), and finally to nitrate (NO_3).

Legume root nodules

Nitrogen Assimilation

Nitrogen is taken up by plants and incorporated into amino acid molecules, which are eventually linked to become protein molecules. The process by which nitrogen is incorporated into organic molecules is called **nitrogen assimilation** (ə sim′ə lā′shən).

We still have not explained how molecular nitrogen is returned to the atmosphere. If nitrogen were fixed only from the atmosphere and cycled through living communities, the atmosphere would eventually run out of atmospheric nitrogen. The process that returns molecular nitrogen to the atmosphere is called **denitrification.** This is done by a variety of bacteria. The job these bacterial populations do is very important because it completes the nitrogen cycle.

The Nitrogen Cycle and Humans

The rapid growth of the human population has made increased food yields from cultivated land a necessity. When it was discovered that plants would grow better with the addition of ammonia or nitrates to the soil, the commercial production of fertilizer began. Soon farmers began buying concentrated fertilizers.

Unfortunately, farmers are not always careful when they spread the new fertilizer and they sometimes use more than is needed. If the fertilizer is applied just before heavy rains, much of it is washed away.

In some areas where ground water is very near the surface, water from shallow wells must be tested for nitrogen compounds as well as for bacteria to make sure it is safe for human use.

In addition to posing a direct danger to human health, fertilizers affect people indirectly. The chemicals that make crops grow so well also increase the growth of algae in lakes and streams. Overfertilized lakes may become entirely clogged by dense concentrations of algae. These algal "blooms," shade out bottom-dwelling plants, which die and decay. The decay process uses up the dissolved oxygen in the water, and the lack of oxygen harms organisms living in the water. Active game fish, such as trout, are especially likely to be harmed.

Another complication exists with the use of ammonium nitrate. Natural gas is used in the industrial production of this fertilizer, and known supplies of natural gas are running out. As countries adopt artificial fertilizers in an attempt to increase food supplies to keep up with their population growth, increased demand is put on remaining supplies of natural gas for fertilizer production.

Excessive algal growth on a lake

Review It

1. Why may there be greater precipitation on one side of a mountain than on the other?
2. What is the value of ozone in the upper atmosphere?
3. How are the carbon and oxygen cycles linked?
4. In what chemical forms may plants utilize nitrogen?

27–2 Succession

Abiotic factors such as water, solar energy, and nitrates determine which populations, especially plant populations, can survive in an area. In general, climate and soil are the main factors that control where plants can survive on land. Remember, plants serve as food for animals and other consumers. The result is that plant distribution will determine which animal populations will be present.

Whereas it is true that the nonliving environment influences populations, it is also true that populations influence the nonliving environment. For example, leaves decaying on the forest floor influence the structure, water-holding capacity, and even the acidity of the soil.

Abiotic factors and biotic factors interact to produce changes in the ecosystem. As conditions change with time, some populations gradually become less capable of surviving while others become more capable. As ecologists have studied selected ecosystems over a period of years, they have learned that there is a successive replacement of one community by another. This sequence of replacement of one community of organisms with another is called ecological succession, or simply **succession.** Succession is an orderly, predictable process, with certain types of organisms always present in a particular stage of development in a given area. For any particular kind of climate and soil, succession leads to a certain community structure that is relatively stable and undergoes little further change. The stable community structure of an ecosystem is called the **climax community.** We will study some climax communities in the last section of this chapter when we examine biomes.

Challenge!
Scientists today are trying to modify the weather in some regions for purposes of warfare and agriculture. Clouds are being seeded in an attempt to increase precipitation. Can you foresee any problems that might occur if they are successful in changing world weather patterns?

Challenge!
Read up on monsoons. What is a monsoon season? How do temperature changes cause seasonal monsoons in an area like eastern Asia?

Succession on a Sand Dune

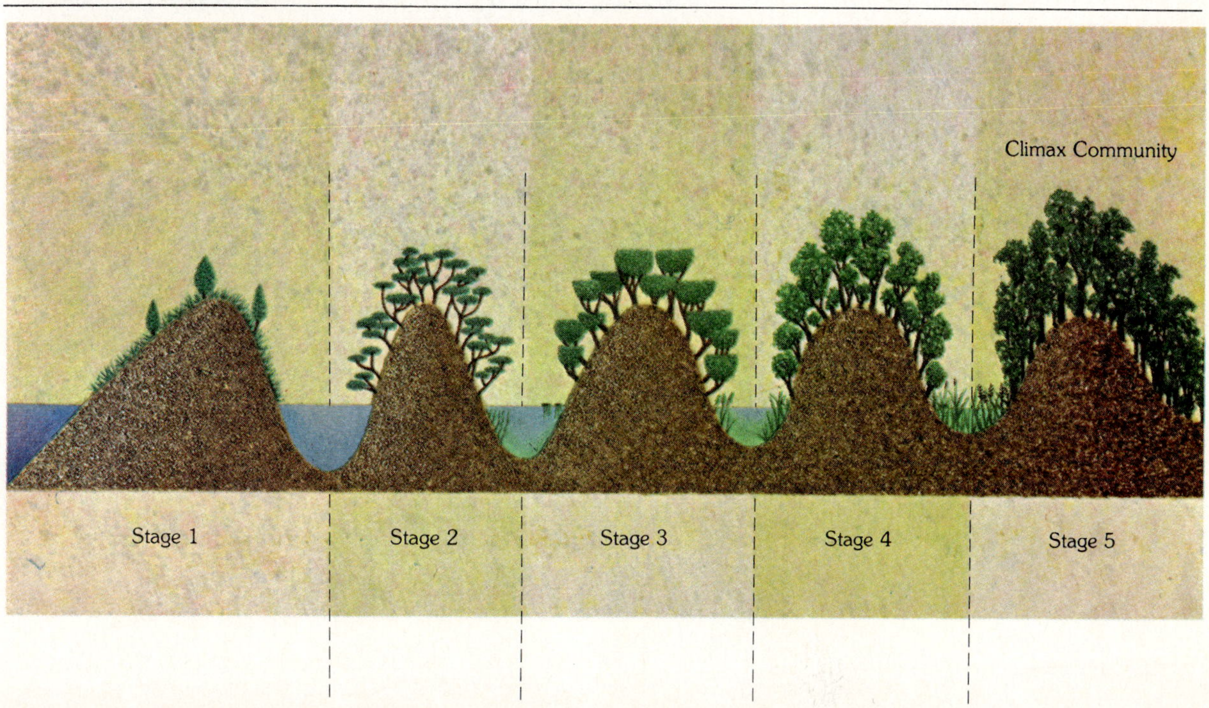

Primary Succession

On August 17, 1959, a severe earthquake struck the Madison River Canyon in southern Montana near Yellowstone Park. One result of the earthquake was an enormous rock slide. In that slide, the U.S. Forest Service estimated that enough rocks and earth to fill the Rose Bowl heaping full 10 times came crashing down into Madison Canyon. One of the many effects was that the entire ecosystem in the immediate area was changed. Plant and animal populations in the area were destroyed. Where an orderly soil structure had existed, there was now a jumble of boulders, rocks, and soil.

Many years later, the slope from which the material came and the valley floor on which it landed are still primarily bare rock. Replacement of the original forested hillside and valley floor cannot simply start with a new forest replacing the old one. There must first be the long process of rebuilding the soil. Organisms such as lichens can survive on bare rock. Gradually, the effects of the lichens together with wind, rain, ice, and temperature changes begin to break rocks into soil. The lichens also help trap soil that is transported by wind and water. As soil formation takes place, other plants such as mosses, grasses, and small flowering forms are able to survive and reproduce. After hundreds or perhaps even thousands of years, succession may result in a climax forest in the area. Such succession from bare rock is called *primary succession.* A lava flow from volcanic action also starts primary succession.

Rock slide

Secondary Succession

Commonly, changes in ecosystem structure are less extreme than in the Madison River Canyon example. Forest fires, clearing of land by humans, severe storms, or the converting of land to agricultural use also change community structure. In these cases, succession may not have to begin with soil development from bare rock. Where soil remains and contains organic matter from previous populations, succession will occur more rapidly. Such succession is called *secondary succession.*

Many examples of secondary succession have been studied in the eastern United States. Before the arrival of European settlers in America, much of this area was covered by forests of beech, maple, oak, and hickory trees. The land was cleared as people settled across the country. Later many fields were abandoned in favor of better farming sites farther west. When the farm fields were abandoned, the plant community changed from agricultural crops to weeds, then to grasses, then shrubs, then conifers (pine trees), and finally back to the original hardwoods. For any particular field, the examples may be different because of soil conditions, drainage, moisture, previous fires in the field, and the kinds of seeds being carried into the field or lying dormant in the soil. However, a climax community with beech and maple trees or oak and hickory trees as the dominant vegetation will commonly result after a succession of many other populations.

Another kind of succession occurs as lakes and ponds age. A young lake usually has very clear water and contains few nutrients. As the years pass, nutrients are carried into the lake by runoff water from the surrounding land. The added nutrients stimulate algae to grow in the water. And plants such as rushes and sedges fringe the shore of the lake. As the plants and animals living in the lake die and decay, their remains are deposited on the bottom of the lake as sediment, and the lake becomes shallower. As bottom-rooted floating plants such as water lilies live and eventually die in the lake, their decaying remains make the lake still more shallow. The rushes and sedges along the lake's shore trap soil washed into the lake, and the lake becomes smaller. Soon, enough soil has been trapped or formed by sediment to support the growth of plants that prefer drier roots than do the aquatic plants. Grasses, woody shrubs, and eventually trees live in the area formerly occupied by the aquatic plants, which have moved closer to the center of the lake. After many, many years, there may be no lake left at all, and the vegetation of the former lakebed will be the climax vegetation of that area.

Pond succession in various stages

Review It

1. Define succession.
2. List some causes of secondary succession.
3. Give at least three examples of abiotic factors that influence succession. Explain *how* an abiotic factor influences succession.

27–3
Biomes

At the beginning of this chapter, you learned that a **biome** (bī′ōm) is a group of ecosystems similar in types of dominant vegetation. From your study of succession, you know that abiotic factors such as temperature, sunlight, soil, and availability of water are important in determining which plant populations can survive in a certain place. Across the surface of the earth, examples of the same biome, a desert, for example, can be found wherever the climate and soil conditions are similar.

Within a biome, patterns of succession are usually similar, but it is the climax community that is characteristic of a biome. This does not mean that all desert climax communities, for example, have exactly the same species. What they do have are species which are all adapted to a similar physical environment. Ecosystems of the same biome from anywhere on earth are similar in many ways.

Biomes of North America

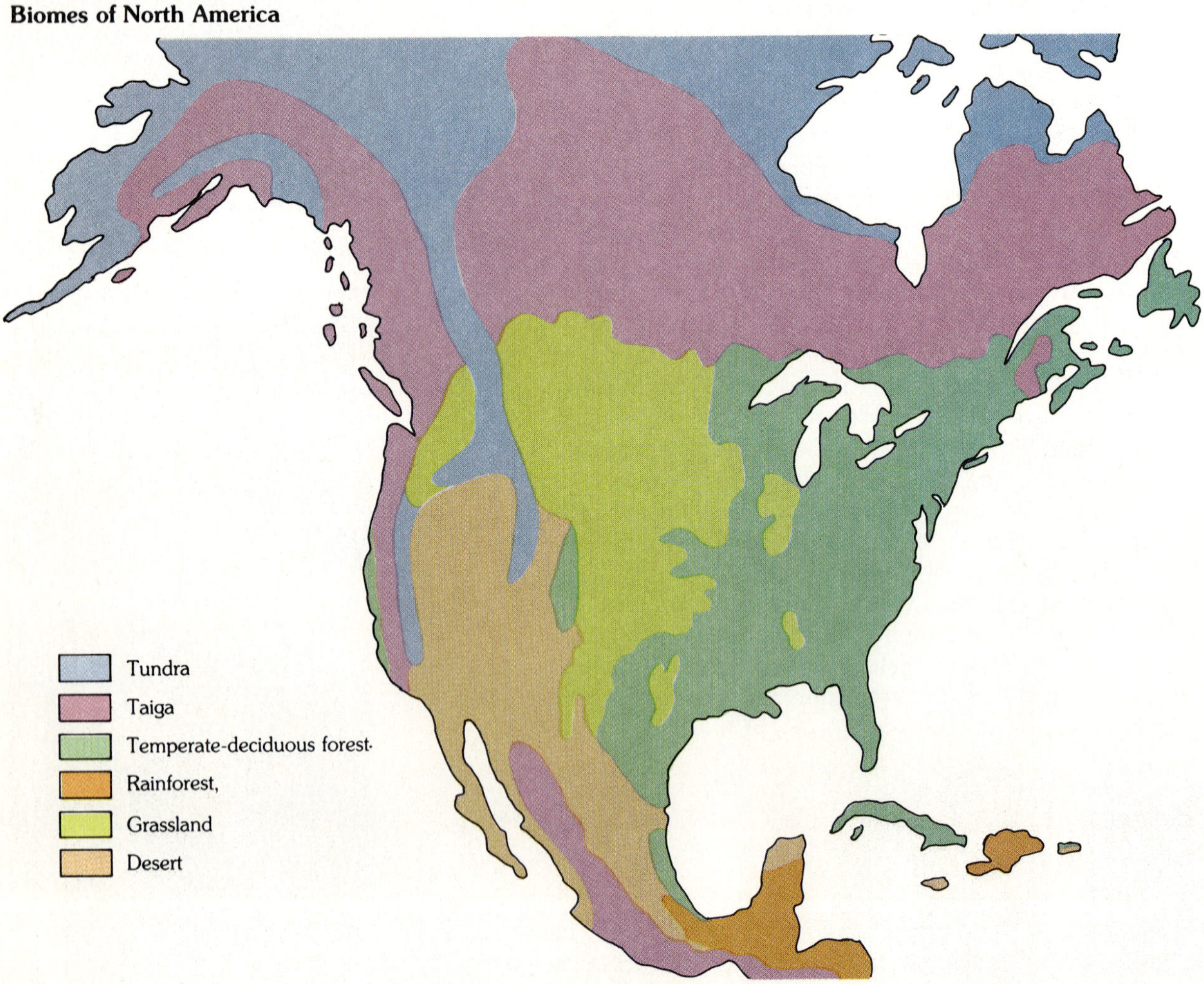

Tundra

The **tundra** (tun′drə) is the almost treeless plain in the far northern parts of Asia, Europe, and North America. Tundra subsoil, which is permanently frozen, is called **permafrost** (pėr′mə frôst′). Winters are very long on the tundra with little daylight. In contrast, summers are short, but there are many daylight hours. Precipitation is low, amounting to only 25 cm or less per year, because cold air can hold relatively little moisture. The ground is soggy in the summer because moisture cannot soak into the permanently frozen ground. Ponds, small lakes, and marshes are abundant due to the nearly flat terrain.

There are no upright trees on the tundra. Only trees such as dwarf willows and birches, which grow low to the ground, can escape the drying effect of the wind which upright trees would experience. The most common plants are mosses, grasses, sedges, and some shrubs.

Caribou, arctic hare, and musk ox are important tundra herbivores. The lemming, an interesting rodent because of its dramatic population fluctuations, is also a herbivore.

Some important carnivores that prey on the herbivores are the arctic fox, arctic wolf, bobcat, and snowy owl. Polar bears live along coastal areas. They prey on seals.

Because of the severe winters, many of the animals are migratory and move from one region to another with the change in seasons. For example, the many shore birds and waterfowl, such as ducks and geese, nest on the tundra during the summer but migrate south for the winter.

Did You Know?

The lemming, a small, furry tundra rodent, has dramatic increases in population. The female can deliver her first young when she is less than six months old and then have as many as eight litters a year. This can result in high increases in population. When the food supply is depleted, the lemmings migrate in huge bands to areas where there is a better food supply. Migration may involve such large numbers of animals that from the air, the ground seems to be moving. Once the migration begins, nothing stops it. Mountains, marshes, rivers, and even the sea do not deter the determined little animals.

Tundra biome

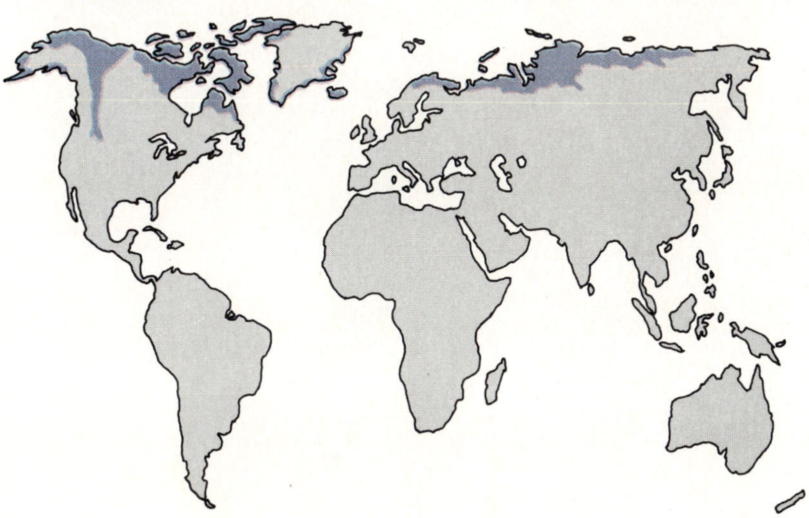

Tundra in summer

Musk oxen in tundra

Taiga

The **taiga** (tī′gə) is a 1300–1450 km wide band south of the tundra. This area also has long, cold winters, but summer temperatures may reach 10–21°C, and the summer and the growing season are longer than in the tundra. Usually the subsoil thaws completely during the summer months. Precipitation is higher than in the tundra, ranging about 38–100 cm annually. The moisture is the combined result of summer rains and winter snows. Lakes, ponds, and bogs are abundant.

The taiga is really a northern forest of coniferous trees such as spruce, fir, and pine. In disturbed areas, deciduous trees such as birch, aspen, willow, and poplar are abundant. In some cases, birch-poplar woods now stand where a genuine coniferous forest once grew. These woods are an intermediate stage in the secondary succession that occurs after logging operations and fires.

Important migratory herbivores include moose, elk, deer, and caribou. Moose and caribou migrate to the taiga for winters and to the tundra for summers. The common smaller mammals are herbivorous squirrels, snowshoe hare, and predatory martens.

Important predators include the timber wolf, grizzly bear, black bear, bobcat, and wolverine. A great many biting insects present during the warmer months can cause humans and other mammals considerable misery. As in the tundra, migratory shore birds and waterfowl are abundant during the summer months.

Taiga

Wolverine

Caribou

Taiga biome

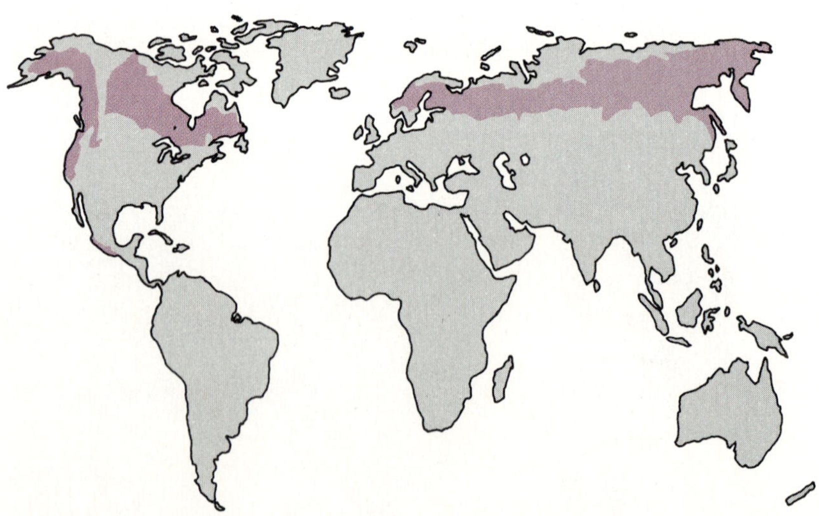

Temperate-Deciduous Forest

The **temperate-deciduous forest** lies south of the taiga in the Northern Hemisphere. The map shows that it does not form as clear a belt around the earth as the taiga does. Temperate-deciduous forests usually occur in regions that have hot summers, cold winters, rich soil, and abundant rain. Annual rainfall is typically around 100 cm per year. Much of the eastern United States was once covered by temperate-deciduous forest.

The dominant vegetation of temperate-deciduous forests is broadleaved trees. The trees lose their leaves in the late fall so the biome has an entirely different appearance in the winter than in the summer. The fallen leaves provide food for a large variety of consumer and decomposer populations including millipedes, snails, and fungi that live in or on the soil. Common deciduous trees are the hardwoods such as beech, maple, oak, hickory, and walnut.

The temperate-deciduous forest produces many flowers, fruits, and seeds that provide a variety of food for animals. Some of the common herbivores are deer, chipmunks, squirrels, rabbits, and beavers. Tree-dwelling birds are abundant in number and diversity. Important predators include black bears, bobcats, and foxes. Predatory birds such as hawks, owls, and eagles are also present.

Another result of the warmer climate of this biome compared to the tundra and taiga is that ectothermic (cold-blooded) animals are much more abundant. Snakes, lizards, frogs, and salamanders are common inhabitants.

The temperate-deciduous forest is a very complex biome. Many changes take place during the year, and a large variety of species inhabits the soil, trees, and air. Food webs in the temperate-deciduous forest are more complex than those of the tundra or taiga.

Temperate-deciduous forest biome

Whitetail deer

Beaver

Tree frog

Bison and antelope

Prairie dogs

Badger

Grassland

Some **grasslands** occur in temperate areas of the earth and some are in tropical regions. Temperate grasslands usually have deep, rich soil. They have hot summers, cold winters, and irregular rainfall. Often they are characterized by high winds. The low irregular rainfall, is the factor which makes the difference between a temperate-deciduous forest and a temperate grassland.

In tall-grass prairies in the United States, important grasses are tall bluestem, Indian grass, and slough grass. Short-grass prairies often have blue grama grass, mesquite grass, and bluegrass. Many grasses have long, well-developed root systems that enable them to survive limited rainfall and the effects of fire.

Historically, important large herbivores were the pronghorn antelope, bison, and the wild horse, though large populations of these animals no longer exist. Smaller herbivores such as the jack rabbit, ground squirrel, and prairie dog are abundant.

Important grassland predators include coyotes, foxes, hawks, and snakes. The small predators have fared better in North America, perhaps because they posed less danger to the domesticated animals of the farmer and rancher than did the larger predators such as wolves and mountain lions.

Grassland biome

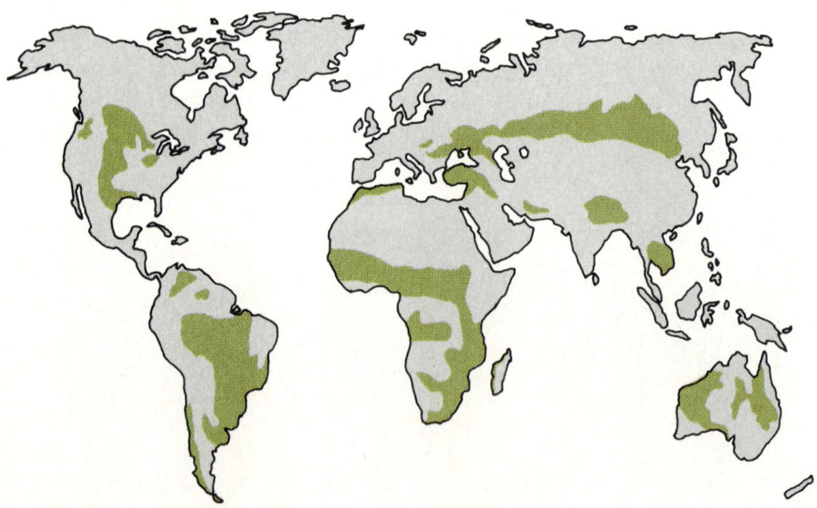

Desert

The **desert** biome is characterized by its very low precipitation, which is usually 25 cm per year or less. Most of this limited moisture comes as short, hard showers. In the United States, most desert area occurs in the West on the leeward side of mountain ranges, the side opposite the main wind direction.

Deserts generally have hot days and cold nights, and they often have high winds. One of the reasons for the dramatic temperature difference between day and night is the lack of water vapor in the air.

Desert plants must meet some initial requirements if they are to survive. They must be able to obtain and conserve water. In order to do this, many adaptations have been made by desert plants. These include: reduced leaf surface area, which reduces evaporation from the plant; loss of leaves during long dry spells; small hairs on the leaf surfaces; and the ability to store large amounts of water. Sagebrush and saltbush are plants commonly found in desert areas where winters are cold. Examples of plants found in desert areas where winters are warm include creosote bush, yucca, and many cacti. Many of the small plants are annuals. They grow, flower, and produce seed quickly during one season, sometimes on only one rainfall.

Animals also must meet the demands of heat, cold, and limited water. Many desert animals are nocturnal. They are active mainly at night. Many reptiles and small mammals burrow to get away from the intense midday sun. In addition to lizards and snakes, common desert animals are the herbivorous kangaroo rat, ground squirrel, and jack rabbit. Important predators include coyotes, badgers, kit fox, eagles, hawks, prairie falcons, and owls.

Desert biome

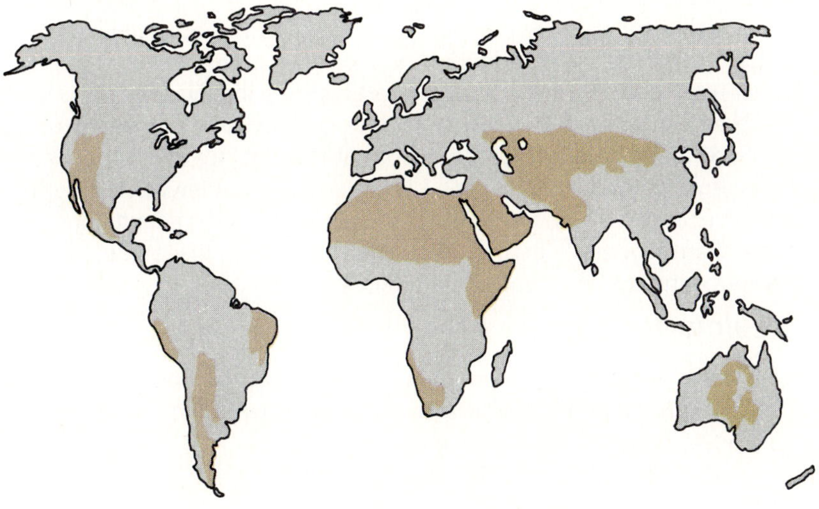

Desert in bloom

Mourning dove

Jackrabbit

Snake eating lizard

Locations of biomes

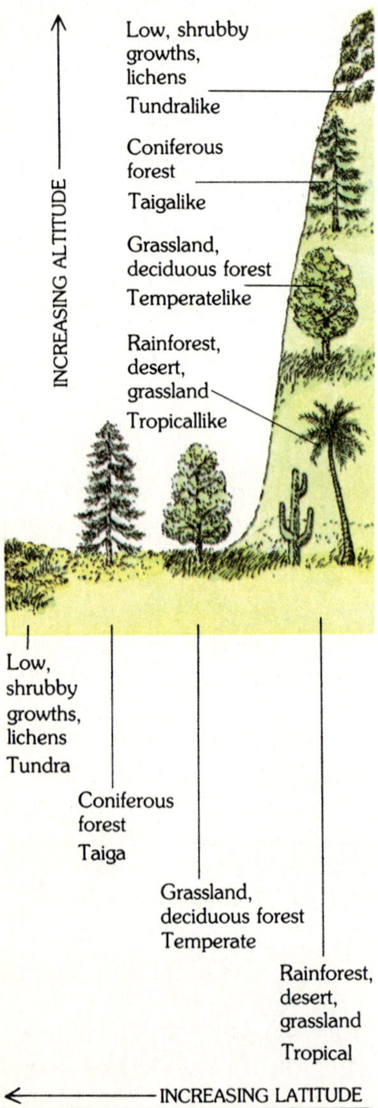

Climate and Biomes

Ecosystems characterized by certain populations occur around the earth as biomes. Their locations are determined mostly by climate. The general pattern from the equator to the North Pole (increasing latitude) is very similar to the pattern upward along mountain slopes (increasing altitude) as shown here. This should not be surprising since climate tends to change with increasing altitude in much the same way that it changes with increasing latitude.

Other Important Biomes

The **tropical rain forest** is an important biome across the earth as a whole, but it is not a major biome in North America. This biome is found in Central America, the Amazon Basin and Orinocan Basin of South America, central Africa, India, and southeast Asia. All tropical rain forests are found near the equator. Tropical rain forests have high rainfall, high temperatures all year, and a great variety of vegetation. An enormous variety of animals lives in the rain forest, including insects, lizards, snakes, monkeys, and colorful birds.

A few areas in Florida and in southern Texas are made up of forests similar to the tropical rain forest, but are commonly called subtropical forests. The temperature is warm year-round, but rainfall varies too much for them to be considered tropical rain forests.

Aquatic biomes occur in ponds, lakes, streams, and rivers. Community structure is likely to be different in still water than in rapidly flowing water. The depth of the water is also important for vegetation, since more light penetrates near the surface of the water than farther down in deep water. Many immature forms of insects and single-celled organisms occur in these biomes in shallow water. Algae are common, as are frogs, salamanders, snails, flatworms, and clams. Water pollution problems caused by humans have become particularly important in aquatic biomes.

Marine biomes are the biomes of the oceans. Factors influencing living things here include depth of the water, currents, tides, light penetration, and temperature. Ocean coasts are fascinating areas in which to observe communities because they are made up of populations which at high tide are under water but at low tide are actually on land. Organisms in this intertidal region live along coasts between the high-tide level and the low-tide level. They have interesting adaptations that allow them to survive both in water and on land between tides, such as burrowing into the sand during low tide.

Review It

1. What is a biome? List five important land biomes. For each, describe an important abiotic factor and list at least two important plants, herbivores, and predators.
2. Describe where tundra and taiga are found.
3. Explain how animals deal with extreme heat in hot deserts.
4. Explain how the pattern of biomes between the equator and polar region resembles the pattern on a mountain slope.

Pioneers in Biology

Rachel Louise Carson: Pioneer ecologist

Rachel Carson's career combined her talent and interest in two fields: science and literature. Two years before she died, she completed a book that was not only a scientific, but a poetic plea for ecological awareness. It focused on the real dangers of the misuse of chemical poisons. This classic work, *Silent Spring,* led to a United States Senate investigation of the use of chemical pesticides.

Rachel was born on May 27, 1907, in Springdale, Pennsylvania. The family homesite included an apple orchard and several acres of woodlands to explore. There were wild animals to listen to and watch, and a few farm animals to be taken care of. Rachel's mother, who had been a teacher, shared with her a love of nature.

Rachel started her writing career at age ten, with a story for a children's magazine. She continued to write for magazines and for school newspapers, and began her college career with a major in literature.

Because she needed science credits for graduation, Rachel took a biology course. A natural interest in life science, a curiosity about the sea, and the enthusiasm of her teacher combined to make the course a turning point in Rachel's life. She switched her major to biology. A summer studying marine biology at

Woods Hole in Cape Cod followed, and then graduate study at Johns Hopkins University for a master's degree in marine biology.

Carson taught biology and zoology after graduation, and might have continued, but in 1935 the country was still in a depression. The death of her father meant that she needed more money to support her mother and herself. At the Bureau of Fisheries (now the U.S. Fish and Wildlife Service) her science and literature background landed her a job as radio script writer for a series on marine life. She became a junior aquatic biologist, and later editor-in-chief of the information division for the Department of the Interior.

In 1948, encouraged by her supervisor at the bureau, she started work on a book that made the *New York Times* best-seller list for 81 weeks. The book, *The Sea Around Us* brought Carson worldwide recognition as a knowledgeable scientist and a brilliant writer. In clear terms she described the mysteries of the sea, conveying her love and respect for the vast oceans and the life they support. She prepared herself for that writing task by exploring the underwater world in a diving suit; and spending hours on the Fish and Wildlife survey ship, working in the laboratory and talking with the scientists. The book, completed in 1950, won several awards, including the National Book Award for the best nonfiction work of 1951.

During the next few years, Carson was busy writing, making public appearances, and giving speeches. Her summers were spent on the coast of Maine where she delighted in the sea and shore life. In 1957, she was to begin a book about the continent, but a letter from a friend changed her mind. The letter told about the spraying of DDT and about dead birds found in the yard for several days following.

Carson had seen the results of the lack of ecological planning. She had seen once lovely areas misused and even destroyed. She had read about environmental problems and took them very seriously. In 1958, she started working on *Silent Spring.* It was published in September of 1962 and became the most talked-about book of the decade. She did not say, as some have misquoted, that pesticides should never be used, but that *misuse* would lead to ecological disaster.

Rachel Carson died of cancer in 1964, two years after the publication of *Silent Spring.*

Chapter Summary

Energy flows from the sun through communities in food webs. Some sunlight also changes to heat when it strikes the surface of the earth, warming ecosystems by day and dissipating into space by night.

Unlike energy, various materials are cycled and conserved by nature on Earth. Examples are the water cycle, oxygen cycle, and nitrogen cycle. These substances, along with many others, are essential to life on Earth. Humans are involved in the cycles of these substances and in some cases have carelessly or accidentally changed the cycles.

Ecosystems around the earth that have similar climates usually are similar in overall structure. Such a group of similar ecosystems is a biome. Particularly important land biomes are tundra, taiga, temperate-deciduous forest, grassland, and desert. In addition, many aquatic and marine biomes are known.

If a drastic change such as a rock slide or fire occurs in an ecosystem, the species destroyed in the change do not usually return directly to populate the area again. Rather, other species such as lichens on bare rock, that are adapted to survive under the conditions after the change will live there. As biotic and abiotic conditions gradually change, a successive replacement of one community for another takes place. This sequence is called succession. For any particular kind of climate and soil, succession leads to a fairly stable community structure called the climax community.

Questions to Think About

1. You have learned that the most severe water pollution takes place during the runoff phase of the water cycle. Even though a river or lake may become polluted and even destroyed, the water cycle itself has continuous potential for purification of water. This means that if pollution were stopped, a river might eventually recover. How does this work?
2. Respiration and burning of fossil fuels continuously place carbon dioxide into the atmosphere. Give at least two ways that carbon dioxide is removed from the atmosphere.
3. Explain how solar energy from the past has been stored in fossil fuel (gas, oil, coal) molecules.
4. Explain the impact on the nitrogen cycle if all denitrification were stopped by destroying all denitrifying populations. What would happen if nitrification to nitrates were stopped?
5. Give two or more reasons why, in a certain area, primary succession would be expected to take longer than secondary succession.
6. Write a paragraph in which you explain why climate is the most important factor that determines the biome of a certain area. (By the way, some scientists think soil may be just as important in this as climate.)

Extra Research

1. Get a physics book and study the First Law of Energy (or Thermodynamics) and the Second Law of Energy (Thermodynamics). Can you harvest more energy as food from a field (or garden) than the total of all forms of energy put into the field? Why or why not?
2. List the possible forms of energy that can be put into an agricultural field. (See question number one.)
3. Talk to officials in your local Department of Health and your State Department of Ecology. What are some of the major water pollution problems in your area?
4. Do some research on the chemical fertilizer industry. What does a fertilizer factory consume? What are some advantages and disadvantages of chemical fertilizers compared with organic fertilizers such as cattle manure?
5. Do some research and talk to local experts about what kind of natural biome you have (or once had) in your community. Which of the major biome types described in this chapter does yours resemble the most? Have studies of primary succession and secondary succession been done in your area?

Chapter Test

Part I

Copy the number of each test item. If the item is *true,* write *T* after the number; if the item is *false,* write *F.*

1. Energy is cycled over and over in ecosystems from food to heat and back to food again. (644)
2. Most of the water of oceans and seas gets there directly from rainfall. (645)
3. Most water pollution problems caused by humans have occurred during the runoff phase of the water cycle. (648)
4. Respiration uses carbon dioxide, and photosynthesis produces carbon dioxide. (647)
5. Nitrogen is an important part of protein molecules in living things. (651)
6. The main form in which nitrogen is used by green plants is free nitrogen (N_2). (651)
7. The living components of an ecosystem are called the biome. (656)
8. Permafrost is a characteristic of tundra. (657)
9. Intertidal marine organisms are adapted to living only in salt water. (662)
10. Secondary succession may occur as a result of human interference with the environment. (655)

Part II

Copy the number of each sentence below. After each number, write the word or words that complete the sentence correctly.

11. The earth's shell of soil, water, and atmosphere and the living things in them make up the _____. (643)
12. In the water cycle, atmospheric water that is blown from above seas to above land is offset or returned to seas by _____. (645)
13. The process of converting atmospheric nitrogen (N_2) into ammonia is _____. (651)
14. Nitrogen fixation, nitrification, and denitrification are all accomplished by populations of _____. (651)
15. When a volcano erupts, it is likely to add a gaseous substance called _____ to the atmosphere. (649)
16. The stable community structure of an ecosystem after succession is complete is called the _____. (653)
17. Succession from bare rock is called _____ succession. (654)
18. Chemical elements and physical elements such as wind and climate are _____ factors of the ecosystem. (644)
19. Probably the main abiotic factor that makes grasslands different from temperate-deciduous forests is that grasslands _____. (660)
20. The most important abiotic factor that determines how deep green plants can survive in aquatic and marine biomes is _____. (662)

Part III

Copy each numbered item from List A. After each item, write the letter and words from List B that best describe it.

List A

21. taiga (658)
22. tundra (657)
23. temperate-deciduous forest (659)
24. desert (661)
25. grassland (660)

List B

a. coniferous forest, very cold winters, moderate summers, moose, bear
b. hot days, cold nights, low precipitation, cacti, saltbush, kangaroo rat, rattlesnake.
c. irregular rainfall, hot summers, cold winters, high winds, herbivorous animals
d. low precipitation, few trees, long, cold winters, short summers, dwarf willow, caribou
e. abundant rain, rich soil, hardwood trees, hot summers, cold winters, black bear

Focus on Careers

Range Manager

"My family hiked and camped in this part of the country every year. I grew to love the vastness of this place and always hoped it would remain the same. As I learned more, however, I realized that a large area of range land like this needs to be protected and managed properly.

"Now as a range manager, I try to do just that: manage, improve, and protect the resources of this area so they can be used without being destroyed. This area is important for recreational purposes as well as commercial purposes.

"Together with the conservation specialist in this area, we are now working to restore an area of the range that has been damaged by mining operations. We have done extensive studies, both on site and in the laboratory, to find out what will grow best here. Now we will plow and plant the area with grasses and other plants that will support livestock. This, in time, will make the area profitable again without hurting it.

"I'm really happy to be doing this kind of work—outside and in a place I love. The work is very productive and far-reaching. Every day I'm glad I completed my college degree in range management."

Range lands of the United States cover more than one billion acres. They contain many natural resources: grass and shrubs for animal grazing, habitats for wildlife, watersheds, minerals, and energy sources. Range managers, also called range ecologists, work to improve and protect these areas. This is done by studying the habits of animals native to the area, providing protected grazing areas for livestock, and conserving the soil and vegetation. A range manager may supervise the controlled burning or biological destruction of undesirable plants. He or she may plan and provide for animal watering facilities, erosion control, and fire prevention. Range managers work closely with foresters and others involved in recreation.

A degree in range management requires courses in biology, chemistry, physics, mathematics, and communication skills. Additional courses in forestry, wildlife, and recreation may be helpful in getting a job. Of course, a love for the outdoors is essential.

Most range managers work for the federal government in the Forest Service, the Soil Conservation Service, or the Bureau of Land Management. There are also jobs working for coal and oil companies. The outlook for job opportunities is good.

Career Information

Forestry Technician: U.S. Department of Agriculture, Forest Service, Washington, DC 20250.

Park Technician: Colorado Mountain College, Glenwood Springs, CO 81601.

Range Manager: Society for Range Management, 2760 W. 5th Avenue, Denver, CO 80204.

Soil Scientist: Soil Society of America, 677 S. Segoe Road, Madison, WI 53711.

Forestry Technician

Duties: Performs many functions to assist foresters in the care and management of forest lands and their resources. These include keeping track of timber production, supervising road crews working in forested areas, leading firefighting crews, supervising planting efforts in burned areas, and enforcing forest rules and regulations for visitors.

Training: Qualifications vary. Work experience as a firefighter or in a tree nursery is helpful. Completing a specialized course which includes surveying, tree identification, and aerial photograph interpretation in a one- or two-year program makes one more valuable. Competition for these jobs is keen.

Opportunities: Many jobs are available in government agencies as well as private industries.

Park Technician

Duties: Helps plan and then maintains the day-to-day operation of a park: caring for the grounds, directing visitors to areas of interest, writing reports concerning park conditions. At a ski resort, for example, a park technician might draw a landscape plan for the area, determining plants to be used; supervise the building of trails and shelters; and make safety checks for people using the ski facilities. A park technician might help plan a golf course or an urban park that features a small lake for fishing.

Training: Requires completion of a two-year program in land management technology.

Opportunities: The need for trained people to plan and manage recreational areas is increasing rapidly.

Soil Scientist

Duties: Collects soil samples and then, in a laboratory, classifies them according to a national system. Using aerial photographs, the soil scientist then prepares a soil-type map, which shows where various soil types are found. The map also shows features such as hills, rivers, lakes, and roads. Some soil scientists conduct experiments on farms to learn what crops grow well in various soils or what effects fertilization, crop rotation, and pollution have on soils. Others investigate building sites to make sure erosion control methods are being followed where areas of vegetation have been removed.

Training: Requires a college degree with a major in soil science.

Opportunities: More than half the soil scientists work for the Soil Conservation Service. About one hundred are hired each year.

Bibliography: Unit Ten

Abbey, Edward. *Cactus Country.* Time Inc., 1973. Beautifully illustrated book describing desert inhabitants.

Allen, Durward. *The Life of Prairies and Plains.* McGraw-Hill, 1967. Vividly illustrated book that describes the variety of creatures that have developed adaptations for life on the plains and prairies.

Buchsbaum, Ralph, and Buchsbaum, Mildred. *Basic Ecology.* Boxwood Press, 1957. Good reference for basic principles of ecology.

Boorer, Michael. *The Living Earth: Forest Life.* The Danbury Press, 1975. Brilliant photographs capture the plant and animal life in the forest.

Cousteau, Jacques-Yves. *Life and Death in a Coral Sea.* Doubleday and Co., 1971. Cousteau and his crew explore and photograph the unique life forms of the great coral areas of the Indian Ocean and the Red Sea.

Darling, Lois, and Darling, Louis. *A Place in the Sun: Ecology and the Living World.* William Morrow and Co., 1968. Discusses cycling of materials, ecosystems, and pollution.

Duffey, Eric. *The Living Earth: Grassland Life.* The Danbury Press, 1975. Beautiful photographs are included in this volume that describes the inhabitants of the grasslands.

Emmel, Thomas. *An Introduction to Ecology and Population Biology.* W. W. Norton and Co., Inc., 1973. Includes information on cycling of materials and energy, interactions of organisms, population growth, succession, and pollution.

Farb, Peter. *Ecology.* Time Inc., 1963. Picture essay volume describing interactions of organisms, biological rhythms and clocks, and the impact of people on nature.

Farb, Peter. *The Forest.* Time Inc., 1963. Absorbing account of the relationships among organisms that live in the forest community.

Friendly, Natalie. *Miraculous Web: The Balance of Life.* Prentice-Hall, Inc., 1968. Describes the different biomes and the animals that inhabit them.

Grossman, Mary Louise, and others. *Our Vanishing Wilderness.* Grosset and Dunlap, Inc., 1969. Picture essay volume that explores the problems that accompany destruction of habitats.

Milne, Lorus, and Milne, Margery. *Patterns of Survival.* Prentice-Hall, 1967. Discusses how adaptations of animals increase each species' chances of survival and reproduction.

Oates, John. *The Living Earth: Web of Life.* The Danbury Press, 1975. Beautifully illustrated encyclopedia volume describing food chains, matter and energy, and ecosystems.

Raskin, Edith. *The Pyramid of Living Things.* McGraw-Hill, 1967. Discusses different biomes and the plants and animals that inhabit them.

Schwartz, George, and Schwartz, Bernice. *Life in a Log.* Natural History Press, 1972. Describes the complex relationships that exist among organisms that make their home in a log.

Shuttlesworth, Dorothy. *Natural Partner-Ships: The Story of Symbiosis.* Doubleday and Co., 1969. Interesting picture essay book that explains partnerships between organisms.

Silverberg, Robert. *The World Within the Tide Pool.* Weybright and Talley, 1972. Describes life in tide pools and along temperate coastlines. Also discusses the dangers posed by oil spills.

Sparks, John. *The Living Earth: The Air Around Us.* The Danbury Press, 1975. Beautiful picture essay volume that describes the importance and cycling of gases that make up the atmosphere. Also discusses adaptations to life in the open air.

Storer, John. *The Web of Life.* The Devin-Adair Co., 1967. Explains in simple terms some of the fundamentals of ecology and describes the natural community in action.

Article:

Young, Gordon, and Blair, James P. "Pollution, Threat to Man's Only Home." *National Geographic,* December 1970.

Acknowledgements

Positions of photographs are shown in abbreviated form as follows: top (t), bottom (b), center (c), left (l), right (r). All natural history photographs credited to Photo Researchers are from the National Audubon Society Collection.

Chapter 1: 2, Animals Animals/Oxford Scientific Films; 4, SF; 5, Nos. 1, 2, 4, 5, 6, 7 Robert D. Allen; 6, 7, 9, NASA; 10, Culver Pictures (1); 14, Courtesy Stanley L. Miller (l), Courtesy Sidney W. Fox, from Fox & Dose, *Molecular Evolution and the Origins of Life,* 2d ed. (c, r); 18, Courtesy Clay Adams, Div. of Becton, Dickinson and Co. (t), SF (c), Courtesy Dr. Leon LeBeau, Univ. of Ill. Med. Ctr., Chicago (b), Lennart Nilsson, from *Behold Man,* Little Brown & Co. (r); 19, Walter Dawn (both); 20, Manfred Kage/Peter Arnold, Inc. (t), Animals Animals/Oxford Scientific Films (c), Walter Dawn (b); 21, Robert D. Allen (tl), Manfred Kage/Peter Arnold, Inc. (tr, bl), Lester V. Bergman & Assoc. (br); 22, Allan Roberts; 23, Patrick Grace (tl), SF (bl), W. H. Hodge/Peter Arnold, Inc. (tr), Walter Dawn (cr), Joy Spurr/Bruce Coleman, Inc. (br); 24, Robert P. Carr/Bruce Coleman, Inc. (l), William H. Amos (c, r); 25, SF (tl), S. J. Krasemann/Peter Arnold, Inc. (tr), W. H. Hodge/Peter Arnold, Inc. (bl), Allan Roberts (bc), Tom Stack (br); 28, Adapted from J. W. Valentine, "The Evolution of Multicellular Plants and Animals." Copyright © 1978 Scientific American, Inc. All rights reserved.; 33, Charles R. S. Pitman, *A Guide to the Snakes of Uganda.* Pub. by Uganda Soc., Kampala, Uganda, 1938.

Chapter 2: 36, SF; 39, Animals Animals/Stouffer Enterprises, (t), D. J. Chivers/Anthro-Photo (c), E. Hanumantha Rao/Photo Researchers (b); 45, Courtesy Clay Adams, Div. of Becton Dickinson and Co. (tl); 47, Reprinted by permission of the Chicago Tribune-New York News Syndicate, Inc.; 50, SF (tl, br), Photri (tr), William F. McMahon (bl); 52–54, Hawaiian Volcano Observatory, U.S. Geological Survey; 58, Ill. Dept. of Law Enforcement, Bureau of Scientific Services; 59, Owen Franken/Stock Boston (l), SF (c, r).

Chapter 3: 62, Manfred Kage/Peter Arnold, Inc.; 64, SF (tl); William H. Amos (tr), Armed Forces Institute of Pathology, #74-6627 (bl), Macmillan Science Co., Inc. (br); 66, SF (t), Manfred Kage/Peter Arnold, Inc. (b); 67, SF (t), Manfred Kage/Peter Arnold, Inc. (b); 69, Courtesy Peter C. Hoppe, from P. C. Hoppe & Karl Illmensee, "Microsurgically Produced Homozygous-Diploid Uniparental Mice," *Proc. Natl. Acad. Sci.,* 74: 5657–5661, 1977; 72, photo by J. David Robertson, from H. Holter, "How Things Get into Cells," *Scientific American,* Sept., 1961; 73, Courtesy Dr. Emma Shelton; 74, Courtesy G. Benjamin Bouck, Univ. of Ill., Chicago; 75, Courtesy Donald Larsen (b); 76, Courtesy Abbott Laboratories (c); 77, Courtesy Robert D. Goldman, Carnegie-Mellon Univ., from R. D. Goldman & D. M. Knipe, "Functions of Cytoplasmic Fibers in Non-Muscle Cell Motility," *Cold Spring Harbor Symposium on Quantitative Biology,* Vol. XXXVII, pp. 523–534, 1973.

Chapter 4: 82, Manfred Kage/Peter Arnold, Inc.; 86–87, Phillip A. Harrington/Peter Arnold, Inc.; 91, Manfred Kage/Peter Arnold, Inc. (bl, bc, tr), Roman Vishniac (br); 100, Courtesy Dr. A. E. Vatter; 102, Robert D. Allen; 103, Courtesy F. K. Fong.

Chapter 5: 106, Roger K. Burnard; 108, Courtesy Dr. A. Solari, *Proc. Natl. Acad. Sci.,* 53: 507, 1965; 109, Courtesy Steven L. McKnight & O. L. Miller, Jr., Dept. of Biol., Univ. of Va., copyright retained; 113, Bettmann Archive; 120, NASA (both); 122, Manfred Kage/Peter Arnold, Inc.; 126, Landrum B. Shettles, M.D.; 127, Thomas Eisner, Cornell Univ.; 130, Animals Animals/Lynn Stone (r); 134, The Intersociety Committee on Pathology Information, Inc.; 135, M. Rotker/Taurus Photos (t), Courtesy Wilson Sporting Goods (bl), D. Ellefsen/FPG/Alpha (br).

Chapter 6: 138, Jack & Betty Cheetham/Magnum; 140, Culver Pictures; 149, SF; 150, Courtesy Dr. Marcus M. Rhoades; 152, Courtesy Walter Plaut; 156, David Linton from *Scientific American;* 157, Robert D. Allen (t), © Carroll H. Weiss, RBP, 1979 (c); 160, Macmillan Science Co., Inc.; 161, from C. B. Bridges, *Journal of Heredity,* 26: 60–64, 1936. By permission. (l), Courtesy Estate of B. P. Kaufmann (r).

Chapter 7: 164, E. Hanumantha Rao/Tom Stack; 170, SF (b), Courtesy Dr. Henry L. Nadler, Genetics Dept., Children's Memorial Hosp., Chicago (tl, tc, tr); 171, Lester V. Bergman & Assoc. (l), Grant Heilman (r); 172–173, Allan Roberts; 176, H. Armstrong Roberts; 179, Courtesy Genentech, Inc., 460 Point San Bruno Blvd., S. S. F., CA 94080; 182, SF; 183, SF (bl, br), Farrell Grehan/Photo Researchers (t).

Chapter 8: 186, Paläotologisches Museum, Museum für Naturkunde der Humboldt-Universität, Berlin DDR; 188, Yoram Kahana/Peter Arnold, Inc. (l), Allan Roberts (r); 189, M. W. Williams, Natl. Park Service; 191, Petrified Forest Museum Assoc. (tl), SF (tr), Allan Roberts (bl, bc), Manfred Kage/Peter Arnold, Inc. (br); 197, Gavin DeBeer, *Charles Darwin: Evolution by Natural Selection,* Thomas Nelson & Sons Ltd.; 201, Constance P. Warner (l), SF (r); 204, Mills Tandy (t), Leonard Lee Rue III/Photo Researchers (c); 207, from Nelson, et al, *Fundamental Concepts of Biology,* 2d ed., 1970, John T. Wiley & Sons, Inc. By permission.; 211, Dr. H. B. D. Kettlewell, Univ. of Oxford, England (both); 212, Courtesy R. F. Johnston, Kansas State Univ., & R. K. Selander, Univ. of Rochester (all).

Chapter 9: 216, William H. Amos; 221, Culver Pictures; 222, Grant Heilman (t, b), John Colwell/Grant Heilman (c); 223, Bureau of Plant Industry, USDA; 224, Ed Park (t, c, bl), Wilford L. Miller (br); 225, Dr. H. B. D. Kettlewell (both); 226, Allan Roberts (tl), SF (tc), Cosmos Blank/Photo Researchers (tr), W. H. Hodge/Peter Arnold, Inc. (bl), Russ Kinne/Photo Researchers (bc), Charlie Ott/Photo Researchers (br); 227, Willis Peterson (both); 228, William H. Amos (tl), Allan Roberts (bl, br); 229, Courtesy Sea World of Ohio (t), Douglas Faulkner (c); 232, William F. McMahon; 233, SF (l), Field Museum of Natural History, Chicago (c), Susan Kuklin/Photo Researchers (r).

Chapter 10: 236, SF; 238, Courtesy C. F. Robinow, Univ. of Western Ontario, Canada (r); 239, SF (tr), Courtesy Dr. R. Wyckoff, Natl. Institutes of Health (cc), Courtesy Dr. Leon LeBeau (cr), Courtesy D. L. Findley, P. L. Walne, and R. W. Holton, Univ. of Tenn., Knoxville (br); 240, Courtesy Stanley W. Watson, Woods Hole Oceanographic Institution (both); 242, Dr. J. G. Zeikus (tl), from E. Klieneberger-Nobel & F. W. Cuckow, "A Study of Organisms of the Pleuropneumonia Group by Electron Microscopy," *J. Gen. Microbiol.,* 12: 95–99, 1955. By permission of Cambridge Univ. Press (br); 245, Courtesy Dr. T. F. Anderson, The Institute for Cancer Research, Phila. (l), Courtesy Maria Costa and George B. Chapman, Georgetown Univ. (r); 247, from R. W. Horne, "The Structure of Viruses," Copyright © 1963 Scientific American, Inc. All rights reserved. (tl), from R. C. Valentine and H. G. Pereira, "Antigens and Structure of the Adenovirus," *J. Mol. Biol.,* 13: 13–20, 1965. Copyright by Academic Press, Inc., (London) Ltd. (bl), Courtesy Michael Moody (bc), Courtesy Gene M. Milbrath (br); 248, Courtesy Lee D. Simon, Waksman Institute of Microbiology, Rutgers Univ.; 250, Courtesy Janet Pisani and George B. Chapman, Georgetown Univ.; 251, Animals Animals/Stewart Halperin; 252, SF.

Chapter 11: 254, Animals Animals/Oxford Scientific Films; 256, Courtesy David Porter, Univ. of Ga., Athens; 258, Walter Dawn; 261, Manfred Kage/Peter Arnold, Inc. (l), Milton Love/Peter Arnold, Inc. (r); 262, Courtesy Dr. I. R. Gibbons and Dr. A. V. Grimstone. By permission of the Editors of the *British Medical Journal;* 263, Robert D. Allen (all); 264, Manfred Kage/Peter Arnold, Inc.; 265, USDA-Forest Service Northeastern Area-FIDM.

Chapter 12: 268, 271, Robert Strindberg; 274, SF (tl), S. J. Krasemann/Peter Arnold, Inc. (bl); 276, Patrick Grace (both); 278, Robert Strindberg (tl), Larry West (bl), John H. Gerard (br); 279, Walter Dawn (tl, br), Patrick Grace (bl, tr), John H. Gerard (cr); 280, Adapted from Walter Litten, "The Most Poisonous Mushrooms," Copyright © 1975 Scientific American, Inc. All rights reserved.; 281, Larry West (l,c), Walter Dawn (r); 282, Patrick Grace; 283, Patrick Grace (bl), Walter Dawn (bc), Larry West (tr), Robert Strindberg (cr,br); 284, W. H. Hodge/Peter Arnold, Inc. (top two), Allan Roberts (third down), John H. Gerard (b); 286, Courtesy Clay Adams, Div. of Becton, Dickinson and Co.; 287, Bettmann Archive; 290, Guy Gillette/Photo Researchers; 291, Paul Moylett/Taurus Photos (l), SF (c,r).

Chapter 13: 294, Grant Heilman; 296, Earth Scenes/Oxford Scientific Films; 300, John MacGregor/Peter Arnold, Inc. (l), Larry West (r); 302, Grant Heilman (bl), John MacGregor/Peter Arnold, Inc. (bc), Allan Roberts (br); 304, John Colwell/Grant Heilman (l), Jeanne Thompson/Grant Heilman (c), Runk/Schoenberger/Grant Heilman (r); 305, John Ebeling (t); 306, SF; 308, Grant Heilman (t,cl,bl,cr), John Colwell/Grant Heilman (c-middle); 309, Grant Heilman.

Chapter 14: 312, Thomas Eisner, Cornell Univ.; 315, Grant Heilman; 318, Alan Pitcairn/Grant Heilman (t), Grant Heilman (c,b); 320, Grant Heilman (both); 323, Runk/Schoenberger/Grant Heilman; 324, Courtesy Dr. J. A. D. Zeevaart, Mich. State Univ.; 325, Grant Heilman; 327, Dr. Hugh H. Iltis, Univ. of Wisc., Madison; 328, Grant Heilman (t,c), SF (b).

Chapter 15: 332, Larry West; 334, William Parker (t), Runk/Schoenberger/Grant Heilman (b); 335, SF; 340, Hugh Spencer; 345, Carl K. Shuman/Grant Heilman (both); 346, David Cavagnaro; 347, Grant Heilman; 348, Courtesy Dr. J. A. D. Zeevaart; 349–350, Runk/Schoenberger/Grant Heilman; 351, Wide World.

Chapter 16: 354, 356, Grant Heilman; 357, SF; 358, John Ebeling (l), SF (r); 359, Allan Roberts; 360, Runk/Schoenberger/Grant Heilman; 362, Courtesy Dr. Walter Preil, from Zeiss Information #75; 364, Library of Congress; 370, S. Rannels/Grant Heilman (t), John Colwell/Grant Heilman (c), William Parker (bl), SF (br); 374, Ellis Herwig/Stock Boston; 375, L. L. T. Rhodes/Taurus Photos (1), H. Armstrong Roberts (c), H. Lank/FPG (r).

Chapter 17: 378, Douglas Faulkner; 382, Al Grotell; 384–385, William H. Amos; 387, Nat Fain; 388, SF; 390, William H. Amos; 392, Armed Forces Institute of Pathology, #78–373 (t); 394–395, William H. Amos; 400, Lionel Atwill/Peter Arnold, Inc.; 402, R. Mariscal/Bruce Coleman, Inc.; 404, Animals Animals/Keith Gillett; 405, William H. Amos; 406, Douglas Faulkner; 407, R. Mariscal/Bruce Coleman, Inc.; 408, William H. Amos (tl,br), Jane Burton/Bruce Coleman, Inc. (both cl), Douglas Faulkner (bl), Al Grotell (bc); 409, UPI.

Chapter 18: 412, Animals Animals/Raymond Mendez; 414, David Waselle (tl), U.S. Naval Photographic Center (cl), Hans Pfletschinger/Peter Arnold, Inc. (bl), S. J. Krasemann/Peter Arnold, Inc. (br); 415, William H. Amos (tl,tr,br), Hans Pfletschinger/Peter Arnold, Inc. (bl); 416, David Scharf/Peter Arnold, Inc.; 417, William H. Amos (tr), William Curtsinger/Photo Researchers (cr), Tom Myers (bl); 421, John MacGregor/Peter Arnold, Inc.; 422, William H. Amos (cl); 423, David Cavagnaro (tr), William H. Amos (cr), Tom Myers (br); 424, John H. Gerard; 425, John MacGregor/Peter Arnold, Inc. (bl,tr,tc); 426, John Ebeling (top two,b), Hans Pfletschinger/Peter Arnold, Inc. (third down); 427, Allan Roberts; 430, S. J. Krasemann/Peter Arnold, Inc.; 432, Animals Animals/D. R. Specker (tl), Animals Animals/R. A. Mendez (cl), Animals Animals/Richard Kolar (bl), Animals Animals/ R. A. Reinhold (bc, br); 433–434, Hans Pfletschinger/Peter Arnold, Inc. (all); 435, Animals Animals/Oxford Scientific Films (tr,cr), Hans Pfletschinger/Peter Arnold, Inc. (bl); 437, USDA Forest Service Northeastern Forest Experiment Station; 440, Dave Woodward/Taurus Photos; 441, Harry Rogers/Photo Researchers (l), SF (c), Peter Menzel/Stock Boston (r).

Chapter 19: 444, Tom Myers; 446, Douglas Faulkner (t), Bo Tallmark (bl), S. J. Krasemann/Photo Researchers (br); 449, R. A. Odum/Peter Arnold, Inc. (l), Allan Roberts (r); 450, Russ Kinne/Photo Researchers (t), Gary G. Gibson/Photo Researchers (c), George Skadding, *Life Magazine*, ©1947, Time Inc. (b); 452, Allan Roberts (t), Douglas Faulkner (l), Tom Myers (r); 453, Douglas Faulkner (l), Tom McHugh/Photo Researchers (r); 454, David Waselle (t); 455, Tom Myers (t); 457, Douglas Faulkner (both); 458, Allan Roberts (r), Field Museum of Natural History, Chicago (l); 460, Animals Animals/Breck P. Kent (l), William H. Amos (r); 466, Hans Pfletschinger/Peter Arnold, Inc. (1st four stages), William H. Amos (last three stages); 469, Tom Myers (top three), Allan Roberts (b); 470, Frederick J. Dodd/Peter Arnold, Inc. (t), Russ Kinne/Photo Researchers (b); 471, Sam Abell.

Chapter 20: 474, Werner H. Mueller/Peter Arnold, Inc.; 476, Townsend Dickinson/Photo Researchers (t), Allan Roberts (b); 478, Animals Animals/R. F. Head; 479, John Ebeling (tl), Grant Heilman (bl,tr); 480, Andreas Feininger, *Life Magazine*, ©1951 Time Inc.; 485, Wilford L. Miller (all); 487, Animals Animals/Oxford Scientific Films (bl,bc,br); 490, Tom McHugh/Photo Researchers (bl), Russ Kinne/Photo Researchers (br); 491, A. B. Joyce/Photo Researchers (t), Allan Roberts (b); 497, Dr. Ronald H. Cohn, The Gorilla Foundation; 500, M. Rosenthal/ATOZ/Van Cleve; 501, SF (b,r), R. L. Williams/Taurus Photos (c).

Chapter 21: 504, dva/Erhard Borcke/Erwin Ranz; 521, Courtesy Pfizer Medical Systems, Inc.

Chapter 22: 534, Manfred Kage/Peter Arnold, Inc.; 537, 539, 540, 541, SF; 542, Courtesy Merck & Co., Inc. (t), Macmillan Science Co., Inc. (b); 545, X ray Dept., Children's Memorial Hospital, Chicago; 547–548, Health Museum, Cologne; 549, Manfred Kage/Peter Arnold, Inc.

Chapter 23: 554, Health Museum, Cologne; 557, © Carroll H. Weiss, RBP, 1979 (l), SF (c), Robert D. Allen (r); 561, SF; 566, Courtesy Pfizer Laboratories; 568, Courtesy Beecham Laboratories; 569, Culver Pictures; 570, Courtesy Virus Research Laboratory, Univ. of Pittsburgh; 571, National Library of Medicine.

Chapter 24: 574, Health Museum, Cologne; 580, Costa Manos/Magnum; 585, Howard Sochurek; 586, Courtesy Cordis Dow Corp.; 589, UPI.

Chapter 25: 592, Quigg Studio; 603, SF; 604–605, Carlo Bevilacqua/CEDRI; 608, Reproduced with permission from the *Birth Atlas*, published and copyright by Maternity Center Assoc.; 609, SF; 610, UPI (tl,bl,bc), Ira Wyman/Sygma (br); 611, Courtesy Kenneth L. Jones, M.D., Univ. of Calif. Med. Ctr., San Diego.

Chapter 26: 618, Joe Branney/Tom Stack; 620, Allan Roberts (both); 621, John Ebeling (both); 622, Peter Arnold (tl), William H. Amos (cl), Victor Englebert (bl), NASA (br); 623, Robert Strindberg (l), Wilford L. Miller (c), Erika Stone/Peter Arnold, Inc. (r); 624, Cosmos Blank/Photo Researchers (tl), John Ebeling (bl), Tom Myers (br); 625, Allan Roberts (b), Bruce Coleman (t), Wilford L. Miller (c); 626, Fred Bruemmer (tl), Dr. George Gerster/Photo Researchers (bl), Al Grotell (bc), SF (br); 627, Allan Roberts (t), Wilford L. Miller (center two); John D. Luke/Tom Stack (b); 628, Helen C. Miller (left four), Allan Roberts (br); 629, Willis Peterson; 630, SF (tl), Robert Strindberg (cl), S. J. Krasemann/Photo Researchers (bl); Harold W. Hoffmann/Photo Researchers (br); 631, William H. Amos (tr), Cindy Ederegger/Peter Arnold, Inc. (br), Animals Animals/Jack Wilburn; 633, M. P. Kahl/Bruce Coleman, Inc. (l), Harry Rogers/Photo Researchers (c), Douglas Faulkner (r); 634, John Ebeling (tl,br), Wilford L. Miller (tc,tr,bl); 635, Animals Animals/Trudy Unverhau (tr), John Ebeling (cl,bl), David Cavagnaro (bc), Allan Roberts (br), Lionel Atwill/Peter Arnold, Inc. (cr); 639, Bruce Roberts/Photo Researchers (l), Patricia Caulfield/Photo Researchers (r).

Chapter 27: 642, R. G. Everts/Photo Researchers/Rapho-Guillumette; 646, Esther Henderson/Photo Researchers (t), Peter Damroth/Photo Researchers (cl); 648, John Ebeling (t,c), Tom Myers (b); 651, Courtesy Dr. J. C. Burton, Nitragin Co.; 652, Jerome Wexler/Photo Researchers; 654, USDA Forest Service; 655, William H. Amos (all); 657, Fred Bruemmer (both); 658, Fred Bruemmer (t), Wilford L. Miller (c), S. J. Krasemann/Peter Arnold, Inc. (b); 659, Animals Animals/Leonard Lee Rue III (t), Wilford L. Miller (c), Allan Roberts (b); 660, Durward L. Allen (t), Wilford L. Miller (c), Willis Peterson (b); 661, Josef Muench (t), David Cavagnaro (second down, b), Willis Peterson (third down); 663, Erich Hartmann/Magnum; 666, Cary Wolinsky/Stock Boston; 667, USDA Forest Service (l,r), Arkansas Dept. of Parks and Tourism/Robyn Hutcheson (c).

Cover: Walter Dawn

Glossary

The pronunciation of each word is shown just after the word, in this way: **ab bre vi ate** (ə brē′vē āt). The letters and signs used are pronounced as in the words below. The mark ′ is placed after a syllable with primary or heavy accent, as in the example above. The mark ′ after a syllable shows a secondary or lighter accent, as in **ab bre vi a tion** (ə brē′vē ā′shən).

a	hat, cap
ā	age, face
ä	father, far
b	bad, rob
ch	child, much
d	did, red
e	let, best
ē	equal, be
ėr	term, learn
f	fat, if
g	go, bag
h	he, how
i	it, pin
ī	ice, five
j	jam, enjoy
k	kind, seek
l	land, coal
m	me, am
n	no, in
ng	long, bring
o	hot, rock
ō	open, go
ô	order, all
oi	oil, voice
ou	house, out
p	paper, cup
r	run, try
s	say, yes
sh	she, rush
t	tell, it
th	thin, both
ŦH	then, smooth
u	cup, butter
u̇	full, put
ü	rule, move
v	very, save
w	will, woman
y	young, yet
z	zero, breeze
zh	measure, seizure
ə	represents: a in about, e in taken, i in pencil, o in lemon, u in circus

abiotic (ā′bī ot′ik), having to do with the non-living parts of the environment, such as rocks, water, air.

absorption (ab sôrp′shən), the taking in of substances, such as digested food and oxygen, across a cell membrane.

active transport, the passage of a substance through a cell membrane to a region of higher concentration; usually requiring the use of energy.

adaptation (ad′ap tā′shən), the evolutionary process by which the descendants of a population become more suited to their environment through genetic change; an inherited feature that helps an organism survive.

ADP (adenosine diphosphate) (ə din′ə sēn′ dī fos′fāt), a molecule that functions in energy storage and transfer; it is formed from ATP.

adrenal (ə drē′nl) **glands,** two endocrine glands located one above each kidney which secrete hormones that control several body processes.

aerobe (er′ōb), a microorganism that uses oxygen.

aerobic (er′ō′bik), living, growing, or taking place where there is oxygen.

algae (al′jē), group of simple photosynthetic organisms, generally one-celled and generally found in ocean water.

allantois (ə lan′tō is), a membrane formed as an outgrowth of the gut in embryos of reptiles, birds, and mammals; it forms part of the placenta in mammals.

alleles (ə lēlz′), genes that code for different forms of the same protein.

allergy (al′ər jē), extreme sensitivity to a substance harmless to most people.

alternation of generations, a life cycle in which a diploid generation alternates with a haploid generation.

alveoli (al vē′ə lī), the air sacs in the lungs.

amino (ə mē′nō) **acid,** any of some twenty organic acids that contain nitrogen and that combine to form proteins in living things; the end products of protein digestion.

amnion (am′nē ən), the fluid-filled sac in which the reptile, bird, or mammal embryo develops.

amphibian (am fib′ē ən), any of the class Amphibia; cold-blooded vertebrates that usually spend their larval life in water and their adult life on land.

anabolism (ə nab′ə liz′əm), the energy-requiring metabolic processes by which simple substances are converted to more complex substances; synthesis.

anaerobe (a ner′ōb), a microorganism that is poisoned by oxygen.

anaerobic (an′ə rō′bik), living, growing, or taking place where there is no oxygen.

analogous (ə nal′ə gəs), having to do with organs that are similar in function but that differ in structure.

anaphase (an′ə fāz), the stage of mitosis during which chromosomes migrate to opposite poles of the cell.

angiosperm (an′jē ə spėrm), one of a group of flowering plants having its seeds enclosed in an ovary, or fruit; for example, grasses, strawberries, and oaks; anthophyte.

annelid (an′l id), any of the phylum Annelida; worms and wormlike animals that have segmented bodies; for example, earthworms and leeches.

annual (an′yü əl), a plant in which the entire life cycle is completed in a single growing season.

anterior (an tir′ē ər), positioned toward the front, or head, end of an animal.

anther (an′thər), the part of the stamen that contains pollen.

antibiotic (an′ti bī ot′ik), a substance made by one microorganism, usually a fungus, that harms another microorganism, often a bacterium or fungus.

antibody (an′ti bod′ē), a protein produced in the blood that becomes attached to foreign substances, making them easier to destroy.

antigen (an′tə jən), any substance that causes the body to produce antibodies.

aorta (ā ôr′tə), the main artery that carries blood from the heart to be distributed through the body.

arachnid (ə rak′nid), any of the class Arachnida; arthropods possessing eight legs but no wings or antennae, a jointed body, and simple eyes; for example, spiders.

arteriole (är tir′ē ōl), a small artery, especially one leading into the capillaries.

artery (är′tər ē), a vessel that carries blood away from the heart.

arthropod (är′thrə pod), any of the phylum Arthropoda; invertebrates having segmented bodies with attached pairs of antennae, wings, and legs; insects, arachnids, and crustaceans.

asexual (ā sek′shü əl) **reproduction,** reproduction that does not involve sex cells or a change in chromosome number, as in fission and budding.

atom (at′əm), the smallest particle of an element that has the chemical properties of that element.

ATP (adenosine triphosphate) (ə din′ə sēn trī fos′fāt), a molecule that stores chemical energy for cell activities; formed when ADP accepts a phosphate group and energy.

atrium (ā′trē əm), either of two chambers of the heart that receive blood from the veins and deliver it to a ventricle.

auricle (ôr′ə kəl), *See* atrium.

autonomic (ô′tə nom′ik) **nervous system,** the portion of the vertebrate nervous system that controls involuntary functions.

autosome (ô′tə sōm), any chromosome other than a sex chromosome.

autotroph (ô′tə trof), an organism that can make its own food from inorganic substances; for example, green plants.

auxin (ək′sən), a plant hormone which stimulates elongation of cells, resulting in the growth of stems.

axial (ak′sē əl) **skeleton,** the bones of the head and trunk, as distinguished from those of the appendages, which make up the appendicular skeleton.

axon (ak′son), a long extension of a nerve cell that carries impulses away from the cell body.

bacteria (bak tir′ē ə), a large group of unicellular, microscopic organisms found in air, water, soil, and the bodies of living plants and animals; monerans.

bacteriophage (bak tir′ē ə fāj), a virus that infects various bacteria.

bark, the outermost tissue on a tree trunk, consisting of cork, sometimes with layers of cortex, and phloem.

bilateral symmetry (bī lat′ər əl sim′ə trē), the condition in which body parts are arranged so that the right and left halves are mirror images of each other.

bile (bīl), a bitter fluid secreted by the liver and stored in the gall bladder. It helps to digest fat.

biome (bī′ōm), the organisms that make up a distinct natural community in a specific climatic region, such as the desert.

biosphere (bī′ə sfir), the parts of the earth, oceans, and air that are inhabited by organisms.

biotic (bī o′tik), having to do with life or living things.

bivalve (bī′valv′), any of a group of mollusks that have two shells, or valves, hinged together; for example, clams.

blastocyst (blas′tə cist), an early stage in the development of a mammal's embryo, consisting of a hollow ball of cells with an off-center cell mass that will become the embryo.

blood (blud), the fluid tissue of the body contained within the circulatory system.

blood type, any of four main groups into which human blood can be placed according to the antigens on the red blood cells.

blood pressure, pressure of the blood against arterial walls, caused by contractions of the heart.

Bowman's (bō′mənz) **capsule,** the cup-shaped structure in a vertebrate nephron that aids in removing wastes from the kidneys.

brain (brān), the mass of nervous tissue within the vertebrate skull; the organ of consciousness, or mind, that furnishes stimulation to the muscles in response to sensory stimulation.

bronchi (brong′kī), the two main branches of the trachea, one going to each lung.

bryophyte (brī′ə fīt), any of a group of nonvascular plants, including mosses and liverworts, that have no true roots, stems, or leaves.

bud (bud), an undeveloped shoot; preliminary stage of a flower.

budding (bud′ing), a form of asexual reproduction in which offspring form directly from a piece, or bud, of the body of the parent, as in yeast and hydra.

Calorie (kal′ərē), the amount of heat needed to raise the temperature of one kilogram of water one degree Celsius.

calyx (kā′liks), all the sepals of a flower, forming a protective outer coat for a flower bud.

cambium (kam′bē əm), a ring of dividing cells in roots and stems that forms secondary tissues, particularly xylem and phloem.

capillary (kap′ə ler′ē), the smallest of blood vessels, through which exchanges occur between blood and tissue fluid.

capillary water, water held in tiny spaces among soil particles.

carbohydrate (kär′bō hī′drāt), compounds composed of chains of carbon atoms to which hydrogen and oxygen atoms are attached in a ratio of $C_1H_2O_1$; made by plants during photosynthesis; for example, sugar and starch.

carbon cycle, the overall circulation of carbon in the biosphere, including photosynthesis, the use of plants by animals for food, and the return of carbon to the atmosphere by respiration and decay.

carcinogen (kär sin'ə jən), any chemical, radiation, or virus that causes cancer.

carnivore (kär'nə vôr), a meat-eating organism.

cartilage (kär'tl ij), a tough, elastic substance that forms parts of the skeleton of vertebrates; gristle.

catabolism (kə tab'ə liz'əm), the energy-yielding metabolic processes, in which foods are broken down into simpler substances.

catalyst (kat'l ist), a substance that influences a chemical reaction but is not itself changed chemically.

cell (sel), the basic unit of living systems.

cell membrane, the outer boundary of a cell; encloses all of the cell's organelles.

cell theory, a set of ideas about the structure of organisms. Ideas include belief that all living things are made of cells and that new cells come from preexisting cells.

cellulose (sel'yə lōs), a carbohydrate that makes up the cell walls of plants.

cell wall, the outer cellulose wall secreted around plant cells.

centriole (sen'trē ōl), an organelle in animal cells that organizes the mitotic spindle.

centromere (sen'trə mir), a region on a chromosome to which a spindle fiber attaches during mitosis.

cephalopod (sef'ə lə pod), any of a group of mollusks that have long tentacles around the mouth, a large head, and a sharp beak; for example, the squid.

cephalothorax (sef'ə lō thôr'aks), the combined head and thorax of some animals, especially arthropods.

cerebellum (ser'ə bel'əm), the region of the vertebrate brain between the cerebrum and the medulla that coordinates muscle actions and helps maintain balance.

cerebrum (sə rē'brəm), the largest region of the human brain; the region that controls thought and voluntary muscular movements.

chemotroph (kē'mə trof), a non-photosynthetic organism that obtains energy from breaking chemical bonds.

chitin (kīt'n), a horny substance forming the exoskeletons of arthropods, some other animals, and the cell walls of fungi.

chlorophyll (klôr'ə fil), the green pigment of plant chloroplasts that functions in photosynthesis.

chloroplast (klôr'ə plast), the organelle of plant cells that contains chlorophyll.

cholesterol (kə les'tə rol'), a fatlike substance contained in animal fats, bile, and blood.

chordate (kôr'dāt), any of the phylum Chordata: animals that have a notochord, gill slits, and a hollow dorsal nerve cord at some time during life.

chorion (kôr'ē on), the outer membrane of the embryonic sac that encloses the amnion.

chromatid (krō'mə tid), one of the two strands that result from the replication of a chromosome during cell division. It forms a new chromosome at anaphase.

chromatin (krō'mə tən), a protein substance found in chromosomes.

chromosome (krō'mə sōm), structure in the eukaryote nucleus that contains DNA.

cilia (sil'ē ə), microscopic organelles of locomotion on certain cells.

class, a taxonomic group that includes one or more families; a primary subdivision of a phylum.

cloaca (klō ā'kə), ending of the gut in birds and other lower vertebrates that receives wastes and reproductive cells.

clone (klōn), several organisms produced asexually from a single parent organism.

cocoon (kə kün'), a protective covering for eggs, pupae, or adult animals, usually insects or annelids.

code, genetic, the sequence of nucleotide bases on a DNA molecule.

codon (kō'don), a sequence of three adjacent nucleotides that specify one amino acid; all organisms use the same codons.

coelenterate (si len'tə rāt), any of the phylum Coelenterata: aquatic invertebrates that have radially symmetrical saclike bodies and a single body opening; for example, hydras, jellyfish, and corals.

coelom (sē'ləm), the body cavity of most multicellular animals, containing organs such as the heart, lungs, and stomach.

cohesion (kō hē'zhən), the attraction of molecules of a compound for one another, as in a column of water.

collagen (kol'ə gən), a strong, elastic protein that gives strength to bone tissue.

commensalism (kə men'sə liz'əm), a symbiotic relationship in which one species benefits and the other is neither helped nor harmed.

community, organisms sharing a common environment and interacting.

compound, a substance formed when two or more elements are combined chemically.

conidium (kə nid'ē əm), a unicellular, asexual spore produced by certain fungi.

conifer (kon'ə fər), a cone-bearing tree or shrub; any of the gymnosperms, including pines, yews, and spruces.

conjugation (kon'jə gā'shən), a kind of sexual reproduction that involves the exchange of nuclear material between two protozoans; also, the injection of DNA from a donor bacterium into a recipient bacterium.

consumer (kən sü'mər), animals and protozoans that eat other organisms; heterotrophs.

contractile vacuole (kən trak'təl vak'yü ōl), an organelle in protozoans that collects fluids before they are discharged from the cell.

cork, the tough, outer bark of roots and stems that gives protection from injury.

corolla (kə rol'ə), all the petals of a flower.

cortex (kôr'teks), the outer portion of an organ or tissue; the region of a root or stem between the epidermis and the xylem and phloem.

cotyledon (kot'l ēd'n), the first leaf or one of the first pair of leaves on the embryo of seed plants.

covalent (kō vā'lənt) **bond,** a chemical bond in which two atoms share a pair of electrons.

cristae (kris'tē), the inner folded membrane of mitochondria.

crop, a muscular swelling of a bird's or insect's food passage where food is stored and prepared for digestion.

crossing-over, the exchange of corresponding sections of homologous chromosomes during meiosis.

crustacean (krus'tā'shən), any of the class Crustacea: mostly aquatic arthropods with hard shells and jointed bodies and appendages; for example, crayfish.

cutin (kyüt'n), a waxy substance covering the leaves and stems of herbaceous plants.

cyclosis (sī klō'sis), the streaming movement of cytoplasm in a cell.

cytokinins (sī tə kin'inz), plant hormones that stimulate cell division.

cytoplasm (sī'tə plaz'əm), in a cell, the clear fluid between the nuclear membrane and the cell membrane; contains all the organelles and aids in transporting materials within a cell.

Darwinism (där'wə niz'əm), a theory of evolution by natural selection developed by Charles Darwin.

deciduous (di sij'ü əs), pertaining to a plant whose leaves fall off at the end of the growing season; the first of two sets of teeth most mammals form.

decomposer (dē'kəm pōz'ər), an organism that causes decay of dead organisms or the wastes of living organisms.

dendrites (den'drītz), short fibers that receive impulses and transmit them to the nerve cell.

dentition (den tish'ən), the shape and arrangement of the teeth.

diaphragm (dī'ə fram), in mammals, the internal muscular layer that separates the chest and abdominal cavities and aids in breathing.

diastolic (dī'ə stol'ik) **pressure,** the lowest pressure that occurs in an artery between heart beats.

dicot (dī'kot), a flowering plant having two cotyledons in the embryo; dicotyledon.

diffusion (di fyü'zhən), the random process by which particles such as atoms and molecules move from an area of higher concentration to one of lower concentration.

digestion (də jes'chən), the mechanical and chemical breakdown of food to compounds that can be absorbed across cell membranes.

dihybrid (dī hī'brid), an organism that is heterozygous for two traits, as in AaBb.

diploid (dip'loid), having two sets of chromosomes per cell.

disaccharide (dī sak'ə rīd), any of a group of sugars formed by the union of two monosaccharides; a double sugar. Sucrose and lactose are disaccharides.

DNA (deoxyribonucleic acid) (dē ok'sə rī'bō nü klē'ik), a molecule found in the nucleus of the cell that carries the genetic information and is the chemical basis of heredity.

DNA replication (rep'lə kā'shən), the process by which a molecule of DNA is duplicated.

dominant (dom'ə nənt) **gene,** a gene that expresses its effect even in the presence of a different gene for the same trait.

dorsal (dôr'səl), pertaining to the upper surface of any animal or organ.

echinoderm (i kī'nə dėrm), any of the phylum Echinodermata: invertebrates with arms arranged around a central disk and a body covered by small plates; for example, starfish.

ecology (ē kol'ə jē), the study of the interactions among organisms and between organisms and their physical environment.

ecosystem (ē'kō sis'təm), the interrelationship of living and nonliving things in a given area.

ectoderm (ek'tə dėrm), the outer layer of cells formed during the development of animal embryos, or tissues that develop from that layer.

ectothermic (ek'tō thėr'mik), unable to maintain a constant body temperature; cold-blooded.

electron (i lek'tron), negatively charged particle of an atom.

element (el'ə mənt), a substance that is composed of atoms of one particular kind and that cannot be broken down by chemical reactions.

embryo (em'brē ō), an organism in the early stages of development.

endocrine (en'dō krən) **gland,** any of several glands that secrete hormones directly into the bloodstream rather than into ducts.

endoderm (en'dō dėrm), the inner layer of cells formed during the dėvelopment of animal embryos, or tissues that develop from that layer.

endodermis (en'dō dėr'mis), a layer of thick cells surrounding the vascular cylinder in plants.

endoplasmic reticulum, ER (en'dō plaz'mik ri tik'yə ləm), the network of tubes and flattened sacs connected to the nuclear membrane that stores, separates, and transports substances within the cell.

endosperm (en'dō spėrm), the food tissue in seeds formed inside the embryo sac but outside the embryo.

endospore (en'dō spôr), a very resistant spore formed inside a cell of certain bacteria and made up of nuclear material condensed and enclosed in a tough covering.

endothelium (en'dō thē'lē əm), in vertebrates, a layer of flat cells that lines the heart, blood vessels, and lymph vessels.

endothermic (en'dō ther'mik), capable of maintaining a constant body temperature; warm-blooded.

environment (en vī'rən mənt), all the surroundings of an organism, living and nonliving.

enzyme (en'zīm), a protein that acts as a catalyst.

epidermis (ep'ə der'mis), a single layer of cells on the outside, as on leaves, roots, stems in plants; the protective layer of skin in animals.

epiglottis (ep'ə glot'is), in mammals, a flap of tissue that covers the entrance to the trachea during swallowing.

erythrocyte (i rith'rō sīt), a red blood cell; red corpuscle.

esophagus (ē sof'ə gəs), the tube that extends from the pharynx to the stomach; gullet.

estrogen (es'trə jən), the hormone secreted by the ovaries that stimulates the expression of female primary and secondary sex characteristics.

eukaryote (yü kar'ē ōt), a kind of cell that has a nucleus surrounded by a membrane, as well as other membrane-bound organelles.

evolution (ev'ə lü'shən), any change in genetic make-up of a population; the process of genetic changes by which species have acquired their distinguishing characteristics.

excretion (ek skrē'shən), the discharge of waste products.

exoskeleton (ek'sō skel'ə tən), the hard external structure of animals such as arthropods.

fallopian (fə lō'pē ən) **tube,** uterine tube; oviduct; the tube by which an ovum is transferred to the uterus.

family, a taxonomic group that includes one or more genera; the main subdivision of an order.

fat, an organic compound formed by a molecule of glycerol and three molecules of fatty acid.

fatty acid, any of a group of organic acids, usually a long chain of carbon atoms with a single acid group attached.

fermentation (fėr'men tā'shən), anaerobic breakdown of carbohydrates and other foods into substances such as carbon dioxide, alcohol, or lactic acid.

fertilization (fėr'tl ə zā'shən), the union of two gametes to form a zygote.

fetus (fē'təs), the unborn young of any animal· in humans, the young from about seven or eight weeks after fertilization.

fibril (fī'brəl), threadlike structure in procaryotes that aids in locomotion.

fibrous (fī'brəs) **root,** secondary roots that quickly outgrow the primary root.

fibrovascular (fī'brō vas'kyə lər) **bundle,** a structure in plants containing xylem and phloem and providing support and transport of materials.

fission (fish'ən), asexual reproduction in which the parent cell divides to form two daughter cells.

flagellum (flə jel'əm), a whiplike organelle of locomotion in some one-celled organisms, sperm cells, and a few animals.

flower (flou'ər), specialized reproduction shoot of higher plants in which the sex organs form.

follicle (fol'ə kəl), cluster of protein-secreting cells from which a hair or a feather grows.

food web, a network of organisms in a community among which nutrients and energy are exchanged, usually including producers, consumers, and decomposers.

fossil (fos'əl), the hardened remains or traces of an organism.

fruit, an enlarged and ripened ovary of a plant

FSH, (follicle-stimulating hormone), a hormone secreted by the pituitary gland that stimulates growth of an ovarian follicle in females and the production of sperm in males.

fungus (fung'gəs), a member of the kingdom Fungi (fun'jī). Fungi are simple multicellular organisms that cannot move about; they get nourishment by decomposing dead or living organisms.

gall (gôl) **bladder,** storage organ for bile.

gametes (gam'ēts), haploid sex cells, known as egg and sperm.

gametophyte (gə mē'tō fīt), the individual haploid plant that forms gametes.

ganglion (gang'glē ən), a group of nerve cells located outside the brain and spinal cord.

gastropod (gas'trə pod), any mollusk that has a single shell, or no shell, and a muscular foot used for locomotion; for example, snails.

gene (jēn), the piece of DNA that codes for a particular protein; location on a chromosome that is associated with a special trait.

genetics (jə net'iks), the branch of science dealing with heredity.

genotype (jen'ə tīp), the genetic make-up of an organism.

genus (jē'nəs), a taxonomic group that includes one or more species; a subdivision of a family.

geotropism (jē ot'rə piz'əm), the response of plants to gravity.

germination (jėr'mə nā'shən), starting to develop; a sprouting.

gestation (je stā'shən), the period of pregnancy, the time from fertilization to delivery.

gibberellins (jib'ə rel'ənz), plant hormones that stimulate the elongation of cells.

gill, an organ adapted for absorbing dissolved oxygen from water; in mushrooms, a platelike structure having reproductive hyphae and spores.

gizzard (giz'ərd), a muscular part of the digestive tract in birds and insects.

glomerulus (glə mer'ə ləs), the mass of capillaries in the tubules of the kidney serving to filter blood.

glucose (glü'kōs), the simple sugar, or monosaccharide that is the main energy source for living things.

glycogen (glī'kə jən), a large carbohydrate molecule that is the storage form of glucose in animal cells.

Golgi (gol'jē) **apparatus,** cytoplasmic membranes that package protein molecules to be secreted from the cell; Golgi body.

guttation (gə tā'shən), loss of water droplets from leaves.

gymnosperm (jim'nə spėrm), one of a group of plants that have exposed ovules, as in conifers.

habitat (hab'ə tat), place where an organism naturally lives.

half-life, the length of time it takes for half the atoms of a particular radioactive substance to disintegrate.

haploid (hap'loid), the reduced number of chromosomes in the gametes; half the diploid number

hemoglobin (hē'mə glō'bən), the iron-containing protein in red blood cells that functions in oxygen transport.

herbivore (hėr'bə vôr), a plant-eating animal.

heterotroph (het'ər ə trof), an organism that cannot make its own food; for example, animals, fungi, and some monerans.

heterozygote (het'ər ə zī'gōt), an organism whose genotype includes two different alleles for a particular gene; a hybrid.

homeostasis (hō'mē ə stā'sis), the capacity of an organism to maintain a stable internal environment by the regulating of its own body processes.

homologous (hō mol'ə gəs) **chromosomes,** pairs of chromosomes that carry genes for the same set of traits.

homologous (hō mol'ə gəs) **organs,** those organs that are similar in origin and structure but not necessarily in function.

homozygote (hō'mə zī'gōt), an organism whose genotype includes identical alleles for a particular trait and, therefore, always breeds true to type for that trait; a purebred.

hormone (hôr'mōn), a chemical substance produced in one part of an organism and carried in the blood to another part where it exerts a specific effect.

hybrid (hī'brid), an offspring from two individuals differing in one or more inherited traits.

hydrologic cycle, the movement of water throughout the biosphere; water cycle.

hypha (hī'fə), one of the filaments that makes up the mycelium of a fungus.

immunity (i myü'nə tē), resistance to disease.
insect (in'sekt), any of the class Insecta: arthropods with a three-part body, three pairs of legs, and one or two pairs of wings; for example, grasshoppers, flies, and bees.
instinct (in'stingkt), an inherited behavior, not dependent on experience.
interferon (in'tər fir'on), a protein that is manufactured by cells infected by a virus and that interferes with the spread of all viruses.
interphase (in'ter fāz), the time in between cell divisions, during which the cell performs its normal functions, including preparation for the next cell division.
intestine (in tes'tən), the part of the digestive tract that extends from the stomach to the anus; small intestine and large intestine.
invertebrate (in vėr'tə brāte), any of a group of animals without a backbone.
ion (ī'ən), an atom or group of atoms having a negative or positive charge.
ionic (ī on'ik) **bond,** a chemical bond in which electrons are transferred from one atom to another.

karyotype (kar'ē ō tīp), the characteristic set of chromosomes of some organism, as defined by their arrangement, number, sizes, and shapes.
kidney (kid'nē), one of the pair of organs in vertebrates that regulate water balance and separate wastes from the blood, passing them into the bladder as urine.
kingdom, in taxonomy, the highest category in a classification system that includes all organisms.

Lamarckism (lə mär'kiz'əm), a theory of evolution developed by Jean Lamarck based on the idea that characteristics acquired by parents are inherited by their offspring.
larva (lär'və), an immature form of an animal that does not resemble the adult and must undergo further development.
larynx (lar'ingks), the voicebox, lying between the pharynx and trachea.
lenticel (len'tə sel'), a small hole on a woody stem through which gases pass between the atmosphere and stem tissues.
leukocyte (lü'kə sīt), in vertebrates, a white blood cell; white corpuscle.
LH (luteinizing hormone) (lü'tē ə niz'ing), the hormone secreted by the pituitary that stimulates ovulation, final maturation of ovarian follicles, formation of corpora lutea in females, and maturation of cells of testes in males.
lichen (lī'kən), a symbiotic association of an alga and a fungus that benefits both organisms.
ligament (lig'ə mənt), a band of strong, flexible connective tissue that connects two bones at a joint.
lipid (lip'id), any of a group of fats and fatlike substances.
liver (liv'ər), a large organ that secretes bile, aids in the absorption of vitamins, minerals, and sugar, and neutralizes many poisons and wastes.
lung, a spongy organ of respiration in land animals and humans in which blood gains oxygen and gives up carbon dioxide.
lymph (limf), fluid in the lymph vessels that contains white blood cells and plasma and that passes through capillary walls and enters the spaces between cells.
lysosome (lī'sə sōm), an organelle that contains digestive enzymes.

mammal (mam'əl), any of the class Mammalia: vertebrates with mammary glands that secrete milk for the young; for example, humans, bats, whales.
mandible (man'də bəl), the lower jawbone in vertebrates; in arthropods, a mouth part for seizing and biting.
mantle (man'tl), in mollusks, a fold of tissue that encloses the soft structures and secretes the shell.
marrow (mar'ō), a soft, fatty substance found in bones. Red marrow produces red blood cells.
marsupial (mär sü'pē əl), a mammal whose young are born as partly developed embryos.
medulla oblongata (mi dul'ə ob'long gä'tə), the lowest segment of the brain stem, responsible for controlling such involuntary functions as respiration, heart rate, and blood pressure.
megaspore (meg'ə spôr), four haploid cells, formed by meiosis from the megaspore mother cell; three die and one develops into a female gametophyte.
meiosis (mī ō'sis), a type of cell division of diploid cells that yields haploid cells such as sex cells; reduction division.
menstruation (men'strü ā'shən), periodic discharge of blood and tissues from the uterus.
meristem (mer'ə stem), in plants, a tissue that is made up of rapidly dividing cells.
mesoderm (mes'ə dėrm'), in animal embryos, the layer of cells between the ectoderm and the endoderm.
mesophyll (mes'ə fil), the green tissue in a leaf, between the upper and lower epidermis.
metabolism (mə tab'ə liz'əm), all the chemical processes by which living things acquire and use energy and materials.
metamorphosis (met'ə môr'fə sis), the developmental changes that take place as a larva is transformed into an adult.
metaphase (met'ə fāz), the state of mitosis during which the chromatids line up across the middle of the spindle.
microbe (mī'krōb), an organism visible only through a microscope; microorganism.
microfilaments (mī'krə fil'ə mənts), submicroscopic protein fibers involved in cell locomotion.
micropyle (mī'krə pīl'), a small opening in an ovule through which the pollen tube enters.
microspores (mī'krə spôrz), a group of four cells, formed from the microspore mother cell, that develop into pollen grains.
microtubules (mī'krə tü'byulz), long, submicroscopic cylinders of protein that provide support for a cell.
migration (mī grā'shən), the movement of individuals into or out of a population.
mitochondria (mī'tə kon'drē ə), rod-shaped organelles that are the principal sites of energy release in the cell.
mitosis (mī tō'sis), process of cell division for non-sex cells, or body cells, in which the two daughter cells receive exactly the same chromosomes as the original cell.

molecule (mol'ə kyül), the smallest particle into which a pure substance can be divided without changing its chemical properties.

mollusk (mol'əsk), any of the phylum Mollusca: invertebrates having a soft, unsegmented body, usually covered with a hard shell; for example, clams, oysters, and octopuses.

molt, to shed an outer layer before new growth occurs.

moneran (mə nir'ən), a member of the kingdom Monera. Monerans are generally unicellular organisms lacking membrane-bound nuclei and specialized organelles. All bacteria are monerans.

monocot (mon'ə kot), a flowering plant having one cotyledon; monocotyledon.

monosaccharide (mon'ə sak'ə rīd), any of a group of simple sugars, such as glucose or fructose.

monotreme (mon'ə trēm), an egg-laying mammal; for example, the duck-billed platypus.

mRNA, messenger RNA, the molecule that carries the coded message from DNA in the nucleus to the ribosomes, the sites of protein synthesis.

mucosa (myü kō'sə), the membrane that lines cavities and tubes of the body that open to the outside. It produces mucus.

mutagen (myü'tə jən), physical or chemical agent that causes a mutation, such as X rays or nitrites.

mutation (myü tā'shən), a change in genetic make-up that results in a new characteristic that can be inherited; a change in the sequence of nucleotides in DNA that changes the amino acid that is coded for.

mutualism (myü'chü ə liz'əm), a symbiotic association in which both organisms benefit.

mycelium (mī sē'lē əm), the mass of hyphae that forms the vegetative body of most fungi.

mycorrhiza (mī'kə rī'zə), a symbiotic association between plant roots and soil fungus that benefits both organisms.

myelin (mī'ə lən), a fatty substance that surrounds certain nerve fibers.

nares (ner'ēz), external openings of the air passages in vertebrates; nostrils.

natural selection, process by which organisms best adapted to their environment tend to survive.

nematode (nem'ə tōd), any of the phylum Nematoda: worms that have unsegmented, cylindrical bodies; roundworms.

nephridium (ni frid'ē əm), an organ of excretion in invertebrates.

nephron (nef'ron), the functional unit of the vertebrate kidney.

nerve, a bundle of nerve fibers, primarily axons, bound together by connective tissue.

nervous system, the network of interconnected nerve cells that together coordinate bodily activities.

neuron (nůr'on), a nerve cell.

neutron (nü'tron), an uncharged particle in the nucleus of an atom.

niche (nich), the life style unique to a species; the role an organism plays in a habitat.

nitrification, conversion by soil bacteria of nitrogen in organic compounds into nitrates that can be used by plants.

nitrogen cycle, the global circulation and re-use of nitrogen and its compounds by organisms.

nitrogen fixation, the process by which atmospheric nitrogen gas is changed into nitrogen compounds by bacteria.

nodule (noj'ül), a small swelling on the root of a legume that contains nitrogen-fixing bacteria.

notochord (nō'tə kôrd), a flexible, rodlike structure of supportive tissue running down the back of the lowest vertebrates and found in the embryos of higher vertebrates.

nuclear (nü'klē ər) **membrane,** a double-layered membrane that surrounds the cell nucleus.

nucleolus (nü klē'ə ləs), a small region, found within the cell nucleus, that is involved in making ribosomes.

nucleotide (nü'klē ə tīd), any of a group of compounds that make up nucleic acids such as DNA and the various RNAs.

nucleus (nü'klē əs), a cellular organelle that contains the chromosomes and that directs all cellular activities; the central part of an atom.

nutrient (nü'trē ənt), nourishing substance; for example, carbohydrates, fats, proteins, vitamins, and minerals.

nymph (nimf), any of certain insects in the second stage of incomplete metamorphosis, between the egg and the adult form.

omnivore (om'nə vôr), an animal that eats both plants and animals.

operculum (ō pėr'kyə ləm), a lidlike part or organ, as the gill-cover of a fish.

opposable (ə pōz'əbəl), capable of being placed opposite something else, as the thumb of a human is opposable to each of the other digits.

optic (op'tik) **nerve,** the nerve of sight; it connects the brain to the retina

order, a subdivision of a class, consisting of one or more families

organ, any part or structure composed of tissues that carry out a specific function; for example, the heart, liver, or kidney.

organelle (ôr'gə nel'), a specialized structure inside the cell that carries out one or more functions; for example, mitochondria.

organic compound, any substance that contains carbon atoms bonded together or contains carbon and hydrogen atoms bonded together.

organism (ôr'gə niz'əm), an individual living thing.

osmosis (oz mō'sis), diffusion of water or any other solvent through a semipermeable membrane.

ovary (ō'vər ē), a female organ of reproduction found in plants and animals; in humans, the female sex gland.

oviduct (ō'və dukt), *See* fallopian tube.

ovulation (ō'vyə lā'shən), the discharge of an ovum from an ovary.

ovule (ō'vyül), in plants, the part that contains the egg, which after fertilization develops into an embryo; an immature seed.

ovum (ō'vəm), a female sex cell; female gamete; egg.

oxidation (ok'sə dā'shən), the burning of a nutrient, by combining with oxygen, for energy and heat.

pacemaker (pās'mā'kər), a cell or a center that sends out rhythmic impulses and thus "drives" other cells, particularly the cells that initiate the heart beat; an electronic device that maintains the heart beat.

pancreas (pan'krē əs), a large gland located behind the stomach that secretes insulin and pancreatic juice, a digestive enzyme.

parasite (par'ə sīt), an organism that lives on or in another (the host) from which it gets its food, always at the expense of the host.

parathyroid (par'ə thī'roid) **glands,** small endocrine glands that secrete parathormone, which enables the body to use calcium and phosphorus.

parenchyma (pə reng'kə mə), soft, loose tissue in higher plants, composed of unspecialized cells; most of the tissue in leaves, the pulp of fruits, the pith of stems, etc.

passive transport, movement of molecules across cell membranes in response to a concentration gradient, not requiring energy.

pathogen (path'ə jən), disease-producing agent.

pedicel (ped'ə səl), the stem that attaches a flower to the main plant body.

peptide (pep'tīd) **bond,** the chemical bond that links one amino acid molecule to another as in a polypeptide or a protein.

peptidoglycan (pep tī'do glī'kan), the organic polymer found only in prokaryotes, which gives strength to the cell wall.

perennial (pə ren'ē əl), a plant that lives for several years.

peristalsis (per'ə stal'sis), wavelike contractions in the wall of a tubular organ by which its contents are moved onward, as in the digestive tract.

phagocyte (fag'ə sīt), a cell, such as a white blood cell, capable of engulfing and destroying microorganisms.

phagocytosis (fag'ə sī tō sis), the process of a cell actively engulfing solids.

pharynx (far'ingks), the part of the digestive tract located between the mouth and the esophagus.

phenotype (fē'nə tīp), the physical appearance of an organism resulting from the interaction of its genotype and its environment.

pheromones (fer'ə mōnz), chemicals that serve as messages between members of a given species.

phloem (flō'em), complex tissue in vascular plants that conducts food downward to the stems and roots.

photoperiodism (fō'tō pir'ē ə diz əm), the response of a plant to varying periods of light, which affects the plant's growth pattern.

photosynthesis (fō'tō sin'thə sis), the process by which plants, using chlorophyll and energy from the sun, produce carbohydrate from carbon dioxide and water.

phototroph (fō'tə trof), an organism that obtains energy from light.

phototropism (fō to'trə piz'əm), the tendency of plants to grow toward light.

phylum (fī'ləm), a taxonomic group that includes one or more classes; one of the main divisions of a kingdom.

pinocytosis (pi'nō sī tō'sis), the process whereby a cell actively engulfs liquids.

pistil (pis'tl), the female reproductive organ of a flower, consisting of an ovary, a style, and a stigma.

pith, the central column of spongy tissue in the stems of most herbaceous plants and some trees.

pituitary (pə tü'ə ter'ē) **gland,** a small endocrine gland, located at the base of the brain, that secretes hormones that promote growth and stimulate other glands.

placenta (plə sen'tə), the round, flat organ in the uterus that is formed during the early stages of a developing embryo, attached to the embryo by the umbilical cord, and that provides nourishment for the embryo.

plasma (plaz'mə), the clear liquid part of blood.

plasmolysis (plaz mol'ə sis), shrinkage of cells due to osmotic loss of water.

plastid (plas'tid), any of a group of organelles in plant cells concerned with special metabolic activity; for example, chloroplasts.

platelet (plāt'lit), one of many small disks that float in plasma and help make blood clot; thrombocyte.

platyhelminth (plat'ə hel'minth), any of the phylum Platyhelminthes: worms that have soft, bilaterally symmetrical bodies and no body cavity; flatworm.

pleiotropy (plī ot'rə pē), the condition in which a single gene affects more than one trait.

pollen (pol'ən), microspores of seed plants.

pollination (pol'ə nā'shən), the transfer of pollen to a pistil or ovule.

polymer, a very large molecule formed by linking many similar smaller units.

polypeptide (pol'ē pep'tīd), a chain of amino acids linked by peptide bonds.

polyploid (pol'ē ploid), having three or more sets of chromosomes per cell.

polysaccharide (pol'ē sak'ə rīd), any of a group of carbohydrate polymers, including starches, glycogen, and cellulose, composed of several simple sugars bonded together.

population, the total number of organisms of the same species inhabiting a place at the same time.

porifera (pə rif'ər ə), members of the phylum Porifera: invertebrates permanently attached to the ocean floor that have porous bodies through which water passes continuously; sponges.

posterior (po stir'ē ər), positioned toward the rear.

predator (pred'ə tər), animal that feeds on other animals; secondary consumers other than parasites.

prokaryote (prō'kar'ē ōt), a kind of cell that does not have a membrane-bound nucleus or any other membrane-bound organelles.

prophase (prō'fāz), the first stage of mitosis during which the spindle forms and the nuclear membrane disappears.

protein (prō'tēn), any of a group of complex chemical compounds made up of long chains of amino acids.

protist (prō'tist), a member of the kingdom Protista. Protists are unicellular organisms with specialized organelles. Some have plantlike characteristics; others have animallike characteristics.

proton (prō'ton), positively charged particle in the nucleus of an atom.

protoplasm (prō'tə plaz'əm), living matter of all cells.

pseudopod (sü'də pod), a temporary protrusion of the protoplasm serving as a means of cellular locomotion and food getting; for example, amebas and white blood cells.

pupa (pyü'pə), the stage between the larva and the adult in insects with complete metamorphosis.

radial (rā'dē əl) **symmetry,** the condition in which similar body parts are arranged around an axis, as in the starfish.

radioactive, having radioactivity: the property exhibited by certain elements of giving off radiation in the form of high-energy particles and being changed into other elements.

receptor (ri sep'tər), a sense organ; the terminal part of a sensory nerve.

recessive (ri ses'iv) **gene,** a gene that does not produce an observable effect when paired with a dominant gene.

reduction division, *See* meiosis.

reflex (rē fleks), a simple behavior in which a stimulus produces a response without involving the brain.

regeneration (ri jen'ə rā'shən), a form of healing in which a structure of an organism is regrown or repaired.

reptile (rep'təl), any of the class Reptilia: cold-blooded chordates having dry skin, lungs, and a four-chambered heart.

respiration (res'pə rā'shən), the chemical reactions in a cell that take food molecules apart to release energy, which is then stored as ATP; taking in oxygen and releasing carbon dioxide; breathing.

rhizome (rī'zom), a thick underground stem that has buds that form new plants.

ribosome (rī'bə sōm), a tiny structure in a cell upon which protein molecules are formed.

RNA, ribonucleic (rī'bō nü klē'ik) acid, nucleic acids found in nuclei and cytoplasm of cells; it plays an important role in protein synthesis; includes messenger RNA, transfer RNA, and ribosomal RNA.

root cap, a mass of cells at the tip of growing roots that protects the root.

root hair, a hairlike outgrowth from a root that absorbs water and dissolved minerals from the soil.

saliva (sə lī'və), a digestive fluid secreted into the mouth by the salivary glands to keep it moist, aid in chewing and swallowing, and start digestion of starches.

seed, a mature, fertilized ovule.

semen (sē'mən), mixture of sperm and seminal fluid produced in the testes.

semipermeable (sem'i pėr'mē ə bəl) **membrane,** a membrane that allows some substances to pass through but not others.

sex chromosomes, a pair of chromosomes that carry genes that determine sex and sex-linked characteristics.

sex-linked trait, a trait controlled by genes located on the sex chromosomes; for example, red-green color blindness.

sexual reproduction, reproduction that involves the fusion of sex cells, or gametes.

spawn (spôn), the eggs of fish, frogs, and shellfish; the newly hatched young from such eggs.

specialization (spesh'ə lə zā'shən), possessing structures adapted to a special function or condition.

speciation (spē shē ā'shən), the evolutionary process whereby one species becomes two species.

species (spē'shēz), a group of organisms that have certain characteristics in common and are able to interbreed; a subdivision of a genus.

sperm, male sex cells.

sphincter (sfingk'tər), a ringlike muscle that surrounds an opening or passage of the body and can contract to close it.

spinal (spī'nl) **cord,** the thick cord of nerve tissue that extends from the medulla oblongata down through most of the backbone and from which nerves branch off to various parts of the body.

spindle (spin'dl), the group of tubelike structures, the ends of which are called poles, along which the chromosomes are arranged during mitosis.

spinneret (spin'ə ret'), an organ by which spiders and certain insect larvae spin their threads.

spiracle (spī'rə kəl), the breathing openings of insects; the first gill slit in cartilaginous fish, such as sharks and rays.

spleen (splēn), an organ, located behind the stomach, that destroys diseased red blood cells and manufactures certain of the white blood cells.

spontaneous generation, the belief that living organisms could be produced from nonliving matter.

sporangium (spə ran'jē əm), a spore case.

spore (spôr), a reproductive cell, usually one-celled, that is capable of developing into a new individual.

sporophyte (spôr'ə fīt), the diploid plant or generation that forms spores.

stamen (stā'mən), the male reproductive organ of a flower that contains the pollen, consisting of a slender filament that supports the anther.

starch, a polysaccharide, formed by the linking of many monosaccharides.

stigma (stig'mə), the part of the pistil having a sticky surface that receives the pollen.

stimulus (stim'yə ləs), anything that produces a response.

stoma (stō'mə), an opening in the epidermis of a leaf through which gases pass.

striated (strī'ā tid), striped, as in skeletal muscles.

style (stīl), the stemlike part of the pistil having the stigma at the top.

succession (sək sesh'ən), the process by which one community or organisms is gradually replaced by another.

sugar, any of a group of carbohydrates with the general formula $(CH_2O)_n$; for example, glucose, sucrose, dextrose.

swimmeret (swim'ə ret'), an abdominal limb or appendage in many crustaceans used for respiration and other functions, such as carrying eggs.

symbiosis (sim'bē ō'sis), the association between organisms of two species living together.

synapse (si naps'), the junction between the tip of an axon and a dendrite where an impulse is transmitted from one neuron to another.

system, a set of organs or parts in the body having the same or similar structure, or serving to perform some function, such as circulation or digestion.

systolic (sis tol'ik) **pressure,** the highest pressure that occurs in an artery when the ventricles of the heart contract.

taproot, the primary root growing downward and sprouting secondary roots.

taxonomy (tak son'ə mē), the classification of organisms according to their characteristics and evolutionary histories.

telophase (tel'ə fāz), the final stage of mitosis, in which the sets of daughter chromosomes separate completely and the nuclei re-form.

tendon (ten'dən), a band of tough connective tissue that attaches a muscle to a bone.

tentacle (ten'tə kəl), a long, flexible organ, usually located on the head or around the mouth of an animal and used in feeding or to touch, hold, or move.

testis (tes'tis), the male reproductive organ; male sex gland.

thromboplastin (throm'bō plas'tən), the substance in platelets that helps blood clot.

thyroid (thī'roid) **gland,** a large endocrine gland in the neck that secretes thyroxin, a hormone that influences metabolic rate.

tissue, a group of similar cells that perform a particular function, such as muscular tissue.

trachea (trā'kē ə), the windpipe, or breathing tube, in vertebrates; tiny air tube of some arthropods.

tracheophyte (trā'kē ə fīt), any of a group of plants that have vessels for transporting nutrients and water.

transcription, synthesis of RNA from a DNA molecule.

translation, synthesis of a protein from the nucleotide sequence in an mRNA molecule.

translocation, the movement of materials within a plant.

transpiration (tran'spə rā'shən), the evaporation of water through leaves and stems.

tRNA, transfer RNA, the molecules that carry amino acids to the ribosomes, the site of protein synthesis. There is a separate tRNA for each amino acid.

tropism (trō'piz əm), movement in a plant in response to an external stimulus.

tuber (tü'bər), a short fleshy underground stem bearing modified axillary buds from which new plants may arise.

turgor (tėr'gər), the condition of a cell swollen with water that gives cells the normal tense condition.

urea (yu̇ rē'ə), a nitrogenous waste product of mammals found in urine.

ureter (yü rē'tər), the duct that transports urine from a kidney to the urinary bladder.

urethra (yu̇ rē'thrə), the duct that transports urine from the bladder out of the body.

urine (yu̇r'ən), the solution of wastes excreted by the kidneys.

uterus (yü'tər əs), the hollow organ in which the fertilized ovum implants and develops.

vaccine (vak'sēn'), the substance, containing weakened or killed viruses or killed bacteria, that stimulates the body to produce antibodies.

vacuole (vak'yü ōl), a tiny saclike organelle inside many cells consisting of a membrane filled with water and other substances and performing various functions, including digestion and storage.

vascular (vas'kyə lər) **tissue,** tissue in a vascular plant, consisting essentially of phloem and zylem, that carries materials throughout the plant.

vegetative (vej'ə tā'tiv) **reproduction,** a form of asexual reproduction in which a new individual is formed from parts of the parent; cloning.

vein (vān), a blood vessel that carries blood toward the heart.

ventral (ven'trəl), the lower surface of an animal or organ; the side opposite the dorsal.

ventricle (ven'trə kəl), either of the two chambers of the heart that receive blood from the atria (auricles) and force it into the arteries.

vertebrate (vėr'tə brit), any of a group of chordates having a segmented spinal column and a brain case, including fish, amphibians, reptiles, birds, and mammals.

vestigial (ve stij'ē əl) **organ,** an organ that is no longer fully developed or useful; for example, the appendix in humans.

villi (vil'ī), tiny fingerlike projections, found in the mucous membrane of the small intestine, that increase the surface available for absorption.

virus (vī'rəs), any of a group of parasites composed of protein and nucleic acid.

vitamin (vī'tə min), an organic molecule, required in small amounts, that an organism cannot make for itself but must obtain from its food.

X chromosome, one of two chromosomes that determine sex. A fertilized human egg cell containing two X chromosomes, one from each parent, develops into a female.

xylem (zī'lem), complex tissue in vascular plants that serves to conduct water upward from the roots and that provides support; wood.

Y chromosome, one of the two chromosomes that determine sex. A fertilized human egg cell containing one X and one Y chromosome develops into a male.

yolk, the part of any animal egg cell that serves as food for the embryo.

zygospore (zī'gə spôr), a spore formed by the union of two gametes, as in certain algae and fungi.

zygote (zī'gōt), a single cell formed by the union of sperm and egg; a fertilized ovum.

Index